Phytotoxicity of Organic POLLUTANTS

Phytotoxicity of Organic POLLUTANTS

Impacts and Management

Edited by

GOLAM JALAL AHAMMED
College of Horticulture and Plant Protection, Henan University of Science and Technology, Luoyang, China

KARINA S.B. MIGLIORANZA
Laboratorio de Ecotoxicología y Contaminación Ambiental, Universidad Nacional de Mar del Plata (UNMdP) & Instituto de Investigaciones Marinas y Costeras (IIMyC-CONICET), Mar del Plata, Argentina

ELSEVIER

ACADEMIC PRESS
An imprint of Elsevier

Academic Press is an imprint of Elsevier
125 London Wall, London EC2Y 5AS, United Kingdom
50 Hampshire Street, 5th Floor, Cambridge, MA 02139, United States

ISBN: 978-0-443-33110-7

For information on all Academic Press publications visit our website at https://www.elsevier.com/books-and-journals

Publisher: Andre Wolff
Acquisitions Editor: Kattie Washington
Editorial Project Manager: Swadha Puri
Production Project Manager: Rajeswari R
Cover Designer: Miles Hitchen

Typeset by TNQ Tech

Contents

Contributors

Ajakaye Abisola Eunice
Department of Public Health, University of Suffolk, Microbiology Programme, Department of Biological Sciences, School of Science, Ipswich, United Kingdom

Bartholomew Saanu Adeleke
Food Security and Safety Focus Area, Faculty of Natural and Agricultural Sciences, North-West University, Mmabatho, South Africa; Agagu University of Science and Technology, Okitipupa, Ondo, Nigeria

Golam Jalal Ahammed
College of Horticulture and Plant Protection, Henan University of Science and Technology, Luoyang, China

Akinlolu Olalekan Akanmu
Food Security and Safety Focus Area, Faculty of Natural and Agricultural Sciences, North-West University, Mmabatho, South Africa

Shamim Akhtar
Department of Botany, Rawalpindi Women University, Rawalpindi, Pakistan

Saheed Adekunle Akinola
Food Security and Safety Focus Area, Faculty of Natural and Agricultural Sciences, North-West University, Mmabatho, South Africa; Department of Microbiology and Parasitology, School of Medicine and Pharmacy, College of Medicine and Health Sciences, University of Rwanda, Butare, Rwanda

Laiba Arif
Department of Environmental Sciences, COMSATS University Islamabad, Abbottabad, Pakistan

Muhammad Arshad
Institute of Environmental Sciences and Engineering, School of Civil and Environmental Engineering, National University of Sciences and Technology (NUST), Islamabad, Pakistan

Modupe Stella Ayilara
Food Security and Safety Focus Area, Faculty of Natural and Agricultural Sciences, North-West University, Mmabatho, South Africa

Belayneh Azene
Yunnan Key Laboratory of Forest Ecosystem Stability and Global Change, Xishuangbanna Tropical Botanical Garden, Chinese Academy of Sciences, Mengla, China

Olubukola Oluranti Babalola
Food Security and Safety Focus Area, Faculty of Natural and Agricultural Sciences, North-West University, Mmabatho, South Africa

Nazneen Bangash
Department of Biosciences, COMSATS University Islamabad, Islamabad, Pakistan

Crislaine Bertoldi
Department of Earth and Environmental Sciences, Faculty of Science, Lund, Sweden

Oussama Bouaicha
Faculty of Agricultural, Environmental and Food Sciences, Free University of Bozen-Bolzano, Bolzano, Italy

Lucia Helena Gomes Coelho
Center for Engineering, Modeling, and Applied Social Sciences, Federal University of ABC, Santo Andre, Brazil

Achlesh Daverey
School of Environment and Natural Resources, Doon University, Dehradun, Uttarakhand, India

Qammar Farooq
Institute of Soil and Environmental Sciences, University of Agriculture, Faisalabad, Pakistan

Andreia Neves Fernandes
Institute of Chemistry, Federal University of Rio Grande do Sul, Porto Alegre, Brazil

Marium Fiaz
School of Interdisciplinary Engineering and Sciences, National University of Sciences and Technology (NUST), Islamabad, Pakistan

Sofía Fideleff
Laboratorio de Ecotoxicología y Contaminación Ambiental, Facultad de Ciencias Exactas y Naturales (FCEyN), Universidad Nacional de Mar del Plata (UNMdP), Mar del Plata, Argentina; Instituto de Investigaciones Marinas y Costeras (IIMyC), Consejo Nacional de Investigaciones Científicas y Tecnológicas (CONICET), Mar del Plata, Argentina

Hardik Giri Gosai
Department of Environmental Studies, Faculty of Science, The Maharaja Sayajirao University of Baroda, Vadodara, Gujarat, India

Malik Tahir Hayat
Department of Environmental Sciences, COMSATS University Islamabad, Abbottabad, Pakistan

Júlia Melo Henrique
Institute of Chemistry, Federal University of Rio Grande do Sul, Porto Alegre, Brazil

Katalin Hubai
University of Pannonia, Centre for Natural Sciences, Veszprém, Hungary

Asma Imtiaz
Department of Environmental Sciences, COMSATS University Islamabad, Abbottabad, Pakistan

Nadeem Iqbal
Department of Environmental Sciences, COMSATS University Islamabad, Abbottabad, Pakistan; State Key Laboratory of Geohazard Prevention and Geoenvironment Protection, College of Ecology and Environment, Chengdu University of Technology, Chengdu, Sichuan, P.R. China

Md Shafiqul Islam
Yunnan Key Laboratory of Forest Ecosystem Stability and Global Change, Xishuangbanna Tropical Botanical Garden, Chinese Academy of Sciences, Mengla, China

Shraddha K. Kanzariya
Department of Environmental Studies, Faculty of Science, The Maharaja Sayajirao University of Baroda, Vadodara, Gujarat, India

Sana Khalid
Department of Environmental Sciences, COMSATS University Islamabad, Vehari, Pakistan

Muhammad Imran Khan
Institute of Soil and Environmental Sciences, University of Agriculture, Faisalabad, Pakistan

Zulqarnain Haider Khan
State Key Laboratory for Ecological Security of Regions and Cities, Ningbo Observation and Research Station, Institute of Urban Environment, Chinese Academy of Sciences, Xiamen, China

Maria Khan
Department of Environmental Sciences, COMSATS University Islamabad, Abbottabad, Pakistan

Nora Kováts
University of Pannonia, Centre for Natural Sciences, Veszprém, Hungary

Zhangtao Li
Key Laboratory of Recycling and Eco-Treatment of Waste Biomass of Zhejiang Province, School of Environment and Natural Resources, Zhejiang University of Science and Technology, Hangzhou, China

Mercedes Lourido
Laboratorio de Ecotoxicología y Contaminación Ambiental, Facultad de Ciencias Exactas y Naturales (FCEyN), Universidad Nacional de Mar del Plata (UNMdP), Mar del Plata, Argentina; Instituto de Investigaciones Marinas y Costeras (IIMyC), Consejo Nacional de Investigaciones Científicas y Tecnológicas (CONICET), Mar del Plata, Argentina

Qaisar Mahmood
Department of Environmental Sciences, COMSATS University Islamabad, Abbottabad, Pakistan; Department of Biology, College of Science, University of Bahrain, Sakhir, Bahrain

Sahil Menghani
Department of Environmental Studies, Faculty of Science, The Maharaja Sayajirao University of Baroda, Vadodara, Gujarat, India

Mirta Luján Menone
Instituto de Investigaciones Marinas y Costeras (IIMYC), Consejo Nacional de Investigaciones Científicas y Técnicas (CONICET), Universidad Nacional de Mar del Plata (UNMDP), Mar del Plata, Provincia de Buenos Aires, Argentina

Karina S.B. Miglioranza
Instituto de Investigaciones Marinas y Costeras (IIMyC), Consejo Nacional de Investigaciones Científicas y Tecnológicas (CONICET), Mar del Plata, Argentina; Laboratorio de Ecotoxicología y Contaminación Ambiental, Facultad de Ciencias Exactas y Naturales (FCEyN), Universidad Nacional de Mar del Plata (UNMdP) & Instituto de Investigaciones Marinas y Costeras (IIMyC-CONICET), Mar del Plata, Argentina

Tanja Mimmo
Faculty of Agricultural, Environmental and Food Sciences, Free University of Bozen-Bolzano, Bolzano, Italy; Competence Centre of Plant Health, Free University of Bozen-Bolzano, Bolzano, Italy

Behzad Murtaza
Department of Environmental Sciences, COMSATS University Islamabad, Vehari, Pakistan

Doris Flore Ngakeu
Center for Engineering, Modeling, and Applied Social Sciences, Federal University of ABC, Santo Andre, Brazil

Minh-Ky Nguyen
Faculty of Environment and Natural Resources, Nong Lam University, Ho Chi Minh City, Viet Nam

Paola M. Ondarza
Laboratorio de Ecotoxicología y Contaminación Ambiental, Facultad de Ciencias Exactas y Naturales (FCEyN), Universidad Nacional de Mar del Plata (UNMdP), Mar del Plata, Argentina; Instituto de Investigaciones Marinas y Costeras (IIMyC), Consejo Nacional de Investigaciones Científicas y Tecnológicas (CONICET), Mar del Plata, Argentina

Jagriti Patel
Department of Environmental Studies, Faculty of Science, The Maharaja Sayajirao University of Baroda, Vadodara, Gujarat, India

Débora Jesabel Pérez
Consejo Nacional de Investigaciones Científicas y Técnicas (CONICET), Universidad Nacional de Mar del Plata, Balcarce, Provincia de Buenos Aires, Argentina; Universidad Fraternidad de Santo Tomás de Aquino, Mar del Plata, Provincia de Buenos Aires, Argentina

Alice Pita-Barbosa
Centre for Coastal, Limnological and Marine Studies, Litoral Norte Campus, Federal University of Rio Grande do Sul, Imbé, Brazil

Md Arifur Rahaman
International Maize and Wheat Improvement Center (CIMMYT), Dhaka, Bangladesh

Midhat Rahim
Department of Environmental Sciences, COMSATS University Islamabad, Abbottabad, Pakistan

Hadia Riaz
Department of Environmental Sciences, COMSATS University Islamabad, Abbottabad, Pakistan

Felipe Klein Ricachenevsky
Institute of Biosciences, Federal University of Rio Grande do Sul, Porto Alegre, Brazil

Palak Savalia
Sophisticated Instrumentation Centre for Applied Research & Testing (SICART), Charutar Vidya Mandal (CVM), Vallabh Vidyanagar, Anand, Gujarat, India

Muhammad Shahid
Department of Environmental Sciences, COMSATS University Islamabad, Vehari, Pakistan

Asha Sharma
Department of Environmental Studies, Faculty of Science, The Maharaja Sayajirao University of Baroda, Vadodara, Gujarat, India

Ricardo Santos Silva
Center for Natural and Human Sciences, Federal University of ABC, Santo Andre, Brazil

Swati Singh
School of Environment and Natural Resources, Doon University, Dehradun, Uttarakhand, India

R. Siranjeevi
Department of Chemistry, Saveetha School of Engineering, Saveetha Institute of Medical and Technical Sciences, Saveetha University, Chennai, Tamil Nadu, India

Thiago Abdo de Souza
Center for Natural and Human Sciences, Federal University of ABC, Santo Andre, Brazil

R. Susmitha
Department of Biotechnology, Saveetha School of Engineering, Saveetha Institute of Medical and Technical Sciences, Saveetha University, Chennai, Tamil Nadu, India

Raphael Tiziani
Faculty of Agricultural, Environmental and Food Sciences, Free University of Bozen-Bolzano, Bolzano, Italy; Competence Centre of Plant Health, Free University of Bozen-Bolzano, Bolzano, Italy

Walter Ruggeri Waldman
Centre for Science and Technology for Sustainability, Sorocaba Campus, Federal University of São Carlos, São Carlos, Brazil

Jianxu Wang
State Key Laboratory of Environmental Geochemistry, Institute of Geochemistry, Chinese Academy of Sciences, Guiyang, China; Institute of Foundation Engineering, Water- and Waste-Management, Laboratory of Soil- and Groundwater-Management, Pauluskirchstraße, University of Wuppertal, School of Architecture and Civil Engineering, Wuppertal, Germany

Muhammad Younis
Department of Environmental Sciences, COMSATS University Islamabad, Vehari, Pakistan

About the editors

Golam Jalal Ahammed is a Distinguished Professor at the College of Horticulture and Plant Protection, Henan University of Science and Technology, Luoyang, China. Dr. Ahammed received his PhD in Olericulture in 2012 from Zhejiang University. Subsequently, he completed two consecutive postdoctoral programs at Zhejiang University, first at the Institute of Pesticide and Environmental Toxicology and then at the Institute of Crop Science. His primary research interests include plant stress physiology, phytohormone-mediated detoxification of organic pollutants, and biological control of plant diseases. Dr. Ahammed has authored over 200 peer-reviewed papers in internationally recognized SCI-indexed journals. He serves as a Senior Editorial Board Member for *Scientific Reports* and as an Associate Editor for the *Journal of Plant Growth Regulation*, *AoB Plants*, *BMC Plant Biology*, and *New Zealand Journal of Crop and Horticultural Science*. Dr. Ahammed was recognized as a Most Cited Chinese Researcher in Crop Science by Scopus (Elsevier) and as a Highly Cited Researcher in Plant and Animal Science by Web of Science (Clarivate) in 2024 and 2025. He has also been recognized among the top 2% of scientists worldwide for four consecutive years, according to Stanford University rankings published by Elsevier.

Karina S.B. Miglioranza received her BS and PhD degrees in Biology from the Mar del Plata University, Argentina. She is a Senior Researcher at the Institute of Marine and Coastal Research (IIMyC) that belongs to the National Scientific and Technical Research Council (CONICET), Argentina. Her expertise and main projects include studies on ecotoxicology and environmental pollution. The research is focused on the environmental dynamics and impacts of persistent organic pollutants, current-use pesticides, and emerging contaminants on biotic (plants, animals, humans) and abiotic (surface waters, rainfall, groundwater, air, soils, sediments) matrices. Moreover, she has expertise in risk analysis, environmental remediation, and phytoremediation of soils. Dr. Miglioranza has authored over 100 peer-reviewed papers in internationally renowned SCI-indexed journals. She serves as a Senior Editorial Board Member for the journals *Environmental Science and Technology*, *Environmental Science and Technology Letters*, and *Bulletin of Environmental Contamination and Toxicology,* and she is an Associate Editor of the *Journal of Hazardous Materials.* In 2018, she received the Capacity-Building Award from the World Council of the Society of Environmental Toxicology and Chemistry (SETAC), and in 2024, the Mention of L'Oreal UNESCO for Women in Science.

Preface

The increasing prevalence of organic pollutants in our environment raises a formidable challenge that compromises not only environmental health but also public safety. Organic pollutants, which include a wide range of carbon-containing compounds such as pesticides, pharmaceuticals, polycyclic aromatic hydrocarbons (PAHs), per- and polyfluoroalkyl substances (PFAS), persistent organic pollutants (POPs), and microplastics, persist in the environment and accumulate through food chains, leading to detrimental effects on human health and the environment as a whole. The complexity of these pollutants, their long half-lives, and their potential for bioaccumulation make addressing organic pollution an urgent and multifaceted concern.

Organic pollution manifests through a variety of pathways, predominantly originating from agricultural practices, industrial activities, urban runoff, and improper waste management. For instance, the overuse of pesticides leads to the leaching of harmful substances into soils, causing soil degradation and adversely affecting plant health and productivity. Organic pollution not only threatens crop production but also diminishes food quality, posing risks to human health through contaminated food sources. Moreover, the presence of residual pharmaceuticals in agricultural systems has been shown to induce antibiotic resistance among soil microorganisms, further complicating the challenges faced by modern agriculture. In this regard, all of these chemicals can reach aquatic environments through runoff or leaching into groundwater.

Plants can detoxify organic pollutants. However, the intricate processes by which organic pollutants affect plant physiology, as well as the innate detoxification mechanisms plants employ, are not fully understood. As we explore more deeply into the complex relationships between organic pollutants, soils, and plants, it becomes increasingly clear that effective remediation strategies, although essential for managing soil contamination, are highly challenging. Various remediation approaches have been developed over recent years, including bioremediation and phytoremediation. These strategies leverage the natural detoxification abilities of microorganisms and plants, thereby restoring the health of contaminated soils. However, critical knowledge gaps remain regarding the mechanisms of toxicity, the fate of organic contaminants in complex environments, and the efficacy of different remediation techniques under varying conditions.

The motivation for compiling this book stems from the urgent need to consolidate knowledge surrounding organic pollutants, their pathways in terrestrial and aquatic environments, and their implications for plant health and agricultural productivity. The book aims to serve as a comprehensive resource for researchers, policymakers, and practitioners alike and covers a broad spectrum of topics related to organic pollutants and phytotoxicity. As a whole, the book is organized into several chapters that systematically explore each aspect of organic pollutants, starting with definitions, then environmental fates, and finally impacts on plants and potential remediation strategies. The discussions extend to emerging contaminants, microplastics, and the potential for remediation through feasible techniques such as biochar application. Each chapter aims to enhance understanding, raise awareness, and inspire further research, ultimately informing policy and practices that mitigate the impacts of organic pollution.

Throughout this journey, it is crucial to recognize the contributions of chapter authors who have supported this work. We would also like to express our sincere gratitude to our fellow researchers and colleagues who provided invaluable insights and assistance in compiling this work. While minor revisions have been made, the authors' original vision has been preserved throughout the chapters. We acknowledge that there may still be areas for improvement, and we welcome feedback from readers to update future editions. Special thanks should also go to the publisher and in-house editors for their expertise and guidance during the production of this manuscript, ensuring high-quality content for readers.

Golam Jalal Ahammed
Luoyang, Henan, China

Karina S. B. Miglioranza
Mar del Plata, Argentina

CHAPTER 1

Introduction to organic pollutants and phytotoxicity

Golam Jalal Ahammed[1] and Karina S.B. Miglioranza[2]
[1]College of Horticulture and Plant Protection, Henan University of Science and Technology, Luoyang, China; [2]Laboratorio de Ecotoxicología y Contaminación Ambiental, Universidad Nacional de Mar del Plata (UNMdP) & Instituto de Investigaciones Marinas y Costeras (IIMyC-CONICET), Mar del Plata, Argentina

1. Introduction

Organic pollutants are a broad class of carbon-based chemical compounds introduced into the environment predominantly through human activities (Pathak et al., 2022; Ahammed and Li, 2022). This definition encompasses a diverse range of substances, including, but not limited to, pesticides, pharmaceuticals, industrial chemicals, personal care products, petroleum hydrocarbons, endocrine disruptors, antibiotics, and microplastics (Awino and Garland, 2025). Organic pollutants are ubiquitous across various ecosystems, including air, water, and soil. They can disrupt a wide range of biological and ecological processes (Zhao et al., 2022). The long-term persistence of some organic pollutants in the environment poses a serious threat since these substances can be accumulated in food chains, leading to a range of human health hazards, including endocrine disruption, reproductive problems, and a higher risk of cancer (Abdul Mutalib and Jaafar, 2022; De-la-Torre, 2019; Du et al., 2024).

The phytotoxicity of organic pollutants is a concern, threatening crop production globally (Haider et al., 2021). In the current context, phytotoxicity refers to the harmful effects of these toxic compounds on plant growth and development (Ahammed and Li, 2022). Organic pollutants, particularly at higher concentrations, have the potential to disrupt fundamental physiological processes of plants, such as photosynthesis, nutrient uptake, and cellular respiration, which may result in decreased agricultural productivity and threaten food security (de Freitas-Silva et al., 2022). Organic pollutants, including pesticides, pharmaceuticals, and microplastics, can alter soil properties, disrupt microbial communities, and thereby influence plant health (Anderson Abel de Souza et al., 2019; Cycoń et al., 2019). Phenotypically, severe phytotoxicity is reflected by reduced seed germination rates, leaf chlorosis, and stunted plant growth (Pathak et al., 2022; Ahammed and Li, 2022; de Freitas-Silva et al., 2022; Haider et al., 2021).

Here, we conducted a comprehensive search of the Web of Science Core Collection database (https://www.webofscience.com; accessed on April 3, 2025) to identify relevant studies. The search used the keywords "organic pollutant" and "phytotoxicity" in the topic field to capture articles on these topics published in the last 10 years, yielding

Phytotoxicity of Organic Pollutants
ISBN 978-0-443-33110-7,
https://doi.org/10.1016/B978-0-443-33110-7.00001-8

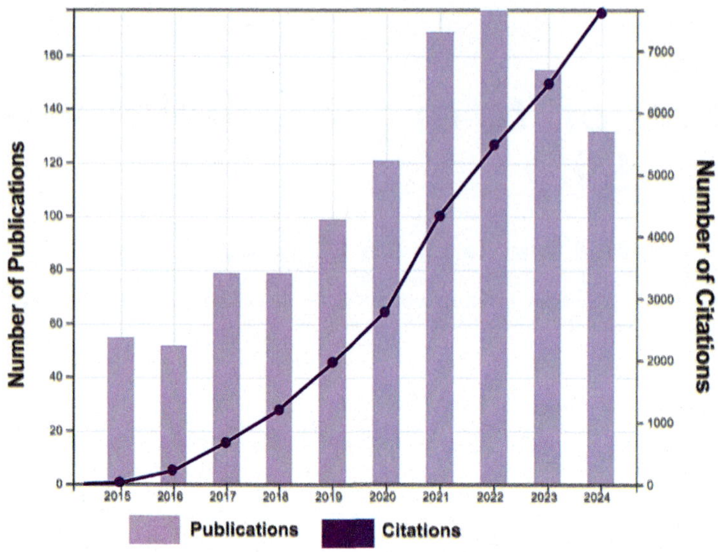

Figure 1.1 Number of publications and citations on "organic pollutant" and "phytotoxicity" from 2015 to 2024 according to the web of science core collection.

1118 articles with 32,575 citations. A steady rise in publications and citations from 2015 to 2024 has been observed (Fig. 1.1), highlighting the expanding research focus on organic pollution and its impact on plant health. Furthermore, a notable increase in both publications and citations over the past 5 years indicates growing academic interest and significance in these topics.

In this introductory chapter, we provide an overview of organic pollutants identified in the scientific literature. We also briefly discuss various types of organic pollutants, their phytotoxicity, detoxification mechanisms, and remediation strategies. Thus, the current chapter serves as a preliminary overview of organic pollutants and associated phytotoxicity, and lays the groundwork for subsequent chapters, which explore deeper into specific pollutants and their interactions with the ecological landscape, focusing on the plants.

2. Classifications of organic pollutants

Organic pollutants can be classified from multiple perspectives, including their sources, chemical structures, functional groups, environmental behavior, and other specific properties relevant to pollution and remediation strategies.

2.1 Classification based on source

Organic pollutants can be divided into two main categories based on their sources or origins, such as natural organic pollutants and anthropogenic organic pollutants.

2.1.1 *Natural organic pollutants*

Natural organic pollutants are substances that occur in the environment without direct human intervention. These compounds can originate from biological processes or natural events, such as volcanic eruptions or the decomposition of organic matter (Tomašek et al., 2021; Wang et al., 2022a). Natural hydrocarbons released during oil seepage and biogenic metabolites produced by microorganisms are examples of natural organic pollutants (Haider et al., 2021; Xu et al., 2021). Notably, many organic compounds produced by living organisms are typically more susceptible to microbial degradation than synthetic alternatives (Pathak et al., 2022; Bose et al., 2021). Although natural organic pollutants often have lower toxicity than synthetic pollutants, elevated concentrations can still be harmful (Tomašek et al., 2021; Lybrand et al., 2020). Nonetheless, natural organic pollutants can significantly impact air and water quality, as well as terrestrial and aquatic ecosystems, mainly when nutrient overloads occur.

2.1.2 *Anthropogenic organic pollutants*

Anthropogenic organic pollutants are synthetic compounds introduced into the environment through human activities (Wang et al., 2022a). This category encompasses a wide variety of substances, including pharmaceuticals, pesticides, plasticizers, and industrial chemicals (Awino and Garland, 2025). Anthropogenic organic pollutants often exhibit high toxicity, persistence, and the potential for bioaccumulation across food webs, leading to long-term ecological consequences (Pathak et al., 2022; Wang et al., 2020). Their interactions with biological organisms can disrupt metabolic pathways, resulting in numerous adverse health effects in both flora and fauna (Hassaan and El Nemr, 2020).

2.2 Classification based on chemical structure

Organic pollutants can also be categorized by their chemical structure, which significantly influences their behavior in environmental contexts.

2.2.1 *Hydrocarbons*

Hydrocarbons are organic compounds consisting solely of hydrogen (H) and carbon (C) (Haider et al., 2021). They can be categorized into aliphatic and aromatic hydrocarbons. Aliphatic hydrocarbons include straight-chain (alkanes) and branched-chain (alkenes) compounds. In contrast, aromatic hydrocarbons contain one or more benzene rings, such as polycyclic aromatic hydrocarbons (PAHs), which are derived from the incomplete combustion of organic matter, fossil fuels, and vehicle exhaust (Kumar et al., 2021). PAHs are among the primary organic pollutants in agricultural soils and can accumulate in edible plant parts, such as cucurbit fruits (Fujita and Inui, 2021; Sun et al., 2018). The hydrophobic nature of hydrocarbons makes them tightly adsorb to sediment particles and persist in the environment, especially in aquatic and sedimentary ecosystems, thereby influencing their bioavailability and potential toxicity to aquatic life (Vijayanand et al., 2023).

2.2.2 Oxygenated organic compounds

Various carbon-based compounds containing oxygen (O), including alcohols, ethers, aldehydes, and carboxylic acids, belong to this category. Natural and anthropogenic sources can contribute to the formation of oxygenated organic compounds. For example, ethanol is widely produced and used in industrial processes. Acetic acid used in consumer products can arise from natural fermentation processes as a byproduct, in addition to anthropogenic sources (Wang et al., 2021). Through their solubility and reactivity, oxygenated organic compounds can influence environmental health by altering the mobility of other pollutants (Bianchi et al., 2019). The atmospheric reaction of oxygenated volatile organic compounds (VOCs) can also lead to ozone formation (Wang et al., 2021).

2.2.3 Nitrogen-containing compounds

Many nitrogen-containing organic compounds, including amines and nitrates, are highly relevant to a range of aquatic and terrestrial environmental processes. They typically emanate from agricultural runoff, especially from fertilizer applications (Ditto et al., 2022). Eutrophication of water bodies can cause substantial ecological disruption. Nitrogen-containing organic compounds and their pathways are also essential to understand interactions with plants, since high concentrations can trigger phytotoxicity, affecting not only species diversity but also the high productivity of terrestrial and aquatic plants (Shilpha et al., 2023).

2.2.4 Halogenated organic compounds

Organic compounds belonging to this category contain halogen atoms (F, Cl, Br, I) (Fig. 1.2). Most halogenated organic pollutants have high persistence, toxicity, and bioaccumulation properties (Wang et al., 2020). Halogen atoms can increase the lipophilicity of halogenated organic pollutants, making them more likely to bioaccumulate in the fatty tissues of organisms, raising concerns about their impacts on food chains and human health (Popek, 2018; Munzel et al., 2023). Polychlorinated biphenyls (PCBs), dioxins, and certain pesticides belong to this category (Popek, 2018).

2.3 By environmental behavior

Organic pollutants can be categorized into two main groups based on their degradability: labile and recalcitrant. Labile organic pollutants are mainly found in wastewater and municipal solid waste, and aquatic environments, whereas recalcitrant organic pollutants typically originate from industrial waste streams or accidental discharges (Wang et al., 2022a). Labile organic pollutants are characterized by their high degradability. They

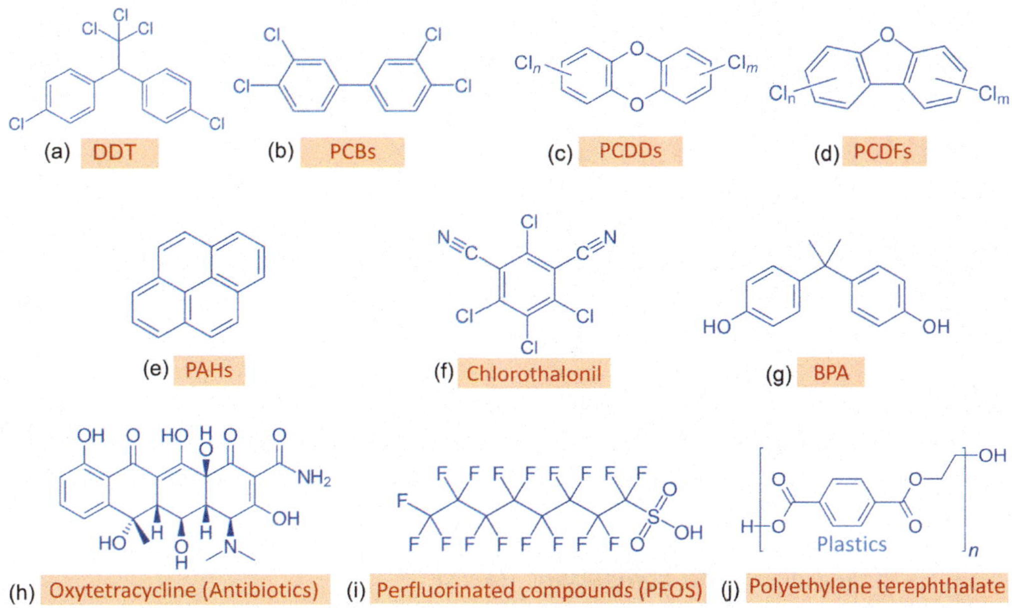

Figure 1.2 ***Chemical structures of various organic pollutants.*** Persistent organic pollutants, such as (a) organochlorine insecticides (dichlorodiphenyltrichloroethanes, DDT), (b) polychlorinated biphenyls (PCBs), (c) polychlorinated dibenzo-p-dioxins (PCDDs), (d) polychlorinated dibenzofurans (PCDFs), (e) polycyclic aromatic hydrocarbons (PAHs), (f) pesticides, (g) endocrine disrupting compounds (bisphenol A, BPA), (h) antibiotics (oxytetracycline), (i) perfluorinated compounds (perfluorooctane sulfonate, PFOS), and (j) plastics (polyethylene terephthalate).

pose a threat to surface water quality by facilitating the proliferation of both aerobic and anaerobic microorganisms. In contrast, recalcitrant organic pollutants are resistant to degradation. They are toxic and pose significant risks to human and ecological health (Sun et al., 2018). Recalcitrant organic pollutants, such as PCBs, can bioaccumulate within the food web, leading to detrimental consequences for organisms and ecosystems alike (Zhao et al., 2022). Thus, the environmental behavior of organic pollutants significantly impacts their fate and associated risks. The following classification is based on how different organic pollutants behave in environmental contexts, beyond their degradability.

2.3.1 Persistent pollutants

Persistent pollutants can resist environmental degradation (Awino and Garland, 2025). The environmental persistence of these pollutants poses serious health risks via the food chain (Arrieta-Cortes et al., 2017). Their characteristics features include low aqueous solubility and high lipophilicity, which facilitate their accumulation in living organisms (Pathak et al., 2022). Dioxins, PCBs, and certain organochlorine pesticides are typical examples of persistent pollutants (Popek, 2018). Depending on concentration and exposure duration, they can substantially affect plant growth and development, thereby reducing crop yield and productivity (Mitton et al., 2018; Haider et al., 2021).

2.3.2 Biodegradable organic pollutants

Biodegradable organic pollutants are chemical compounds that can be broken down by biological processes (Raffa and Chiampo, 2021; Su et al., 2019). These include many agricultural chemicals and household products, such as common pesticides and food waste, that are designed to decompose effectively (Bhandari et al., 2021). Because of their biodegradability, they pose a lower risk of long-term accumulation in the environment. However, improper disposal and excessive application can still lead to significant environmental impacts (Raffa and Chiampo, 2021; Liu and Zhu, 2020). Notably, certain environmental conditions can promote microbial communities capable of degrading these organic pollutants, and thus microbial degradation plays a vital role in their detoxification (Bose et al., 2021; Boudh et al., 2019).

2.3.3 Endocrine-disrupting compounds (EDCs)

Endocrine-disrupting compounds are organic pollutants that interfere with hormonal systems in animals and plants (Ishtiaq et al., 2021; Kanwar et al., 2020). A variety of substances, such as certain pesticides, pharmaceuticals, and plastic additives, such as bisphenol A (BPA) and phthalates, belong to this group. EDCs have the potential to disrupt reproductive and developmental processes in organisms, leading to harmful effects across ecosystems (Molina-Lopez et al., 2023). Understanding the mechanisms through which EDCs exert their effects is crucial for addressing their implications on plant and animal health (Molina-Lopez et al., 2023; Kanwar et al., 2020).

2.4 By physicochemical properties

Physicochemical properties of organic pollutants play a critical role in determining their behavior and fate in the environment. Solubility and volatility are two key physicochemical properties that are used to classify organic pollutants.

2.4.1 Solubility

The solubility of organic pollutants in water significantly influences their distribution and impact potential (Popek, 2018). Pollutants that are highly soluble (hydrophilic pollutants) can be readily transported in water bodies, thereby affecting aquatic ecosystems and terrestrial plants receiving polluted water (Pathak et al., 2022). Conversely, hydrophobic pollutants tend to adhere to sediments and particles, influencing their bioavailability for organisms (Boudh et al., 2019). Previous studies have shown that the solubility of pollutants can affect their adsorption to sediment and microplastics, further complicating their environmental fate (Pulitika et al., 2024).

2.4.2 Volatility

Organic pollutants can be categorized into two main groups based on their boiling points: volatile organic compounds (VOCs) and semivolatile organic compounds (SVOCs) (Popek, 2018). In qualitative terms, VOCs are organic substances that readily vapourize at room temperature, contributing to air pollution and potential health impacts through inhalation (Liu, 2023). The distinction between VOCs and SVOCs is not entirely clear-cut, as several substances can be classified within both categories (Popek, 2018). SVOCs include a wide variety of industrial chemicals, including pesticides, phenolic compounds, ketones, ethers, phthalate esters, anilines, pyridine derivatives, and some persistent organic pollutants (POPs) (Sun et al., 2018; Pathak et al., 2022). VOCs including some pesticides, benzene, formaldehyde, and various solvents that are used in industrial processes, present unique challenges for regulatory management and requires strategies that address both air quality and potential impacts on soil, plant, and human health (Backer et al., 2018; Haider et al., 2021).

3. Organic pollutant pathways into soil and aquatic ecosystems

Organic pollutants infiltrate soil and aquatic systems through multiple pathways, primarily driven by human activities, with significant ramifications for ecological health and human safety. Agricultural practices are one of the most prominent sources of organic pollutants in soil and water, particularly through the use of pesticides and fertilizers (Sun et al., 2018; Shilpha et al., 2023). They are introduced via direct application or through indirect routes, such as atmospheric deposition via wind dispersal (Hassaan and El Nemr, 2020) or leaching and runoff, thereby affecting soil health, microbial diversity, and aquatic ecosystems (Shilpha et al., 2023). Moreover, veterinary medicines (antibiotics and hormonal growth promoters) used in livestock production also enter soils and aquatic ecosystems, predominantly through manure application or improper disposal of unused drugs (Minden et al., 2017), leading to the development of antibiotic-resistant bacteria (Cycoń et al., 2019). Notably, antibiotics can be incorporated by edible plants (Minden et al., 2017). Industrial activities are another critical pathway for the entry of organic pollutants into the soil and aquatic ecosystems (Gautam et al., 2023). Industrial hazardous waste, including organic solvents and VOCs can contaminate soils and water through accidental spills, leaks, and improper waste-disposal practices (Haider et al., 2021; Gautam et al., 2023).

Urban runoff is a prevalent source of organic pollutants in many cities, as stormwater collects contaminants as it flows over impervious surfaces such as roads and pavements. This runoff can transport oils, heavy metals, and organic compounds into nearby soils and waterways (Modabberi et al., 2018). Wastewater discharge also plays a vital role in introducing organic pollutants into soil and water systems (Hechmi et al., 2024). Treated or untreated wastewater often contains a cocktail of organic chemicals, including pharmaceuticals and personal care products, which can be absorbed by agricultural soils if used for irrigation (Bauer et al., 2020). Another emerging concern is the contribution of microplastics to soil and water pollution (Tang, 2023; Hasan et al., 2024). Microplastics can enter soil and aquatic environments through various pathways, including the degradation of larger plastic debris and urban runoff (Ramanayaka et al., 2024). Studies indicate that microplastics can absorb organic contaminants from the environment, subsequently affecting plant health through direct uptake (Zhou et al., 2024).

4. Uptake of organic pollutants by plants

Organic pollutants can enter plants through various routes. One primary pathway is the transfer of pollutants from contaminated soil to the plant roots, where they are then transported through the xylem to different parts of the plant (Fig. 1.3). The xylem transports water from the roots to the leaves via transpiration. The degree to which an organic pollutant migrates into a plant's roots from contaminated soil is primarily influenced by the water solubility of the compound, its Henry's law constant, and its octanol-water partitioning coefficient (Kow) (Simonich and Hites, 1995; Popek, 2018). Other contributing factors include the organic matter present in the soil and the specific plant species (Fujita and Inui, 2021). For hydrophilic compounds, the predominant mechanism of accumulation is soil uptake. Following uptake from the soil, these compounds move from the outer to inner root, rapidly drawn into the plant via the xylem, and their lipophilicity influences their distribution within the plant. Moreover, as hydrophilic compounds are translocated throughout the plant, they are often subject to significant metabolic processes (Zhu et al., 2022).

The Cucurbitaceae family is known for its capacity to accumulate hydrophobic organic pollutants such as organochlorine insecticides, PCBs, and polychlorinated dibenzo-p-dioxins and dibenzofurans (PCDD/Fs), resulting in significantly elevated concentrations found in their fruits (Fujita and Inui, 2021). The mechanisms by which cucurbitaceous plants uptake organic pollutants encompass four key steps: (1) solubilization of organic pollutants, (2) absorption of organic pollutants into the root cells, (3) translocation of organic pollutants into xylem vessels, and (4) transport of organic pollutants to the above-ground plant parts and fruits (Fig. 1.4). In the rhizosphere, pollutants tend to bind to soil organic matter (SOM), but organic acids released from roots as exudates can disrupt these interactions, thereby enhancing pollutant desorption and bioavailability (Ren et al., 2018). Once inside the roots, the pollutants diffuse through the plasma membrane of root cells and move toward the endodermis and pericycle via plasmodesmata. Importantly, major latex-like proteins (MLPs) play a crucial role in the translocation of the absorbed organic pollutants into the xylem vessels (Fujita and Inui, 2021). MLPs bind to organic pollutants at the plasma membrane of the endodermis and pericycle, followed by translocation of MLP-organic pollutant complexes

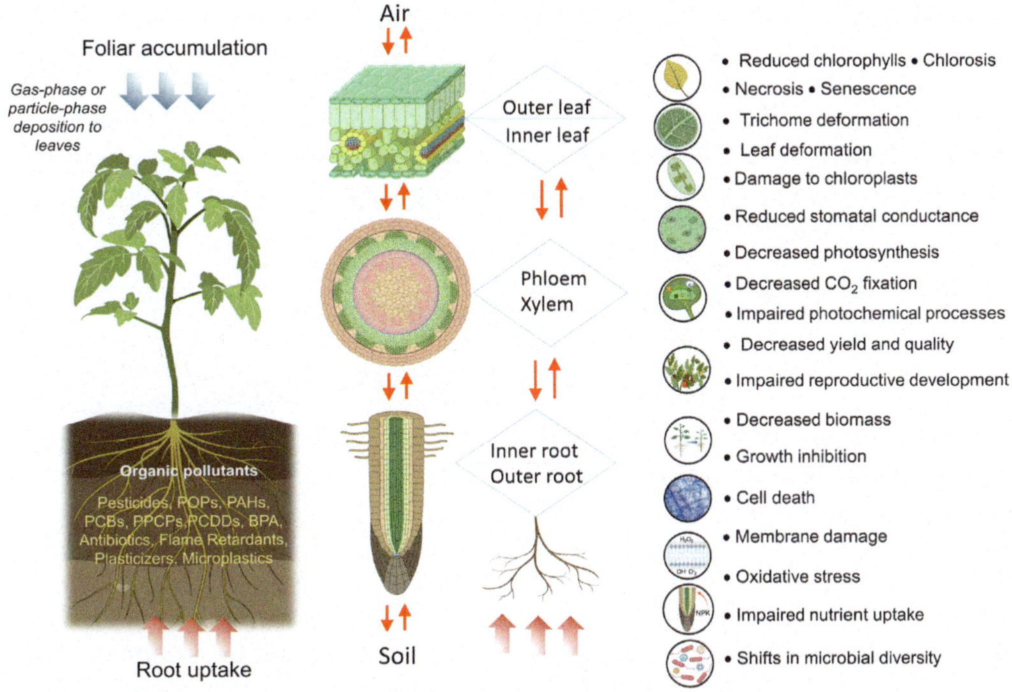

Figure 1.3 Uptake, accumulation and phytotoxicity of organic pollutants. *(Created with Biorender.com.)*

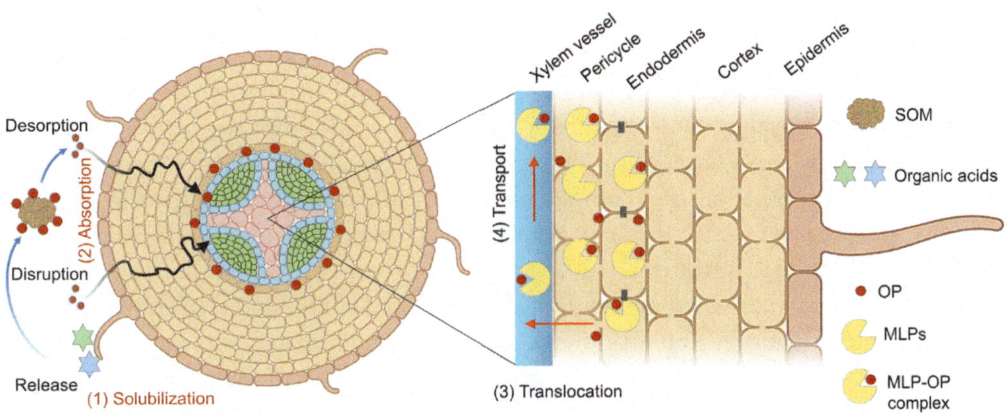

Figure 1.4 Root absorption of organic pollutants from the soil and mechanisms of translocation and transport to the aerial plant parts. *(Modified and redrawn from Fujita, K., Inui, H., 2021. How does the Cucurbitaceae family take up organic pollutants (POPs, PAHs, and PPCPs)? Rev. Environ. Sci. Biotechnol. 20 (3):751–779, https://doi.org/10.1007/s11157-021-09578-w, Created with Biorender.com.)*

into xylem vessels, ultimately facilitating the transport of organic pollutants to the fruits and other aerial parts of the plant. Additionally, organic pollutants can enter terrestrial plants from the atmosphere through gas–or particle-phase deposition onto cuticular wax on leaves, or via absorption through stomata, and are subsequently transported by the phloem, which distributes photosynthates to the roots and other plant tissues (Fig. 1.3). For lipophilic organic pollutants, the predominant accumulation route is from the atmosphere to the leaf surface.

5. Phytotoxicity and impacts of organic pollutants

Phytotoxicity, the toxic effect of pollutants on plants, is a major impact factor. Exposure to organic pollutants can inhibit root development and nutrient uptake, resulting in stunted growth and leaf chlorosis (Haider et al., 2021). Many organic pollutants disrupt essential physiological processes, particularly photosynthesis and respiration, through non-specific interference with cellular functions (Fig. 1.3). For example, certain pesticides, such as glyphosate, inhibit the shikimic acid pathway, affecting amino acid biosynthesis, which is essential for plant growth and development, leading to reduced biomass and vitality (de Freitas-Silva et al., 2022). Organic pollutants such as PAHs, PCBs and BPA can induce oxidative stress by generating reactive oxygen species (ROS) that damage cellular components, including proteins, lipids, and nucleic acids (Ahammed et al., 2020; Kanwar et al., 2020; Pathak et al., 2022). Furthermore, pollutants can reduce stomatal conductance, impairing gas exchange and reducing photosynthetic efficiency (de Freitas-Silva et al., 2022).

In addition to direct toxicity, organic pollutants can also affect plant health by altering soil microbial communities (Anderson Abel de Souza et al., 2019, Cycoń et al., 2019). For instance, the presence of certain antibiotics in soil can alter microbial diversity, affecting nutrient cycling and availability, ultimately impacting plant growth (Cycoń et al., 2019; Minden et al., 2017). The interplay between organic pollutants and soil biota introduces another layer of complexity to understanding phytotoxicity, as a healthy soil microbiome plays a crucial role in helping plants acquire nutrients and resist diseases (Munzel et al., 2023; Pathak et al., 2022).

Furthermore, widely used agrochemicals may have unintended consequences when they leach into surrounding environments, often harming non-target plant species (Nazish et al., 2022). Research indicates that herbicide application can lead to biodiversity loss within natural habitats when residues persist in soils, affecting susceptible plant populations and altering ecological dynamics (Pathak et al., 2022; Ren et al., 2018). Contaminated plants, whether through direct uptake from soil, water, or air, or via bioaccumulation of pollutants within the food chain, pose risks to consumers (Munzel et al., 2023). Recent studies have highlighted the presence of pesticide residues and pharmaceuticals in food supplies, magnifying the need for monitoring and regulatory measures to safeguard public health (Zhang and Yang, 2021; Awino and Garland, 2025; Yee et al., 2021).

6. Mechanisms of detoxification of organic pollutants

Plants have evolved sophisticated innate detoxification mechanisms to mitigate the impacts of organic pollutants. Detoxification of organic pollutants, particularly pesticides, typically occurs in plants through three key phases: Phase I, Phase II, and Phase III (Zhang and Yang, 2021). In Phase I, oxidative enzymes such as cytochrome P450 monooxygenases modify the pollutants by adding reactive groups, which enhances their solubility and reduces toxicity. This is followed by Phase II, during which enzymes such as glutathione S-transferases (GSTs) conjugate reactive metabolites to endogenous compounds, further increasing their solubility and facilitating sequestration (Yu et al., 2022a). Glutathione (GSH) plays a critical role in phase II reaction (Yu and Ahammed, 2025). Upon exposure to organic pollutants, plants typically enhance GSH accumulation through biosynthesis from cysteine and regeneration from its oxidized form (GSSG) (Yu et al., 2013). Then GSH makes conjugates with organic pollutants (Kanwar et al., 2020; Yu et al., 2022b). Finally, Phase III involves ATP-binding cassette (ABC) transporters that actively transport these conjugated products to the vacuole or out of the plant cells, thus attenuating the toxic effects of organic pollutants (Zhang and Yang, 2021; Ahammed and Li, 2022).

Mechanisms of enhanced tolerance to organic pollutants also involve the synthesis of various secondary metabolites and the activation of antioxidant defense systems, which scavenge ROS generated during pollutant exposure, thereby protecting against oxidative stress (Cesco et al., 2021). The induction of these detoxification pathways highlights plants' resilience to environmental stressors; however, the capacity to detoxify varies widely among plant species, depending on genetic and physiological traits (Zhang and Yang, 2021). Nonetheless, certain plants can tolerate high levels of specific organic pollutants through adaptive physiological changes. Moreover, some plant species alter root exudation patterns, thereby promoting the growth of beneficial microbial communities that can assist in degrading pollutants (Fujita and Inui, 2021).

The physiological responses and adaptations of plants to organic pollutants involve intricate interactions across multiple biochemical pathways (Ahammed et al., 2017). One significant aspect is the alteration of phytohormone levels, which are crucial regulators of plant growth and stress response (Cesco et al., 2021). Certain phytohormones participate in the detoxification of organic pollutants, particularly pesticides, across various plant species (Wang et al., 2022b). For example, brassinosteroids notably enhance detoxification processes of organophosphorus, organochlorine, and carbamate pesticides, reducing pesticide residues by 30%–70% in crops like *Solanum lycopersicum* (tomato) and *Cucumis sativus* (cucumber), by promoting metabolic activity and bolstering antioxidant defenses (Zhou et al., 2015; Hou et al., 2019). Likewise, salicylic acid and jasmonic acid have been linked to improved detoxification mechanisms, enabling plants to better cope with pesticide stress (Yu et al., 2021, 2022b). Melatonin also enhances plants' antioxidant capacity and detoxification potential, thereby reducing phytotoxicity (Peng et al., 2023). These findings illustrate the critical role of phytohormones not only in mitigating the harmful effects of organic pollutants but also in reducing pesticide residues, highlighting their potential to ensure food safety.

7. Potential remediation strategies

Remediation strategies for organic pollutants can broadly be classified into physical, chemical, and biological techniques (Sun et al., 2018). Physical remediation methods often involve removing pollutants from contaminated sites, primarily through excavation, soil washing, or thermal desorption (Haider et al., 2021). However, these techniques can be costly and logistically challenging, especially for large-scale applications. Chemical remediation, on the other hand, uses chemical reactions to stabilize or neutralize pollutants (Abdul Mutalib and Jaafar, 2022). This may include oxidation, reduction, or precipitation processes designed to detoxify contaminants before they are released back into the environment. While effective in certain scenarios, these methods may also pose risks associated with introducing new hazardous substances and potentially incomplete degradation of the original contaminants (Morsi et al., 2021).

In nature, organic pollutants can be attenuated via degradation and humification (Wang et al., 2022a). As a coexisting process with humification in natural environments, biodegradation of organic pollutants may yield a range of intermediates and final products that serve as precursors for humification. Many pathways have been identified for the generation of humic substances (HS), in which microbes play key roles in providing HS precursors and in subsequent polymerization. Humic substances, such as humic acids, fulvic acids, and humins, have been shown to enhance plant growth, stress tolerance and soil microbial activity. Bioremediation is increasingly recognized as a promising approach to remove organic pollutants from the environment, leveraging biological processes mediated by microorganisms, plants, or their interactions (Pathak et al., 2022; Raffa and Chiampo, 2021). Microbial bioremediation has been widely applied, leveraging the metabolic activities of bacteria and fungi to degrade contaminants in soil and water (Boudh et al., 2019). For instance, the bacterium *Pseudomonas putida* has shown remarkable capabilities in degrading a wide range of organic solvents and hydrocarbons, making it a valuable candidate for bioremediation applications (Han et al., 2024). Additionally, certain fungi exhibit efficient lignin-degrading capabilities through enzyme production, enabling the breakdown of complex organic pollutants like phenols (Athamneh et al., 2022). The adaptation of these microorganisms provides a flexible and effective means of addressing pollution across a variety of settings, from contaminated industrial sites to agricultural areas.

While microbial approaches provide significant advantages, phytoremediation, the use of plants to absorb, accumulate, and detoxify pollutants, offers complementary benefits (Haider et al., 2021). Phytoremediation involves using plants' natural abilities to take up, degrade, or stabilize contaminants in soil and water (An et al., 2020). Plant species capable of accumulating organic pollutants can be employed in contaminated sites to mitigate environmental risks (Haider et al., 2021). However, the effectiveness of phytoremediation can be influenced by various environmental factors, including soil conditions and climatic conditions. Plant species that show enhanced tolerance to organic pollutants and also can absorb contaminants from the soil and transport them to their aerial parts, can be used for phytoextraction. This strategy not only removes pollutants but also improves soil health and increases biodiversity, offering ecological and esthetic benefits to degraded landscapes. Additionally, plants can enhance microbial activity in the rhizosphere, promoting synergistic interactions that can further accelerate the degradation of harmful substances (Pathak et al., 2022).

Despite the promise of bioremediation and phytoremediation, challenges remain. The time required for biological processes can be significantly longer than for physical and chemical methods, making it crucial to optimize conditions for microbial activity and plant growth (Haider et al., 2021; Raffa and Chiampo, 2021). The integration of genetic engineering techniques, such as isolating and enhancing the traits of microorganisms or plants that facilitate pollutant degradation, can provide avenues for accelerating remediation processes (Su et al., 2019; Bhandari et al., 2021). Moreover, the scale of these methods can range from laboratory trials to full-field applications, necessitating careful evaluation of results before widespread implementation.

8. Summary and outlook

The impacts of organic pollutants are profound, as evidenced by the adverse effects on plant physiology. They interfere with essential processes such as photosynthesis and nutrient uptake, leading to diminished growth and crop yields, and in turn, threatening food security. As noted, the uptake mechanisms for organic pollutants rely heavily on their physicochemical properties, such as lipophilicity and water solubility, as well as on environmental conditions and the specific traits of plant species. The persistence of these chemicals in the environment raises concerns about their role in disrupting ecosystems and affecting biodiversity.

Given the tenacious need to address the challenges posed by organic pollutants, this book explores various types of organic pollutants, their environmental fate, and their specific implications for phytotoxicity. This book is compiled to provide a comprehensive update on organic pollutant research, with an emphasis on their impact not only on plant health but also on broader environmental health and food safety. Each chapter explores specific types of organic pollutants, from POPs to emerging organic pollutants, including microplastics and pharmaceuticals. The integration of insights from recent studies helps elucidate the intricacies of how these pollutants function within ecosystems and associate with different biota, particularly plants, ultimately providing valuable information for addressing environmental pollution and sustainable crop production. The insights gathered throughout this book aim to contribute to a greater understanding of the environmental issues related to organic pollutants, fostering not only academic discussions but also practical approaches to mitigate their negative impacts. We invite readers to explore the subsequent chapters, which offer richer, more detailed discussions of specific classes of organic pollutants and their impacts on plant and human health.

References

Abdul Mutalib, A.A., Jaafar, N.F., 2022. ZnO photocatalysts applications in abating the organic pollutant contamination: a mini review. Total Environ. Res. Themes 3–4. https://doi.org/10.1016/j.totert.2022.100013.

Ahammed, G.J., He, B.B., Qian, X.J., Zhou, Y.H., Shi, K., Zhou, J., Yu, J.Q., Xia, X.J., 2017. 24-Epibrassinolide alleviates organic pollutants-retarded root elongation by promoting redox homeostasis and secondary metabolism in *Cucumis sativus* L. Environ. Pollut. 229, 922–931. https://doi.org/10.1016/j.envpol.2017.07.076.

Ahammed, G.J., Li, X., 2022. Melatonin-induced detoxification of organic pollutants and alleviation of phytotoxicity in selected horticultural crops. Horticulturae 8 (12), 1142.

Ahammed, G.J., Wang, Y., Mao, Q., Wu, M., Yan, Y., Ren, J., Wang, X., Liu, A., Chen, S., 2020. Dopamine alleviates bisphenol A-induced phytotoxicity by enhancing antioxidant and detoxification potential in cucumber. Environ. Pollut. 259. https://doi.org/10.1016/j.envpol.2020.113957.

An, Y., Wang, Y., Tan, S.N., Yusof, M.L.M., Ghosh, S., Chen, Z., 2020. Phytoremediation: a promising approach for revegetation of heavy metal-polluted land. Front. Plant Sci. 11. https://doi.org/10.3389/fpls.2020.00359.

Anderson Abel de Souza, M., Lau, C.W., Kloas, W., Bergmann, J., Bachelier, J., Faltin, E., Becker, R., Görlich, A.S., Rillig, M.C., 2019. Microplastics can change soil properties and affect plant performance. Environ. Sci. Technol. 53 (10), 6044–6052. https://doi.org/10.1021/acs.est.9b01339.

Arrieta-Cortes, R., Farias, P., Hoyo-Vadillo, C., Kleiche-Dray, M., 2017. Carcinogenic risk of emerging persistent organic pollutant perfluorooctane sulfonate (PFOS): a proposal of classification. Regul. Toxicol. Pharmacol. 83, 66–80. https://doi.org/10.1016/j.yrtph.2016.11.021.

Athamneh, K., Alneyadi, A., Alsadik, A., Wong, T.S., Ashraf, S.M., 2022. Efficient degradation of various emerging pollutants by wild type and evolved fungal DyP4 peroxidases. PLoS One 17 (1), e0262492. https://doi.org/10.1371/journal.pone.0262492.

Awino, F.B., Garland, G., 2025. Occurrence of emerging and persistent organic pollutants in dumpsite environments: a review. Environ. Chall. 18. https://doi.org/10.1016/j.envc.2025.101094.

Backer, R., Ghidotti, M., Schwinghamer, T., Saeed, W., Grenier, C., Dion-Laplante, C., Fabbri, D., Dutilleul, P., Séguin, P., Smith, D.L., 2018. Getting to the root of the matter: water-soluble and volatile components in thermally-treated biosolids and biochar differentially regulate maize (*Zea mays*) seedling growth. PLoS One 13 (11), e0206924. https://doi.org/10.1371/journal.pone.0206924.

Bauer, S., Linke, H.J., Wagner, M., 2020. Optimizing water-reuse and increasing water-saving potentials by linking treated industrial and municipal wastewater for a sustainable urban development. Water Sci. Technol. 81 (9), 1927–1940. https://doi.org/10.2166/wst.2020.257.

Bhandari, S., Poudel, D.K., Marahatha, R., Dawadi, S., Khadayat, K., Phuyal, S., Shrestha, S., Gaire, S., Basnet, K., Khadka, U., Parajuli, N., 2021. Microbial enzymes used in bioremediation. J. Chem. 2021, 1–17. https://doi.org/10.1155/2021/8849512.

Bianchi, F., Kurten, T., Riva, M., Mohr, C., Rissanen, M.P., Roldin, P., Berndt, T., Crounse, J.D., Wennberg, P.O., Mentel, T.F., Wildt, J., Junninen, H., Jokinen, T., Kulmala, M., Worsnop, D.R., Thornton, J.A., Donahue, N., Kjaergaard, H.G., Ehn, M., 2019. Highly oxygenated organic molecules

(HOM) from gas-phase autoxidation involving peroxy radicals: a key contributor to atmospheric aerosol. Chem. Rev. 119 (6), 3472–3509. https://doi.org/10.1021/acs.chemrev.8b00395.

Bose, S., Kumar, P.S., Vo, D.N., 2021. A review on the microbial degradation of chlorpyrifos and its metabolite TCP. Chemosphere 283, 131447. https://doi.org/10.1016/j.chemosphere.2021.131447.

Boudh, S., Singh, J.S., Chaturvedi, P., 2019. Microbial resources mediated bioremediation of persistent organic pollutants. In: New and Future Developments in Microbial Biotechnology and Bioengineering, pp. 283–294. https://doi.org/10.1016/b978-0-12-818258-1.00019-4.

Cesco, S., Lucini, L., Miras-Moreno, B., Borruso, L., Mimmo, T., Pii, Y., Puglisi, E., Spini, G., Taskin, E., Tiziani, R., Zangrillo, M.S., Trevisan, M., 2021. The hidden effects of agrochemicals on plant metabolism and root-associated microorganisms. Plant Sci. 311, 111012. https://doi.org/10.1016/j.plantsci.2021.111012.

Cycoń, M., Mrozik, A., Piotrowska-Seget, Z., 2019. Antibiotics in the soil environment—degradation and their impact on microbial activity and diversity. Front. Microbiol. 10. https://doi.org/10.3389/fmicb.2019.00338.

De-la-Torre, G.E., 2019. Microplastics: an emerging threat to food security and human health. J. Food Sci. Technol. 57 (5), 1601–1608. https://doi.org/10.1007/s13197-019-04138-1.

de Freitas-Silva, L., Araujo, H.H., Meireles, C.S., da Silva, L.C., 2022. Plant exposure to glyphosate-based herbicides and how this might alter plant physiological and structural processes. Botany 100 (6), 473–480. https://doi.org/10.1139/cjb-2021-0033.

Ditto, J.C., Machesky, J., Gentner, D.R., 2022. Analysis of reduced and oxidized nitrogen-containing organic compounds at a coastal site in summer and winter. Atmos. Chem. Phys. 22 (5), 3045–3065. https://doi.org/10.5194/acp-22-3045-2022.

Du, J., Huang, W., Pan, Y., Xu, S., Li, H., Jin, M., Liu, Q., 2024. Ecotoxicological effects of microplastics combined with antibiotics in the aquatic environment: recent developments and prospects. Environ. Toxicol. Chem. 43 (9), 1950–1961. https://doi.org/10.1002/etc.5950.

Fujita, K., Inui, H., 2021. How does the Cucurbitaceae family take up organic pollutants (POPs, PAHs, and PPCPs)? Rev. Environ. Sci. Biotechnol. 20 (3), 751–779. https://doi.org/10.1007/s11157-021-09578-w.

Gautam, K., Sharma, P., Dwivedi, S., Singh, A., Gaur, V.K., Varjani, S., Srivastava, J.K., Pandey, A., Chang, J.S., Ngo, H.H., 2023. A review on control and abatement of soil pollution by heavy metals: emphasis on artificial intelligence in recovery of contaminated soil. Environ. Res. 225, 115592. https://doi.org/10.1016/j.envres.2023.115592.

Haider, F.U., Ejaz, M., Cheema, S.A., Khan, M.I., Zhao, B., Liqun, C., Salim, M.A., Naveed, M., Khan, N., Nunez-Delgado, A., Mustafa, A., 2021. Phytotoxicity of petroleum hydrocarbons: sources, impacts and remediation strategies. Environ. Res. 197, 111031. https://doi.org/10.1016/j.envres.2021.111031.

Han, S., Kim, D., Kim, Y.-S., Yoon, S.H., 2024. Genome-scale metabolic network model and phenome of solvent-tolerant *Pseudomonas putida* S12. BMC Genom. 25 (1). https://doi.org/10.1186/s12864-023-09940-y.

Hasan, M.K., Jahan, I., Suravi, T.I., Al Hasib Imon, M.A., Shopan, J., Ahammed, G.J., 2024. Heavy metals and microplastics derived from laboratory effluents enhance toxicological risks to the ecosystems of canals in Bangladesh. Environ. Sci. Pollut. Res. Int. https://doi.org/10.1007/s11356-024-35758-6.

Hassaan, M.A., El Nemr, A., 2020. Pesticides pollution: classifications, human health impact, extraction and treatment techniques. Egypt. J. Aquat. Res. 46 (3), 207–220. https://doi.org/10.1016/j.ejar.2020.08.007.

Hechmi, S., Bhat, M.A., Kallel, A., Khiari, O., Louati, Z., Khelil, M.N., Zoghlami, R.I., Cherni, Y., Melki, S., Trabelsi, I., Jedidi, N., 2024. Soil contamination with microplastics (MPs) from treated wastewater and sewage sludge: risks and sustainable mitigation strategies. Disc. Environ. 2 (1). https://doi.org/10.1007/s44274-024-00135-0.

Hou, J., Sun, Q., Li, J., Ahammed, G.J., Yu, J., Fang, H., Xia, X., 2019. Glutaredoxin S25 and its interacting TGACG motif-binding factor TGA2 mediate brassinosteroid-induced chlorothalonil metabolism in tomato plants. Environ. Pollut. https://doi.org/10.1016/j.envpol.2019.113256.

Ishtiaq, A., Ali, T., Bakhtiar, A., Bibi, R., Bibi, K., Mushtaq, I., Li, S., Khan, W., Khan, U., Anis, R.A., Anees, M., Sultan, A., Murtaza, I., 2021. Melatonin abated bisphenol A-induced neurotoxicity via p53/PUMA/Drp-1 signaling. Environ. Sci. Pollut. Res. Int. 28 (14), 17789–17801. https://doi.org/10.1007/s11356-020-12129-5.

Kanwar, M.K., Xie, D., Yang, C., Ahammed, G.J., Qi, Z., Hasan, M.K., Reiter, R.J., Yu, J.Q., Zhou, J., 2020. Melatonin promotes metabolism of bisphenol A by enhancing glutathione-dependent detoxification in *Solanum lycopersicum* L. J. Hazard Mater. 388, 121727. https://doi.org/10.1016/j.jhazmat.2019.121727.

Kumar, M., Bolan, N.S., Hoang, S.A., Sawarkar, A.D., Jasemizad, T., Gao, B., Keerthanan, S., Padhye, L. P., Singh, L., Kumar, S., Vithanage, M., Li, Y., Zhang, M., Kirkham, M.B., Vinu, A., Rinklebe, J., 2021. Remediation of soils and sediments polluted with polycyclic aromatic hydrocarbons: to immobilize, mobilize, or degrade? J. Hazard Mater. 420, 126534. https://doi.org/10.1016/j.jhazmat.2021.126534.

Liu, N., Zhu, L., 2020. Metabolomic and transcriptomic investigation of metabolic perturbations in *Oryza sativa* L. triggered by three pesticides. Environ. Sci. Technol. 54 (10), 6115–6124. https://doi.org/10.1021/acs.est.0c00425.

Liu, Y., 2023. In-situ online detection of atmospheric volatile organic compounds based on laser induced breakdown spectroscopy: a review. Atom. Spectros 44 (03), 178–190. https://doi.org/10.46770/as.2023.117.

Lybrand, D.B., Xu, H., Last, R.L., Pichersky, E., 2020. How plants synthesize pyrethrins: safe and biodegradable insecticides. Trends Plant Sci. 25 (12), 1240–1251. https://doi.org/10.1016/j.tplants.2020.06.012.

Minden, V., Deloy, A., Volkert, A.M., Leonhardt, S.D., Pufal, G., 2017. Antibiotics impact plant traits, even at small concentrations. Aob Plants 9 (2). https://doi.org/10.1093/aobpla/plx010.

Mitton, F.M., Gonzalez, M., Monserrat, J.M., Miglioranza, K.S., 2018. DDTs-induced antioxidant responses in plants and their influence on phytoremediation process. Ecotoxicol. Environ. Saf. 147, 151–156.

Modabberi, S., Tashakor, M., Sharifi Soltani, N., Hursthouse, A.S., 2018. Potentially toxic elements in urban soils: source apportionment and contamination assessment. Environ. Monit. Assess. 190 (12), 715. https://doi.org/10.1007/s10661-018-7066-8.

Molina-Lopez, A.M., Bujalance-Reyes, F., Ayala-Soldado, N., Mora-Medina, R., Lora-Benitez, A., Moyano-Salvago, R., 2023. An overview of the health effects of bisphenol A from a one health perspective. Animals (Basel) 13 (15). https://doi.org/10.3390/ani13152439.

Morsi, R., Al-Maqdi, K.A., Bilal, M., Iqbal, H.M., Khaleel, A., Shah, I., Ashraf, S.M., 2021. Immobilized soybean peroxidase hybrid biocatalysts for efficient degradation of various emerging pollutants. Biomolecules 11 (6), 904. https://doi.org/10.3390/biom11060904.

Munzel, T., Hahad, O., Daiber, A., Landrigan, P.J., 2023. Soil and water pollution and human health: what should cardiologists worry about? Cardiovasc. Res. 119 (2), 440–449. https://doi.org/10.1093/cvr/cvac082.

Nazish, T., Huang, Y.J., Zhang, J., Xia, J.Q., Alfatih, A., Luo, C., Cai, X.T., Xi, J., Xu, P., Xiang, C.B., 2022. Understanding paraquat resistance mechanisms in *Arabidopsis thaliana* to facilitate the development of paraquat-resistant crops. Plant Commun. 3 (3), 100321. https://doi.org/10.1016/j.xplc.2022.100321.

Pathak, V.M., Verma, V.K., Rawat, B.S., Kaur, B., Babu, N., Sharma, A., Dewali, S., Yadav, M., Kumari, R., Singh, S., Mohapatra, A., Pandey, V., Rana, N., Cunill, J.M., 2022. Current status of pesticide effects on environment, human health and it's eco-friendly management as bioremediation: a comprehensive review. Front. Microbiol. 13, 962619. https://doi.org/10.3389/fmicb.2022.962619.

Peng, X., Wang, N., Sun, S., Geng, L., Guo, N., Liu, A., Chen, S., Ahammed, G.J., 2023. Reactive oxygen species signaling is involved in melatonin-induced reduction of chlorothalonil residue in tomato leaves. J. Hazard Mater. 443. https://doi.org/10.1016/j.jhazmat.2022.130212.

Popek, E., 2018. Environmental chemical pollutants. In: Sampling and Analysis of Environmental Chemical Pollutants, pp. 13–69. https://doi.org/10.1016/b978-0-12-803202-2.00002-1.

Pulitika, A., Karamanis, P., Kovačić, M., Božić, A.L., Kušić, H., 2024. An atomic-level perspective on the interactions between organic pollutants and PET particles: a comprehensive computational investigation. ChemPhysChem 25 (5). https://doi.org/10.1002/cphc.202300854.

Raffa, C.M., Chiampo, F., 2021. Bioremediation of agricultural soils polluted with pesticides: a review. Bioengineering (Basel) 8 (7). https://doi.org/10.3390/bioengineering8070092.

Ramanayaka, S., Zhang, H., Semple, K.T., 2024. Environmental fate of microplastics and common polymer additives in non-biodegradable plastic mulch applied agricultural soils. Environ. Pollut. 363 (Pt. 2), 125249. https://doi.org/10.1016/j.envpol.2024.125249.

Ren, X., Zeng, G., Tang, L., Wang, J., Wan, J., Liu, Y., Yu, J., Yi, H., Ye, S., Deng, R., 2018. Sorption, transport and biodegradation—an insight into bioavailability of persistent organic pollutants in soil. Sci. Total Environ. 610–611, 1154–1163. https://doi.org/10.1016/j.scitotenv.2017.08.089.

Shilpha, J., Song, J., Jeong, B.R., 2023. Ammonium phytotoxicity and tolerance: an insight into ammonium nutrition to improve crop productivity. Agronomy 13 (6), 1487.

Simonich, S., Hites, R., 1995. Organic pollutant accumulation in vegetation. Environ. Sci. Technol. 29, 2905–2914. https://doi.org/10.1021/es00012a004.

Su, X.N., Zhang, J.J., Liu, J., Zhang, N., Li, Y., Lu, F., Chen, Z.J., Shi, Z., Jing, W., Liu, C., Yang, H., 2019. Biodegrading two pesticide residues in paddy plants and the environment by a genetically engineered approach. J. Agric. Food Chem. 67 (17), 4947–4957. https://doi.org/10.1021/acs.jafc.8b07251.

Sun, J., Pan, L., Tsang, D.C.W., Zhan, Y., Zhu, L., Li, X., 2018. Organic contamination and remediation in the agricultural soils of China: a critical review. Sci. Total Environ. 615, 724–740. https://doi.org/10.1016/j.scitotenv.2017.09.271.

Tang, K.H.D., 2023. Microplastics in agricultural soils in China: sources, impacts and solutions. Environ. Pollut. 322, 121235. https://doi.org/10.1016/j.envpol.2023.121235.

Tomašek, I., Damby, D.E., Andronico, D., Baxter, P.J., Boonen, I., Claeys, P., Denison, M.S., Horwell, C. J., Kervyn, M., Kueppers, U., Romanias, M.N., Elskens, M., 2021. Assessing the biological reactivity of organic compounds on volcanic ash: implications for human health hazard. Bull. Volcanol. 83 (5). https://doi.org/10.1007/s00445-021-01453-4.

Vijayanand, M., Ramakrishnan, A., Subramanian, R., Issac, P.K., Nasr, M., Khoo, K.S., Rajagopal, R., Greff, B., Wan Azelee, N.I., Jeon, B.H., Chang, S.W., Ravindran, B., 2023. Polyaromatic hydrocarbons (PAHs) in the water environment: a review on toxicity, microbial biodegradation, systematic biological advancements, and environmental fate. Environ. Res. 227, 115716. https://doi.org/10.1016/j.envres.2023.115716.

Wang, H., Xue, S., Hao, R., Fang, L., Nie, L., 2021. Emission characteristics and ozone formation potential assessment of VOCs from typical metal packaging plants. Atmosphere 13 (1), 57. https://doi.org/10.3390/atmos13010057.

Wang, S., Wang, Y., He, X., Lu, Q., 2022a. Degradation or humification: rethinking strategies to attenuate organic pollutants. Trends Biotechnol. 40 (9), 1061–1072. https://doi.org/10.1016/j.tibtech.2022.02.007.

Wang, T., Wang, L., Chen, Q., Kalogerakis, N., Ji, R., Ma, Y., 2020. Interactions between microplastics and organic pollutants: effects on toxicity, bioaccumulation, degradation, and transport. Sci. Total Environ. 748. https://doi.org/10.1016/j.scitotenv.2020.142427.

Wang, X.D., Zhang, C.Y., Yuan, Y., Hua, Y.F., Asami, T., Qin, Y., Xiong, X.H., Zhu, J.L., Lu, Y.C., 2022b. Molecular responses and degradation mechanisms of the herbicide Diuron in rice crops. J. Agric. Food Chem. https://doi.org/10.1021/acs.jafc.2c05142.

Xu, D., Xue, M., Shen, Z., Jia, X., Hou, X., Lai, D., Zhou, L., 2021. Phytotoxic secondary metabolites from fungi. Toxins 13 (4). https://doi.org/10.3390/toxins13040261.

Yee, M.S., Hii, L.W., Looi, C.-K., Lim, W.M., Wong, S.F., Kok, Y.-Y., Tan, B.-K., Wong, C.-Y., Leong, C.O., 2021. Impact of microplastics and nanoplastics on human health. Nanomaterials 11 (2), 496. https://doi.org/10.3390/nano11020496.

Yu, G., Chen, Q., Chen, F., Liu, H., Lin, J., Chen, R., Ren, C., Wei, J., Zhang, Y., Yang, F., Sheng, Y., 2022a. Glutathione promotes degradation and metabolism of residual fungicides by inducing UDP-glycosyltransferase genes in tomato. Front. Plant Sci. 13. https://doi.org/10.3389/fpls.2022.893508.

Yu, G.B., Ahammed, G.J., 2025. Detoxification of pesticide residue for safe vegetable production. In: Ahammed, G.J., Zhou, J. (Eds.), Growth Regulation and Quality Improvement of Vegetable Crops. Springer, Singapore. https://doi.org/10.1007/978-981-96-0169-1_14.

Yu, G.B., Chen, R.N., Chen, Q.S., Chen, F.Q., Liu, H.L., Ren, C.Y., Zhang, Y.X., Yang, F.J., Wei, J.P., 2022b. Jasmonic acid promotes glutathione assisted degradation of chlorothalonil during tomato growth. Ecotoxicol. Environ. Saf. 233, 113296. https://doi.org/10.1016/j.ecoenv.2022.113296.

Yu, G.B., Zhang, Y., Ahammed, G.J., Xia, X.J., Mao, W.H., Shi, K., Zhou, Y.H., Yu, J.Q., 2013. Glutathione biosynthesis and regeneration play an important role in the metabolism of chlorothalonil in tomato. Chemosphere 90 (10), 2563–2570. https://doi.org/10.1016/j.chemosphere.2012.10.100.

Yu, Q.Q., Lu, F., Ya, L.I., Yang, H., Song, N.H., 2021. Residues of reduced herbicides terbuthylazine, ametryn, and atrazine and toxicology to maize and the environment through salicylic acid. ACS Omega 6 (41), 27396–27404. https://doi.org/10.1021/acsomega.1c04315.

Zhang, J.J., Yang, H., 2021. Metabolism and detoxification of pesticides in plants. Sci. Total Environ. 790, 148034. https://doi.org/10.1016/j.scitotenv.2021.148034.

Zhao, W., Teng, M., Zhang, J., Wang, K., Zhang, J., Xu, Y., Wang, C., 2022. Insights into the mechanisms of organic pollutant toxicity to earthworms: advances and perspectives. Environ. Pollut. 303, 119120. https://doi.org/10.1016/j.envpol.2022.119120.

Zhou, J., Xu, H., Xiang, Y., Wu, J., 2024. Effects of microplastics pollution on plant and soil phosphorus: a meta-analysis. J. Hazard Mater. 461, 132705. https://doi.org/10.1016/j.jhazmat.2023.132705.

Zhou, Y., Xia, X., Yu, G., Wang, J., Wu, J., Wang, M., Yang, Y., Shi, K., Yu, Y., Chen, Z., Gan, J., Yu, J., 2015. Brassinosteroids play a critical role in the regulation of pesticide metabolism in crop plants. Sci. Rep. 5, 9018. https://doi.org/10.1038/srep09018.

Zhu, M., Tang, J., Shi, T., Ma, X., Wang, Y., Wu, X., Li, H., Hua, R., 2022. Uptake, translocation and metabolism of imidacloprid loaded within fluorescent mesoporous silica nanoparticles in tomato (*Solanum lycopersicum*). Ecotoxicol. Environ. Saf. 232. https://doi.org/10.1016/j.ecoenv.2022.113243.

CHAPTER 2

Phytotoxicity of Persistent Organic Pollutants (POPs)

Karina S.B. Miglioranza[1, 2], Paola M. Ondarza[1, 2], Mercedes Lourido[1, 2] and Sofía Fideleff[1, 2]

[1]Laboratorio de Ecotoxicología y Contaminación Ambiental, Facultad de Ciencias Exactas y Naturales (FCEyN), Universidad Nacional de Mar del Plata (UNMdP), Mar del Plata, Argentina; [2]Instituto de Investigaciones Marinas y Costeras (IIMyC), Consejo Nacional de Investigaciones Científicas y Tecnológicas (CONICET), Mar del Plata, Argentina

1. Introduction

Persistent organic pollutants (POPs) are a group of toxic chemicals characterized by their resistance to environmental degradation, their ability to bioaccumulate in organisms and their potential for long-range transport. These properties allow POPs to persist in the environment for long periods, constituting a significant threat to ecosystems and human health. POPs include a wide variety of industrial chemicals, unintentional by-products and pesticides. Their toxicity and persistence in the environment have raised a global concern, driving international efforts to regulate and mitigate their impact (UNEP, 2001). POPs are prohibited worldwide and regulated by the Stockholm Convention, an international treaty whose objective is to eliminate or restrict their production and use. At present, POPs constitute a list of 35 compounds and others are in the process of revision for inclusion.

Among these contaminants, it can be mentioned.

- **Polychlorinated Biphenyls (PCBs)**: They were widely employed in industrial applications due to their thermal stability, such as dielectric fluids in transformers or in lubricants, plasticizers, printing, and many other applications and capacitors. PCBs were produced commercially as mixtures of congeners such as Aroclor 1221,1254, 1260. The PCBs degrade very slowly in the environment and are lipophilic compounds. Despite their ban in many countries, and due to the high persistence in the environment, continuous leaching from old equipment and waste sites is commonly found. Their toxic effects include endocrine disruption, immune suppression, and carcinogenicity, among others (ATSDR, 2000).
- **Organochlorine Pesticides (OCPs)**: This group includes pesticides such as DDT, endosulfan, aldrin, dieldrin, dicofol, hexachlorocyclohexanes (HCHs), heptachlor, methoxychlor, mirex, chlordanes and their metabolites. These chemicals were extensively used for agricultural and public health purposes to control pests and vector-borne diseases. However, their bioaccumulative nature and harmful effects on nontarget species led to a continuous concern at worldwide level (UNEP, 2019).

Phytotoxicity of Organic Pollutants
ISBN 978-0-443-33110-7
https://doi.org/10.1016/B978-0-443-33110-7.00002-X

- **Dioxins and Furans**: These compounds are unintentional by-products of industrial processes such as waste incineration, metal smelting, and the production of chlorinated chemicals, due to incomplete combustion. They are found at very low levels in the environment. Both dioxins and furans are highly toxic, even at very low concentrations, with adverse effects ranging from developmental disorders to cancer. There are 75 different dioxins, of which seven are considered to be of concern. The one considered most toxic is referred to as 2,3,7,8-tetrachlorodibenzo-p-dioxin (TCDD). They are also classified as endocrine disruptors (WHO, 2010).
- **Per- and Polyfluoroalkyl Substances (PFAS)**: They are a group of over 4700 heterogeneous compounds with amphipathic properties and very high stability to chemical and thermal degradation. Often referred to as "forever chemicals", due to their extreme persistence and bioaccumulative potential. Some PFAS are a class of POPs with widespread use in firefighting foams, nonstick coatings, photo imaging, hydraulic fluids, leather, paper, textiles and water-resistant textiles, and many others (Panieri et al., 2022). Their presence in drinking water has raised concerns about developmental and immune effects in humans (Giesy and Kannan, 2001). In 2009, perfluorooctane sulfonic acid (PFOS), its salts and perfluorooctane sulfonyl fluoride (PFOSF) were listed in Annex B to the Stockholm Convention. Nowadays, other PFAS are being considered for inclusion on the list of POPs (UNEP, 2009) (Figs. 2.1 and 2.2).
- **Hexachlorobenzene (HCB)**: Initially, HCB was used as a pesticide and fungicide, but now it is largely produced unintentionally as a by-product of chemical manufacturing. It is highly toxic to aquatic life and poses risks to human health, including liver damage and reproductive effects.
- **Polybrominated Diphenyl Ethers (PBDEs)**: They have been used since the 1970s as additive flame retardants in a wide range of mainly consumer products, such as furniture, electronics, and building materials. PBDEs are known to disrupt thyroid hormones and affect neurodevelopment, particularly in children (Costa and Giordano, 2007).
- **Short-Chain Chlorinated Paraffins (SCCPs)**: Chlorinated paraffins (CPs) are complex mixtures of certain organic compounds containing chloride: polychlorinated n-alkanes. The chlorination degree of CPs can vary between 30 and 70 wt%. SCCPs can be used as a plasticizer in rubber, paints, adhesives, flame retardants for plastics as well as an extreme pressure lubricant in metal working fluids. SCCPs are persistent and bioaccumulative. They are toxic to aquatic organisms and may disrupt endocrine functions (UNEP, 2017).
- **Pentachlorobenzene (PeCB)**: PeCB was used in PCB products, in dyestuff carriers, as a fungicide, a flame retardant and as a chemical intermediate. It also presents as impurities in products such as solvents or pesticides. PeCB is also produced unintentionally during combustion, thermal and industrial processes.

- **Polychlorinated Naphthalenes (PCNs)**: Commercial PCNs are mixtures of up to 75 chlorinated naphthalene congeners plus byproducts. These compounds were used in lubricants and electrical insulators, as wood preservatives, as rubber and plastic additives, and for capacitor dielectrics. They are persistent, bioaccumulative, and toxic to both aquatic life and humans (Falandysz, 1998).
- **Hexachlorobutadiene (HCBD)**: It is a byproduct of industrial processes, such as the manufacture of chlorinated aliphatic compounds. HCBD is toxic to the liver, kidneys, and respiratory system (UNEP, 2015).

1.1 Chemical properties of POPs

POPs share several properties that contribute to their persistence and toxicity. Among these characteristics, POPs are halogenated hydrocarbons, with chlorine, fluorine or bromine atoms bonded to carbon skeletons. This halogenation enhances their stability making them less susceptible to enzymatic and microbial degradation (Muir and Howard, 2006). Moreover, many POPs contain aromatic rings, which confer high chemical stability and resistance to breakdown processes. These compounds are lipophilic, leading to their accumulation in the lipid tissues of organisms and, the organic carbon of soils and sediments (Gobas, 1993).

POPs exhibit remarkable environmental persistence, owing to their resistance to chemical, biological, and photolytic degradation, and can persist in the environment for years or even decades. The strong carbon-halogen bonds in POPs make them chemically inert and resistant to breakdown processes, such as hydrolysis, oxidation, and microbial action (Sinkkonen and Paasivirta, 2000).

1.2 Environmental behavior of POPs

The distribution and fate of POPs in environmental compartments are mainly controlled by the physicochemical properties of the compound, the environmental conditions (e.g., temperature, pH, solar radiation, and wind speed), physicochemical properties of abiotic components (e.g., organic carbon content, particle size distribution of soil or sediment), and physiological, anatomical and behavioral characteristics of the species inhabiting exposed ecosystems.

These pollutants can enter aquatic environments through both point and nonpoint sources, depending on land use and how contaminants are transported (Fig. 2.1). Due to their hydrophobic nature, they tend to bind to sediment organic matter and/or accumulate in the lipid tissues of organisms. Although the toxicity of weathered organic compounds is generally considered to decrease over time in the environment, their persistence and potential for bioaccumulation and biomagnification through food webs pose a serious environmental risk. Agricultural practices, urban development,

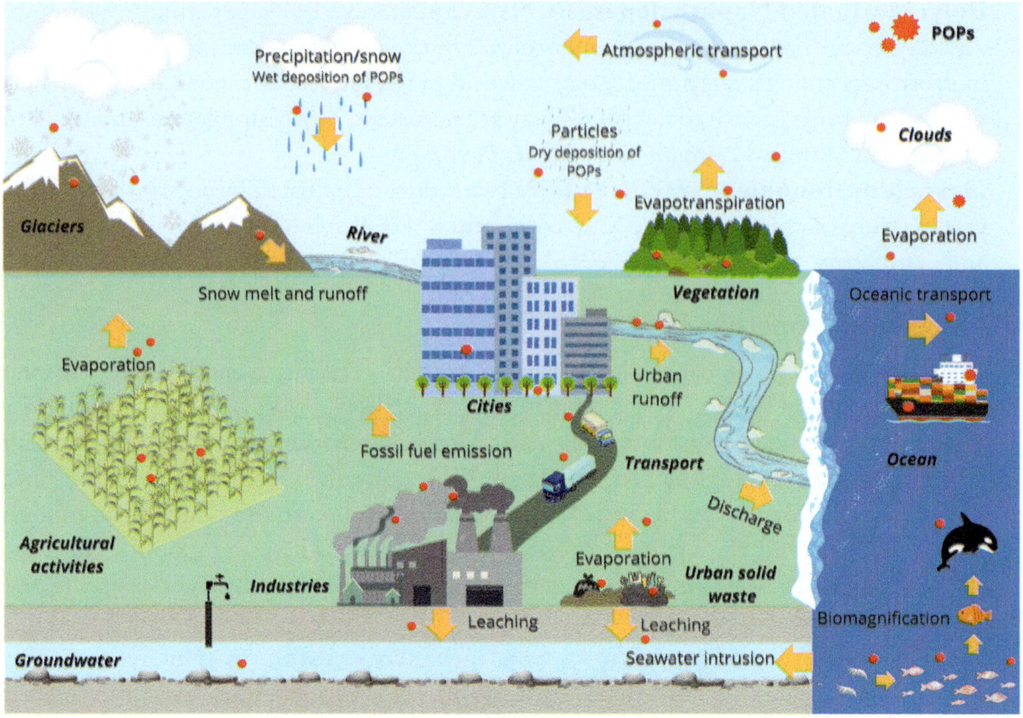

Figure 2.1 Environmental behavior of POPs.

industrial activities, and the presence of dams are recognized as potential contributors of POPs in large watersheds.

Moreover, POPs pose a global environmental challenge due to their long-range transport potential. In this sense, POPs can evaporate into the atmosphere under warmer conditions and condense under cooler conditions (Dietz, 2013). This cycle, known as the "grasshopper effect," allows POPs to travel great distances from their original sources. They often accumulate in remote, pristine regions such as the Arctic and Antarctic, impacting also those extreme ecosystems (Wania and Mackay, 1993; Miglioranza et al., 2021).

Once deposited in the ecosystems, POPs accumulate in biota, disrupting food chains and biodiversity (Miglioranza et al., 2013a; Quadri-Adrogué et al., 2021). Additionally, POPs contribute to the degradation of soil and water quality, further compromising the resilience of ecosystems already under pressure from climate change and other anthropogenic stressors (Jones and de Voogt, 1999).

1.3 Pathways of POPs uptake in plants

Plants are directly exposed to POPs through soil, water, and air, in both terrestrial and aquatic ecosystems (Miglioranza et al., 2013b). Vegetation covers about 80% of the Earth's terrestrial surface and constitutes a key component of the land-atmosphere interface. Plant species are highly organized and structured, compose large areas of vegetation canopies that link the soil and air environments. Terrestrial vegetation may either take up pollutants from the air through the surfaces of above ground parts, such as leaves, stems, or bark, or directly from the soil through the roots. Consumption of plants or plant products such as vegetables, fruits, and grains constitutes a major part of the food consumption of human beings and cattle. Therefore, contamination of plants can have a great influence on the total daily intake of a substance.

Root uptake of POPs can occur from both soils and sediments (Fig. 2.2). Due to their hydrophobic nature, these compounds tend to bind strongly to organic matter. Plants can absorb POPs through their root systems, and although they exhibit low water solubility, their association with dissolved organic matter can enhance their availability

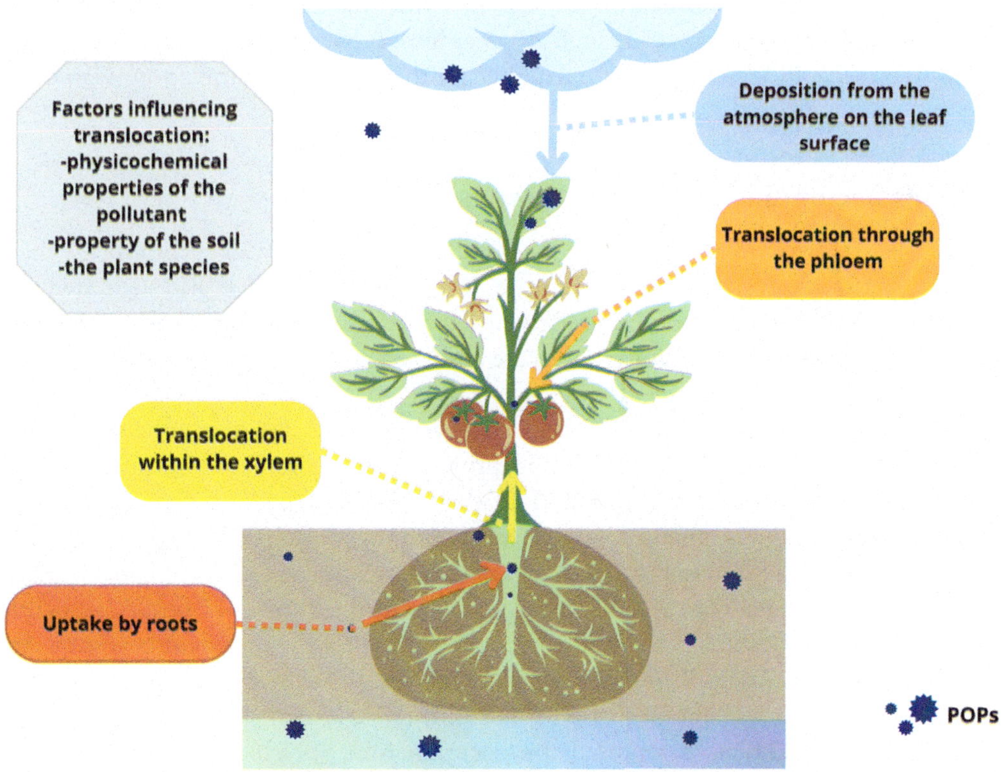

Figure 2.2 Pathways of POPs uptake by plants.

for root uptake (Mitton et al., 2016a,b). Once inside the plant, certain POPs are capable of entering the xylem and being translocated to aerial parts, including shoots and leaves (Mitton et al., 2016a,b; Miglioranza et al., 2013a). The incorporation of POPs through leaf implies the uptake of contaminants from atmospheric deposition because POPs in the atmosphere can settle on leaf surfaces as particulates or aerosols. Gas-phase POPs can diffuse into plant tissues through stomatal openings, leading to internal contamination (Simonich and Hites, 1995). Moreover, it can occur by cuticular absorption because the waxy cuticle of leaves can absorb lipophilic POPs, which may then penetrate deeper into plant tissue (Reyes-Jurado et al., 2020). Once inside the plant, POPs can be transported through the xylem and phloem to various tissues, including fruits and seeds (Li et al., 2018). Besides, POPs may be sequestered in vacuoles or lipid-rich tissues within the plant, reducing their immediate toxicity but prolonging their presence within the ecosystem (Barra Caracciolo et al., 2015).

Some plants can metabolize POPs into less toxic forms, although this process is often slow and incomplete (Zhang et al., 2021).

In the case of aquatic plants, the incorporation of POPs is by direct exposure to water, sediments or air. Uptake occurs through submerged roots and leaves, with some plants showing higher accumulation capacities (Ghisi et al., 2019). It has been demonstrated that certain aquatic species, such as water hyacinths and duckweed, have the potential to remove POPs from contaminated water through bioaccumulation (Gao et al., 2022).

2. POPs accumulation in plant tissues

Qian et al. (2023) found that ferns could accumulate PFAS from water, immobilize them in roots, and store them in harvestable tissue. Moreover, the authors also suggested that long-chain hydrophobic compounds tend to be adsorbed and retained on the root epidermis, while short-chain compounds are absorbed and quickly translocated upward (Qian et al., 2023).

It has been reported that *Oryza sativa* L. (rice) accumulates and metabolizes CB-61 to para-substituted hydroxylated CB-61 as its primary metabolite, in addition to a minor product of para-substituted methoxy CB-61 (Sun et al., 2016).

In a hydroponic experiment, UV-328 was significantly absorbed by the roots of lettuce (*Lactuca sativa* L.) and radish (*Raphanus sativus* L.), with average root concentration factors (RCFs) ranging from 58.5 to 400 mL/g in lettuce and 84.4–154 mL/g in radish. However, its translocation from roots to shoots was limited, with a translocation factor (TF) below 0.055. Additionally, UV-328 underwent transformation and metabolism within the plant, identifying 14 phase I metabolites of UV-328 (Li et al., 2024).

Dong et al. (2022) have reported the occurrence of polychlorinated naphthalenes (PCNs) in plants from the Arctic. Total PCN concentrations ranged from 77 to

870 pg/g dw in plants. The higher levels of PCNs near the research stations highlighted the significant influence of local anthropogenic emission sources. The composition of PCNs was dominated by lower chlorinated homologs, especially mono-to trichlorinated naphthalenes, which accounted for over 80% of total PCNs in the soil and plant samples. The use of pine needles as a biomonitor is a practical tool for understanding the occurrence of organic contaminants in areas with different impacts. A study carried out in needles of two conifers (*Pinus nigra* and *Cedrus* atlantica), detected levels of organochlorine pesticides (OCPs) and polychlorinated biphenyls in an urban area in Strasbourg (France). Eleven PCBs, 11 OCPs, were detected. The urban and rural sites were more exposed to PCBs than the suburban site. The highest concentration of PCBs was found at the urban site, but the largest number of congeners (10) was detected at the rural site. For OCPs, the rural site displayed the highest concentrations (up to 22.9 ng g−1) (Al Dine et al., 2015).

The epiphyte *Tillandsia bergeri* is feasible for use as a biomonitor of POPs. A study performed in the southeast region of Buenos Aires Province, Argentina, reported a predominance of endosulfan in leaves of *T. bergeri* collected in an agricultural watershed up to 300 ng/g dry weight. Endosulfan sulfate showed higher concentrations than α- and β-isomers predominantly during the winter period (Silva Barni et al., 2019).

3. Mechanisms of POPs toxicity in plants

POPs can disrupt vital physiological processes in plants, including photosynthesis, nutrient uptake, and growth regulation (Mitton et al., 2018). These disruptions can cascade through ecosystems, affecting biodiversity and the services ecosystems, such as food production, carbon sequestration, and water purification. Furthermore, plants often act as a sink for POPs, transferring these pollutants to higher trophic levels, by the biomagnification process through the food webs leading to potential ecological and health risks (Windsor et al., 2019).

The uptake of POPs by plants occurs predominantly through their leaves and roots. Following internalization within the cells, POPs have been observed to interact with biological macromolecules, due to their structural similarity with small organic molecules, such as proteins, nucleic acids, lipids, and carbohydrates. The biotransformation of pesticides in plants occurs primarily through enzymatic pathways. This process has the potential to reduce the toxicity of pesticides to both target and nontarget organisms, and enhance pesticide vulnerability to other biological, chemical, or physical degradative transformations. Hydrolases, dehalogenases, transferases, peroxidases, oxygenases and oxidoreductases are the most representative enzyme classes in the remediation of polluted environments (Mousavi et al., 2021).

The POPs in plants are metabolized in three phases. In the first phase, called transformation, the plants absorb organic pollutants through their roots with the help of enzymes that catalyze transformations. Then, the organic products resulting from the transformation react with endogenous enzymes, generating metabolites derivatized with carbohydrates and other molecules. The third phase of metabolism is compartmentalization, which leads to the sequestration of conjugated organic compounds.

There are different processes related to the toxicity of POPs. It could be related to a direct interaction with cellular processes, such as damage to cell membranes. POPs have the potential to get into the plants, inhibiting several biological processes such as growth, photosynthesis, seed germination and elongation of pollen tubes (Tomar et al., 2019). In addition, numerous studies mention the alteration of hormone levels in the plants, disturbed in cell biosynthesis, damage in cell ultrastructure, membrane stability and DNA (Zhang et al., 2017). For example, according to Zhang et al. (2012), tobacco cell growth was inhibited by PCDD/Fs. In some cells, exposed to concentration of 1000 ng/g of PCDD/Fs, lead to an unusual aspect in the surface of cell wall known as plasmolysis, which change the shape of chloroplast and a rupture in the cytoplasm.

POPs can also affect membrane stability by generating reactive oxygen species (ROS). Moreover, the unsaturated lipids in cells deteriorate upon contact with oxygen. This process is known as lipid peroxidation and usually alters the integrity, fluidity and permeability of the membrane triggering a loss of functionality. In addition, ROS can react with lipoproteins and modify them into potentially toxic products (Zhang et al., 2017). MDA is used to determine membrane damage by lipid peroxidation. Some results reported 2.7 times higher MDA values than control in tobacco cells exposed at high PCDD/Fs levels (Zhang et al., 2012). *Vallisneria natans* exposed to environmentally relevant concentrations of PFAS exhibited an increase in ROS production and stress tolerance. Furthermore, the study demonstrated the upregulation of gene expression, resulting in the activation of aquaporins and the slow-type anion channels (Xiao et al., 2024).

DDT, a widely used organochlorine pesticide, has left significant residues in agricultural soils despite being banned in many countries decades ago. Studies have shown that plants grown in DDT-contaminated soils exhibit reduced germination rates, stunted growth, and oxidative stress due to ROS production. For instance, rice plants exposed to DDT-contaminated soils demonstrated decreased chlorophyll content and increased lipid peroxidation, indicating cellular membrane damage (Kumar et al., 2020). These effects not only reduce crop yields but also pose a risk to food safety due to bioaccumulation in edible tissues.

On the other hand, the damage to nucleic acids is another effect of plants exposed to POPs. Numerous studies have demonstrated that organic pollutants possess the potential to bind to nucleic acids, which can consequently lead to a disruption of their biological function. It was observed that in tobacco plants exposed to soil contaminated by PCBs, there has been a significant increase in DNA damage associated with toxic effects such as growth inhibition, brownish and distorted leaves.

Qin et al. (2022) proposed that PFAAs induce DNA deformation in plant seedlings, being the main mechanisms employed related to the predominant interaction forces in the PFAAS-DNA binding process, such as hydrogen bonds and van der Waals forces.

The alterations in photosynthesis and respiration are key effects as a consequence of POPs exposure. Photosynthesis is the process by which green organisms convert light energy into chemical energy, which is subsequently used to fix carbon for cellular processes. It has been demonstrated that the photosynthetic oxygen production of *Ceratophyllum demersum* was inhibited (33%–66%) by 3-chlorobiphenyl (CB 3) at different exposure levels. This inhibition process may be associated with the potential breakdown of the detoxification enzymes in chloroplast cells, suggesting that an acute stress condition, in which a possible abrupt generation of high levels of free radicals can overload the antioxidant capacity and exert acute effects (Menone and Pflugmacher, 2005).

Several studies have reported that POPs inhibit the photosynthetic electron transport chain in chloroplasts. Studies across various plant and algal species emphasize the inhibition that compounds such as DDT or DDE produce in photosynthesis (Tomar et al., 2019).

It has been demonstrated that PCBs, hydroxy-PCBs and PBDEs, have the capacity to reduce the biomass of rice (*Oryza sativa* L.). Thus, the inhibition of the photosynthetic rate by these compounds has been shown to result in a reduction in the enzymatic carbon fixation pathway (Zhang et al., 2022).

Dioxins and furans, by-products of industrial processes, have been detected in forest soils near industrial sites. These compounds adversely affect the growth of tree seedlings, such as pine and spruce. Studies have demonstrated that dioxins impair photosynthesis by inhibiting the activity of key enzymes involved in carbon fixation. In a controlled study, spruce seedlings exposed to dioxin-contaminated soils showed significant reductions in leaf area and biomass, emphasizing the vulnerability of forest ecosystems to POP contamination (Ivanova et al., 2018).

Per- and polyfluoroalkyl substances (PFAS), have been detected in agricultural soils irrigated with contaminated water. Studies on lettuce and spinach revealed that PFAS compounds are taken up by the roots and translocated to leaves. This accumulation has been linked to reduced photosynthetic efficiency and growth inhibition. Moreover, PFAS contamination poses a dual threat by affecting plant health and entering the human food chain (Brendel et al., 2018).

Li et al. (2020) reported that lettuce employed multiple changes in metabolic pathways in roots to increase its tolerance to PFOA and PFOS, under hydroponic conditions. The results offer insights into the molecular reprogramming of plant roots in response to PFAS exposure and provide important information for the risk assessment of PFASs in the environment.

Organic pollutants reduced CO_2 capture efficiency in rice (*Oryza sativa* L.). Under laboratory conditions, it has been demonstrated that the mechanisms of disturbed carbon fixation in rice (*Oryza sativa* L.) by 2,3,4,5-tetrachlorobiphenyl (CB 61), 4′-hydroxy-2,3,4,5-tetrachlorobiphenyl (4′–OH–CB 61), and 2,2′,4,4′-tetrabromo diphenyl ether (BDE #47). The biomass of rice exposed to 4′–OH–CB 61 and BDE #47 was 20% lower than control. Moreover, an inhibition of the net photosynthetic rate (60%) of rice exposed to 4′–OH–CB 61 was also observed. These results were attributed to the decrease in ribulose-1,5-bisphosphate carboxylase/oxygenase (Rubisco) activity (6%–33%). Molecular docking and simulation suggested that these pollutants potentially competed with CO_2 for binding to the active sites in Rubisco, leading to reduced CO_2 capture efficiency (Zhang et al., 2022).

Oxidative stress has been demonstrated to play a significant role in the toxic action of POPs. This process can be induced by two main mechanisms: the first is the direct production of reactive oxygen species (ROS), and the second is the indirect inhibition of antioxidant enzymes or nonenzymatic antioxidants, resulting in a modification of the cellular redox balance. In the presence of reactive oxygen intermediates (ROI), organisms produce a defense system to prevent serious or irreversible damage. Superoxide dismutase (SOD) is an enzyme that protects cells from free radicals. It is the first line of defense that is activated against ROI to prevent the formation of lipid peroxidation by eliminating free radicals. Peroxidase (POD) is an enzyme that has various functions in plants, including the breakdown of H_2O_2, in a series of reactions closely related to the activity of catalase (CAT) and superoxide dismutase (SOD). Similar features to GPX that consumes H_2O_2 to generate phenoxy compounds (Reddy et al., 2005). For example, tobacco cells showed an increase in enzyme activity of SOD, POD, GPX and GR in responses for ROI produced by PCDD/Fs (Zhang et al., 2012).

In aquatic plants, the oxidative metabolism of POPs is typically catalyzed by cytochrome P450 enzymes, peroxidases, or mixed-function oxidases (Chroma et al., 2002). In this sense, the involvement of peroxidases, phase I (cytochrome P450 systems) and phase II (i.e., glutathione-S-transferase GST-) enzymes in the detoxification of PCBs in plants has been demonstrated (Chroma et al., 2002; Menone and Pflugmacher, 2005). GST is an enzyme that catalyzes the binding of reduced glutathione (GSH) to electrophilic substances. GSH is the main thiol compound that acts as a buffer, maintaining adequate levels of the thiol-disulfide redox potential.

Aquatic plants, such as duckweed (*Lemna minor*), are particularly sensitive to organochlorine pesticides such as endosulfan. Exposure to endosulfan-contaminated water caused a reduction in frond number, chlorophyll degradation, and increased activity of antioxidant enzymes, indicating oxidative stress. These findings highlight the potential of using aquatic plants as bioindicators for monitoring POPs contamination in aquatic ecosystems (Ghisi et al., 2019).

Polybrominated diphenyl ethers (PBDEs), widely used as flame retardants, have been detected in wetland sediments. Research on *Typha latifolia* (cattails) shows that PBDEs accumulate in roots and leaves, leading to impaired photosynthetic efficiency and oxidative stress. Wetland plants exposed to high PBDEs concentrations also exhibit reduced reproductive success, affecting the wetland ecosystem services (Chen et al., 2020).

Regarding edible vegetables, a study was conducted on the antioxidant response of soybean and alfalfa plants growing in DDT-contaminated soil. Protein content, antioxidant capacity, GST activity and thiol-groups on roots, stems and leaves, were altered by DDTs. In particular, soybean exhibited higher susceptibility to DDT than alfalfa plants, a consequence of its lower antioxidant capacity and GST activity in leaves.

The insecticide DDT and its metabolites (DDE and DDD) also induced enzymatic and nonenzymatic antioxidant responses in tomato and zucchini plants growing in DDT-contaminated soils. Changes in GST, glutathione reductase (GR), and glutathione peroxidase (GPx) activities were identified in tomato plants. These findings suggest that glutathione (nonenzymatic mechanisms) plays a role in the oxidative stress response in this species. Conversely, zucchini plants presented a lack of antioxidant enzyme activity (Mitton et al., 2018).

Moreover, the pesticide Endosulfan, has been shown to accumulate in tomato plants grown in contaminated soils. Exposure leads to reduced fruit yield and quality, as well as physiological stress marked by increased production of antioxidant enzymes (Rani et al., 2017).

Mixed crop systems, such as intercropping maize and legumes, have been shown to experience differential POPs uptake and toxicity. While maize tends to accumulate higher levels of POPs such as DDT, legumes such as beans show more pronounced oxidative stress and enzyme inhibition.

Endogenous hormones play a fundamental role in plant responses to diverse environmental stimuli. Abiotic stress can induce hormonal changes in plants, leading to interactions between different types of hormones (Zhang et al., 2012). There is a wide variety of phytohormones, among which auxin, gibberellin, and cytokinins are the most important, as they are responsible for regulating aspects such as development, growth, and gene expression in plants. Auxins are plant hormones that promote cell elongation, inhibit the growth of main roots, prevent organ abscission, regulate phototropism, and induce the development of fruits and flowers. Gibberellins are a family of hormones that promote stem growth, facilitate the mobilization of reserves, seed germination, and flower and fruit development. On the other hand, cytokinins play an important role in gene expression, cell division, bud sprouting, organogenesis, and senescence. Compounds such as PCDD/Fs are capable of modifying the endogenous hormonal balance in tobacco cells after 144 h of exposure. In this sense, tobacco cells showed a decrease in jasmonates content and an increase in indole-3-acetic acid content, resulting in a new endogenous hormone balance that could be favorable for PCDD/Fs tolerance (Zhang et al., 2012).

Eco-physiological impacts include other effects related to alterations in growth and development. A study on *Brassica juncea* (Indian mustard) revealed that PCB-contaminated soils significantly inhibited root elongation and biomass accumulation. The phytotoxicity was attributed to disruptions in cellular metabolism and the impairment of nutrient uptake mechanisms. Additionally, PCBs were found to accumulate in root tissues, suggesting limited translocation to above-ground parts, which might reduce the risk of entry into the food chain but also highlights their impact on plant development (Zhang et al., 2019).

Hexachlorobenzene (HCB), an organochlorine fungicide, has been reported to persist in soil, causing toxic effects on cereal crops such as wheat and maize. Studies indicate that HCB exposure reduces seed germination rates, inhibits root elongation, and disrupts enzymatic activity critical for plant metabolism. Maize grown in HCB-contaminated soils exhibited reduced chlorophyll content and lower biomass production, as a result of its phytotoxic nature (Wang et al., 2021).

A study reported the combined effects of POPs on rice and wheat crops. In agricultural areas with mixed POPs contamination, such as DDT, PCBs, and HCB, combined phytotoxic effects on rice and wheat crops included synergistic impacts on germination rates, root elongation, and photosynthetic efficiency. For example, rice plants exposed to a mixture of DDT and PCBs exhibited more severe reductions in chlorophyll content and biomass compared with exposure to individual compounds. Similarly, wheat exposed to HCB and PCBs demonstrated significant oxidative stress, as evidenced by increased malondialdehyde (MDA) levels and reduced enzymatic activity (Li et al., 2020).

4. Gaps and perspectives

The phytotoxicity of POPs in plants is essential for predicting their environmental and ecological impacts. However, significant research gaps remain, particularly in the context of long-term effects, interactions with multiple stressors, and the influence of climate change on the behavior of POPs. It is clear that there is a need for long-term studies, since most of the available studies are focused on the phytotoxicity of POPs in the short-term on acute toxicity endpoints, such as growth inhibition, photosynthetic disruption, or oxidative stress. While these findings are very important, there is a gap concerning chronic and sublethal effects of prolonged POP exposure, which may include cumulative damage. In this fact, over time, POPs can alter plant physiology and metabolism, leading to reduced fitness, lower reproductive success, and impaired resilience to environmental stressors. Another possible outcome is related to epigenetic effects, because long-term exposure to POPs may induce heritable epigenetic changes that affect plant populations and their ability to adapt to contaminated environments.

Studies of impacts at the ecosystem level are also highly required, because chronic phytotoxicity can disrupt plant community dynamics, reduce biodiversity, and impair ecosystem services, such as carbon sequestration and nutrient cycling. This kind of studies are recommended to investigate the impacts of POPs across multiple plant generations, focusing on both physiological and ecological endpoints.

The combined effects of multiple contaminants should be considered, since plants are exposed to complex mixtures of POPs and other pollutants. In this sense, many of these pollutants can exhibit synergistic, antagonistic, or additive effects, complicating risk assessments. Some POPs may enhance or inhibit the uptake and metabolism of other pollutants in plants. For example, coexposure to PCBs and heavy metals may exacerbate oxidative stress and impair enzymatic detoxification pathways. Processes of interactive phytotoxicity involve mixtures of POPs and other chemicals, leading to simultaneous alterations in photosynthesis, water uptake, and hormonal regulation.

References

Al Dine, E.J., Mokbel, H., Elmoll, A., Massemin, S., Vuilleumier, S., Toufaily, J., Hanieh, T., Millet, M., 2015. Concomitant evaluation of atmospheric levels of polychlorinated biphenyls, organochlorine pesticides, and polycyclic aromatic hydrocarbons in Strasbourg (France) using pine needle passive samplers. Environ. Sci. Pollut. Control Ser. 22 (22), 17850–17859. https://doi.org/10.1007/s11356-015-5030-5.

ATSDR, Agency for Toxic Substances and Disease Registry, 2002. Toxicological Profile for Polychlorinated Biphenyls (PCBs). U.S. Department of Health and Human Services.

Barra Caracciolo, A., Topp, E., van Elsas, J.D., 2015. Microbial responses to persistent organic pollutants in soil: Ecotoxicology effects and bioremediation. Front. Microbiol. 6, 1401.

Brendel, S., Fetter, É., Staude, C., Vierke, L., Biegel-Engler, A., 2018. Short-chain perfluoroalkyl acids: environmental concerns and a regulatory strategy under REACH. Environ. Sci. Eur. 30 (1). https://doi.org/10.1186/s12302-018-0134-4.

Chen, X., Zhu, Y., Wang, J., 2020. Effects of polybrominated diphenyl ethers on wetland plants: a case study on *Typha latifolia*. Ecotoxicol. Environ. Saf. 193, 110330.

Chroma, L., Mackova, M., Kucerova, P., in der Wiesche, C., Burkhard, J., Macek, T., 2002. Enzymes in plant metabolism of PCBs and PAHs. Acta Biotechnol. 22 (1–2), 35–41. https://doi.org/10.1002/1521-3846(200205)22:1/2<35::aid-abio35>3.0.co;2-u.

Costa, L.G., Giordano, G., 2007. Developmental neurotoxicity of PBDEs: mechanisms and implications. Environ. Health Perspect. 115.

Dietz, R., 2013. Spatial and temporal trends of persistent organic pollutants in Greenlandic biota: an updated review. Sci. Total Environ. 454–455.

Dong, C., Xiong, S., Yang, R., Pei, Z., Li, Y., Zhang, Q., Jiang, G., 2022. Polychlorinated naphthalenes (PCNs) in soils and plants from Svalbard, Arctic: levels, distribution, and potential sources. Sci. Total Environ. 849, 157883. https://doi.org/10.1016/j.scitotenv.2022.157883.

Falandysz, J., 1998. Chlorinated naphthalenes as persistent organic pollutants. Environ. Pollut. 101 (1).

Gao, Y., Fang, X., Tsang, D.C., 2022. Phytoremediation of persistent organic pollutants in water and soil: progress, challenges, and future perspectives. Environ. Res. 210, 112963.

Ghisi, R., Buschini, A., Lori, V., 2019. Phytotoxicity and bioconcentration of persistent organic pollutants in aquatic plants. Chemosphere 229, 353–364.

Giesy, J.P., Kannan, K., 2001. Global distribution of perfluorooctane sulfonate in wildlife. Environ. Sci. Technol. 35 (7), 1339–1342. https://doi.org/10.1021/es001834k.

Gobas, F.A., 1993. Bioaccumulation of hydrophobic chemicals in aquatic food-webs: a dynamic model. Environ. Sci. Technol. 12.

Ivanova, T., Zhelev, Z., Dobreva, I., 2018. Effects of dioxins on forest ecosystems: a case study of coniferous seedlings. J. Environ. Sci. Health Part B 53 (7), 445–452.

Jones, K.C., de Voogt, P., 1999. Persistent organic pollutants (POPs): state of the science. Environ. Pollut. 100 (1–3), 209–221. https://doi.org/10.1016/s0269-7491(99)00098-6.

Kumar, V., Chopra, A.K., Singh, J., 2020. Effects of DDT contamination on soil health and crop productivity: a case study in agricultural regions. Environ. Monit. Assess. 192 (3), 161.

Li, J., Zhang, G., Liu, C., 2018. Mechanisms of bioaccumulation and translocation of organic pollutants in plants. Sci. Total Environ. 634, 1685–1695.

Li, P., Xiao, Z., Sun, J., Oyang, X., Xie, X., Li, Z., Tian, X., Li, J., 2020. Metabolic regulations in lettuce root under combined exposure to perfluorooctanoic acid and perfluorooctane sulfonate in hydroponic media. Sci. Total Environ. 726, 138382. https://doi.org/10.1016/j.scitotenv.2020.138382.

Li, B., Yao, Z., Qi, L., Guo, L., Ma, Z., Li, C., 2024. Insights into the uptake, accumulation, and metabolism of UV-328 in lettuce (*Lactuca sativa* L.) and radish (*Raphanus sativus* L.). Chemosphere 366, 143488. https://doi.org/10.1016/j.chemosphere.2024.143488.

Menone, M.L., Pflugmacher, S., 2005. Effects of 3-chlorobiphenyl on photosynthetic oxygen production, glutathione content and detoxication enzymes in the aquatic macrophyte *Ceratophyllum demersum*. Chemosphere 60 (1), 79–84.

Miglioranza, K.S., Gonzalez, M., Ondarza, P.M., Shimabukuro, V.M., Isla, F.I., Fillmann, G., et al., 2013a. Assessment of Argentinean Patagonia pollution: PBDEs, OCPs and PCBs in different matrices from the Río Negro basin. Sci. Total Environ. 452, 275–285.

Miglioranza, K.S., Gonzalez, M., Ondarza, P.M., Mitton, F., Grondona, S., Shimabukuro, V., et al., 2013b. Assessment of persistent organic pollutants in the atmosphere of Latin America. In: In Occurrence, Fate and Impact of Atmospheric Pollutants on Environmental and Human Health. American Chemical Society, pp. 183–199.

Miglioranza, K.S.B., Ondarza, P.M., Costa, P.G., de Azevedo, A., Gonzalez, M., Shimabukuro, V.M., Grondona, S.I., Mitton, F.M., Barra, R.O., Wania, F., Fillmann, G., 2021. Spatial and temporal distribution of Persistent Organic Pollutants and current use pesticides in the atmosphere of Argentinean Patagonia. Chemosphere 266. https://doi.org/10.1016/j.chemosphere.2020.129015. Available from: www.elsevier.com/locate/chemosphere.

Mitton, F.M., Ribas Ferreira, J.L., Gonzalez, M., Miglioranza, K.S.B., Monserrat, J.M., 2016a. Antioxidant responses in soybean and alfalfa plants grown in DDTs contaminated soils: useful variables for selecting plants for soil phytoremediation. Pestic. Biochem. Physiol. 130, 17–21. https://doi.org/10.1016/j.pestbp.2015.12.005.

Mitton, F.M., Gonzalez, M., Monserrat, J.M., Miglioranza, K.S., 2016b. Potential use of edible crops in the phytoremediation of endosulfan residues in soil. Chemosphere 148, 300–306.

Mitton, F.M., Gonzalez, M., Monserrat, J.M., Miglioranza, K.S.B., 2018. DDTs-induced antioxidant responses in plants and their influence on phytoremediation process. Ecotoxicol. Environ. Saf. 147, 151–156. https://doi.org/10.1016/j.ecoenv.2017.08.037.

Mousavi, S.M., Hashemi, S.A., Iman Moezzi, S.M., Ravan, N., Gholami, A., Lai, C.W., Chiang, W.H., Omidifar, N., Yousefi, K., Behbudi, G., 2021. Recent advances in enzymes for the bioremediation of pollutants. Biochemistry Research International 2021. https://doi.org/10.1155/2021/5599204. Available from: http://www.hindawi.com/journals/bcri/.

Muir, D.C.G., Howard, P.H., 2006. Are there other persistent organic pollutants? A challenge for environmental chemists. Environ. Sci. Technol. 40 (23), 7157–7166. https://doi.org/10.1021/es061677a.

Panieri, E., Baralic, K., Djukic-Cosic, D., Buha Djordjevic, A., Saso, L., Molecules, P.F.A.S., 2022. A major concern for the human health and the environment. Toxics 10 (2), 44. https://doi.org/10.3390/toxics10020044.

Qian, S., Lu, H., Xiong, T., Yue, Z., Munoz, G., Zhang, C., Li, Z., Liu, C., Li, W., Wang, X., He, Q., 2023. Bioaccumulation of per- and Polyfluoroalkyl substances (PFAS) in ferns: effect of PFAS molecular structure and plant root characteristics. Environ. Sci. Technol. 57 (11), 4443–4453. https://doi.org/10.1021/acs.est.2c06883.

Qin, C., Lu, Y.X., Borch, T., Yang, L.L., Li, Y.W., Zhao, H.M., Hu, X., Gao, Y., Xiang, L., Mo, C.H., Li, Q.X., 2022. Interactions between extracellular DNA and perfluoroalkyl acids (PFAAs) decrease the bioavailability of PFAAs in Pakchoi (*Brassica chinensis* L.). J. Agric. Food Chem. 70 (46), 14622–14632. Available from: http://pubs.acs.org/journal/jafcau. doi: 10.1021/acs.jafc.2c04597.

Quadri-Adrogué, A., Seco Pon, J.P., García, G.O., Castano, M.V., Copello, S., Favero, M., Beatriz Miglioranza, K.S., 2021. Chlorpyrifos and persistent organic pollutants in feathers of the near threatened Olrog's Gull in southeastern Buenos Aires Province, Argentina. Environ. Pollut. 272. https://doi.org/10.1016/j.envpol.2020.115918. Available from: https://www.journals.elsevier.com/environmental-pollution.

Rani, L., Thapa, K., Singh, S., 2017. Impact of endosulfan on crop plants and its remediation strategies. Environ. Sci. Pollut. Control Ser. 24 (15), 13672–13687.

Reddy, A.M., Kumar, S.G., Jyothsnakumari, G., Thimmanaik, S., Sudhakar, C., 2005. Lead induced changes in antioxidant metabolism of horsegram (Macrotyloma uniflorum (Lam.) Verdc.) and bengal-gram (Cicer arietinum L.). Chemosphere 60 (1), 97–104. https://doi.org/10.1016/j.chemosphere.2004.11.092.

Reyes-Jurado, F., Cabrera, M., Trejo, A., 2020. Cuticular permeability of persistent organic pollutants in crop plants. Environ. Toxicol. Chem. 39 (9), 1738–1746.

Silva-Barni, M.F., Gonzalez, M., Miglioranza, K.S.B., 2019. Comparison of the epiphyte *Tillandsia bergeri* and the XAD-resin based passive air sampler for monitoring airborne pesticides. Atmos. Pollut. Res. 10 (5), 1507–1513.

Simonich, S.L., Hites, R.A., 1995. Organic pollutant accumulation in vegetation. Environ. Sci. Technol. 29 (12), 2905–2914. https://doi.org/10.1021/es00012a004.

Sinkkonen, S., Paasivirta, J., 2000. Degradation half-life times of PCDDs, PCDFs and PCBs for environmental fate modeling. Chemosphere 40 (9–11), 943–949. https://doi.org/10.1016/s0045-6535(99)00337-9.

Sun, J., Pan, L., Su, Z., Zhan, Y., Zhu, L., 2016. Interconversion between methoxylated and hydroxylated polychlorinated biphenyls in rice plants: an important but overlooked metabolic pathway. Environ. Sci. Technol. 50 (7), 3668–3675. https://doi.org/10.1021/acs.est.6b00266.

Tomar, R.S., Singh, B., Jajoo, A., 2019. Effects of organic pollutants on photosynthesis. In: Photosynthesis, Productivity and Environmental Stress, pp. 1–26.

UNEP, 2001. Stockholm Convention on Persistent Organic Pollutants.

UNEP, 2009. Stockholm Convention on Persistent Organic Pollutants. United Nations Environment Programme.

UNEP, 2015. Hexachlorobutadiene Risk Profile.

UNEP, 2017. Report of the Conference of the Parties to the Stockholm Convention on Persistent Organic Pollutants on the Work of its Eighth Meeting.

UNEP, 2019. Guidelines on Best Available Techniques and Best Environmental Practices for POPs.

Wang, X., Liu, H., Zhao, Y., 2021. Hexachlorobenzene toxicity and its effects on cereal crop productivity. Sci. Total Environ. 771, 145212.

Wania, F., Mackay, D., 1993. Global fractionation and cold condensation of low volatility organochlorine compounds in polar regions. Ambio 22 (1).

WHO, 2010. Health Effects of Dioxins and Furans.

Windsor, F.M., Pereira, M.G., Tyler, C.R., Ormerod, S.J., 2019. Biological traits and the transfer of persistent organic pollutants through river food webs. Environ. Sci. Technol. 53 (22), 13246–13256. Available from: http://pubs.acs.org/journal/esthag. doi: 10.1021/acs.est.9b05891.

Xiao, Y., Qi, L., Yang, Y., Zhang, Y., Shen, Y., Liu, J., Lei, N., Zhang, W., Wang, Q., 2024. Unravelling the mechanisms of PFAS toxicity to submerged macrophytes and epiphytic biofilms at metabolic and molecular levels. Sci. Total Environ. 952, 175726. https://doi.org/10.1016/j.scitotenv.2024.175726.

Zhang, B., Zhang, H., Jin, J., Ni, Y., Chen, J., 2012. PCDD/Fs-induced oxidative damage and antioxidant system responses in tobacco cell suspension cultures. Chemosphere 88 (7), 798–805. https://doi.org/10.1016/j.chemosphere.2012.03.085.

Zhang, C., Feng, Y., Liu, Y.w., Chang, H.q., Li, Z.j., Xue, J.m., 2017. Uptake and translocation of organic pollutants in plants: a review. J. Integr. Agric. 16 (8), 1659–1668. https://doi.org/10.1016/S2095-3119(16)61590-3.

Zhang, X., Hu, R., Sun, Y., 2019. PCB phytotoxicity and accumulation in Brassica species: implications for phytoremediation. Environ. Sci. Pollut. Control Ser. 26 (10), 9781–9789.

Zhang, X., Hu, R., Sun, Y., 2021. Plant metabolic pathways and detoxification mechanisms for persistent organic pollutants. Environ. Sci. Pollut. Control Ser. 28 (12), 14567–14581.

Zhang, X., Liu, N., Lu, H., Zhu, L., 2022. Molecular mechanism of organic pollutant-induced reduction of carbon fixation and biomass yield in Oryza sativa L. Environ. Sci. Technol. 56 (7), 4162–4172. https://doi.org/10.1021/acs.est.1c07835.

CHAPTER 3

Phytotoxicity of organic air pollutants

Nora Kováts and Katalin Hubai
University of Pannonia, Centre for Natural Sciences, Veszprém, Hungary

1. Introduction

Air pollution is a growing environmental problem, affecting not only human health but the nonhuman biota as well (Harrison and Yin, 2017). Air pollutants are a wide range of chemicals (Carter and Atkinson, 2019), including nitrogen oxides (NO_x), sulfur oxides (SO_x), ozone (O_3), volatile and semivolatile organic compounds (VOCs and SVOCs), heavy metals, particulate matter (PM), and even nanoplastics (Joksimovic et al., 2022). These pollutants are emitted by human activities and natural processes and significantly impact air quality, ecosystems, and climate.

Anthropogenic sources practically involve all kinds of human activity: agriculture, industry, traffic, biomass burning, or waste disposal. Amongst natural sources, wildfires, which emit particulate matter (PM), volatile organic compounds (VOCs), nitrogen oxides (NO_x), and carbon monoxide (CO), contribute significantly to air pollution, especially in forested areas (Vicente et al., 2025).

Air pollutants can be classified into primary and secondary (Anttila and Tuovinen, 2010). Primary pollutants are directly emitted into the atmosphere from sources such as transportation, industry, and agriculture. Secondary pollutants are formed by chemical reactions in the atmosphere.

1.1 Organic pollutants in the atmosphere

Volatile organic compounds (VOCs) are a group of organic chemicals with high vapor pressure at room temperature, enabling them to easily evaporate into the atmosphere (Mangotra and Singh, 2024). VOCs are emitted from natural and anthropogenic sources (Duan et al., 2023), and they interact with other atmospheric constituents and contribute to the formation of secondary pollutants.

The occurrence and quantity of endocrine-disrupting chemicals (EDCs) have become a specific target of investigations. Teil et al. (2016) examined the spatial and temporal patterns of 55 EDCs in France, comparing three sites (urban, suburban, and forest) and two climatic periods (warm vs. cold). Phthalate diesters (PAEs) were the most abundant compounds, followed by alkylphenols (APs) and polycyclic aromatic hydrocarbons (PAHs). Polychlorinated biphenyls (PCBs) and bisphenol A (BPA) were measured at considerably lower concentrations. As expected, these chemicals occurred

Phytotoxicity of Organic Pollutants
ISBN 978-0-443-33110-7
https://doi.org/10.1016/B978-0-443-33110-7.00013-4

at the highest rate in the urban areas, followed by suburban and, finally, forest sites. Seasonal analysis revealed that most EDCs are most abundant in summer, except for PAHs, which reach the highest concentrations during the winter heating period. PAH emissions were higher in the cold period, due to combustion processes. Teil et al. (2016) found that the EDCs measured in France were largely distributed in the gaseous phase (>80% in the summer).

Organic pollutants are ubiquitous in practically all environmental media—soil, water, air, and even living organisms—due to their diverse emission sources. Their occurrence and potential exposure with higher plants, however, are discussed in the literature at varying depths. Except for PAHs, mostly soil-to-plant or water-to-plant exposure pathways have been examined, providing much less information about aerial contact. This fact, unfortunately, leaves many questions unanswered about how their atmospheric occurrence would influence plants' health.

Despite all difficulties, the present report tries to give a summary of the potential phytotoxic effects of these compounds based on available literature. Readers, however, should bear in mind that these data have to be interpreted very carefully, as atmospheric pollution will differ from soil-to-plant or water-to-plant exposure pathways in many ways. First, reported studies generally use concentrations that are higher than predicted atmospheric occurrences. In addition, ecotoxicological tests by nature use acute exposure, treating test plants at relatively higher concentrations for a short time, while atmospheric exposure is more of a chronic nature: relatively low concentrations influence the life quality of the plants, but for a prolonged period or repeatedly.

1.2 Atmospheric particulate matter

Atmospheric particulate matter (PM) is grouped as coarse, fine, and ultrafine particles (UFPs) with aerodynamic diameters of 2.5–10 μm (PM10), <2.5 μm (PM2.5), and <0.1 μm (PM0.1), respectively. The sources of different compounds could be both anthropogenic and natural, including vehicular emissions, industrial activities (Caumo et al., 2023), biomass burning (Cipoli et al., 2023), secondary atmospheric reactions, as well as natural sources (e.g., dust storms, volcanic activity, forest fires, and sea spray) (Filonchyk and Peterson, 2024). PM exerts its harmful effects through physical processes, as these particles may directly block stomatal openings. However, the major risk is the presence of potentially harmful contaminants bound on the surface, such as metals (Kelly and Fussel, 2012), organic compounds, e.g., primary organic carbon (POC), secondary organic carbon (SOC) (Saarikoski et al., 2008) and polycyclic aromatic hydrocarbons (PAHs) (Kim et al., 2013). The smaller the more hazardous: particles with decreasing diameter will have increasing relative surface area, implying they can bind relatively higher amounts of potentially toxic pollutants (Samek et al., 2017). Zhang et al. (2025) estimated the concentrations of different PAHs in the PM2.5 and PM > 2.5

fractions and found that concentrations in the PM2.5 fraction were roughly 10-fold higher than in the bigger (coarse) particle fractions. The chemical composition and toxicity of PM show high temporal and spatial variability, and the primary source is also a critical factor. The season or the time of year can affect the chemical composition of PM due to seasonal variations in emissions and atmospheric conditions (Moura et al., 2024). For example, in the winter, heating activities contribute more to PM, while in the summer, photochemical reactions may lead to an increase in secondary organic aerosols.

1.3 Phytotoxic responses

Bioindication studies have long been in use to examine the behavior of plants in polluted air. Plants, being sessile organisms, involuntarily tolerate the presence of atmospheric pollutants. This tolerance, however, will depend on species-specific features; even cultivars of the same species might show varying sensitivity (Nafula et al., 2025).

A wide range of indicators has been investigated, which reflect the general stress of the plants. These indicators (or endpoints) include biochemical markers, and morphological, physiological symptoms and can even be extended to community-level responses (Bayouli et al., 2021). Leaf functional traits include reliable and sensitive endpoints such as morphological traits, leaf water status, leaf ultrastructure (Mukherjee and Agrawal, 2018), stomatal characteristics such as stomatal conductance (Bell et al., 2011), or stomatal density (Kardel et al., 2010). Symptoms of visible leaf injury, such as chlorotic and necrotic spots, were early reported in the 1940's–50's (Roth and Swenson, 1947; Middleton et al., 1950). Visual assessments have recently been substituted by measuring injured leaf area reflectance (e.g., Khavaninzadeh et al., 2014).

Inhibition of the photosynthetic activity is a reliable measure of plant stress. Decreased levels of photosynthetic pigments chlorophyll *a* and *b* were reported by Rabe and Kreeb as early as 1979. In addition to direct measurement of photosynthetic pigments, reduction in the levels of PSI and PSII photochemical activity (Kreslavski et al., 2017), reduction in net photosynthetic rate and of Rubisco activity (Tomar et al., 2019) also reflect the damage in the photosynthetic machinery. Protein content is also an easy-to-measure and useful endpoint, as air pollutants may cause an inhibitory effect on protein synthesis (Kumar and Dubey, 1998), which is reflected in the decrease in the foliar protein content (Verma and Singh, 2006).

Most studies agree that atmospheric Persistent Organic Pollutants (POPs) exert their phytotoxic effects—partially or fully—through reactive oxygen species generation. Reactive oxygen species (ROS) are defined as reactive molecules with free radicals and highly reactive oxygen molecules that may impact cells (Bailey-Seress and Mittler, 2006) and play a key role in several biological processes in plants (Bardaweel et al., 2018). In chloroplasts, mitochondria, and peroxisomes, ROS are produced in plant cells in normal metabolic processes, for example, during photosynthesis and respiration. The activation or reduction of oxygen gives different reactive oxygen species in plant cells, listed below.

Superoxide anion (O_2−): Formed by the one-electron reduction of molecular oxygen, which has limited activity and can accumulate in plant cells, particularly in meristems. In the cell, the superoxide reacts with different proteins, ascorbate, and nitric oxide (Karpinska and Foyer, 2024).

Hydrogen peroxide (H_2O_2): comes from a two-electron reduction of O2 and is produced by the dismutation of superoxide anions or through enzymatic reactions (e.g., ascorbate peroxidase) and is generated during normal processes in photosynthetic cells (Mhamdi, 2023). It can enter cell membranes, interacts with thiol-containing proteins, and activates different signalling pathways as well as transcription factors, which in turn regulate gene expression and cell-cycle processes (Slesak et al., 2007).

Hydroxyl radical (OH): is the most "damaging" ROS type, it has extreme reactivity and the ability to oxidize all cell components (Sharma et al., 2012). It is generated through the Fenton reaction or Haber–Weiss mechanism from H_2O_2 and O_2 in the presence of transition metals (e.g., copper and iron) (Gill and Tuteja, 2010).

Singlet oxygen ($^1O_2$): Produced continuously in chloroplasts under light conditions by PSII, but during light stress and photoinhibition of PSII, the production increases (Apel and Hirt, 2004). It is highly reactive and can be toxic to cells because it causes lipid peroxidation and damages proteins and DNA.

ROS has several roles in plant cells; it is essential for response to biotic and abiotic stresses (Su et al., 2024). For example, during pathogen infection, plants start to generate ROS as part of the response, which is a signal for programmed cell death, so the cell can limit the spread of pathogens and help to limit the growth of the pathogen (Mukhtar et al., 2016). ROS also has a secondary messenger function in cell signalling pathways. The increasing amount of ROS can alter the growth together with hormones (Tognetti et al., 2017). ROS modifies the activity of proteins and processes; the synthesis and action of plant hormones such as abscisic acid (ABA), auxins, and ethylene, which mediate plant responses to stress (Suzuki et al., 2012). In the study of Su et al. (2024), ROS regulated root growth and morphology. In primary root cell division, expansion, and cell wall remodelling were modulated.

During stress conditions such as drought, salinity, extreme temperatures, or pathogen attack, the production of ROS can significantly increase, and inequality between the production of ROS and the ability of detoxification of ROS may arise in what we call "oxidative stress" (Tripathy and Oelmüller, 2012).

Plants have developed experienced antioxidant systems to keep balance in ROS homeostasis (Nadarajah, 2020). These systems include two main groups of antioxidants. 1. Enzymatic antioxidants: Superoxide dismutase (SOD), catalase (CAT), and peroxidases (POX) (Joshi and Chinnusamy, 2014). 2. Nonenzymatic antioxidants: Ascorbic acid (vitamin C), glutathione, and flavonoids (Waskiewicz et al., 2014a,b). Enzyme levels have long been used to indicate toxic effects exerted by air pollution (Keller, 1974; Rabe and Kreeb, 1979; Sarkar et al., 1986).

Superoxide dismutase is one of the first defense enzymes that neutralize superoxide radicals (O_2^-). There are different forms of SOD based on the metal cofactor: Cu/Zn-SOD, Mn-SOD, and Fe-SOD. The main function of SOD is to convert superoxide radicals into hydrogen peroxide (H_2O_2) and molecular oxygen (O_2), which can then be further detoxified (Asada, 2006).

Catalase's function is to break down hydrogen peroxide (H_2O_2) into water and oxygen, reducing oxidative damage. It can be found only in peroxisomes, but it has an obligatory role in the detoxification of high-level ROS during stress (Mittler, 2002).

Peroxidases are familiar enzymes with catalase that use H_2O_2 as an electron acceptor and convert it into water and oxygen during the detoxifying process (Passardi et al., 2005). They play a central role in various plant tissues and are involved in the lignification process and response to pathogens. The nonanimal peroxidases have three classes, namely, group I, II, and III peroxidases. Group I. and Group III. are distributed in plant cells (Mathé et al., 2010). Group I. includes cytochrome peroxidase (CCP; EC 1.11.1.5), ascorbate peroxidase (APX; EC1.11.1.11), and catalase peroxidase (CP; EC 1.11.1.6). Group III. peroxidases include horseradish peroxidases (HRP), peanut peroxidase (PNP), soybean peroxidase (SBP), etc. (Pandey et al., 2017).

Ascorbate peroxidase reduces hydrogen peroxide using ascorbic acid (vitamin C) as an electron donor (Nakano and Asada, 1981); it is responsible for the fine modulation of ROS (Mittler, 2002). Ascorbate peroxidases are specific enzymes in those organisms that contain chloroplasts (Mathé et al., 2010). One important function of this enzyme is to block cell damage caused by H_2O_2, generally in chloroplasts.

Glutathione peroxidase also catalyzes the reduction of hydrogen peroxide using glutathione as an electron donor (Attacha et al., 2017). It helps to prevent oxidative damage in plant cells and is particularly important during stress conditions such as drought and high salinity (Waskiewicz et al., 2014a,b).

Glutathione S-transferases are detoxifying enzymes that belong to multifunctional proteins. They are catalyzing glutathione conjugation to reactive molecules, making them more water-soluble and easier to detoxify (Estévez and Hernandez, 2020).

Glutathione reductase is a flavoprotein that is responsible for reducing oxidized glutathione (GSSG) to its reduced form (GSH), which is vital for the proper functioning of glutathione peroxidases and other ROS scavenging enzymes (Sharma and Dubey, 2005).

Thioredoxins are small proteins that help in the reduction of ROS and are found in chloroplasts, cytosol, and mitochondria (Dos Santos and Ray, 2006). Another important function of this enzyme is to decrease disulfide bridges of oxidized target proteins and also to play an important role in the reduction of other proteins with the cut of disulfide bonds (Balmer and Buchanan, 2002). Thereby, By the maintenance of cellular redox balance, they are essential in protecting plant cells from oxidative stress.

To compare the performance of plant species in potentially polluted environments, the air pollution tolerance index (APTI) has been suggested (Singh et al., 1991). APTI is based on four biochemical parameters: total chlorophyll content, relative water content, pH of leaf extract, and ascorbic acid (Karmakar et al., 2021). APTI, however, is a measure of general stress and is not pollutant-specific (Anand et al., 2022) as it has been suggested to assess plant health affected by traffic-imposed heavy metal pollution (Nadgórska-Socha et al., 2017).

2. Chemicals of interest

2.1 Polycyclic aromatic hydrocarbons

2.1.1 PAHs in the atmosphere

Polycyclic aromatic hydrocarbons (PAHs) are ubiquitous organic pollutants. PAHs were classified as persistent organic pollutants (POPs) by the Stockholm Convention, which entered into force in 2004. They contain two or more fused benzene rings. These compounds are characterized by low water solubility, which decreases with increasing molecular mass.

Over 200 congeners have been described, of which 16 have been designated by the United States Environmental Protection Agency (US EPA) as being of greatest concern, due to their environmental hazard (Table 3.1). Although this list has been questioned (Andersson and Achten, 2015), most scientific reports still address them.

PAHs are generated during the incomplete combustion of biomass and fossil fuels. Emission sources include natural processes such as fires, however, anthropogenic activities provide the main inputs (Zhang et al., 2023). These sources include the combustion of fossil fuels, biomass burning, industrial activities, or waste incineration.

In the global context, input from anthropogenic sources shows strong spatial and temporal patterns. Most studies agree that when assessing the contribution of the combustion of fuels for heating, it significantly increases winter pollution (Siudek, 2022). This seasonal pattern, naturally, will also influence the exposure of plants to airborne PAHs. Traffic and industry will provide whole-year-round sources, impacting practically all plants: natural coniferous and deciduous forests, annuals, perennials, crop species, etc. On the other hand, winter heating-generated PAH emissions mostly damage coniferous species.[1]

[1] It should be noted that the growing season will most likely be extended due to climate change (Anav et al., 2019).

Table 3.1 The physicochemical properties of EPA 16 PAHs.

Compound name	Chemical formula (H)	Molecular weight (g mol^{-1})
Naphthalene	$C_{10}H_8$	128.2
Acenaphthylene	$C_{12}H_8$	152.2
Acenaphthene	$C_{12}H_{10}$	154.2
Fluorene	$C_{13}H_{10}$	166.2
Phenanthrene	$C_{14}H_{10}$	178.2
Anthracene	$C_{14}H_{10}$	178.2
Fluoranthene	$C_{16}H_{10}$	202.3
Pyrene	$C_{16}H_{10}$	202.3
Benzo[a]anthracene	$C_{12}H_{12}$	228.3
Chrysene	$C_{18}H_{12}$	228.3
Benzo[b]fluoranthene	$C_{20}H_{12}$	252.3
Benzo[k]fluoranthene	$C_{20}H_{12}$	252.3
Benzo[e]pyrene	$C_{20}H_{12}$	252.3
Benzo[a]pyrene	$C_{20}H_{12}$	252.3
Indeno[1,2,3-cd]pyrene	$C_{22}H_{12}$	276.3
Dibenzo[a,h]anthracene	$C_{22}H_{14}$	278.3
Benzo[g,h,i]perylene	$C_{22}H_{12}$	276.3

2.1.2 Exposure pathways and uptake mechanisms

In the atmosphere, PAHs can occur in the atmosphere in gaseous form (mainly PAHs with low molecular weight) or bound to particulate matter (PAHs with higher molecular weight) (Liu et al., 2022). Plants are exposed to both phases. Gaseous PAHs normally diffuse through the wax and cuticular lamellae into the interior parts of the leaves; also, they can enter directly through the stomata (Lehndorff and Schwark, 2004). PAHs adsorbed on atmospheric particles are first primarily deposited on the surface of the leaves and desorbed from the particles (De Nicola et al., 2017). However, depending on the size of the particle, PM can directly pass through the stoma (Fig. 3.1).

The uptake from the gas phase represents a more important exposure route than the particle deposition onto plant surfaces (Ortiz et al., 2023). It is generally assumed that foliar uptake is the major PAH exposure pathway for higher plants (Tao et al., 2004). However, PAHs from the atmosphere can be transported to the soil, providing an additional exposure route for plants (Kulhánek et al., 2005). Accumulation of PAHs in soils has been widely addressed, as a significant proportion of PAHs remains in the soil, depending on the type of the compound (Sehili and Lammel, 2007). Being lipophilic molecules, PAHs generally tend to bind strongly to soil particles (Yang et al., 2013), and soils will then provide a sink for these compounds.

Behavior of low molecular weight (LMW) and high molecular weight (HMW) PAHs differs: 4–6-ring PAHs are most likely deposited close to emission sources, while the lower molecular weight compounds are more capable of long-range transport (Nadal

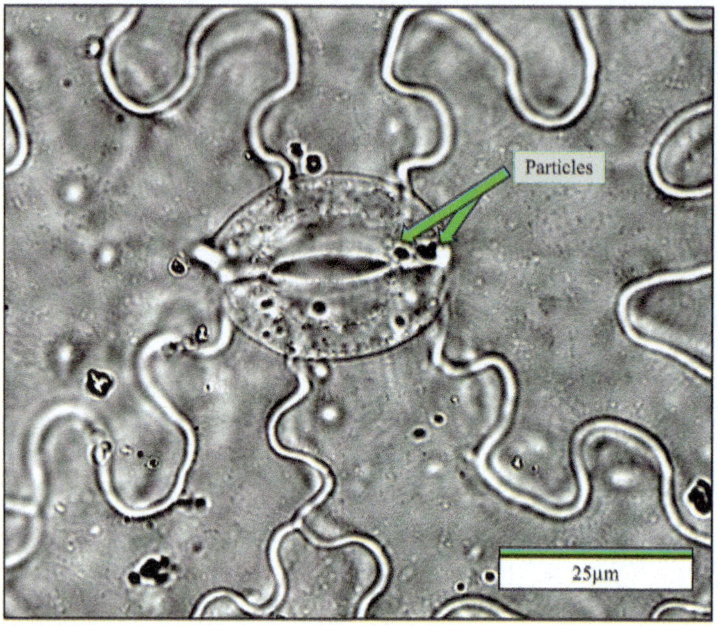

Figure 3.1 $PM_{2.5}$ particles around stomata of tomato (*Lycopersicum esculentum*) leaves. These leaves were experimentally treated with diesel exhaust PM. The photo was taken by Thermo Scientific *Talos F200X* scanning transmission electron microscope.

et al., 2004). Also, after reaching the ground, LMW PAHs will be degraded or returned to the atmosphere due to volatilization or microbial biodegradation (Al-Nasir et al., 2022); HMW PAHs will remain in the soil bound to particles (Biache et al., 2014). As PAHs are poorly soluble in water, once deposited in the soil, their downward migration is limited. They mostly accumulate in the upper layers, according to Clément et al. (2015), the upper 10 cm layer is the most affected.

Zhang et al. (2018) showed a close relationship between PAHs in dust fall and surface soil. In the study of Jia et al. (2019), soil absorption contributed to 9.4% of total PAH uptake in tested vegetables, demonstrating that atmospheric dust fall serves as an important source for increasing soil PAH concentrations. Once uptaken, compounds are assumed to be transported within the plant through the plant vascular system, or by diffusion through the cells (Watts et al., 2008).

Aerial tissues of winter wheat samples were analyzed in the study of Tian et al. (2018). Samples were collected in areas affected by coal combustion, and PAH concentrations were measured. Results showed that HMW (6-ring) PAHs originated from the atmosphere, while LMW (3-4-ring) species from root uptake. Xiong et al. (2017) investigated the relationships between bioconcentration factors and the physicochemical properties of PAHs and found that solubility and the octanol-water partition coefficient were strongly correlated with the soil-to-root bioconcentration factors, while vapor pressure and the octanol–air partition coefficient were proven to be good predictors for the air-to-leaf accumulation of PAHs.

In addition to the direct deposition of particles on soil or plant surfaces, both gas and particle phase PAHs can be washed out by rain or snow, defined as wet deposition (Birgül et al., 2011). Wet deposition provides an important pathway for PAHs removal from the atmosphere; however, it depends on precipitation amount, showing seasonal patterns (Wang et al., 2016). The process has been proven efficient for all particle-associated PAHs (Škrdlíková et al., 2011).

Figure 3.2 illustrates possible exposure routes for both LMW and HMW PAHs.

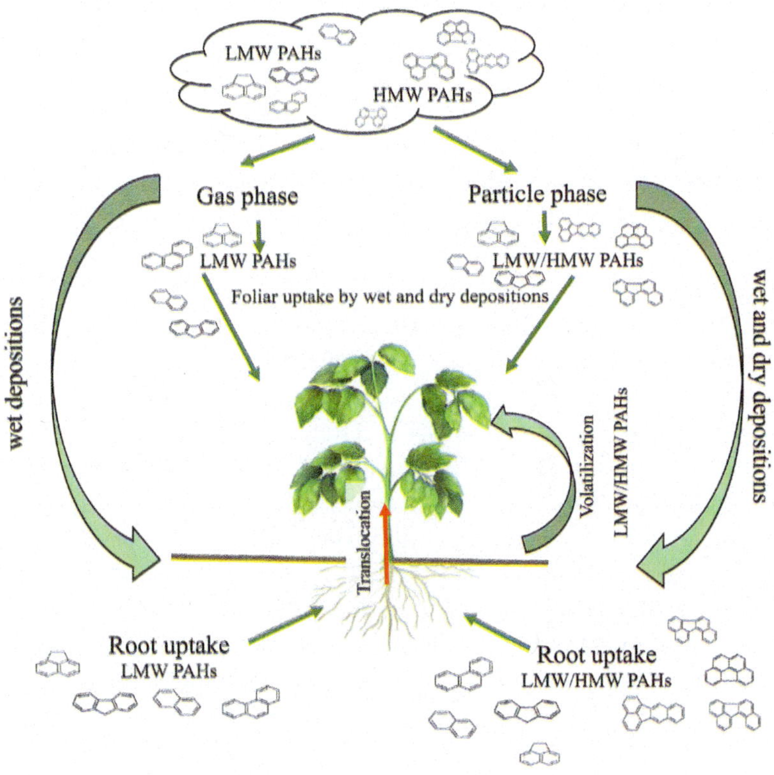

Figure 3.2 Typical exposure routes for LMW and HMW PAHs: uptake through roots versus foliar uptake.

2.1.3 Assessment of complex environmental samples

Incomplete combustion of fossil fuels, biomass, or even plastic materials results in the release of a wide range of PAHs. In order to assess the deleterious impacts of traffic, especially diesel emissions, "traditional" bioindication studies can establish a cause-and-effect relationship by investigating the spatial pattern of symptoms recorded. In simple words, if symptoms are less and less severe as we move away from the hypothesized source of pollution, then there is a real causality: the effect is purportedly linked to the cause.

However, traffic-related emissions are complex mixtures of possibly phytotoxic compounds, both organic and inorganic. A wide range of studies have concentrated on nitric oxide (NO) and nitrogen dioxide (NO_2) as they occur in relatively high concentrations in exhaust emissions and have proven phytotoxicity.

Viskari et al. (2000a,b) designed a fumigation chamber for the analysis of exhaust emissions. These emissions were generated by a 4-stroke spark-ignition motor, which was fueled with lead-free gasoline. Abundant VOCs were in the air of the exposure chamber: toluene, xylenes, ethylbenzene, methyl, and propyl-benzenes, as well as naphthalene. Exposed Norway spruce seedlings (*Picea abies* (L.) Karst) showed stress symptoms as concentrations of proline, glutamine, threonine, aspartic acid, glycine, and phenylalanine increased, but concentrations of arginine, serine, alanine, and glycine decreased in young needles (Viskari et al., 2000a). Deterioration of the epistomatal wax structure and mesophyll ultrastructure in needles was also experienced (Viskari et al., 2000b).

Under laboratory conditions, Honour et al. (2009) and Bell et al. (2011) tested the phytotoxic effects of simulated diesel exhaust emissions in a Solardome fumigation facility, setting prolonged exposure, varying between 12 and 45 weeks. Native herbaceous vegetation was represented by 12 species such as: great plantain (*Plantago major*) (Plantaginaceae), black knapweed (*Centaurea nigra*) (Asteraceae), sorrel (*Rumex acetosa*) (Polygonaceae), autumn hawkbit (*Leontodon autumnalis*) (Asteraceae), common bird's-foot trefoil (*Lotus corniculatus*) (Fabaceae), white clover (*Trifolium repens*) (Fabaceae), perennial ryegrass (*Lolium perenne*) (Poaceae), common sowthistle (*Sonchus oleraceus*) (Asteraceae), white goosefoot (*Chenopodium album*), Oxford ragwort (*Senecio squalidus*) (Asteraceae), annual meadow grass (*Poa annua*) (Poaceae) and Timothy grass (*Phleum pratense*) (Poaceae). The study was based on the measurement of a wide range of endpoints, including growth (based on plant height and leaf number), flower development, leaf senescence, and leaf surface wax characteristics.

These studies intended to follow a relatively long exposure, also comparing the sensitivities of younger plants to older ones. The generalized conclusion was that older plants proved less sensitive than younger plants. However, in terms of reliability or sensitivity of individual endpoints, no specific trend could be experienced, as most of the symptoms were strongly species-specific. Premature leaf senescence, however, seemed to be a general response across species.

Unfortunately, the PAH content of the emission was not monitored; pollution levels were assessed by following NOx, benzene, and toluene concentrations. It should be noted that the ecotoxicity of the volatile organic compound, toluene, has been much less addressed. According to Paterson et al. (1990), the main source could be the volatilization from hazardous waste sites; however, toluene was detected in the aqueous extract of winter urban aerosol samples in a concentration of 0.84 μg/L (Kováts et al., 2017). Toluene was also detected in diesel exhaust emissions (Mondal et al., 2014). Jen et al. (1995) used experimental chambers (both flow-through and air-tight systems) to assess the foliar uptake and potential phytotoxicity of this compound using Soybean [*Glycine max* (L.) Merr. cv. Davis]. Not only was a dose-dependent effect established, but accumulation was experienced.

Diesel emissions were tested on Guabiroba (*Campomanesia xanthocarpa* O. Berg.) plants using an extended exposure of 120 days (Storch-Böhm et al., 2020). During the exposure, emission intensity was changed, simulating actual traffic conditions. Damage was assessed by leaf injury (calculating an index based on chlorosis and necrosis), plant biomass, and physiological/biochemical parameters (chlorophyll, peroxidase enzyme activity, and protein concentrations). These end-points showed different sensitivities, in the order of peroxidase > biomass/chlorophyll > protein > leaf injury.

In a glasshouse experiment by Paull et al. (2018), hybrid diesel fuel candles were used to simulate traffic-related exposure conditions. The sensitivities of eight commonly used green wall plants were compared based on multiple end-points such as leaf area, leaf pH, relative water content, chlorophyll concentrations, stomatal conductance, and Maximum Quantum Efficiency of Photosystem II. The sensitivity of these end-points varied by species; however, stomatal conductance significantly decreased in all cases, proving to be an ultimate response.

Testing of atmospheric samples is rather difficult to standardize. Kováts et al. (2017) adopted and modified the No. 227 OECD Guideline for the Testing of Chemicals: Terrestrial Plant Test: Vegetative Vigor Test (OECD, 2006) to simulate wet deposition implied phytotoxicity. The Guideline was first designed to assess the phytotoxicity of general chemicals, biocides, and crop protection products by spraying the test substance on the above-ground parts of the test plants (Boutin et al., 2012; Carpenter et al., 2013).

According to the OECD protocol, phytotoxicity is assessed 3–4 weeks after the plants receive the treatment, based on growth impairment (reduced biomass or length in comparison to the control) and visible symptoms. In order to get a more comprehensive insight into the possible effects of the atmospheric samples, these endpoints were supplemented by biochemical markers such as concentration of photosynthetic pigments, peroxidase, and total protein content.

Aqueous extract prepared from PM1-10 emitted by operating a diesel-powered vehicle was used to compare the sensitivity of different, common European roadside plants including wild parsnip (*Pastinaca sativa* L.), common daisy (*Bellis perennis* L.), common thistle (*Cirsium arvense* (L.) Scop.), common sowthistle (*Sonchus oleraceus* L.), common dandelion (*Taraxacum officinale* F.H. Wigg.), soapwort (*Saponaria officinalis* L.), goosefoot (*Chenopodium album* L.), white clover (*Trifolium repens* L.), Mediterranean sage (*Salvia aethiopis* L.), hoary plantain (*Plantago media* L.), common sorrel (*Rumex acetosa* L.), meadow buttercup (*Ranunculus acris* L.) and wood avens (*Geum urbanum* L.) (Kováts et al., 2021). Biomass reduction in comparison to the control reflected the deleterious effects of traffic-related emissions; however, these common roadside plants showed varying sensitivities, and many of them proved resistant to road environments. A similar diesel exhaust sample was applied in the study of Hubai et al. (2021) to test if bioactive compound production in essential oil-bearing plants can be used as a measure of chemical stress. Linalool production in sweet basil (*Ocimum basilicum* L.) was shown to be significantly impacted.[2,3]

It should be noted that the $\sum$PAH content of the diesel extract used in these works was 0.818 μg/L, with the dominance of phenanthrene (0.264 μg/L) and fluoranthene (0.220 μg/L). Birgül et al. (2011) reported dissolved and particle phase rainwater concentrations of PAHs amounting to 0.290 μg/L and 0.665 μg/L in a Turkish study, implying that the PAHs content of the simulated diesel deposition represented actual environmental exposure.

In these studies, particles emitted by diesel-powered vehicles were collected on glass-fibre filters, and an aqueous extract was further made of these filters. A similar extraction procedure was applied for emission samples generated during the controlled burning of plastic waste. Open burning of plastic waste is a global problem (Velis and Cook, 2021), releasing significant amounts of persistent organic compounds (Cogut, 2016). Plastic waste burning samples elucidated measurable plant stress indicated by the elevated levels of peroxidase enzyme activity (Hubai et al., 2022).

Direct toxicity of particles can also be addressed. Originally developed for testing soil phytotoxicity, the international standard "ISO/DIS 18,763: Soil quality—Determination of the toxic effects of pollutants on germination and early growth of higher plants" assesses the level of soil pollution by calculating the ratio of germinated seeds after 72 h exposure.

Mondal et al. (2014) followed this principle to assess plant germination on black carbon (BC) samples collected directly from the emission pipes of a diesel-powered car. Reduced germination rate was reported; in young plants, the abnormal structure of stomata was detected. Daresta et al. (2015) assessed the germination rate of tomatoes grown on quartz fibre filters containing PM10 samples, which were collected in urban

[2] Blande (2021) hypothesize that air pollution-induced changes in the volatile oil production of plants will in turn plant-insect interactions, triggering community-level alterations.

[3] Paul et al. (2024) reported that essential oil production of Psidium guajava was negatively affected in sites with heavy traffic in Kolkata city (India).

background monitoring sites of Canosa (Apulia, Southern Italy). After emergence, plants were grown for a further 18 days. In the roots, oxidative stress was demonstrated. The phytotoxic effect of PM10 on photosynthetic activity was also reported, as two stress indicators, Chl a/Chl b and Chl(a + b)/carotenoid content, showed a decrease in all filters. In the samples, significant PAH contamination was measured.

2.1.4 Phytotoxicity of selected PAHs

The above-mentioned fumigation chamber-based experiments agree that an artificially generated emission is used as a sample. It will, in fact, represent or approximate real-world traffic conditions; however, it will still be difficult to link phytotoxic responses to a certain pollutant. On the other hand, laboratory-based experiments have been reported where a single PAH was chosen as a test material.

The following PAHs have been investigated in most of the cases: naphthalene (NAP), phenanthrene (PHE), and fluoranthene (FLT). They are lighter (two to three-ring PAHs). According to Fang et al. (2021), gasoline exhaust was the largest source of NAP emission in the Pearl River Delta region (China), followed by biomass/coal burning and diesel exhaust. NAP has been widely used as a combustion tracer for vehicle emissions (Uria-Tellaetxe et al., 2016).

FLT has been commonly applied in phytotoxicity studies as a model PAH. Babula et al. (2012) studied the cytotoxic effects of FLT on an experimental plant cell model, tobacco BY-2 cell suspension culture. FLT was found to generate ROS, causing irreversible damage to cell integrity. It also accumulated in the cells. Freshly cut leaves of pea plants (*Pisum sativum* L.) were immersed in a solution containing FLT concentrations of 0.1, 1, and 10 mg/L (Kummerová and Vánová, 2007). The inhibitory effect of photosynthesis was demonstrated, indicated by chlorophyll fluorescence parameters ($F0$, FV/FM, and ΦII). In a following study (Vánová et al., 2009), more specific parameters were measured: levels of phytohormones such as abscisic acid (ABA) and cytokinins. These hormone concentrations proved to be sensitive indicators of plant stress, ABA showing concentration-effect response to FLT treatments.

In the study of Wang et al. (2022), a mixed solution of three PAHs (naphthalene, phenanthrene, and benzo(b)fluoranthene) was sprayed on the leaves of winter wheat (*Triticum aestivum* L.) during different stages of its life-cycle. The concentrations applied were 0.75 and 4.5 mg/L, but only the highest concentration elucidated a measurable effect, resulting in stem biomass reduction.

Wieczorek et al. (2015) applied 3-ring anthracene (ANT) and 5-ring benzo[*k*]fluoranthene (BkF), and the combination of ANT and BkF on the above-ground parts of celery plants for 45 days. Plants were treated thrice a day, the concentration applied was 4 mg/L in each treatment. ANT applied alone reduced ascorbic acid levels. As a response to all treatments, the total phenolic content of leaves increased, indicating stress. However, no synergistic effect was detected.

Desalme et al. (2011a,b) tested PHE in a concentration of 160 μg/m^3 in a controlled chamber study. In the chambers, the air was enriched with PHE by passing through an evaporator filled with PHE pills. The test species was red clover (*Trifolium pratense* L., var. Jonas, Caillard), and the exposure used was 1 month. Ultimate biomass reduction was observed, most possibly caused by reduced net assimilation rate, lower carbon assimilation rate, and chlorophyll content. Biomass reduction was highest (25%) in leaves. In addition, disturbances in carbon allocation were also observed. The authors compared PHE concentrations applied in the study with reported atmospheric loads, claiming that the study represented actual atmospheric exposure. A similar device was used to compare the sensitivities of red clover and ryegrass (*Lolium perenne* L.), setting PHE exposure of 150 μg/m^3 for 1 month. In clover, a significant reduction of biomass (30%) was measured in the shoots, but root biomass remained stable. The other test species, ryegrass, showed practically negligible sensitivity.

Oguntimehin et al. (2008) applied a growth chamber to study the effect of two PAHs, FLT and PHE, on the evergreen conifer, Japanese red pine (*Pinus densiflora* Sieb. et Zucc.), using prolonged exposure of 3 months. 2-year-old seedlings were sprayed with the test solution containing 10 mM FLT or PHE. Both PAHs had negative effects on net photosynthesis, chlorophyll fluorescence, stomatal conductance, and total chlorophyll content. Also, visible decoloration (yellowing) of the test plants was demonstrated after the exposure. The two PAHs, however, had different phytotoxic potentials, with FLT exerting more severe effects. The difference can be attributed to their different lipophilicity, as the log K_{OW} of FLU is 5.16 while that of PHE is 4.48. Higher K_{OW} indicates higher bioaccumulation capacity. Phytotoxic effects were explained by the generation of OH radicals, which could have multiple inhibitory mechanisms, for example, reducing the synthesis of enzymes in the C3 pathway. Ribulose 1,5-bisphosphate carboxylase (Rubisco) content in current-year needles showed reduced levels after PAH exposure. The importance of OH radicals in phytotoxic effects was demonstrated by adding mannitol to PHE or FLT-containing solutions. Mannitol, which has known OH-radical scavenger capacity, mitigated negative effects of PAHs.

Higher toxicity of FLT fluoranthene was also supported by Khpalwak et al. (2018) when individual and combined effects of FLT, PHE, mannitol, and sulfuric acid were assessed using marigold (*Calendula officinalis*) as a model species.

The same experimental setup was used by Oguntimehin and Sakugawa (2008) to determine the relationship between exposure time and damage caused. FLU was applied in the same concentration as in the previously reported study (10 mM FLU), but phytotoxic responses were assessed after 5, 10, and 14 days of exposure. Symptoms such as gas exchange measurements appeared only after the 10-day exposure period; no negative effects could be measured during the first 5 days. Mannitol was reported to mitigate phytotoxic effects.

Cherry tomato plants (*Lycopersicon esculentum* Mill) were also subjected to fluoranthene treatment (Oguntimehin et al., 2010a). Seedlings were sprayed with FLT inat a concentration of 10 μM and a mixture of FLT and mannitol solutions during the 30-

day test period. Phytotoxicity was assessed based on the following parameters: CO_2 uptake and water vapor exchange, chlorophyll fluorescence, leaf pigment contents, visual symptoms, and biomass allocation. Treated plants all showed inhibition in these endpoints; in addition, visible injury symptoms were seen on the foliage. Similar to other studies, the mitigating effect of mannitol was demonstrated.

Other test plants, the annual herbaceous sunpatiens (*Impatiens* spp.) were exposed separately or in combination to FLT (10 μM), ozone gas (130 ppb), and sulfuric acid mists (pH 3) using a 21-day exposure time (Oguntimehin et al., 2010b). Chlorophyll fluorescence and gas exchange were negatively affected by each treatment; however, FLT and acid mist seemed to exert a synergistic effect when applied in combination. A similar experience was conducted on juvenile Japanese red pine (*P. densiflora* Sieb. et Zucc.) (Oguntimehin et al., 2013). In this case, relatively long exposures, 20 and 60 days were used. Also, effects were assessed under different pH conditions. Endpoints included gas exchange measurement, chlorophyll fluorescence analysis, and analysis of needle chemical content. Effects were pronounced by acid mist at pH 2 and 3, and even visible damage appeared. These effects were enhanced by the addition of FLT to the mist.

The most important aspect of this study was that the authors compared the estimated deposition rate of FLT to doses applied during the experiment, taking the average value of 2.81 $\mu g/m^2$/day. In comparison, the amount of FLT applied in a 21-week treatment period was calculated as 113 $\mu g/m^2$/day, considerably exceeding realistic exposure.

Foliar treatment was followed in a study of Ahammed et al. (2012a), spraying PHE solution in 30, 100 and 300 μM concentrations on the following test plants: pakchoi (*Brassica rapa* cv. chinensis), cucumber (*Cucumis sativus* cv. Jinyan4), flowering Chinese cabbage (*Brassica campestris* L. ssp. chinensis var. purpurea Bailey), tomato (*Solanum lycopersicum* L. cv.Hezuo903) and lettuce (*Lactuca sativa* L.). Concentration-dependent decrease in growth, photosynthesis, and chlorophyll contents was found; also, antioxidant enzymes (superoxide dismutase, guaiacol peroxidase, catalase, ascorbate peroxidase, and glutathione reductase) were upregulated. However, a decrease in the antioxidant enzymes was experienced in higher concentrations, indicating severe damage to plants. 24-epibrassinolide (EBR), a bioactive plant steroid, was proven to counteract the negative effects of PHE through the enhancement of antioxidant and detoxification enzymes, inducing the transcripts of antioxidant genes and enhancing the activities of antioxidant and detoxification enzymes (Ahammed et al., 2015). It was also shown to induce the expression of detoxification genes (CYP90b3, GSH1, GST1) in tomato (Ahammed et al., 2012b) and to mitigate growth inhibition in tomato seedlings (Ahammed et al., 2012c).

Othman et al. (2012) compared the toxicity of benz(a)anthracene and fluoranthene on seven species of marine algae belonging to pico-, nano-, and microphytoplankton. Concentrations applied were in the range of up to 2 mg/L. FLT was more toxic than benz(a)anthracene; LOEC was 50.6 μg/L, while that of benz(a)anthracene was

186 μg/L. After 24 h exposure, toxicity was assessed based on photosynthetic parameters such as the maximum quantum yield Fv/Fm. After 72 h exposure, the endpoint used was cell density. In both cases, concentration-dependent negative effects were experienced; however, the damage was also dependent on the cell size of the algal species tested, with larger species showing lower sensitivity.

Wieczorek and Wieczorek (2007) examined the phytotoxicity of anthracene (ANT) by spraying the test solution separately onto the leaves of crop species (*Lactuca sativa* L. and *Raphanus sativus* L.) and onto the soil surface. Treated plants received a total ANT dose of approximately 140 μg in the first group and 280 μg in the second group. Exposure was 21 days for lettuce and 12 days for radish plants. Both foliar and soil spraying resulted in lower biomass production; however, only foliar application caused decreased photosynthetic rate.

2.1.5 Phytotoxicity of photomodified PAHs

PAHs strongly absorb UV-B radiation both in the atmosphere and in the aquatic environments, resulting in the formation of several products. Mallakin et al. (2000) examined the photomodification pathway of ANT under simulated solar radiation, detecting over 20 compounds such as anthraquinones, hydroxyanthraquinones, benzoic acids, and phenols. Mallakin et al. (1999) compared the toxicities of ANT and its photomodified products on *Lemna gibba* and found higher toxicities for the photomodified compounds (anthraquinone and hydroxyanthraquinones).

Photomodification of three PAHs: anthracene (ANT), benzo[*a*]pyrene (BaP), and fluoranthene (FLT) resulted in higher toxicity measured as biomass reduction in *Brassica napus* (canola) (Ren et al., 1996). Higher toxicity of photoinduced ANT and BaP compounds was also reported by Huang et al. (1996) when canola and Cucumber were foliar exposed. The toxicity of FLT and its photomodified form was studied by using acute (12, 24, and 48 h) and prolonged exposure (up to 25 days) (Kummerová and Vánová, 2007; Kummerová et al., 2008). Previously cut leaves of pea plants (*Pisum sativum* L.) were immersed in FLT solution in increasing concentrations (0.1, 1, and 10 mg/L. Based on chlorophyll fluorescence parameters (*F*0, *FV*/*FM*, and ΦII), enhanced toxicity of photomodified FLT (phFLT) was suggested. Higher phytotoxicity of photomodified forms of phenanthrene (Huang et al., 1993, 1995), pyrene (Ren et al., 1994; Huang et al., 1995), and naphthalene (Ren et al., 1994) was also reported.

Tomar and Jajoo (2015) investigated what end-points reflect best the increased phytotoxicity of photomodified FLT in wheat. Primary photochemistry and the photosystem II system were the most affected, resulting in reduced growth. No difference was experienced, however, in the pattern of antioxidant enzymes.

3. Bisphenol A (BPA)

Bisphenol A [2,2-bis (4-hydroxyphenyl) propane, BPA; CAS No. 80-05-7] is an industrial chemical used as a monomer in the production of polycarbonates and epoxy resins. BPA-based plastic polymers are commonly applied as an additive to various products such as thermal paper, paints, food containers, flame retardants, electronic products, etc. Due to its widespread use, BPA has become one of the world's most productive chemicals. Its estimated global market was 6.4 million tons in 2023 (https://www.researchandmarkets.com).

Due to its high production and consumption, BPA has become ubiquitous in all environmental compartments. In Europe, reported soil concentrations reached the ~140 μg/kg level (Kwak et al., 2018). As high as 1400 μg/ L concentration was detected in landfill leachate in Germany (Tappert et al., 2024). Considering aquatic environments, as high as 922 ng/L concentration was detected in surface water in China and 546 ng/L in India (reviewed by Dogra et al., 2024).

The half-life of BPA is rather short, approximately 4.5 days in water and soil, and less than 1 day in the air. Therefore, in classical terms, it is not persistent, though it is often added to other POPs due to its sustained release into the environment.

3.1 Atmospheric sources and occurrence

Fu and Kawamura (2010) mapped global atmospheric concentrations of BPA. The highest occurrence was detected in Chennai, India (17,400 pg/m3), followed by another Indian city, Mumbai (9820 pg/m3). These high episodes were recorded at nighttime, showing temporal coincidence with the open burning of municipal waste. A typical tracer of plastic burning, 3,5-triphenylbenzene (Simoneit et al., 2005), was also measured in these aerosol samples, giving indirect evidence that burning of plastic-enriched municipal waste could be the primary source of BPA emission. Direct measurements of open-air barrel burn samples revealed the occurrence of BPA in the emissions (Sidhu et al., 2005). In Chinese cities, BPA was found to occur in fine particles, due to its lower volatility (Li and Wang, 2015).

Munyaneza et al. (2022) measured the occurrence of BPA and its analogues in fine particulate matter (PM2.5) samples in Shanghai (China). BPA dominated the samples and was identified in 88% of the cases, followed by bisphenol F (BPF) (82%) and bisphenol S (BPS) (75%). The highest average BP concentrations were found in winter samples (BPA: 3.47 ng/m^3; BPF: 2.46 ng/m^3; BPS: 0.58 ng/m^3, and bisphenol AF (BPAF): 1.14 ng/m^3). Direct emission of BPA and four analogues were detected among the gaseous compounds released from petrochemical complexes that produce polypropylene (Fernández et al., 2022).

Based on perceived human health risk, Directive 2011/8/EU was issued in 2011 on "Prohibition of the BPA use in baby bottles and drinking cups." In order to avoid potential risks posed by BPA, bisphenol analogues such as bisphenol S (BPS), bisphenol D (BPD) or bisphenol AF (BPAF) have been introduced and can be found in "BPA-free" products. These compounds, however, are structurally similar to BPA, which questions their safety (Thoene et al., 2018). Da Silva Santos et al. (2023) compared the toxicities of BPA, bisphenol F, and bisphenol S exposing *Salvinia biloba* to environmentally relevant and sublethal concentrations (1, 10, 50, 100, and 150 μM). Similar toxicities were recorded, affecting lipid peroxidation, free-radical generation, photosynthetic pigments, and chlorophyll fluorescence.

Bisphenols (BPA and analogues) have already been detected in indoor air (Vasiljevic and Harner, 2021). The main sources, in addition to infiltration of outdoor air, are locally used plasticware. When they are exposed to high temperatures (heat sources, sunlight through windows, etc.), bisphenols can be emitted and become airborne. Physical abrasion can also result in the release of bisphenols (Moorchilot et al., 2025).

3.2 Phytotoxic effects

Dos Santos et al. (2024) give an extensive review of the comparative (aquatic) toxicity of a wide range of bisphenols on different taxa (algae, crustaceans, fish, and others). The review concludes that "analogues were many times more toxic than BPA, which applied to acute and chronic toxicity, regardless of the trophic level."

At present, practically no studies are available that address the atmospheric uptake of BPA by either lower or higher plants. On the other hand, a wide range of reports suggests its phytotoxicity using other—soil-to-plant or water-to-plant pathways. In terms of test layout, two types of experiments can be distinguished: test plants are either cultivated in BPA-spiked soil or in BPA solution.

Abdelmoneim et al. (2024) conducted a pot study, cultivating tomato plants in soil spiked with BPA in concentrations of 250, 500, 750, 1,000, and 1250 mg/kg. Biomass inhibition and a decrease in all measured photosynthetic pigments content (Chl-*a*, Chl-*b*, and carotenoids) were found in a concentration-dependent manner. BPA-induced osmotic stress was also recorded, as the content of soluble sugars and soluble proteins decreased. Moreover, genetic analyses revealed that BPA elucidated an increase in relative gene expression of POD, reaching maximum expression at 1250 mg/kg soil concentration.

Liang et al. (2022) exposed duckweed (*Lemna minor*) to BPA at 1, 5, 20, and 50 mg/L. At a low (5 mg/L) concentration, a higher growth rate and an increase in the concentration of photosynthetic pigments (chlorophyll *a*, *b*, and total Chl) was experienced, but BPA at the highest concentration applied (50 mg/L) had a significant inhibitory effect on plant growth, the Chl content, and the maximal photochemical efficiency. Chl content showed a positive and significant correlation with the fresh weight of plants, indicating that changes in the Chl content might influence the growth of plants.

Photochemical efficiency of PSII expressed as Fv/Fm decreased at 50 mg/L, showing that this BPA concentration would destroy the Chl fluorescence.

In addition, the 50 mg/L concentration induced ROS accumulation, which in turn increased the activities of antioxidant enzymes. Stimulating effects of BPA at low concentrations were experienced in other plants such as *Arabidopsis thaliana* (Rapala et al., 2017) and soybean (Jiao et al., 2017). Zhang et al. (2015) carried out a comparative study on seedlings of five plants, such as tomato (*Lycopersicum esculentum*), lettuce (*Lactuca sativa*), soybean (*Glycine max*), maize (*Zea mays*), and rice (*Oryza sativa*), using chlorophyll fluorescence techniques. Low concentrations, 1.5 and 3.0 mg/L BP,A had a stimulating effect, improving photosystem II efficiency. 1.5 mg/L was chosen, representing the limit set for drinking water by the U.S. Environmental Protection Agency. On the other hand, concentrations of 10.0 and 17.2 mg/L inhibited the photochemical reaction and damaged the photosystem II reaction centre. The five plants showed different sensitivity, lettuce being the most and rice the least sensitive.

A similar approach was used in the study of Jiao et al. (2017) when soybean seedlings were directly treated with BPA solutions. The lowest concentration was also 1.5 mg/L, but two relatively high concentrations, 17.2 and 50.0 mg/L, were set representing landfill leachates and contaminated sites. The test demonstrated that higher concentrations inhibited Chl synthesis by decreasing the concentrations of key enzymes. Treatment also resulted in lower concentrations of Chl in a dose-response manner. Qiu et al. (2013) reported that concentrations of 7.0, 12.0, 17.2, and 50.0 mg/L BPA negatively affected Chl concentrations and the chlorophyll fluorescence in soybean seedlings.

Aquatic tests also showed the harmful effects of BPA. Esperanza et al. (2020) reported that in the bioassay using *Chlamydomonas reinhardtii* microalga, the 96 h-EC50 value for growth was 30 mg/L. Biomarkers, including oxidative stress markers and cytoplasmic membrane potential, were also affected in exposed cells. Accumulation reaching 0.16 pg BPA/cell was also measured.

Lu et al. (2015) applied foliar treatment with a 50 μg/L BPA solution using tomato and lettuce. An application volume of 5–200 mL/day was used for lettuce and 2–30 mL/day for tomato. The treatment and the concentration applied were intended to simulate realistic environmental exposure when crops are irrigated with reclaimed water. Tomato fruits accumulated 8.3 ± 3.5, lettuce leaves 128.9 ± 17.4 μg/kg.

The above studies generally concluded the deleterious effects of BFA on plant growth and photosynthesis. Also, the level of antioxidant enzymes is a very often applied endpoint. Other studies, however, try to investigate other mechanisms that can reflect the health of exposed plants and their strategy to cope with chemical stress. Li et al. (2018) assessed exposure to BPA on endogenous plant hormones in soybean (*Glycine max* L.). A low concentration of BPA (1.5 mg L^{-1}) increased indole-3-acetic acid (IAA) and zeatin (ZT) content but decreased gibberellic acid (GA3) content at the seedling stage. On the contrary, moderate and high concentrations (6.0 and 12.0 mg L^{-1}) decreased IAA, GA3 contents and increased abscisic acid (ABA) content at all growth stages examined.

Most studies conclude the Janus-faced acute effects of BPA are defined as "low promotion and high inhibition" on growth and photosynthesis. Effective concentrations were, in most cases, defined as 10 mg/L and above. Unfortunately, atmospheric transfer of either BPA or any other analogue has not been investigated. Also, comparing the 10 mg/L concentration to reported atmospheric occurrence, it seems very unlikely that plants will be exposed to harmful atmospheric concentrations.

4. Phthalates

Phthalates (PAEs) are the esters of phthalic acid (1,2-benzenedicarboxylic acid). These compounds are characterized by low volatility (semivolatile), and high octanol-to-water coefficient (K_{ow}). They have been recognized as having an endocrine-disrupting nature and have become an increasingly important environmental issue worldwide. The United States Environmental Protection Agency listed these compounds as priority pollutants (US EPA, 2013).

Phthalates are plasticizer compounds widely used in plastic and nonpolymer products (Zhang et al., 2015) to improve their flexibility and durability (Nagorka and Koschorreck, 2020). Polyvinyl chloride is the major source of phthalate contamination especially in water as it is generally used in water supply systems or hydroponic cultivation systems (Mondal et al., 2022). PAEs are also common in personal care products such as shampoos, soaps, colognes, and perfumes as fragrances (Khalil et al., 2022).

PAEs have weak bonding with plastic matrices, therefore they can be easily released during ageing or decomposition (Gao et al., 2019; Tang et al., 2021), potentially polluting different environmental compartments. PAEs occur both in surface waters and groundwater in the μg/L range, diethyl phthalate (DEP) and di-2-ethylhexyl phthalate (DEHP) being the most abundant congeners. In soils, diverse sources include the use of agricultural plastic films, irrigation with contaminated water, and leaching from landfills.

4.1 Atmospheric sources and occurrence

PAEs have also been released into the atmosphere. Salapasidou et al. (2011) analyzed urban air samples in Thessaloniki (Greece) and found a wide range of PTEs, including dimethyl phthalate (DMP); diethyl phthalate (DEP); di-n-butyl phthalate (DBP), butylbenzyl phthalate (BBP), di-2-ethylhexyl phthalate (DEHP) and di(n-octyl) phthalate (DNOP). In this study, DEHP was one of the dominant EDC compounds. According to a Chinese urban study (Li and Wang, 2015), bis(2-ethylhexyl) phthalate (BEHP) was the predominant congener, amounting to 23%–79% of the total phthalates' concentration. It was also reported that diisobutyl and dibutyl phthalates were mainly found in coarse particles because of their high volatilities. BEHP, on the contrary, was dominant in fine particles.

Chen et al. (2018) compared indoor and outdoor phthalate concentrations in PM2.5 samples in Beijing, China. DBP and DEHP were the most abundant compounds. Their concentrations showed significant differences between indoor and outdoor samples, suggesting that indoor sources (such as PVC products) contributed to PAE contamination in the air.

A study conducted in Delhi (India) found diisobutyl phthalate and di-n-butyl phthalate in organic aerosols collected at a landfill site after waste-burning episodes (Agarwal et al., 2020). The main source was supposed to be commonly used plasticizers, due to the high share of plastics in municipal waste. Emissions generated during the grilling of different meals (e.g., salmon or rump cap steak) were analyzed. Diverse plasticizers were observed in these samples, explained as food contact with packaging materials (Alves et al., 2021).

4.2 Phytotoxicity

Uptake has been mainly investigated from soil or from water in hydroponic cultures (Cai et al., 2008; Liao et al., 2006). Ma et al. (2015) compared the sensitivities of seven crop species, such as wheat (*Triticum aestivum*), alfalfa (*Medicago sativa*), perennial ryegrass (*Lolium perenne*), radish (*Raphanus sativus*), cucumber (*Cucumis sativus*), oat (*Avena sativa*) and onion (*Allium cepa*) to di-n-butyl phthalate (DnBP) and bis (2-ethylhexyl) phthalate (DEHP). Plants were exposed to 0–500 mg/kg soil concentrations. In general, endpoints such as root elongation, seedling growth, and biomass indicated lower phytotoxicity of DEHP than DnBP. On the contrary, both chemicals had significant inhibitory effects on the photosynthetic machinery, as reduced levels of chlorophyll *a* and *b* were measured.

Yao et al. (2022) cultivated pakchoi (*Brassica campestris*) in soils with concentrations of 2, 20, and 200 mg/kg of dibutyl phthalate (DBP). To assess the magnitude and effects of oxidative stress induced by DBP, both levels of antioxidant enzymes (SOD, POD, and CAT) and molecular markers were determined. These markers revealed that the three antioxidant enzyme genes (sod, pod, and cat) were significantly downregulated. Antioxidant enzyme activities were, in addition, directly affected as DBP could stably bind near the SOD/POD active centre, indicated by molecular docking. Oxidative stress, in turn, had a negative effect on the plants' performance, reducing the content of photosynthetic pigments (chlorophyll *a* and *b* and carotenoids) and ultimately, inhibiting growth.

Xiao et al. (2023) cultivated cucumber plants in soil artificially spiked with di (2-ethylhexyl) phthalate (DEHP) in a concentration of 40 mg/kg. After 3 weeks of exposure, a significant reduction of the membrane stability index (MSI) and of photosynthetic pigments of cucumber seedlings s found, also, antioxidant enzyme activities such as SOD, POD, CAT, and APX increased.

Gao et al. (2021) cultivated lettuce (*L. sativa*) seedlings in a 5 mg/L di-n-butyl phthalate (DBP) solution. After 28 days of exposure, the treatment significantly reduced shoot biomass and the height of the plants, and the level of antioxidant enzymes SOD, CAT, and GSH-Px decreased in the leaves and roots of the plants. TEM images indicated cell membrane rupture and cytoplasm leakage in treated plants. Hu et al. (2023) experimentally proved that DEHP had significant negative effects on genes responsible for antioxidative component synthesis. In the study of Yuan et al. (2020), soil was spiked with DEHP, setting a concentration series ranging from 0.432 mg/kg to 50 mg/kg. In *Brassica chinensis* test plants, low DEHP concentrations (in the range of 0.432 mg/kg to 8.81 mg/kg) had a stimulating effect on total superoxide dismutase, peroxidase, and catalase activities. On the contrary, high DEHP contamination (21.1 mg/kg and above) elucidated an inhibitory effect.

Extensive reviews (Kumari and Kaur, 2020; Sharma and Kaur, 2020) have been published on PAE phytotoxicity. Cited works agree that these compounds have inhibitory effects on seed germination, growth (both in roots and shoots), photosynthesis, and induce oxidative stress. However, the question arises to what extent these observations can be extrapolated to atmospheric exposure, both in terms of uptake and in realistic concentrations. An experimental study by Wang et al. (2024) suggests that atmospheric uptake of PAEs does occur, mainly through particles, and PAEs will be stored in the epicuticular wax.

5. Polychlorinated organic compounds

Polychlorinated biphenyl (PCB) molecules have two linked benzene rings (C_6H_5–C_6H_5) connected by a single carbon-carbon bond with various numbers of chlorine atoms attached at different positions on the rings. The chemical formula for PCBs is C12H10-xClx, where x can range from 1 to 10, indicating the number of chlorine atoms (Montano et al., 2022). Based on the number and position of the chlorine atoms, 209 different PCBs congeners are often classified into groups based on the number of chlorine atoms they contain: mono-, di-, tri-, tetra-, etc., chlorinated biphenyls (Mills et al., 2007). The number and position of chlorine atoms significantly affect the chemical properties and toxicity of individual PCB congeners (Pajurek et al., 2024).

The chlorine atoms in the PCB molecule affect the molecule's polarity, making the PCBs lipophilic (fat-soluble) and resistant to biodegradation. This gives them a long environmental half-life and high bioaccumulation potential (Grimm et al., 2015).

The manufacture and commercial use of PCBs started in 1929 for use in dielectric fluids. They were also used as plasticizers and flame retardants. According to the estimation of Breivik et al. (2002), over 1.3 million tonnes of pure PCBs were produced between 1930 and 1993. Its production and use resulted in app. 17 million tonnes of PCB-containing materials and waste (UNEP, 2016). Recognizing its high environmental risk, many manufacturers stopped or reduced manufacturing PCBs back in the 1970s. PCBs are classified and regulated as persistent organic pollutants (POPs) under the Stockholm Convention, which came into force in 2004. The Convention currently has 185 parties (184 countries plus the European Union) that ban the production of PCBs and aim to phase out in-use PCBs by 2025 (Melymuk et al., 2022).

The use of PCB-containing products and materials still provides an important source of PCB emission, however, major sources are secondary such as polluted soil or waste disposal sites (Mulder et al., 2015).

Polychlorinated dibenzo-p-dioxins (PCDDs) and polychlorinated dibenzofurans (PCDFs) are collectively known as PCDD/Fs. They are persistent semivolatile organic contaminants that also fall under the Stockholm Convention. This group consists of 75 PCDD and 135 PCDF isomers, 2,3,7,8-TCDD (20,30,70,80-tetrachlorinated dibenzo-p-dioxin), having the highest reported toxicity (Li et al., 2024).

5.1 Atmospheric sources and occurrence

PCDD/Fs are produced by industrial processes, incineration of medical and municipal solid waste, or burning of biomass and fossil fuels. Of industrial processes, metal production activities have been the dominant sources (Deng et al., 2020). Hu et al. (2014) documented a maximum concentration of 178 pg/m3 PCDD/F in the ambient air of secondary copper and aluminium metallurgical facilities. Oil field activities were also reported to emit considerable amounts of PCDD/Fs (Martínez-Guijarro et al., 2017). Atmospheric deposition provides an important pathway to both soil and water contamination (Wu et al., 2009). Kishida (2013) measured the maximum concentration of 120 pg/L of PCDDs and 60 pg/L of PCDFs in river samples in the Osaka urban area (Japan).

Municipal waste incineration serves as a major secondary source of PCBs, PBDEs, and PCDD/Fs emissions to the atmosphere (Li et al., 2021). In addition, e-waste recycling processes have also been identified as important emission sources of PCBs and PBDEs (Liu et al., 2019).

Atmospheric occurrence of PCBs and PBDEs has even been demonstrated in remote areas such as Antarctica (Wang et al., 2017). Lower chlorinated PCBs and PCDD/Fs predominantly exist in the gas phase; higher chlorinated compounds will have a higher tendency to bind with particles (Mila et al., 2022), most typically with fine PM (Degrendele et al., 2014). Being semivolatile compounds, PCBs are capable of long-range atmospheric transport (Wania and Westgate, 2008). Nellier et al. (2015) demonstrated that atmospheric deposition of PCBs provided a significant input for mountain lakes in the French Alps.

5.2 Phytotoxicity (Consider that this chapter should be focused on atmospheric contaminants)

Phytotoxicity studies on polychlorinated organic compounds are rather rare, and they address either aquatic (Chen et al., 2024; Pan et al., 2023) or terrestrial environments (Hanano et al., 2014; Jia et al., 2018). In addition, relatively higher concentrations are used in comparison to reported atmospheric concentrations (Wang et al., 2021). On the other hand, bioaccumulation measurements clearly prove that plants can be impacted by PCBs from the atmosphere. Concentrations of PBDEs and PCBs were determined in native and transplanted specimens of the moss *Pleurozium schreberi* in Upper Silesia (Poland) (Kosior et al., 2017). ΣPBDE and ΣPCB levels amounted to as high as 63.6 μg/kg and 4.47 ng/kg in native moss, respectively.

6. Combined effects of micro-, nanoplastics, and organic pollutants

The presence of micro- and nanoplastic particles in different environmental compartments is closely associated with the growing production and usage of plastics for various purposes. Microplastics are defined as plastic particles $\leq$5 mm in size, while nanoplastics are particles with diameters under 0.1 mm.

Micro- and nanoplastic pollution in terrestrial ecosystems has gained increasing attention. Airborne particles can undergo long-range transport and are subjected to dry and wet deposition (Allen et al., 2019), exposing both soil and plant surfaces. Their presence in bulk deposition was confirmed in cities such as Paris (Dris et al., 2016) or London (Wright et al., 2020); however, remote places also seem contaminated. Bergmann et al. (2019) e.g, reported microplastics in remote (Arctic, Swiss Alps) areas.

Abbasi and Turner (2021) found that the most abundant polymers were polypropylene, polyethylene, polystyrene, and polyethylene terephthalate in atmospheric deposition. As they have a low rate of degradation, long-term ecological risk can be anticipated (Chen et al., 2020). Plant leaves, due to their large and rough surfaces, can act as temporary sinks of airborne microplastics. Liu et al. (2020) directly measured plastic particles deposited on plant leaves and found a wide range of plastics, including polyethylene terephthalate (PET), polyethylene, polypropylene, cellophane, polyacrylonitrile, polybutyl acrylate, polyacrylic acid, polyamide, and epoxy resin, PET was amongst the dominant plastic types. However, the study revealed that these particles easily resuspend from leaf surfaces.

Sun et al. (2021) exposed maize plants to foliar treatment with polystyrene nanoplastics. Particles were shown to enter through the stomatal opening, and they also moved down to the roots through the vascular bundle. Some studies have already indicated that plastic particles elucidate toxic effects: Shi et al. (2022) e.g., demonstrated that polystyrene, polyethylene, and polypropylene microplastics caused oxidative stress in tomato plants, and they also inhibited seed germination. Plastic particles in addition, might cause physical damage: Gao et al. (2021) reported that polystyrene (PS) particles adhered to the root surfaces, causing physical blockage of the root pores.

In terms of organic pollutants, plastic particles pose additional stressors coexisting with chemicals: they might act as carriers, binding and transferring these contaminants. A wide range of studies have demonstrated that plastic particles enhanced the toxic effects of organic pollutants tested; they even had a synergistic effect. Gao et al. (2021) used lettuce test plants to investigate the potential effects of small and large polystyrene particles (SPS and LPS) alone and in combination with di-butyl-phthalate (DBP). All treatments induced $O_2^{\bullet-}$ and H_2O_2 generation in lettuce leaves and roots, while combined pollution of plastic particles and DBP significantly increased the activities of antioxidant enzymes (SOD, CAT, and GSH-Px).

Pan et al. (2024) experimentally demonstrated that PS microplastics acted as carriers for DBP by efficiently adsorbing DBP molecules. Microplastics not only facilitated chemical translocation in choy sum (*Brassica parachinensis*) but also synergistic toxic effects appeared, causing disruptions in osmoregulation and photosynthetic function, enhancing oxidative stress. Dong et al. (2021) examined the effects of nano- and micro-sized PS and DBP on the photosynthesis of lettuce. Combined treatment resulted in a higher decrease in the photosynthetic parameters and total chlorophyll content compared with DBP used alone. Alterations in the crystalline structure of the chlorophyll protein were also demonstrated. Higher oxidative stress was also reported. These effects, however, were size-dependent, with nano-sized PS being more hazardous.

In the study of Li et al. (2023), the individual and combined effects of PS NPs and DBP were examined on dandelion (*Taraxacum officinale*) seedlings. Coexposure to PS NPs and DBP had a more intense effect compared with individually treated plants. Symptoms were reduced growth, oxidative stress, and altered enzymatic and nonenzymatic antioxidant levels. Metabolic analysis revealed that some of the metabolism pathways (e.g., galactose metabolism) were also affected.

Sun et al. (2022) used foliar treatment to assess the environmental impacts of two phthalate esters (PAEs), di-n-butyl phthalate (DBP) and di-(2-ethylhexyl) phthalate (DEHP), in corn (*Zea mays*) seedlings. Treatments were done separately and in combination with amino-functionalized polystyrene nanoplastics (PSNPs-NH2), in the following concentrations: DBP alone (1 mg/L), DEHP alone (1 mg/L), PSNPs-NH2 (1 mg/L) + DBP (1 mg/L), and PSNPs-NH2 (1 mg/L) + DEHP (1 mg/L). Nanoplastics acted as carriers of the PAE compounds, increasing DBP and DEHP accumulation in the leaves by 1.36 and 1.32 times, respectively. The study also demonstrated that PSNPs-NH2 accumulated on the leaf surfaces, followed by internalization into the inner compartments. In addition to enhancing accumulation, the copresence of PSNPs-NH2 and PAEs elucidated a higher negative effect on photosystem II efficiency than PAE treatment alone. As a result, a greater inhibitory effect on plant growth was observable. Association with the leaf surfaces differed, being higher for positively charged PS-NH2 than negatively charged PS-COOH, most possibly due to electrostatic forces (Sun et al., 2021).

Although it is known that MPs can bind PAHs (Cui et al., 2024), especially through hydrophobic and electrostatic interactions (Wisniowska and Wlodarczyk-Makula, 2022), little is known about their combined effects. Studies available exclusively report cocontamination of MP and NP with individual PAHs such as phenanthrene in soil environments (Liu et al., 2021; Xu et al., 2021).

7. Concluding remarks

Despite all efforts (technological improvements or legislative means), the emission of persistent organic pollutants cannot be fully avoided, and even worse, they still remain in environmental compartments in considerable concentrations. Atmosphere receives serious emissions, both primary and secondary. Plants are practically constantly exposed to these emissions, enduring a lifetime stress.

Exact stressor-effect relationships are hard to determine. First, plants living in a polluted habitat are exposed to polluted air, which is a complex cocktail of chemicals. The composition of such a cocktail will show very big spatial and temporal differences, which also means that a "unified" response is absolutely impossible to define.

Secondly, atmospheric exposure is very difficult to mimic under laboratory conditions. Most of the studies cited here use concentrations that are well above predicted atmospheric exposure. Even in cases of known and reported atmospheric concentrations, how would we calculate the required concentration(s) or exposure duration that could best represent these real-world conditions?

A careful conclusion of this study is that we have very limited information about phytotoxic effects on plants exposed to air pollutants. Considering the increasing tendency in plastic and plastic waste production (a primary source of many POPs), a gap has been identified here, necessitating the targeted evaluation of these contaminants.

References

Abbasi, S., Turner, A., 2021. Dry and wet deposition of microplastics in a semi-arid region. Shiraz, Iran. Sci. Total Environ. 786, 147358. https://doi.org/10.1016/j.scitotenv.2021.147358.

Abdelmoneim, M.S., Hafez, E.E., Dawood, M.F., Hammad, S.F., Ghazy, M.A., 2024. Toxicity of bisphenol A and p-nitrophenol on tomato plants: morpho-physiological, ionomic profile, and antioxidants/defense-related gene expression studies. Biomol. Concepts 15 (1), 20220049. https://doi.org/10.1515/bmc-2022-0049.

Agarwal, R., Shukla, K., Kumar, S., Aggarwal, S.G., Kawamura, K., 2020. Chemical composition of waste burning organic aerosols at landfill and urban sites in Delhi. Atmos. Pollut. Res. 11 (3), 554–565. https://doi.org/10.1016/j.apr.2019.12.004.

Ahammed, G.J., Wang, M.M., Zhou, Y.H., Xia, X.J., Mao, W.H., et al., 2012a. The growth, photosynthesis and antioxidant defense responses of five vegetable crops to phenanthrene stress. Ecotoxicol. Environ. Saf. 80, 132–139. https://doi.org/10.1016/j.ecoenv.2012.02.015.

Ahammed, G.J., Gao, C.J., Ogweno, J.O., Zhou, Y.H., Xia, X.J., Mao, W.H., et al., 2012b. Brassinosteroids induce plant tolerance against phenanthrene by enhancing degradation and detoxification in *Solanum lycopersicum* L. Ecotoxicol. Environ. Saf. 80, 28–36. https://doi.org/10.1016/j.ecoenv.2012.02.004.

Ahammed, G.J., Zhang, S., Shi, K., Zhou, Y.H., Yu, J.Q., 2012c. Brassinosteroid improves seed germination and early development of tomato seedling under phenanthrene stress. Plant Growth Regul. 68, 87–96. https://doi.org/10.1007/s10725-012-9696-0.

Ahammed, G.J., Li, X., Xia, X.-J., Shi, K., Zhou, Y.-H., Yu, J.-Q., 2015. Enhanced photosynthetic capacity and antioxidant potential mediate brassinosteriod-induced phenanthrene stress tolerance in tomato. Environ. Pollut. 201, 58–66. https://doi.org/10.1016/j.envpol.2015.02.024.

Allen, S., Allen, D., Phoenix, V.R., Le Roux, G., Durantez Jimenez, P., et al., 2019. Atmospheric transport and deposition of microplastics in a remote mountain catchment. Nat. Geosci. 12, 339–344. https://doi.org/10.1038/s41561-019-0335-5.

Al-Nasir, F., Hijazin, T.J., Al-Alawi, M.M., Jiries, A., Al-Madanat, O.Y., Mayyas, A., 2022. Accumulation, source identification, and cancer risk assessment of polycyclic aromatic hydrocarbons (PAHs) in different Jordanian vegetables. Toxics 10 (11), 643. https://doi.org/10.3390/toxics10110643, 2022.

Alves, C.A., Vicente, E.D., Evtyugina, M., Vicente, A.M., Sainnokhoi, T.A., Kováts, N., 2021. Cooking activities in a domestic kitchen: chemical and toxicological profiling of emissions. Sci. Total Environ. 772, 145412. https://doi.org/10.1016/j.scitotenv.2021.145412.

Anand, P., Mina, U., Khare, M., Kumar, P., Kota, S.H., 2022. Air pollution and plant health response-current status and future directions. Atmos. Pollut. Res. 13 (8), 101508. https://doi.org/10.1016/j.apr.2022.101508.

Anav, A., De Marco, A., Friedlingstein, P., Savi, F., Sicard, P., et al., 2019. Growing season extension affects ozone uptake by European forests. Sci. Total Environ. 669, 1043–1052. https://doi.org/10.1016/j.scitotenv.2019.03.020.

Andersson, J.T., Achten, C., 2015. Time to say goodbye to the 16 EPA PAHs? Toward an up-to-date use of PACs for environmental purposes. Polycycl. Aromat. Compd. 35, 330–354. https://doi.org/10.1080/10406638.2014.991042.

Anttila, P., Tuovinen, J.P., 2010. Trends of primary and secondary pollutant concentrations in Finland in 1994–2007. Atmos. Environ. 44 (1), 30–41. https://doi.org/10.1016/j.atmosenv.2009.09.041.

Apel, K., Hirt, H., 2004. Reactive oxygen species: metabolism, oxidative stress, and signal transduction. Annu. Rev. Plant Biol. 55 (1), 373–399. https://doi.org/10.1146/annurev.arplant.55.031903.141701.

Asada, K., 2006. Production and scavenging of reactive oxygen species in chloroplasts and their functions. Plant Physiol. 141 (2), 391–396. https://doi.org/10.1104/pp.106.082040.

Attacha, S., Solbach, D., Bela, K., Moseler, A., Wagner, S., et al., 2017. Glutathione peroxidase-like enzymes cover five distinct cell compartments and membrane surfaces in *Arabidopsis thaliana*. Plant Cell Environ. 40 (8), 1281–1295. https://doi.org/10.1111/pce.12919.

Babula, P., Vodicka, O., Adam, V., Kummerova, M., Ladislav Havel, L., Hosek, J., 2012. Effect of fluoranthene on plant cell model: tobacco BY-2 suspension culture. Environ. Exp. Bot. 78, 117–126. https://doi.org/10.1016/j.envexpbot.2011.12.024.

Bailey-Serres, J., Mittler, R., 2006. The roles of reactive oxygen species in plant cells. Plant Physiol. 141 (2). https://doi.org/10.1104/pp.104.900191, 311-311.

Balmer, Y., Buchanan, B.B., 2002. Yet another plant thioredoxin. Trends Plant Sci. 7 (5), 191–193. https://doi.org/10.1016/S1360-1385(02)02263-X.

Bardaweel, S.K., Gul, M., Alzweiri, M., Ishaqat, A., Salamat, H.A., Bashatwah, R.M., 2018. Reactive oxygen species: the dual role in physiological and pathological conditions of the human body. The Eurasian Journal of Medicine 50 (3), 193. https://doi.org/10.5152/eurasianjmed.2018.17397.

Bayouli, I.T., Bayouli, H.T., Dell'Oca, A., Meers, E., Sun, J., 2021. Ecological indicators and bioindicator plant species for biomonitoring industrial pollution: eco-based environmental assessment. Ecol. Indic. 125, 107508. https://doi.org/10.1016/j.ecolind.2021.107508.

Bell, J.N.B., Honour, S.L., Power, S.A., 2011. Effects of vehicle exhaust emissions on urban wild plant species. Environ. Pollut. 159, 1984–1990. https://doi.org/10.1016/j.envpol.2011.03.006.

Bergmann, M., Mutzel, S., Primpke, S., Tekman, M.B., Trachsel, J., Gerdts, G., 2019. White and wonderful? Microplastics prevail in snow from the Alps to the Arctic. Sci. Adv. 5, 8. https://doi.org/10.1126/sciadv.aax1157.

Biache, C., Mansuy-Huaulta, L., Faure, P., 2014. Impact of oxidation and biodegradation on the most commonly used polycyclic aromatic hydrocarbon (PAH) diagnostic ratios: implications for the source identifications. J. Hazard Mater. 267 (28), 31–39. https://doi.org/10.1016/j.jhazmat.2013.12.036.

Birgül, A., Tasdemir, Y., Cindoruk, S.S., 2011. Atmospheric wet and dry deposition of polycyclic aromatic hydrocarbons (PAHs) determined using a modified sampler. Atmos. Res. 101 (1–2), 341–353. https://doi.org/10.1016/j.atmosres.2011.03.012.

Blande, J.D., 2021. Effects of air pollution on plant–insect interactions mediated by olfactory and visual cues. Curr. Opin. Environ. Sci. Health 19, 100228. https://doi.org/10.1016/j.coesh.2020.100228.

Boutin, C., Aya, K.L., Carpenter, D., Thomas, P.J., Rowland, O., 2012. Phytotoxicity testing for herbicide regulation: shortcomings in relation to biodiversity and ecosystem services in agrarian systems. Sci. Total Environ. 415, 79–92. https://doi.org/10.1016/j.scitotenv.2011.04.046.

Breivik, K., Sweetman, A.J., Pacyna, J.M., Jones, K.C., 2002. Towards a global historical emission inventory for selected PCB congeners–a mass balance approach. 1. Global production and consumption. Sci. Total Environ. 290, 181–198. https://doi.org/10.1016/S0048-9697(01)01075-0.

Cai, Q.Y., Mo, C.H., Wu, Q.T., Zeng, Q.Y., 2008. Polycyclic aromatic hydrocarbons and phthalic acid esters in the soil-radish (*Raphanus sativus*) system with sewage sludge and compost application. Bioresour. Technol. 99, 1830–1836. https://doi.org/10.1016/j.biortech.2007.03.035.

Carpenter, C., Boutin, C., Allison, J.E., 2013. Effects of chlorimuron ethyl on terrestrial and wetland plants: levels of, and time to recovery following sublethal exposure. Environ. Pollut. 172, 275–282. https://doi.org/10.1016/j.envpol.2012.09.007.

Carter, W.P.L., Atkinson, R., 2019. The impacts of air pollution on ecosystems and biodiversity: a global assessment. Environ. Sci. Technol. 53 (9), 5171–5183. https://doi.org/10.1021/acs.est.9b01535.

Caumo, S., Yera, A.B., Alves, C., Rienda, I.C., Kováts, N., et al., 2023. Assessing the chemical composition, potential toxicity and cancer risk of airborne fine particulate matter (PM2. 5) near a petrochemical industrial area. Environ. Toxicol. Pharmacol. 101, 104170. https://doi.org/10.1016/j.etap.2023.104170.

Chen, Y., Lv, D., Li, X., Zhu, T., 2018. PM2. 5-bound phthalates in indoor and outdoor air in Beijing: seasonal distributions and human exposure via inhalation. Environ. Pollut. 241, 369–377. https://doi.org/10.1016/j.envpol.2018.05.081.

Chen, G., Feng, Q., Wang, J., 2020. Mini review of microplastics in the atmosphere and their risks to humans. Sci. Total Environ. 703, 135504. https://doi.org/10.1016/j.scitotenv.2019.135504.

Chen, G., Pan, T., Gao, D., Liao, H., Wang, J., 2024. Enhanced competitiveness of Spirodela polyrhiza in co-culture with Salvinia natans under combined exposure to polystyrene nanoplastics and polychlorinated biphenyls. Sci. Total Environ. 955, 176870. https://doi.org/10.1016/j.scitotenv.2024.176870.

Cipoli, Y.A., Alves, C., Rapuano, M., Evtyugina, M., Rienda, I.C., et al., 2023. Nighttime–daytime PM10 source apportionment and toxicity in a remoteness inland city of the Iberian Peninsula. Atmos. Environ. 303, 119771. https://doi.org/10.1016/j.atmosenv.2023.119771.

Clément, N., Muresan, B., Hedde, M., François, D., 2015. PAH dynamics in roadside environments: influence on the consistency of diagnostic ratio values and ecosystem contamination assessments. Sci. Total Environ. 538, 997–1009. https://doi.org/10.1016/j.scitotenv.2015.08.072.

Cogut, A., 2016. Open Burning of Waste: A Global Health Disaster. R20 Regions of Climate Action.

Cui, L., Liang, R., Zhang, C., Zhang, R., Wang, H., Wang, X.X., 2024. Coupling polyethylene microplastics with other pollutants: exploring their combined effects on plant health and technologies for mitigating toxicity. Sci. Total Environ. 176657. https://doi.org/10.1016/j.scitotenv.2024.176657.

Daresta, B.E., Italiano, F., de Gennaro, G., Trotta, M., Tutino, M., Veronico, P., 2015. Atmospheric particulate matter (PM) effect on the growth of *Solanum lycopersicum* cv. Roma plants. Chemosphere 119, 37–42. https://doi.org/10.1016/j.chemosphere.2014.05.054.

da Silva Santos, J., da Silva Pontes, M., de Souza, M.B., Fernandes, S.Y., Azevedo, R., et al., 2023. Toxicity of bisphenol A (BPA) and its analogues BPF and BPS on the free-floating macrophyte Salvinia biloba. Chemosphere 343, 140235. https://doi.org/10.1016/j.chemosphere.2023.140235.

De Nicola, F., Graña, E.C., Mahía, P.L., 2017. Evergreen or deciduous trees for capturing PAHs from ambient air? A case study. Environ. Pollut. 221, 276–284. https://doi.org/10.1016/j.envpol.2016.11.074.

Degrendele, C., Okonski, K., Melymuk, L., Landlová, L., Kukučka, P., et al., 2014. Size specific distribution of the atmospheric particulate PCDD/Fs, dl-PCBs and PAHs on a seasonal scale: implications for cancer risks from inhalation. Atmos. Environ. 98, 410–416. https://doi.org/10.1016/j.atmosenv.2014.09.001.

Deng, Y., Peng, P., Jia, L., Yin, H., Hu, J., Mao, W., 2020. Atmospheric bulk deposition of polychlorinated dibenzo-p-dioxins and dibenzofurans (PCDD/Fs) in the vicinity of MSWI in Shanghai, China. Ecotoxicol. Environ. Saf. 196, 110493. https://doi.org/10.1016/j.ecoenv.2020.110493.

Desalme, D., Binet, P., Epron, D., Bernard, N., Gilbert, D., et al., 2011a. Atmospheric phenanthrene pollution modulates carbon allocation in red clover (*Trifolium pratense* L.). Environ. Pollut. 159, 2759–2765. https://doi.org/10.1016/j.envpol.2011.05.015.

Desalme, D., Binet, P., Bernard, N., Gilbert, D., Toussaint, M.L., Chiapusio, G., 2011b. Atmospheric phenanthrene transfer and effects on two grassland species and their root symbionts: a microcosm study. Environ. Exp. Bot. 71, 146–151. https://doi.org/10.1016/j.envexpbot.2010.11.009.

Dogra, K., Lalwani, D., Dogra, S., Panday, D.P., Raval, N., et al., 2024. Indian and global scenarios of bisphenols A distribution and its new analogues: prevalence and probability exceedance. J. Hazard Mater. 135128. https://doi.org/10.1016/j.jhazmat.2024.135128.

Dong, Y., Song, Z., Liu, Y., Gao, M., 2021. Polystyrene particles combined with di-butyl phthalate cause significant decrease in photosynthesis and red lettuce quality. Environ. Pollut. 278, 116871. https://doi.org/10.1016/j.envpol.2021.116871.

dos Santos, C.R., Arcanjo, G.S., Araújo, A.A.D., de Souza Santos, L.V., Amaral, M.C.S., 2024. Occurrence, environmental risks, and removal of bisphenol A and its analogues by membrane bioreactors. Chem. Eng. J. 494, 153278. https://doi.org/10.1016/j.cej.2024.153278.

Dos Santos, C.V., Rey, P., 2006. Plant thioredoxins are key actors in the oxidative stress response. Trends Plant Sci. 11 (7), 329–334. https://doi.org/10.1016/j.tplants.2006.05.005.

Dris, R., Gasperi, J., Saad, M., Mirande, C., Tassin, B., 2016. Synthetic fibers in atmospheric fallout: a source of microplastics in the environment? Mar. Pollut. Bull. 104, 290–293. https://doi.org/10.1016/j.marpolbul.2016.01.006.

Duan, C., Liao, H., Wang, K., Ren, Y., 2023. The research hotspots and trends of volatile organic compound emissions from anthropogenic and natural sources: a systematic quantitative review. Environ. Res. 216, 114386. https://doi.org/10.1016/j.envres.2022.114386.

Esperanza, M., Seoane, M., Servia, M.J., Cid, Á., 2020. Effects of Bisphenol A on the microalga Chlamydomonas reinhardtii and the clam Corbicula fluminea. EES (Ecotoxicol. Environ. Saf.) 197, 110609. https://doi.org/10.1016/j.ecoenv.2020.110609.

Estévez, I.H., Hernández, M.R., 2020. Plant glutathione S-transferases: an overview. Plant Gene 23, 100233. https://doi.org/10.1016/j.plgene.2020.100233.

Fernández, J.H., Guerra, Y., Cano, H., 2022. Detection of bisphenol A and four analogues in atmospheric emissions in petrochemical complexes producing polypropylene in South America. Molecules 27 (15), 4832. https://doi.org/10.3390/molecules27154832.

Fang, H., Luo, S., Huang, X., Fu, X., Xiao, S., et al., 2021. Ambient naphthalene and methylnaphthalenes observed at an urban site in the Pearl River Delta region: sources and contributions to secondary organic aerosol. Atmos. Environ. 252, 118295. https://doi.org/10.1016/j.atmosenv.2021.118295.

Filonchyk, M., Peterson, M.P., 2024. Changes in aerosol properties at the El Arenosillo site in Southern Europe as a result of the 2023 Canadian forest fires. Environ. Res. 260, 119629. https://doi.org/10.1016/j.envres.2024.119629.

Fu, P., Kawamura, K., 2010. Ubiquity of bisphenol A in the atmosphere. Environ. Pollut. 158 (10), 3138–3143. https://doi.org/10.1016/j.envpol.2010.06.040.

Gao, M., Xu, Y., Liu, Y., Wang, S., Wang, C., et al., 2021. Effect of polystyrene on di-butyl phthalate (DBP) bioavailability and DBP-induced phytotoxicity in lettuce. Environ. Pollut. 268, 115870. https://doi.org/10.1016/j.envpol.2020.115870.

Gill, S.S., Tuteja, N., 2010. Reactive oxygen species and antioxidant machinery in abiotic stress tolerance in crop plants. Plant Physiol. Biochem. 48 (12), 909–930. https://doi.org/10.1016/j.plaphy.2010.08.016.

Grimm, F.A., Hu, D., Kania-Korwel, I., Lehmler, H.-J., Ludewig, G., et al., 2015. Metabolism and metabolites of polychlorinated biphenyls. Crit. Rev. Toxicol. 45, 245–272. https://doi.org/10.3109/10408444.2014.999365.

Hanano, A., Almousally, I., Shaban, M., 2014. Phytotoxicity effects and biological responses of *Arabidopsis thaliana* to 2, 3, 7, 8-tetrachlorinated dibenzo-p-dioxin exposure. Chemosphere 104, 76–84. https://doi.org/10.1016/j.chemosphere.2013.10.060.

Harrison, R.M., Yin, J., 2017. Atmospheric pollution and its effects on ecosystems: a review of key findings. Environ. Sci. Pollut. Control Ser. 24 (6), 5357–5369. https://doi.org/10.1007/s11356-017-9185-1.

Honour, S.L., Bell, J.N.B., Ashenden, T.W., Cape, J.N., Power, S.A., 2009. Responses of herbaceous plants to urban air pollution: effects on growth, phenology and leaf surface characteristics. Environ. Pollut. 157, 1279–1286. https://doi.org/10.1016/j.envpol.2008.11.049.

Hu, J., Zheng, M., Liu, W., Nie, Z., Li, C., et al., 2014. Characterization of polychlorinated dibenzo-p-dioxins and dibenzofurans, dioxin-like polychlorinated biphenyls, and polychlorinated naphthalenes in the environment surrounding secondary copper and aluminum metallurgical facilities in China. Environ. Pollut. 193, 6–12. https://doi.org/10.1016/j.envpol.2014.06.007.

Huang, X.D., Dixon, D.G., Greenberg, B.M., 1993. Impacts of UV radiation and photomodification ont he toxicity of PAHs to the higher plant *Lemna gibba* (Duckweed). Environ. Toxicol. Chem. 12, 1067–1077. https://doi.org/10.1002/etc.5620120613.

Huang, X.D., Dixon, D.G., Greenberg, B.M., 1995. Increased polycyclic aromatic hydrocarbon toxicity following their photomodification in natural sunlight: impacts on the duckweed *Lemna gibba* L. G-3. EES (Ecotoxicol. Environ. Saf.) 32 (2), 194–200. https://doi.org/10.1006/eesa.1995.1102.

Huang, X.D., Zeiler, L.F., Dixon, D.G., Greenberg, B.M., 1996. Photoinduced toxicity of PAHs to the foliar regions of *Brassica napus* (Canola) and *Cucumbis sativus* (Cucumber) in simulated solar radiation. Ecotoxicol. Environ. Saf. 35, 190–197. https://doi.org/10.1006/eesa.1996.0099.

Hubai, K., Székely, O., Teke, G., Kováts, N., 2021. Is essential oil production influenced by air pollution in *Ocimum basilicum* L. Biochem. Systemat. Ecol. 96, 104248. https://doi.org/10.1016/j.bse.2021.104248.

Hubai, K., Kováts, N., Sainnokhoi, T.A., Eck-Varanka, B., Hoffer, A., et al., 2022. Phytotoxicity of particulate matter from controlled burning of different plastic waste types. Bull. Environ. Contam. Toxicol. 109 (5), 852–858. https://doi.org/10.1007/s00128-022-03581-9.

Jen, M.S., Hoylman, A.M., Edwards, N.T., Walton, B.T., 1995. Experimental method to measure gaseous uptake of 14C-Toluene by foliage. Environ. Exp. Bot. 35 (3), 389–398. https://doi.org/10.1016/0098-8472(95)00007-4.

Jia, J., Bi, C., Zhang, J.J., Chen, Z., 2019. Atmospheric deposition and vegetable uptake of polycyclic aromatic hydrocarbons (PAHs) based on experimental and computational simulations. Atmos. Environ. 204, 135–141. https://doi.org/10.1016/j.atmosenv.2019.02.030.

Jia, Y., Li, H., Qu, Y., Chen, W., Song, L., 2018. Phytotoxicity, bioaccumulation and potential risks of plant irrigations using cyanobloom-loading freshwater. Sci. Total Environ. 624, 704–712. https://doi.org/10.1016/j.scitotenv.2017.12.164.

Jiao, L., Ding, H., Wang, L., Zhou, Q., Huang, X., 2017. Bisphenol A effects on the chlorophyll contents in soybean at different growth stages. Environ. Pollut. 223, 426–434. https://doi.org/10.1016/j.envpol.2017.01.042.

Joksimovic, N., Selakovic, D., Jovicic, N., Jankovic, N., Pradeepkumar, P., et al., 2022. Nanoplastics as an invisible threat to humans and the environment. J. Nanomater. 2022 (1), 6707819. https://doi.org/10.1155/2022/6707819.

Joshi, R., Chinnusamy, V., 2014. Antioxidant enzymes: defense against high temperature stress. In: Ahmad, P. (Ed.), Oxidative Damage to Plants. Academic Press, pp. 369–396. https://doi.org/10.1016/B978-0-12-799963-0.00012-5.

Kardel, F., Wuyts, K., Babanezhad, M., Vitharana, U.W.A., Wuytack, T., et al., 2010. Assessing urban habitat quality based on specific leaf area and stomatal characteristics of *Plantago lanceolata* L. Environ. Pollut. 158, 788–794. https://doi.org/10.1016/j.envpol.2009.10.006.

Karmakar, D., Deb, K., Padhy, P.K., 2021. Ecophysiological responses of tree species due to air pollution for biomonitoring of environmental health in urban area. Urban Clim. 35, 100741. https://doi.org/10.1016/j.uclim.2020.100741.

Karpinska, B., Foyer, C.H., 2024. Superoxide signalling and antioxidant processing in the plant nucleus. J. Exp. Bot. 75 (15), 4599–4610. https://doi.org/10.1093/jxb/erae090.

Keller, T., 1974. The use of peroxidase activity for monitoring and mapping air pollution areas. Eur. J. For. Path. 4, 11–19. https://doi.org/10.1111/j.1439-0329.1974.tb00407.x.

Kelly, F.J., Fussell, J.C., 2012. Size, source and chemical composition as determinants of toxicity attributable to ambient particulate matter. Atmos. Environ. 60, 504–526. https://doi.org/10.1016/j.atmosenv.2012.06.039.

Khalil, M., Iqbal, M., Turan, V., Tauqeer, H.M., Farhad, M., et al., 2022. Chapter 11—household chemicals and their impact. In: Hashmi, M.Z., Wang, S., Ahmed, Z. (Eds.), Advances in Pollution Research. Environmental Micropollutants, pp. 201–232. https://doi.org/10.1016/B978-0-323-90555-8.00022-2.

Khavaninzadeh, A.R., Veroustraete, F., Buytaert, J.A.N., Samson, R., 2014. Leaf injury symptoms of *Tilia* sp. as an indicator of urban habitat quality. Ecol. Indic. 41, 58–64. https://doi.org/10.1016/j.ecolind.2014.01.014.

Khpalwak, W., Abdel-Dayem, S.M., Sakugawa, H., 2018. Individual and combined effects of fluoranthene, phenanthrene, mannitol and sulfuric acid on marigold (*Calendula officinalis*). Ecotoxicol. Environ. Saf. 148, 834–841. https://doi.org/10.1016/j.ecoenv.2017.11.065.

Kim, K.H., Jahan, S.A., Kabir, E., Brown, R.J., 2013. A review of airborne polycyclic aromatic hydrocarbons (PAHs) and their human health effects. Environ. Int. 60, 71–80. https://doi.org/10.1016/j.envint.2013.07.019.

Kishida, M., 2013. Distribution characteristics and source identification of polychlorinated dibenzo-p-dioxin and dibenzofurans, and dioxin-like polychlorinated biphenyls in the waters from River Kanzaki, running through Osaka urban area, Japan. J. Environ. Sci. 25 (3), 441–451. https://doi.org/10.1016/S1001-0742(12)60091-6.

Kosior, G., Přibylová, P., Vaňková, L., Kukučka, P., Audy, O., et al., 2017. Bioindication of PBDEs and PCBs by native and transplanted moss *Pleurozium schreberi*. Ecotoxicol. Environ. Saf. 143, 136–142. https://doi.org/10.1016/j.ecoenv.2017.05.025.

Kováts, N., Horváth, E., Eck-Varanka, B., Csajbók, E., Hoffer, A., 2017. Adapting the vegetative Vigour terrestrial plant test for assessing ecotoxicity of aerosol samples. Environ. Sci. Pollut. Control Ser. 24, 15291–15298. https://doi.org/10.1007/s11356-017-9103-5.

Kováts, N., Hubai, K., Diósi, D., Sainnokhoi, T.A., Hoffer, A., et al., 2021. Sensitivity of typical European roadside plants to atmospheric particulate matter. Ecol. Indic. 124, 107428. https://doi.org/10.1016/j.ecolind.2021.107428.

Kreslavski, V.D., Brestic, M., Zharmukhamedov, S.K., Lyubimov, V.Y., Lankin, A.V., et al., 2017. Mechanisms of inhibitory effects of polycyclic aromatic hydrocarbons in photosynthetic primary processes in pea leaves and thylakoid preparations. Plant Biol. 19 (5), 683–688. https://doi.org/10.1111/plb.12598.

Kulhánek, A., Trapp, S., Sismilich, M., Jankl, J., Zimová, M., 2005. Crop-specific human exposure assessment for polycyclic aromatic hydrocarbons in Czech soils. Sci. Total Environ. 339 (1–3), 71–80. https://doi.org/10.1016/j.scitotenv.2004.08.003.

Kumar, G.S., Dubey, P.S., 1998. Differential response and detoxifying mechanism of *Cassia siamea* Lam. and *Dalbergia sissoo* Roxb. of different ages to SO2 treatment. J. Environ. Biol. 9 (3), 243–249.

Kumari, A., Kaur, R., 2020. A review on morpho-physiological traits of plants under phthalates stress and insights into their uptake and translocation. Plant Growth Regul. 91 (3), 327–347. https://doi.org/10.1007/s10725-020-00625-0.

Kummerová, M., Váňová, L., 2007. Chlorophyll fluorescence as an indicator of fluoranthene phototoxicity. Plant Soil Environ. 53 (10), 430–436.

Kummerová, M., Váňová, L., Krulová, J., Zezulka, S., 2008. The use of physiological characteristics for comparison of organic compounds phytotoxicity. Chemosphere 71, 2050–2059. https://doi.org/10.1016/j.chemosphere.2008.01.060.

Kwak, J.I., Moon, J., Kim, D., Cui, R., An, Y.J., 2018. Determination of the soil hazardous concentrations of bisphenol A using the species sensitivity approach. J. Hazard Mater. 344, 390–397. https://doi.org/10.1016/j.jhazmat.2017.10.048.

Lehndorff, E., Schwark, L., 2004. Biomonitoring of air quality in the Cologne Conurbation using pine needles as a passive sampler—part II: polycyclic aromatic hydrocarbons (PAH). Atmos. Environ. 38 (23), 3793–3808. https://doi.org/10.1016/j.atmosenv.2004.03.065.

Liao, C.S., Yen, J.H., Wang, Y.S., 2006. Effects of endocrine disruptor di-n-butyl phthalate on the growth of Bok choy (Brassica rapa subsp. chinensis). Chemosphere 65, 1715–1722. https://doi.org/10.1016/j.chemosphere.2006.04.093.

Li, J., Wang, G., 2015. Airborne particulate endocrine disrupting compounds in China: compositions, size distributions and seasonal variations of phthalate esters and bisphenol A. Atmos. Res. 154, 138–145. https://doi.org/10.1016/j.atmosres.2014.11.013.

Li, M., Zhou, Y., Wang, G., Zhu, G., Zhou, X., et al., 2021. Evaluation of atmospheric sources of PCDD/Fs, PCBs and PBDEs around an MSWI plant using active and passive air samplers. Chemosphere 274, 129685. https://doi.org/10.1016/j.chemosphere.2021.129685.

Li, X., Wang, L., Wang, S., Yang, Q., Zhou, Q., Huang, X., 2018. A preliminary analysis of the effects of bisphenol A on the plant root growth via changes in endogenous plant hormones. Ecotox. Environ. Safe. 150, 152–158. https://doi.org/10.1016/j.ecoenv.2017.12.031.

Li, X., Zhang, Y., Wang, J., Zeng, G., Tong, X., et al., 2023. Revealing the metabolomics and biometrics underlying phytotoxicity mechanisms for polystyrene nanoplastics and dibutyl phthalate in dandelion (*Taraxacum officinale*). Sci. Total Environ. 905, 167071. https://doi.org/10.1016/j.scitotenv.2023.167071.

Li, Y., Sidikjan, N., Huang, L., Chen, Y., Zhang, Y., et al., 2024. Multi-media environmental fate of polychlorinated dibenzo-p-dioxins and dibenzofurans in China: a systematic review of emissions, presence, transport modeling and health risks. Environ. Pollut. 124970. https://doi.org/10.1016/j.envpol.2024.124970.

Liang, J., Li, Y., Xie, P., Liu, C., Yu, L., Ma, X., 2022. Dualistic effects of bisphenol A on growth, photosynthetic and oxidative stress of duckweed (*Lemna minor*). Environ. Sci. Pollut. Control Ser. 29 (58), 87717–87729. https://doi.org/10.1007/s11356-022-21785-8.

Liu, K., Wang, X., Song, Z., Wei, N., Li, D., 2020. Terrestrial plants as a potential temporary sink of atmospheric microplastics during transport. Sci. Total Environ. 742, 140523. https://doi.org/10.1016/j.scitotenv.2020.140523.

Liu, S., Wang, J., Zhu, J., Wang, J., Wang, H., Zhan, X., 2021. The joint toxicity of polyethylene microplastic and phenanthrene to wheat seedlings. Chemosphere 282, 130967. https://doi.org/10.1016/j.chemosphere.2021.130967.

Liu, J., Jia, J., Grathwohl, P., 2022. Dilution of concentrations of PAHs from atmospheric particles, bulk deposition to soil: a review. Environ. Geochem. Health 44 (12), 4219–4234. https://doi.org/10.1007/s10653-022-01216-w.

Liu, R., Ma, S., Li, G., Yu, Y., An, T., 2019. Comparing pollution patterns and human exposure to atmospheric PBDEs and PCBs emitted from different e-waste dismantling processes. J. Hazard Mater. 369, 142–149. https://doi.org/10.1016/j.jhazmat.2019.02.029.

Lu, J., Wu, J., Stoffella, P.J., Wilson, P.C., 2015. Uptake and distribution of bisphenol A and nonylphenol in vegetable crops irrigated with reclaimed water. J. Hazard Mater. 283, 865–870. https://doi.org/10.1016/j.jhazmat.2014.10.018.

Ma, T., Teng, Y., Christie, P., Luo, Y., 2015. Phytotoxicity in seven higher plant species exposed to di-n-butyl phthalate or bis (2-ethylhexyl) phthalate. Front. Environ. Sci. Eng. 9, 259–268. https://doi.org/10.1007/s11783-014-0652-2.

Mallakin, A., McConkey, B.J., Miao, G., McKibben, B., Snieckus, V., Dixon, D.G., Greenberg, B.M., 1999. Impacts of structural photomodification on the toxicity of environmental contaminants: anthracene photooxidation products. Ecotoxicol. Environ. Saf. 43, 204–212. https://doi.org/10.1006/eesa.1999.1764.

Mallakin, A., Dixon, D.G., Greenberg, B.M., 2000. Pathway of anthracene modification under simulated solar radiation. Chemosphere 40 (12), 1435–1441. https://doi.org/10.1016/S0045-6535(99)00331-8.

Mangotra, A., Singh, S.K., 2024. Volatile organic compounds: a threat to the environment and health hazards to living organisms—A review. J. Biotechnol. 382, 51–69. https://doi.org/10.1016/j.jbiotec.2023.12.013.

Martínez-Guijarro, K., Ramadan, A., Gevao, B., 2017. Atmospheric concentration of polychlorinated dibenzo-p-dioxins, polychlorinated dibenzofurans (PCDD/Fs) and dioxin-like polychlorinated biphenyls (dl-PCBs) at Umm-Al-Aish oil field-Kuwait. Chemosphere 168, 147–154. https://doi.org/10.1016/j.chemosphere.2016.10.036.

Mathé, C., Barre, A., Jourda, C., Dunand, C., 2010. Evolution and expression of class III peroxidases. Arch. Biochem. Biophys. 500 (1), 58–65. https://doi.org/10.1016/j.abb.2010.04.007.

Melymuk, L., Blumenthal, J., Sanka, O., Shu-Yin, A., Singla, V., et al., 2022. Persistent problem: global challenges to managing PCBs. Environ. Sci. Technol. 56 (12), 9029–9040. https://doi.org/10.1021/acs.est.2c01204.

Mhamdi, A., 2023. Hydrogen peroxide in plants. In: Advances in Botanical Research, vol 105, pp. 43–75. https://doi.org/10.1016/bs.abr.2022.11.002.

Middleton, J.T., Kendrick Jr., J.B., Schwalm, H.W., 1950. Injury to herbaceous plants by smog or air pollution. Plant Dis.Reporter. 34, 245–252.

Mila, A., Cao, R., Geng, N., Zhu, X., Chen, J., 2022. Characteristics of PAHs, PCDD/Fs, PCBs and PCNs in atmospheric fine particulate matter in Dalian, China. Chemosphere 288, 132488. https://doi.org/10.1016/j.chemosphere.2021.132488.

Mills, S.A., Thal, D.I., Barney, J., 2007. A summary of the 209 PCB congener nomenclature. Chemosphere 68, 1603–1612. https://doi.org/10.1016/j.chemosphere.2007.03.052.

Mittler, R., 2002. Oxidative stress, antioxidants, and stress tolerance. Trends Plant Sci. 7 (9), 405–410.

Mondal, N.K., Panja, D., Das, C., Dey, U., Das, K., 2014. Impacts of vehicle exhaust black and soot on germination of gram seed (*Cicer arietinum* L.). Communications in Plant Sciences 4 (1–2), 01–09.

Mondal, T., Mondal, S., Ghosh, S.K., Pal, P., Soren, T., et al., 2022. Phthalates-A family of plasticizers, their health risks, phytotoxic effects, and microbial bioaugmentation approaches. Environ. Res. 214, 114059. https://doi.org/10.1016/j.envres.2022.114059.

Montano, L., Pironti, C., Pinto, G., Ricciardi, M., Buono, A., et al., 2022. Polychlorinated biphenyls (PCBs) in the environment: occupational and exposure events, effects on human health and fertility. Toxics 10 (7), 365. https://doi.org/10.3390/toxics10070365.

Moorchilot, V.S., Louis, H., Haridas, A., Praveena, P., Arya, S.B., et al., 2025. Bisphenols in indoor dust: a comprehensive review of global distribution, exposure risks, transformation, and biomonitoring. Chemosphere 370, 143798. https://doi.org/10.1016/j.chemosphere.2024.143798.

Moura, B.B., Zammarchi, F., Manzini, J., Hoshika, Y., Brilli, L., et al., 2024. Assessment of seasonal variations in particulate matter accumulation and elemental composition in urban tree species. Environ. Res. 252, 118782. https://doi.org/10.1016/j.envres.2024.118782.

Mukherjee, A., Agrawal, M., 2018. Use of GLM approach to assess the responses of tropical trees to urban air pollution in relation to leaf functional traits and tree characteristics. Ecotoxicol. Environ. Saf. 152, 42–54. https://doi.org/10.1016/j.ecoenv.2018.01.038.

Mukhtar, M.S., McCormack, M.E., Argueso, C.T., Pajerowska-Mukhtar, K.M., 2016. Pathogen tactics to manipulate plant cell death. Curr. Biol. 26 (13), R608–R619.

Mulder, M.D., Heil, A., Kukučka, P., Kuta, J., Přibylová, P., et al., 2015. Long-range atmospheric transport of PAHs, PCBs and PBDEs to the central and eastern Mediterranean and changes of PCB and PBDE congener patterns in summer 2010. Atmos. Environ. 111, 51–59. https://doi.org/10.1016/j.atmosenv.2015.03.044.

Munyaneza, J., Qaraah, F.A., Jia, Q., Cheng, H., Zhen, H., Xiu, G., 2022. Seasonal trends, profiles, and exposure risk of PM2. 5-bound bisphenol analogs in ambient outdoor air: a study in Shanghai, China. Aerosol Air Qual. Res. 22 (9), 210324. https://doi.org/10.4209/aaqr.210324.

Nadal, M., Schuhmacher, M., Domingo, J.L., 2004. Levels of PAHs in soil and vegetation samples from Tarragona County Spain. Environ. Pollut. 132, 1–11. https://doi.org/10.1016/j.envpol.2004.04.003.

Nadarajah, K.K., 2020. ROS homeostasis in abiotic stress tolerance in plants. Int. J. Mol. Sci. 21 (15), 5208. https://doi.org/10.3390/ijms21155208.

Nadgórska-Socha, A., Kandziora-Ciupa, M., Trzęsicki, M., Barczyk, G., 2017. Air pollution tolerance index and heavy metal bioaccumulation in selected plant species from urban biotopes. Chemosphere 183, 471–482. https://doi.org/10.1016/j.chemosphere.2017.05.128.

Nafula, W.S., Orlóci, L., Kisvarga, S., Boroknay, G., Tóthmérész, B., et al., 2025. Rose Cultivars for Biomonitoring of Air Quality. https://doi.org/10.1016/j.heliyon.2025.e41978.

Nagorka, R., Koschorreck, J., 2020. Trends for plasticizers in German freshwater environments–evidence for the substitution of DEHP with emerging phthalate and non-phthalate alternatives. Environ. Pollut. 262, 114237. https://doi.org/10.1016/j.envpol.2020.114237.

Nakano, Y., Asada, K., 1981. Hydrogen peroxide is scavenged by ascorbate-specific peroxidase in spinach chloroplasts. Plant Cell Physiol. 22 (4), 867–880. https://doi.org/10.1093/oxfordjournals.pcp.a076232.

Nellier, Y.M., Perga, M.E., Cottin, N., Fanget, P., Malet, E., Naffrechoux, E., 2015. Mass budget in two high altitude lakes reveals their role as atmospheric PCB sinks. Sci. Total Environ. 511, 203–213. https://doi.org/10.1016/j.scitotenv.2014.12.052.

Oguntimehin, I., Nakatani, N., Sakugawa, H., 2008. Phytotoxicities of fluoranthene and phenanthrene deposited on needle surfaces of the evergreen conifer, Japanese red pine (*Pinus densiflora* Sieb. et Zucc.). Environ. Pollut. 154 (2), 264–271. https://doi.org/10.1016/j.envpol.2007.10.039.

Oguntimehin, I., Eissa, F., Sakugawa, H., 2010a. Negative effects of fluoranthene on the ecophysiology of tomato plants (*Lycopersicon esculentum* Mill) Fluoranthene mists negatively affected tomato plants. Chemosphere 78, 877–884. https://doi.org/10.1016/j.chemosphere.2009.11.030.

Oguntimehin, I., Kondo, H., Sakugawa, H., 2010b. The use of Sunpatiens (Impatiens spp.) as a bioindicator of some simulated air pollutants—using an ornamental plant as bioindicator. Chemosphere 81, 273–281. https://doi.org/10.1016/j.chemosphere.2010.05.052.

Oguntimehin, I., Bandai, S., Sakugawa, H., 2013. Mannitol can mitigate negative effects of simulated acid mist and fluoranthene in juvenile Japanese red pine (*P. densiflora* Sieb. et Zucc.). Environ. Pollut. 174, 78–84. https://doi.org/10.1016/j.envpol.2012.10.023.

Ortiz, J.E., Sánchez-Palencia, Y., Gallego, J.L.R., Borrego, Á.G., Baragaño, D., Torres, T., 2023. Deposition of atmospheric polycyclic aromatic hydrocarbons in rural areas: current data and historical record from an ombrotrophic peatland. Int. J. Coal Geol. 268, 104199. https://doi.org/10.1016/j.coal.2023.104199.

Othman, H.B., Leboulanger, C., Le Floch, E., Mabrouk, H.H., Hlaili, A.S., 2012. Toxicity of benz (a) anthracene and fluoranthene to marine phytoplankton in culture: does cell size really matter? J. Hazard Mater. 243, 204–211. https://doi.org/10.1016/j.jhazmat.2012.10.020

Pajurek, M., Warenik-Bany, M., Mikolajczyk, S., 2024. Feed materials–levels and characteristic profiles of dioxins and PCBs. Sci. Total Environ. 916, 170227. https://doi.org/10.1016/j.scitotenv.2024.170227.

Pan, T., Chen, X., Kong, C., Gao, D., Liu, W., et al., 2023. Single and combined toxicity of polystyrene nanoplastics and PCB-52 to the aquatic duckweed *Spirodela polyrhiza*. Sci. Total Environ. 901, 166482. https://doi.org/10.1016/j.scitotenv.2023.166482.

Pan, B., Pan, B., Lu, Y., Cai, K., Zhu, X., et al., 2024. Polystyrene microplastics facilitate the chemical journey of phthalates through vegetable and aggravate phytotoxicity. J. Hazard Mater. 480, 135770. https://doi.org/10.1016/j.jhazmat.2024.135770.

Pandey, V.P., Awasthi, M., Singh, S., Tiwari, S., Dwivedi, U.N., 2017. A comprehensive review on function and application of plant peroxidases. Biochem. Anal. Biochem. 6 (1), 308. https://doi.org/10.4172/2161-1009.1000308.

Passardi, F., Cosio, C., Penel, C., Dunand, C., 2005. Peroxidases have more functions than a Swiss army knife. Plant Cell Rep. 24, 255–265. https://doi.org/10.1007/s00299-005-0972-6.

Paterson, S., MacKay, D., Tam, D., Shin, W.Y., 1990. Uptake of organic chemicals by plants: a review of processes, correlations and models. Chemosphere 21, 297–331. https://doi.org/10.1016/0045-6535(90)90002-B.

Paul, S., Biswas, A., Roy, A., Chatterjee, A., Mukherjee, S., 2024. Impact of automobile induced air pollution on the quality of *Psidium guajava* fruit. Biocatal. Agric. Biotechnol. 103230. https://doi.org/10.1016/j.bcab.2024.103230.

Paull, N.J., Irga, P.J., Torpy, F.R., 2018. Active green wall plant health tolerance to diesel smoke exposure. Environ. Pollut. 240, 448–456. https://doi.org/10.1016/j.envpol.2018.05.004.

Qiu, Z., Wang, L., Zhou, Q., 2013. Effects of bisphenol A on growth, photosynthesis and chlorophyll fluorescence in above-ground organs of soybean seedlings. Chemosphere 90 (3), 1274–1280. https://doi.org/10.1016/j.chemosphere.2012.09.085.

Rabe, R., Kreeb, K.H., 1979. Enzyme activities and chlorophyll and protein content in plants as indicators of air pollution. Environ. Pollut. 19, 119–137. https://doi.org/10.1016/0013-9327(79)90143-5.

Rapala, M., Plucinski, B., Jedynak, P., 2017. The effect of bisphenol A on growth, pigment composition and photosystem II activity of *Arabidopsis thaliana*. Acta Biochim. Pol. 64, 407–413. https://doi.org/10.18388/abp.2017_1626.

Ren, L., Huang, X.D., McConkey, B.J., Dixon, D.G., Greenberg, B.M., 1994. Photoinduced toxicity of three polycyclic aromatic hydrocarbons (fluoranthene, pyrene, and naphthalene) to the duckweed *Lemna gibba* L. G-3. Ecotoxicol. Environ. Saf. 28 (2), 160–171. https://doi.org/10.1006/eesa.1994.1042.

Ren, L., Zeiler, L.F., Dixon, D.G., Greenberg, B.M., 1996. Photoinduced effects of polycyclic aromatic hydrocarbons on *Brassica napus* (Canola) during germination and early seedling development. Ecotoxicol. Environ. Saf. 33 (1), 73–80. https://doi.org/10.1006/eesa.1996.0008.

Roth, H.P., Swenson, E.A., 1947. Physiological Studies of Irritant Aspects of Atmospheric Pollution. Report to Los Angeles County Department of Air Pollution.

Saarikoski, S., Timonen, H., Saarnio, K., Aurela, M., Järvi, L., et al., 2008. Sources of organic carbon in fine particulate matter in northern European urban air. Atmos. Chem. Phys. 8, 6281–6295. https://doi.org/10.5194/acp-8-6281-2008, 2008.

Salapasidou, M., Samara, C., Voutsa, D., 2011. Endocrine disrupting compounds in the atmosphere of the urban area of Thessaloniki, Greece. Atmos. Environ. 45 (22), 3720–3729. https://doi.org/10.1016/j.atmosenv.2011.04.025.

Samek, L., Furman, L., Mikrut, M., Regiel-Futyra, A., Macyk, W., et al., 2017. Chemical composition of submicron and fine particulate matter collected in Krakow, Poland. Consequences for the APARIC project. Chemosphere 187, 430–439. https://doi.org/10.1016/j.chemosphere.2017.08.090.

Sarkar, R.K., Banerjee, A., Mukherji, S., 1986. Acceleration of peroxidase and catalase activities in leaves of wild dicotyledonous plants, as an indication of automobile exhaust pollution. Environ. Pollut. 42, 289–295. https://doi.org/10.1016/0143-1471(86)90013-9.

Sehili, A.M., Lammel, G., 2007. Global fate and distribution of polycyclic aromatic hydrocarbons emitted from Europe and Russia. Atmos. Environ. 41, 8301–8315. https://doi.org/10.1016/j.atmosenv.2007.06.050.

Sharma, P., Dubey, R.S., 2005. Drought-induced oxidative stress and antioxidant responses in plants. Plant Growth Regul. 46 (3), 233–241. https://doi.org/10.1002/9783527675265.ch09.

Sharma, P., Jha, A.B., Dubey, R.S., Pessarakli, M., 2012. Reactive oxygen species, oxidative damage, and antioxidative defense mechanism in plants under stressful conditions. J. Bot. 217037. https://doi.org/10.1155/2012/217037.

Sharma, R., Kaur, R., 2020. Physiological and metabolic alterations induced by phthalates in plants: possible mechanisms of their uptake and degradation. Environ. Sustain. 3 (4), 391–404. https://doi.org/10.1007/s42398-020-00141-x.

Shi, R., Liu, W., Lian, Y., Wang, Q., Zeb, A., Tang, J., 2022. Phytotoxicity of polystyrene, polyethylene and polypropylene microplastics on tomato (*Lycopersicon esculentum* L.). J. Environ. Manag. 317, 115441. https://doi.org/10.1016/j.jenvman.2022.115441.

Sidhu, S., Gullett, B., Striebich, R., Klosterman, J., Contreras, J., DeVito, M., 2005. Endocrine disrupting chemical emissions from combustion sources: diesel particulate emissions and domestic waste open burn emissions. Atmos. Environ. 39 (5), 801–811. https://doi.org/10.1016/j.atmosenv.2004.10.040.

Simoneit, B.R.T., Medeiros, P.M., Didyk, B.M., 2005. Combustion products of plastics as indicators for refuse burning in the atmosphere. Environ. Sci. Technol. 39, 6961–6970. https://doi.org/10.1021/es050767x.

Singh, S.K., Rao, D.N., Agrawal, M., Pandey, J., Naryan, D., 1991. Air pollution tolerance index of plants. J. Environ. Manag. 32 (1), 45–55. https://doi.org/10.1016/S0301-4797(05)80080-5.

Siudek, P., 2022. Seasonal distribution of PM2. 5-bound polycyclic aromatic hydrocarbons as a critical indicator of air quality and health impact in a coastal-urban region of Poland. Sci. Total Environ. 827, 154375. https://doi.org/10.1016/j.scitotenv.2022.154375.

Škrdlíková, L., Landlová, L., Klánová, J., Lammel, G., 2011. Wet deposition and scavenging efficiency of gaseous and particulate phase polycyclic aromatic compounds at a central European suburban site. Atmos. Environ. 45 (25), 4305–4312. https://doi.org/10.1016/j.atmosenv.2011.04.072.

Slesak, I., Libik, M., Karpinska, B., Karpinski, S., Miszalski, Z., 2007. The role of hydrogen peroxide in regulation of plant metabolism and cellular signalling in response to environmental stresses. Acta Biochim. Pol. 54 (1), 39–50.

Storch-Böhm, R.F., Somensi, C.A., Cotelle, S., Deomar-Simões, M.J., Poyer-Radetski, L., et al., 2020. Sensitivity of different parameters for selection of higher plants in urban afforestation: exposure of Guabiroba (*Campomanesia xanthocarpa* O. Berg.) to diesel engine exhaust. Environ. Pollut. 265, 114675. https://doi.org/10.1016/j.envpol.2020.114675.

Su, J., Liu, Y., Han, F., Gao, F., Gan, F., et al., 2024. ROS, an important plant growth regulator in root growth and development: functional genes and mechanism. Biology 13 (12), 1033. https://doi.org/10.3390/biology13121033.

Sun, H., Lei, C., Xu, J., Li, R., 2021. Foliar uptake and leaf-to-root translocation of nanoplastics with different coating charge in maize plants. J. Hazard Mater. 416, 125854. https://doi.org/10.1016/j.jhazmat.2021.125854.

Sun, H., Lei, C., Yuan, Y., Xu, J., Han, M., 2022. Nanoplastic impacts on the foliar uptake, metabolism and phytotoxicity of phthalate esters in corn (*Zea mays* L.) plants. Chemosphere 304, 135309. https://doi.org/10.1016/j.chemosphere.2022.135309.

Tang, X.Y., Liu, H.P., Naila, R.S.L., Dai, Y., Zhang, X.M., et al., 2021. Irrigation using hybrid constructed wetland treated domestic sewage: uptake of phthalic acid esters and antibiotics by *Ipomoea aquatica* forssk. J. Hazard Mater. 405, 10. https://doi.org/10.1016/j.jhazmat.2020.124025.

Tappert, L., Bunge, M., Hoehne, D., Dlugi, I., Fetters, K., et al., 2024. Bisphenol A in surface waters in Germany: Part I. Reassessment of sources and emissions pathways for FlowEQ modeling. Integrated Environ. Assess. Manag. 20 (1), 211–225. https://doi.org/10.1002/ieam.4805.

Tao, S., Cui, Y.H., Xu, F.L., Li, B.G., Cao, J., Liu, W.X., 2004. Polycyclic aromatic hydrocarbons (PAHs) in agricultural soil and vegetables from Tianjin. Sci. Total Environ. 320, 11–24. https://doi.org/10.1016/S0048-9697(03)00453-4.

Teil, M.J., Moreau-Guigon, E., Blanchard, M., Alliot, F., Gasperi, J., et al., 2016. Endocrine disrupting compounds in gaseous and particulate outdoor air phases according to environmental factors. Chemosphere 146, 94–104. https://doi.org/10.1016/j.chemosphere.2015.12.015.

Tian, K., Bao, H.Y., Liu, X.P., Wu, F.Y., 2018. Accumulation and distribution of PAHs in winter wheat from areas influenced by coal combustion in China. Environ. Sci. Pollut. Control Ser. 25, 23780–23790. https://doi.org/10.1007/s11356-018-2456-6.

Thoene, M., Rytel, L., Nowicka, N., Wojtkiewicz, J., 2018. The state of bisphenol research in the lesser developed countries of the EU: a mini-review. Toxicol. Res. 7, 371–380. https://doi.org/10.1039/c8tx00064f.

Tognetti, V.B., Bielach, A., Hrtyan, M., 2017. Redox regulation at the site of primary growth: auxin, cytokinin and ROS crosstalk. Plant Cell Environ. 40, 2586–2605. https://doi.org/10.1111/pce.13021.

Tomar, R.S., Jajoo, A., 2015. Photomodified fluoranthene exerts more harmful effects as compared to intact fluoranthene by inhibiting growth and photosynthetic processes in wheat. Ecotoxicol. Environ. Saf. 122, 31–36. https://doi.org/10.1016/j.ecoenv.2015.07.002.

Tomar, R.S., Singh, B., Jajoo, A., 2019. Effects of organic pollutants on photosynthesis. Photosynthesis, productivity and environmental stress 1–26. https://doi.org/10.1002/9781119501800.ch1.

Tripathy, B.C., Oelmüller, R., 2012. Reactive oxygen species generation and signaling in plants. Plant Signal. Behav. 7 (12), 1621–1633. https://doi.org/10.4161/psb.22455.

UNEP, 2016. Consolidated Assessment of Efforts Made towards the Elimination of Polychlorinated Biphenyls: Geneva.

Uria-Tellaetxe, I., Navazo, M., de Blas, M., Durana, N., Alonso, L., Iza, J., 2016. Gas-phase naphthalene concentration data recovery in ambient air and its relevance as a tracer of sources of volatile organic compounds. Atmos. Environ. 131, 279–288. https://doi.org/10.1016/j.atmosenv.2016.01.047.

Vasiljevic, T., Harner, T., 2021. Bisphenol A and its analogues in outdoor and indoor air: properties, sources and global levels. Sci. Total Environ. 789, 148013. https://doi.org/10.1016/j.scitotenv.2021.148013.

Vánová, L., Kummerová, M., Klemš, M., Zezulka, S., 2009. Fluoranthene influences endogenous abscisic acid level and primary photosynthetic processes in pea (Pisum sativum L.) plants in vitro. Plant Growth Regul. 57, 39–47. https://doi.org/10.1007/s10725-008-9318-z.

Velis, C.A., Cook, E., 2021. Mismanagement of plastic waste through open burning in the Global South: a systematic review of risks to occupational and public health. ES T (Environ. Sci. Technol.) 55 (19), 7186–7207. https://doi.org/10.1021/acs.est.0c08536.

Verma, A., Singh, S.N., 2006. Biochemical and ultrastructural changes in plant foliage exposed to auto-pollution. Environ. Monit. Assess. 120, 585–602. https://doi.org/10.1007/s10661-005-9105-5.

Vicente, E.D., Figueiredo, D., Gonçalves, C., Kováts, N., Hubai, K., et al., 2025. Toxicological screening of PM2. 5 from wildfires involving different biomass fuels. Environ. Pollut. 125887. https://doi.org/10.1016/j.envpol.2025.125887.

Viskari, E.L., Surakka, J., Pasanen, P., Mirme, A., Kössi, S., et al., 2000a. Responses of spruce seedlings (*Picea abies*) to exhaust gas under laboratory conditions—I plant–insect interactions. Environ. Pollut. 107 (1), 89–98. https://doi.org/10.1016/S0269-7491(99)00128-1.

Viskari, E.L., Holopainen, T., Kärenlampi, L., 2000b. Responses of spruce seedlings (*Picea abies*) to exhaust gas under laboratory conditions—II ultrastructural changes and stomatal behaviour. Environ. Pollut. 107 (1), 99–107. https://doi.org/10.1016/S0269-7491(99)00127-X.

Wang, Q., Liu, M., Li, Y., Liu, Y., Li, S., Ge, R., 2016. Dry and wet deposition of polycyclic aromatic hydrocarbons and comparison with typical media in urban system of Shanghai, China. Atmos. Environ. 144, 175–181. https://doi.org/10.1016/j.atmosenv.2016.08.079.

Wang, J., Bao, H., Cai, J., Li, J., Li, J., Wu, F., 2022. Uptake and accumulation of naphthalene, phenanthrene, and benzo (b) fluoranthene in winter wheat affected by foliar exposure at different growth stages. Environ. Sci. Pollut. Control Ser. 29 (31), 47617–47628. https://doi.org/10.1007/s11356-022-19263-2.

Wang, C., Sun, Y., Ruan, H., Yang, J., 2021. Toxic effects of 2, 4, 4′-trichlorobiphenyl (PCB-28) on growth, photosynthesis characteristics and antioxidant defense system of *Lemna minor* L. Plant Physiol. Biochem. 166, 505–511. https://doi.org/10.1016/j.plaphy.2021.06.031.

Wang, Y., Gao, F., Xu, Y., Rodgers, T.F., Tan, F., 2024. Field study on the uptake pathways and their contributions to the accumulation of organophosphate esters, phthalates, and polycyclic aromatic hydrocarbons in upland rice. Sci. Total Environ. 174205. https://doi.org/10.1016/j.scitotenv.2024.174205.

Wang, P., Li, Y., Zhang, Q., Yang, Q., Zhang, L., et al., 2017. Three-year monitoring of atmospheric PCBs and PBDEs at the Chinese Great Wall Station, West Antarctica: levels, chiral signature, environmental behaviors and source implication. Atmos. Environ. 150, 407–416. https://doi.org/10.1016/j.atmosenv.2016.11.036.

Wania, F., Westgate, J.N., 2008. On the mechanism of mountain cold-trapping of organic chemicals. Environ. Sci. Technol. 42, 9092–9098. https://doi.org/10.1021/es8013198.

Waśkiewicz, A., Beszterda, M., Goliński, P., 2014a. Nonenzymatic antioxidants in plants. In: Ahmad, P. (Ed.), Oxidative Damage to Plants. Academic Press, pp. 369–396. https://doi.org/10.1016/B978-0-12-799963-0.00007-1.

Waśkiewicz, A., Gładysz, O., Szentner, K., Goliński, P., 2014b. Role of glutathione in abiotic stress tolerance. In: Ahmad, P. (Ed.), Oxidative Damage to Plants. Academic Press, pp. 149–181. https://doi.org/10.1016/B978-0-12-799963-0.00005-8.

Watts, A.W., Ballestero, T.P., Gardner, K.H., 2008. Soil and atmospheric inputs to PAH concentrations in salt marsh plants. Water Air Soil Pollut. 189, 253–263. https://doi.org/10.1007/s11270-007-9572-0.

Wieczorek, J., Sienkiewicz, S., Pietrzak, M., Wieczorek, Z., 2015. Uptake and phytotoxicity of anthracene and benzo[k]fluoranthene applied to the leaves of celery plants (*Apium graveolens* var. secalinum L.). Ecotoxicol. Environ. Saf. 115, 19–25. https://doi.org/10.1016/j.ecoenv.2015.01.032.

Wieczorek, J.K., Wieczorek, Z.J., 2007. Phytotoxicity and accumulation of anthracene applied to the foliage and sandy substrate in lettuce and radish plants. Ecotoxicol. Environ. Saf. 66, 369–377. https://doi.org/10.1016/j.ecoenv.2005.10.002.

Wisniowska, E., Wlodarczyk-Makula, M., 2022. Evaluation of the adsorption efficiency of carcinogenic PAHs on microplastic (polyester) fibers-preliminary results. Appl. Water Sci. 12, 119. https://doi.org/10.1007/s13201-022-01654-y.

Wright, S.L., Ulke, J., Font, A., Chan, K.L.A., Kelly, F.J., 2020. Atmospheric microplastic deposition in an urban environment and an evaluation of transport. Environ. Int. 136, 105411. https://doi.org/10.1016/j.envint.2019.105411.

Wu, Y.L., Lin, L.F., Hsieh, L.T., Wang, L.C., Chang-Chien, G.P., 2009. Atmospheric dry deposition of polychlorinated dibenzo-p-dioxins and dibenzofurans in the vicinity of municipal solid waste incinerators. J. Hazard Mater. 162 (1), 521–529. https://doi.org/10.1016/j.jhazmat.2008.05.075.

Xiao, H., Liu, Y., Yu, H., Yadav, N., He, J., et al., 2023. Combined toxicity influence of polypropylene microplastics and di-2-ethylhexyl phthalate on physiological-biochemical characteristics of cucumber (Cucumis sativus L.). Plant Physiol. Biochem. 201, 107811. https://doi.org/10.1016/j.plaphy.2023.107811.

Xiong, G.N., Zhang, Y.H., Duan, Y.H., Cai, C.Y., Wang, X., et al., 2017. Uptake of PAHs by cabbage root and leaf in vegetable plots near a large coking manufacturer and associations with PAHs in cabbage core. Environ. Sci. Pollut. Control Ser. 24, 18953–18965. https://doi.org/10.1007/s11356-017-9548-6.

Xu, G., Liu, Y., Yu, Y., 2021. Effects of polystyrene microplastics on uptake and toxicity of phenanthrene in soybean. Sci. Total Environ. 783, 147016. https://doi.org/10.1016/j.scitotenv.2021.147016.

Yang, L., Jin, M., Tong, C., Xie, S., 2013. Study of dynamic sorption and desorption of polycyclic aromatic hydrocarbons in silty-clay soil. J. Hazard Mater. 244–245, 77–85. https://doi.org/10.1016/j.jhazmat.2012.11.029.

Yao, X., Zhang, J., Wang, C., Wang, Q., Li, X., et al., 2022. Toxicity of dibutyl phthalate to pakchoi (*Brassica campestris* L.): evaluation through different levels of biological organization. Sci. Total Environ. 849, 157943. https://doi.org/10.1016/j.scitotenv.2022.157943.

Yuan, L., Cheng, J., Wang, Y., Wang, W., Gao, R., Yu, X., 2020. Uptake and toxicity of di-(2-ethylhexyl) phthalate in *Brassica chinensis* L. Chemosphere 252, 126640. https://doi.org/10.1016/j.chemosphere.2020.126640.

Zhang, J., Wang, L., Li, M., Jiao, L., Zhou, Q., Huang, X., 2015. Effects of bisphenol A on chlorophyll fluorescence in five plants. Environ. Sci. Pollut. Control Ser. 22, 17724–17732. https://doi.org/10.1007/s11356-015-5003-8.

Zhang, Y., Hou, D., Xiong, G., Duan, Y., Cai, C., Wang, X., 2018. Structural equation modeling of PAHs in ambient air, dust fall, soil, and cabbage in vegetable bases of Northern China. Environ. Pollut. 239, 13–20. https://doi.org/10.1016/j.envpol.2018.03.084.

Zhang, X., Zhang, H., Wang, Y., Bai, P., Zhang, L., et al., 2025. Estimation of gaseous polycyclic aromatic hydrocarbons (PAHs) and characteristics of atmospheric PAHs at a traffic site in Kanazawa, Japan. J. Environ. Sci. 149, 57–67. https://doi.org/10.1016/j.jes.2023.09.009.

CHAPTER 4

Phytotoxicity of pesticides

Hardik Giri Gosai[1], Asha Sharma[1], Jagriti Patel[1], Palak Savalia[2], Shraddha K. Kanzariya[1] and Sahil Menghani[1]

[1]Department of Environmental Studies, Faculty of Science, The Maharaja Sayajirao University of Baroda, Vadodara, Gujarat, India; [2]Sophisticated Instrumentation Centre for Applied Research & Testing (SICART), Charutar Vidya Mandal (CVM), Vallabh Vidyanagar, Anand, Gujarat, India

1. Introduction

Pesticides have steadily become an essential component of agricultural systems throughout the world. This has enabled a significant increase in crop yields and the quantity of food that is produced (Schleiffer and Speiser, 2022). The term pesticide refers to any substance or combination of compounds that are used to prevent or eliminate pests. Some potential pests include insects, fungus, rodents, and undesired plant species that cause damage to crop while they are being grown or stored. The phrase pesticide refers to a wide range of chemicals that are used to eradicate certain pests, including insecticides, herbicides, fungicides, and rodenticides (Abubakar et al., 2019). The origin of pesticides is the primary factor that determines whether they are classified as chemical pesticides or biopesticides. There is a host-specificity associated with biological pesticides. As opposed to chemical pesticides, which are generic and influence a wide variety of organisms that are not the intended target, they have a high level of specificity because they target the desired pest as well as critters that are closely related to it. As a result of their decreased toxicity, fast disintegration, and limited amount requirements, biopesticides are environmentally friendly. Because of their toxicity and the fact that they do not biodegrade, chemical pesticides are a contributor to the contamination of the environment (Maino et al., 2023). Figure 4.1 depicts different types of plant protective products used in the agricultural land.

Since the beginning of time, several types of pesticides have been utilized for the purpose of protecting crops. However, even though pesticides are beneficial to crops, they also have a significant and detrimental effect on the ecosystem. It is possible that the degradation of biodiversity could be caused by the overuse of pesticides. For the sake of their continued existence, a great number of animals, aquatic species, and birds are in danger of being killed by hazardous chemicals. Pesticides are a cause for concern with regard to the durability of the environment and the stability of the world. The most practical and widespread basis for classifying pesticides is the chemical composition of the active components and the properties of those components (Hassaan and El Nemr, 2020). It is possible to draw conclusions about the efficiency of the substances based on

Phytotoxicity of Organic Pollutants
ISBN 978-0-443-33110-7,
https://doi.org/10.1016/B978-0-443-33110-7.00003-1

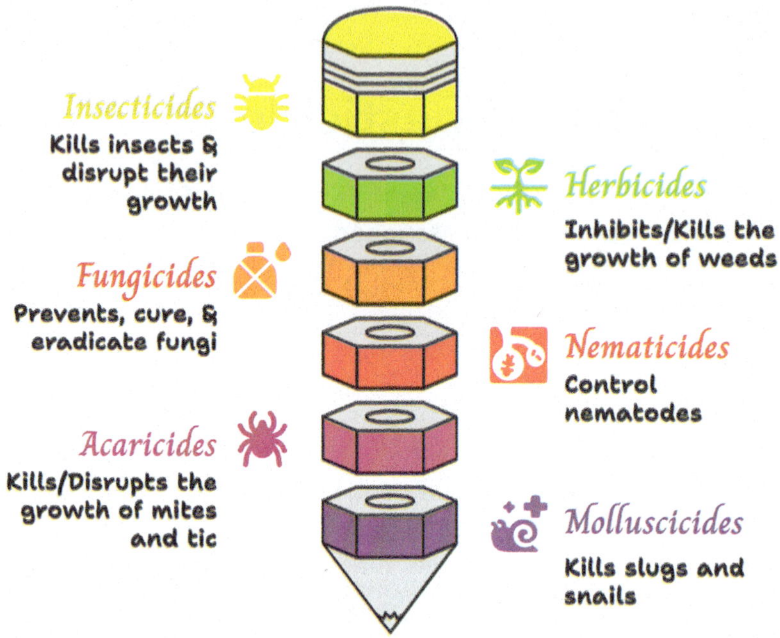

Figure 4.1 Types of pesticides used in the agricultural field. The figure illustrates various PPP utilized in agricultural practices to enhance crop yield.

this form of categorization, as well as their physical and chemical properties. When determining how to apply pesticides, how much to apply, and what safety measures to take, it is essential to have a comprehensive understanding of the chemical and physical properties of the pesticides. Organochlorines, carbamates, organophosphorus, and pyrethroids are the four basic categories that pesticides fall into regarding their chemical composition. Organochlorines are the most common type of pesticide. The method of categorizing pesticides according to the chemical composition of their compounds is a challenging one. Most pesticides that are currently available are organic chemicals (Liguori et al., 2022).

The application of pesticides in agricultural settings can be accomplished through a variety of approaches, ranging from the manual spraying of personnel to the employment of various vehicles, such as trucks and airplanes. Several of these approaches, or maybe all of them, have been utilized in different regions of the world at different times and in different places. Numerous studies on poisoning and the effects of synthetic chemicals on human health provide evidence that pesticide applications did, in fact, cause intoxication among farmers, rural workers, and their families (Pathak et al., 2022). These studies were conducted in the United States. Due to the use of crop protection pesticides, it was immediately recognized that the use of synthetic chemicals was contaminating both local and global scales. This pollution was caused by the use of synthetic chemicals. As a result of the rising usage of these chemicals on a global basis, the negative impacts that pesticides have on the environment and on human health have become even more damaging. When pesticides are used in an irresponsible manner or in excessive amounts, they can gradually cause harm to both the environment and human health, particularly in countries that are less developed. This chapter examines the detrimental impacts of pesticides on environmental integrity and food safety. It further elucidates the significant risks associated with prolonged pesticide use and explores alternative sustainable practices aimed at promoting healthier consumption (Ahmed et al., 2023).

Chemical agents have been employed in pest control for centuries. The history of pesticides was influenced in the mid-20th century, particularly by the rise of DDT (dichlorodiphenyltrichloroethane) after the second World War. Prominent entomologist, Robert van den Bosch critiqued the pesticide treadmill, which he argued was driven by vested interest that alternative pest control methods and early pesticide use included the mineral compounds and plant extracts, evolving with copper fungicides and arsenic compounds by the late 19th century, despite advancements in pesticide application, there were concern about the rapid adoption of chemical methods as the expense of biological controls (Thellier et al., 2024). Although pesticides could be divided into biopesticides and chemical pesticides. Among the most widely used chemical pesticides are organophosphates, carbamates, and pyrethroids. Organophosphates include active ingredients like chlorpyrifos, dimethoate, glyphosate and profenofos. Carbamate contain mancozeb, propineb, fenobucarb and carbosulfan. Pyrethroid made with ingredients like cypermethrin, permethrin, emamectin, deltamethrin and beta-cyfluthrin. Depend on the population, the effects of pesticides on factors like exposure, toxicity, life history, timing of application, population structure and landscape characteristics. Insects are targeted through various neuroactive mechanisms, including acetylcholinesterase inhibition for organophosphates and methylcarbamates, nicotinic acetylcholine receptors for neonicotinoids, gamma aminobutyric acid (GABA) receptor channels for polychlorocyclohexanes and fiproles, Voltage-gated sodium channels for pyrethroids and DDT (Aqeel et al., 2023; Casida and Durkin, 2013; Das et al., 2022). Application of these substances from manual pumps to aerial spraying has also advanced and changed agricultural practices. Aerial spraying was introduced in the 1920s, for large-scale applications in agriculture and public health campaigns. The wide use of pesticides, especially DDT, led to significant environmental concerns. These concerns focus on the ecological impacts of pesticides, setting the stage for modern debates on their safety and alternatives in pest management (Bertomeu-Sánchez, 2019). Figure 4.2 shows the timeline of pesticide use.

In the 1960s, the government of Canada conducted a study to collect and analyze wildlife species, such as mammals, fish, migratory and non-migratory birds, and invertebrates, to assess pesticide residue levels in these organisms. In the 1970s, the use of

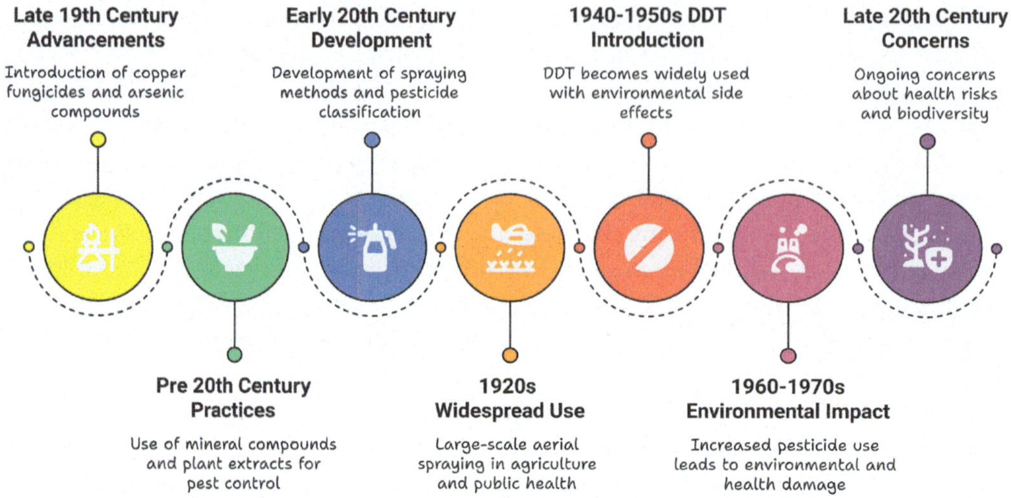

Figure 4.2 Timeline of pesticide use, illustrating key developments, regulatory changes, and emerging concerns over time.

organophosphate and carbamate insecticides rose significantly, they were more environmentally friendly due to their reduced stability and lower likelihood of widespread transport and bioaccumulation. However, research such as Carbofuran spill site on a remote island of the Hawaiian national wildlife refuge, and chemical and biochemical evaluation of Swainson's-hawk mortalities in Argentina, highlighted the exposure to these insecticides still posed risks to non-target wildlife. In the 1980s, Pyrethroids, a new class of Insecticides, were introduced and quickly became more popular (Johnston, 2009). Brazil is currently the world's leading consumer of pesticides, with increasing use of pesticides, alongside the advancement in agriculture production. The awareness of the risks of pesticides has grown globally since the publication of Silent Spring by Rachel Carson in 1962. The book highlighted the effects of pesticide use in the 1940s and 1950s in the United States, which led to large-scale environmental damage. This widespread use of pesticides aimed to eliminate pests but has unintended consequences for the environment as well as human health (Clasen et al., 2019).

1.1 Overview of pesticides action

Pesticides play a very important role in crop protection and enhancing agricultural productivity or yield. By managing pests, weeds, pesticides, they help safeguard crops both before and after harvest to increase yields. These products enable farmers to grow healthier, more abundant food, which is essential for meeting the growing global demand. With continued advancements in agricultural technology, pesticides are key tools in ensuring affordable, safe and sustainable food production, ultimately supporting a healthier population. Various technologies for crop protection, including pesticides, are essential for both increasing the quantity and improving the quality of food production (Kalogiannidis et al., 2022). It helps farmers boost crop yields and efficiency, preventing up to 40% of potential crop losses caused by weeds, pests. Additionally, pesticides help ensure that consumers receive high quality produce, free from insect damage and contamination. Without these kinds of technologies, the production of food would decrease, leading to a shortage and higher prices. Thus, crop protection plays a very important role in both food supplies and supporting economic stability (García-Díez et al., 2021). The mode of action of pesticides is a subject to intersection with fields of biology and chemistry, and with many practical implications. The advent of synthetic pesticides DDT, just after World War 2. At that time, understanding of biochemical and physiological processes within organisms was not sufficiently clarified to understand their mechanism of action. This lack of knowledge made it challenging to know the mechanisms through which pesticides acted at the target sites, as well as how chemicals were absorbed, distributed, and degraded in the environment. Today we are better equipped to address and understand the challenges posed by pesticides, but the historical context underscores the importance of continually refining our understanding of how these chemicals interact with living systems (Clarke and Lean, 2022; Das, 2013).

Insecticides act as nerve poisons by altering the permeability of nerve axons to Na^+ and K^+ ions. Leading to disruption of ionic balance, thus the imbalance causes repetitive firing of nerve axons. This imbalance results in nerve axons firing repetitively (Toshio, 1992). The repetitive firing of nerve axons leads to tremors, convulsions and eventually death. Insecticides are toxic to mammals. They are chemically stable, making them persistent in the environment. They are lipophilic, meaning they are fat soluble. They are readily absorbed by fish. Environmental impact due to their ability to be absorbed by fish, they can enter the food chain, posing further environmental and health risks. Insecticides are not safe, especially because of their persistence and ability to accumulate in ecosystems (Eldefrawi and Eldefrawi, 2020; Sparks and Nauen, 2015). ***Organophosphate and Carbamate*** inhibit cholinesterase (ChE) in which organophosphates phosphorylate the enzyme and carbamates carbamylate it. The binding to the enzyme is irreversible, especially when the insecticide is attached to the phosphorous moiety. The inhibition causes accumulation of acetylcholine (Ach) at synapses (neuron to neuron or neuron to muscle). This leads to rapid twitching of voluntary muscles, followed by paralysis (Das, 2013; De Bleecker, 2008).

Fungicides having sulfur, which was first reported for pest control in 1000 BCE that acts as an H-acceptor in metabolic systems, hindering hydrogenation and dehydrogenation. Bordeaux mixture used for controlling *Plasmopara viticola*. Causes precipitation and denaturation of proteins/enzymes, leading to their inactivation. Organometallic fungicides, organomercury, tin (Sn) and arsenic (As) inhibit biochemical reactions in the electron transport chain of photosynthesis. Inhibit oxidative phosphorylation in mitochondria. Anilinopyrimidines (AP) inhibit secretion of hydrolyzing enzymes and fungal methionine biosynthesis. Benzimidazole fungicides involved inhibition of fungal mitosis by binding to fungal tubulin or microtubules. It blocks microtubule polymerization, disrupting cell division. Melanin biosynthesis inhibitors have a fused bicyclic structure that acts as competitors for intermediates in fungal melanin biosynthesis, disrupting fungal activity rapidly. Dicarboxamides cause disorder in fungal cell structure during growth and multiplication. Leads to morphological changes in hyphae and leakage of fungal cell contents (Davidse, 1986; Kuplińska and Rząd, 2021; Yang et al., 2024).

Herbicides, known as weed killers, are used to kill unwanted plants and to clear ground, industrial sites, railways and railway embankments, which are not selective and kill all plant material with which they come into contact (Alghuthaymi et al., 2019). Some plants, like walnuts and tree of heaven, produce natural herbicides and this chemical action is called allelopathy. Pyruvate binds to thiamine pyrophosphate (TPP) at the active site, decarboxylating to form an enzyme substrate complex and release CO_2. The second pyruvate molecule is prevented from adding due to herbicide binding, halting the process. Induces the activity of 1-amino-cyclopropane-1-carboxylic acid (ACC). Promotes ethylene biosynthesis, which triggers ABA accumulation in susceptible dicots, leading to growth inhibition, Epinasty (downward leaf curling), senescence (aging of the plant). 4-Hydroxyphenylpyruvate dioxygenase (HPPD) inhibition leads to increased tyrosine levels in plants, depletion of plastoquinone, a critical cofactor for phytoene desaturase, resulting in reduced carotenoid production and causing bleaching symptoms in the plant (Duke, 1990; He et al., 2022; Khamare et al., 2022).

2. Phytotoxicity of pesticide on crops

The insecticide reduces the possibility of harm during post-harvest storage while also protecting crops on agricultural land. It is highly successful and useful in managing a variety of human and plant diseases, including typhoid and malaria. However, it also degrades the soil quality of agricultural land, which is why its adverse effects are taken into consideration (Pathak et al., 2022). Phytotoxic pesticides typically disrupt the physiological and biochemical functions of plants that are not their intended targets. Measuring these reactions can reveal the degree of physiological harm brought on by pesticides, as many of these impacts are connected to photosynthetic function (Shahid et al., 2024). Key photosynthetic responses, such as A/Ci curves, which can reveal information about photosynthetic reactions and potentially serve as early warning signs of phytotoxic reactions, are now easier to evaluate because to advancements in portable photosynthesis equipment. The plant taxa that are irrigated, the pesticide concentration in the water, and the length of time that the pesticide is applied can all affect the phytotoxicity of isoxaben, oxyfluorfen, and chlorpyrifos. Growers are not well-informed about the long-term phytotoxicity these pesticides cause in runoff water. Additionally, fewer people are aware of how seriously these herbicides harm typical landscaping plants through phytotoxicity (Poudyal et al., 2019). The use of pesticides causes numerous enzymatic and non-enzymatic changes in biochemical and physiological antioxidants, and these changes can initially impact plant growth from germination and ultimately impact plant yield output (Yengkokpam and Mazumder, 2021).

2.1 Pesticide residues in food

Pesticides are a major contributor to ecosystem pollution because they build up in crops and can be carried for great distances by wind, water, and soil (Khan et al., 2023). Air pollution caused by agrochemicals is generally caused by spraying techniques, such as those used in the air or on the ground. Particulate semi-volatile pesticides have half-lives that can range anywhere from a few days to a month and are even capable of withstanding gas-phase interactions with hydroxyl radicals that are present in the environment. The majority of these insecticides are attracted to particles that are found in aerosols. Pesticides that are disseminated in the air have the capacity to travel tremendous distances, and chemicals that are stable in gaseous form have the potential to be transported all over the environment (Zaller et al., 2022). The most common cause of soil contamination is the direct application of pesticides to the soil for the aim of protecting crops. It is common practice in Southeast Asia for agricultural practitioners who do not possess specialist knowledge to rely on the recommendations of sales workers working at retail establishments that sell pesticides (Schreinemachers et al., 2015). As a consequence of this, it is possible that they will apply an excessive number of pesticides in order to manage the pests that are present in their agricultural areas. As a result of the application of pesticides to crops, these substances are transported downhill through the flow of water. As they move through the soil matrix, they may go through processes of adsorption, desorption, and destruction. In addition to the features of the pesticide, the qualities of the soil, and the environmental elements that are already present, the rate at which chemicals move through the soil is also affected by these factors (Cui et al., 2023; Rasool et al., 2022).

The selection of pesticide application methods is determined by analyzing the behaviors of the target pests, the specific attributes of the application site, and the chemical properties of the pesticides involved. The prevalent methods for pesticide application include broadcast spraying, band spraying, space treatment, spot treatment, foliar application, drench and soil injection (Tudi et al., 2022). Band spraying and broadcast spraying are two techniques that are frequently utilized to treat crop regions in a regular manner. There is a technique known as band spraying, which involves the use of non-selective herbicides along fence rows and borders in order to eradicate all plants. In broadcast spraying, pesticides are uniformly applied to a large area of turf grass either on foot or using motorized equipment. There are a number of critical parameters that contribute to the effectiveness of these application methods, including spray pressure, walking speed, nozzle tip, and height (Tudi et al., 2022; Chaudhary et al., 2020).

People are exposed to pesticides in a direct manner through a variety of application methods in a variety of situations, including agricultural, occupational, and residential environments (Figueiredo et al., 2021). In addition, there is a pathway for indirect exposure through polluted environmental media, which includes the environments of the air, water, soil, and food. The level of toxicity that is associated with pesticides is affected by the many modes of exposure that are available. There are three basic routes that pesticides take to interact with the human body: the cutaneous route, the oral route, and the respiratory route (Vandana et al., 2022; Singh et al., 2017). Dermal exposure represents the predominant and most efficient route of exposure for agricultural workers encountering pesticides, primarily due to factors such as splashing, spillage, or spray drift, particularly during the application of pesticides in agricultural practices or domestic tasks (Mazlan et al., 2016; Chester, 2010). An investigation aimed at evaluating the hazards associated with pesticide exposure among workers utilizing hand-held sprayers in greenhouse environments. This study investigates the accumulation of pesticides on the bodies of operators, evaluates the efficacy of protective measures, and explores possible strategies for mitigation. Individuals engaged in greenhouse operations utilizing hand-held sprayers encounter considerable dermal exposure, particularly on areas of the body that are not covered. The enclosed greenhouse setting exacerbates this risk owing to the absence of natural airflow and the close proximity to cultivated plants that have undergone treatment (Cerruto et al., 2018; Naik et al., 2024).

Observations indicated that differences in pesticide deposition were influenced by the type of sprayer used, with certain devices leading to irregular distribution and elevated losses to the soil, thereby exacerbating environmental contamination (Felsot et al., 2010). Furthermore, a risk assessment was carried out on individuals who had been directly dermally exposed to Trinexapac-ethyl (TE) by the utilization of absorbent paper patches. When the exposure intensity (EI) and potential dermal exposure (PDE) for the body sections of workers were measured, it was discovered that the EI for each body section varied from 7.22 to 73.0 pg cm^{-2} (Tudi et al., 2022). Among workers who were mixing/loading, hand-held power sprayer, and manual sprayer workers. The hands of manual sprayers were found to have the greatest exposure index of total exposure, whereas the upper arms of workers who were responsible for mixing and loading were found to have the lowest exposure index of total exposure. The chests and backs of mixing and loading people had the highest recorded exposure levels, which were 29%. Meanwhile, the chests and backs of hand-held power sprayers had the greatest exposure levels, which were 40%. The thighs of manual sprayers had the highest exposure levels, which were 32% (Choi et al., 2013).

2.2 Pesticide degradation

The extensive insecticidal efficacy, chemical resilience, high operational efficiency, and economical production costs position organophosphates as a leading category of pesticides utilized across agricultural, veterinary, and urban environments. Nonetheless, due to their significant impact on human health, their utilization has emerged as a major anthropogenic contributor to environmental pollution (Raj et al., 2023). Most of the pesticides cannot be easily degraded, they remain persistent in the soil and leak into nearby surface and groundwater bodies contaminating the surrounding environment. Two majorly used degradation techniques for degradation includes biotic and abiotic degradation strategies. Biotic techniques include degradation using microorganisms i.e., bioremediation, by plants (phyoremediation), by algae (mycoremediation) or enzymatic degradation. Abiotic strategies includes chemical breakdown of pesticides into non or less toxic forms (Akhtar et al., 2021; Sinam et al., 2023). A study provides evidence that nitrogen-fixing bacteria have the capacity to breakdown pesticides, hence reducing the amount of contamination that is found in the environment. In addition to this, the review investigates the ways in which biotic stress factors, such as interactions with other species, and abiotic stress variables, such as temperature, pH, and salinity, influence the efficacy of these bacteria in decomposing pesticides (Ridwan et al., 2022; Shahid et al., 2023). An investigation was carried out to examine the environmental dynamics of exianliumi, a newly identified pesticide, within three agricultural soils representative of China. The varying adsorption capacities of different soil types influence the mobility and leaching potential of xianliumi, highlighting the significance of these differences in soil characteristics. The decomposition of this substance is affected by both biological and environmental factors, with the rate of degradation and the pathways involved differing based on the characteristics of the soil (Liu et al., 2022).

Photodegradation refers to the process by which pesticides undergo degradation through the influence of sunlight. The majority of insecticides exhibit a capacity for photodegradation to varying degrees, with the degradation rate influenced by factors such as light intensity, exposure duration, and the specific properties of the insecticide (Tudi et al., 2022). Molecules can undergo photosensitizer-mediated photodegradation when they absorb light during a photochemical reaction. When electrons or atoms are initially transferred during redox processes in photosensitization, free radicals such as the hydroxyl radical (OH) are formed. But there are a lot of steps for sensitizers that have been oxidized or reduced to get back to their original state (Liao et al., 2024). Phytoremediation is an efficient technique for the degradation of pesticides. The absorption of pesticides by plants is contingent upon the physicochemical characteristics of the compound, the method of application, soil composition, environmental conditions, and the species of the plant. Plants employ many methods to eliminate pesticides from soil, such as phytodegradation, phytoextraction, rhizodegradation, rhizofiltration, and phytostabilization (Bhalla et al., 2022). A study examined the degradation of two organophosphorus insecticides, malathion, and dimethoate, using four plant species: *Neriumoleander*, *Ricinuscommunis*, *Rhazyastricta*, *Conocarpuserectus*. The researchers found that *Ricinuscommunis* and *Neriumoleander* were particularly effective in decomposing dimethoate and malathion residues in plant tissues. This signifies that these plants possess significant potential for the phytoremediation of organophosphorus pesticides (Jia et al., 2023). A recent study investigated the capacity of Plantago major to remediate soil contaminated with cypermethrin and its protective effects on tomato plants. The findings revealed that Plantago major exhibited a significant ability to absorb cypermethrin residues from contaminated soil, surpassing the absorption capabilities of tomato plants. This suggests that Plantago major can be effectively employed for the phytoremediation of cypermethrin-polluted soils. Additionally, exposure to cypermethrin increased the activity of antioxidant enzymes, such as superoxide dismutase, peroxidase, and catalase, in the roots of both Plantago major and tomato plants. This response indicates the activation of a defensive mechanism against oxidative damage induced by cypermethrin (Aioub et al., 2021).

3. Impacts of pesticide

3.1 Pesticides impacts on the environment

The potential hazards associated with pesticide application surpass the advantages observed. The application of pesticides exerts a considerable influence on non-target species, thereby impacting animal and plant biodiversity, as well as aquatic and terrestrial food webs and overall ecosystem dynamics. Pesticides are utilized to eliminate pests and control weeds due to their chemical properties; however, they may pose risks to various organisms, including birds, fish, beneficial insects, non-target plants, and essential natural resources like crops, water, and soil. Pesticides can spread through several means, including air, wind, water, runoff, and contact with plants, animals, and humans (Elgueta et al., 2016; He et al., 2023; Tudi et al., 2021). Pesticides are predominantly detected in aquatic environments, including both surface water and groundwater systems. Pesticides infiltrate surface and groundwater via several pathways, including direct application aimed at managing aquatic weeds and insects, runoff originating from agricultural fields, drift associated with industrial wastewater, and discharge from pesticide formulation and application apparatus. Additionally, atmospheric deposition and air-water exchange contribute to this process (AbuQamar et al., 2024). Pesticides have the potential to contaminate groundwater as they can seep from agricultural fields, mixing and washing areas, or waste disposal locations. Surface water systems, including rivers, lakes, streams, reservoirs, and estuaries, exhibit a susceptibility to the accumulation of pesticides and chemicals, attributable to their limited ability to assimilate byproducts resulting from anthropogenic activities. Surface water systems are interconnected with groundwater and atmospheric water through the hydrologic cycle. The migration of pesticides from surface water to groundwater can occur via soil seepage mechanisms. They additionally infiltrate the atmosphere through the processes of evaporation and transpiration (Peluso et al., 2023).

A study investigated the occurrence of pesticide residues in public drinking water systems across Brazil and evaluated the related cancer risks. The research reveals extensive pollution of potable water sources with pesticides, encompassing both currently sanctioned and previously banned substances acknowledged for their potential to induce cancer. The presence of pesticide contaminants in drinking water impacts approximately five million individuals. The findings of the study indicate that this exposure could lead to an increase in cancer incidence, thereby highlighting significant public health implications (Panis et al., 2022). A study was undertaken on the Thamirabarani River to evaluate the occurrence of pesticide residues. The study identified the presence of pesticides in water samples, with concentrations varying from non-detectable levels to 31.69 μg/L. In sediment samples, the detected levels ranged from non-detectable to 14.77 μg/kg, while fish samples exhibited concentrations between 0.02 and 26.05 μg/kg. The organochlorine insecticides that were most frequently utilized include endosulfan, aldrin, and endrin. The research indicated that both infants and adults who come into contact with pesticides through ingestion and dermal exposure to contaminated water and sediments may face elevated cancer risks (cancer risk $>10^{-4}$) compared to those exposed via inhalation (Arisekar et al., 2021).

Soil quality encompasses its capacity to filter, decompose, and neutralize pesticides. The process of pesticide decomposition results in residues that persist and transform within both aquatic and terrestrial ecosystems over extended periods, thereby posing a threat to environmental integrity (Barron et al., 2017). The contamination of soil and sediment by pesticides has adversely affected the quality of food and the sustainability of agricultural practices. Soils serve as the principal reservoir for pesticides in terrestrial environments, facilitating their retention via adsorption and subsequently releasing them into the atmosphere, groundwater, and biota as a secondary source. This has a crucial impact on the worldwide dispersal and eventual outcome of pollutants. Pesticide contamination can induce sub-lethal effects on terrestrial plants and adversely affect non-target plant species (Al-Wabel et al., 2016; Mahmood et al., 2016). The persistence of pesticide residues in soil is intricately linked to various characteristics of the pesticides, including their water solubility, soil sorption constant (K_{oc}), octanol/water partition coefficient (K_{ow}), and half-life in soil (DT_{50}) (Zhang et al., 2013). Pesticides including organochlorine DDT, endosulfan, endrin, heptachlor, and lindane are highly persistent in soil particles and have been banned in many countries, including China. The behavior of pesticide transformation in soil is influenced by the interactions among various soil matrices, including organic matter content, pH, temperature, humidity, types of microorganisms, irrigation methods, and the presence of grass hedges, as well as the inherent physical and chemical properties of the pesticides themselves. The adsorption of pesticides, for instance, is positively correlated with the quantity of organic matter available (Yadav et al., 2015).

The exposure to pesticides may induce sub-lethal effects on terrestrial plant species, alongside detrimental impacts on non-target flora. The terrestrial animal population has been adversely affected by pesticides. The application of broad-spectrum pesticides such as carbamates, organophosphates, and pyrethroids has the potential to significantly diminish populations of beneficial insects, including bees and beetles. Observations indicate that insect populations tend to be more abundant on organic farms compared to their non-organic counterparts. The co-application of pyrethroids alongside triazole or imidazole fungicides is lethal to honeybee populations (Mahmood et al., 2016). Since the era preceding agricultural development, there has been a documented decline of 20%–25% in avian populations. A major factor contributing to this notable decline is the application of pesticides, which were not previously recognized before 1962. The accumulation of pesticides in the tissues of avian species leads to increased mortality rates. The population of bald eagles in the United States has significantly decreased primarily as a result of exposure to DDT and its metabolites (Padayachee et al., 2023). Fungicides have the potential to reduce populations of avian and terrestrial fauna by eliminating earthworms, a primary food source. Birds may ingest granular insecticides that are camouflaged as food grains. Organophosphate pesticides are notably toxic to avian species and have demonstrated detrimental effects on raptors in natural environments. Even sublethal doses of these pesticides can adversely affect the nervous system, leading to behavioral anomalies (Costas-Ferreira et al., 2022; Dudley et al., 2017). Herbicides exert considerable adverse effects on fungal species within the soil ecosystem, with insecticides such as trifluralin and oryzalin recognized for their capacity to inhibit the proliferation of symbiotic mycorrhizal fungi, which play a crucial role in nutrient acquisition. Oxadiazon has been documented to reduce the quantity of fungal spores, whereas triclopyr exhibits toxicity toward specific varieties of mycorrhizal fungi (Ali et al., 2020).

3.2 Pesticides impact on human health

Pesticides have been linked to a wide range of diseases, including cancer, leukemia, and asthma, according to the findings of many studies. There are health hazards associated with pesticide exposure due to the presence of toxic components as well as exposure levels (Alengebawy et al., 2021). It is possible that certain populations, such as youngsters, pregnant women, and the elderly, are more susceptible to the adverse effects of pesticides. There are around 385 million people around the world who suffer from acute pesticide poisoning as a result of accidental exposure, and approximately 11,000 people lose their lives as a result. It is estimated that around 44% of the world's farming population is poisoned by pesticides (Zhou et al., 2025). Organochlorines (OCs) and organophosphates (OPs) are extremely hazardous to the nervous system and have the potential to bring about a variety of major health problems, including cancer and damage to organs. The cytotoxic effects and organ damage that might result from exposure can be caused by both diet and environment (Chuang et al., 2019). Pesticide poisoning represents a significant global public health concern, leading to nearly 300,000 fatalities annually. The exposure to pesticides is an omnipresent phenomenon, with various pathways through which individuals come into contact with these substances (Sabarwal et al., 2018). Pesticides are frequently taken up by the body's circulatory system, yet they can also be expelled through exhalation, dermal pathways, or urinary excretion. Pesticides can infiltrate the human body via four distinct pathways: ingestion through the mouth, absorption through the skin, exposure through the eyes, or inhalation through the respiratory tract. The toxicity of pesticides is contingent upon the mode of exposure, which may occur through oral, dermal, or respiratory routes (Alengebawy et al., 2021). The major damages caused by pesticides in human health are given in Fig. 4.3.

Numerous studies have found a link between pesticide use and cancer (Bonner et al., 2017; de Graaf et al., 2022; Matich et al., 2021; Navarrete-Meneses and Pérez-Vera, 2019; Silva Pinto et al., 2020; VoPham et al., 2017). This research investigates multiple

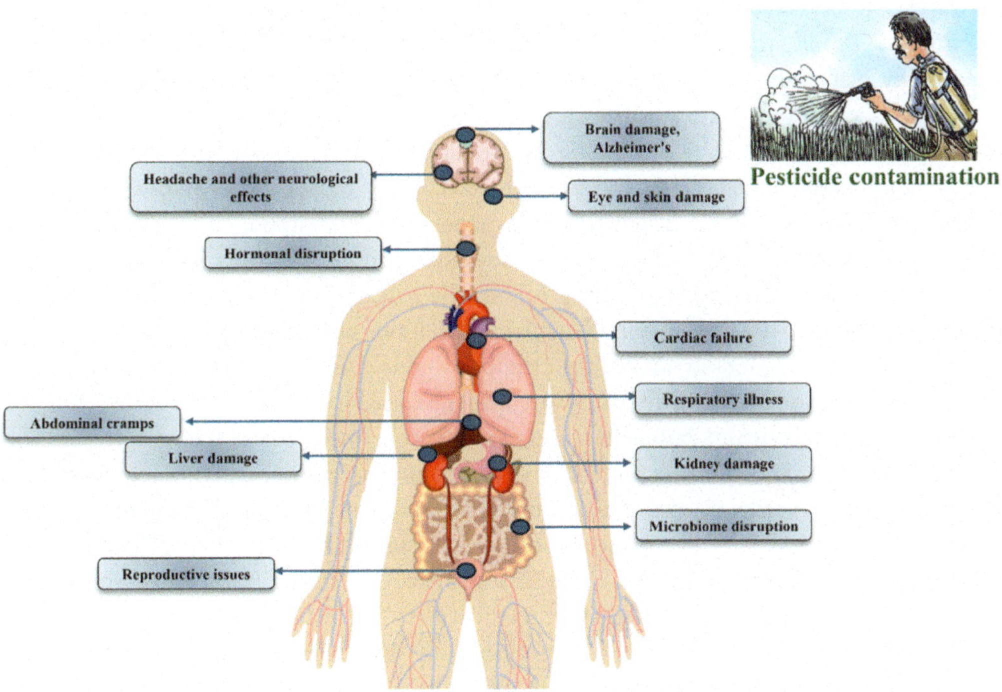

Figure 4.3 Illustrative representation of the impact of pesticides on human health.

epidemiological studies to assess the potential correlation between pesticide exposure, especially in occupational settings, and an increased risk of colon cancer. Occupational exposure to pesticides has been associated with minimal or absent correlations with colon and rectal cancer. This suggests a potential association between pesticide exposure and elevated colorectal cancer risk, especially concerning specific chemicals and in scenarios of high exposure (Matich et al., 2021). A separate investigation demonstrated the correlation between the application of particular pesticides and the occurrence of aggressive prostate cancer in individuals who apply these substances. Research indicates that individuals with a history of exposure to the organodithioate insecticide dimethoate exhibit an increased likelihood of developing aggressive prostate cancer, as evidenced by a hazard ratio of 1.37 (95% confidence interval [CI]: 1.04–1.80). The herbicide triclopyr exhibited a negative correlation with the incidence of aggressive prostate cancer. In particular, each user exhibited a hazard ratio of 0.68 (95% CI: 0.48–0.95), with the most significant inverse correlation observed in individuals who reported 4 years or more of usage (HR = 0.44; 95% CI: 0.26–0.77) (Pardo et al., 2020). The fundamental process responsible for the inhibition of cell growth induced by CPF involves an elevation in p-ERK1/2 levels, which is facilitated by H_2O_2 in breast cancer cells. Certain OCs have been associated with breast cancer as a result of their estrogenic effects on mammary cells (Rivero et al., 2015). A research investigation conducted in Tunisia identified a significant association between serum levels of organochlorine pesticides, specifically polychlorinated biphenyls, and the manifestation of xenoestrogenic effects, which correspondingly elevates the risk of breast cancer (Kim et al., 2017). The application of pyrethroid insecticides has been associated with hematological cancers, particularly in pediatric populations. Pyrethroids have been recognized for their genotoxic properties, indicating their potential to damage genetic material. This process entails the generation of genetic rearrangements, alterations in gene expression, and modifications to DNA, all of which may play a role in the carcinogenic mechanisms within hematopoietic (blood-forming) cells (Navarrete-Meneses and Pérez-Vera, 2019).

A study was conducted involving children aged 7 to 12 residing in low-income, multi-family urban housing, focusing on the relationship between pesticide exposure and asthma-related health outcomes. Researchers recruited individuals diagnosed with asthma from urban, economically disadvantaged, multi-family residential settings. The evaluation of pesticide exposure was performed via the examination of urinary biomarkers, focusing specifically on metabolites that signify the presence of organophosphate and pyrethroid pesticides. The assessment of asthma morbidity involved a comprehensive analysis of healthcare utilization data, including emergency department visits and hospital admissions, in conjunction with symptom questionnaires completed by caregivers. The findings suggest that while low-level pesticide exposure may not have a uniform impact on asthma morbidity, certain pesticide biomarkers and cumulative exposures could correlate with increased healthcare utilization associated with asthma (Chetty-Mhlanga et al., 2021). Environmental pollutants may affect diabetes, according to new research in the scientific community. An increased risk of developing type 2 diabetes and its consequences may be associated with exposure to pesticides, especially organochlorines and metabolites (Miranda et al., 2023). Multiple organochlorine pesticides, including DDT, DDE, oxychlordane, trans-nonachlor, hexachlorobenzene, and hexachlorocyclohexane, as well as polychlorinated dibenzodioxins and furans (PCDD/Fs), PCBs, and a systematic review of the literature found a positive correlation between diabetes and serum levels of these pollutants (Jaacks and Staimez, 2015). A significant number of pesticides, especially those comprising organophosphorus compounds, have demonstrated adverse effects on the male reproductive system via multiple mechanisms. A decline in sperm parameters is observed, including counts, motility, viability, and density. Additionally, there is an inhibition of spermatogenesis, a reduction in testis weights, damage to sperm DNA, and an increase in aberrant sperm morphology (Kim et al., 2017). There is evidence that pesticide exposure may play a role in the development of neurological illnesses such as Parkinson's disease and Alzheimer's disease (Chetty-Mhlanga et al., 2021; Cresto et al., 2023; Jokanović et al., 2023).

4. Overview of detection and quantification techniques for pesticide detection

Since the introduction of pesticides on fruits and vegetables by farmers, individuals have employed traditional methods to check for pesticide presence before consumption. In the beginning, pesticides were removed manually or washed off with water. However, these methods were not the most efficient at eliminating pesticide residues. Nowadays, various methods like cleaning, scrubbing, using baking soda and water, and saltwater mixed with vinegar, can potentially eliminate up to 70% of pesticide residue (Thorat et al., 2023). Pesticides and their byproducts can be analyzed using several conventional analytical methods, including non-selective techniques of UV liquid chromatography (LC), gas chromatography (GC) with electron capture detection, diode array, electrochemical detection, fluorescence, and mass ionization detection (Hercegová et al., 2007). Pesticides can also be identified through Effective machine learning (ML) and artificial intelligence (AI) techniques are available (Wakchaure et al., 2023). AI and ML techniques are rarely used because it takes time to teach and render the model with multiple options. Additionally, there is an error case (Thorat et al., 2023). Since the approaches can help us identify pesticide residue on the spot in less time, it is necessary to comprehend them well by considering time, cost, portability, and ease of use. The most often-used methods combine gas and chromatography liquid chromatography with mass spectrometry (MS). These techniques have great sensitivity and selectivity at low detection limits, but they are constrained by employing complex, costly, and time-consuming equipment that requires expertise to operate (Samsidar et al., 2018).

4.1 Gas chromatography (GC)

The procedure of gas chromatography (GC) is used to separate volatile substances. One of the primary analytical techniques used in food analysis to find and measure pesticide residues in intricate matrices is gas chromatography. GC distinguishes between the pesticides according to their thermal stability and volatilities. Since gas is examined in gas chromatography (GC), its operation is different from that of LC (Petsas and Vagı, 2017). For pesticides like N-methyl carbamate, the methods are restricted since they either remain on the chromatographic column or break down into their phenols. Derivatization techniques, which include first hydrolysis to the corresponding phenols or amines and reaction with halogen-rich reagents, can also restrict the sensitivity and usefulness of fragrant carbamates ECD (Munir and Badri, 2020). Innovations in mass spectrometry and gas chromatography are desperately needed. The developments have been tested and summarized using GC-MS in conjunction with a variety of equipment and sample extraction techniques (Shinde et al., 2021). The improvements in GC-MS techniques simplify analysis and reduce the amount of time needed for sensitivity testing. highlighting one benefit when used in conjunction with extraction by sorbents that hold onto the analytes before they are thermally absorbed by the GC indicator. For rapid GC-MS-MS, the LOD ranges from 0.07 to 4.7×10^{-3} ppb. One significant benefit of the GC technique is its high resolution and ability to identify volatile chemicals. This method's drawback is that mass spectroscopy is necessary to confirm the peak. Rather than being a replacement for the LC technique, the GC technique might be viewed as complementary (Pico et al., 2020). Depending on the type of pesticide being measured, gas chromatography can be used in conjunction with various detection techniques (LeDoux, 2011). Pesticide residues in cereal samples are identified using techniques including mass selective detection (MSD), flame ionization detection (FID), nitrogen–phosphorous detection (NPD), and flame photometric detection (FPD) (Grimalt and Dehouck, 2016). For a specific pesticide, the detectors offer sensitivity and selectivity. ECDs, or electron capture detectors, are particularly useful for halogenated substances like herbicides that contain organic chlorine (Lian et al., 2010). The FPD is sensitive to phosphorus and sulfur pesticides, while the NPD is sensitive to nitrogenous and organophosphate pesticides (Aruna and Baskaran, 2010).

4.2 Liquid chromatography (LC)

Initially detectors—UV or diode array detectors (DAD) were used in Liquid Chromatography (LC) (Esquinas-Requena et al., 2020). But later pesticides are detected using a variety of methods, including ultraviolet (UV), refractive, fluorescent, and electrochemical detectors. Pesticide residue on fruits and vegetables can be examined using a variety of additional methods based on liquid chromatography. The machine in the LC uses chemicals and detectors to identify pesticide residue. The machine is connected to several detectors, including UV, IR, Refractive index, mass spectrometry (MS), electrochemical, and others. For a triple quadrupole linear ion trap mass analyzer, the stated limit of detection for LC-MS-MS is 0.001 ppb (Narenderan et al., 2020). Due to its excellent sensitivity, the LC-MS and LC-MS/MS technology uses the electrospray ionization method (Savant et al., 2010). Compared to other methods, the LC-MS/MS methodology can detect a broad spectrum of pesticide residues and is helpful for chemicals having polar characteristics. Additionally, LC-MS/MS mostly uses high-performance liquid chromatography (HPLC) due to developments in its equipment (Stachniuk and Fornal, 2016). LC when combined with mass spectrometry (MS), QuEChERS, and solid-phase extraction helps in achieving superior findings, for pesticide residue measurement in biological samples of humans (Prodhan et al., 2015). Quadrupole(Q), triple quadrupole (QqQ), time of fight (TOF), hybrid quadrupole ion trap (QTrap), and Orbitrap are the MS analyzers coupled with LC for analysis of pesticides in food (Celeiro et al., 2014).

4.3 High-pressure liquid chromatography (HPLC)

High-performance Liquid Chromatography (HPLC) is frequently used and compatible with a variety of detectors. It can identify triazines and organophosphorus in a variety of matrices when combined with DAD and/or UV. HPLC is an expensive and time-consuming conventional technology that highlights the need for quick and easy methods to be developed (Meyers, 2006). The usage of HPLC is growing, especially in the investigation of pesticides that GC is unable to identify directly because of the compounds' poor thermal stability, polarity, or volatility (Sannino et al., 2004). Diode array detection (DAD) in high-quality liquid chromatography may precisely identify pesticides in intricate combinations. These methods have made it possible to identify and measure pesticide residues in a variety of foods and atmospheres (Spizzirri and Cirillo, 2016). The most often used type of HPLC is reversed-phase liquid chromatography (RPLC), which can successfully separate and detect polar and non-polar pesticides, which we cannot detect from GC directly (Meyers, 2006). In 2018, an integrated analysis technique coupled mass spectrometry (MS), ultra-performance liquid chromatography (UPLC), and solid-phase extraction (SPE) techniques to quickly and sensitively identify 50 pesticides. Out of 50 compounds, the combined approaches showed LOQs less than 1 ng/mL, and the greatest linear response for 48 substances, with acceptable accuracy, reached 200 ng/mL in 10 minutes of total analysis time (Castro et al., 2018).

4.4 Spectroscopic methods

Chromatographic techniques have some drawbacks, including a drawn-out evaluation process, a lengthy detection cycle, and delayed detection results. Therefore, it's critical to create quick, dependable methods. NIR reflectance and NIR transmission spectroscopy are two subcategories of near-infrared spectroscopy (Sánchez et al., 2010). Unlike chromatographic procedures, this approach does not require sample preparation and is inexpensive, safe, easy to use, and ecologically friendly. It also does not require the use of organic solvents. A convenient method for both quantitative and qualitative analysis in fields including chemistry, medicine, and agriculture is near-infrared (NIR) spectroscopy (Jamshidi, 2017). Pesticide absorption and soil composition can be predicted using ultraviolet visible–NIR spectroscopy (Levasseur-Garcia, 2018).

5. Strategies to mitigate pesticide residues

In contrast to synthetic pesticides, microbial pesticides are environmentally sustainable, have no lasting effects, are easily obtained, and are particular in their action. Biopesticides are inexpensive, environmentally benign, selective in their method of action, long-lasting, residue-free, and unrelated to greenhouse gas emissions (Borges et al., 2021; Hummadi et al., 2021). In addition to having a wide range of phytochemical components that enable them to demonstrate different mechanisms of action, phytopesticides pose fewer health concerns to humans and are not linked to the generation of greenhouse gases (Nurmansyah et al., 2022), microbial pesticides (Harish et al., 2022), nano biopesticides (Pan et al., 2023), and phytopesticides (Nurmansyah et al., 2022) are some of the several forms of these biopesticides. Unlike synthetic pesticides, nanobiopesticides exhibit superior biocompatibility, biodegradability, targeted or controlled release, and better pesticidal action (Pan et al., 2023). The retardation and destruction of diseases' and pests' plasma membranes and protein translation are two of the various ways that biopesticides work. In contrast to synthetic pesticides, biopesticides are extremely selective in their target, have a short shelf life, are less persistent in the soil environment, and are made from sustainable raw materials, despite a few disadvantages that have decreased their acceptance and commercial use (Kumar et al., 2021).

5.1 Integrated pest management (IPM) approaches

Research on host plant resistance involves a complex network of interactions, mediated by morphological and chemical traits that influence the extent of damage caused by pests (Mouden et al., 2017). Antixenosis and antibiosis resistance are two examples of the various resistance traits that may be present (Angon et al., 2023). This strategy exploits cultivars that have morphological, biochemical, or physical traits produced by genetic engineering that reduce their appeal to pests or hinder their ability to grow, develop, or reproduce (Han et al., 2018). Certain plants are rewarded with traits that help with indirect defense, such as nectar that attracts and rewards natural foes of pests and scents (often created by bugs) that attract natural enemies to afflicted plants (Ostrem et al., 2016; Stenberg et al., 2015). Isobutyl amide is associated with host plant resistance in several plants in the chrysanthemum genus. The chemicals enabling constitutive resistance to western flower thrips were discovered and confirmed in later in vitro bioassays by using an ecometabolomic approach and incorporating the metabolomic profiles of susceptible and resistant plants (Leiss et al., 2009). Most transgenic crops are resistant to lepidopteran and/or coleopteran pests, or sometimes a combination of both, including glyphosate, dicamba, glufosinate ammonium, and 2–4 D (Anderson et al., 2019). Until recently, insecticidal proteins derived from *Bacillusthuringiensis* (Bt) (lepidopteran and coleopteran pests) were the primary source of insect defenses for cotton, soybeans (lepidopteran pests), and maize. Aubergines in Bangladesh have been found to possess a Bt trait for a lepidopteran pest since 2014 (Fahad et al., 2021). Plant regulators like salicylic acid (SA), jasmonic acid (JA) and ethylene (ET), coupled with other phytochemicals defend plants from infections and insects (Bruce et al., 2017). Biological control and crop rotation are the two main aspects of integrated pest management (IPM), which reduce the concentration of pesticides (Baker et al., 2020).

5.2 Biological control

Biological control is the application of natural enemies to control insects, diseases, and weeds (Baker et al., 2020). Augmentative biological control with repeated releases only temporarily reduces pest outbreaks, and few programs achieve permanent control, despite its restrictions based on how it is used. Few programs can permanently stop pest outbreaks, although the efficiency of biological management utilizing evolved extractions varies depending on how it is applied (Torres and Bueno, 2018). Natural enemies like parasitic wasps and predatory arthropods can sometimes drastically lower pest populations (Hajek, 2004). Common methods for managing endemic pests include releasing commercially available natural enemies regularly and protecting natural enemy populations by offering them refuges or avoiding actions that could endanger them (Dara, 2019). Entomopathogenic fungi (EPF) have been shown in numerous studies to be efficient against mite and insect infestations. Additionally, they improve the plant's capacity to react to additional biotic stressors (Abd-Elgawad, 2019). The production of chemicals such as insecticides, herbicides, antifungals, and antivirals during various biocidal actions results in the induction of systemic resistance. Predatory insects such as lacewings, ground beetles, centipedes, dragonflies, rove, midges, ladybirds, pirate bugs, and aphids are used in integrated pest management systems (Angon et al., 2023). When aphids occur in large numbers, lady beetles are great predators; but, when pest consistency is low, they are believed to be less. EPF, a crucial class of biological control agents for pests with chewing mouthparts, such as whiteflies and other sap-sucking pests, can directly infect insects through their cuticle, which is why they are crucial to the natural fall of whitefly populations (Sani et al., 2020).

5.3 Crop rotation

Crop rotation is a strategic strategy to mitigate pest pressure by altering the planting of crops within a defined sequence. This disturbs the life cycle of pests that are specific to crops, hence reducing their population and lessening the need for chemical interventions (Shah et al., 2021). Crop rotation can help reduce a lot of illnesses and pests. Numerous agricultural and vegetable crops are susceptible to infection by the common fungal disease *Fusarium* sp. (van Bruggen et al., 2016). The main strategies for controlling these infections include fungicides, crop rotation, and genetic resistance. Using a non-host crop can also assist lower the likelihood of disease development, even when resistant types are used (McLaughlin et al., 2023). Rotating 2-year cycles of corn or winter wheat with soybeans significantly reduced the quantity of apothecia produced by the sclerotia of *Sclerotiniasclerotium* in soybean fields (Kora et al., 2025). Crop rotations lowered the germination rates of sclerotia compared to continuous soybean cultivation. Alternating between wheat and soybeans has the potential to significantly diminish Fusarium head blight in wheat (Sumner, 2018). As noted by (Stetina et al., 2007), cultivating corn for 2 years was more effective in managing reniform nematode populations than continuously planting cotton. However, reniform nematode populations tended to rise to damaging levels toward the end of the cotton growing season after only 1 year of corn. Crops such as rye and triticale, which exhibit allelopathic properties, facilitate sustainable weed management while reducing agriculture's environmental footprint (Singh et al., 2003; Knodel and Shrestha, 2018) report that wireworms and soil-dwelling insects are significant pests that inflict considerable damage in the Pacific Northwest region of the United States. Furthermore, managing wireworms might be improved by implementing a rotation of winter wheat and summer fallow along with continuous spring wheat.

5.4 Bioremediation

Eco-friendly methods, such as bioremediation, are available for the removal of hazardous pollutants to maintain a sustainable environment with a strong and healthy ecosystem (Desisa et al., 2022). Currently, pesticide remediation includes a variety of environmentally friendly techniques, such as bacterial degradation of pesticides, myco-remediation, bioremediation with microalgae, and phytoremediation (Singh et al., 2020). Plants can eliminate pollutants through several mechanisms, including pollutant transpiration (phytovolatilization), cleanup of the rhizosphere microbiome (rhizo-degradation), enzymatic breakdown (phytodegradation), and the accumulation of pesticides in different plant parts (phytoextraction) (Koranteng et al., 2018; Pérez-Lucas et al., 2019). A few species of microalgae, including *Chlamydomonasreinhardtii*, *Chlamydomonasmexicana*, and *Dunaliella* sp., have been identified as capable of removing the pesticides prometryne, atrazine, and mirex, respectively (Torres et al., 2017). The bioremediation of pesticides using microalgae demonstrates a dual process involving both the accumulation of pesticides and the conversion of harmful chemicals into less toxic forms. The degradation process is influenced by the presence of resilient microalgae, optimal conditions, and the chemical properties of the pesticides involved (Monteiro et al., 2012). Additionally, key factors such as molecular weight, functional groups, concentration, and solubility in water can impact the degradation of pesticides (Mir et al., 2017). Microalgae have an edge over bacteria and fungi in the biodegradation process because they utilize a mixotrophic mode of metabolism under stress, deriving energy from both light and organic carbon (Singh et al., 2020).

Another biological method for managing pesticide waste is myco-remediation, in which fungus use the pollutants as a source of carbon to change them into less harmful substances that are then removed from the soil and water system (Wu et al., 2015). A range of extracellular enzymes that help break down resistant chemical compounds are secreted by ligninolytic fungus. The most enzymes are expelled by saprotrophic fungi, which are followed by brown rot, white rot, and soft rot (Maqbool et al., 2016). The extracellular enzyme complex of white rot fungus (*Pleurotus ostreatus*, *Trameteshirsutus*, and *Cyathusbulleri*) which includes laccase, manganese peroxidase, and lignin peroxidase is well known for its ability to biodegrade pesticides (Janusz et al., 2017). Many developing nations now must adopt mycoremediation or other bioremediation techniques to reduce pesticide contamination concurrently because they cannot afford biopesticides or avoid using chemical pesticides (Singh et al., 2020). Unlike other remediation methods, there is substantial evidence that bacteria can decompose and remove pesticides. The primary bacterial genera utilized for the removal of pesticides from affected environments include *Arthrobacter*, *Flavobacterium*, *Azotobacter*, and *Pseudomonas* (Singh et al., 2020). These microorganisms mitigate the environmental threats posed by pesticides by utilizing them as nutrients and producing CO_2 and H_2O as byproducts. Pesticides accumulate in the soil ecosystem and act as carbon sources and electron donors for soil bacteria (Rani and Dhania, 2014). Advanced strategies such as bioaugmentation, bio-stimulation, and natural attenuation are employed to enhance the biodegradability of pesticides. These strategies involve introducing indigenous species to the contaminated site, adding nutrients, and employing effective bacteria (Omokhagbor Adams et al., 2015).

6. Future trends and research direction

- Future food safety greatly depends on scientific innovation and cross-sector cooperation as the world's population continues to rise and climate change changes the dynamics of food systems. The way that agricultural methods can be made more efficient, safe, and sustainable is changing due to new trends and research avenues.

These developments seek to satisfy strict food safety regulations around the world in addition to minimizing environmental impacts (Ashraf et al., 2021).

- Although traditional pesticide use has proven successful in protecting crops, it has caused serious issues with food safety, environmental contamination, and the toxicity of non-target species. The development of bio-based, biodegradable, and target-specific pesticide formulations that are safer for ecosystems and people is the main goal of future research. The application of nano-formulated pesticides, which enable controlled release and enhanced adherence to plant surfaces, is one promising field. This minimizes residues and uses less chemical. The study of botanical and microbial pesticides, which use natural plant components or advantageous microbes to repel pests without endangering beneficial species or leaving hazardous residues, is also growing (Ali et al., 2024).
- Through the introduction of characteristics that improve insect resistance, lessen reliance on pesticides, and boost yields in harsh environmental circumstances, genetic engineering is still transforming agriculture. Bt (*Bacillusthuringiensis*) genes, for example, are used to create crops that produce proteins that are harmless for humans and other species but harmful to certain insect pests. More accurate and effective changes are now possible thanks to more recent gene-editing technologies like CRISPR-Cas9. Scientists can use these technologies to improve stress tolerance, increase natural pest resistance, and even eliminate allergenic or antinutritional elements from crops (Anjanappa and Gruissem, 2021).
- A lot of research is being done on residue-free crop protection technology to better guarantee food safety. By removing or reducing chemical residues on harvested produce, these advances guarantee adherence to consumer expectations and food safety laws. Non-chemical methods for protecting crops before and after harvest include plasma-activated water, ozone sprays, and ultraviolet (UV–C) light treatments. These technologies are appropriate for organic and high-value agricultural (Lengai et al., 2022).
- Strong international cooperation is necessary to ensure consistent food safety across borders as the food economy becomes more globalized. Harmonizing standards, exchanging best practices, and funding cooperative research are essential for future advancements in sustainable food safety. International rules on pesticide residues, maximum residue levels (MRLs), and sustainable agricultural inputs are being developed in collaboration with national governments by agencies such as the Food and Agriculture Organization (FAO), the World Health Organization (WHO), and Codex Alimentarius (Ong et al., 2021)

7. Conclusion

Although pesticides were originally employed to improve human lives by raising crop yields and combating illness, their negative consequences have now outweighed their advantages. The increasing reliance on pesticides in modern agriculture presents a twin challenge: maximize crop output while limiting negative effects on food safety, human health, and the environment. This chapter evaluated the difficult balance between pesticide benefits and risks, emphasizing the significance of rigorous regulations, public awareness, and sustainable alternatives. Agricultural systems may develop toward safer and more sustainable food production by using integrated pest management, organic practices, and biopesticides. Moving forward, additional research and innovation will be necessary to develop environmentally friendly pest control strategies that safeguard both agricultural efficiency and public health. To bridge the productivity and sustainability gap, academics, legislators, and industry stakeholders must collaborate to keep food safety and environmental integrity at the forefront of global agriculture practices.

AI disclosure

During the preparation of this work the author(s) used Grammarly/Quilbot in order to check the Grammer and language. After using this tool/service, the author(s) reviewed and edited the content as needed and take (s) full responsibility for the content of the publication.

References

Abd-Elgawad, M.M.M., 2019. Towards optimization of entomopathogenic nematodes for more service in the biological control of insect pests. Egypt. J. Biol. Pest Control 29 (1). https://doi.org/10.1186/s41938-019-0181-1.

Abubakar, Y., Tijjani, H., Egbuna, C., Adetunji, C.O., Kala, S., Kryeziu, T.L., Patrick-Iwuanyanwu, K.C., 2019. Pesticides, history, and classification. In: Natural Remedies for Pest, Disease and Weed Control. Elsevier, Nigeria, pp. 29–42. https://doi.org/10.1016/B978-0-12-819304-4.00003-8.

AbuQamar, S.F., El-Saadony, M.T., Alkafaas, S.S., Elsalahaty, M.I., Elkafas, S.S., Mathew, B.T., Aljasmi, A.N., Alhammadi, H.S., Salem, H.M., Abd El-Mageed, T.A., Zaghloul, R.A., Mosa, W.F. A., Ahmed, A.E., Elrys, A.S., Saad, A.M., Alsaeed, F.A., El-Tarabily, K.A., 2024. Ecological impacts and management strategies of pesticide pollution on aquatic life and human beings. Mar. Pollut. Bull. 206. https://doi.org/10.1016/j.marpolbul.2024.116613.

Ahmed, M., Asim, M., Ahmad, S., Aslam, M., 2023. Climate change, agricultural productivity, and food security. In: Global Agricultural Production: Resilience to Climate Change. Springer International Publishing, Pakistan, pp. 31–73. https://doi.org/10.1007/978-3-031-14973-3_2.

Aioub, A.A.A., Zuo, Y., Aioub, A.A.A., Hu, Z., 2021. Biochemical and phytoremediation of Plantago major L. to protect tomato plants from the contamination of cypermethrin pesticide. Environ. Sci. Pollut. Control Ser. 28 (32), 43992–44001. https://doi.org/10.1007/s11356-021-13853-2.

Akhtar, N., Syakir Ishak, M.I., Bhawani, S.A., Umar, K., 2021. Various natural and anthropogenic factors responsible for water quality degradation: a review. Water (Switzerland) 13 (19). https://doi.org/10.3390/w13192660.

Al-Wabel, M., El-Saeid, M.H., El-Naggar, A.H., Al-Romian, F.A., Osman, K., Elnazi, K., Sallam, A.S., 2016. Spatial distribution of pesticide residues in the groundwater of a condensed agricultural area. Arabian J. Geosci. 9 (2). https://doi.org/10.1007/s12517-015-2122-y.

Alengebawy, A., Abdelkhalek, S.T., Qureshi, S.R., Wang, M.Q., 2021. Heavy metals and pesticides toxicity in agricultural soil and plants: ecological risks and human health implications. Toxics 9 (3), 1–34. https://doi.org/10.3390/toxics9030042.

Alghuthaymi, M., Amal, A.-, Mostafa, M., Abd-Elsalam, K.A., 2019. Carbon nanotubes: an efficient sorbent for herbicide sensing and remediation. In: Carbon Nanomaterials for Agri-Food and

Environmental Applications. Elsevier, Saudi Arabia, pp. 429–457. https://doi.org/10.1016/B978-0-12-819786-8.00019-0.

Ali, S., Ullah, M.I., Sajjad, A., Shakeel, Q., Hussain, A., 2020. Environmental and Health Effects of Pesticide Residues. Springer Science and Business Media LLC, pp. 311–336. https://doi.org/10.1007/978-3-030-54719-6_8.

Ali, S., Ahmad, N., Dar, M.A., Manan, S., Rani, A., Alghanem, S.M.S., Khan, K.A., Sethupathy, S., Elboughdiri, N., Mostafa, Y.S., Alamri, S.A., Hashem, M., Shahid, M., Zhu, D., 2024. Nano-agrochemicals as substitutes for pesticides: prospects and risks. Plants 13 (1). https://doi.org/10.3390/plants13010109.

Anderson, J.A., Ellsworth, P.C., Faria, J.C., Head, G.P., Owen, M.D.K., Pilcher, C.D., Shelton, A.M., Meissle, M., 2019. Genetically engineered crops: importance of diversified integrated pest management for agricultural sustainability. Front. Bioeng. Biotechnol. 7. https://doi.org/10.3389/fbioe.2019.00024.

Angon, P.B., Mondal, S., Jahan, I., Datto, M., Antu, U.B., Ayshi, F.J., Shafiul Islam, M., 2023. Integrated pest management (IPM) in agriculture and its role in maintaining ecological balance and biodiversity. Adv. Agric. 2023. https://doi.org/10.1155/2023/5546373.

Anjanappa, R.B., Gruissem, W., 2021. Current progress and challenges in crop genetic transformation. J. Plant Physiol. 261. https://doi.org/10.1016/j.jplph.2021.153411.

Aqeel, M., Ran, J., Hu, W., Irshad, M.K., Dong, L., Akram, M.A., Eldesoky, G.E., Aljuwayid, A.M., Chuah, L.F., Deng, J., 2023. Plant-soil-microbe interactions in maintaining ecosystem stability and coordinated turnover under changing environmental conditions. Chemosphere 318. https://doi.org/10.1016/j.chemosphere.2023.137924.

Arisekar, U., Jeya Shakila, R., Shalini, R., Jeyasekaran, G., 2021. Pesticides contamination in the Thamirabarani, a perennial river in peninsular India: the first report on ecotoxicological and human health risk assessment. Chemosphere 267, 129251. https://doi.org/10.1016/j.chemosphere.2020.129251.

Aruna, G., Baskaran, V., 2010. Comparative study on the levels of carotenoids lutein, zeaxanthin and β-carotene in Indian spices of nutritional and medicinal importance. Food Chem. 123 (2), 404–409. https://doi.org/10.1016/j.foodchem.2010.04.056.

Ashraf, S.A., Siddiqui, A.J., Elkhalifa, A.E.O., Khan, M.I., Patel, M., Alreshidi, M., Moin, A., Singh, R., Snoussi, M., Adnan, M., 2021. Innovations in nanoscience for the sustainable development of food and agriculture with implications on health and environment. Sci. Total Environ. 768. https://doi.org/10.1016/j.scitotenv.2021.144990.

Baker, B.P., Green, T.A., Loker, A.J., 2020. Biological control and integrated pest management in organic and conventional systems. Biol. Control 140. https://doi.org/10.1016/j.biocontrol.2019.104095.

Barron, M.G., Ashurova, Z.J., Kukaniev, M.A., Avloev, H.K., Khaidarov, K.K., Jamshedov, J.N., Rahmatullova, O.S., Atolikshoeva, S.S., Mamadshova, S.S., Manzenyuk, O., 2017. Residues of organochlorine pesticides in surface soil and raw foods from rural areas of the Republic of Tajikistan. Environ. Pollut. 224, 494–502. https://doi.org/10.1016/j.envpol.2017.02.031.

Bertomeu-Sánchez, J.R., 2019. Introduction. Pesticides: past and present. J. Hist. Sci. Technol. 13 (1), 1–27. https://doi.org/10.2478/host-2019-0001.

Bhalla, G., Bhalla, B., Kumar, V., Sharma, A., 2022. Bioremediation and phytoremediation of pesticides residues from contaminated water: a novel approach. In: Pesticides Remediation Technologies from Water and Wastewater. Elsevier, India, pp. 339–363. https://doi.org/10.1016/B978-0-323-90893-1.00016-7.

De Bleecker, J.L., 2008. Chapter 13 Organophosphate and carbamate poisoning. Handb. Clin. Neurol. 91, 401–432. https://doi.org/10.1016/S0072-9752(07)01513-8.

Bonner, M.R., Beane Freeman, L.E., Hoppin, J.A., Koutros, S., Sandler, D.P., Lynch, C.F., Hines, C.J., Thomas, K., Blair, A., Alavanja, M.C.R., 2017. Occupational exposure to pesticides and the incidence of lung cancer in the agricultural health study. Environ. Health Perspect. 125 (4), 544–551. https://doi.org/10.1289/EHP456.

Borges, S., Alkassab, A.T., Collison, E., Hinarejos, S., Jones, B., McVey, E., Roessink, I., Steeger, T., Sultan, M., Wassenberg, J., 2021. Overview of the testing and assessment of effects of microbial pesticides on bees: strengths, challenges and perspectives. Apidologie 52 (6), 1256–1277. https://doi.org/10.1007/s13592-021-00900-7.

Bruce, T.J.A., Smart, L.E., Birch, A.N.E., Blok, V.C., MacKenzie, K., Guerrieri, E., Cascone, P., Luna, E., Ton, J., 2017. Prospects for plant defence activators and biocontrol in IPM—concepts and lessons learnt so far. Crop Prot. 97, 128–134. https://doi.org/10.1016/j.cropro.2016.10.003.

van Bruggen, A.H.C., Gamliel, A., Finckh, M.R., 2016. Plant disease management in organic farming systems. Pest Manag. Sci. 72 (1), 30–44. https://doi.org/10.1002/ps.4145.

Casida, J.E., Durkin, K.A., 2013. Neuroactive insecticides: targets, selectivity, resistance, and secondary effects. Annu. Rev. Entomol. 58, 99–117. https://doi.org/10.1146/annurev-ento-120811-153645.

Castro, G., Pérez-Mayán, L., Rodríguez-Cabo, T., Rodríguez, I., Ramil, M., Cela, R., 2018. Multianalyte, high-throughput liquid chromatography tandem mass spectrometry method for the sensitive determination of fungicides and insecticides in wine. Anal. Bioanal. Chem. 410 (3), 1139–1150. https://doi.org/10.1007/s00216-017-0724-9.

Celeiro, M., Llompart, M., Lamas, J.P., Lores, M., Garcia-Jares, C., Dagnac, T., 2014. Determination of fungicides in white grape bagasse by pressurized liquid extraction and gas chromatography tandem mass spectrometry. J. Chromatogr. A 1343, 18–25. https://doi.org/10.1016/j.chroma.2014.03.057.

Cerruto, E., Manetto, G., Santoro, F., Pascuzzi, S., 2018. Operator dermal exposure to pesticides in tomato and strawberry greenhouses from hand-held sprayers. Sustainability 10 (7), 2273. https://doi.org/10.3390/su10072273.

Chaudhary, N., Choudhary, K.K., Agrawal, S.B., Agrawal, M., 2020. Pesticides usage, uptake and mode of action in plants with special emphasis on photosynthetic characteristics. In: Pesticides in Crop Production: Physiological and Biochemical Action. wiley, Israel, pp. 159–180. https://doi.org/10.1002/9781119432241.ch9.

Chester, G., 2010. Worker exposure: methods and techniques. In: Hayes' Handbook of Pesticide Toxicology. Elsevier Inc., United Kingdom, pp. 1127–1138. https://doi.org/10.1016/B978-0-12-374367-1.00052-5.

Chetty-Mhlanga, S., Fuhrimann, S., Basera, W., Eeftens, M., Röösli, M., Dalvie, M.A., 2021. Association of activities related to pesticide exposure on headache severity and neurodevelopment of school-children in the rural agricultural farmlands of the Western Cape of South Africa. Environ. Int. 146, 106237. https://doi.org/10.1016/j.envint.2020.106237.

Choi, H., Moon, J.K., Kim, J.H., 2013. Assessment of the exposure of workers to the insecticide imidacloprid during application on various field crops by a hand-held power sprayer. J. Agric. Food Chem. 61 (45), 10642–10648. https://doi.org/10.1021/jf403169t.

Chuang, C.S., Yang, K.W., Yen, C.M., Lin, C.L., Kao, C.H., 2019. Risk of seizures in patients with organophosphate poisoning: a nationwide population-based study. Int. J. Environ. Res. Publ. Health 16 (17). https://doi.org/10.3390/ijerph16173147.

Clarke, S., Lean, T., 2022. Turning DDT into 'Didimac': making insecticide products and consumers in British farming after 1945. Hist. Technol. 38 (1), 31–61. https://doi.org/10.1080/07341512.2022.2085492.

Clasen, B., Murussi, C., Storck, T., 2019. Pesticide contamination in southern Brazil. In: Pollution of Water Bodies in Latin America: Impact of Contaminants on Species of Ecological Interest. Springer International Publishing, Brazil, pp. 43–54. https://doi.org/10.1007/978-3-030-27296-8_3.

Costas-Ferreira, C., Durán, R., Faro, L.R.F., 2022. Toxic effects of glyphosate on the nervous system: a systematic review. Int. J. Mol. Sci. 23 (9). https://doi.org/10.3390/ijms23094605.

Cresto, N., Forner-Piquer, I., Baig, A., Chatterjee, M., Perroy, J., Goracci, J., Marchi, N., 2023. Pesticides at brain borders: impact on the blood-brain barrier, neuroinflammation, and neurological risk trajectories. Chemosphere 324, 138251. https://doi.org/10.1016/j.chemosphere.2023.138251.

Cui, S., Qi, Y., Zhu, Q., Wang, C., Sun, H., 2023. A review of the influence of soil minerals and organic matter on the migration and transformation of sulfonamides. Sci. Total Environ. 861, 160584. https://doi.org/10.1016/j.scitotenv.2022.160584.

Dara, S.K., 2019. The new integrated pest management paradigm for the modern age. J. Integr. Pest Manag. 10 (1). https://doi.org/10.1093/jipm/pmz010.

Das, S.K., 2013. Mode of action of pesticides and the novel trends-a critical review. Int. Res. J. Agric. Sci. Soil Sci. 3 (11), 393–401.

Das, P.P., Singh, K.R., Nagpure, G., Mansoori, A., Singh, R.P., Ghazi, I.A., Kumar, A., Singh, J., 2022. Plant-soil-microbes: a tripartite interaction for nutrient acquisition and better plant growth for sustainable agricultural practices. Environ. Res. 214. https://doi.org/10.1016/j.envres.2022.113821.

Davidse, L., 1986. Benzimidazole fungicides: mechanism of action and biological impact. Annu. Rev. Phytopathol. 24 (1), 43–65. https://doi.org/10.1146/annurev.phyto.24.1.43.

Desisa, B., Getahun, A., Muleta, D., 2022. Advances in biological treatment technologies for some emerging pesticides. In: Pesticides Bioremediation. Springer International Publishing, Ethiopia, pp. 259–280. https://doi.org/10.1007/978-3-030-97000-0_10.

Dudley, N., Attwood, S.J., Goulson, D., Jarvis, D., Bharucha, Z.P., Pretty, J., 2017. How should conservationists respond to pesticides as a driver of biodiversity loss in agroecosystems? Biol. Conserv. 209, 449–453. https://doi.org/10.1016/j.biocon.2017.03.012.

Duke, S.O., 1990. Overview of herbicide mechanisms of action. Environ. Health Perspect. 87, 263–271. https://doi.org/10.1289/ehp.9087263.

Eldefrawi, M.E., Eldefrawi, A.T., 2020. Nervous-System-Based Insecticides Safer Insecticides, pp. 155–207.

Elgueta, S., Moyano, S., Sepúlveda, P., Quiroz, C., Correa, A., 2016. Pesticide residues in leafy vegetables and human health risk assessment in North Central agricultural areas of Chile. Food Addit. Contam. B 10 (2), 105–112. https://doi.org/10.1080/19393210.2017.1280540.

Esquinas-Requena, J.L., Lozoya-Moreno, S., García-Nogueras, I., Atienzar-Núñez, P., Sánchez-Jurado, P. M., Abizanda, P., 2020. La anemia aumenta el riesgo de mortalidad debido a fragilidad y discapacidad en mayores: Estudio FRADEA. Atención Primaria 52 (7), 452–461. https://doi.org/10.1016/j.aprim.2019.07.001.

Fahad, S., Saud, S., Akhter, A., Bajwa, A.A., Hassan, S., Battaglia, M., Adnan, M., Wahid, F., Datta, R., Babur, E., Danish, S., Zarei, T., Irshad, I., 2021. Bio-based integrated pest management in rice: an agro-ecosystems friendly approach for agricultural sustainability. J. Saudi Soc. Agric. Sci. 20 (2), 94–102. https://doi.org/10.1016/j.jssas.2020.12.004.

Felsot, A.S., Unsworth, J.B., Linders, J.B.H.J., Roberts, G., Rautman, D., Harris, C., Carazo, E., 2010. Agrochemical spray drift; assessment and mitigation—a review. J. Environ. Sci. Health B 46 (1), 1–23. https://doi.org/10.1080/03601234.2010.515161.

Figueiredo, D.M., Krop, E.J.M., Duyzer, J., Gerritsen-Ebben, R.M., Gooijer, Y.M., Holterman, H.J., Huss, A., Jacobs, C.M.J., Kivits, C.M., Kruijne, R., Mol, H.J.G.J., Oerlemans, A., Sauer, P.J.J., Scheepers, P.T.J., van de Zande, J.C., van den Berg, E., Wenneker, M., Vermeulen, R.C.H., 2021. Pesticide exposure of residents living close to agricultural fields in The Netherlands: protocol for an observational study. JMIR Res. Protoc. 10 (4). https://doi.org/10.2196/27883.

García-Díez, J., Gonçalves, C., Grispoldi, L., Cenci-Goga, B., Saraiva, C., 2021. Determining food stability to achieve food security. Sustainability 13 (13). https://doi.org/10.3390/su13137222.

de Graaf, L., Boulanger, M., Bureau, M., Bouvier, G., Meryet-Figuiere, M., Tual, S., Lebailly, P., Baldi, I., 2022. Occupational pesticide exposure, cancer and chronic neurological disorders: a systematic review of epidemiological studies in greenspace workers. Environ. Res. 203, 111822. https://doi.org/10.1016/j.envres.2021.111822.

Grimalt, S., Dehouck, P., 2016. Review of analytical methods for the determination of pesticide residues in grapes. J. Chromatogr. A 1433, 1–23. https://doi.org/10.1016/j.chroma.2015.12.076.

Hajek, A.E., 2004. Frontmatter.

Han, P., Zhang, Y.N., Lu, Z.Z., Wang, S., Ma, D.Y., Biondi, A., Desneux, N., 2018. Are we ready for the invasion of Tuta absoluta? Unanswered key questions for elaborating an Integrated Pest Management package in Xinjiang, China. Entomol. Gen. 38 (2), 113–125. https://doi.org/10.1127/entomologia/2018/0739.

Harish, S., Murugan, M., Kannan, M., Parthasarathy, S., Prabhukarthikeyan, S.R., Elango, K., 2022. Entomopathogenic viruses. In: Microbial Approaches for Insect Pest Management. Springer Singapore, India, pp. 1–57. https://doi.org/10.1007/978-981-16-3595-3_1.

Hassaan, M.A., El Nemr, A., 2020. Pesticides pollution: classifications, human health impact, extraction and treatment techniques. Egypt. J. Aquat. Res. 46 (3), 207–220. https://doi.org/10.1016/j.ejar.2020.08.007.

He, B., Hu, Y., Wang, W., Yan, W., Ye, Y., 2022. The progress towards novel herbicide modes of action and targeted herbicide development. Agronomy 12 (11). https://doi.org/10.3390/agronomy12112792.

He, R., Yuan, L., Jiang, Y., Jia, Z., Ding, W., Yang, Z., 2023. Pollution status, distribution, source analysis, and risk assessment of OCPs in soil from the Hexi Corridor in Northwest China. J. Environ. Expos. Assess. 2 (4). https://doi.org/10.20517/jeea.2023.22.

Hercegová, A., Dömötörová, M., Matisová, E., 2007. Sample preparation methods in the analysis of pesticide residues in baby food with subsequent chromatographic determination. J. Chromatogr. A 1153 (1–2), 54–73. https://doi.org/10.1016/j.chroma.2007.01.008.

Hummadi, E.H., Dearden, A., Generalovic, T., Clunie, B., Harrott, A., Cetin, Y., Demirbek, M., Khoja, S., Eastwood, D., Dudley, E., Hazir, S., Touray, M., Ulug, D., Gulsen, S.H., Cimen, H., Butt, T., 2021. Volatile organic compounds of Metarhizium brunneum influence the efficacy of entomopathogenic nematodes in insect control. Biol. Control 155, 104527. https://doi.org/10.1016/j.biocontrol.2020.104527.

Jaacks, L.M., Staimez, L.R., 2015. Association of persistent organic pollutants and non-persistent pesticides with diabetes and diabetes-related health outcomes in Asia: a systematic review. Environ. Int. 76, 57–70. https://doi.org/10.1016/j.envint.2014.12.001.

Jamshidi, B., 2017. Non-destructive safety assessment of agricultural products using Vis/NIR spectroscopy. NIR News 28 (1), 4–8. https://doi.org/10.1177/0960336016687519.

Janusz, G., Pawlik, A., Sulej, J., Świderska-Burek, U., Jarosz-Wilkołazka, A., Paszczyński, A., 2017. Lignin degradation: microorganisms, enzymes involved, genomes analysis and evolution. FEMS (Fed. Eur. Microbiol. Soc.) Microbiol. Rev. 41 (6), 941–962. https://doi.org/10.1093/femsre/fux049.

Jia, F., Li, Y., Hu, Q.n., Zhang, L., Mao, L.g., Zhu, L.z., Jiang, H.y., Liu, X.g., Sun, Y., 2023. Factors impacting the behavior of phytoremediation in pesticide-contaminated environment: a meta-analysis. Sci. Total Environ. 892. https://doi.org/10.1016/j.scitotenv.2023.164418.

Johnston, J.J., 2009. Introduction to pesticides and wildlife. Am. Chem. Soc. 1–5. https://doi.org/10.1021/bk-2001-0771.ch001.

Jokanović, M., Oleksak, P., Kuca, K., 2023. Multiple neurological effects associated with exposure to organophosphorus pesticides in man. Toxicology 484, 153407. https://doi.org/10.1016/j.tox.2022.153407.

Kalogiannidis, S., Kalfas, D., Chatzitheodoridis, F., Papaevangelou, O., 2022. Role of crop-protection technologies in sustainable agricultural productivity and management. Land 11 (10). https://doi.org/10.3390/land11101680.

Khamare, Y., Chen, J., Marble, S.C., 2022. Allelopathy and its application as a weed management tool: a review. Front. Plant Sci. 13. https://doi.org/10.3389/fpls.2022.1034649.

Khan, B.A., Nadeem, M.A., Nawaz, H., Amin, M.M., Hassan Abbasi, G., Nadeem, M., Ali, M., Ameen, M., Javaid, M.M., Maqbool, R., Ikram, M., Ayub, M.A., 2023. Pesticides: Impacts on Agriculture Productivity, Environment, and Management Strategies. Springer Science and Business Media LLC, pp. 109–134. https://doi.org/10.1007/978-3-031-22269-6_5.

Kim, K.H., Kabir, E., Jahan, S.A., 2017. Exposure to pesticides and the associated human health effects. Sci. Total Environ. 575, 525–535. https://doi.org/10.1016/j.scitotenv.2016.09.009.

Knodel, J.J., Shrestha, G., 2018. Pulse crops: pest management of wireworms and cutworms in the northern great plains of United States and Canada. Ann. Entomol. Soc. Am. 111 (4), 195–204. https://doi.org/10.1093/aesa/say018.

Kora, C., McDonald, M.R., Boland, G.J., 2025. New Progress in the Integrated Management of Sclerotinia Rot of Carrot. Springer Science and Business Media LLC. https://doi.org/10.1007/978-1-4020-8571-0_13.

Koranteng, S.S., Darko, D.A., Nukpezah, D., Ameka, G.K., 2018. Pesticides bioconcentration potential of aquatic plants in the volta lake. West Afr. J. Appl. Ecol. 26, 193–202.

Kumar, J., Ramlal, A., Mallick, D., Mishra, V., 2021. An overview of some biopesticides and their importance in plant protection for commercial acceptance. Plants 10 (6), 1185. https://doi.org/10.3390/plants10061185.

Kuplińska, A., Rząd, K., 2021. Molecular targets for antifungals in amino acid and protein biosynthetic pathways. Amino Acids 53 (7), 961–991. https://doi.org/10.1007/s00726-021-03007-6.

LeDoux, M., 2011. Analytical methods applied to the determination of pesticide residues in foods of animal origin. A review of the past two decades. J. Chromatogr. A 1218 (8), 1021–1036. https://doi.org/10.1016/j.chroma.2010.12.097.

Leiss, K.A., Choi, Y.H., Abdel-Farid, I.B., Verpoorte, R., Klinkhamer, P.G.L., 2009. NMR metabolomics of thrips (frankliniella occidentalis) resistance in senecio hybrids. J. Chem. Ecol. 35 (2), 219–229. https://doi.org/10.1007/s10886-008-9586-0.

Lengai, G.M.W., Fulano, A.M., Muthomi, J.W., 2022. Improving access to export market for fresh vegetables through reduction of phytosanitary and pesticide residue constraints. Sustainability 14 (13). https://doi.org/10.3390/su14138183.

Levasseur-Garcia, C., 2018. Updated overview of infrared spectroscopy methods for detecting mycotoxins on cereals (corn, wheat, and barley). Toxins 10 (1). https://doi.org/10.3390/toxins10010038.

Lian, Y.J., Pang, G.F., Shu, H.R., Fan, C.L., Liu, Y.M., Feng, J., Wu, Y.P., Chang, Q.Y., 2010. Simultaneous determination of 346 multiresidue pesticides in grapes by PSA-MSPD and GC-MS-SIM. J. Agric. Food Chem. 58 (17), 9428–9453. https://doi.org/10.1021/jf1019592.

Liao, Z., He, H., Wang, Y., Liu, F., Cui, D., Cui, J., Guo, Z., Lai, C., Huang, B., Sun, H., Pan, X., 2024. Algal extracellular organic matter induced photochemical oxidation of Mn(II) to solid Mn oxide: role of Mn(III)-EOM complex and its ability to remove 17α-ethinylestradiol. Environ. Sci. Technol. 58 (13), 5832–5843. https://doi.org/10.1021/acs.est.3c07970.

Liguori, J., Trübswasser, U., Pradeilles, R., Le Port, A., Landais, E., Talsma, E.F., Lundy, M., Béné, C., Bricas, N., Laar, A., Amiot, M.J., Brouwer, I.D., Holdsworth, M., 2022. How do food safety concerns affect consumer behaviors and diets in low- and middle-income countries? A systematic review. Global Food Secur. 32. https://doi.org/10.1016/j.gfs.2021.100606.

Liu, L., Xu, M., Ye, Y., Zhang, B., 2022. On the degradation of (micro) plastics: degradation methods, influencing factors, environmental impacts. Sci. Total Environ. 806, 151312.

Mahmood, I., Imadi, S.R., Shazadi, K., Gul, A., Hakeem, K.R., 2016. Effects of pesticides on environment. In: Plant, Soil and Microbes: Volume 1: Implications in Crop Science. Springer International Publishing, Pakistan, pp. 253–269. https://doi.org/10.1007/978-3-319-27455-3_13.

Maino, J.L., Thia, J., Hoffmann, A.A., Umina, P.A., 2023. Estimating rates of pesticide usage from trends in herbicide, insecticide, and fungicide product registrations. Crop Prot. 163. https://doi.org/10.1016/j.cropro.2022.106125.

Maqbool, Z., Hussain, S., Imran, M., Mahmood, F., Shahzad, T., Ahmed, Z., Azeem, F., Muzammil, S., 2016. Perspectives of using fungi as bioresource for bioremediation of pesticides in the environment: a critical review. Environ. Sci. Pollut. Control Ser. 23 (17), 16904–16925. https://doi.org/10.1007/s11356-016-7003-8.

Matich, E.K., Laryea, J.A., Seely, K.A., Stahr, S., Su, L.J., Hsu, P.C., 2021. Association between pesticide exposure and colorectal cancer risk and incidence: a systematic review. Ecotoxicol. Environ. Saf. 219. https://doi.org/10.1016/j.ecoenv.2021.112327.

Mazlan, A.Z., Hussain, H., Zawawi, M.A.M., 2016. Potential dermal exposure assessment of farmers to herbicide imazapic in an agriculture area. Proced. Soc. Behav. Sci. 234, 144–153. https://doi.org/10.1016/j.sbspro.2016.10.229.

McLaughlin, M.S., Roy, M., Abbasi, P.A., Carisse, O., Yurgel, S.N., Ali, S., 2023. Why do we need alternative methods for fungal disease management in plants? Plants 12 (22). https://doi.org/10.3390/plants12223822.

Meyers, R.A., 2006. High-Performance Liquid Chromatography Methods in Pesticide Residue Analysis Encyclopedia of Analytical Chemistry: Applications, Theory, and Instrumentation. Wiley. https://doi.org/10.1002/9780470027318.a1712.

Mir, Z.A., Bharose, R., Malik, Z.A., e, A., 2017. Review on phytoremediation: an ecofriendly and green technology for removal of heavy metals. Crop Res. 52, 74–82.

Miranda, R.A., Silva, B.S., de Moura, E.G., Lisboa, P.C., 2023. Pesticides as endocrine disruptors: programming for obesity and diabetes. Endocrine 79 (3), 437–447. https://doi.org/10.1007/s12020-022-03229-y.

Monteiro, C.M., Castro, P.M.L., Malcata, F.X., 2012. Metal uptake by microalgae: underlying mechanisms and practical applications. Biotechnol. Prog. 28 (2), 299–311. https://doi.org/10.1002/btpr.1504.

Mouden, S., Sarmiento, K.F., Klinkhamer, P.G.L., Leiss, K.A., 2017. Integrated pest management in western flower thrips: past, present and future. Pest Manag. Sci. 73 (5), 813–822. https://doi.org/10.1002/ps.4531.

Munir, M.A., Badri, K.H., 2020. The importance of derivatizing reagent in chromatography applications for biogenic amine detection in food and beverages. J. Anal. Methods Chem. 2020. https://doi.org/10.1155/2020/5814389.

Naik, M.A., Kumar, A., Gupta, S., Kushwaha, D.K., Kushwaha, H.L., Singh, A.K., Vaidhyanathan, R., Hasan, M., 2024. Potential dermal exposure and ergonomic assessment in greenhouse spraying. J. Occup. Environ. Hyg. 21 (11), 817–830. https://doi.org/10.1080/15459624.2024.2400241.

Narenderan, S.T., Meyyanathan, S.N., Babu, B., 2020. Review of pesticide residue analysis in fruits and vegetables. Pre-treatment, extraction and detection techniques. Food Res. Int. 133, 109141. https://doi.org/10.1016/j.foodres.2020.109141.

Navarrete-Meneses, M.D.P., Pérez-Vera, P., 2019. Pyrethroid pesticide exposure and hematological cancer: epidemiological, biological and molecular evidence. Rev. Environ. Health 34 (2), 197–210. https://doi.org/10.1515/reveh-2018-0070.

Nurmansyah, H.I., Suryani, E., Gustia, H., Ilmar Ramadhan, A., 2022. The effect of various essential oil and solvent additives on the botanical pesticide of Piper Aduncum essential oil on formulation antifungal activity. Results Eng. 16, 100644. https://doi.org/10.1016/j.rineng.2022.100644.

Omokhagbor Adams, G., Tawari Fufeyin, P., Eruke Okoro, S., Ehinomen, I., 2015. Bioremediation, biostimulation and bioaugmention: a review. Int. J. Environ. Bioremed. Biodegrad. 3 (1), 28–39. https://doi.org/10.12691/ijebb-3-1-5.

Ong, K.J., Johnston, J., Datar, I., Sewalt, V., Holmes, D., Shatkin, J.A., 2021. Food safety considerations and research priorities for the cultured meat and seafood industry. Compr. Rev. Food Sci. Food Saf. 20 (6), 5421–5448. https://doi.org/10.1111/1541-4337.12853.

Ostrem, J.S., Pan, Z., Flexner, J.L., Owens, E., Binning, R., Higgins, L.S., 2016. Monitoring susceptibility of western bean cutworm (Lepidoptera: Noctuidae) field populations to bacillus thuringiensis Cry1F protein. J. Econ. Entomol. 109 (2), 847–853. https://doi.org/10.1093/jee/tov383.

Padayachee, K., Reynolds, C., Mateo, R., Amar, A., 2023. A global review of the temporal and spatial patterns of DDT and dieldrin monitoring in raptors. Sci. Total Environ. 858 (Pt 1), 159734. https://doi.org/10.1016/j.scitotenv.2022.159734.

Pan, X., Guo, X., Zhai, T., Zhang, D., Rao, W., Cao, F., Guan, X., 2023. Nanobiopesticides in sustainable agriculture: developments, challenges, and perspectives. Environ. Sci. Nano 10 (1), 41–61. https://doi.org/10.1039/d2en00605g.

Panis, C., Candiotto, L.Z.P., Gaboardi, S.C., Gurzenda, S., Cruz, J., Castro, M., Lemos, B., 2022. Widespread pesticide contamination of drinking water and impact on cancer risk in Brazil. Environ. Int. 165. https://doi.org/10.1016/j.envint.2022.107321.

Pardo, L.A., Beane Freeman, L.E., Lerro, C.C., Andreotti, G., Hofmann, J.N., Parks, C.G., Sandler, D.P., Lubin, J.H., Blair, A., Koutros, S., 2020. Pesticide exposure and risk of aggressive prostate cancer among private pesticide applicators. Environ. Health Glob. Access Sci. Sour. 19 (1). https://doi.org/10.1186/s12940-020-00583-0.

Pathak, V.M., Verma, V.K., Rawat, B.S., Kaur, B., Babu, N., Sharma, A., Dewali, S., Yadav, M., Kumari, R., Singh, S., Mohapatra, A., Pandey, V., Rana, N., Cunill, J.M., 2022. Current status of pesticide effects on environment, human health and it's eco-friendly management as bioremediation: a comprehensive review. Front. Microbiol. 13. https://doi.org/10.3389/fmicb.2022.962619.

Peluso, J., Chehda, A.M., Olivelli, M.S., Ivanic, F.M., Pérez Coll, C.S., Gonzalez, F., Valenzuela, L., Rojas, D., Cristos, D., Butler, M., Candal, R.J., Aronzon, C.M., 2023. Metals, pesticides, and emerging contaminants on water bodies from agricultural areas and the effects on a native amphibian. Environ. Res. 226. https://doi.org/10.1016/j.envres.2023.115692.

Pérez-Lucas, G., Vela, N., El Aatik, A., Navarro, S., 2019. Environmental Risk of Groundwater Pollution by Pesticide Leaching through the Soil Profile. IntechOpen. https://doi.org/10.5772/intechopen.82418.

Petsas, A., Vagı, M., 2017. Determination of pesticides residues in food of vegetal origin: sample preparation, chromato-graphic techniques and applications. In: 15th International Conference on Environmental Science and Technology Rhodes.

Pico, Y., Alfarhan, A.H., Barcelo, D., 2020. How recent innovations in gas chromatography-mass spectrometry have improved pesticide residue determination: an alternative technique to be in your radar. TrAC, Trends Anal. Chem. 122, 115720. https://doi.org/10.1016/j.trac.2019.115720.

Poudyal, S., Fernandez, R.T., Owen, J., Cregg, B., 2019. Dose-dependent phytotoxicity of pesticides in simulated nursery runoff on landscape nursery plants. Water 11 (11), 2354. https://doi.org/10.3390/w11112354.

Prodhan, M.D.H., Papadakis, E.N., Papadopoulou-Mourkidou, E., 2015. Determination of multiple pesticide residues in eggplant with liquid chromatography-mass spectrometry. Food Anal. Methods 8 (1), 229–235. https://doi.org/10.1007/s12161-014-9898-3.

Raj, A., Dubey, A., Malla, M.A., Kumar, A., 2023. Pesticide pestilence: global scenario and recent advances in detection and degradation methods. J. Environ. Manag. 338. https://doi.org/10.1016/j.jenvman.2023.117680.

Rani, K., Dhania, G., 2014. Bioremediation and biodegradation of pesticide from contaminated soil and water—a noval approach. Int. J. Curr. Microbiol. Appl. Sci. 3 (10), 23–33.

Rasool, S., Rasool, T., Gani, K.M., 2022. A review of interactions of pesticides within various interfaces of intrinsic and organic residue amended soil environment. Chem. Eng. J. Adv. 11, 100301. https://doi.org/10.1016/j.ceja.2022.100301.

Ridwan, Q., Anjum, N., Rashid, S., Akhter, F., Hanief, M., 2022. Plant-Microbe Interactions in Agro-Ecological Perspectives and Degradation of Pesticides. Informa UK Limited, pp. 129–158. https://doi.org/10.1201/9781003281207-6.

Rivero, J., Luzardo, O.P., Henríquez-Hernández, L.A., Machín, R.P., Pestano, J., Zumbado, M., Boada, L.D., Camacho, M., Valerón, P.F., 2015. In vitro evaluation of oestrogenic/androgenic activity of the serum organochlorine pesticide mixtures previously described in a breast cancer case-control study. Sci. Total Environ. 537, 197–202. https://doi.org/10.1016/j.scitotenv.2015.08.016.

Sabarwal, A., Kumar, K., Singh, R.P., 2018. Hazardous effects of chemical pesticides on human health–Cancer and other associated disorders. Environ. Toxicol. Pharmacol. 63, 103–114. https://doi.org/10.1016/j.etap.2018.08.018.

Samsidar, A., Siddiquee, S., Shaarani, S.M., 2018. A review of extraction, analytical and advanced methods for determination of pesticides in environment and foodstuffs. Trends Food Sci. Technol. 71, 188–201. https://doi.org/10.1016/j.tifs.2017.11.011.

Sánchez, M.T., Flores-Rojas, K., Guerrero, J.E., Garrido-Varo, A., Pérez-Marín, D., 2010. Measurement of pesticide residues in peppers by near-infrared reflectance spectroscopy. Pest Manag. Sci. 66 (6), 580–586. https://doi.org/10.1002/ps.1910Spain.

Sani, I., Izera Ismail, S., Abdullah, S., Jalinas, J., Jamian, S., Saad, N., 2020. A review of the biology and control of whitefly, bemisia tabaci (Hemiptera: aleyrodidae), with special reference to biological control using entomopathogenic fungi. Insects 11 (9), 619. https://doi.org/10.3390/insects11090619.

Sannino, A., Bolzoni, L., Bandini, M., 2004. Application of liquid chromatography with electrospray tandem mass spectrometry to the determination of a new generation of pesticides in processed fruits and vegetables. J. Chromatogr. A 1036 (2), 161–169. https://doi.org/10.1016/j.chroma.2004.02.078.

Savant, R.H., Banerjee, K., Utture, S.C., Patil, S.H., Dasgupta, S., Ghaste, M.S., Adsule, P.G., 2010. Multiresidue analysis of 50 pesticides in grape, pomegranate, and mango by gas chromatography-Lon trap mass spectrometry. J. Agric. Food Chem. 58 (3), 1447–1454. https://doi.org/10.1021/jf903398f.

Schleiffer, M., Speiser, B., 2022. Presence of pesticides in the environment, transition into organic food, and implications for quality assurance along the European organic food chain—a review. Environ. Pollut. 313, 120116. https://doi.org/10.1016/j.envpol.2022.120116.

Schreinemachers, P., Afari-Sefa, V., Heng, C.H., Dung, P.T.M., Praneetvatakul, S., Srinivasan, R., 2015. Safe and sustainable crop protection in Southeast Asia: status, challenges and policy options. Environ. Sci. Pol. 54, 357–366. https://doi.org/10.1016/j.envsci.2015.07.017.

Shah, K.K., Modi, B., Pandey, H.P., Subedi, A., Aryal, G., Pandey, M., Shrestha, J., 2021. Diversified crop rotation: an approach for sustainable agriculture production. Adv. Agric. 2021. https://doi.org/10.1155/2021/8924087.

Shahid, M., Khan, M.S., Singh, U.B., 2023. Pesticide-tolerant microbial consortia: potential candidates for remediation/clean-up of pesticide-contaminated agricultural soil. Environ. Res. 236. https://doi.org/10.1016/j.envres.2023.116724.

Shahid, M., Shafi, Z., Ilyas, T., Singh, U.B., Pichtel, J., 2024. Crosstalk between phytohormones and pesticides: insights into unravelling the crucial roles of plant growth regulators in improving crop resilience to pesticide stress. Sci. Hortic. 338, 113663. https://doi.org/10.1016/j.scienta.2024.113663.

Shinde, R., Pardeshi, A., Dhanshetty, M., Anastassiades, M., Banerjee, K., 2021. Development and validation of an analytical method for the multiresidue analysis of pesticides in sesame seeds using liquid- and gas chromatography with tandem mass spectrometry. J. Chromatogr. A 1652, 462346. https://doi.org/10.1016/j.chroma.2021.462346.

Silva Pinto, B.G., Marques Soares, T.K., Azevedo Linhares, M., Castilhos Ghisi, N., 2020. Occupational exposure to pesticides: genetic danger to farmworkers and manufacturing workers—a meta-analytical review. Sci. Total Environ. 748. https://doi.org/10.1016/j.scitotenv.2020.141382.

Sinam, V., Kumar, P., Singh, J., 2023. Bioremediation and ecorestoration strategies of aquatic environment. In: Spatial Modeling of Environmental Pollution and Ecological Risk. Elsevier, India, pp. 483–499. https://doi.org/10.1016/B978-0-323-95282-8.00025-0.

Singh, H.P., Batish, D.R., Kohli, R.K., 2003. Allelopathic interactions and allelochemicals: new possibilities for sustainable weed management. Crit. Rev. Plant Sci. 22 (3–4), 239–311. https://doi.org/10.1080/713610858.

Singh, N.S., Sharma, R., Parween, T., Patanjali, P.K., 2017. Pesticide contamination and human health risk factor. In: Modern Age Environmental Problems and Their Remediation. Springer International Publishing, India, pp. 49–68. https://doi.org/10.1007/978-3-319-64501-8_3.

Singh, D.V., Ali, R., Anita, A., Kulsum, M., Bhat, R.A., 2020. Ecofriendly approaches for remediation of pesticides in contaminated environs. Bioremediation and Biotechnology, Vol 3: Persistent and Recalcitrant Toxic Substances 3, 173–194. https://doi.org/10.1007/978-3-030-46075-4_8.

Sparks, T.C., Nauen, R., 2015. IRAC: mode of action classification and insecticide resistance management. Pestic. Biochem. Physiol. 121, 122–128. https://doi.org/10.1016/j.pestbp.2014.11.014.

Spizzirri, U.G., Cirillo, G., 2016. Analytical Methods for Pesticide Detection in Foodstuffs Food Safety: Innovative Analytical Tools for Safety Assessment. Wiley, pp. 221–269. https://doi.org/10.1002/9781119160588.ch7.

Stachniuk, A., Fornal, E., 2016. Liquid chromatography-mass spectrometry in the analysis of pesticide residues in food. Food Anal. Methods 9 (6), 1654–1665. https://doi.org/10.1007/s12161-015-0342-0.

Stenberg, J.A., Heil, M., Åhman, I., Björkman, C., 2015. Optimizing crops for biocontrol of pests and disease. Trends Plant Sci. 20 (11), 698–712. https://doi.org/10.1016/j.tplants.2015.08.007.

Stetina, S.R., Young, L.D., Pettigrew, W.T., Bruns, H.A., 2007. Effect of corn-cotton rotations on reniform nematode populations and crop yield. Nematropica 37 (2), 237–248.

Sumner, D.R., 2018. Crop rotation and plant productivity. Handbk. Agric. Product. 1, 273–313.

Thellier, M., Gemegah, A.A.J., Tantaoui, I., 2024. Global fight against malaria: goals and achievements 1900–2022. J. Clin. Med. 13 (19). https://doi.org/10.3390/jcm13195680.

Thorat, T., Patle, B.K., Kumar Kashyap, S., 2023. Intelligent insecticide and fertilizer recommendation system based on TPF-CNN for smart farming. Smart Agric. Technol. 3, 100114. https://doi.org/10.1016/j.atech.2022.100114.

Torres, J.B., Bueno, A.d.F., 2018. Conservation biological control using selective insecticides—a valuable tool for IPM. Biol. Control 126, 53–64. https://doi.org/10.1016/j.biocontrol.2018.07.012.

Torres, E.M., Hess, D., McNeil, B.T., Guy, T., Quinn, J.C., 2017. Impact of inorganic contaminants on microalgae productivity and bioremediation potential. Ecotoxicol. Environ. Saf. 139, 367–376. https://doi.org/10.1016/j.ecoenv.2017.01.034.

Toshio, N., 1992. Nerve membrane Na+ channels as targets of insecticides. Trends Pharmacol. Sci. 13 (6), 236–241. https://doi.org/10.1016/0165-6147(92)90075-h.

Tudi, M., Ruan, H.D., Wang, L., Lyu, J., Sadler, R., Connell, D., Chu, C., Phung, D.T., 2021. Agriculture development, pesticide application and its impact on the environment. Int. J. Environ. Res. Publ. Health 18 (3), 1–24. https://doi.org/10.3390/ijerph18031112.

Tudi, M., Li, H., Li, H., Wang, L., Lyu, J., Yang, L., Tong, S., Yu, Q.J., Ruan, H.D., Atabila, A., Phung, D.T., Sadler, R., Connell, D., 2022. Exposure routes and health risks associated with pesticide application. Toxics 10 (6). https://doi.org/10.3390/toxics10060335.

Vandana, M.P., Mahto, U., Das, S., 2022. Mechanism of toxicity and adverse health effects of environmental pollutants. In: Microbial Biodegradation and Bioremediation: Techniques and Case Studies for Environmental Pollution. Elsevier, India, pp. 33–53. https://doi.org/10.1016/B978-0-323-85455-9.00024-2.

VoPham, T., Bertrand, K.A., Hart, J.E., Laden, F., Brooks, M.M., Yuan, J.M., Talbott, E.O., Ruddell, D., Chang, C.C.H., Weissfeld, J.L., 2017. Pesticide exposure and liver cancer: a review. Cancer Causes Control 28 (3), 177–190. https://doi.org/10.1007/s10552-017-0854-6.

Wakchaure, M., Patle, B.K., Mahindrakar, A.K., 2023. Application of AI techniques and robotics in agriculture: a review. Artif. Intell. Life Sci. 3, 100057. https://doi.org/10.1016/j.ailsci.2023.100057.

Wu, S., Zhang, X., Sun, Y., Wu, Z., Li, T., Hu, Y., Su, D., Lv, J., Li, G., Zhang, Z., Zheng, L., Zhang, J., Chen, B., 2015. Transformation and immobilization of chromium by arbuscular mycorrhizal fungi as revealed by SEM–EDS, TEM–EDS, and XAFS. Environ. Sci. Technol. 49 (24), 14036–14047. https://doi.org/10.1021/acs.est.5b03659.

Yadav, I.C., Devi, N.L., Syed, J.H., Cheng, Z., Li, J., Zhang, G., Jones, K.C., 2015. Current status of persistent organic pesticides residues in air, water, and soil, and their possible effect on neighboring countries: a comprehensive review of India. Sci. Total Environ. 511, 123–137. https://doi.org/10.1016/j.scitotenv.2014.12.041.

Yang, S., Zhang, J.R., Li, M., Feng, Y.M., Yang, L.L., Long, Z.Q., Zhou, X., Wu, Z.B., Liu, L.W., Yang, S., 2024. Expanding the structural diversity of tubulin-targeting agents: development of highly potent benzimidazoles for treating fungal diseases. J. Agric. Food Chem. 72 (28), 15541–15551. https://doi.org/10.1021/acs.jafc.4c01578.

Yengkokpam, P., Mazumder, P.B., 2021. Antioxidant enzymatic activities and profiling of gene expression associated with organophosphate stress tolerance in Solanum melongena L.cv. Longai. 3 Biotech 11 (12). https://doi.org/10.1007/s13205-021-03061-7.

Zaller, J.G., Kruse-Plaß, M., Schlechtriemen, U., Gruber, E., Peer, M., Nadeem, I., Formayer, H., Hutter, H.P., Landler, L., 2022. Pesticides in ambient air, influenced by surrounding land use and weather, pose a potential threat to biodiversity and humans. Sci. Total Environ. 838. https://doi.org/10.1016/j.scitotenv.2022.156012.

Zhang, F., He, J., Yao, Y., Hou, D., Jiang, C., Zhang, X., Di, C., Otgonbayar, K., 2013. Spatial and seasonal variations of pesticide contamination in agricultural soils and crops sample from an intensive horticulture area of Hohhot, North-West China. Environ. Monit. Assess. 185 (8), 6893–6908. https://doi.org/10.1007/s10661-013-3073-y.

Zhou, W., Li, M., Achal, V., 2025. A comprehensive review on environmental and human health impacts of chemical pesticide usage. Emerging Contam. 11 (1), 100410. https://doi.org/10.1016/j.emcon.2024.100410.

CHAPTER 5

Phytotoxicity of antibiotics

Marium Fiaz[1] and Muhammad Arshad[2]
[1]School of Interdisciplinary Engineering and Sciences, National University of Sciences and Technology (NUST), Islamabad, Pakistan; [2]Institute of Environmental Sciences and Engineering, School of Civil and Environmental Engineering, National University of Sciences and Technology (NUST), Islamabad, Pakistan

1. Introduction

1.1 Definition of phytotoxicity

Phytotoxicity refers to the harmful effects of chemical substances, such as agro-contaminants, on plant growth and development (Ahmed et al., 2022). Agro-contaminants, encompassing pesticides, heavy metals, and other pollutants, can induce biochemical and physiological disturbances that impede vital plant activities, including nutrient absorption, photosynthesis, and root development (Kumar et al., 2021). These pollutants frequently build in agricultural soil and water, adversely impacting plant health and productivity over time (Apreja et al., 2022).

Among these agro-contaminants, antibiotic residues have emerged as a significant problem due to their extensive application in agricultural practices, including livestock farming and crop treatment (Table 5.1). Antibiotic-induced phytotoxicity arises from the disturbance of plant cell structure and function, encompassing the compromise of membrane integrity, enzyme activity, and hormone control (Khan et al., 2021). Recent research indicates that antibiotics, such as tetracyclines, sulfonamides, and fluoroquinolones, intensify oxidative stress in plants by promoting the overproduction of reactive oxygen species (ROS). Oxidative stress results in the impairment of cellular membranes, proteins, and DNA (Fiaz et al., 2023; Riaz et al., 2017). The resultant cellular damage hinders plant processes, including photosynthesis and respiration, ultimately resulting in stunted development, decreased agricultural yields, and compromised ecosystem stability (Chauhan et al., 2023; Timm, 2020). Moreover, antibiotics lead to soil pollution, disrupting microbial populations vital for nitrogen fixation and nutrient cycling, hence exacerbating negative impacts on plant health (Zhao et al., 2023; Fang et al., 2023).

1.2 Overview of antibiotics as organic pollutants

Antibiotics represent a major class of organic pollutants, attributable to their extensive application in human healthcare, veterinary practices, and agricultural settings. Numerous antibiotics remain unmetabolized and are excreted in active forms through human and animal waste, resulting in environmental contamination through wastewater and agricultural runoff (Bansal, 2022; Martinez, 2009). Antibiotics are utilized in agriculture for the

Phytotoxicity of Organic Pollutants
ISBN 978-0-443-33110-7,
https://doi.org/10.1016/B978-0-443-33110-7.00010-9

Table 5.1 Physicochemical properties of selected antibiotics.

Antibiotic class	Antibiotic	Usage	pKa	Log Kow	Kd (L/kg^{-1})	Half-life in soil (days)
Fluroquinolones	Ciprofloxacin	Human health, veterinary	6.09, 6.82	0.28	61,000	2310 ± 1155
	Enrofloxacin	Veterinary	6.27, 8.3	1.1	260–6000	n/a
	Ofloxacin	Human health	5.97, 8.28	0.35	310	1386 ± 434
	Norfloxacin	Human health	6.40–8.68	−1.0	n/a	1155
Tetracycline	Oxytetracycline	Human health, veterinary	3.3, 7.3, 9.1	1.22	680–1030	30.2–41.3
	Chlortetracycline	Human health, veterinary	3.3, 7.44, 9.27	−0.36	794	25.9–30.8
	Tetracycline	Human health, veterinary	3.32, 7.78, 9.58	1.30	400–1620	578
	Doxycycline	Human health, veterinary	3.02, 7.97, 9.15	0.02	n/a	533 ± 23
Sulfonamide	Sulfamethazine	Veterinary	2.65, 7.65	0.89	0.6–3.1	18.6
	Sulfamethazole	Human health	1.97,5.70	0.98		9–18.3
	Sulphadiazine	Human health, veterinary	2.01, 6.15	−0.092	2.0	12–18
	Sulfachloropyridazine	Veterinary	1.87, 5.45	0.31	09–1.8	21.3

Data from Zhou, Z., Huang, F., Chen, L., Liu, F., Wang, B., Tang, J., 2024. Effects of antibiotics on microbial nitrogen cycling and N_2O emissions: a review. Chemosphere 357, 142034, https://doi.org/10.1016/j.chemosphere.2024.142034; Yang, J.-F., Ying, G.-G., Zhao, J.-L., Ran, T., Su, H.-C., Liu, Y.-S., 2011. Spatial and seasonal distribution of selected antibiotics in surface waters of the Pearl Rivers, China. J. Environ. Sci. Health B 46 (3), 272–280, https://doi.org/10.1080/03601234.2011.540540; DeVries, S. L., Zhang. P., 2016.

control of livestock infections and as growth promoters, resulting in contamination of soil and water (Khan et al., 2013). Runoff from agricultural fields introduces antibiotic residues into aquatic environments, where they persist and affect ecosystems (Kümmerer, 2010). The chemical stability of antibiotics and their potential for bioaccumulation in the food chain increase environmental risks (Zhou et al., 2024; Yang et al., 2011) (Table 5.1), impacting both plant and animal health (Zafar et al., 2021; Saima et al., 2019). These substances interfere with microbial communities essential for nutrient cycling, particularly nitrogen-fixing bacteria, thereby adversely affecting plant growth and agricultural productivity (Rasheela et al., 2025; Chen et al., 2024; Cycoń et al., 2019).

1.3 Importance of studying antibiotic phytotoxicity

Understanding antibiotic phytotoxicity is crucial for evaluating its impact on plant health, crop yields, and food safety. Antibiotics interfere with plant metabolic processes, including nitrogen fixation, photosynthesis, and nutrient absorption (Khan et al., 2022). Nutrient deficiencies, stunted growth, and reduced yields adversely affect food security (Fiaz et al., 2023; Minden et al., 2018). Additionally, antibiotic residues in crops may infiltrate the food chain (Fig. 5.1), leading to concerns regarding the emergence of antibiotic-resistant bacteria (Geng et al., 2022; Saima et al., 2019).

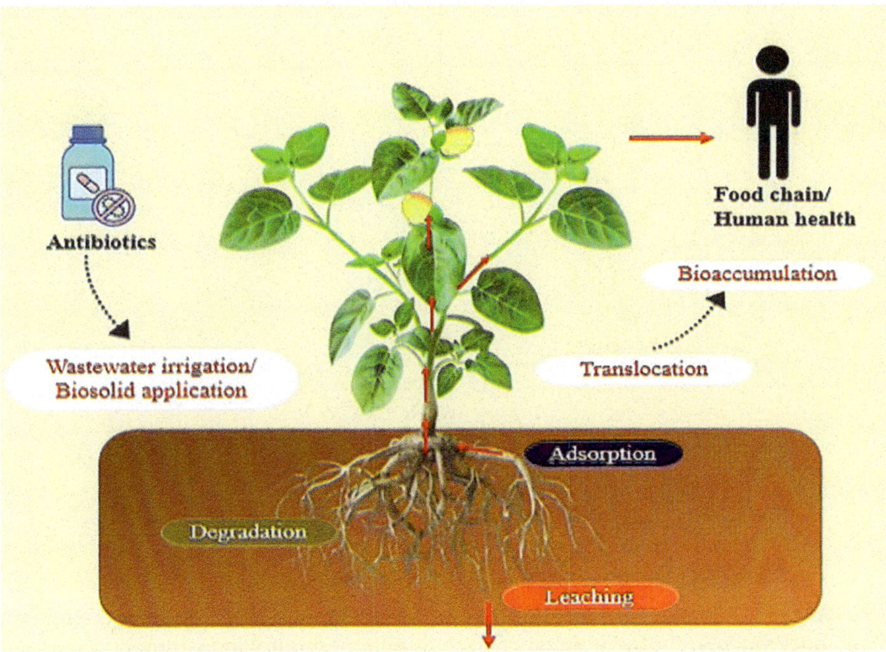

Figure 5.1 ***Pathways for antibiotic uptake and associated risks.*** This figure illustrates the pathways through which antibiotics from contaminated water sources can be taken up by plants, potentially leading to human exposure through the food chain and contributing to antibiotic resistance.

Extended exposure to subtherapeutic concentrations of antibiotics within the food chain may intensify the global challenge of antibiotic resistance (Wu et al., 2023; Mathé-Hubert et al., 2022). Comprehending these phytotoxic effects is essential for safeguarding plant health and averting the dissemination of resistance. Optimized fertilizer use and enhanced wastewater treatment are sustainable agricultural practices that can mitigate associated risks. Additionally, bioremediation techniques may effectively reduce antibiotic residues in the environment (Visca et al., 2024; Ren et al., 2023).

2. Sources and pathways of antibiotic contamination

2.1 Agricultural practices

Agricultural practices significantly contribute to contamination in the environment (Barnett et al., 2025; Manzoor et al., 2024; Zafar et al., 2024). Antibiotics are frequently utilized in livestock production for both therapeutic and growth-promoting objectives, with an estimated annual usage of around 63,000 tons globally (Zafar et al., 2021; Van Boeckel et al., 2015). Antibiotics frequently remain unmetabolized in animals and are excreted in manure, commonly utilized as fertilizer (Fig. 5.2). Manure has the potential

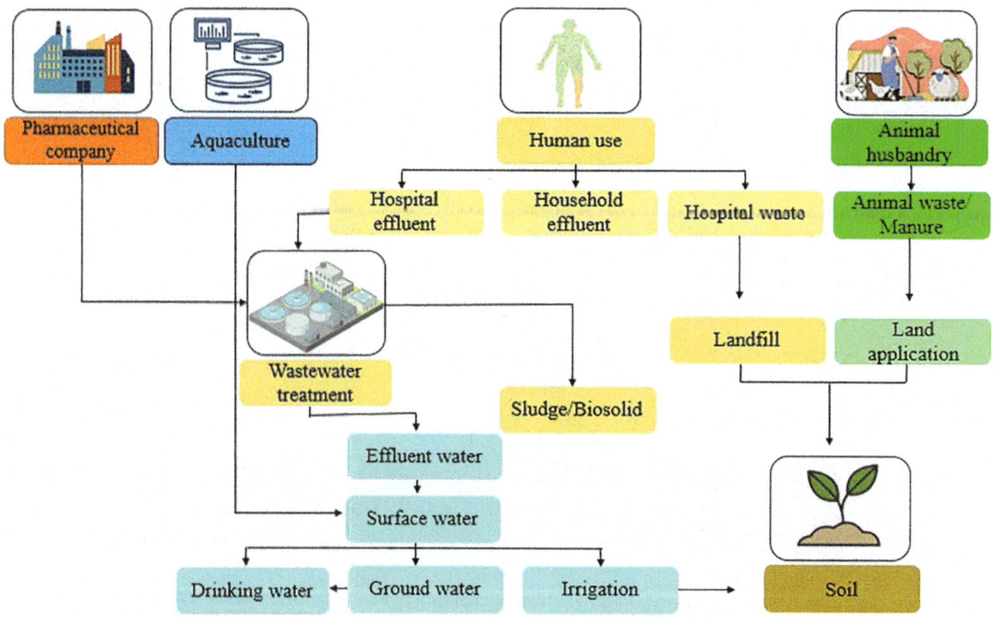

Figure 5.2 ***Sources and fate of antibiotics in the environment.*** This figure outlines the primary sources of antibiotics entering the environment—such as pharmaceutical industries, hospitals, livestock, and agriculture—and traces their movement through various pathways, including wastewater, soil, surface water, and ultimately into the food chain, contributing to environmental contamination and antimicrobial resistance.

to contaminate soil, which can adversely affect microbial diversity and disrupt essential processes such as nitrogen fixation (Arshad and Zafar, 2020; Cycoń et al., 2019). Antibiotics such as streptomycin, commonly used in agricultural practices, exacerbate soil contamination and disrupt ecological balance (Petre et al., 2020). The presence of antibiotics in soil results in enduring environmental issues, such as antibiotic resistance, highlighting the need for improved management and alternative practices to mitigate agricultural contamination (Rasheela et al., 2025; Chen et al., 2024).

2.2 Wastewater and runoff

Wastewater and agricultural runoff represent major routes for antibiotic contamination. Pharmaceutical manufacturing and municipal wastewater treatment facilities discharge effluents containing antibiotics, which frequently retain biological activity and infiltrate aquatic ecosystems (Kümmerer, 2010). Runoff from agricultural fields, which contain antibiotics from manure and fertilizers, exacerbates this issue by contaminating rivers and lakes (Fig. 5.2). The persistence of these compounds in aquatic environments presents risks to both aquatic organisms and human health (Vestel et al., 2022). Inefficient wastewater treatment techniques perpetuate antibiotic pollution, particularly in areas without sophisticated filtration systems (Zafar et al., 2021; Chen et al., 2020).

2.3 Industrial discharges

Pharmaceutical production facilities significantly contribute to antibiotic contamination. Improperly managed industrial discharges frequently introduce elevated concentrations of antibiotics into the environment, especially in areas characterized by insufficient regulatory oversight (Kawai, 2022). Antibiotics in industrial effluents contribute to soil and water contamination, facilitating the growth of antibiotic-resistant bacteria (Su et al., 2023). Inadequate waste management and insufficient regulatory enforcement aggravate these issues, highlighting the necessity for stricter regulations and enhanced monitoring systems to address antibiotic levels in industrial waste (Kumar et al., 2021).

2.4 Soil and water accumulation

The concentrations of different antibiotics reported in wastewater and river water used for irrigation are presented in Table 5.2. The presence of antibiotics in soil and water represents a significant environmental concern, as their persistence and mobility adversely affect ecosystem health and agricultural productivity (Fig. 5.2). The chemical properties of antibiotics, such as solubility, adsorption characteristics, and stability, influence their duration of activity in the environment (Martin-Laurent et al., 2019). Numerous widely utilized antibiotics, including tetracyclines and fluoroquinolones, exhibit high adsorption rates, especially to organic matter and clay particles in soil, enabling them to maintain biological activity for prolonged durations

Table 5.2 Reported concentrations of antibiotics in the environment.

Antibiotics	Predicted No. effect concentration- ENV (mg L^{-1})	Reported concentration (mg L^{-1})	References
Ciprofloxacin	0.00045	0.157–332.155	Zafar et al. (2021)
Levofloxacin	0.0015	0.012–6.638	
Ofloxacin	0.01	0.005–2.558	
Sulfamethoxazole	0.0066	0.318–16.009	
Ampicillin	0.0006	0.012–32.570	
Oxytetracycline	0.047	0.027	Khan et al. (2013)
Trimethoprim	0.928	0.028	
Erythromycin	0.0005	0.001	
Lincomycin	0.00081	0.004	

From; Zafar, R., Bashir, S., Nabi, D., Arshad, M., 2021. Occurrence and quantification of prevalent antibiotics in wastewater samples from Rawalpindi and Islamabad, Pakistan. Sci. Total Environ. 764, 142596, https://doi.org/10.1016/j.scitotenv.2020.142596; Khan, G.A., Berglund, B., Khan, K.M., Lindgren, P.E., Fick, J., 2013. Occurrence and abundance of antibiotics and resistance genes in rivers, canal and near drug formulation facilities—a study in Pakistan. PLoS One 8 (6), https://doi.org/10.1371/journal.pone.0062712; Alliance, A.I., 2023. AMR alliance science-based PNEC targets for risk assessments. AMR Ind. Allian.

(Conde-Cid et al., 2020). The extended presence of antibiotics in soil leads to various ecological disruptions. Antibiotics disrupt soil microbial communities, which are crucial for nutrient cycling, organic matter decomposition, and disease suppression. Antibiotics inhibit nitrogen-fixing bacteria, resulting in decreased nitrogen availability for plants. This can lead to nutrient deficiencies, reduced crop yields, and overall poorer soil health (Rasheela et al., 2025; Chen et al., 2024). Antibiotics can disrupt the equilibrium of soil microorganisms, promoting antibiotic-resistant strains, which may lead to enduring effects on microbial diversity and ecosystem stability (Xiao et al., 2023). The mobility of antibiotics in soil is influenced by factors including soil texture, pH, and moisture content. In soils with elevated clay content, antibiotics exhibit a stronger binding affinity to particles, which diminishes their mobility and extends their residence time. Sandy soil exhibits reduced adsorption capacities, resulting in increased leaching of antibiotics into groundwater and surface waters (Wu et al., 2023). Leaching can lead to the contamination of drinking water sources, presenting significant risks to human and environmental health.

The presence of antibiotics in aquatic environments intensifies the issue. Antibiotics are introduced into aquatic ecosystems through runoff from agricultural fields and untreated wastewater discharges, where they can persist and adversely affect aquatic organisms. Antibiotics in aquatic environments can disrupt microbial populations, alter food webs, and promote the emergence of antibiotic-resistant bacteria, which may disseminate to other ecosystems and potentially enter the human food chain (Arshad and Zafar, 2020; Kumar et al., 2020). The intricate relationships among antibiotics, soil characteristics, and microbial populations require a thorough strategy for addressing antibiotic pollution. To address the challenges of soil and water accumulation, it is essential to improve waste management, enhance wastewater treatment technologies, and implement sustainable agricultural practices aimed at reducing antibiotic use and minimizing environmental contamination (Chen et al., 2024).

3. Mechanisms of phytotoxicity

3.1 Cellular and molecular responses in plants

Antibiotics adversely affect plant growth by interfering with essential cellular processes, such as membrane stability, enzymatic activity, and signal transduction. These disruptions lead to impeded growth and increased susceptibility to further environmental stressors (Minden et al., 2018). Antibiotics including tetracyclines, sulfonamides, and fluoroquinolones interfere with membrane functions in plant cells (Cheong et al., 2020). Research demonstrates that antibiotics can trigger lipid peroxidation, resulting in cell membrane damage and disruption of cellular integrity (Yang et al., 2024). Membrane damage impairs nutrient transport and ion-homeostasis, thereby reducing nutrient absorption efficiency and hindering overall plant growth. Antibiotics disrupt membranes and inhibit enzymatic activities critical for plant function. Antibiotics inhibit enzymes essential for cellular respiration, photosynthesis, and antioxidant defense, thus further compromising plant health (Fiaz et al., 2023). Studies demonstrate that sulfonamides and tetracyclines can impede the function of critical enzymes in the tricarboxylic acid (TCA) cycle and other metabolic pathways (Zhao et al., 2023). As a result, plants demonstrate diminished metabolic efficiency, leading to impaired growth and lower overall vitality (Javornik et al., 2023). Antibiotics interfere with cellular signaling pathways, affecting hormone regulation crucial for growth, stress response, and reproductive development. Antibiotic exposure in plants results in abnormal growth patterns, reduced disease resistance, and increased vulnerability to abiotic stress (Khan et al., 2022).

3.2 Disruption of photosynthesis and respiration

Antibiotics negatively affect photosynthesis and respiration, indicating considerable phytotoxicity that directly impacts energy production and plant health (Jin et al., 2022). Antibiotics such as tetracyclines and sulfonamides are known to harm chloroplasts, reduce chlorophyll levels, and disrupt photosynthesis. Research demonstrates that antibiotic exposure diminishes chlorophyll biosynthesis, resulting in yellowing or chlorosis of plant tissues (Chauhan et al., 2023). The reduction in chlorophyll content impairs the plant's ability to absorb sunlight and produce energy through photosynthesis, which is crucial for growth and reproduction (Bashir and Cho, 2016).

Antibiotics induce damage to photosynthetic machinery (Fig. 5.3), leading to the degradation of photosystem proteins and disruption of electron transport chains, thereby

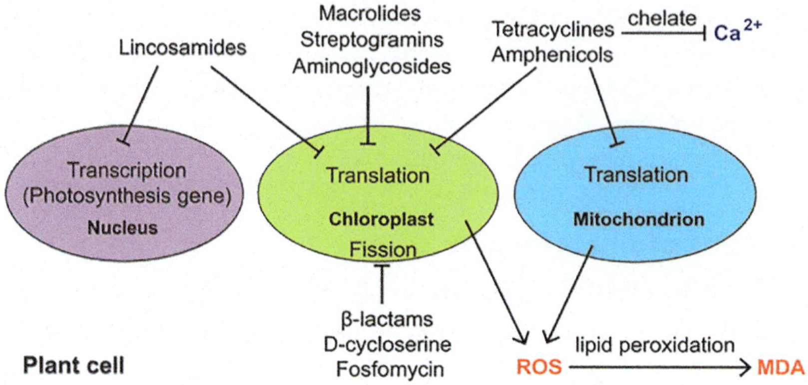

Figure 5.3 ***Mechanism of antibiotic stress in plants.*** **This figure presents how different antibiotics affect plant cells by interfering with transcription in the nucleus, translation in chloroplasts and mitochondria, and causing oxidative stress, which can lead to cell damage.** *(From Wang, X., Ryu, D., Houtkooper, R.H., Auwerx, J., 2015. Antibiotic use and abuse: a threat to mitochondria and chloroplasts with impact on research, health, and environment. Bioessays 37 (10), 1045–1053, https://doi.org/10.1002/bies.201500071.)*

reducing the energy efficiency of the plant (Jin et al., 2022). Studies demonstrate that tetracyclines hinder the function of essential enzymes involved in both the light and dark reactions of photosynthesis, leading to a diminished overall photosynthetic rate (Yang et al., 2024). In addition to photosynthesis, antibiotics interfere with respiratory pathways in plants. Respiratory processes, including mitochondrial function and energy production through oxidative phosphorylation, are essential for plant growth (Wang et al., 2015). Tetracyclines, a class of antibiotics, disrupt mitochondrial function, reducing the metabolic capacity of plants and impairing their energy generation for growth and reproduction (Minden et al., 2018). Another study conducted by Khan et al. (2021) explored the phytotoxicity of antibiotics and observed that antibiotic exposure in plants results in reduced growth, decreased productivity, and impaired reproductive success.

3.3 Impact on nutrient uptake and metabolism

Antibiotics disrupt essential nutrient absorption and metabolic functions in plants, significantly affecting nitrogen metabolism (Minden et al., 2018).

Nitrogen is a crucial nutrient for plants, playing a significant role in the synthesis of amino acids, proteins, and nucleic acids. Antibiotics interfere with critical nitrogen processes, such as nitrogen fixation in legumes and nitrate assimilation in non-leguminous plants, both essential for plant growth and agricultural productivity (Fiaz et al., 2023).

Nitrification and denitrification are critical processes in the nitrogen cycle that can be affected by antibiotic exposure. A previous study indicated that antibiotics impact the activity of nitrogen-fixing bacteria and also influence the broader microbial communities engaged in the conversion of ammonia to nitrites and nitrates (nitrification) or the reduction of nitrate to nitrogen gas (denitrification) (DeVries and Zhang, 2016). Antibiotics can inhibit microbial populations involved in these processes, disrupting nitrogen turnover in the soil and causing imbalances that affect plant nutrient availability (Roose-Amsaleg and Laverman, 2016).

Antibiotics reduce the diversity of soil microbes, thereby affecting soil health and its capacity to sustain agricultural productivity. Microbial diversity is crucial for maintaining healthy soil, as it supports nutrient cycling, organic matter decomposition, and disease suppression (Cycoń et al., 2019). Antibiotics reduce microbial diversity by favoring resistant strains and inhibiting sensitive ones, leading to altered soil functions and diminished agricultural productivity (Fiaz et al., 2021). The disruption of microbial communities diminishes the soil's ability to decompose organic materials, recycle nutrients, and support healthy plant growth.

The disruption of nitrogen cycling due to antibiotics impacts nitrification and denitrification processes, which are critical for maintaining soil fertility and preventing nutrient imbalances (Lucas et al., 2021). Antibiotics inhibit ammonia-oxidizing bacteria (AOB), disrupting nitrification and leading to ammonium accumulation, potentially causing toxicity in plants. Tetracycline and other antibiotics have been shown to reduce the effectiveness of denitrification, a process that converts excess nitrates into gaseous nitrogen, thereby preventing the accumulation of harmful nitrogen compounds in soil and water (Yang et al., 2023). The inhibition of this process leads to increased nitrogen runoff, contributing to water pollution and eutrophication (Chen et al., 2024). The impact of antibiotics on nitrogen metabolism is significant in agricultural systems, affecting plant growth, nutrient availability, and soil health.

Antibiotics, including tetracyclines and sulfonamides, reduce the activity of nitrogen-fixing bacteria in the soil, which are crucial for converting atmospheric nitrogen into bioavailable forms for plant uptake (DeVries and Zhang, 2016). Research demonstrates that antibiotics negatively affect the activity of nitrogen-fixing microorganisms, including *Rhizobium* in legumes, leading to reduced nitrogen availability in soils (Ostermann et al., 2019). Antibiotics, including tetracyclines and sulfonamides, have demonstrated a reduction in nitrification rates, leading to the accumulation of ammonium (NH_4^+) and a subsequent decrease in nitrate (NO_3^-) availability for plants (Rasheela et al., 2025; Chen et al., 2024). This disruption in the nitrogen cycle leads to decreased nitrogen availability, thereby impeding plant growth and productivity. Studies show that tetracyclines reduce the effectiveness of the nitrification process by inhibiting nitrifying bacteria such as *Nitrosomonas* and *Nitrobacter*. The inhibition results in ammonium accumulation in the soil, potentially becoming toxic to plants at elevated levels (Katipoglu-Yazan et al., 2015).

Antibiotics, alongside nitrogen, hinder the absorption of essential nutrients such as phosphorus and potassium. Exposure to antibiotics results in reduced root growth and altered nutrient uptake mechanisms, consequently hindering the plant's ability to absorb essential nutrients. Zhai et al. (2023) conducted a study that provides significant insights into the subject matter. This nutrient deficiency significantly affects the synthesis of amino acids, proteins, and other essential compounds necessary for plant health (Romera et al., 2021). Research demonstrates that antibiotics, including sulfadiazine and oxytetracycline, impede root development in crops such as wheat and rice, leading to reduced yields and impaired plant health (Zhao et al., 2023). The disruptions in nutrient metabolism underscore the importance of addressing antibiotic contamination in agricultural soils and its broader implications for crop production (Youssef and Bashour, 2017).

3.4 Induction of oxidative stress

Exposure to antibiotics induces oxidative stress in plants, which acts as a primary mechanism underlying the toxic effects of these substances (Wahid et al., 2021). Antibiotics such as tetracyclines, sulfonamides, and fluoroquinolones have been shown to generate reactive oxygen species (ROS) in plant cells. Reactive oxygen species (ROS) are highly reactive molecules that can induce damage to cellular components, including lipids, proteins, and DNA (Jin et al., 2022).

Exposure to antibiotics leads to the generation of reactive oxygen species (ROS), which causes oxidative damage to plant tissues and reduces their ability to endure further environmental stresses, such as drought and pathogen attacks (Chauhan et al., 2023). Plants react to oxidative stress through the activation of antioxidant defense mechanisms, which include the upregulation of enzymes like superoxide dismutase (SOD), catalase (CAT), and peroxidase (POD). These enzymes remove reactive oxygen species (ROS) and reduce their harmful effects, thus maintaining cellular integrity and function (Sahoo and Tiwari, 2022). Extended exposure to antibiotics may surpass the antioxidant defenses of plants, leading to irreversible oxidative damage (Jin et al., 2023). Studies demonstrate that the buildup of reactive oxygen species (ROS) from antibiotic exposure results in lipid peroxidation, DNA mutations, and cell membrane disruption, consequently compromising plant health and growth (Zhao et al., 2023). The restricted ability of plants to alleviate oxidative damage under prolonged antibiotic stress may lead to significant reductions in vitality, growth, and reproductive success. Furthermore, the induction of oxidative stress by antibiotics affects various physiological processes, including photosynthesis and respiration, by impairing the relevant enzymes (Guillouzo and Guguen-Guillouzo, 2020). Oxidative damage to the photosynthetic apparatus reduces the energy production capacity of the plant, while damage to respiratory enzymes undermines the overall metabolic efficiency. This sequence of events leads to inhibited growth, decreased yield, and, in severe cases, plant mortality.

4. Effects of antibiotics on plant growth and development

4.1 Germination rates

Antibiotics adversely affect seed germination, an essential stage in plant development. Antibiotics present in soil interfere with metabolic processes critical for seedling emergence, such as water absorption and enzyme activity. A previous study conducted by Liu et al. (2009) provides significant insights into the subject matter. Water absorption during seed imbibition is essential for germination, and antibiotics can alter this process by affecting the seed's hydration capacity. Studies demonstrate that antibiotic exposure, particularly tetracyclines and sulfonamides, reduces seed germination rates by inhibiting critical enzymes involved in cellular functions, including water absorption and the mobilization of stored reserves in seeds (Minden et al., 2017; Saleem et al., 2022).

Recent studies indicate that antibiotics significantly impact the germination of crop species, with evidence suggesting that exposure to these substances may delay or hinder seedling emergence. Tetracyclines inhibit the germination of species such as wheat and lettuce by disrupting enzyme activities essential for nutrient breakdown and the initiation of cellular growth necessary for sprout formation (Yang et al., 2024). This disruption results in reduced crop establishment, which may lead to lower plant survival rates and decreased productivity. Furthermore, delayed germination may affect the agricultural cycle, leading to diminished yields and posing a threat to food security. The presence of antibiotics in soil, particularly from agricultural runoff or manure application, presents a considerable challenge for crop production in contaminated environments.

4.2 Root and shoot development

Antibiotic exposure markedly affects root and shoot development, essential for overall plant growth and nutrient acquisition. Roots are essential for the absorption of water and nutrients, whereas shoots enable photosynthesis, thus promoting plant growth. Antibiotics such as tetracyclines and sulfonamides impede root elongation by interfering with cell division and differentiation processes (Minden et al., 2017). The reduction in root growth limits the plant's ability to acquire essential nutrients, including nitrogen and phosphorus, required for cellular growth and metabolic functions (Minden et al., 2018).

Studies demonstrate that antibiotics disrupt nutrient transport by impairing the functionality of plant root systems. Tetracyclines influence the activity of transporters involved in nutrient uptake, resulting in reduced efficiency of mineral absorption in plants (Tong et al., 2019). Antibiotics, including fluoroquinolones, tetracyclines and sulfonamides, impair the integrity of cell walls in root tissues, thereby undermining the structural support necessary for root elongation (Zhao et al., 2023). Exposure to antibiotics in the shoot system can negatively affect chlorophyll production, reduce photosynthetic efficiency, and hinder overall shoot growth. This results in reduced biomass production and compromised plant health (Fiaz et al., 2023).

The inhibition of root and shoot growth leads to stunted development, significantly reducing agricultural productivity and biomass, with lasting impacts on crop yields and the ecosystem services offered by plants (Mukhtar et al., 2020).

4.3 Reproductive success and seed viability

Antibiotics negatively affect plant reproduction, which is crucial for the sustainability and genetic diversity of plant populations. Reproductive success, which includes flower formation, pollen viability, and seed production, is notably influenced by disturbances caused by antibiotics. Sulfonamides and tetracyclines negatively affect flower development and pollen germination, thereby diminishing plant fertility. Minden et al. (2017) indicate that antibiotics interfere with hormonal signaling pathways essential for flower formation, leading to abnormal or incomplete development of reproductive organs (Yang et al., 2024).

The impact of antibiotics on seed production and viability presents important issues, affecting both individual plants and the broader plant population. Research demonstrates that antibiotic exposure may impair seed viability by interfering with the physiological processes linked to seed maturation. Tetracycline exposure reduces the viability of seeds from plants like wheat and tomatoes, impacting their reproductive capacity and subsequent generations (Zhao et al., 2023). The reduction in seed viability substantially affects plant biodiversity and the stability of plant communities. A decrease in genetic diversity may increase the vulnerability of plant populations to diseases, pests, and environmental stressors.

4.4 Long-term consequences on plant populations

Extended exposure to antibiotics can considerably affect plant populations, reducing their resilience and altering species composition. Antibiotics persist in the environment, particularly in agricultural soils that receive continuous applications of manure or wastewater containing these substances, which may lead to cumulative toxicity over time (Zhao and Xiao, 2019). Extended exposure to sub-lethal concentrations of antibiotics leads to enduring changes in plant community dynamics, often causing a decrease in sensitive species and an increase in antibiotic-tolerant species. Studies show that antibiotics modify microbial communities in soil (Fig. 5.4), which may negatively affect plant health and ecosystem functionality (Bansal, 2022). Chronic exposure to antibiotics like tetracycline and sulfonamides can reduce plant species diversity in agricultural ecosystems by facilitating the growth of species that demonstrate increased resistance to antibiotic stress. Changes in species composition can reduce the overall biodiversity of an ecosystem (Fig. 5.4), consequently heightening its susceptibility to invasive species and diseases. The disruption of plant growth and reproduction may result in cascading effects on the food web, as plants are essential to numerous terrestrial ecosystems.

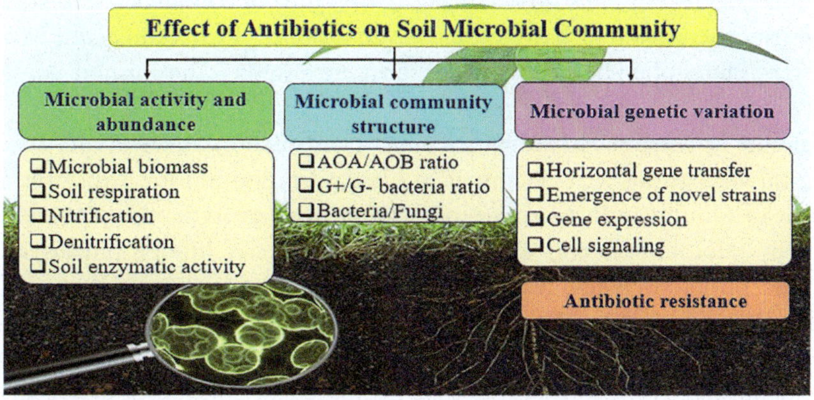

Figure 5.4 ***Effect of antibiotics on soil microbial community.*** This figure illustrates how antibiotics impact the soil microbial community by altering microbial activity, community structure, and genetic variation, which can lead to antibiotic resistance in the environment.

Reduced plant populations, especially those critical for ecosystem functions such as nitrogen fixation or pollination, can disrupt ecosystem processes, leading to lasting ecological imbalances (Rasheela et al., 2025; Chen et al., 2024). Antibiotic exposure has long-term effects on plant populations, impacting on individual crops and entire ecosystems. Antibiotic contamination diminishes the resilience of plant communities by hindering their recovery from disturbances such as drought and pest outbreaks (Minden et al., 2017).

Decreased resilience can lead to reduced ecosystem stability and a decline in ecosystem services such as soil fertility, carbon sequestration, and water filtration (Radu et al., 2017). Investigating the long-term effects of antibiotic exposure on plant populations is essential for formulating strategies to alleviate its impacts and promote the sustainability of agricultural and natural ecosystems.

5. Plant stress responses to antibiotic exposure

5.1 Physiological adaptations

Plants exposed to antibiotics demonstrate various physiological adaptations to counteract the negative impacts of these agents. Antibiotics cause stress in plants, impacting essential functions including water uptake, transpiration, and stomatal conductance. Plants can alter their physiological processes to conserve water and limit the uptake of harmful substances (Kumari et al., 2021; Mathews and Reinhold, 2013). Changes in water uptake mechanisms constitute the primary response. Antibiotics compromise the integrity of cell membranes in plant roots, resulting in diminished water retention capacity and reduced efficiency in water absorption (Zhao et al., 2023). Plants often adjust their transpiration rates by closing stomata or reducing stomatal conductance to conserve water and restrict the uptake of harmful antibiotic residues. Plants subjected to antibiotics such as tetracyclines often demonstrate a response marked by diminished stomatal opening. This functions as a protective mechanism designed to reduce water loss and restrict the infiltration of detrimental substances into plant tissues (Zhao et al., 2023). These physiological adaptations reduce immediate harm from antibiotic exposure; however, they may lead to long-term negative impacts on plant growth and development. The reduction in transpiration and altered water uptake can hinder photosynthesis and nutrient absorption efficiency, leading to stunted growth, decreased biomass, and lower plant productivity over time (Fiaz et al., 2023). Prolonged antibiotic exposure in plants can lead to diminished root growth and overall physiological decline, as the energy and resources dedicated to stress responses may limit growth and reproductive success.

5.2 Biochemical changes and secondary metabolite production

Antibiotic-induced stress triggers biochemical changes in plants, significantly impacting the production of secondary metabolites. Plants synthesize compounds such as flavonoids, phenolics, and alkaloids as defense mechanisms to combat oxidative stress and mitigate damage from environmental stressors, including antibiotics (Ghorai et al., 2021; Zia ur Rehman et al., 2021). Secondary metabolites play a crucial role in plant defense by neutralizing reactive oxygen species (ROS) and other detrimental compounds generated during stress responses.

Research demonstrates that plants such as lettuce and oilseed rape significantly increase the production of flavonoids and phenolics when exposed to antibiotics (Tasho et al., 2020). These compounds demonstrate antioxidant properties and are capable of scavenging reactive oxygen species (ROS), which are byproducts of oxidative stress caused by antibiotics (Chauhan et al., 2023). The enhancement of secondary metabolite production contributes to the reduction of oxidative damage, thereby lessening negative impacts on photosynthesis, respiration, and overall plant health. In addition to their antioxidant properties, these secondary metabolites improve the plant's resistance to pathogens and herbivores, providing an additional layer of defense against environmental threats (Bano et al., 2023). The synthesis of secondary metabolites reduces oxidative damage caused by antibiotics; however, it diverts resources from other metabolic pathways, which may impede long-term plant growth and development. Prolonged exposure to antibiotics may surpass the plant's ability to produce sufficient protective compounds, leading to reduced plant health and overall fitness.

5.3 Role of antioxidants in mitigating stress

The activation of antioxidant defense mechanisms is a crucial response of plants to oxidative stress caused by antibiotics. Antioxidant enzymes such as superoxide dismutase

(SOD), catalase (CAT), and peroxidase (POD) play a crucial role in neutralizing reactive oxygen species (ROS), which can cause considerable cellular damage if not regulated (Guillouzo and Guguen-Guillouzo, 2020; Alscher et al., 2002). These enzymes aid plants in reducing oxidative stress by converting reactive oxygen species (ROS) into less harmful compounds, thereby protecting cellular components such as lipids, proteins, and DNA from damage (Chauhan et al., 2023). Studies demonstrate that antibiotic exposure markedly increases the activity of antioxidant enzymes in plants. Antibiotic exposure, specifically to sulfonamides and tetracyclines, significantly enhances the activities of superoxide dismutase (SOD), catalase (CAT), and peroxidase (POD) in plants such as lettuce, wheat, and oilseed rape. This response functions as a protective mechanism against oxidative stress caused by these antibiotics (Riaz et al., 2017; Fiaz et al., 2023; Chauhan et al., 2023). Increased enzymatic activity reduces oxidative damage to vital cellular components, allowing the plant to survive and continue growing despite exposure to antibiotics. However, while these antioxidant responses help mitigate oxidative damage, prolonged exposure to antibiotics may eventually exceed the plant's antioxidant capabilities. Chronic antibiotic stress leads to persistent production of reactive oxygen species (ROS), potentially surpassing the plant's ability to neutralize these molecules, resulting in sustained oxidative damage and reduced plant vitality. The increased requirement for antioxidant enzymes during extended stress can deplete the plant's resources, diverting energy away from growth and reproduction.

5.4 Genetic and epigenetic responses

Exposure to antibiotics may induce genetic and epigenetic responses that influence plant adaptability and stress tolerance. Genetic responses involve the activation of stress-responsive genes that aid plants in mitigating antibiotic-induced damage (Khan et al., 2021).

The genes involved regulate the production of antioxidant enzymes, the synthesis of stress hormones, and DNA repair mechanisms, allowing the plant to mitigate the effects of antibiotic exposure (Zhao et al., 2023). The expression of genes related to flavonoid, phenolic, and other secondary metabolite synthesis is often upregulated in response to antibiotic stress, which enhances plant resistance to oxidative damage.

Antibiotics can induce epigenetic modifications that alter gene expression without changing the DNA sequence, alongside triggering genetic responses. The modifications may influence the plant's ability to respond to environmental stressors, such as antibiotic exposure, by changing the function of stress-responsive genes (D'Aquila et al., 2023; Lu and Hyun, 2021; Kumar, 2018).

Epigenetic modifications, such as DNA methylation, histone modification, and non-coding RNA regulation, can lead to lasting changes in gene expression that affect plant growth, development, and stress tolerance (Zhao et al., 2023). Epigenetic responses enable plants to cope with short-term stress; however, prolonged exposure to antibiotics can lead to irreversible genetic and epigenetic changes that adversely affect plant fitness and reproductive success (Li et al., 2023). Research demonstrates that extended exposure to antibiotics may lead to enduring changes in stress-responsive gene expression, which could impair the plant's ability to recover from future stressors and reduce its long-term resilience (Zhao et al., 2023).

Genetic and epigenetic responses to antibiotic stress are critical for plant adaptation to environmental challenges. These adaptive mechanisms may involve trade-offs, particularly with extended antibiotic exposure (Sharma et al., 2023).

Chronic antibiotic stress induces genetic and epigenetic alterations that compromise plant health, reducing the plant's ability to endure subsequent stressors and negatively impacting its reproductive success. The long-term effects can substantially influence plant populations, leading to alterations in species composition and disruptions in ecological balances.

6. Assessment methods for phytotoxicity

6.1 Laboratory techniques

The phytotoxicity of antibiotics is primarily evaluated in laboratory settings through the measurement of various plant growth parameters that reflect overall health and vitality. Essential parameters encompass germination rates, root elongation, chlorophyll content, biomass production, and enzymatic activities. The laboratory experiments create a controlled environment, allowing for direct attribution of observed changes in plant health to antibiotic exposure (Minden et al., 2017; Hillis et al., 2011). Seed germination is a frequently evaluated parameter, given that antibiotics may adversely affect seedling emergence. Root elongation exhibits heightened sensitivity to environmental stressors, such as antibiotic exposure, thereby serving as a critical early indicator of phytotoxicity. Chlorophyll content, crucial for photosynthesis, frequently declines in species such as wheat and lettuce upon exposure to antibiotics, indicating possible disturbances in photosynthetic mechanisms (Su et al., 2023). Low concentrations of antibiotics can induce significant growth alterations, highlighting the efficacy of these compounds (Minden et al., 2017; Thiele-Bruhn and Beck, 2005).

6.2 Field studies and ecotoxicological assessments

Laboratory studies isolate the effects of antibiotics in controlled settings, whereas field studies provide essential insights into the impact of antibiotics on plants and ecosystems in natural environments. These studies are crucial for comprehending the wider ecological implications of antibiotic contamination, especially in agricultural systems where antibiotic use is prevalent. Antibiotics may accumulate in soil and water, potentially impacting plants, soil microbes, invertebrates, and herbivores within the food chain.

Ecotoxicological assessments examine the bioaccumulation and trophic transfer of antibiotics, analyzing the movement of these substances through ecosystems and their effects on various organisms, which raises concerns regarding antibiotic residues in the food chain (Fang et al., 2023; Martin-Laurent et al., 2019). Field studies indicate that antibiotics can disrupt soil microbial communities critical for plant nutrition, including nitrogen-fixing bacteria, resulting in nutrient deficiencies and diminished crop yields. Changes in microbial diversity can adversely impact soil health and ecosystem stability, which are essential for the sustainability of agriculture in the long term (Fang et al., 2023).

6.3 Biomarkers for stress responses

Biochemical and molecular biomarkers are essential for evaluating plant responses to antibiotic exposure. These biomarkers serve as early indicators of plant stress prior to the manifestation of visible symptoms. Common biomarkers encompass the activity of antioxidant enzymes, including superoxide dismutase (SOD), catalase (CAT), and peroxidase (POD), along with the expression of stress-responsive genes. Exposure to antibiotics results in the production of reactive oxygen species (ROS), which contributes to oxidative stress. Plants increase antioxidant enzyme activity to reduce damage from reactive oxygen species (ROS), positioning enzyme activity as a sensitive indicator of oxidative stress (Fiaz et al., 2023; Banerjee et al., 2019). Field studies assessing antioxidant enzyme levels in plants subjected to antibiotic-contaminated environments provide important insights into plant responses to oxidative stress. The expression of stress-responsive genes associated with secondary metabolite production, such as flavonoids and phenolics, as well as environmental stress hormones like abscisic acid, may function as molecular biomarkers. Quantitative PCR techniques have been employed to assess alterations in gene expression, indicating early plant responses to antibiotics (Fang et al., 2023; Piña et al., 2007). These biomarkers enhance the understanding of the mechanisms underlying phytotoxicity and aid in the identification of strategies to alleviate the detrimental effects of antibiotics on plant health.

7. Remediation strategies for antibiotic pollution

7.1 Phytoremediation techniques

Phytoremediation provides a sustainable approach for the removal or degradation of antibiotics from contaminated soils and water using plants that can tolerate and interact with these pollutants. Specific plant species, including *Triticum aestivum* (wheat), *Lupinus* (legumes), and *Vetiveria zizanoides* (vetiver grass), demonstrate the capacity to absorb antibiotics such as fluoroquinolones, tetracycline and sulfonamides via their roots and leaves (Visca et al., 2024). Certain plants possess the ability to degrade antibiotics via metabolic processes, transforming these harmful compounds into less toxic forms, thereby diminishing their persistence in the environment. This method is particularly beneficial in agricultural contexts, where the use of antibiotics is prevalent, as it aids in reducing environmental contamination by improving soil and water quality. The effectiveness of phytoremediation is influenced by various factors, such as the chosen plant species, soil conditions, and the specific types of antibiotics involved. This underscores the necessity for additional research to enhance these techniques (Sidhu, O'Connor and Kruse, 2019; Paz-Alberto and Sigua, 2013).

7.2 Microbial and plant interactions

Microbial-plant interactions are acknowledged as a viable approach to improve the remediation of antibiotic contamination (Fig. 5.5). Beneficial microorganisms, including plant growth-promoting rhizobacteria (PGPR) and mycorrhizal fungi, significantly

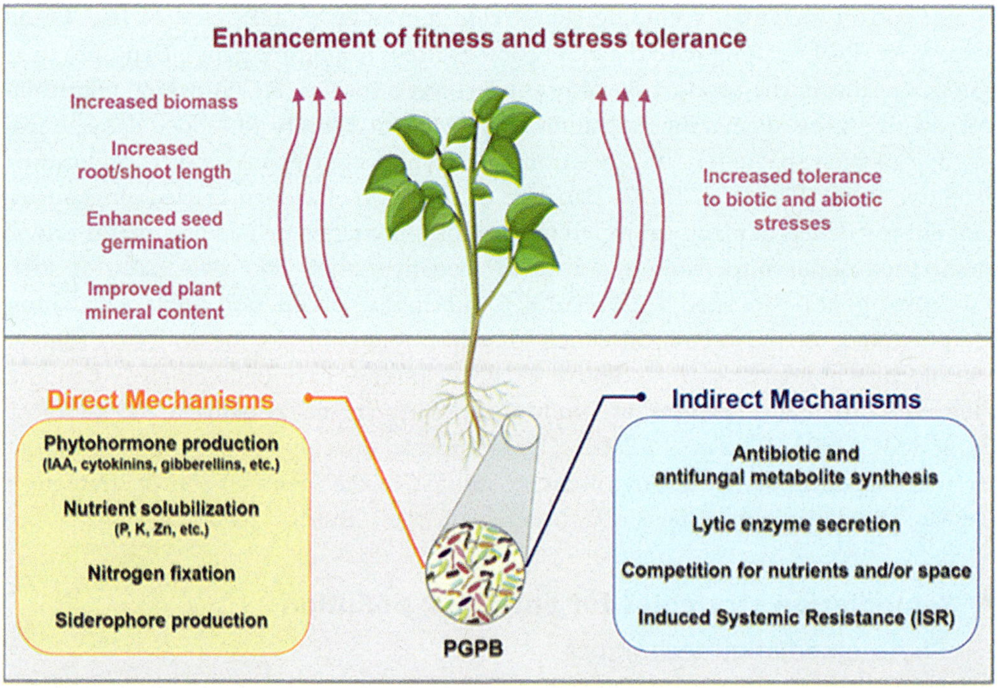

Figure 5.5 ***Mechanism of stress tolerance and plant growth improvement by PGPR.*** This figure explains how plant growth-promoting bacteria (PGPB) enhance plant fitness and stress tolerance through direct mechanisms like hormone production, nutrient solubilization, nitrogen fixation, and siderophore production, as well as indirect mechanisms such as antimicrobial compound synthesis, enzyme secretion, competition, and induced systemic resistance. *(From Vuolo, F., Novello, G., Bona, E., Gorrasi, S., Gamalero, E., 2022. Impact of plant-beneficial bacterial Inocula on the resident bacteriome: current knowledge and future perspectives. Microorganisms 10 (12), 2462, https://doi.org/10.3390/microorganisms10122462.)*

enhance plant resilience to stressors such as antibiotic exposure and facilitate the degradation of soil contaminants (Vuolo et al., 2022). Plant growth-promoting rhizobacteria (PGPR), including *Pseudomonas* and *Bacillus*, enhance plant growth through improved nutrient absorption, suppression of soil pathogens, and activation of plant defense mechanisms. These bacteria degrade antibiotics, thereby diminishing their bioavailability and detrimental effects on the environment (Sinha et al., 2021). Mycorrhizal fungi establish symbiotic associations with plant roots, enhancing nutrient acquisition and water absorption, while also facilitating antibiotic degradation. These fungi enhance soil health by improving soil structure and promoting microbial diversity, essential for the long-term sustainability of agricultural systems (Fig. 5.5). The integration of PGPR and mycorrhizal fungi in phytoremediation enhances antibiotic degradation, presenting a promising and comprehensive approach to contamination (Arun et al., 2023; Saleem et al., 2022).

8. Implications for food security and environmental sustainability

Antibiotic contamination adversely affects crop yield and quality, posing a significant threat to global food security. Antibiotic exposure negatively affects plant growth, hinders photosynthesis, and disrupts nutrient absorption, resulting in decreased biomass and diminished crop yields. This has a direct effect on food availability, particularly in areas already experiencing environmental stressors such as climate change (Yang et al., 2024). Furthermore, antibiotic residues in plants may infiltrate the food chain, thereby posing risks to human health through the promotion of antibiotic-resistant bacteria. This establishes a cycle of resistance that complicates the treatment of infections in humans and animals. The transfer of antibiotic residues from plants to herbivores, and subsequently to humans, elevates public health risks related to food consumption (Mirza et al., 2019).

Effective risk management strategies are necessary to address these concerns, which include researching alternative farming methods that minimize antibiotic reliance. Enhanced regulatory measures and monitoring systems are essential to restrict antibiotic usage in agriculture, especially in crop production and livestock farming. Sustainable farming practices, including crop rotation, organic fertilizers, and integrated nutrient management, contribute to decreased antibiotic reliance, enhanced soil health, and increased crop productivity. These practices mitigate environmental contamination and enhance ecosystem resilience and long-term food security. Emerging technologies, including nanomaterials (Ashraf et al., 2025) and bioremediation methods, demonstrate potential for mitigating antibiotic pollution in agricultural systems. Further investigation is required to evaluate the safety, feasibility, and effectiveness of these technologies. Integrating sustainable antibiotic management practices into agriculture, alongside comprehensive policy and research initiatives, is essential for ensuring environmental health, public safety, and the future sustainability of global food systems (Singh et al., 2020; Zhao et al., 2023).

9. Conclusions

Antibiotic contamination in agricultural systems presents significant risks to plant health, ecosystem stability, and food security. The rising use of antibiotics in agriculture, particularly in livestock production, has led to considerable contamination of soil, water, and crops. Antibiotics remain in the environment and adversely affect plant growth by interfering with critical physiological processes. The disruption of nitrogen cycling, essential for plant nutrition, represents a notable phytotoxic effect of antibiotics. Antibiotics interfere with nitrogen fixation in soil, hinder nutrient uptake, and alter microbial communities responsible for nutrient cycling, resulting in nutrient deficiencies in plants and reduced agricultural productivity (Zhao et al., 2023).

Antibiotics interfere with nutrient cycling and hinder essential processes like photosynthesis by compromising chloroplast function and reducing chlorophyll levels, leading to stunted growth and decreased biomass production. Antibiotic-induced phytotoxicity reduces overall biodiversity in agricultural systems by favoring antibiotic-tolerant plant species and weakening plant resilience to disease and environmental stress. Disruptions cause ecological imbalances, leading to long-term effects on plant populations, food webs, and ecosystem services (Yang et al., 2024). The presence of antibiotics in agricultural ecosystems presents considerable threats to crop yields, biodiversity, and ecosystem stability, highlighting the need for immediate research and policy measures to tackle these challenges.

Addressing this challenge necessitates a thorough understanding of the effects of antibiotics on plants and soil ecosystems. Research should concentrate on elucidating the mechanisms of antibiotic-induced phytotoxicity, evaluating long-term impacts on agricultural productivity, and investigating advanced remediation technologies, including bioremediation and nanotechnology. It is essential to develop sustainable agricultural practices that reduce antibiotic use, improve soil health, and foster biodiversity. Immediate action is essential, as the future of agriculture relies on effectively addressing these critical challenges. Ongoing research and the implementation of sustainable farming practices can mitigate antibiotic pollution, enhance plant health, and promote the long-term stability of food systems, thereby ensuring food security and preserving ecosystems for future generations.

References

Ahmed, H.M., Amiri-Ardekani, E., Ebadi, S., 2022. Phytotoxicity of natural molecules derived from cereal crops as a means to increase yield productivity. Int. J. Agron. 2022. https://doi.org/10.1155/2022/4223853.

Alscher, R.G., Erturk, N., Heath, L.S., 2002. Role of superoxide dismutases (SODs) in controlling oxidative stress in plants. J. Exp. Bot. 53, 372–1331. https://doi.org/10.1093/jxb/53.372.1331.

Apreja, M., Sharma, A., Balda, S., Kataria, K., Capalash, N., Sharma, P., 2022. Antibiotic residues in environment: antimicrobial resistance development, ecological risks, and bioremediation. Environ. Sci. Pollut. Control Ser. 29 (3), 3355–3371. https://doi.org/10.1007/s11356-021-17374-w.

Arshad, M., Zafar, R., 2020. Antibiotics, AMRs, and ARGs: Fate in the Environment. Elsevier BV, pp. 138–154. https://doi.org/10.1016/b978-0-12-818882-8.00009-7.

Arun, K.B., Madhavan, A., Tarafdar, A., Sirohi, R., Anoopkumar, A.N., Kuriakose, L.L., Awasthi, M.K., Binod, P., Varjani, S., Sindhu, R., 2023. Filamentous fungi for pharmaceutical compounds degradation in the environment: a sustainable approach. Environ. Technol. Innovat. 31. https://doi.org/10.1016/j.eti.2023.103182.

Ashraf, K., Liu, Z., Zaman, Q.u., Arshad, M., Zaman, W.Q., Shan, A., Yu, J., Rehman, T.u., Zhuang, Y., Guo, M., Mohsin, A., 2025. De Novo synthesis of selenium-doped CeO2@Fe3O4 nanoparticles for improving secondary metabolite biosynthesis in Carthamus tinctorius cell suspension culture. Chem. Eng. J. 505. https://doi.org/10.1016/j.cej.2025.159705.

Banerjee, S., Nandy, S.K., Chakraborti, S., 2019. Oxidative Stress as a Determinant of Antimicrobial Action, Resistance, and Treatment Oxidative Stress in Microbial Diseases. Springer Singapore, India, pp. 111–124. https://doi.org/10.1007/978-981-13-8763-0_7.

Bano, A., Qadri, T.A., Mahnoor, Khan, N., 2023. Bioactive metabolites of plants and microbes and their role in agricultural sustainability and mitigation of plant stress. South Afr. J. Bot. 159, 98–109. https://doi.org/10.1016/j.sajb.2023.05.049.

Bansal, O.P., 2022. Antibiotics in the environment: a review. Int. J. Front. Biol. Pharm. 2 (1), 57–083.

Barnett, A.N., Irum, A., Nabi, D., Khawar, M.I., Arshad, M., 2025. Occurrence of some heavy metals in drinking water, vegetables, and urine in Rawalpindi and Islamabad, Pakistan—human health risk assessment. Int. J. Environ. Res. 19 (1). https://doi.org/10.1007/s41742-024-00686-3.

Bashir, K.M.I., Cho, M.G., 2016. The effect of Kanamycin and tetracycline on growth and photosynthetic activity of two chlorophyte algae. BioMed Res. Int. 2016. https://doi.org/10.1155/2016/5656304.

Chauhan, J., Prathibha, M.D., Singh, P., Choyal, P., Mishra, U.N., Saha, D., Kumar, R., Anuragi, H., Pandey, S., Bose, B., Mehta, B., Dey, P., Dwivedi, K.K., Gupta, N.K., Singhal, R.K., 2023. Plant photosynthesis under abiotic stresses: damages, adaptive, and signaling mechanisms. Plant Stress 10. https://doi.org/10.1016/j.stress.2023.100296.

Chen, X., Song, Y., Chen, L., Shen, Y., Zhan, X., Xing, B., 2024. Fate of emerging antibiotics in soil-plant systems: a case on fluoroquinolones. Sci. Total Environ. 951, 175487. https://doi.org/10.1016/j.scitotenv.2024.175487.

Chen, Y., Shen, W., Wang, B., Zhao, X., Su, L., Kong, M., Li, H., Zhang, S., Li, J., 2020. Occurrence and fate of antibiotics, antimicrobial resistance determinants and potential human pathogens in a wastewater treatment plant and their effects on receiving waters in Nanjing, China. Ecotoxicol. Environ. Saf. 206, 111371. https://doi.org/10.1016/j.ecoenv.2020.111371.

Cheong, M.S., Seo, K.H., Chohra, H., Yoon, Y.E., Choe, H., Kantharaj, V., Lee, Y.B., 2020. Influence of sulfonamide contamination derived from veterinary antibiotics on plant growth and development. Antibiotics 9 (8), 1–18. https://doi.org/10.3390/antibiotics9080456.

Conde-Cid, M., Núñez-Delgado, A., Fernández-Sanjurjo, M., Álvarez-Rodríguez, E., Fernández-Calviño, D., Arias-Estévez, M., 2020. Tetracycline and sulfonamide antibiotics in soils: presence, fate and environmental risks. Processes 8 (11), 1479. https://doi.org/10.3390/pr8111479.

Cycoń, Mrozik, Piotrowska-Seget, 2019. Antibiotics in the soil environment—degradation and their impact on microbial activity and diversity. Front. Microbiol. 10.

D'Aquila, P., De Rango, F., Paparazzo, E., Passarino, G., Bellizzi, D., Josenhans, C., 2023. Epigenetic-Based regulation of transcriptome in *Escherichia coli* adaptive antibiotic resistance. Microbiol. Spectr. 11 (3). https://doi.org/10.1128/spectrum.04583-22.

DeVries, S.L., Zhang, P., 2016. Antibiotics and the terrestrial nitrogen cycle: a review. Curr. Pollut. Rep. 2 (1), 51–67. https://doi.org/10.1007/s40726-016-0027-3.

Fang, L., Chen, C., Li, S.Y., Ye, P., Shi, Y., Sharma, G., Sarkar, B., Shaheen, S.M., Lee, S.S., Xiao, R., Chen, X., 2023. A comprehensive and global evaluation of residual antibiotics in agricultural soils: accumulation, potential ecological risks, and attenuation strategies. Ecotoxicol. Environ. Saf. 262, 115175. https://doi.org/10.1016/j.ecoenv.2023.115175.

Fiaz, M., Ahmed, I., Hassan, S.M.U., Niazi, A.K., Khokhar, M.F., Zeshan, Farooq, M.A., Arshad, M., 2023. Antibiotics induced changes in nitrogen metabolism and antioxidative enzymes in mung bean (Vigna radiata). Sci. Total Environ. 873. https://doi.org/10.1016/j.scitotenv.2023.162449.

Fiaz, M., Ahmed, I., Riaz, R., Nawaz, U., Arshad, M., 2021. Prevalence of antibiotic-resistant bacterial strains in wastewater streams: molecular characterization and relative abundance. Folia Microbiol. 66 (6), 1023–1037. https://doi.org/10.1007/s12223-021-00902-z.

Geng, J., Liu, X., Wang, J., Li, S., 2022. Accumulation and risk assessment of antibiotics in edible plants grown in contaminated farmlands: a review. Sci. Total Environ. 853, 158616. https://doi.org/10.1016/j.scitotenv.2022.158616.

Ghorai, A.K., Patsa, R., Jash, S., Dutta, S., 2021. Microbial Secondary Metabolites and Their Role in Stress Management of Plants. Elsevier BV, pp. 283–319. https://doi.org/10.1016/b978-0-12-822919-4.00012-0.

Guillouzo, A., Guguen-Guillouzo, C., 2020. Antibiotics-induced oxidative stress. Curr. Opin. Toxicol. 20–21, 23–28. https://doi.org/10.1016/j.cotox.2020.03.004.

Hillis, D.G., Fletcher, J., Solomon, K.R., Sibley, P.K., 2011. Effects of ten antibiotics on seed germination and root elongation in three plant species. Arch. Environ. Contam. Toxicol. 60 (2), 220–232. https://doi.org/10.1007/s00244-010-9624-0.

Javornik, T., Poljak, M., Carović-Stanlo, K., Lazarević, B., 2023. Common bean (Phaseolus vulgaris L.) gas exchange capacity under nutrient deficiency. J. Cent. Eur. Agric. 24 (1), 216–224. https://doi.org/10.5513/jcea01/24.1.3667.

Jin, J., Xu, L., Zhang, S., Jin, M.K., Zhang, P., Shen, L., Chen, J., Li, Z., Zhao, W., Liu, H., 2023. Oxidative response of rice (Oryza sativa L.) seedlings to quinolone antibiotics and its correlation with phyllosphere microbes and antibiotic resistance genes. Sci. Total Environ. 867, 161544. https://doi.org/10.1016/j.scitotenv.2023.161544.

Jin, M.K., Yang, Y.T., Zhao, C.X., Huang, X.R., Chen, H.M., Zhao, W.L., Yang, X.R., Zhu, Y.G., Liu, H.J., 2022. ROS as a key player in quinolone antibiotic stress on *Arabidopsis thaliana*: from the perspective of photosystem function, oxidative stress and phyllosphere microbiome. Sci. Total Environ. 848. https://doi.org/10.1016/j.scitotenv.2022.157821.

Katipoglu-Yazan, T., Merlin, C., Pons, M.N., Ubay-Cokgor, E., Orhon, D., 2015. Chronic impact of tetracycline on nitrification kinetics and the activity of enriched nitrifying microbial culture. Water Res. 72, 227–238. https://doi.org/10.1016/j.watres.2014.12.041.

Kawai, M., 2022. Environmental monitoring in a pharmaceutical manufacturing facility using a culture independent approach. Yakugaku Zasshi 142 (1), 33–37. https://doi.org/10.1248/yakushi.21-00161-4.

Khan, G.A., Berglund, B., Khan, K.M., Lindgren, P.E., Fick, J., 2013. Occurrence and abundance of antibiotics and resistance genes in rivers, canal and near drug formulation facilities—a study in Pakistan. PLoS One 8 (6). https://doi.org/10.1371/journal.pone.0062712.

Khan, K.Y., Ali, B., Zhang, S., Stoffella, P.J., Yuan, S., Xia, Q., Qu, H., Shi, Y., Cui, X., Guo, Y., 2021. Effects of antibiotics stress on growth variables, ultrastructure, and metabolite pattern of Brassica rapa ssp. chinensis. Sci. Total Environ. 778. https://doi.org/10.1016/j.scitotenv.2021.146333.

Khan, K.Y., Ali, B., Zhang, S., Stoffella, P.J., Shi, S., Xia, Q., Cui, X., Ali, Z., Guo, Y., 2022. Phytotoxic effects on chloroplast and UHPLC-HRMS based untargeted metabolomic responses in Allium tuberosum Rottler ex Sprengel (Chinese leek) exposed to antibiotics. Ecotoxicol. Environ. Saf. 234. https://doi.org/10.1016/j.ecoenv.2022.113418.

Kumar, G., Singh, S., Singh, R., Mishra, R., 2021. Role of Physical Agents in Inducing Genotoxicity and Oxidative Stress in Plants Induced Genotoxicity and Oxidative Stress in Plants. Springer Nature, India, pp. 65–102. https://doi.org/10.1007/978-981-16-2074-4_3.

Kumar, S.B., Arnipalli, S.R., Ziouzenkova, O., 2020. Antibiotics in food chain: the consequences for antibiotic resistance. Antibiotics 9 (10), 1–26. https://doi.org/10.3390/antibiotics9100688.

Kumar, S., 2018. Epigenomics of plant responses to environmental stress. Epigenomes 2 (1), 6. https://doi.org/10.3390/epigenomes2010006.

Kumari, A., Sharma, B., Singh, B.N., Hidangmayum, A., Jatav, H.S., Chandra, K., Singhal, R.K., Sathyanarayana, E., Patra, A., Mohapatra, K.K., 2021. Physiological Mechanisms and Adaptation Strategies of Plants under Nutrient Deficiency and Toxicity Conditions Plant Perspectives to Global

Climate Changes: Developing Climate-Resilient Plants. Elsevier, India, pp. 173–194. https://doi.org/10.1016/B978-0-323-85665-2.00012-1.

Kümmerer, K., 2010. Pharmaceuticals in the environment. Annu. Rev. Environ. Resour. 35 (1), 57–75. https://doi.org/10.1146/annurev-environ-052809-161223.

Li, L., Li, T., Liu, Y., Li, L., Huang, X., Xie, J., 2023. Effects of antibiotics stress on root development, seedling growth, antioxidant status and abscisic acid level in wheat (*Triticum aestivum* L.). Ecotoxicol. Environ. Saf. 252, 114621. https://doi.org/10.1016/j.ecoenv.2023.114621.

Liu, F., Ying, G.G., Tao, R., Zhao, J.L., Yang, J.F., Zhao, L.F., 2009. Effects of six selected antibiotics on plant growth and soil microbial and enzymatic activities. Environ. Pollut. 157 (5), 1636–1642. https://doi.org/10.1016/j.envpol.2008.12.021.

Lu, X., Hyun, T.K., 2021. The role of epigenetic modifications in plant responses to stress. Botanica Serbica 45 (1), 3–12. https://doi.org/10.2298/BOTSERB2101003L.

Lucas, J.M., Sone, B.S., Whitmore, D., Strickland, M.S., 2021. Antibiotics and temperature interact to disrupt soil communities and nutrient cycling. Soil Biol. Biochem. 163. https://doi.org/10.1016/j.soilbio.2021.108437.

Manzoor, S., Abbas, S., Zulfiqar, S., Wang, H.-C., Xiao, M., Li, W.-J., Arshad, M., Ahmed, I., 2024. Functional genomics and taxonomic insights into heavy metal tolerant novel bacterium Brevibacterium metallidurans sp. nov. NCCP-602T isolated from tannery effluent in Pakistan. Antonie Leeuwenhoek 117 (1). https://doi.org/10.1007/s10482-024-02006-3.

Martinez, J.L., 2009. Environmental pollution by antibiotics and by antibiotic resistance determinants. Environ. Pollut. 157 (11), 2893–2902. https://doi.org/10.1016/j.envpol.2009.05.051.

Martin-Laurent, F., Topp, E., Billet, L., Batisson, I., Malandain, C., Besse-Hoggan, P., Morin, S., Artigas, J., Bonnineau, C., Kergoat, L., Devers-Lamrani, M., Pesce, S., 2019. Environmental risk assessment of antibiotics in agroecosystems: ecotoxicological effects on aquatic microbial communities and dissemination of antimicrobial resistances and antibiotic biodegradation potential along the soil-water continuum. Environ. Sci. Pollut. Control Ser. 26 (18), 18930–18937. https://doi.org/10.1007/s11356-019-05122-0.

Mathé-Hubert, H., Amia, R., Martin, M., Gaffé, J., Schneider, D., 2022. Evolution of bacterial persistence to antibiotics during a 50,000-generation experiment in an antibiotic-free environment. Antibiotics 11 (4), 451. https://doi.org/10.3390/antibiotics11040451.

Mathews, S., Reinhold, D., 2013. Biosolid-borne tetracyclines and sulfonamides in plants. Environ. Sci. Pollut. Control Ser. 20 (7), 4327–4338. https://doi.org/10.1007/s11356-013-1693-y.

Minden, V., Schnetger, B., Pufal, G., Leonhardt, S.D., 2018. Antibiotic-induced effects on scaling relationships and on plant element contents in herbs and grasses. Ecol. Evol. 8 (13), 6699–6713. https://doi.org/10.1002/ece3.4168.

Minden, V., Deloy, A., Volkert, A.M., Leonhardt, S.D., Pufal, G., 2017. Antibiotics impact plant traits, even at small concentrations. AoB Plants 9 (2). https://doi.org/10.1093/aobpla/plx010.

Mirza, S.A., Afzaal, M., Begum, S., Arooj, T., Almas, M., Ahmed, S., Younus, M., 2019. Uptake Mechanism of Antibiotics in Plants Antibiotics and Antimicrobial Resistance Genes in the Environment: Volume 1 in the Advances in Environmental Pollution Research Series. Elsevier, Pakistan, pp. 183–188. https://doi.org/10.1016/B978-0-12-818882-8.00011-5.

Mukhtar, A., Manzoor, M., Gul, I., Zafar, R., Jamil, H.I., Niazi, A.K., Ali, M.A., Park, T.J., Arshad, M., 2020. Phytotoxicity of different antibiotics to rice and stress alleviation upon application of organic amendments. Chemosphere 258. https://doi.org/10.1016/j.chemosphere.2020.127353.

Ostermann, A., Mortimer, P.E., Huang, R., Ma, L., Bu, D., Xu, J., 2019. Symbiotic nitrogen fixation in soil contaminated with the veterinary antibiotics oxytetracycline and sulfamethazine . American Society of Agronomy, Crop Science Society of America, Soil Science. J. Environ. Qual. 48 (4), 1067–1073. https://doi.org/10.2134/jeq2019.01.0021.

Paz-Alberto, A.M., Sigua, G.C., 2013. Phytoremediation: a green Technology to remove environmental pollutants. Am. J. Clim. Change 02 (01), 71–86. https://doi.org/10.4236/ajcc.2013.21008.

Petre, A.A., Nenciu, F., Vanghele, N.A., Stanciu, M.M., Mihalache, D.B., Grigore, I.A., Vladutoiu, L., 2020. EDP Sciences Romania Study on the effects and changes of soil degradation under the influence of antibiotics. In: E3S Web of Conferences. https://doi.org/10.1051/e3sconf/202018003018, 180. www.e3s-conferences.org/.

Piña, B., Casado, M., Quirós, L., 2007. Analysis of gene expression as a new tool in ecotoxicology and environmental monitoring. TrAC, Trends Anal. Chem. 26 (11), 1145–1154. https://doi.org/10.1016/j.trac.2007.09.009.

Radu, E., Woegerbauer, M., Oismuller, M., Kreuzinger, N., 2017. Impact of antibiotics of anthropogenic origin on bacterial soil communities in agricultural ecosystems. Int. Sympos. Environ. Ind.

Rasheela, A.R.P., Khalid, M.F., Abumaali, D.A., Alatalo, J.M., Ahmed, T., 2025. Impact of abiotic stressors on soil microbial communities: a focus on antibiotics and their interactions with emerging pollutants. Soil Syst. 9 (1), 2. https://doi.org/10.3390/soilsystems9010002.

Ren, S., Wang, S., Liu, Y., Wang, Y., Gao, F., Dai, Y., 2023. A review on current pollution and removal methods of tetracycline in soil. Separ. Sci. Technol. 58 (14), 2578–2602. https://doi.org/10.1080/01496395.2023.2259079.

Riaz, L., Mahmood, T., Coyne, M.S., Khalid, A., Rashid, A., Hayat, M.T., Gulzar, A., Amjad, M., 2017. Physiological and antioxidant response of wheat (*Triticum aestivum*) seedlings to fluoroquinolone antibiotics. Chemosphere 177, 250–257. https://doi.org/10.1016/j.chemosphere.2017.03.033.

Romera, F.J., Lucena, C., García, M.J., Alcántara, E., Angulo, M., Aparicio, M.Á., Pérez-Vicente, R., 2021. Plant Hormones and Nutrient Deficiency Responses. Springer Science and Business Media LLC, pp. 29–65. https://doi.org/10.1007/978-3-030-77477-6_2.

Roose-Amsaleg, C., Laverman, A.M., 2016. Do antibiotics have environmental side-effects? Impact of synthetic antibiotics on biogeochemical processes. Environ. Sci. Pollut. Control Ser. 23 (5), 4000–4012. https://doi.org/10.1007/s11356-015-4943-3.

Sahoo, A., Tiwari, S., 2022. Antioxidants and antioxidative enzymes as potential biomarkers for assessing stress in plants. Int. J. Plant Environ. 8 (02), 95–105. https://doi.org/10.18811/ijpen.v8i02.01.

Saima, S., Fiaz, M., Zafar, R., Ahmed, I., Arshad, M., 2019. Dissemination of Antibiotic Resistance in the Environment Antibiotics and Antimicrobial Resistance Genes in the Environment: Volume 1 in the Advances in Environmental Pollution Research Series. Elsevier, Pakistan, pp. 99–116. https://doi.org/10.1016/B978-0-12-818882-8.00006-1.

Saleem, A., Zulfiqar, A., Alamoudi, S.A., Ashkan, M.F., Alhomaidi, E., 2022. Plant growth-promoting bacteria improve *Triticum aestivum* L. growth and photosynthetic activity in sulfonamide-contaminated soil. Turk. J. Agric. For. 46 (6), 822–837. https://doi.org/10.55730/1300-011X.3045.

Sharma, R., Thakur, A., Saini, A., Giri, S.K., Kumar, A., Priya, K., Singh, G., 2023. Antibiotics stress response of bacteria as mechanism of development of drug resistance. ACS (Am. Chem. Soc.) Symp. Ser. 1434, 23–42. https://doi.org/10.1021/bk-2023-1434.ch002.

Sidhu, H., George, O., Connor, Kruse, J., 2019. Plant toxicity and accumulation of biosolids-borne ciprofloxacin and azithromycin. Sci. Total Environ. 648, 1219–1226. https://doi.org/10.1016/j.scitotenv.2018.08.218.

Singh, T.B., Ali, A., Prasad, M., Yadav, A., Shrivastav, P., Goyal, D., Dantu, P.K., 2020. Role of Organic Fertilizers in Improving Soil Fertility Contaminants in Agriculture: Sources, Impacts and Management. Springer International Publishing, India, pp. 61–77. https://doi.org/10.1007/978-3-030-41552-5_3.

Sinha, D., Mukherjee, S., Mahapatra, D., 2021. Multifaceted Potential of Plant Growth Promoting Rhizobacteria (PGPR). IGI Global, pp. 205–268. https://doi.org/10.4018/978-1-7998-7062-3.ch008.

Su, Z., Wen, D., Gu, A.Z., Zheng, Y., Tang, Y., Chen, L., 2023. Industrial effluents boosted antibiotic resistome risk in coastal environments. Environ. Int. 171, 107714. https://doi.org/10.1016/j.envint.2022.107714.

Tasho, R.P., Ryu, S.-H., Cho, J.-Y., 2020. Effect of sulfadimethoxine, oxytetracycline, and streptomycin antibiotics in three types of crop plants—root, leafy, and fruit. Appl. Sci. 10 (3), 1111. https://doi.org/10.3390/app10031111.

Thiele-Bruhn, S., Beck, I.C., 2005. Effects of sulfonamide and tetracycline antibiotics on soil microbial activity and microbial biomass. Chemosphere 59 (4), 457–465. https://doi.org/10.1016/j.chemosphere.2005.01.023.

Timm, S., 2020. The impact of photorespiration on plant primary metabolism through metabolic and redox regulation. Biochem. Soc. Trans. 48 (6), 2495–2504. https://doi.org/10.1042/bst20200055.

Tong, X., Wang, X., He, X., Sui, Y., Shen, J., Feng, J., 2019. Effects of antibiotics on nitrogen uptake of four wetland plant species grown under hydroponic culture. Environ. Sci. Pollut. Control Ser. 26 (11), 10621–10630. https://doi.org/10.1007/s11356-019-04184-4.

Van Boeckel, T.P., Brower, C., Gilbert, M., Grenfell, B.T., Levin, S.A., Robinson, T.P., Teillant, A., Laxminarayan, R., 2015. Global trends in antimicrobial use in food animals. Proc. Natl. Acad. Sci. U. S. A 112 (18), 5649–5654. https://doi.org/10.1073/pnas.1503141112.

Vestel, J., Caldwell, D., Tell, J., et al., 2022. Default predicted no-effect target concentrations for antibiotics in the absence of data for the protection against antibiotic resistance and environmental toxicity. Integr. Environ. Assess. Manag. 18 (4), 863–867. https://doi.org/10.1002/ieam.4560.

Visca, A., Di Gregorio, L., Clagnan, E., Bevivino, A., 2024. Sustainable strategies: nature-based solutions to tackle antibiotic resistance gene proliferation and improve agricultural productivity and soil quality. Environ. Res. 248, 118395. https://doi.org/10.1016/j.envres.2024.118395.

Vuolo, F., Novello, G., Bona, E., Gorrasi, S., Gamalero, E., 2022. Impact of plant-beneficial bacterial inocula on the resident bacteriome: current knowledge and future perspectives. Microorganisms 10 (12), 2462. https://doi.org/10.3390/microorganisms10122462.

Wahid, F., Baig, S., Bhatti, M.F., Manzoor, M., Ahmed, I., Arshad, M., 2021. Growth responses and rubisco activity influenced by antibiotics and organic amendments used for stress alleviation in Lactuca sativa. Chemosphere 264 (Pt 1), 128433. https://doi.org/10.1016/j.chemosphere.2020.128433.

Wang, X., Ryu, D., Houtkooper, R.H., Auwerx, J., 2015. Antibiotic use and abuse: a threat to mitochondria and chloroplasts with impact on research, health, and environment. Bioessays 37 (10), 1045–1053. https://doi.org/10.1002/bies.201500071.

Wu, J., Wang, J., Li, Z., Guo, S., Li, K., Xu, P., Ok, Y.S., Jones, D.L., Zou, J., 2023. Antibiotics and antibiotic resistance genes in agricultural soils: a systematic analysis. Crit. Rev. Environ. Sci. Technol. 53 (7), 847–864. https://doi.org/10.1080/10643389.2022.2094693.

Xiao, R., Huang, D., Du, L., Song, B., Yin, L., Chen, Y., Gao, L., Li, R., Huang, H., Zeng, G., 2023. Antibiotic resistance in soil-plant systems: a review of the source, dissemination, influence factors, and potential exposure risks. Sci. Total Environ. 869, 161855. https://doi.org/10.1016/j.scitotenv.2023.161855.

Yang, J.-F., Ying, G.-G., Zhao, J.-L., Ran, T., Su, H.-C., Liu, Y.-S., 2011. Spatial and seasonal distribution of selected antibiotics in surface waters of the Pearl Rivers, China. J. Environ. Sci. Health B 46 (3), 272–280. https://doi.org/10.1080/03601234.2011.540540.

Yang, K., Zhao, F., Yang, L., Huang, Y., Shen, L., Liu, H., Feng, Q., Chen, L., 2024. Occurrence and dissipation of antibiotics in manure-amended vegetable greenhouse soils under sprinkling irrigation. Environ. Res. Commun. 6 (12), 125024. https://doi.org/10.1088/2515-7620/ad9e8d.

Yang, L., Yuan, C., Chen, X., Xue, W., Cao, G., Meng, S., Bai, L., 2023. The effect of nitrification inhibitors on the aerobic biodegradation of tetracycline antibiotics in swine wastewater. Chemosphere 311 (Pt 1), 136849. https://doi.org/10.1016/j.chemosphere.2022.136849.

Youssef, S.A., Bashour, I.I., 2017. Uptake of Antibiotics by Plants. Springer Nature, pp. 221–237. https://doi.org/10.1007/978-3-319-66260-2_12.

Zafar, R., Bashir, S., Nabi, D., Arshad, M., 2021. Occurrence and quantification of prevalent antibiotics in wastewater samples from Rawalpindi and Islamabad, Pakistan. Sci. Total Environ. 764, 142596. https://doi.org/10.1016/j.scitotenv.2020.142596.

Zafar, R., Zafar Paracha, R., Fahad Alrefaei, A., Albeshr, M.F., Baloch, M.Y.J., Jean, K., Arshad, M., 2024. Removal of four antibiotics from the aqueous phase using Polyethylene adsorbent: mechanistic understanding and reusability potential. Environ. Technol. Innovat. 36, 103831. https://doi.org/10.1016/j.eti.2024.103831.

Zhai, W., Jiang, W., Guo, Q., Wang, Z., Liu, D., Zhou, Z., Wang, P., 2023. Existence of antibiotic pollutant in agricultural soil: exploring the correlation between microbiome and pea yield. Sci. Total Environ. 871, 162152. https://doi.org/10.1016/j.scitotenv.2023.162152.

Zhao, X.D., Qiao, Q.Q., Qin, X.R., Li, X.J., Li, Y.T., 2023. Characteristics of antibiotic contamination of soil in China in past fifteen years and the bioremediation Technology: a review. Environ. Sci. 44 (7), 4059–4076. https://doi.org/10.13227/j.hjkx.202208244.

Zhao, Y., Xiao, X., 2019. Environmental Antibiotics: Exposure Monitoring and Health Endpoints Emerging Chemicals and Human Health. Springer Singapore, China, pp. 165–178. https://doi.org/10.1007/978-981-32-9535-3_10.

Zhou, Z., Huang, F., Chen, L., Liu, F., Wang, B., Tang, J., 2024. Effects of antibiotics on microbial nitrogen cycling and N2O emissions: a review. Chemosphere 357, 142034. https://doi.org/10.1016/j.chemosphere.2024.142034.

Zia ur Rehman, M., Ayub, M.A., Umair, M., Ahmad, Z., Khalid, H., Ali, S., Rizwan, M., 2021. Synthesis and Regulation of Secondary Metabolites in Plants in Conferring Tolerance against Pollutant Stresses Approaches to the Remediation of Inorganic Pollutants. Springer Singapore, Pakistan, pp. 57–76. https://doi.org/10.1007/978-981-15-6221-1_4.

CHAPTER 6

MeHg phytotoxicity and detoxification

Muhammad Shahid[1], Jianxu Wang[2,3], Muhammad Younis[1], Sana Khalid[1] and Behzad Murtaza[1]

[1]Department of Environmental Sciences, COMSATS University Islamabad, Vehari, Pakistan; [2]State Key Laboratory of Environmental Geochemistry, Institute of Geochemistry, Chinese Academy of Sciences, Guiyang, China; [3]Institute of Foundation Engineering, Water- and Waste-Management, Laboratory of Soil- and Groundwater-Management, Pauluskirchstraße, University of Wuppertal, School of Architecture and Civil Engineering, Wuppertal, Germany

1. Introduction

During recent decades, the rapid global economic surge and industrial revolution have caused significant environmental contamination (Shahid et al., 2025). The ever-increasing environmental contamination by different types of toxic pollutants, especially heavy metals, is a serious and widespread global issue (Rafiq et al., 2024). These heavy metals generally induce serious concerns to environmental safety and human health (Khalid et al., 2024). Heavy metals, characterized by high persistence, toxicity, destructiveness, and bioaccumulation potential, are capable of provoking numerous health issues in plants and animals (Shahid, 2021). The elements defined as "heavy metal" vary from 60–70 in number, depending on their definition. However, the most commonly reported heavy metals include mercury (Hg), arsenic (As), cadmium (Cd), lead (Pb), chromium (Cr), and nickel (Ni).

Among the heavy metals, Hg (atomic number 80) is considered of growing public health concern at a global scale (Wang et al., 2012). This metal was previously famous as "quicksilver or hydrargyrum." Mercury is characterized to be a globally priority-controlled pollutant owing to its severe toxic effects on human health and the ecosystem (Hu et al., 2024). Mercury is naturally present in the Earth's crust, mainly found in rocks and minerals (USEPA, 2025). Mercury is a rare earth metal (an average abundance of 0.08 ppm) and is ranked the 66th most abundant element of the Earth's crust (Meyer et al., 2023).

Mercury is emitted into the ecosystem both by natural and anthropogenic sources (Wang et al., 2012; Maqsood et al., 2024; USEPA, 2025). Mining activities are considered the major contributor to Hg environmental contamination. High Hg concentration in surface soil is linked with the atmospheric deposition of elemental Hg and rock weathering (Rizwan et al., 2024). Mercury has high mobility and can be easily transported through the atmosphere and is directly deposited into different environmental compartments such as soil, water, and sediments.

Owing to these sources, the environmental levels of Hg may reach toxic levels, thereby provoking numerous noxious effects on living organisms. Numerous reports

Phytotoxicity of Organic Pollutants
ISBN 978-0-443-33110-7,
https://doi.org/10.1016/B978-0-443-33110-7.00011-0

revealed the toxic levels of this metal in the ecosystem (Santos et al., 2024). Therefore, this metal is listed among the top 10 compounds of greatest concern by the World Health Organization. Moreover, the Agency for Toxic Substances and Disease Registry (ATSDR) has registered Hg as the third priority pollutant. Similarly, Hg is classified as a group 1 environmental pollutant (Zhang et al., 2024).

Plants have the potential to uptake Hg from both soil and the atmosphere (Natasha et al., 2020). Plants can accumulate low to high levels of Hg in different plant parts (Maqsood et al., 2024). The Hg accumulation and compartmentation inside plants vary with the plant type, Hg form, and pathway of entrance (root vs. foliar). There is no known Hg biological role in living organisms, and it is regarded as a nonessential element. Therefore, exposure to even a low level of Hg can harmfully affect the normal physiological, morphological, and biochemical aspects of plants (Chakraborty and Choudhury, 2023). In most cases, the phytotoxicity of Hg is attributed to its forms. High-level build-up of Hg in plants induces negative effects on growth, biomass, photosynthesis, and water/nutritional balance (Chakraborty and Choudhury, 2023). Additionally, Hg is capable of mediating oxidative damage, genotoxicity, and lipid peroxidation. The genotoxic effects of Hg in plants include aberrations to chromosomes and mitotic division, DNA base damage, and effects on DNA repair machinery.

Similar to some other heavy meta(loid)s such as As and Cr, the toxicity of Hg to plants is also governed by its chemical forms (Khalid et al., 2023). Methylmercury (MeHg) is considered the most toxic form of Hg, while inorganic mercury (IHg) is the most predominant form in the environment (Hu et al., 2024). It is reported that the IHg is converted to MeHg, mainly by microbial activity under optimal biochemical conditions in aquatic systems (Zhang et al., 2024). This highly toxic form of Hg can be taken up by plants and thereby induce numerous toxic effects on plants.

MeHg has been the focus of research because of its high potential phytotoxicity and strong phytoaccumulation (Kabata-Pendias, 2011; Maqsood et al., 2024). Mainly, consumption of fish is considered the main human exposure route of MeHg globally. However, the latest data revealed that rice intake also greatly contributes to human exposure to MeHg. This exposure of humans to MeHg through rice consumption is not limited to Hg-contaminated regions but also to non-Hg-contaminated regions (Ran et al., 2022).

Owing to its potential toxicity, MeHg induces numerous toxic effects in plants after being accumulated in plants. Mercury can mediate retardation in plant growth, imbalance of some essential nutrients, decrease in pigment contents, enhanced accumulation of reactive oxygen species (ROS), decrease in photosynthesis, damage to DNA, and even cell death under severe toxicity (Hu et al., 2024; Maqsood et al., 2024).

Under severe Hg or other heavy metal toxicity, plants possess a well-organized tolerance and defense system (Shahid, 2021; Rafiq et al., 2024). This system helps plants to withstand stressful conditions, including Hg exposure and toxicity. The plant defense/tolerance system generally consists of enzymatic and nonenzymatic antioxidants

(Natasha et al., 2022). These antioxidants are capable of neutralizing the reactive radicals produced after Hg exposure to plants. In this way, certain plant species are capable of tolerating Hg toxicity (Santos et al., 2024). However, the Hg-mediated phytotoxicity and the activation of plant tolerance mechanisms are greatly governed by Hg forms inside plants (MeHg vs. IHg). In this book chapter, the main focus is therefore to discuss the recent findings related to Hg (especially MeHg) uptake by plants, its phytotoxicity and tolerance.

2. Occurrence of different Hg forms in the environment

Mercury may exist in the environment in various forms, including organic (mainly as MeHg), inorganic (such as $HgCl_2$, HgS), and elemental (Hg^0) (Natasha et al., 2020). Among these various Hg forms, Hg^0 occurs in liquid form and is the most identified Hg form. The inorganic forms of Hg mainly occur in the salts of HgCl, HgS, and Hg (OH) (O'Connor et al., 2019). The organic Hg forms, such as phenyl mercury, methyl-mercuric chloride (CH_3HgCl), methyl-mercury ion ($HgCH^{3+}$), dimethyl mercury and methyl-mercuric hydroxide (CH_3HgOH), are recognized as the most toxic Hg forms (O'Connor et al., 2019).

Among these organic Hg forms, MeHg is the most predominant in the ecosystem. MeHg constitutes only 2% of the total soil Hg (Ran et al., 2022; Tang and Zhong, 2025). MeHg is highly lipophilic and possesses outstanding binding affinity with sulfur-containing proteins (Zhang et al., 2024). Therefore, MeHg complexed with sulfur ligands has a high potential for plant uptake and thereby transfer to the aquatic food chain. MeHg is the most toxic, having numerous noxious effects on human health, including Parkinson's and Alzheimer's diseases (Mohamadi and Saeedi, 2024). MeHg can provoke extreme neurological damage. The WHO has suggested a maximum tolerable Hg ingestion of 2 μg/kg of body weight per day.

3. Global uses and reserves of Hg

Since the beginning of human civilization, Hg has been categorized as an important metal and is used for numerous human activities. Mercury was used as a pigment by ancient Egyptian, Chinese, and Hindu civilizations (Mahbub et al., 2017). Mercury was also used for centuries in numerous medicines because of its healing properties. Since Hg can contract and expand regularly, it is used in thermostats, blood pressure devices, and thermometers.

Other uses of Hg include dental fillings, fluorescent lighting, batteries, soaps, and paints. The amalgams of Hg are employed for gold extraction from rocks (Usgs, 2025). In the United States, about 65% of the Hg is used for sensors, relays, switches, and valves (Usgs, 2025). About 27% Hg was used for dental amalgam, followed by

7% for formulated products. The manufacturing of catalysts or as cathodes also consumes a huge amount of Hg^0 (about 163 tons).

The USGS data show that the world mine production of Hg was 3700 metric tons in 2020, which decreased to 1200 metric tons in 2024 (Usgs, 2025). In fact, Hg is not produced as a principal mineral commodity in many countries, such as the United States, since 1992. Mercury is mainly recovered as a byproduct from the processing of other metals, such as gold-silver ore. Mercury deposition from the atmosphere in the USA due to power plants reduced by 91% during 2010–2020 (from 6.4–0.55 Mg) (Dai et al., 2023). The reserves of Hg are mainly found in Ukraine, China, Mexico, Kyrgyzstan, Russia, Peru, Spain, and Slovenia. Most of the world's Hg reserves are found in Ukraine (about 600,000 tons) (Usgs, 2025).

4. Environmental sources of Hg

Mercury is emitted into the environment by both natural and human sources (Singh et al., 2023). High levels of Hg radiation (varying from >6000–8000 Mg Hg annually) are considered as a global pollutant that is released in the air by different anthropogenic and natural processes (Maqsood et al., 2024). The key natural sources of Hg include weathering of rocks, volcanic activity, forest fires, vegetable decomposition, leaching, and fossil fuels. In fact, Hg is found naturally in the environment in the form of Hg-bearing minerals found in rocks (Aleku et al., 2024). The commonly found Hg-bearing minerals are meta-cinnabar (β-HgS), cinnabar (HgS), and native Hg. HgS may contain Hg up to 86%. Mercury also exists in several igneous rocks such as dacites, diabases, and basalts, with a mean value ranging from 80 to 100 μg/kg (Usgs, 2025). The sedimentary ores of Mn and Fe also contain Hg (Mahbub et al., 2017). Low levels of Hg (~0.5 mg/kg) occur naturally in the Earth's crust. However, Hg levels in certain regions termed as the "hot spots" may be very high. The natural processes, such as hot springs or volcanic explosions, may release Hg from these "hot spots" (Fig. 6.1).

In addition to natural sources, anthropogenic sources also release Hg into the environment. It is revealed that the anthropogenic activities, primarily since the industrial

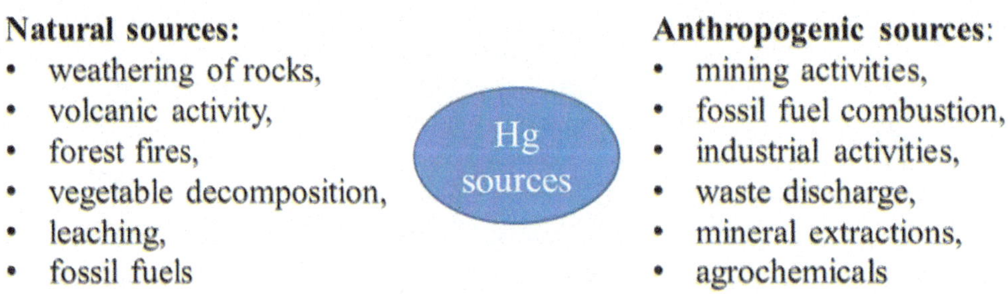

Figure 6.1 Environmental sources of Hg.

revolution, have significantly enhanced Hg environmental release at a global scale (Tang et al., 2023; Santos et al., 2024). About 2000 tons of Hg were released to the environment in 2015 by different human activities (Rizwan et al., 2024). The major anthropogenic Hg sources include mining activities, fossil fuel combustion, industrial activities, waste discharge, mineral extractions, and agrochemicals (Tang et al., 2023). Coal or fossil fuel combustion is considered a major anthropogenic source of Hg. It is proposed that coal-burning plants contribute 65% of Hg emissions and depositions in the environment. Even in the future (by 2050), global Hg release can be up to 96% (Rizwan et al., 2024). Gold mining is reported to contribute 37% of Hg emissions. Moreover, about 50 tons of Hg is emitted by the global electronic waste every year (Tang et al., 2023; Santos et al., 2024).

5. Soil levels of Hg

Mercury is ubiquitous in the environment. Soil acts as "sink or storage" of various pollutants, including Hg. Once released into the atmosphere, Hg can be transferred over long distances before being deposited on land or water surfaces (Aleku et al., 2024). Mercury can linger in the atmosphere from 0.5 to 2 years (Ran et al., 2022). Mercury deposition on soil can occur through wet and dry deposition. It is revealed that about 60% of atmospheric Hg finally deposits on the terrestrial surface and 40% into the aquatic systems (Mahbub et al., 2017). Some studies even estimated that about 75%–80% of Hg deposited from the atmosphere is retained in the soil. The Hg level in soil after atmospheric deposition depends on soil type, contents of organic matter, pH and soil use type.

Numerous studies have reported Hg levels in soil. Mercury is reported to persist and stay in soil for a longer time than in air or water. Soil is the largest sink of Hg with 250–1000 Gg, of which 86 Gg is added by anthropogenic activities (Obrist et al., 2018). This reveals that the human contribution to Hg release and buildup in soil cannot be ignored. High levels of Hg have been reported in different soils of the world. For example, Rizwan et al. (2024) reviewed and reported Hg ranges in different countries as follows: 3.8–11in the USA, 0.35–244 in Slovenia (Rizwan et al., 2024). Owing to high chemical risk, Hg-polluted soils are considered a severe concern for human health (Hu et al., 2023). It is reported that the concentration of MeHg in paddy soil may reach up to 0.27–12.1 ng/g (Guo et al., 2024).

6. Methylation of Hg in paddy soil

In soil, Hg may transform from one form to another, hence it exists in different chemical forms. Mercury biogeochemical behavior (such as mobility and complexation in soil, uptake by plants, compartmentation inside plants and toxicity to plants) mainly depends on its form/speciation (Rizwan et al., 2024). Mercury chemical speciation refers to its

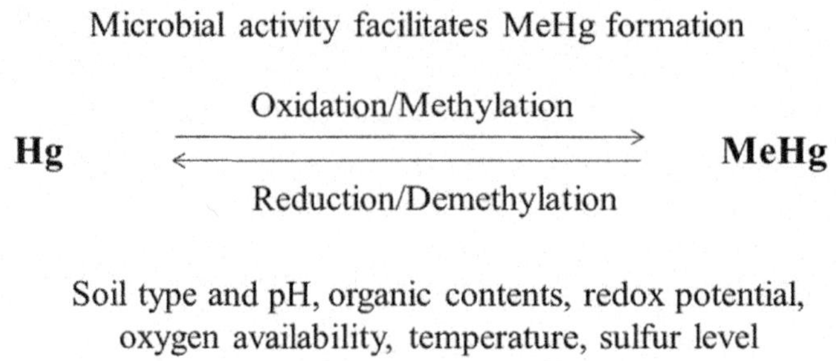

Figure 6.2 Factors affecting methylation of Hg under paddy soil conditions.

existence in soil, mainly in inorganic and organic forms. In soil, especially paddy soil systems having an anoxic (oxygen-poor) environment, the methylation of IHg to MeHg is regarded as the key pathway of Hg speciation transformation (Zhang et al., 2024) (Fig. 6.2).

This Hg speciation transformation can occur through abiotic or biological means. The biotic pathway of Hg methylation is the most important process. Mercury is generally biomethylated by the activity of different types of microorganisms, especially sulfate-reducing bacteria (SRB), iron-reducing bacteria (FeRB), and methanogens. Recently, Tang and Zhong (2025) revealed that FeRB contains *hgcA,* which mediates a key role in MeHg generation in paddy soils. Using data of 67 paddy soils (3600 km), they showed that *hgcA*-containing FeBR exhibited the strongest correlation with Hg methylation capacity, the highest relative abundance (50%), and a strong correlation with iron reduction (Tang and Zhong, 2025).

Microbial activity in soil, especially bacteria, converts Hg^{2+} into CH_3Hg (Rizwan et al., 2024). In fact, sulfur and iron act as precursors for Hg-methylating through interactions between Hg and sulfur. MeHg is highly lipophilic and possesses high binding strength for sulfur-containing proteins such as glutathione and metallothionein (Zhang et al., 2024). Recently in California–USA, Windham-Myers et al. (2014) assessed the live vegetation role on MeHg production under varied agricultural wetlands (wild rice, domesticated or white rice and fallow fields) along with and neighboring managed natural wetlands (cattail- and bulrush or tule-dominated). It was observed that the microbial-assisted formation of MeHg increased by 20%–669% in the presence of live plants for each wetland during the active growing season than devegetated fields. Similarly, Heim et al. (2023) revealed that microbial-assisted MeHg generation was enhanced by decomposing vegetation from 3.41–56.8 ng/g dw. Moreover, 17 weeks after flooding, MeHg in vegetation enhanced from 2.78–45.4 ng/g dw (Heim et al., 2023).

However, not all the species of SRB, FeRB, or methanogens are capable of methylating IHg to MeHg. Moreover, the different types of microorganisms greatly differ in their ability to methylate IHg to MeHg. Even some studies reported a decrease in MeHg concentration after inoculation with sulfate-reducing bacteria. For example, Guo et al. (2024) reported that MeHg production was decreased by 22% and 17% in the two soils by the presence of sulfate-reducing bacteria.

The MeHg formation in the paddy soil system not only depends on the activity of Hg-methylating microbes but also total soil Hg level. It has been shown that soil MeHg level positively correlates with total Hg level. Based on literature data of 51 studies, Rothenberg et al. (2014) found that total Hg (THg) and MeHg levels were higher in polluted areas than nonpolluted areas, confirming higher MeHg production under higher THg conditions. Moreover, the percentage of MeHg to THg remained the same in all the soils (Rothenberg et al., 2014). This shows that only the total level of MeHg increases in soils under increasing THg. Similarly, Guo et al. (2021) assessed the role of varying Hg levels (0.3, 3, 30 mg/kg) on MeHg formation. Under the Hg soil level of 3 mg/kg, there was a great increase in MeHg/THg (1.69%). At 30 mg/kg Hg level, there was a significant reduction in MeHg/THg (0.29%).

Moreover, some Hg forms are more prone to methylation by Hg-methylating microorganisms. In addition to Hg form, different soil characteristics such as soil pH, organic contents and temperature also control MeHg formation. Generally, the process of Hg methylation is favored under acidic and neutral-to-alkaline soil conditions. The management practices may also affect Hg methylation. For example, Rothenberg et al. (2014) performed alternate drying and wetting (interrupting nonstop flooding) and observed spiking in MeHg levels in soil. They revealed increased MeHg pulse after reflooding of fields, thereby confirming variations in MeHg levels due to interrupted drying and flooding. Xie et al. (2023) evaluated the transformation of nano-HgSe and micro-HgSe to MeHg by rice plants under hydroponic conditions. They reported that nano-HgSe was more methylated than micro-HgSe in the rice plant.

7. Uptake and accumulation of MeHg in plants

The threshold level of Hg in soil varies between 0.5–5 mg/kg (Kabata-Pendias, 2011). When Hg is present at high levels in soil, it can mediate severe phytotoxicity. The absorption/uptake of Hg by plants has been reported in numerous studies (Wang et al., 2018; Rizwan et al., 2024). Once inside the root tissues, Hg is transferred to plant shoot tissues through the transpiration stream, thereby causing food-chain contamination. In this way, Hg buildup in edible plant tissues imparts a potential risk to human health. Although Hg fish contamination is considered a major route of human exposure to Hg, the MeHg uptake by rice plants under paddy conditions and consumption by humans is also a major rout of Hg food contamination (Liu et al., 2019). It was predicted that rice plants produced about 180 metric tons of Hg, which is 14-times higher than that exported from the aquatic system through fisheries (Liu et al., 2019).

Mercury is taken up by plants in different forms, such as IHg and MeHg (Hu et al., 2023). Different studies have revealed that uptake and distribution trends of MeHg and IHg in rice seedlings differ greatly. The IHg level in rice plants mainly depends on its concentrations in soil and air (Zhao et al., 2020). In fact, rice has been reported to accumulate IHg from the atmosphere through its leaves and stems. Moreover, rice plants are also able to accumulate MeHg from paddy soil conditions. Hu et al. (2023) revealed that the highest MeHg level was found in rice grains, followed by other parts such as roots > husks > stems > leaves. It was also noted that the level of MeHg in rice grains was much more than IHg.

Under paddy soil conditions, the mobility and uptake of MeHg increase, which causes an increase in MeHg contents in rice grains (Zhang et al., 2024). It is revealed that MeHg is the predominant form in rice. This is because of the fact that MeHg has a high affinity for cysteine and forms MeHg-cysteine complexes (Feng et al., 2021). These MeHg complexes are easily transferred to rice grains (Feng et al., 2021). In this way, high MeHg levels are concentrated in rice grains. However, different studies have reported different trends. For example, the highest IHg and MeHg levels were reported in rice bran than the rice grains and husk (Meng et al., 2014). Nevertheless, the pattern was different for IHg and MeHg. In the case of IHg, rice bran and husk were the main sink, while for MeHg, the accumulation was the highest in white rice grains. This shows that a major part of MeHg was sequestered in the edible tissues of white rice (Meng et al., 2014) and imparts a serious health risk to humans. In fact, MeHg is transported inside rice plants through membrane transport pathways in the endosperm, and therefore has higher mobility inside plants than IHg. Recently, Zhang et al. (2024) reported that Hg concentration in rice was as follows: Roots > third leaf > second leaf - > first leaf > stem > grains. They reported that presence of selenium (Se) in growth medium decreased Hg transfer to aerial tissues of rice.

Reports revealed that MeHg uptake by rice roots and transfer to aerial tissues are highly dynamic. In a hydroponic study, Li et al. (2023) reported that the bioaccumulation factor of Hg was >1, while the transfer factor to aerial parts was <1 in *Arundo donax*. It is shown that rice roots can efficiently absorb MeHg from soil. Thus, revealing that MeHg is more readily absorbed by rice roots than IHg. This varied uptake of MeHg and IHg can be due to their varied complexations. IHg has been reported to be sequestered by phytochelatins but not the MeHg. Moreover, iron plaque in plant roots acts as a key physical barrier against metal transfer toward the shoot. It is revealed that MeHg can easily penetrate iron plaque, signifying high mobility of MeHg inside plants (Zhang et al., 2010). This is a possible reason for increased MeHg accumulation in aerial tissues.

It is reported that the uptake of MeHg in plants, especially rice, is several times higher than IHg. In a study conducted in the Hg mining areas of China, Zhang et al. (2010) collected soil and rice grain samples from 59 sites. They reported that the bioaccumulation factors of MeHg were 0.71–50, which were about 800-folds higher than IHg (ranging from 0.00014 to 0.51). they concluded that rice grains mainly accumulate MeHg in grains, while IHg may get sequestered in roots.

The uptake and compartmentation of MeHg in plants also vary with applied Hg level. Guo et al. (2021) reported that varying Hg levels (0, 0.3, 3, and 30 mg/kg) in paddy soil had different effects on MeHg accumulation in rice. The presence of 3 mg Hg in 1 kg of soil greatly increased MeHg levels in rice seeds, and significantly decreased MeHg contents in roots, stalks, and leaves. Increasing soil Hg level to 30 mg/kg slightly sequestered MeHg in seeds, with high MeHg levels in the husk. Moreover, the presence of microbes greatly affects MeHg uptake by roots and transfer to aerial tissues. Li et al. (2023) reported that the inoculation of arbuscular mycorrhizal fungi greatly decreased Hg levels in rice, varying from 53% to 96%. It was reported that MeHg accumulation in edible parts was reduced by triggering its buildup in nonedible rice tissues such as stems and leaves. The MeHg level in rice grains decreased by 10%–28%. Guo et al. (2024) reported that sulfate-reducing bacteria reduced MeHg production by 22% and 17% in two types of soil. However, MeHg concentration in rice grains increased by 55% and 99% in two soils (Guo et al., 2024).

Application of sulfur also decreases MeHg buildup in rice plants (Huang et al., 2024). Furthermore, the presence of Se can affect MeHg formation in soil and thereafter its mobility, uptake and compartmentation inside plants. For example, Xie et al. (2023) assessed phytouptake of MeHg by exposing rice to nano-HgSe and micro-HgSe under hydroponic conditions results revealed a higher level of nano-HgSe in aerial tissues of rice than micro-HgSe. Moreover, a small percentage (1.2%) of MeHg to THg was detected in rice stem, signifying a possible decomposition of nano-HgSe (Xie et al., 2023). Tian et al. (2023) reported biochar-mediated enhanced MeHg generation (19.01%) in paddy soil. The available and dissolved MeHg concentration in paddy soil decreased, respectively, by 75.69% and 88.64%. Tian et al. (2023) also revealed that the presence of biochar in soil decreased (61.10%) MeHg accumulation in rice. Application of biochar derived from rice hull immobilizes MeHg by decreasing its level in pore water (Xing et al., 2020). However, an increase in Hg mobilization and methylation in soil was reported by Xing et al. (2022). Moreover, MeHg accumulation in rice was decreased by biochar. Similarly, Boie et al. (2024) reported biochar mediated methylation of Hg in paddy soil.

8. Food chain contamination by MeHg

One of the biggest challenges the world is facing today is food security. During the past few decades, food-chain contamination by different types of pollutants, including MeHg, has become common. The above-mentioned studies revealed that MeHg can be easily up-taken by plants, especially by rice. Most importantly, MeHg is mainly accumulated in rice grains, thereby posing a serious threat to human exposure to MeHg and associated health hazards. It is revealed that rice seedlings have a high capacity to absorb and accumulate MeHg from paddy soil conditions. Rice is considered a major food source for >3 billion people globally. Rice is consumed as a staple food by 50% of the world's habitants (Li et al., 2010). In Central Asia where rice is the major food source, the human exposure to MeHg through rice consumption contributes about 98% (15a). Because of growing international rice trade, there is an increased risk of MeHg transfer and exposure between different countries, especially in Central Asia, Africa and Europe (Liu et al., 2019).

In most cases, the Hg level in foodstuffs is <20 ng/g, and primarily occurs as IHg (Li et al., 2010). Wang et al. (2020) determined THg and MeHg level in 172 rice samples in Bangladesh. They reported 0.42–14.4 ng/g of THg, while 0.026–7.47 ng/g of MeHg in all rice samples. However, recent literature reveals that MeHg is the main Hg form reported in rice. The MeHg level in rice grains may vary from 0.69 to 144 ng/g. Even some reports showed higher MeHg levels in rice grains near the Wanshan Hg mining area (174 ng/g) (Feng et al., 2021). The threshold limit set for MeHg in rice grains is (20 ng/g). This reveals that rice grain can accumulate several-fold higher concentrations of MeHg than the threshold limit. Liu et al. (2019) revealed that MeHg was mainly generated in rice grain by East, South, and Southeast Asia, respectively by 0.45, 0.77, and 0.42 Mg, which accounts for 91% of global MeHg production. Moreover, the generation of MeHg in rice plants has greatly enhanced in many parts of the world, including South and Southeast Asia, Africa, North America, and South America.

Hence, consumption of MeHg through rice grains can be a major route of human exposure to MeHg, especially in South and Southeast Asia. In China, 26% of human exposure to MeHg is through rice ingestion. The trade of rice across borders has caused human exposure to MeHg about 9.9% (Liu et al., 2019). There was also a significant increase in MeHg human exposure through rice consumption in Central Asia (98%), Africa (62%), and Europe (42%) (Liu et al., 2019). In 2013, India was listed as a major country exporting MeHg through international rice trade, and was recognized as a key net source of global MeHg exports (Liu et al., 2019). However, the major pathway of human exposure to MeHg is through fish ingestion (approximately 56%) in China. MeHg is highly toxic to humans. MeHg to children can interrupt long-term development and can provoke cardiovascular issues in adults (Guo et al., 2024).

9. Toxicity of MeHg to plants

Inside plants, MeHg can induce several toxic effects. The toxic effects of MeHg to plants include reduced biomass production and stunted growth, disruption of photosynthetic processes, interference with water and essential nutrient phytouptake and chlorophyll synthesis, induction of oxidative stress, damage to cellular components, binding to DNA and causing genetic mutation, and altering antioxidant enzyme activity (Natasha

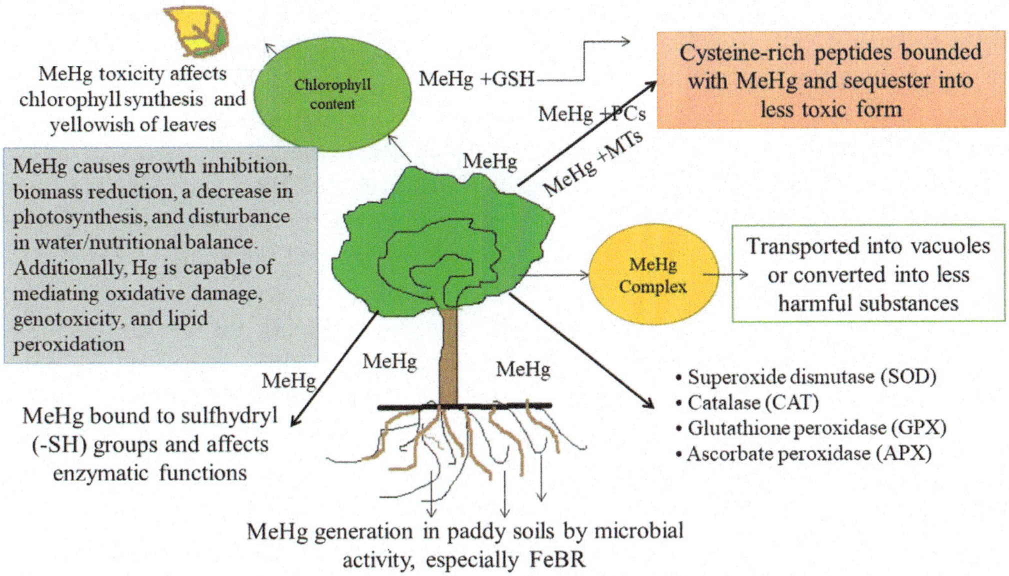

Figure 6.3 MeHg toxicity and detoxification in plants.

et al., 2022). MeHg can cause chlorosis (yellow leaves). Exposure to MeHg has been reported to induce a decrease in growth rate, enhanced ROS production, chlorophyll degradation and effects on plasma membrane permeability (Santos et al., 2024). These toxic effects of MeHg to plants have been reported in numerous studies (Fig. 6.3).

MeHg has been well-reported to affect the growth of plants through various mechanisms and processes. For example, recently, Rajendran et al. (2024) reviewed the following toxic effects of Hg on plants: decreased photosynthetic rate, plant heights, root and shoot growth, biomass, relative water contents, pigment contents, stomatal conductance, and carbon assimilation. Xie et al. (2023) reported that rice growth was reduced after exposure to nano-HgSe and micro-HgSe, which were methylated (MeHg) inside plants. In fact, MeHg is capable of interfering with pigment content formation and photosynthesis processes. In this way, MeHg affects the plant growth. A decrease in chlorophyll contents of rice plants after exposure to Hg has been reported (Xie et al., 2023). MeHg caused a decrease in the light-saturated CO_2 assimilation rate of rice leaves (Mao et al., 2021). Beauvais-Flück et al. (2018) reported that MeHg exposure caused up-regulation of contigs associated with the photosystem II biogenesis.

Exposure to MeHg has been reported to cause oxidative damage, mainly through enhanced production of ROS. For example, Rajendran et al. (2024) reported Hg-mediated increased ROS and malondialdehyde concentration in plants. Xie et al. (2023) evaluated the toxic effects of nano-HgSe and micro-HgSe on rice plants under hydroponic conditions. They reported that nano-HgSe exposure caused oxidative damage, mainly due to enhanced MeHg generation than nano-HgSe. The presence of Se in the growth medium minimizes the toxic effects of MeHg on plants. For example, Zhang et al. (2024) exposed high and low Se genotypes to varied levels of Hg and Se and reported a significant decrease in malondialdehyde (MDA) contents and oxidative damage in high Se genotypes. Huang et al. (2024) revealed that sulfur caused a decrease in Hg-mediated ROS generation and lipid peroxidation in rice. MeHg exposure also reduced GSH-Px activity in rice plants (Xie et al., 2023).

Studies also reported Hg-mediated up-regulation or down-regulation of genes. Beauvais-Flück et al. (2018) exposed *Elodea nuttallii* to MeHg for 2 h and reported the up-regulation of 4389 genes, while down-regulating 16853 genes. These genes were related to amino acid, sugar, and secondary metabolism. After MeHg exposure, there was a regulation of up to 18′557 contigs for 1.54 μg/g dw and 16′853 for 18.3 μg/g dw (Beauvais-Flück et al., 2018). These contigs were mainly involved in gene regulation (protein, signaling, RNA, and DNA). MeHg can bind with p62 of DNA, thereby causing DNA mutation (Kiyono et al., 2023). İskil et al. (2022) evaluated the effect of Hg (5, 10, and 20 μM) on biochemical and molecular alterations in *Triticum aestivum* after 24 and 48 h of exposure. They reported a significant augmentation in ROS generation, MDA and proline levels after Hg exposure.

However, some studies have revealed that the toxic effects of MeHg may increase under extreme environmental conditions. For example, Tang et al. (2023) reported that water deficit conditions aggravated Hg stress to plants by enhancing Hg contents in plants, enhancing ROS levels, provoking thylakoid membrane damage, and reducing pigment contents. Moreover, the majority of the down-regulated genes were related to the pathways of chlorophyll, photosynthesis and (PsbS1, PSBQ1 and FDX1 etc.). Overall, Hg exposure with water-deficient conditions reduced photosynthetic efficacy and severely affected electron transport and light absorption.

Some studies reported no effect or even an increase in plant growth under low Hg exposure. For example, pigment contents of *Elodea nuttallii* remained unaffected after MeHg exposure (Beauvais-Flück et al., 2018). Microbial activity affected plant health and growth by mediating Hg methylation in paddy soil (Guo et al., 2023). There was an increase in plant development from 48%–50% (Guo et al., 2023). Li et al. (2023) performed a hydroponic study to assess the individual and combined effects of Hg and Cd stress on *Arundo donax*. Exposure to individual Hg marginally increased fresh biomass, stem height, and pigment contents, while combined Hg–Cd exposure decreased these plant growth variables. Moreover, Hg–Cd exposure caused oxidative stress by enhancing ROS and MDA contents.

Recent studies have also reported that the presence of some amendments in the growth medium decreases the toxic effects of MeHg to plants. For example (Zhang et al., 2024) reported Se-mediated decrease in MeHg toxicity to rice plants in term of MDA accumulation and superoxide dismutase (SOD) antioxidant activity. When Se was applied 1.0 mg/kg soil, there was a significant decrease in MeHg contents inside plants and improvements in physiological and agronomic characteristics of rice. Guo et al. (2021) reported that exposure to MeHg caused variations in chlorophyll contents, ROS, MDA and antioxidant enzyme activities in young and old leaves. An increase in MDA contents under Hg exposure in rice was also reported by Ren et al. (2014).

10. Detoxification of MeHg in plants

Plants are exposed to various stresses naturally and anthropogenically. Because of their sessile nature, plants have to cope with all types of stresses for their survival. Plants are naturally equipped with various defense and tolerance strategies (including biochemical, morphological, and genetic in nature) to cope with different types of biotic and abiotic stresses. Plant exposure to high levels of MeHg, they can survive its toxic effects up to a certain level using their defense mechanisms (Singh et al., 2023). Generally, the defense and tolerance strategies of plants include extrusions, chelations, regulatory disseminations, and vacuolar compartmentations (Podar and Maathuis, 2022) (Fig. 6.4).

These defense and tolerance strategies are operated inside plants by different types of transporter proteins and enzymatic and nonenzymatic antioxidants. The key antioxidant enzymes responsible to detoxify heavy metals (such as MeHg) include glutathione peroxidase (GPX), peroxidase (POD), glutathione reductase (GR), catalase (CAT), SOD, and ascorbate peroxidase (APX). Similarly, different types of nonenzymatic antioxidants such as phytochelatins (PCs), glutathione (GSH), carotenoids, ascorbic acid (AsA), prolines, and α-tocopherol also play key role in metal tolerance. Additionally, various Hg-transporter proteins such as ABC transporters, ATP-binding cassette (ABC) transporters and other bacteria-derived transporter proteins are involved in Hg transport and compartmentalization inside plants. Beauvais-Flück et al. (2018) showed MeHg-mediated up- and down-regulation of ATPases. Huang et al. (2024) reported that the presence of sulfur alleviated Hg-mediated oxidative stress by enhancing glutathione content in leaves. Therefore, the antioxidant enzymes get down-regulated.

The activation of antioxidant enzymes inside plants controls the level of free radicals such as ROS. Plant exposure to heavy metals, including MeHg, causes enhanced

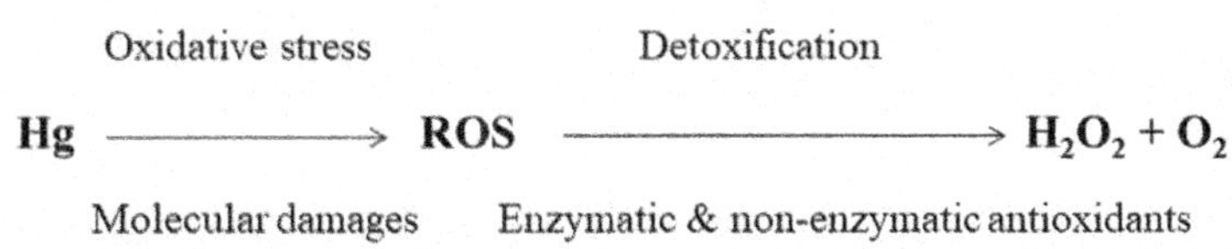

Figure 6.4 Hg-induced oxidative stress through enhanced ROS production and its detoxification by enzymatic and nonenzymatic antioxidants.

production of superoxide radical ($O_2^{\cdot-}$) (Shahid et al., 2017). The SOD enzyme causes dismutation of $O_2^{\cdot-}$ and transforms it to hydrogen peroxide (H_2O_2) (Natasha et al., 2022). This form of ROS (H_2O_2) is scavenged by different enzymes such as CAT, APX, POD, and GPX, thereby converting them to H_2O (Natasha et al., 2020). The pathways and mechanisms of H_2O_2 conversion to H_2O vary for different enzymes. For example, Ascorbate is used as a substrate by APX to produce H_2O and O_2 from H_2O_2. While GSH is employed by GPX for this conversion. In this way, antioxidant enzymes alleviate ROS-induced oxidative damage in plants.

Various studies have reported the activation of antioxidant enzymes in plants under MeHg stress. For example, Zhang et al. (2024) reported Hg-mediated increase in SOD antioxidant activity in the presence of Se. Santos et al. (2024) revealed that activation in the antioxidant metabolites of glutathione was reduced after MeHg treatments. İskil et al. (2022) exposed *Triticum aestivum* for 24 and 48 h to varying levels of Hg (5, 10, and 20 μM) and observed a decrease in SOD and CAT activity. Guo et al. (2021) revealed MeHg-mediated alterations in antioxidant enzyme activities in young and old leaves of rice. Zhang et al. (2024) reported enhanced SOD activity in high Se-genotypes than low Se-genotypes to mitigate MeHg stress.

(İskil et al., 2022) showed that the expression of Fe-SOD, Cu/Zn-SOD, TaDREB1 and TaWRKY19 genes was declined by Hg in *Triticum aestivum*. However, the expression of the CAT gene showed a significant increase in leaves of *Triticum aestivum*. In contrast, a varying pattern of decrease and increase was observed of TaWRKY2 and Mn-SOD genes. They concluded that *Triticum aestivum* can withstand Hg toxicity by altering the induction of various genes encoding. Huang et al. (2024) showed that Hg exposure upregulated genes associated with biological processes. Using physiological and transcriptomic evidence, they concluded that sulfur helps plants cope with Hg toxicity. In fact, sulfur is capable of improving the synthesis of thiol-containing molecules in rice plants, including PCs, GSH, metallothioneins, and nonprotein thiols (NPTs) (Huang et al., 2024).

11. Conclusions

The literature data of the current book chapter reveal that mercury (Hg) is greatly released into the environment by various natural and anthropogenic. In the soil, Hg is methylated to the organic form of methylmercury (MeHg). This methylation is mainly attributed to microbial activity under paddy soil conditions. Therefore, abundant MeHg is generated globally inside rice plants. This organic form of Hg is reported to be highly phytoavailable in soil and mobile inside plants. MeHg is mainly accumulated in rice grains compared with other rice tissues. The uptake and compartmentation mechanisms of MeHg are different from inorganic Hg (IHg). MeHg is also highly toxic and induces numerous toxic effects in plants. The key toxic effects of MeHg to plants include low biomass production and stunted growth, disturbance in photosynthetic processes, interference with water and nutrient uptake, and chlorophyll synthesis reduction. MeHg can induce oxidative stress through enhanced generation of ROS, thereby damaging key cellular components such as RNA, DNA, lipids and proteins. MHg can also cause genetic mutation and alter antioxidant enzyme activity. However, plants can cope with MeHg stress through various tolerance and detoxification mechanisms. These tolerance mechanisms are mainly governed by enzymatic and nonenzymatic antioxidants. In contrast to other heavy metals such as Pb, Cd, As, and Ni, very limited data are available related to biogeochemical behavior (mobility, phytoavailability, compartmentation, toxicity and tolerance) of Hg, especially MeHg, in the soil-plant system.

Acknowledgments

Prof. Dr. Jianxu Wang acknowledges ANSO project "Funded by CAS-ANSO Fellowship. Grant No.CAS-ANSO-FS-2024-06."

References

Aleku, D.L., Lazareva, O., Pichler, T., 2024. Mercury in groundwater—source, transport and remediation. Appl. Geochem. 170.

Beauvais-Flück, R., Slaveykova, V.I., Cosio, C., 2018. Effects of two-hour exposure to environmental and high concentrations of methylmercury on the transcriptome of the macrophyte Elodea nuttallii. Aquat. Toxicol. 194, 103–111. https://doi.org/10.1016/j.aquatox.2017.11.010.

Boie, F., Ducey, T.F., Xing, Y., Wang, J., Rinklebe, J., 2024. Field-aged rice hull biochar stimulated the methylation of mercury and altered the microbial community in a paddy soil under controlled redox condition changes. J. Hazard Mater. 472, 134446. https://doi.org/10.1016/j.jhazmat.2024.134446.

Chakraborty, D., Choudhury, B., 2023. Toxic effects of mercury on crop plants and its physiological and biochemical responses—a review. Int. J. Adv. Res. 11 (02), 168–174. https://doi.org/10.21474/ijar01/16236.

Dai, M.Q., Geyman, B.M., Hu, X.C., Thackray, C.P., Sunderland, E.M., 2023. Sociodemographic disparities in mercury exposure from United States coal-fired power plants. Environ. Sci. Technol. Lett. 10 (7), 589–595. https://doi.org/10.1021/acs.estlett.3c00216.

Feng, L., Li, P., Feng, X., 2021. Methylmercury bioaccumulation in rice and health effects: a systematic review. Curr. Opin. Environ. Sci. Health 23, 100285. https://doi.org/10.1016/j.coesh.2021.100285.

Guo, P., Rennenberg, H., Du, H., Wang, T., Gao, L., Flemetakis, E., Hänsch, R., Ma, M., Wang, D., 2023. Bacterial assemblages imply methylmercury production at the rice-soil system. Environ. Int. 178, 108066. https://doi.org/10.1016/j.envint.2023.108066.

Guo, P., Du, H., Wang, D., Ma, M., 2021. Effects of mercury stress on methylmercury production in rice rhizosphere, methylmercury uptake in rice and physiological changes of leaves. Sci. Total Environ. 765, 142682. https://doi.org/10.1016/j.scitotenv.2020.142682.

Guo, P., Du, H., Mao, Q., Deng, Y., Wang, X., Li, J., Xiong, B., Fan, X., Wang, D., Agathokleous, E., Ma, M., 2024. The duality of sulfate-reducing bacteria: reducing methylmercury production in rhizosphere but enhancing accumulation in rice plants. J. Hazard Mater. 476, 135049. https://doi.org/10.1016/j.jhazmat.2024.135049.

Heim, W.A., Bosworth, D., DiGiorgio, C., Stephenson, M., Gill, G., 2023. Effects of vegetation on methylmercury concentrations and loads in a mercury contaminated floodplain. Sci. Total Environ. 901, 165864. https://doi.org/10.1016/j.scitotenv.2023.165864.

Hu, H., Gao, Y., Yu, H., Xiao, H., Chen, S., Tan, W., Tang, J., Xi, B., 2023. Mechanisms and biological effects of organic amendments on mercury speciation in soil–rice systems: a review. Ecotoxicol. Environ. Saf. 251, 114516. https://doi.org/10.1016/j.ecoenv.2023.114516.

Hu, W.Y., Mao, H.T., Yin, X.Y., Chen, J.Y., He, A.Q., Huang, L.Y., Zhang, Z.W., Yuan, S., Yuan, M., Su, Y.Q., Chen, Y.E., 2024. Melatonin alleviates Hg toxicity by modulating redox homeostasis and the urea cycle in moss. Sci. Total Environ. 907. https://doi.org/10.1016/j.scitotenv.2023.167958. Available from. www.elsevier.com/locate/scitotenv.

Huang, Y., Yi, J., Li, X., Li, F., 2024. Transcriptomics and physiological analyses reveal that sulfur alleviates mercury toxicity in rice (Oryza sativa L.). J. Environ. Sci. 135, 10–25. https://doi.org/10.1016/j.jes.2023.01.001.

İskil, R., Surgun-Acar, Y., Çatav, S.S., Zemheri-Navruz, F., Erden, Y., 2022. Mercury toxicity affects oxidative metabolism and induces stress responsive mechanisms in wheat (*Triticum aestivum* L.). Physiol. Mol. Biol. Plants 28 (4), 911–920. https://doi.org/10.1007/s12298-022-01171-x.

Kabata-Pendias, A., 2011. Trace Elements in Soils and Plants. CRC Press.

Khalid, S., Shahid, M., Bibi, I., Nadeem, H.M., Younis, M., Natasha, N., Murtaza, B., Niazi, N.K., 2024. Assessing the arsenic contents and associated risks in groundwater of Vehari and Lodhran districts, Pakistan. Water (Switzerland) 16 (21). https://doi.org/10.3390/w16213055. Available from: http://www.mdpi.com/journal/water.

Khalid, S., Shahid, M., Alothman, Z.A., Al-Kahtani, A.A., Murtaza, B., Dumat, C., 2023. Predicting chemical speciation of metals in soil using Visual Minteq. Soil Ecol. Lett. 5 (3). https://doi.org/10.1007/s42832-022-0162-2. Available from: https://www.springer.com/journal/42832.

Kiyono, M., Ohshiro, Y., Nakamura, R., Uraguchi, S., Takanezawa, Y., 2023. Molecular toxicology of methylmercury and phytoremediation of toxic metals for human health. BPB Rep. 6 (4), 126–132. https://doi.org/10.1248/bpbreports.6.4_126.

Li, P., Feng, X., Qiu, G., 2010. Methylmercury exposure and health effects from rice and fish consumption: a review. Int. J. Environ. Res. Publ. Health 7 (6), 2666–2691. https://doi.org/10.3390/ijerph7062666.

Li, X., Zhao, L., Teng, Y., Luo, Y., Zhao, Q., 2023. Effects of cadmium on mercury accumulation and transformation by Arundo donax L. Environ. Sci. Pollut. Control Ser. 30 (22), 62461–62469. https://doi.org/10.1007/s11356-023-26516-1.

Liu, M., Zhang, Q., Cheng, M., He, Y., Chen, L., Zhang, H., Cao, H., Shen, H., Zhang, W., Tao, S., Wang, X., 2019. Rice life cycle-based global mercury biotransport and human methylmercury exposure. Nat. Commun. 10 (1). https://doi.org/10.1038/s41467-019-13221-2.

Mahbub, K.R., Krishnan, K., Naidu, R., Andrews, S., Megharaj, M., 2017. Mercury toxicity to terrestrial biota. Ecol. Indic. 74, 451–462. https://doi.org/10.1016/j.ecolind.2016.12.004.

Mao, Q., Tang, L., Ji, W., Rennenberg, H., Hu, B., Ma, M., 2021. Elevated CO2 and soil mercury stress affect photosynthetic characteristics and mercury accumulation of rice. Ecotoxicol. Environ. Saf. 208, 111605. https://doi.org/10.1016/j.ecoenv.2020.111605.

Maqsood, M., Zahra, N., Kausar, A., Shahzad, S., Batool, A., Naseer, R., 2024. Mercury Contamination and It's Dynamics in Soil–Plant Systems. Springer Science and Business Media LLC, pp. 45–63. https://doi.org/10.1007/978-3-031-48817-7_2.

Meng, B., Feng, X., Qiu, G., Anderson, C.W.N., Wang, J., Zhao, L., 2014. Localization and speciation of mercury in Brown rice with implications for Pan-Asian public health. Environ. Sci. Technol. 48 (14), 7974–7981. https://doi.org/10.1021/es502000d.

Meyer, L., Guyot, S., Chalot, M., Capelli, N., 2023. The potential of microorganisms as biomonitoring and bioremediation tools for mercury-contaminated soils. Ecotoxicol. Environ. Saf. 262, 115185. https://doi.org/10.1016/j.ecoenv.2023.115185.

Mohamadi, S., Saeedi, M., 2024. Review and assessment of mercury contamination and risk in sediment and food chain near a chlor-alkali plant. Emerging Contam. 10 (4), 100384. https://doi.org/10.1016/j.emcon.2024.100384.

Natasha, M.S., Khalid, S., Bibi, I., Bundschuh, J., Khan Niazi, N., Dumat, C., 2020. A critical review of mercury speciation, bioavailability, toxicity and detoxification in soil-plant environment: ecotoxicology and health risk assessment. Sci. Total Environ. 711. https://doi.org/10.1016/j.scitotenv.2019.134749. Available from. www.elsevier.com/locate/scitotenv.

Natasha, N., Shahid, M., Bibi, I., Iqbal, J., Khalid, S., Murtaza, B., Bakhat, H.F., Farooq, A.B.U., Amjad, M., Hammad, H.M., Niazi, N.K., Arshad, M., 2022. Zinc in soil-plant-human system: a data-analysis review. Sci. Total Environ. 808. https://doi.org/10.1016/j.scitotenv.2021.152024. Available from: www.elsevier.com/locate/scitotenv.

Obrist, D., Kirk, J.L., Zhang, L., Sunderland, E.M., Jiskra, M., Selin, N.E., 2018. A review of global environmental mercury processes in response to human and natural perturbations: changes of emissions, climate, and land use. Ambio 47 (2), 116–140. https://doi.org/10.1007/s13280-017-1004-9.

O'Connor, D., Hou, D., Ok, Y.S., Mulder, J., Duan, L., Wu, Q., Wang, S., Tack, F.M.G., Rinklebe, J., 2019. Mercury speciation, transformation, and transportation in soils, atmospheric flux, and implications for risk management: a critical review. Environ. Int. 126, 747–761. Available from: www.elsevier.com/locate/envint. doi: 10.1016/j.envint.2019.03.019.

Podar, D., Maathuis, F.J.M., 2022. The role of roots and rhizosphere in providing tolerance to toxic metals and metalloids. Plant Cell Environ. 45 (3), 719–736. https://doi.org/10.1111/pce.14188.

Rafiq, M., Shahid, M., Bibi, I., Khalid, S., Tariq, T.Z., Al-Kahtani, A.A., Alothman, Z.A., Murtaza, B., Niazi, N.K., 2024. Role of organic and inorganic amendments on physiological attributes of germinating pea seedlings under arsenic stress. Int. J. Phytoremediation 26 (8), 1243–1252. https://doi.org/10.1080/15226514.2024.2305684.

Rajendran, S., Kamaludeen, S.P.B., Subramanian, A., 2024. Antioxidant Defense Mechanism in Plants Exposed to Mercury Toxicity: Response, Tolerance and Remediation. Springer Science and Business Media LLC, pp. 225–245. https://doi.org/10.1007/978-3-031-48817-7_10.

Ran, S., He, T., Zhou, X., Yin, D., 2022. Effects of fulvic acid and humic acid from different sources on Hg methylation in soil and accumulation in rice. J. Environ. Sci. 119, 93–105. https://doi.org/10.1016/j.jes.2022.02.023.

Ren, J.H., Sun, H.J., Wang, S.F., Luo, J., Ma, L.Q., 2014. Interactive effects of mercury and arsenic on their uptake, speciation and toxicity in rice seedling. Chemosphere 117 (1), 737–744. https://doi.org/10.1016/j.chemosphere.2014.10.035.

Rizwan, M., Murtaza, G., Zulfiqar, F., Moosa, A., Iqbal, R., Ahmed, Z., Khan, I., Siddique, K.H.M., Leng, L., Li, H., 2024. Tuning active sites on biochars for remediation of mercury-contaminated soil: a comprehensive review. Ecotoxicol. Environ. Saf. 270, 115916. https://doi.org/10.1016/j.ecoenv.2023.115916.

Rothenberg, S.E., Windham-Myers, L., Creswell, J.E., 2014. Rice methylmercury exposure and mitigation: a comprehensive review. Environ. Res. 133, 407–423. https://doi.org/10.1016/j.envres.2014.03.001.

Santos, J.P., Li, W., Keller, A.A., Slaveykova, V.I., 2024. Mercury species induce metabolic reprogramming in freshwater diatom Cyclotella meneghiniana. J. Hazard Mater. 465. https://doi.org/10.1016/j.jhazmat.2023.133245. Available from: https://www.sciencedirect.com/science/journal/03043894.

Shahid, M., Dumat, C., Khalid, S., Niazi, N.K., Antunes, P.M.C., 2017. Cadmium bioavailability, uptake, toxicity and detoxification in soil-plant system. Rev. Environ. Contam. Toxicol. 241, 73–137. https://doi.org/10.1007/398_2016_8.

Shahid, M., Khalid, S., Alothman, Z.A., Al-Kahtani, A.A., Bibi, I., Naz, R., Natasha, N., Niazi, N.K., Iqbal, J., Han, C., Shah, N.S., Murtaza, B., 2025. Trace element removal from wastewater by agricultural biowastes: a data analysis on removal efficacy and optimized conditions. Sci. Total Environ. 975. https://doi.org/10.1016/j.scitotenv.2025.179235. Available from: https://www.sciencedirect.com/science/journal/00489697.

Shahid, M., 2021. Effect of soil amendments on trace element-mediated oxidative stress in plants: meta-analysis and mechanistic interpretations. J. Hazard Mater. 407, 124881. https://doi.org/10.1016/j.jhazmat.2020.124881.

Singh, A.D., Khanna, K., Kour, J., Dhiman, S., Bhardwaj, T., Devi, K., Sharma, N., Kumar, P., Kapoor, N., Sharma, P., Arora, P., Sharma, A., Bhardwaj, R., 2023. Critical review on biogeochemical dynamics of mercury (Hg) and its abatement strategies. Chemosphere 319, 137917. https://doi.org/10.1016/j.chemosphere.2023.137917.

Tang, W., Zhong, H., 2025. Developing methylmercury-targeted strategies to safeguard rice consumers. Environ. Health 3 (3), 213–217. https://doi.org/10.1021/envhealth.4c00257.

Tang, Y., Zhang, J., Wang, L., Wang, H., Long, H., Yang, L., Li, G., Guo, J., Wang, Y., Li, Y., Yang, Q., Shi, W., Shao, R., 2023. Water deficit aggravated the inhibition of photosynthetic performance of maize under mercury stress but is alleviated by brassinosteroids. J. Hazard Mater. 443 (Pt B), 130365. https://doi.org/10.1016/j.jhazmat.2022.130365.

Tian, X., Chai, G., Xie, Q., Li, G., 2023. Response of methylmercury in paddy soil and paddy rice to pristine biochar: a meta-analysis and environmental implications. Ecotoxicol. Environ. Saf. 257, 114933. https://doi.org/10.1016/j.ecoenv.2023.114933.

USEPA, :~:text=Mercury%20is%20a%20naturally%2Doccurring,Elemental%20(metallic)%20mercury Assessed Online, 2025.

Usgs, Mercury Statistics and Information, 2025.

Wang, J., Feng, X., Anderson, C.W.N., Xing, Y., Shang, L., 2012. Remediation of mercury contaminated sites—a review. J. Hazard Mater. 221–222, 1–18. https://doi.org/10.1016/j.jhazmat.2012.04.035.

Wang, J., Anderson, C.W.N., Xing, Y., Fan, Y., Xia, J., Shaheen, S.M., Rinklebe, J., Feng, X., 2018. Thiosulphate-induced phytoextraction of mercury in Brassica juncea: spectroscopic investigations to define a mechanism for Hg uptake. Environ. Pollut. 242, 986–993. https://doi.org/10.1016/j.envpol.2018.07.065.

Wang, Y., Habibullah-Al-Mamun, Md, Han, J., Wang, Le, Zhu, Y., Xu, X., Li, N., Qiu, G., 2020. Total mercury and methylmercury in rice: exposure and health implications in Bangladesh. Environ. Pollut. 265 (Pt A), 114991. https://doi.org/10.1016/j.envpol.2020.114991.

Windham-Myers, L., Marvin-DiPasquale, M., Stricker, C.A., Agee, J.L., Kieu, L.H., Kakouros, E., 2014. Mercury cycling in agricultural and managed wetlands of California, USA: experimental evidence of vegetation-driven changes in sediment biogeochemistry and methylmercury production. Sci. Total Environ. 484 (1), 300–307. https://doi.org/10.1016/j.scitotenv.2013.05.028.

Xie, H., He, L., Tian, X., Zhang, W., Cui, L., Shang, L., Zhao, J., Bai, L., Li, Y.-F., 2023. Nano mercury selenide as a source of mercury for rice. Environ. Pollut. 318, 120918. https://doi.org/10.1016/j.envpol.2022.120918.

Xing, Y., Wang, J., Kinder, C.E.S., Yang, X., Slaný, M., Wang, B., Song, H., Shaheen, S.M., Leinweber, P., Rinklebe, J., 2022. Rice hull biochar enhances the mobilization and methylation of mercury in a soil under changing redox conditions: implication for Hg risks management in paddy fields. Environ. Int. 168. https://doi.org/10.1016/j.envint.2022.107484.

Xing, Y., Wang, J., Shaheen, S.M., Feng, X., Chen, Z., Zhang, H., Rinklebe, J., 2020. Mitigation of mercury accumulation in rice using rice hull-derived biochar as soil amendment: a field investigation. J. Hazard Mater. 388, 121747. https://doi.org/10.1016/j.jhazmat.2019.121747

Zhang, H., Feng, X., Larssen, T., Shang, L., Li, P., 2010. Bioaccumulation of methylmercury versus inorganic mercury in rice (Oryza sativa L.) grain. Environ. Sci. Technol. 44 (12), 4499–4504. https://doi.org/10.1021/es903565t.

Zhang, Y., Ahsan, M.Z., Luo, D., Hussain Panhwar, F., Li, L., Su, Y., Jia, X., Ye, X., Chen, R., Lihua, L., Zhu, J., 2024. Dynamics of selenium-mercury interaction under mercury stress in high and low selenium rice genotypes. Environ. Exp. Bot. 224, 105822. https://doi.org/10.1016/j.envexpbot.2024.105822.

Zhao, L., Meng, B., Feng, X., 2020. Mercury methylation in rice paddy and accumulation in rice plant: A review. Ecotoxicol. Environ. Saf. 195, 110462.

CHAPTER 7

Perfluoroalkyl and polyfluoroalkyl substances (PFAS) and phytotoxicity in aquatic plants

Débora Jesabel Pérez[1,2] and Mirta Luján Menone[3]

[1]Consejo Nacional de Investigaciones Científicas y Técnicas (CONICET), Universidad Nacional de Mar del Plata, Balcarce, Provincia de Buenos Aires, Argentina; [2]Universidad Fraternidad de Santo Tomás de Aquino, Mar del Plata, Provincia de Buenos Aires, Argentina; [3]Instituto de Investigaciones Marinas y Costeras (IIMYC), Consejo Nacional de Investigaciones Científicas y Técnicas (CONICET), Universidad Nacional de Mar del Plata (UNMDP), Mar del Plata, Provincia de Buenos Aires, Argentina

1. PFAS, from contaminants of emerging concern to "forever chemicals"

Perfluoroalkyl and polyfluoroalkyl substances (PFAS) are a group of synthetic chemicals that have been manufactured globally since the 1950s. Due to their extreme persistence and widespread environmental distribution and mobility, high bioaccumulation and biomagnification potential (Abusallout et al., 2022; George et al., 2023) they are known as "forever chemicals," i.e., persistent organic pollutants considered to be contaminants of emerging concern (Brunn et al., 2023). Concerns over PFAS toxicity have received global public attention due to recent evidence about adverse effects on microorganisms, plankton, invertebrates, vertebrates and terrestrial plants (e.g., Ma et al., 2022, 2023; Zhao et al., 2011; Yang et al., 2015).

PFAS comprise a diverse chemical group with great economic value, which includes residential, commercial, and industrial applications, such as surfactants in fluoropolymer production, aqueous film-forming foams for firefighting, metal plating, paper, textile, cosmetics, household products, semiconductors, lubricants, coating additives, agricultural applications, pesticides, and aviation industry (Ahrens and Bundschuh., 2014; Kurwadkar et al., 2022). Their broad spectrum uses and applications are associated with their ubiquitous occurrence in all environmental compartments, and they have been detected all over the globe, including at remote locations distant from the sources of pollution. The long-range transport of PFAS occurs through the sorption to aerosol particles in the atmosphere (Faust, 2023).

PFAS encompass $\geq$4,000 substances, which can be divided into three classes: perfluoroalkyl substances (PerFAS), polyfluoroalkyl substances (PolyFAS), and fluorinated polymers (Fig. 7.1). The chemical structure of PFAS consists of a carbon–fluorine backbone that may be partially or fully fluorinated and can be linear, branched, or cyclic (Fig.

Phytotoxicity of Organic Pollutants
ISBN 978-0-443-33110-7
https://doi.org/10.1016/B978-0-443-33110-7.00012-2

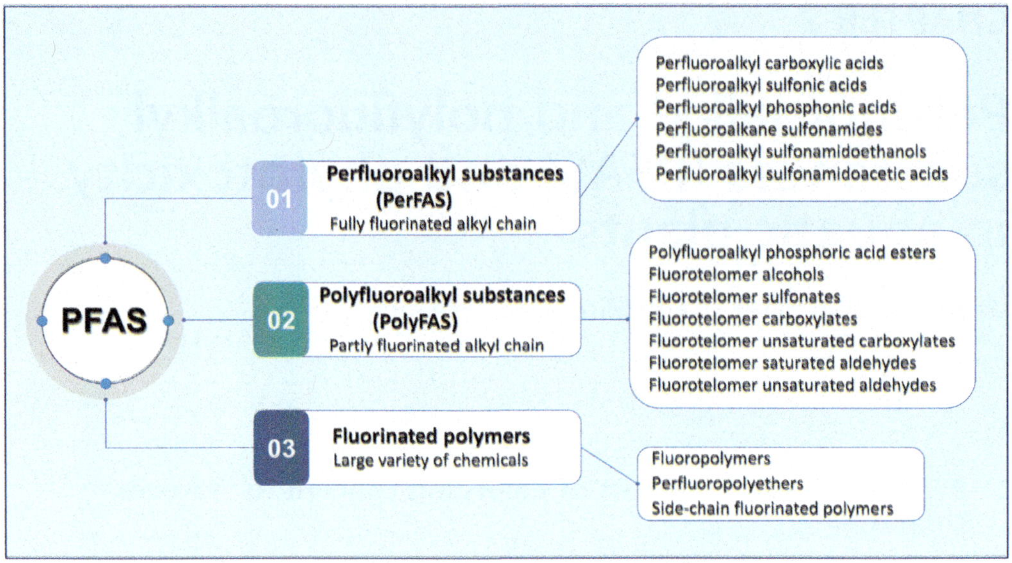

Figure 7.1 General classification of PFAS, following (Ahrens and Bundschuh, 2014).

7.2). The chain length of PFAS is defined by the number of perfluorinated carbons ($C \geq 6$–7 or $C < 6$), including their precursor structures.

Regarding their technical properties, PFAS exhibit high thermal and chemical stability, which is attributed to the exceptional strength of the C–F bond and the high electronegativity of fluorine atoms. The lack of biological or abiotic degradation mechanisms leading to mineralization results in their extreme environmental persistence (Brunn et al., 2023).

PFAS are aliphatic compounds with high water solubility and high $LogK_{OW}$ values. This is because many PFAS contain polar, nonfluorinated functional groups with hydrophilic properties and are negatively charged (e.g., carboxylic acids, sulfonic acids, and phosphonic acids). Consequently, many PFAS, especially long-chain compounds, exhibit amphiphilic properties (Ghisi et al., 2019). For example, perfluorooctanesulfonic acid (PFOS) has a water solubility of 3300 mg/L at 25°C and a $LogK_{OW}$ of 6.3 (Pubchem, 2025). Environmental physicochemical properties, such as persistence, mobility, bioaccumulation, and ecotoxicity, vary depending on chain length and functional groups (Ahrens and Bundschuh, 2014; Muñoz et al., 2022).

Global legislation has been particularly interested in long-chain PFAS, especially PFOS ($C_nF_{2n+1}SO_3H$, $n \geq 6$) and perfluorooctanoic acid (PFOA) ($C_nF_{2n+1}COOH$, $n \geq 7$) and their related compounds, which are the most frequently reported and discussed in the scientific literature (Buck et al., 2011; Wee and Aris, 2023). The massive global production and synthesis, widespread use, extreme persistence and recalcitrance, bioaccumulation and

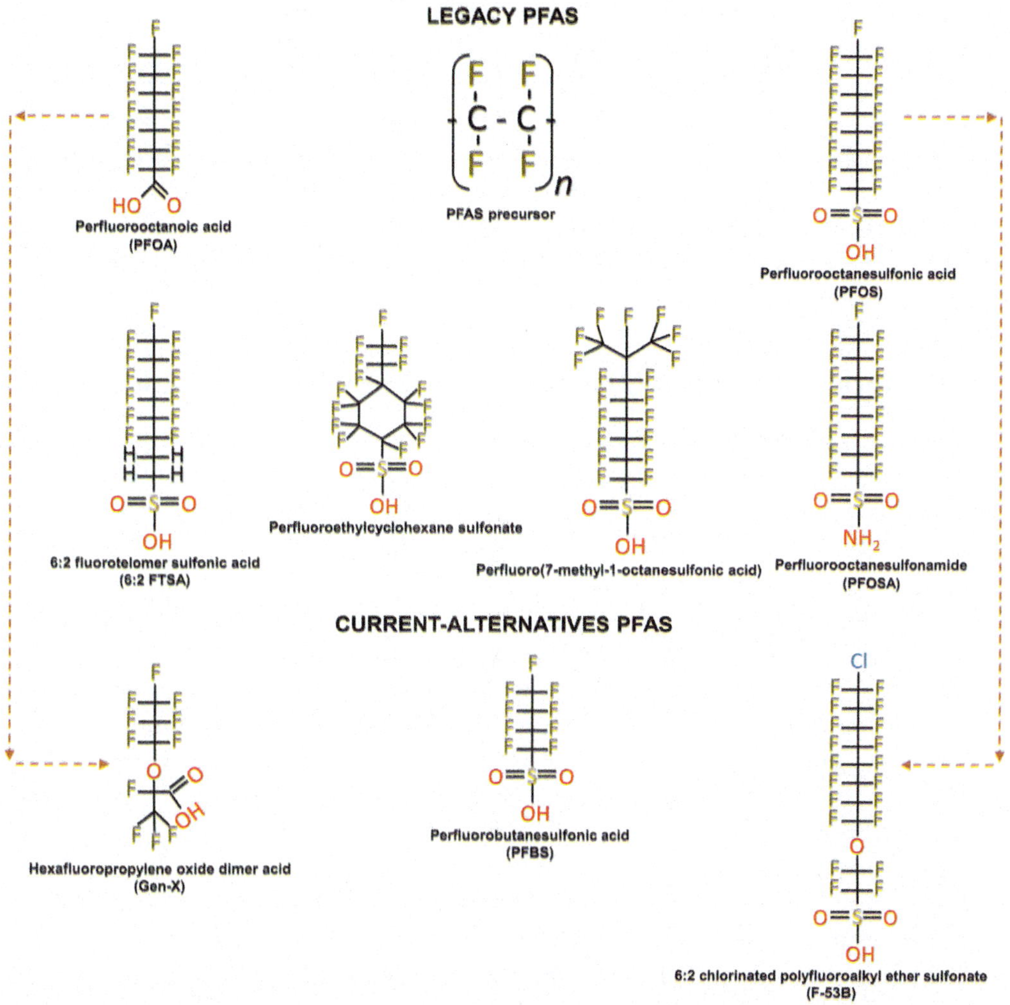

Figure 7.2 Chemical structures, lineal, branched and cyclic, of some legacy and current-alternative PFAS. Arrows connect the corresponding legacy to current-alternative PFAS.

biomagnification potential of PFAS led to list PFOS at the Conference of the Parties to the Stockholm Convention in 2009 and its salts and perfluorooctane sulfonyl fluoride in Annex B to the Convention (Stockholm Convention on Persistent Organic Pollutants, United Nations Environment Program, Decision SC-4/17, 2025). Furthermore, in 2019, following the recommendations of the Persistent Organic Pollutants Review Committee, PFOA, its salts and related compounds were added to Annex A of the Stockholm Convention (Stockholm Convention on Persistent Organic Pollutants, United Nations Environment Program, Decision SC-9/12., 2025). Thus, PFOS, PFOA and their related compounds are commonly referred to as legacy PFAS. In recent years, with the development of the fluoridation industry and strict regulatory controls on the production and use of long-chain PFAS, current-alternatives to PFOS and PFOA, including PFAS with fluoroalkyl ether linkages ($-CF_2-O-CF_2-$) and lower production costs, have become popular and are gradually replacing the market share of legacy PFAS (Munoz et al., 2022). Among these, Gen-X and F–53B are alternatives to PFOA and PFOS, respectively (USEPA, 2022). As shorter-chain PFAS alternatives have been designed to be less persistent in the environment (Yan et al., 2020), the easy degradability of GenX is desirable compared to PFOA (Yang et al., 2022).

2. Exposure, uptake and bioaccumulation of PFAS in aquatic plants

Aquatic macrophytes community constitutes an assemblage taxonomically diverse, which includes bryophytes and vascular plants, as ferns and flowering plants, which are characterized by a life cycle that takes place completely or partially in the freshwater ecosystems. Following Menone et al. (2023) the aquatic macrophytes can be classified depending on their growth forms as:

- Free-floating: those that typically float on or under the water surface but are not attached to the bottom (e.g., *Eichhornia crassipes*, *Lemna spp.*);
- Submerged: those that grow completely submerged under the water, with roots attached to, or closely associated with the substrate (e.g., *Hydrilla verticillata*, *Vallisneria natans*);
- Wetland or emergent: those that are rooted in sediments or soils that are periodically inundated, but with aerial leaves (e.g., *Cyperus alternifolius*, *Juncus spp.*, *Phragmites* spp., *Salix spp.*) (Fig. 7.3).

Aquatic macrophytes are used as in situ bioindicators of water quality and as an important green tool in phytoremediation processes for water purification (Pérez et al., 2023; Pi et al., 2017; Pérez and Menone, 2024). Hydrophytes are a component of benthic primary production in wetlands, acting as primary producers at the base of both herbivorous and detritivorous food chains, as photosynthesis driving primary production provides energy flow to the food webs of a range of freshwater ecosystems. Aquatic macrophytes also provide physical structure to ecosystems, increase habitat complexity and heterogeneity, influence oxygen and nutrient concentrations and cycling, provide shelter and refuge from predation, and release dissolved organic carbon that can be utilized by microbial complexes in the periphyton or plankton (Bakker et al., 2016). Nowadays, water purification is one of the most important ecosystem services provided by macrophytes (Thomaz, 2021). This process occurs through the remotion, uptake, tissue bioaccumulation and potential biotransformation process inorganic and organics pollutants that occur in the water body (Bertrand et al., 2019; Pérez et al., 2017).

The main uptake pathways for PFAS into roots and leaves of aquatic macrophytes are the water column and the pore water in sediments for those submerged or wetland species

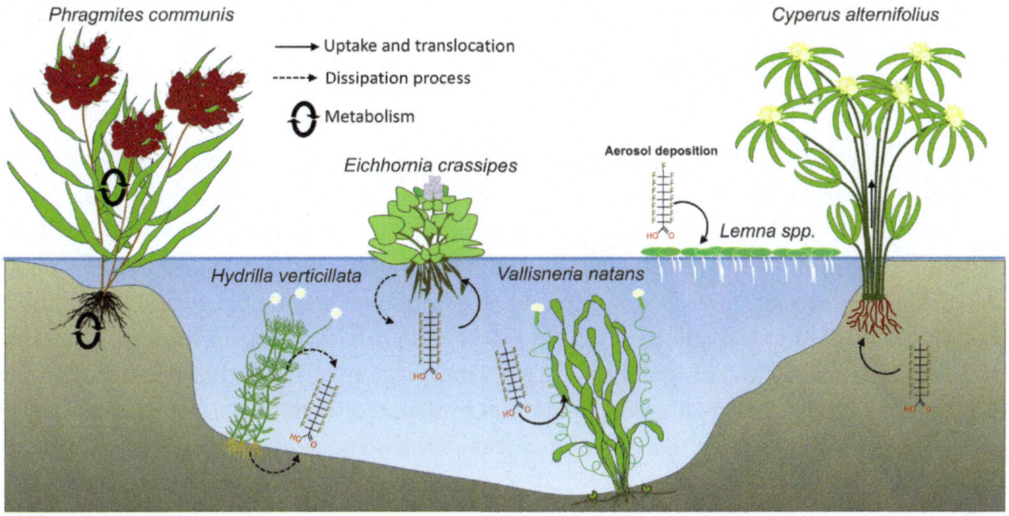

Figure 7.3 Pathways of PFAS uptake, distribution, metabolism, and dissipation in aquatic macrophytes of different growth forms.

(Ghisi et al., 2019; Pi et al., 2017) studied the differential uptake and elimination kinetics of PFAS between the submerged plant *Echinodorus horemanii* and the free-floating plant *Eichhornia crassipes*. They found that in *E. horemanii* leaves exhibited higher k_u (uptake constant) compared to *E. crassipes*, which is likely the result of the partitioning of PFAS via leaf-water exchange in the submerged macrophyte. Moreover, in both species, it was shown that the uptake and elimination capacity was correlated with the chain length. The *ku* for short-chain PFAS was lower than those longer PFAS, for example the *ku* ranged between 5 L/kg per day for perfluoropentanoic acid (PFPeA; $C–F_n = 4$) to 160 L/kg per day for perfluorotetradecanoic acid (PFTeDA; $C–F_n = 13$). The elimination process in *E. crassipes* was negatively correlated with the chain length. The k_e (elimination constant) values ranged from approximately 0.03 d^{-1} for PFTeDA to 0.24 d^{-1} to perfluorobutanesulfonic acid (PFBS; $C–F_n = 4$), indicating relatively slower elimination/dissipation of long-chain PFAS.

PFAS, specifically perfluorinated sulfonates and carboxylates, are generally not volatile, with very low Henry's Laws constants, which makes partitioning to the atmosphere unlikely (Abusallout et al., 2022). However, the sorption of PFAS to aerosol particles in the atmosphere (Faust, 2023) could be considered as an exposure and uptake pathway to aquatic plants. Nearby point sources, such as a fluorochemical industry, the potential exposure of PFAS via air or deposition of particles on above-ground parts can increase (Lesmeister et al., 2021).

PFOA and PFOS are the most studied and detected PFAS in the environment. They have been detected in levels typically <100 ng/L to >1000 ng/L in groundwater, surface water, and drinking water or in the order of μg/L in areas nearby of point sources of PFAS production. Levels upper than mg/L have been detected in groundwater and surface water close to fire-training locations and airports have also been measured (Ahrens, 2011).

On the other hand, levels of PFAS in sediments from Jucar River in Spain ranged between < 10 μg/kg to 100 μg/kg (Campo et al., 2016), similar to those detected in the Pensacola Bay, USA (Ahmadireskety et al., 2021). Moreover, in the sediments of the Qing River in China, the proportion of long-chain PFAS was higher than those of short-chain, being the dominant component in sediments PFOA followed by PFOS (Zhou et al., 2017). However, the PFAS sediment bioavailability could change depending on the sediment composition. It was demonstrated that the montmorillonite clay decreases the bioavailability of PFOA and PFOS due the hydrophobic and electrostatic interactions at the terminal fluorinated carbon chains of PFAS that interact with the clay charged surface (Hearon et al., 2022). In constructed wetlands, the presence of submerged macrophytes, particularly *Ceratophyllum demersum*, reduced PFAS adsorption to sediments. Presumably, this fact occurs because uptake of PFAS from overlying water into shoots could reduce their levels in the pore water phase and then root uptake of PFAS from pore water could compete with sediment adsorption (Li et al., 2021).

The root uptake and shoot translocation mechanisms of PFAS into plants are influenced by intrinsic plant properties. The uptake mechanisms of PFOS and PFOA have been studied in several wetland species, including *Canna indica, Thalia dealbata, Cyperus alternifolius, and Phragmites australis* among others. The uptake process involves first the adsorption of PFAS on the root surface, then the transport to root epidermal cells, and later the radial transport to the cortex. In some species, radial transport from the cortex to the vascular bundle was hindered by the selective Casparian band strips. Depending on the root structure, PFAS movement into the root could follow either or both apoplastic and symplastic routes, and to the vascular bundle via the symplastic route. In *C. indica* and *C. alternifolius*, movement of PFAS into the root is mainly by both apoplastic and symplastic routes, whereas in *T. dealbata* and *P. australis* it is mainly by the symplastic route. The large development of root cortex aerenchyma reduces the movement of PFAS in the root thought the apoplastic pathway, due to fewer intercellular routes. The less developed root cortex aerenchyma in *C. indica* and *C. alternifolius* may have less effect on the apoplastic pathway, resulting in greater transport of PFAS from the cortex to the vascular bundle than in *T. dealbata* and *P. australis* (Wang et al., 2020).

Then, to be taken up by root and/or leaves, PFAS could cross cell membranes through aquaporins and anion channels. This mode of enter was proposed in *V. natans* because PFOS and PFOA exposure increased the expression of genes encoding aquaporins and slow-type anion channels (Xiao et al., 2024). Moreover, in the wetland specie *Alisma orientale*, it was demonstrated that the uptake mechanisms involved the facilitate active process by water and anion channels proteins in the root cell (Wang et al., 2020). After that, PFAS can be bioaccumulated into the plant tissues. Although the information about the bioaccumulation behavior of PFAS in relation to the physico-chemical properties of these compounds in aquatic plants there is a consensus that the long-chain PFAS with high LogK_{ow} has greater bioaccumulation in the root, and those of short-chain compounds exhibit low adsorption affinity to the root, and are easier translocated to shoot and bioaccumulated in leaf (e.g., Li et al., 2021; Pi et al., 2017; Zhang et al., 2019; Zhou et al., 2017). Thus, as the chain length increases, the translocation factors decrease, while the whole-body bioaccumulation factor increases mainly at the expense of root bioaccumulation (Pi et al., 2017; Zhang et al., 2020). In addition, the PFAS bioaccumulation in roots of *Typha angustifolia* exhibited a positive concentration dependence, being PFAS accumulation in roots increased with increasing initial PFAS concentrations in soil (Zhang et al., 2020).

After exposure and bioaccumulation of PFAS no apparent negative effects could be detected, however, growth reduction in *Juncus effusus* and *E. horemani* has been reported, indicating phytotoxic (Zhang et al., 2019; Pi et al., 2017). Therefore, the phytotoxicity of PFAS in aquatic plants was reviewed in the present work. The main goal of this chapter was to compile, revise, and summarize the available information about phytotoxicity (e.g., morphological, physiological, molecular and biochemical biomarkers) of PFAS in aquatic macrophytes. Data on uptake, remotion, tissue bioaccumulation, metabolism, biofilm-related effects, and phytotoxicity of mixtures of PFAS with other contaminants were discussed.

3. Search methodology and data analysis

Data were collected from original research articles available from Science Direct published between 2000 and 2024. Search criteria were performed using combinations of the following keywords: aquatic macrophyte OR hydrophyte AND effect OR toxicity OR phytotoxicity OR growth inhibition OR toxicity endpoint OR non-observed effect concentration (NOEC) OR lowest observed effect concentration (LOEC) AND PFAS OR Polyfluoroalkyl substance OR Perfluoroalkyl substance. The selected articles represent laboratory or greenhouse scale studies evaluating individual PFAS exposure bioassays reporting data about phytotoxicity and their associated effects (i.e., morphological, physiological, biochemical responses, among others). The data were classified according to (1) species; (2) growth form; (3) PFAS individual compound or mixture, concentration and time of exposure; (4) phytotoxicity endpoints; (5) additional information, as uptake, removal efficiency, metabolization. The data were combined for: growth form * species * PFAS phytotoxicity (individual compound, as legacy or current-use alternative, or PFAS mixture), to identify the most commonly studied aquatic plant as a possible model for PFAS phytotoxicity, and also to identify the most studied PFAS and types of effects described. Papers reporting uptake/translocation/bioaccumulation data without toxicity effects were not considered.

All the articles selected have used purified active compound of PFAS in the exposure bioassays. This fact is relevant because commercial formulations of PFAS contain excipients that can modify the bioavailability and toxicity profile of the active compounds, influencing the interpretation of the results.

In the review of phytotoxicity data, it was observed that the most studied PFAS were PFOS and PFOA, and physiological and biochemical enzymatic and nonenzymatic biomarkers were the most analyzed as phytotoxicity endpoints. Therefore, to characterize a potential phytotoxicity profile of PFAS in aquatic plants, total chlorophyll content, superoxide dismutase (SOD), guaiacol peroxidase (POD) and catalase (CAT) activities, and malondialdehyde (MDA) levels were used in several macrophytes.

4. Effects of PFAS on aquatic plants

4.1 Data collection results

A total of 17 original papers were reviewed. All of them show legacy or current-alternative PFAS, as individual compounds or mixtures, as pollutants inducing phytotoxicity in different aquatic plant species (Table 7.1). The most studied species were the submerged macrophyte *V. natans* and the free-floating species *E. crassipes* and *Lemna spp.*, with seven, five and four combinations of growth form * species * PFAS phytotoxicity, respectively. Several studies on *V. natans* have been carried out by Asian authors because of its ecological relevance, being a native species of temperate and tropical Asia; and because of its tolerance to environmental stress in this region (World Flora Online (2025). Vallisneria natans (Lour.) H. Hara, 2025; Yang et al., 2023). In second place *E. crassipes* is generally used but this is an introduced species inhabiting tropical Asia, native from Amazonas to Mato Grosso do Sul, Brazil (World Flora Online. Eichhornia crassipes (Mart.) Solms., 2025). On the other hand, *L. minor* and *L. gibba* are aquatic plant models to test the toxicity of inorganic and organic pollutants (Menone and Pérez, 2023).

In terms of PFAS types, 16 compounds were identified (Table 7.1), including legacy and current–alternative PerFAS and PolyFAS. As individual toxicity compounds the legacy PerFAS such as PFOS and PFOA were the most studied pollutants, and only one current-alternative PerFAS was identified as Gen X (Hexafluoropropylene oxide dimer acid [HFPO–DA]). Within PolyFAS only two compounds were identified, the legacy 6:2 FTSA (6:2 fluorotelomer sulfonic acid) and the current–alternative of PFOS, F–53B (6:2 chlorinated polyfluoroalkyl ether sulfonate [6:2 Cl–PFAES]). In addition, PerFAS single or complex mixtures were the most frequently studied (Table 7.1, Fig. 7.4). The submerged *V. natans* have been studied for PFOS and PFOA as individual compounds, and for their mixture in phytotoxicity hydroponic bioassays (Table 7.1, Fig. 7.4). Likewise, *L. gibba* was studied for PFOS, while *L. minor* was studied with both PerFAS, as PFOA and PFOSA (perfluorooctanesulfonamide), for the PolyFAS 6:2

Table 7.1 Selected PFAS phytotoxicity studies in aquatic macrophytes available in literature.

Plant species	PFAS (exposure concentration-time)[a]	Phytotoxicity endpoints	Uptake, remotion efficiency, metabolism	References
Free-floating				
Eichhornia crassipes	PFOS 1, 100, 10,000 μg/L 42 days	Growth rate, root fresh weight, aerial parts fresh weight, total fresh weight Chlorophyll *a*, *b*, carotenoids Activity of SOD, POD, CAT Levels of MDA Root cell ultrastructure Transcription, gene expression		Li et al. (2020)
	F–53B Gen X PFOA PFOS 10 μg/L 14 days	Shoot biomass, root biomass, total biomass, growth rate, root length, Cotyledon or foliage number Chlorophyll *a*, *b*, total Total soluble protein and carbohydrate Levels of 8-OHdG Activity of SOD, POD, CAT Levels of MDA Root activity Root cell ultrastructure	Uptake and translocation	Chen et al. (2024)

Continued

Table 7.1 Selected PFAS phytotoxicity studies in aquatic macrophytes available in literature.—cont'd

Plant species	PFAS (exposure concentration-time)[a]	Phytotoxicity endpoints	Uptake, remotion efficiency, metabolism	References
Free-floating				
Lemna gibba	PFOS 0, 10, 20, 40, 80, and 160 mg/L 7 days	Inhibition of growth 50% (IC_{50})		Boudreau et al. (2003)
Lemna minor	PFOA 0, 2, 20, 200 μg/L 7 days	Mean frond area, frond number, multiplication rate, average specific growth Chlorophyll total Chlorophyll fluorescence parameters: PSII, F_v/F_m, ΦPSII, NPQ, ETR	Uptake and removal efficiency	Pietrini et al. (2019)
	6:2 FTSA PFOSA 10, 200 μg/L 7, 14 days	Activity of SOD, POD, CAT	Uptake and removal efficiency	Zhang and Liang (2021)
	PFBA PFBS PFHpA PFHxA PFHxS PFOA PFOS PFPeA 200 μg/L of each one in the mixture 7 h	Biomass	Uptake and removal efficiency	Zhang and Liang (2020)
Submerged				
Elodea canadensis	PFBA, 110 ± 0.5 μg/L PFBS, 78 ± 1.2 μg/L	Activity of LAC and POD for metabolization process	Uptake and removal efficiency Metabolites	Greger and Landberg (2024)

Table 7.1 Selected PFAS phytotoxicity studies in aquatic macrophytes available in literature.—cont'd

Plant species	PFAS (exposure concentration-time)[a]	Phytotoxicity endpoints	Uptake, remotion efficiency, metabolism	References
Free-floating				
	PFHpA, 63 ± 1.2 μg/L PFHxS, 92 ± 0.9 μg/L PFOA, 172 ± 0.5 μg/L PFOS, 126 ± 1.5 μg/L, in the mixture 96 h			
Hydrilla verticillata	PFBA PFBS PFDA PFDoDA PFHpA PFHxA PFHxS PFNA PFOA PFOS PFPeA PFUnDA Toxicity bioassay: 1, 10, 100 μg/L in the mixture of each one 0, 10, 20 days	Biomass, relative growth rate Chlorophyll *a*, *b*, total, chlorophyll autofluorescence Activity of SOD, POD, CAT Levels of MDA, H_2O_2 Shifts of microbial community in water and biofilm	Removal efficiency	Li et al. (2022)
Vallisneria natans	PFOS 1, 10, 100 μg/L 14 days	Shoot length Total soluble protein Activity of SOD, CAT Levels of MDA, GSH Shifts of		Wang et al. (2023)

Continued

Table 7.1 Selected PFAS phytotoxicity studies in aquatic macrophytes available in literature.—cont'd

Plant species	PFAS (exposure concentration-time)[a]	Phytotoxicity endpoints	Uptake, remotion efficiency, metabolism	References
Free-floating				
		microbial community in leaf biofilm Transcription, gene expression		
	PFOA PFOS PFOS + PFOA mixture 0.1, 1, 10, 100 μg/L 14 days	Chlorophyll *a*, *b*, total, carotenoids ROS contents Total soluble protein Levels of polysaccharides Activity of SOD, POD, CAT, GST Levels of MDA, GSH Transcription, gene expression Shifts of microbial community in leaf biofilm	Removal efficiency Metabolites	Xiao et al. (2024)
	PFOA 0.1, 1, 10 μg/L 14 days	Growth, root length, plant height Chlorophyll *a*, *b* total, carotenoids Total soluble protein Activity of SOD, POD Levels of MDA, GSH Leaf cell ultrastructure Shifts of		Li et al. (2023)

Table 7.1 Selected PFAS phytotoxicity studies in aquatic macrophytes available in literature.—cont'd

Plant species	PFAS (exposure concentration-time)[a]	Phytotoxicity endpoints	Uptake, remotion efficiency, metabolism	References
Free-floating				
		microbial community in leaf biofilm		
	PFOA 0.01, 0.1, 1, 10 μg/L 14 days	Biomass Chlorophyll *a*, *b*, total, carotenoids Total soluble protein Activity of SOD, POD, CAT Levels of MDA, levels of GSH Leaf cell ultrastructure Transcription, gene expression Shifts of microbial community in leaf biofilm	Metabolites	Yang et al. (2023)
	PFOA 1, 10, 100 μg/L 14 days	Chlorophyll *a*, *b*, total, carotenoids Activity of SOD, POD, GST Rubisco concentration Levels of MDA, levels of GSH Levels of fatty acids Transcription, gene expression Shifts of microbial community in leaf biofilm		Zhang et al. (2023)

Continued

Table 7.1 Selected PFAS phytotoxicity studies in aquatic macrophytes available in literature.—cont'd

Plant species	PFAS (exposure concentration-time)[a]	Phytotoxicity endpoints	Uptake, remotion efficiency, metabolism	References
Free-floating				
Wetland				
Acorus calamus	PFOS 100, 1000, 10,000, 50,000 μg/L 48 days	Chlorophyll total Total soluble protein Activity of SOD, POD, CAT Levels of MDA, levels of H_2O_2		Qian et al. (2019)
Carex rostrata	PFBA, 110 ± 0.5 μg/L PFBS, 78 ± 1.2 μg/L PFHpA, 63 ± 1.2 μg/L PFHxS, 92 ± 0.9 μg/L PFOA, 172 ± 0.5 μg/L PFOS, 126 ± 1.5 μg/L, in the mixture 96 h	Activity of LAC and POD for metabolization process	Uptake and removal efficiency Metabolites	Greger and Landberg (2024)
Cyperus alternifolius	PFOS 1, 100, 10,000 μg/L 42 days	Growth rate, root fresh weight, aerial parts fresh weight, total fresh weight Chlorophyll *a*, *b*, carotenoids Activity of SOD, POD, CAT Levels of MDA Root ultrastructure Transcription, gene expression		Li et al. (2020)

Table 7.1 Selected PFAS phytotoxicity studies in aquatic macrophytes available in literature.—cont'd

Plant species	PFAS (exposure concentration-time)[a]	Phytotoxicity endpoints	Uptake, remotion efficiency, metabolism	References
Free-floating				
Eriophorum angustifolium	PFBA, 110 ± 0.5 μg/L PFBS, 78 ± 1.2 μg/L PFHpA, 63 ± 1.2 μg/L PFHxS, 92 ± 0.9 μg/L PFOA, 172 ± 0.5 μg/L PFOS, 126 ± 1.5 μg/L, in the mixture 96 h	Activity of LAC and POD for metabolization process	Uptake and removal efficiency Metabolites	Greger and Landberg (2024)
Juncus effusus	PFBS: 110, 1100 μg/L PFHpA: 75, 750 μg/L PFHxA: 120, 1200 μg/L PFHxS: 290, 2900 μg/L PFOA: 250, 2500 μg/L PFOS: 4300, 43,000 μg/L PFPA: 66, 660 μg/L 7, 14, 21 days	Activity of SOD, CAT Levels of H_2O_2	Uptake and removal efficiency	Zhang et al. (2019)
Juncus sarophorus	PFHxS PFOA PFOS 10, 100 μg/L of each one in the irrigation water mixture 190 days	Shoot height	Uptake and translocation	Zhu et al. (2022)

Continued

Table 7.1 Selected PFAS phytotoxicity studies in aquatic macrophytes available in literature.—cont'd

Plant species	PFAS (exposure concentration-time)[a]	Phytotoxicity endpoints	Uptake, remotion efficiency, metabolism	References
Free-floating				
Phragmites communis	PFOS 100, 1000, 10,000, 50,000 μg/L 48 days	Chlorophyll total Total soluble protein Activity of SOD, POD, CAT Levels of MDA, levels of H_2O_2		Qian et al. (2019)
Salix spp. S. eleagnos S. purpurea S. triandra	PFBA PFBS PFDA PFDoDA PFHpA PFHxA PFNA PFOA PFOS PFPeA PFUnA 10 μg/L of each one in the mixture 8 days	Relative growth rate, dry weight Chlorophyll fluorescence parameters: NPQ, F_v/F_m, ETR Photosynthetic gas exchange parameters: Net CO_2 assimilation, intracellular CO_2, stomatal conductance to water vapor, vapor pressure deficit	Uptake and removal efficiency	Sharma et al. (2020)

6:2 FTSA, 6:2 fluorotelomer sulfonic acid; *F–53B*, 6:2 chlorinated polyfluoroalkyl ether sulfonate [6:2 Cl-PFAES]; *8-OHdG*, 8-hydroxy-deoxy-guanosine; *ΦPSII*, Quantum efficiency of PSII photochemistry; *CAT*, Catalase; *ETR*, Photosynthetic electron transport rate through PSII; F_v/F_m, photochemistry; *GSH*, Glutathione; *GST*, glutathione S-transferase; H_2O_2, Hydrogen peroxide; *LAC*, Laccase; *MDA*, Malondialdehyde; *NPQ*, Nonphotochemical quenching; *PerFAS*, Perfluoroalkyl substances; *Gen-X*, Hexafluoropropylene oxide dimer acid [HFPO–DA]; *PFBA*, Perfluorobutanoic acid; *PFBS*, Perfluorobutanesulfonic acid; *PFDA*, Perfluorodecanoic acid; *PFDoDA*, Perfluorododecanoic acid; *PFHpA*, Perfluoroheptanoic acid; *PFHxS*, Perfluorohexanesulfonic acid; *PFHxA*, Perfluorohexanoic acid; *PFNA*, Perfluorononanoic acid; *PFOA*, Perfluorooctanoic acid; *PFOS*, Perfluorooctanesulfonic acid (commonly Perfluorooctane sulfonate); *PFOSA*, Perfluorooctanesulfonamide; *PFPeA*, Perfluoropentanoic acid; *PFUnDA*, Perfluoroundecanoic acid; *PolyFAS*, Polyfluoroalkyl substances; *POD*, Peroxidase; *PSII*, Maximum quantum yield of photosystem II; *Rubisco*, ribulose-1, 5-diphosphate carboxylase/oxygenase; *SOD*, Superoxide dismutase.

[a]High purity active compounds of each PFAS (>90%) were used in the phytotoxicity bioassays.

FTSA, and for the eight PerFAS mixture, that included PFOS and PFOA (Table 7.1). *E. crassipes* was studied for PFOS and PFOA, and for F–53B and Gen X (Table 7.1, Fig. 7.4). In particular, the single mixture of PFOS and PFOA was studied in *V. natans*, while complex mixtures have been studied in three growth form species, such as *Lemna spp.*, *Elodea canadensis*, *Hydrilla verticillata*, *Carex rostrata*, *Eriophorum angustifolium*, *Juncus spp.*, and *Salix spp.* (Table 7.1). Concluding, several types of studies using individual or complex mixtures of PFAS have been observed in the literature. As mentioned by Ofoegbu et al. (2022), combined exposure to a mixture of PFAS reflects environmental conditions but it is also important to conduct studies with individual PFAS members such as PFOS to understand PFOS specific toxicity.

In general, most studies have been conducted to test the toxicity of environmentally relevant concentrations of PFAS, both individual and mixtures, ranging from low concentration such as 0.01 μg/L of PFOA to as high as > 10,000 μg/L of PFOS, in hydroponic solution or irrigation water (Table 7.1). This wide range of concentrations correlates well with both nonpoint sources of pollution detected in the surface water and near-point sources of pollution (as described in section 8.2). The bioassay exposure time ranged from acute, with only a few hours or days of exposure to chronic ranging from 21 to 190 days. The exposure time may depend on the objective of each study, whether it was only to assess toxicity using early responses or late responses such as morphological changes or biofilm formation, quantification of removal efficiency, bioaccumulation and/or metabolic processes (as described in section 8.2).

4.2 Phytotoxicity of PFAS

Considering the recent concerns about PFAS toxicity, all the selected papers, except the one of Boudreau et al. (2003) were novel and timely from 2019 to 2024 (Table 7.1). The pioneering paper by Boudreau et al. (2003) deals with the 7 days-acute toxicity test,

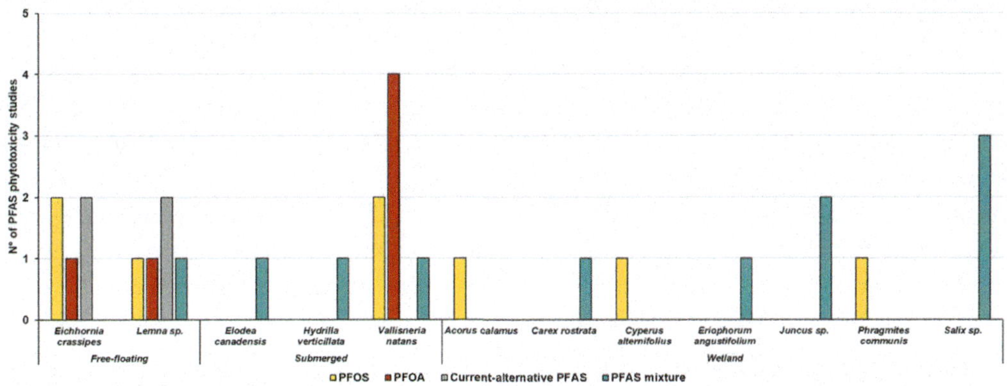

Figure 7.4 Distribution of published studies per PFAS type in different aquatic macrophyte species.

using toxicity endpoints in *L. gibba* to determine the IC_{50} and NOEC of PFOS. That work showed that in this species the IC_{50} of PFOS was 59.1 mg/L in fronds number and 31.1 mg/L total wet weight while the NOEC was 6.6 mg/L wet weight. The highest concentrations used in the range of 80–160 mg/L induced chlorosis and necrosis, root growth reduction and at 160 mg/L the plants died. These results indicate the low acute toxicity of PFOS in that model species compared with other organic pollutants with similar physico-chemical and environmental properties, such as the insecticide chlorpyriphos ($EC_{50} = 0.53$ mg/L in *L. minor*) or the fungicide azoxystrobin ($EC_{50} = 3.2$ mg/L in *L. gibba*) (Pesticide Properties DataBase, 2025).

The phytotoxicity endpoints studied in aquatic plants to determine the adverse effects of PFAS include morphological (i.e., growth rate, biomass, root and shoot length and weight, frond area, root ultrastructure, among others), physiological (i.e., related to the photosynthesis and mitochondrial respiration process, such as chlorophyll and carotenoid contents, chlorophyll fluorescence, functioning of PSII, CO_2 assimilation, gas exchange, mitochondrial electron transport), biochemical (i.e., enzymatic antioxidant activities such as SOD, POD, CAT, GST, levels of MDA, H_2O_2 and GSH, among others), and molecular (i.e., differential gene expression). In addition, indirect effects such as shifts in the microbiota and periphyton diversity, and in the structure of the associated biofilm are highlighted as novel tools to assess the ecological relevance of macrophytes (Table 7.1).

Although plant growth rate and morphological parameters can be considered as the most direct index of phytotoxicity these endpoints have shown some controversial results in macrophytes to PFOS and PFOA. For example, PFOS and PFOA induced a dose-dependent decrease in growth and biomass production in *V. natans* (Li et al., 2023; Yang et al., 2023; Wang et al., 2023) but increases of root fresh weight and length were observed in *E. crassipes* exposed to a range from 1 to 10,000 μg/L of PFOS (Li et al., 2021; Chen et al., 2024) (Table 7.1). Likewise, increased frond area was observed in *L. minor* exposed to PFOA, without changes in the biomass production (Pietrini et al., 2019). In the case of PerPFAS mixtures, macrophytes also showed variable responses of growth and morphological changes. While *H. verticillata* and *J. sarophorus* showed decreases in biomass and shoot height, respectively (Li et al., 2020; Zhu et al., 2022), no changes in *L. minor* and *Salix spp.* were observed (Zhang and Liang, 2020; Sharma et al., 2020).

In this section, the effects of PFOS and PFOA on the different phytotoxicity endpoints will be summarized. We have found that the main effects evaluated were those related to photosynthesis (mainly chlorophyll contents), and oxidative stress enzymatic systems like SOD, POD, CAT and MDA levels.

In short, it shows that both PFOS and PFOA tended to reduce chlorophyll content in aquatic macrophytes, and that *V. natans* was the most sensitive species for this phytotoxic endpoint, showing this effect at low concentrations (from 0.01 μg/L PFOA and from 0.1 μg/L PFOS (Fig. 7.5).). Regarding photosynthesis, the gas exchange parameters were affected more than those related to chlorophyll fluorescence in *Salix* species, with significant increase of the net CO_2 assimilation rate and parameters related to stomatal conductance (Sharma et al., 2020).

In terms of antioxidant enzyme systems, in general, all the enzymes tested showed changes. SOD responded to PFOS exposure by increasing or decreasing its activity depending on the macrophyte species (Fig. 7.6). In *V. natans* an increase of SOD activity was observed at low concentrations while the same behavior was registered in *A. calamus* and *P. communis* at high concentrations. In the case of PFOA, *V. natans* was the only species that showed changes in SOD activity being increased at low concentrations (0.1–1 μg/L). POD activity increased in *V. natans* at low concentrations of PFOS (1 μg/L) and at concentrations of 1, 10 and 100 μg/L PFOA (Fig. 7.7). In addition, the activity of this enzyme decreased at 10 μg/L PFOS in *E. crassipes* and at 1000–˃1000 μg/L in *C. calamus* and *P. communis*. In the case of CAT, *V. natans* showed increased activity at low concentrations of PFOS while this behavior occurred in other species like *A. calamus* and *P. communis* at high concentrations (Fig. 7.8). In this sense, it is noteworthy that both POD and CAT share the same substrate and, therefore, they then showed a concordant response when *V. natans* was exposed to PFOS indicating the sensitivity and usefulness of these biomarkers of oxidative stress. MDA levels, an indicator of oxidative damage, increased only in *V. natans* exposed to 1 μg/L PFOS and at higher concentrations of 1000 μg/L in *C. calamus, C. alternifolius, E. crassipes* and *P.*

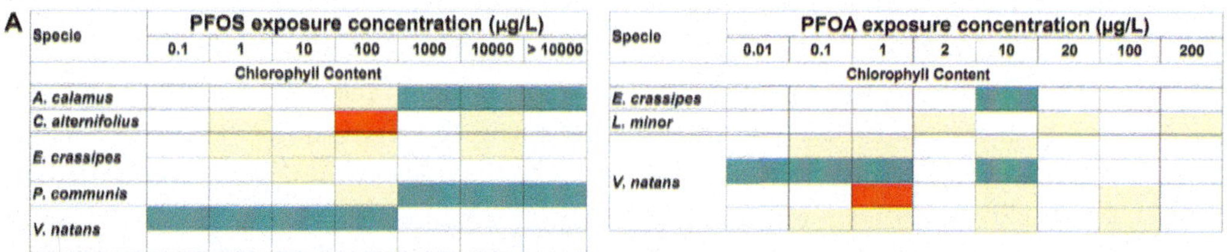

Figure 7.5 Phytotoxicity endpoint of PFOS and PFOA in different aquatic macrophyte species. A: Chlorophyll content.

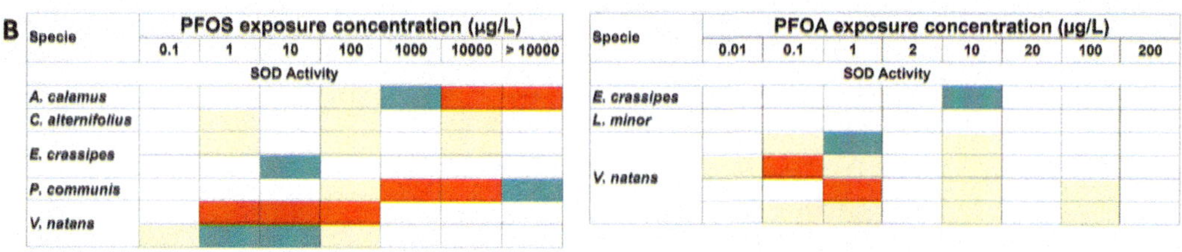

Figure 7.6 Phytotoxicity endpoint of PFOS and PFOA in different aquatic macrophyte species. B: SOD activity.

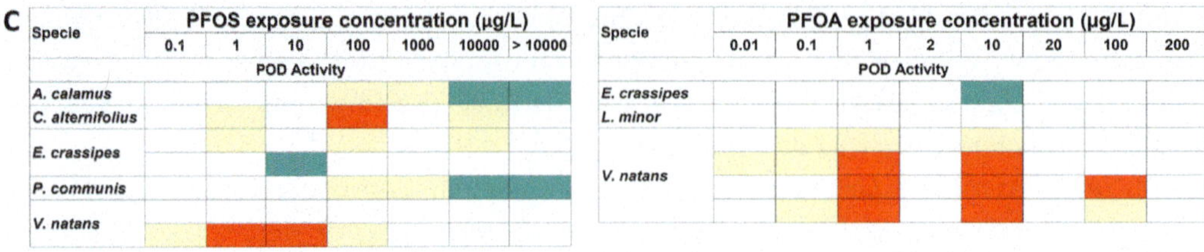

Figure 7.7 Phytotoxicity endpoint of PFOS and PFOA in different aquatic macrophyte species. C: POD activity.

communis whereas no changes in this product of lipid peroxidation were observed in macrophytes exposed to PFOA (Fig. 7.9). Therefore, this endpoint showed that although oxidative stress may occur in macrophytes exposed to PFAS, no oxidative damage would be expected. On the other hand, like the chlorophyll content, the responses of oxidative stress were conclusive in *V. natans*, making it the most sensitive species among those tested in different papers and for that we propose it as a model species to study effects of PFAS.

In general, PFOS may be more toxic than PFOA in aquatic macrophytes (Xiao et al., 2024). Since all endpoints summarized in show more changes (increase/decrease) for PFOS than for PFOA in the present work, the same conclusion can be drawn. In particular, the observed inhibition of antioxidant enzymes may indicate oxidative damage to proteins, which in the case of enzymes, results in enzyme inactivity. This difference in toxicity between PFOS and PFAS in aquatic macrophytes is consistent with previous reports on invertebrates and vertebrates. For example, PFOS causes developmental and reproductive toxicity in the cladocera *Daphnia carinata* at concentrations of 1 and 10 mg/L PFOS and was more toxic than PFOA in terms of days to first brood, average number of offspring per brood, and total live offspring (Logeshwaran et al., 2021). These authors also tested concentrations in the order of µg/L. They found that no mortality (%) of *D. carinata* was recorded at exposures of 1 µg/L PFOS or PFOA, but animals exposed to 10 µg/L PFOS showed 2.5% and no mortality with PFOA; and at higher concentrations, the % was always higher for PFOS than for PFOA. In the fish *Oryzias latipes*, Ji

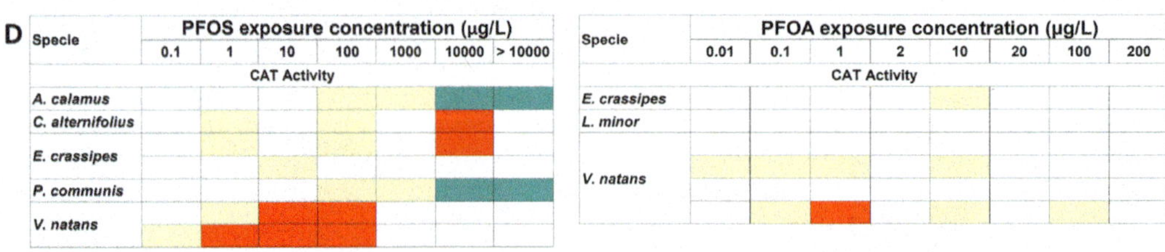

Figure 7.8 Phytotoxicity endpoint of PFOS and PFOA in different aquatic macrophyte species. D: CAT activity.

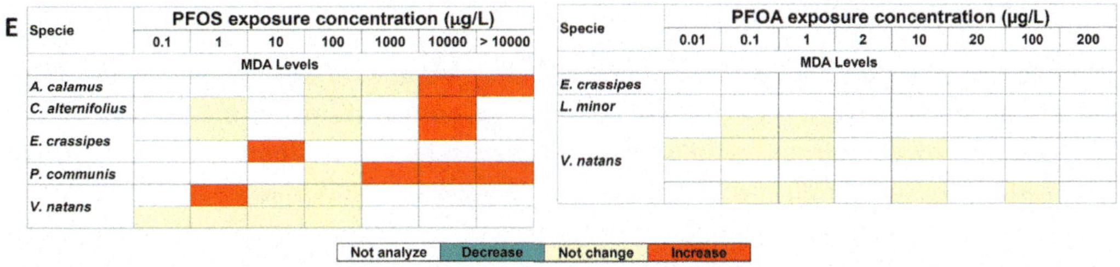

Figure 7.9 Phytotoxicity endpoint of PFOS and PFOA in different aquatic macrophyte species. E: MDA levels.

et al. (2008) demonstrated that PFOS was approximately 10 times more toxic than PFOA.

5. Detoxification of PFAS

Although most studies have been carried out to understand the uptake and potential effects of PFAS, some work has been done on their detoxification. Looking at Table 7.1, we can see that the presence of PFAS metabolites has been described in some species, suggesting that once absorbed, plants try to eliminate PFAS. Greger and Landberg (2024) measured and identified PFOS and PFOA metabolites in roots and shoots of *Elodea canadensis*, *Carex rostrata* and *Eriophorum angustifolium*. In addition, as the laccase and peroxidase activities are related to PFOS and PFOA degradation, the authors found that both enzymatic activities increased in plants exposed to PFAS, indicating a possible metabolization of these pollutants. In *V. natans* metabolites of PFOS and PFOA were identified (Yang et al., 2023; Xiao et al., 2024) (Table 7.1). Moreover, Xiao et al. (2024) showed that PFAS bind to functional proteins for detoxification. In fact, these authors showed that PFOA and PFOS can bind to GSTF10 via channel proteins after entering plant cells. The mechanism involved appears to be the interfacial interactions of GSTF10 with amino acid residues of PFOA and PFOS, with their binding mainly maintained by hydrogen and halogen bond interactions. In the same study, the metabolic response to PFOA, PFOS and their combination was investigated in *V. natans*. It was observed that there was a higher degree of down-regulation than up-regulation of metabolites in pollutant-exposed plants compared to those in the control group, with an increase in differentially expressed metabolites (DEMs) observed following combined exposure. The DEMs identified in macrophytes exposed to PFAS were mainly malic acid, sucrose, citric acid, cyclopentanol and harmaline, among which sucrose was particularly reduced, in agreement with previous reports showing this response under exposure to other organic pollutants, as a strategy to provide more energy for stress adaptation (Xiao et al., 2024).

6. Biofilm-related effects

Among the publications that have addressed the potential effects of the PFAS found, several have also investigated the effects of these contaminants in the biofilm (Table 7.1). As microbial aggregates in water, periphytic biofilms adhere to aquatic plants and participate in nutrient cycling in aquatic ecosystems. Previous studies have shown that the biofilm is an important reservoir for contaminants and that the microorganisms, and the extracellular polysaccharides (EPS) involved in biofilms generation, contribute to the bioaccumulation of PFOS (Wang et al., 2023). Moreover, under PFAS exposure, microbial communities on leaf biofilms of plants changed at the genus level (Zhang et al., 2023). In the present work we can see that biofilm in *V. natans* and *H. verticillata*. Wang et al. (2023) have conducted a study using *V. natans* as a model species exposed to PFOS and Cu. The authors have found that PFOS, Cu, and the mixed treatment changed the structural features of EPS and induced alterations in abundances and structures of the microbial community in biofilms. Similarly, the structure and function of biofilm in *V. natans* have changed after exposure to PFOA and the antibiotic sulfadiazine (Zhang et al., 2023), to PFOA and the cyanotoxin microcystin-LR (Yang et al., 2023) and to PFOA and microplastics (Li et al., 2023). In the three cases, some antagonistic effects were induced by the combination of PFOA and the other pollutants studied, but the underlying mechanisms remain unclear and require further research.

7. Gaps of information and future perspectives

This chapter focused on aquatic macrophytes because they are of high ecological relevance in aquatic ecosystems but data about the effects of PFAS on them is scarce in the literature. The literature cited in this chapter is a compilation of original articles that complement the seminal review by Huang et al. (2021). However, this paper emphasized the role of plants in the PFAS remediation process and included little data on aquatic macrophyte species. Except for the work by Boudreau et al. (2003), the present chapter in particular shows very recent work on PFAS phytotoxicity in aquatic macrophytes, as most of the work was done from 2019 onwards, with 9 out of 17 from 2023 to 2024.

Although the uptake via and mechanisms are already known and are described by Huang et al. (2021) and Xiao et al. (2024) there is still a lack of information on the mechanisms underlying PFAS phytotoxicity. In this sense, promoting the use of molecular biomarkers of toxicity may be a valuable approach. The techniques of genomics and transcriptomics have great potential in ecotoxicology as the basis for biomarker assays. The pattern of genomic response may give critical evidence about the modes of action and these molecular biomarkers have great potential for screening for toxic effects in the environment (Menone and Pérez, 2023).

As can be seen here, the main processes investigated are photosynthesis and oxidative stress. Therefore, we would like to recommend that biomarkers related to other processes potentially affected by PFAS exposure should also be investigated. For example, several authors have used *V. natans* to study environmentally relevant concentrations of PFAS, but in all cases plants were exposed for 14 days and the effects were mostly on oxidative stress and chlorophylls, as well as shifts in microbial communities in the leaf biofilm. In fact, this submerged macrophyte was the most frequently selected species in the literature analyzed. The same pattern is seen for wetland species, where 10 different species were analyzed, but only effects on growth, photosynthesis and oxidative stress were considered. In addition, PFOS and PFOA have been studied more extensively. Therefore, we recommend to expand the knowledge of others like PerFAS and PolyFAS, as well as the current-alternatives because their environmental occurrence in water and various matrices is expected to increase as they replace older substances.

References

Abusallout, I., Holton, C., Wang, J., Hanigan, D., 2022. Henry's Law constants of 15 per- and polyfluoroalkyl substances determined by static headspace analysis. J. Hazard. Mater. Lett. 3, 100070. https://doi.org/10.1016/j.hazl.2022.100070.

Ahmadireskety, A., Silva, Awkerman, J.A., Aufmuth, J., Yost, R.A., Bowden, J.A., 2021. Per-and polyfluoroalkyl substances (PFAS) in sediments collected from the Pensacola Bay System watershed. Environ. Adv. 5.

Ahrens, L., 2011. Polyfluoroalkyl compounds in the aquatic environment: a review of their occurrence and fate. J. Environ. Monit. 13 (1), 20–31. https://doi.org/10.1039/c0em00373e.

Ahrens, L., Bundschuh, M., 2014. Fate and effects of poly- and perfluoroalkyl substances in the aquatic environment: a review. Environ. Toxicol. Chem. 33 (9), 1921–1929. https://doi.org/10.1002/etc.2663.

Bakker, E.S., Wood, K.A., Pagès, J.F., Veen, G.F., Christianen, M.J.A., Santamaría, L., Nolet, B.A., Hilt, S., 2016. Herbivory on freshwater and marine macrophytes: a review and perspective. Aquat. Bot. 135, 18–36. https://doi.org/10.1016/j.aquabot.2016.04.008.

Bertrand, L., Monferrán, M.V., Valdés, M.E., Amé, M.V., 2019. Usefulness of a freshwater macrophyte (Potamogeton pusillus) for an environmental risk assessment in a multi-source contaminated basin. Chemosphere 222, 1003–1016. https://doi.org/10.1016/j.chemosphere.2019.02.018.

Boudreau, T.M., Sibley, P.K., Mabury, S.A., Muir, D.G.C., Solomon, K.R., 2003. Laboratory evaluation of the toxicity of perfluorooctane sulfonate (PFOS) on Selenastrum capricornutum , Chlorella vulgaris , Lemna gibba , Daphnia magna , and *Daphnia pulicaria*. Arch. Environ. Contam. Toxicol. 44 (3), 307–313. https://doi.org/10.1007/s00244-002-2102-6.

Brunn, H., Arnold, G., Körner, W., Rippen, G., Steinhäuser, K.G., Valentin, I., 2023. PFAS: forever chemicals—persistent, bioaccumulative and mobile. Reviewing the status and the need for their phase out and remediation of contaminated sites. Environ. Sci. Eur. 35 (1), 1–50.

Buck, R.C., Franklin, J., Berger, U., Conder, J.M., Cousins, I.T., Voogt, P.D., Jensen, A.A., Kannan, K., Mabury, S.A., van Leeuwen, S.P.J., 2011. Perfluoroalkyl and polyfluoroalkyl substances in the environment: terminology, classification, and origins. Integrated Environ. Assess. Manag. 7 (4), 513–541. https://doi.org/10.1002/ieam.258.

Campo, J., Lorenzo, M., Pérez, F., Picó, Y., Farré, M.l., Barceló, D., 2016. Analysis of the presence of perfluoroalkyl substances in water, sediment and biota of the Jucar River (E Spain). Sources, partitioning and relationships with water physical characteristics. Environ. Res. 147, 503–512. https://doi.org/10.1016/j.envres.2016.03.010.

Chen, Z.w., Hua, Z.l., Guo, P., 2024. The bioaccumulation and ecotoxicity of co-exposure of per(poly) fluoroalkyl substances and polystyrene microplastics to *Eichhornia crassipes*. Water Res. 260. https://doi.org/10.1016/j.watres.2024.121878. Available from: https://www.sciencedirect.com/science/journal/00431354.

Faust, J.A., 2023. PFAS on atmospheric aerosol particles: a review. Environ. Sci. Process. Impacts 25 (2), 133–150. https://doi.org/10.1039/d2em00002d.

George, S.E., Baker, T.R., Baker, B.B., 2023. Nonlethal detection of PFAS bioaccumulation and biomagnification within fishes in an urban- and wastewater-dominant Great Lakes watershed. Environ. Pollut. 321. https://doi.org/10.1016/j.envpol.2023.121123. Available from: https://www.journals.elsevier.com/environmental-pollution.

Ghisi, R., Vamerali, T., Manzetti, S., 2019. Accumulation of perfluorinated alkyl substances (PFAS) in agricultural plants: a review. Environ. Res. 169, 326–341. https://doi.org/10.1016/j.envres.2018.10.023.

Greger, M., Landberg, T., 2024. Removal of PFAS from water by aquatic plants. J. Environ. Manag. 351, 119895. https://doi.org/10.1016/j.jenvman.2023.119895.

Hearon, S.E., Orr, A.A., Moyer, H., Wang, M., Tamamis, P., Phillips, T.D., 2022. Montmorillonite clay-based sorbents decrease the bioavailability of per- and polyfluoroalkyl substances (PFAS) from soil and their translocation to plants. Environ. Res. 205. https://doi.org/10.1016/j.envres.2021.112433. Available from: http://www.elsevier.com/inca/publications/store/6/2/2/8/2/1/index.htt.

Huang, D., Xiao, R., Du, L., Zhang, G., Yin, L., Deng, R., Wang, G., 2021. Phytoremediation of poly- and perfluoroalkyl substances: a review on aquatic plants, influencing factors, and phytotoxicity. J. Hazard Mater. 418.

Ji, K., Kim, Y., Oh, S., Ahn, B., Jo, H., Choi, K., 2008. Toxicity of perfluorooctane sulfonic acid and perfluorooctanoic acid on freshwater macroinvertebrates (Daphnia magna and Moina macrocopa) and fish (*Oryzias latipes*). Environ. Toxicol. Chem. 27 (10), 2159–2168. https://doi.org/10.1897/07-523.1.

Kurwadkar, S., Dane, J., Kanel, S.R., Nadagouda, M.N., Cawdrey, R.W., Ambade, B., 2022. Wilkin, Per- and polyfluoroalkyl substances in water and wastewater: a critical review of their global occurrence and distribution. Sci. Total Environ. 809.

Lesmeister, L., Lange, F.T., Breuer, J., Biegel-Engler, A., Giese, E., Scheurer, M., 2021. Extending the knowledge about PFAS bioaccumulation factors for agricultural plants – a review. Sci. Total Environ. 766, 142640. https://doi.org/10.1016/j.scitotenv.2020.142640.

Li, Q., Xiao, Y., Zhang, W., Li, S., Liu, J., Yu, Y., Wen, Y., Zhang, Y., Lei, N., Wang, Q., 2023. Single and combined toxicity effects of microplastics and perfluorooctanoic acid on submerged macrophytes and biofilms. Sci. Total Environ. 897, 165370. https://doi.org/10.1016/j.scitotenv.2023.165370.

Li, R., Tang, T., Qiao, W., Huang, J., 2020. Toxic effect of perfluorooctane sulfonate on plants in vertical-flow constructed wetlands. J. Environ. Sci. 92, 176–186. https://doi.org/10.1016/j.jes.2020.02.018.

Li, X.q., Hua, Z.l., Zhang, J.y., Gu, L., 2022. Ecotoxicological responses and removal of submerged macrophyte Hydrilla verticillate to multiple perfluoroalkyl acid (PFAA) pollutants in aquatic environments. Sci. Total Environ. 825. https://doi.org/10.1016/j.scitotenv.2022.153919. Available from: www.elsevier.com/locate/scitotenv.

Li, X.q., Hua, Z.l., Wu, J.y., Gu, L., 2021. Removal of perfluoroalkyl acids (PFAAs) in constructed wetlands: considerable contributions of submerged macrophytes and the microbial community. Water Res. 197. https://doi.org/10.1016/j.watres.2021.117080. Available from: www.elsevier.com/locate/watres.

Logeshwaran, P., Kenday Sivaram, A., Surapaneni, A., Kannan, K., Naidu, R., Megharaj, M., 2021. Exposure to perfluorooctanesulfonate (PFOS) but not perflurooctanoic acid (PFOA) at ppb concentration induces chronic toxicity in Daphnia carinata. Sci. Total Environ. 769, 144577. https://doi.org/10.1016/j.scitotenv.2020.144577.

Ma, T., Ye, C., Wang, T., Li, X., Luo, Y., 2022. Toxicity of per-and polyfluoroalkyl substances to aquatic invertebrates, planktons, and microorganisms. Int. J. Environ. Res. Publ. Health 19 (24).

Ma, T., Wu, P., Wang, L., Li, Q., Li, X., Luo, Y., 2023. Toxicity of per-and polyfluoroalkyl substances to aquatic vertebrates. Front. Environ. Sci. 11.

Menone, M.L., Humeniuk, B.W., Metcalfe, C.D., 2023. The Ecotoxicology of Aquatic Macrophytes: An Overview. Springer Science and Business Media LLC, pp. 1–10. https://doi.org/10.1007/978-3-031-27833-4_1.

Menone, M.L., Pérez, D.J., 2023. Biomarkers in Aquatic Macrophytes: Traditional and Novel Approaches for Monitoring Responses to Exposure to Pollutants. Springer Science and Business Media LLC, pp. 11–52. https://doi.org/10.1007/978-3-031-27833-4_2.

Munoz, G., Mercier, L., Vo Duy, S., Liu, J., Sauvé, S., Houde, M., 2022. Bioaccumulation and trophic magnification of emerging and legacy per- and polyfluoroalkyl substances (PFAS) in a St. Lawrence River food web. Environ. Pollut. 309, 119739. https://doi.org/10.1016/j.envpol.2022.119739.

Ofoegbu, P.C., Wagner, D.C., Abolade, O., Clubb, P., Dobbs, Z., Sayers, I., Zenobio, J.E., Adeleye, A.S., Rico, C.M., 2022. Impacts of perfluorooctanesulfonic acid on plant biometrics and grain metabolomics of wheat (*Triticum aestivum* L.). J. Hazard. Mater. Adv. 7, 100131. https://doi.org/10.1016/j.hazadv.2022.100131.

Pérez, D.J., Menone, M.L., 2024. Effects and Translocation of Pharmaceuticals and Personal Care Products Into Aquatic Plants. Springer Science and Business Media LLC. https://doi.org/10.1007/698_2024_1187.

Pérez, D.J., Okada, E., Menone, M.L., Costa, J.L., 2017. Can an aquatic macrophyte bioaccumulate glyphosate? Development of a new method of glyphosate extraction in Ludwigia peploides and watershed scale validation. Chemosphere 185, 975–982. https://doi.org/10.1016/j.chemosphere.2017.07.093.

Pérez, D.J., Lombardero, L.R., Doucette, W.J., 2023. Influence of exposure time, physicochemical properties, and plant transpiration on the uptake dynamics and translocation of pharmaceutical and personal care products in the aquatic macrophyte Typha latifolia. Sci. Total Environ. 896. https://doi.org/10.1016/j.scitotenv.2023.165107. Available from: www.elsevier.com/locate/scitotenv.

Pesticide Properties DataBase, 2025. Available from: https://sitem.herts.ac.uk/aeru/ppdb/en/.

Pi, N., Ng, J.Z., Kelly, B.C., 2017. Uptake and elimination kinetics of perfluoroalkyl substances in submerged and free-floating aquatic macrophytes: results of mesocosm experiments with Echinodorus horemanii and *Eichhornia crassipes*. Water Res. 117, 167–174. https://doi.org/10.1016/j.watres.2017.04.003.

Pietrini, F., Passatore, L., Fischetti, E., Carloni, S., Ferrario, C., Polesello, S., Zacchini, M., 2019. Evaluation of morpho-physiological traits and contaminant accumulation ability in Lemna minor L. treated with increasing perfluorooctanoic acid (PFOA) concentrations under laboratory conditions. Sci. Total Environ. 695, 133828. https://doi.org/10.1016/j.scitotenv.2019.133828.

Pubchem, (2025), Available from: https://pubchem.ncbi.nlm.nih.gov/compound/Perfluorooctanesulfonic-acid.

Qian, J., Lu, B., Chen, H., Wang, P., Wang, C., Li, K., Tian, X., Jin, W., He, X., Chen, H., 2019. Phytotoxicity and oxidative stress of perfluorooctanesulfonate to two riparian plants: Acorus calamus and Phragmites communis. Ecotoxicol. Environ. Saf. 180, 215–226. https://doi.org/10.1016/j.ecoenv.2019.04.078.

Sharma, N., Barion, G., Shrestha, I., Ebinezer, L.B., Trentin, A.R., Vamerali, T., Mezzalira, G., Masi, A., Ghisi, R., 2020. Accumulation and effects of perfluoroalkyl substances in three hydroponically grown Salix L. species. Ecotoxicol. Environ. Saf. 191. https://doi.org/10.1016/j.ecoenv.2019.110150. Available from. http://www.elsevier.com/inca/publications/store/6/2/2/8/1/9/index.htt.

Stockholm Convention on Persistent Organic Pollutants, 2025. United Nations Environment Program. Available from:, Decision SC-4/17. https://chm.pops.int/Implementation/IndustrialPOPs/PFAS/Decisions/tabid/5222/Default.aspx.

Stockholm Convention on Persistent Organic Pollutants, 2025. United Nations Environment Program. Available from:, Decision SC-9/12. https://chm.pops.int/Implementation/IndustrialPOPs/PFAS/Decisions/tabid/5222/Default.aspx.

Thomaz, S.M., 2021. Ecosystem services provided by freshwater macrophytes. Hydrobiologia 1.

USEPA, 2022. Drinking Water Health Advisories for GenX Chemicals and PFBS. US EPA. Available from: https://www.epa.gov/sdwa/drinking-water-health-advisories-genx-chemicals-and-pfbs.

Wang, T.T., Ying, G.G., Shi, W.J., Zhao, J.L., Liu, Y.S., Chen, J., Ma, D.D., Xiong, Q., 2020. Uptake and translocation of perfluorooctanoic acid (PFOA) and perfluorooctanesulfonic acid (PFOS) by wetland plants: tissue-A nd cell-level distribution visualization with desorption electrospray ionization mass spectrometry (DESI-MS) and transmission electron microscopy equipped with energy-dispersive spectroscopy (TEM-EDS). Environ. Sci. Technol. 54 (10), 6009–6020. https://doi.org/10.1021/acs.est.9b05160.

Wang, Y., Xiao, Y., Wu, H., Zhang, K., Tian, X., Cao, X., Wang, Q., Lei, N., Zheng, Z., Zhang, W., Gu, P., Qi, L., 2023. Ecological effects and molecular mechanisms of single and coexisting PFOS and Cu exposure on submerged macrophytes and periphytic biofilms in aquatic environments. Environ. Exp. Bot. 213, 105435. https://doi.org/10.1016/j.envexpbot.2023.105435.

Wee, S.Y., Aris, A.Z., 2023. Revisiting the "forever chemicals," PFOA and PFOS exposure in drinking water. Npj Clean Water 6 (1). https://doi.org/10.1038/s41545-023-00274-6. Available from: https://www.nature.com/npjcleanwater/.

World Flora Online, 2025a. Vallisneria Natans (Lour.) H. Hara. Available from: http://53 https://www.worldfloraonline.org/taxon/wfo-0000770251#distributionMap.

World Flora Online, 2025b. *Eichhornia crassipes* (Mart.) Solms. Available from: https://www.worldfloraonline.org/taxon/wfo-0000401771.

Xiao, Y., Qi, L., Yang, Y., Zhang, Y., Shen, Y., Liu, J., Lei, N., Zhang, W., Wang, Q., 2024. Unravelling the mechanisms of PFAS toxicity to submerged macrophytes and epiphytic biofilms at metabolic and molecular levels. Sci. Total Environ. 952, 175726. https://doi.org/10.1016/j.scitotenv.2024.175726.

Yan, N., Ji, Y., Zhang, B., Zheng, X., Brusseau, M.L., 2020. Transport of GenX in saturated and unsaturated porous media. Environ. Sci. Technol. 54 (19), 11876–11885. https://doi.org/10.1021/acs.est.9b07790.

Yang, L.H., Yang, W.J., Lv, S.H., Zhu, T.T., Adeel Sharif, H.M., Yang, C., Du, J., Lin, H., 2022. Is HFPO–DA (GenX) a suitable substitute for PFOA? A comprehensive degradation comparison of PFOA and GenX via electrooxidation. Environ. Res. 204. https://doi.org/10.1016/j.envres.2021.111995. Available from: http://www.elsevier.com/inca/publications/store/6/2/2/8/2/1/index.htt.

Yang, X., Ye, C., Liu, Y., Zhao, F.J., 2015. Accumulation and phytotoxicity of perfluorooctanoic acid in the model plant species *Arabidopsis thaliana*. Environ. Pollut. 206, 560–566. https://doi.org/10.1016/j.envpol.2015.07.050.

Yang, Y., Li, Q., Shen, Y., Wei, R., Lan, Y., Wang, Q., Lei, N., Xie, Y., 2023. Combined toxic effects of perfluorooctanoic acid and microcystin LR on submerged macrophytes and biofilms. J. Hazard Mater. 459, 132193. https://doi.org/10.1016/j.jhazmat.2023.132193.

Zhang, D.Q., Wang, Mo, He, Q., Niu, X., Liang, Y., 2020. Distribution of perfluoroalkyl substances (PFASs) in aquatic plant-based systems: from soil adsorption and plant uptake to effects on microbial community. Environ. Pollut. 257, 113575. https://doi.org/10.1016/j.envpol.2019.113575.

Zhang, D., Zhang, W., Liang, Y., 2019. Distribution of eight perfluoroalkyl acids in plant-soil-water systems and their effect on the soil microbial community. Sci. Total Environ. 697, 134146. https://doi.org/10.1016/j.scitotenv.2019.134146.

Zhang, W., Liang, Y., 2021. Interactions between Lemna minor (common duckweed) and PFAS intermediates: perfluorooctanesulfonamide (PFOSA) and 6:2 fluorotelomer sulfonate (6:2 FTSA). Chemosphere 276, 130165. https://doi.org/10.1016/j.chemosphere.2021.130165.

Zhang, W., Liang, Y., 2020. Removal of eight perfluoroalkyl acids from aqueous solutions by aeration and duckweed. Sci. Total Environ. 724, 138357. https://doi.org/10.1016/j.scitotenv.2020.138357.

Zhang, W., Li, Q., Yang, Y., Yu, Y., Li, S., Liu, J., Xiao, Y., Wen, Y., Wang, Q., Lei, N., Gu, P., 2023. Joint toxicity mechanisms of perfluorooctanoic acid and sulfadiazine on submerged macrophytes and periphytic biofilms. J. Hazard Mater. 458, 131910. https://doi.org/10.1016/j.jhazmat.2023.131910.

Zhao, H., Chen, C., Zhang, X., Chen, J., Quan, X., 2011. Phytotoxicity of PFOS and PFOA to Brassica chinensis in different Chinese soils. Ecotoxicol. Environ. Saf. 74 (5), 1343–1347. https://doi.org/10.1016/j.ecoenv.2011.03.007.

Zhou, Y., Wang, T., Jiang, Z., Kong, X., Li, Q., Sun, Y., Wang, P., Liu, Z., 2017. Ecological effect and risk towards aquatic plants induced by perfluoroalkyl substances: bridging natural to culturing flora. Chemosphere 167, 98–106. https://doi.org/10.1016/j.chemosphere.2016.09.146.

Zhu, J., Wallis, I., Guan, H., Ross, K., Whiley, H., Fallowfield, H., sarophorus, J., 2022. A native Australian species, tolerates and accumulates PFOS, PFOA and PFHxS in a glasshouse experiment. Sci. Total Environ. 826, 154184. https://doi.org/10.1016/j.scitotenv.2022.154184.

CHAPTER 8

Phytotoxicity of pharmaceuticals and personal care products

Swati Singh and Achlesh Daverey
School of Environment and Natural Resources, Doon University, Dehradun, Uttarakhand, India

1. Introduction

Pharmaceuticals and personal care products (PPCPs), are a group of chemicals that constitutes medical drugs (prescriptive and nonprescriptive drugs) such as antibiotics, analgesics, antiseptics, diuretics, β-blockers as pharmaceuticals (Zeng et al., 2022; Raina et al., 2024) and personal care products including cosmetics, preservative, shampoos, toothpaste and toiletries (soaps and detergents), and fragrances (Al-Baldawi et al., 2021; Mukherjee et al., 2024; Singh et al., 2024). Due to overconsumption and improper disposal of these compounds, they get transported indiscriminately to different environmental matrices including the terrestrial and aquatic ecosystem, and ultimately get transferred into the wastewater treatment plants (WWTPs). Conventional wastewater treatment facilities are not designed for the elimination of these micropollutants or their harmful metabolites as their treatment goal is to provide treated water free of organic matter, nutrient load, solids and pathogens, and therefore does not remove PPCPs entirely (Singh et al., 2024). Biosolids are nutrient rich organic residuals that can be applied as fertilizer after treatment and processing. Land application of biosolids as fertilizer is a common practice because it is inexpensive resource which provides nutrients back to the soil and its disposal provide waste management route of these abundant material from treatment facilities (Clarke and Cummins, 2015). Some of the PPCPs therefore, gets introduced into the soil matrix via application of biosolids (treated sewage sludge) for agriculture (Jayampathi et al., 2019). A number of PPCPs have been found in biosolids which may range from ng. kg^{-1} to mg. kg^{-1} (Clarke and Cummins, 2015). For example, in a study, high concentration of ciprofloxacin and triclocarban (1590 μg. kg^{-1} and 1400 μg. kg^{-1}) was found in biosolids (Kumar et al., 2023). Reclaimed water, manure and compost application may potentially release PPCPs into the agricultural soil systems (Keerthanan et al., 2021) causing ecotoxicity. For instance, gemfibrozil, a lipid regulator was found in the wastewater influent (63.8 μg. L^{-1}) as well as in the treated water (19.4 μg. L^{-1}), suggesting that the drug could reach groundwater following the land application of treated water (Fang et al., 2012). Similarly, several researchers have emphasized on the concerns related to discharge of PPCPs in environment and its harmful impacts to humans, animals and plants (Fang et al., 2012; Clarke and Cummins, 2015; Carter et al., 2018; Xu et al., 2021; Wei et al., 2023; Singh et al., 2024). In order to establish the

Phytotoxicity of Organic Pollutants
ISBN 978-0-443-33110-7
https://doi.org/10.1016/B978-0-443-33110-7.00007-9

relationship and dynamics between PPCPs and their uptake by plants, it is important to understand the properties of these organic pollutants and their distribution across environmental matrices (Carter et al., 2014). The chemical structures of PPCPs are stable, they are often resistant to biodegradation which makes them bioavailable to both aquatic and terrestrial ecosystems (Hena et al., 2021) thus giving rise to potential uptake by plants via root and further accumulation in plants tissues (Al-Farsi et al., 2017; Bartrons and Peñuelas, 2017). PPCPs uptake and accumulation by plants may pose potential threat to human beings and other feeding organisms such as herbivores. PPCPs may get accumulated in the plant tissues in the range of ng.kg^{-1} to mg. kg^{-1}, their metabolism and bioaccumulation may lead to potential toxicity in plants (Carter et al., 2014; Wei et al., 2023). Phytotoxicity refers to any adverse effect on plants such as delay in seed germination, decreased photosynthesis and respiration that inhibits plant growth and development. The concerns related to the possible toxic effects of these emerging contaminants on the plants have increased immensely in recent years due to dearth of knowledge in this area. The objective of this chapter is to discuss the physicochemical factors affecting the PPCPs uptake in various sections of plants and the phytotoxic responses such as inhibitory effects on seed germination, photosynthesis, respiration and plant growth, triggered due to the accumulation of these emerging contaminants.

2. PPCPs in water and soil (sources and discharge)

Increased awareness of maintaining health and hygiene over the past few decades had led to incorrect and sometimes overconsumption of PPCPs, but incorrect disposal has transported them in different matrices including water and soil leading to contamination (Zind et al., 2021). PPCPs and their residues enter the environment through various routes (Fig. 8.1).

In water, PPCPs have been detected in river water, seawater, sewage effluents, ground water through leaching, and their concentrations have been found to range from ng. L^{-1} to mg. L^{-1} (Fang et al., 2012). Wastewater discharge is the main source of pharmaceuticals to enter the surface water. Another route for drugs to enter the water circle is either by direct discharge from hospitals, industries, improper disposal of unused medicines by flushing into the toilets and drains or dumping in household garbage (Diaz-Cruz et al., 2003; Zhang et al., 2008). Upon consuming the pharmaceuticals, parent compound undergoes biotic or abiotic transformations . They break down into more soluble and polar forms with the help of variety of conjugative and oxidative enzymes, for example acetylation, glutathione conjugation, amino acid conjugation and methylation etc. (Zhao et al., 2021). Depending on the half-life of the drugs the percentage of drug metabolized often varies significantly, but in general about 70% of the administered drugs or their active metabolites are passed through urine and feces (Fang et al., 2012; Veiga-Gómez et al., 2019) and ultimately to WWTPs, then into other water bodies. In surface waters for example, Zind et al. (2021) reported a highly consumed drugs like carbamazepine (CBZ) and

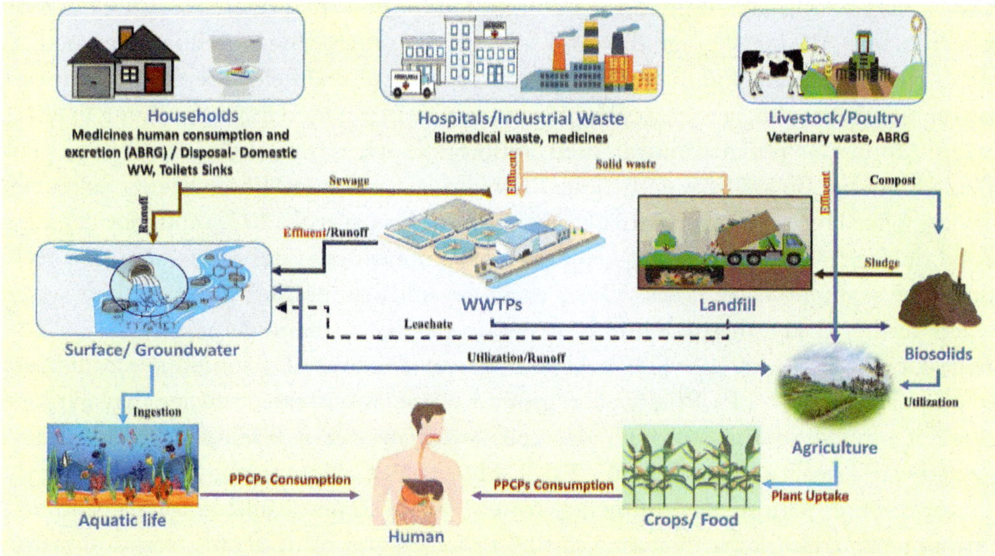

Figure 8.1 ***Sources and distribution of PPCPs in environment.*** Schematic representation of the major sources of PPCPs and their subsequent distribution across various environmental compartments including water, wastewater, soil and biota. *(From Singh, S., Pant, A., Dutta, K., Rani, R., Vithanage, M., Daverey, A., 2024. Environ. Chem. Ecotoxicol. 6, 104–116. https://doi.org/10.1016/j.enceco.2024.04.001.)*

diclofenac (DCF) in water, sediment and biofilms of three French rivers with mean concentrations of 0.52 μg. L^{-1}, 229 ng. g^{-1} and 2153 ng. g^{-1} respectively. In another study, Sharma et al. (2019) reported the PPCPs in the range approximately from 54 ng. L^{-1}–826 ng. L^{-1} in the river Ganges, India, among which stimulant caffeine was found to be highest which is also used in creams and other cosmetic product. Similarly personal care products, like surfactants and antimicrobial agent triclosan have been reported by many studies in the past (Yuan et al., 2019; Wang et al., 2024). Surfactants in the sewage have been reported to be in the range 3 mg. L^{-1}–20 mg. L^{-1} (Jena et al., 2023). In addition, the surfactant pollution in rivers, lakes, open and bore well have been evidently reported in the concentrations (2.5 mg. L^{-1}, 9.5 mg. L^{-1}, 50 g. L^{-1} and 720 g. L^{-1}) (Chaturvedi and Kumar, 2010; Seth et al., 2013). In a study, triclosan levels were found in the range of 55 ng. L^{-1}–134 ng. L^{-1} at Tone Canal, Chiba, Japan (Nishi et al., 2008). According to Zhang et al. (2015), approximately 91% of the triclosan was getting discharged in water bodies and 9% in the soil in China. PPCPs gets introduced into the soil metric in two ways from direct and indirect emission sources. Point sources include the identifiable sources from which PPCPs may enter in soil including municipal and industrial discharge, use of effluents from the WWTPs for irrigation in agricultural fields, sewer overflows, sewage sludge, dumping of solid waste in landfills containing expired medications or their bio-transformed drugs among others (Dey et al., 2019; Okoye et al., 2022). Karnjanapiboonwong et al. (2011) reported the presence of PPCPs in the concentration 319 ng. g^{-1} in the land application site receiving effluents treated for past 70 years. Biosolids and manure applied to the agricultural field as fertilizer often contains antibiotics and analgesics, as pharmaceuticals are being used for maintaining livestock health. Among the pharmaceuticals used, antibiotics such as trimethoprim-sulfa and penicillin administration are frequently done for veterinary purposes. Pharmaceuticals remains unmetabolized by cattle (approximately 90%–95%) (Mojiri et al., 2022) and they get transferred during field application of manure. Consumption of water containing PPCPs by cattle, for example, dairy cows, allows these organic compounds to enter their system which further leads pharmaceuticals to enter the food chain (Moreno et al., 2017; Veiga-Gómez et al., 2019). The leakage in sewer conduits also sometimes contributes to soil pollution with PPCPs as well as ground water, as untreated sewage may exfiltrate into unconfined aquifers through decrepit sewer networks leading to compromised ground water quality (Kuroda et al., 2012). Hence, PPCPs enters the soil and water dynamics through various environmental routes and becomes available for plant uptake. Further to understand the dynamics of PPCPs in soil and plant uptake, sorption process and factors influencing mobility and uptake of these micropollutants by plants needs to be explored.

3. Sorption, translocation, and bioaccumulation of the PPCPs within plants

The fate and transport of PPCPs is influenced by different processes and factors, such as sorption which includes both adsorption and absorption in soil and rhizosphere; root uptake and foliar uptake, governed by diffusion through cell membrane in case of roots and via stomata in leaf (Briggs et al., 1982); as well as translocation of the organic contaminants to the above ground tissues and their subsequent accumulation in the plant tissues (Limmer and Burken, 2014; Miller et al., 2016). Several physicochemical factors such as molecular size of the PPCPs, ionic properties and hydrophobicity also play a crucial role in the uptake and translocation of these organic compounds in the different parts of the plants due to the transpirational pull and root pressure.

3.1 Sorption in soil

Sorption is an important process that affects the fate and transport (mobility) of the organic pollutants in the soil (Yu et al., 2013). In some cases, the pollutants are not retained in the soil and show higher mobility while others show less mobility as they are strongly retained in the soil. One of the factors on which persistence of the PPCPs depends on is its half-life ($t_{1/2}$), which may be hours, weeks, months or even years (Fang et al., 2012; Pérez-Lucas and Navarro, 2024). Half-life ($t_{1/2}$) refers to the time required for the drug to reduce to half of its initial concentrations (Smith et al., 2017). The other factor for sorption of PPCPs on soil is the composition and type of soil. The mobility of PPCPs is also based on the sorption capacity of soil for these organic contaminants. For instance, under aerobic conditions sorption capacity of the gemfibrozil drug in soil was in the order silt loam > sandy loam > sand (Fang et al., 2012). Soil can be divided into three sorption categories with different sorption capacity: (1) strong sorption capacity e.g., montmorillonite and humic acids (HAs), (2) medium sorption capacity, e.g., iron oxide, (3) poor sorption capacity e.g., sand and scree (Scheidegger and Sparks, 1996; Chang et al., 2016). Soil properties like pH, cation exchange capacity, organic carbon content, texture and environmental temperature also play an important role in sorption of PPCP in soil. For instance, Kočárek et al. (2016) evaluated the impact of soil properties on sorption of pharmaceuticals via sorption isotherms and reported that sorption of ionizable pharmaceuticals was highly affected by soil pH. A low sorption coefficient (*Kd*) of organic substances often indicates that they have lower affinity for soil particles due to development of negative charge on both PPCPs and the soil surface with higher pH. For example, surface charge of the soil (clay, minerals and organic matter) and most of the PPCPs tend to be negatively charged at higher pH (Xu et al., 2021). PPCPs emerge as cationic, anionic, nonionic and zwitterionic modalities individually or simultaneously depending on different *pKa* values. Sorption behavior of ionizable PPCPs is strongly influenced by pH-dependent molecule speciation (Schaffer and Licha, 2015). Depending on the surface charge on PPCPs, its interactions with the soil components differ for each organic compound. For example, in ionic compounds, electrostatic interactions including cation bridging, cation exchange and surface complexation play an important role in adsorption of organic compounds. In cationic molecules, sorption occurs with negatively charged soil (clay surface) whereas anionic molecules get attached to the positive charges on soil surface. In case of nonionic moieties, they get attracted to soil organic matter (SOM) by hydrophobic interactions and via electron donor–acceptor and van der Waals interactions (Xu et al., 2021; Alhalabi et al., 2024). Sorption behavior is also influenced by the texture of the soil i.e., soil minerals and SOM, therefore understanding the components of soil is important. For instance, clay being the mineral component of the soil is divided according to the different stacking ratios, sequence of silicon wafers and aluminum sheet (1:1 type, 2:1 type and 2:1:1 type). Montmorillonite has 2:1 layered silicate structure which imparts a permanent negative charge to the soil, it has large surface area and higher cation exchange capacity which aids in the adsorption of organic pollutants.

In the case of humus when the organic matter content is high the surface charge varies significantly. Humus has various active groups, such as carboxyl (-COOH), amino ($-NH_2$), hydrosulfide (-SH), ketone (-C=O), phenolic hydroxyl ($-C_6H_5OH$) and enol (-COH=CH), which contribute to its overall surface charge (Kočárek et al., 2016) The dissociation of hydroxyl or carboxyl functional groups provides the negative surface charge to the humus whereas protonation of amino groups leads to the positive charge to the surface. Chefetz et al. (2008) studied the mobility, sorption and desorption pattern of pharmaceutical drugs CBZ, diclofenac and naproxen in soil irrigated with reclaimed wastewater at different soil depth (0–5 cm, 5–15 cm and 15–25 cm). They reported that the quantity and physicochemical properties of SOM affected the mobility of PPCPs. Mobility reduced with decreasing organic matter at higher depth. CBZ had slower mobility in SOM rich soil depth i.e., 0–5 cm. CBZ was possibly adsorbed due to hydrogen bonding and hydrophobic interaction as it is nonionic. However, in another report by Williams and McLain (2012), they found low levels of CBZ (0.18 ng. g^{-1} in soil) at depth 0–5 cm when compared with higher depth (>10 cm) suggesting possible degradation or leaching, whereas caffeine accumulation was found higher in soil. Hence, the sorption capacity and mobility of different PPCPs on soil depends on various factors including soil components, different chemical interaction with the soil surface and soil depth etc.

3.2 Uptake, translocation and accumulation of PPCPs by plant

Organic pollutants enter the plant tissues by two pathways namely: by root uptake and through leaves (foliar uptake). The most prominent pathway for PPCPs uptake via plant is through roots. Plants roots may uptake PPCPs adsorbed on soil particles or from soil water interphase in similar manner in which roots absorb nutrients and water (Kurade et al., 2021; Wei et al., 2023). Plant roots can absorb the PPCPs which are bound to the soil particles either by direct contact of roots with the high concentration of PPCPs in the soil, or absorption of desorbed PPCPs in soil water interphase making the contaminant available for root uptake. For example, the ability of plants to uptake pharmaceuticals from soil have been advocated in a study conducted by Carter et al. (2014) in which the CBZ, diclofenac, and sulfamethazine were detected in plant tissues of *Lolium perenne* (ryegrass) and *Raphanus sativa* (radish).

The organic pollutant enters the plant root largely by passive diffusion (less through active transport), as plants lack specific transporters for PPCPs on their cell membranes. PPCPs enter the plants first via root hair followed by cortex, endodermis and casparian strip before entering to the xylem or phloem vessel (Miller et al., 2016; Al-Farsi et al., 2017). The root cortex comprises loosely packed parenchyma cells with intercellular spaces for water and solutes movement which helps in the apoplastic movement. The translocation of water and minerals through xylem, and organic solutes by phloem is greatly influenced and facilitated by transpirational pull, root pressure, water potential gradient and cohesive and adhesives forces (Dodgen et al., 2015). The soil–plant–atmosphere continuum (SPAC) is maintained by the movement of water from soil and then plant and further to atmosphere. This occurs due to water movement that by convention occurs from higher water potential area to lower water potential area.

The organic pollutants during intercellular transport from soil to root and then to casparian strip may follow three pathways for short distances to reach vascular tissues: (1) symplastic movement (2) apoplastic movement and, (3) via vacuolar transport. In the symplastic pathway, movement occurs through cell cytoplasm interconnected through plasmodesmata (cell to cell transport) (Al-Farsi et al., 2017; Kurade et al., 2021). PPCPs primarily follows symplastic pathway, although it depends on plant species and PPCPs type. In apoplastic route, the movement occurs through the intercellular spaces and through cell walls i.e., outside the cytoplasm. Water following the apoplastic pathway is blocked at the endodermis by casparian strip that consists of lignin and suberin which acts as hydrophobic barrier and seals off the route of water and impels water to enter the vascular tissue through symplastic pathway. Sometimes water and minerals take vacuolar pathway i.e., through one vacuole to another facilitated by plasmodesmata (Fig. 8.2). Some compounds may get transported through one or more pathways (alternate symplastic or apoplastic transport) within the plant roots (Zhang et al., 2017). In general, hydrophobic PPCPs tend to follow the apoplastic pathway as cell wall has some hydrophobic properties and similarly, symplastic pathway is preferred by the hydrophilic PPCPs (Wei et al., 2023). However, PPCPs not always follow this trend as at endodermis all PPCPs are driven by symplastic pathway into the vascular tissue resulting in hydrophobic substances to follow the cell-to-cell transport. The properties of PPCPs like water solubility, lipophilicity and charge among others define that from which vascular tissue (xylem or phloem) does organic pollutant shall be transported. For example, more water-soluble PPCPs get transported through xylem and reach plant parts like stem and leaves through transpiration (Zhang et al., 2017). The organic pollutants may also get transferred to plant parts like fruit through the phloem transport due to pressure gradient which is least affected by the transpiration (Miller et al., 2016) (Fig. 8.2). Foliar transport is another pathway for PPCPs uptake which occurs in leaf surface through stomata or cuticle. In stomatal route, the PPCPs in vapor phase such as fragrances enters the stomata during the gaseous exchange along with CO_2 and gets subsequently transferred to the other plant compartments via phloem including roots (Bartrons and Peñuelas, 2017). The absorption of PPCPs from air may be affected by various factors like vapor pressure, hydrophobicity, and gas concentrations.

Carter et al. (2018) reported high concentration of benzodiazepines in leaves than in radish bulb, suggesting PPCPs accumulation in plant parts also depends on the octanol–-water partition of the organic pollutants. The quantity of water required by each plant species is different and it contributes to the accumulation of PPCPs in crops when irrigated with recycled water. Environmental conditions like higher atmospheric temperature influences speed of transpiration stream and pull which in turn affects the accumulation in different plant parts (Pérez et al., 2023). For example, in summers leafy and fruity crops have higher evapotranspiration rate (190 mm in some cases) which declines in winters.

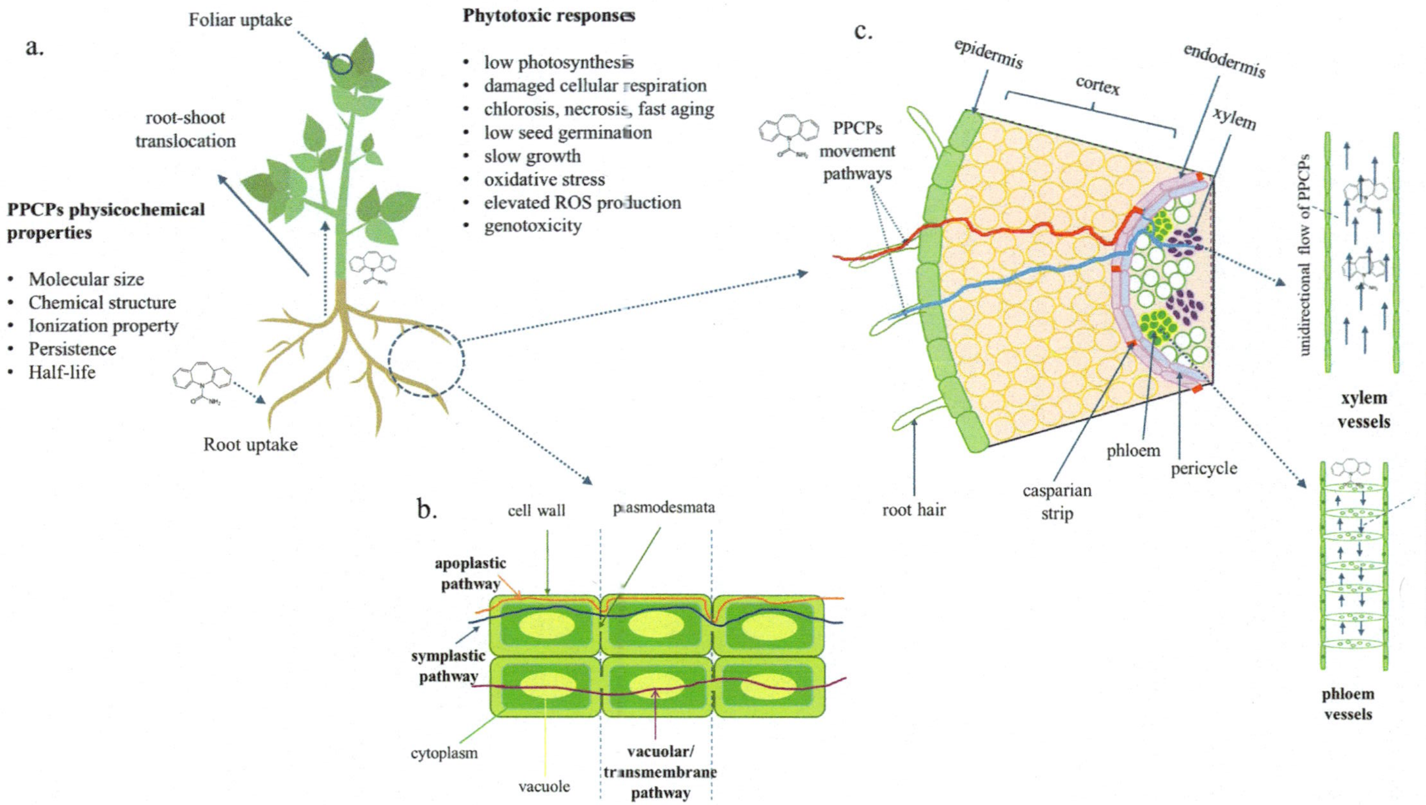

Figure 8.2 ***Plant uptake and transport mechanisms of PPCPs.*** Physicochemical factors of PPCPs affecting plant uptake and potential phytotoxic responses by plants against organic pollutants (a). Different routes for potential PPCPs uptake by root hair (b). Transportation of PPCPs via xylem and phloem vascular bundles (c).

In a study conducted by Azad and Liu (2024), they reported higher accumulation of DCF (among eight studied PPCPs in hydroponic system) in six leafy vegetable crops (lettuce, cabbage, cauliflower, spinach, collard and Wisconsin fast plant) when compared with three fruity vegetable crops (pepper, pea and cucumber) in water availability conditions, and vice versa in water limiting condition suggesting higher accumulation of PPCPs in leaves in the direction of transpiration stream (root to edible part). Hence, in general a higher transpiration rate may lead to the accumulation of solutes and nutrients in plants.

3.3 Physicochemical factors affecting plant uptake, translocation and bioaccumulation

3.3.1 Physicochemical properties of PPCPs

The uptake of PPCPs is influenced by their physicochemical properties such as environmental conditions, hydrophobicity, pH and ionization properties of chemical compound and molecular size.

3.3.1.1 Hydrophobicity and water solubility of PPCPs

The hydrophobicity and water solubility of PPCPs are important factors to consider while accessing the plant uptake and potential toxicity caused by PPCPs. The *Kow* or the octanol–water partition coefficient defines the lipophilicity (*Kow* >1) and hydrophilicity (*Kow* <1) of an organic compound. The uptake of PPCPs by plants is highly dependent on *Kow*, as the plant uptake is better for PPCPs with higher hydrophobicity, log *Kow* > 4, and can be easily transported from root to shoot (Vo et al., 2018) than hydrophilic PPCPs (*Kow* < 1). PPCPs with high hydrophilicity and polarity are likely to move across slower resulting in plant uptake to occur relatively at slower rate. Chemical compounds with high hydrophilicity potentially get translocated and absorbed to the root (Hu et al., 2021a). The PPCPs with log *Kow* values from 1-4, are easily translocated from root to shoot of the plants (Kinney and Heuvel, 2020). However, hydrophobicity alone is not responsible for the uptake of PPCPs. In some cases, studies have revealed that the slight change in the chemical structure and pH of a compound changes its solubility, *pKa* and polarity. This suggests that hydrophobicity of the chemical compound is not the only parameter that determines the uptake of the pollutant in the plants.

3.3.1.2 Molecular size and chemical structure of chemical compound

The molecular size and chemical structure of the PPCPs affects the transfer of organic pollutants across varied plant parts. Integral proteins on bio-membranes may allow ionic PPCPs to enter plant cells, primarily due to concentration gradients (Miller et al., 2016). Ionic PPCPs with small molecular weight of MW $< 1000\ g\ mol^{-1}$ may diffuse prominently across plant cell membrane and show higher accumulation in roots than PPCPs with higher MW (Wei et al., 2023). For instance, PPCPs with MW $< 300\ g\ mol^{-1}$ (CBZ and caffeine) may diffuse in lettuce roots and further transfer to shoot effectively (Keerthanan et al., 2021). Large sized PPCPs may take apoplastic pathway to enter the roots with slower mobility due to slow diffusion rate and may result in less accumulation in plant shoot. However, more contact time may result in higher uptake of large sized PPCPs.

Chemical structure of the PPCPs like alkyl chain length, number of aromatic rings and different groups plays a crucial role in determining if the organic compound shall be accumulated or transpire back into the environment. Toxic effects on plants due to accumulation of PPCPs with different structure is potentially different as they may disrupt plant cell structures and hinder cellular pathways while directly interacting with plant cellular components (Singh et al., 2025). The shape and functional groups of the PPCPs may affect its ability to bind with cell receptors and enzymes which might inhibit plants metabolic activities and growth pattern.

3.3.1.3 pH and ionization properties

PPCPs exhibits diverse ionization properties due to presence of varied functional groups and has different acid dissociation constants which dissociates at specific pH. In the case of ionic PPCPs their sorption and accumulation is greatly influenced by the electrostatic interaction (attraction and repulsion) of anionic, cationic, zwitterionic species and its dissociated ions with the cell membrane (Franklin et al., 2022). As stated earlier, PPCPs with charge tends to move slower than neutral species as lipids are major constituent in plant cell membranes (Trapp et al., 2023). Different pH in various plant cell organelles may change charge of PPCPs while ion trapping. For instance, at low pH (acidic conditions) PPCPs with acidic groups may become negatively charged due to lose of proton (H^+). Cationic and neutral species tends to move upwards (above ground tissues) whereas anionic species tends to remain in the root tissues due to electrostatic repulsion with plant cell membrane (Miller et al., 2016). In case of weak acids or bases, partial dissociation may result in neutral or ionized species. The uptake of weak acids by plants increases with the pH difference, when the pH outside the cell compartment decreases as compared to inside of the cell. During the shift in pH the equilibrium tends to shift toward neutral form of weak acid, as more H^+ ions in acidic environment keep weak acid in its undissociated or neutral form, and these molecules cross cell membrane barrier easily (Goldstein et al., 2014).

3.3.1.4 Persistence and half-life of PPCPs

The uptake of organic pollutants in plants depends on half-life and subsequent persistence of PPCPs in environment. Persistence means the time for which the PPCP remains in the environment and half-life ($t_{1/2}$), as stated earlier is the time taken by PPCPs to degrade half of its initial concentration. PPCPs with short half-lives are less likely to be available for plant uptake due to their faster breakdown whereas emerging pollutants with longer half-lives are more likely to persist in the environment. Fang et al. (2012) in their study reported that dissipation half-lives of the gemfibrozil drug in sandy and silt soil were 17.8 and 20.6 days. In another study, the $t_{1/2}$ of the same organic pollutant in biosolids-amended soil was 224–231 days (Walters et al., 2010) suggesting difference in persistence was dependent on the medium i.e., biosolids and soil. Similarly, the dissipation of pharmaceuticals such as atenolol and amitriptyline, follow first-order exponential decay with $t_{1/2}$ in sediments ranging from 9.5 days for atenolol and 78.8 days for amitriptyline (Al-Khazrajy, Bergström and Boxall, 2018). Highly persistent PPCPs tends to be available for longer time and may get adsorbed onto the soil surface and become available for plant uptake. According to Ying et al. (2007), triclosan and triclocarban persist longer in anaerobic soil than aerobic soil with half lives ($t_{1/2}$) of 108 and 18 days, respectively. Likewise, CBZ is recognized as one of the persistent organic compounds in surface waters, biosolids–soil mixtures (Walters et al., 2010; Du et al., 2023) and surrounding factors like water depth, latitude and solar irradiance significantly affects the $t_{1/2}$ and contributes in the persistence of the organic pollutant as it delays the process of PPCPs degradation. Other PPCPs like norfloxacin, ofloxacin and doxycycline have also shown persistence in biosolid matrix in the past.

3.3.2 Plant-specific factors

Plant specific factors such as transpiration rate of different species, lipid and carbohydrate content of the cell membrane, protein channels, presence of rhizosphere bacteria, leaf morphology etc., also contributes in the uptake, translocation and accumulation of PPCPs in plants (Goldstein et al., 2014; Dodgen et al., 2015). The extent of translocation of PPCPs is majorly defined by the rate of transpiration. For example, in a study conducted by Ben Mordechay et al. (2018), on uptake of CBZ and its metabolites in tomato, wheat and lettuce plants, they declared transpiration as a major factor for the accumulation of nonionic CBZ and suggested that younger leaves accumulate less pollutant than outer older leaves in lettuce (higher transpiration occurs in the matured leaves). The environmental factors like wind speed, temperature, humidity, water content and growth stage, also contributes in transpiration rate and thereby influence the accumulation, translocation and bioconcentration of PPCPs in the leaves (Singh et al., 2025) and fruits. For instance, Dodgen et al. (2015) reported that warm and dry environment led to 3–4-fold increase in the rate of transpiration than cold-dry condition for the lettuce and tomato (same plant species), leading to higher translocation in PPCPs in higher rate of transpiration. They found positive correlation between bioconcentration factor (BCF) for leaf and transpiration of ionic PPCPs.

The BCF is defined as the ratio of concentration of pollutant in plant tissues to its concentration in soil/water. According to the literature, BCF >1 indicates accumulation of chemical in the plant tissues whereas BCF<1 suggests that plant has not accumulated the pollutant. The translocation of PPCPs from root to above ground tissues is measured by translocation factor (TF), given as the ratio of concentration of pollutant in aerial part to the concentration of pollutant in roots (Al-Farsi et al., 2017). Few studies have suggested higher accumulation of PPCPs in roots as compared to shoot areas in the order anionic > neutral ≥ cationic. However, it is not always the case, for example, one of the study reported higher accumulation of CBZ in leaves (BCF >36) of cabbage than in roots (BCF <16) when grown in solution spiked with mixture of PPCPs. Likewise, Singh et al. (2025) reported higher accumulation of sodium dodecyl sulfate (SDS), an anionic surfactant in the aerial part of ornamental plant *Canna indica* than in roots (as compared with *Acorus calamus*) in the vertical flow constructed wetland systems (VFCWs) with higher translocation of SDS (TF = 4.7–17.8), and suggested high transpiration due to higher surface area of leaves as one of the contributing factors. Ravichandran and Philip (2022) suggested that uptake, accumulation and translocation of drug CBZ was dependent on variety of parameters such as plant biomass, transpiration rate, chlorophyll content, lipid content among others. Therefore, the uptake potential and translocation of PPCPs by plants is a complex process and is not dependent on just one factor; they may also be influenced by plant-specific factors such as root lipid and carbohydrate content.

The lipid content is the contributing factor as it varies in cells for different plant species, and influences the intensity of hydrophobic PPCPs uptake by different plants parts. The hydrophobic compounds may retain in the root as they may get partitioned into lipids. Nonionic PPCPs may get accumulated in leaves (in the direction of transpiration stream) (Bartrons and Peñuelas, 2017). According to Azad and Liu (2024), root lipid content was positively related to the concentration of the antibiotics in the root, suggesting lipid content as an important indicator of accessing concentration of PPCPs.

4. Phytotoxic responses in plants caused due to PPCPs

In general, PPCPs are bioactive compounds (except for few personal care products) that may alter plant physiology and hinder biochemical processes (Carter et al., 2015; Amy-Sagers et al., 2017). Generally, the level of toxicity on the different plants species depends on the types of PPCPs and their concentration. In some plant species, the transformed products of the parent compounds or the initial compound itself, may result in toxicity. The long-term exposure of PPCPs in plants, their bioaccumulation and transformation may trigger plant physiological responses like increased oxidative stress, hyperaccumulation of reactive oxygen species leading to cell death, plant hormonal imbalances, impairment in photosynthesis etc. (Pérez and Menone, 2024). These pollutants in higher concentrations may disrupt the cell membrane structure while interacting with the lipids layer leading to potential damage and subsequently hinder the uptake and transport of essential nutrients. These chemical compounds sometimes alter the ion channels' chemistry and does not allow the entry of essential proteins/enzymes and nutrients that also hinders plant growth (Al-Farsi et al., 2017). PPCPs may inhibit soil respiration, soil microbial activities, seed germination and growth rate of plant species (Pandey et al., 2022). The phytotoxic responses in plants differs greatly among terrestrial and aquatic plants due to different physiological and environmental adaptations including exposure routes, uptake mechanisms, tolerance, sensitivity and species-to-species variation. PPCPs exposure in terrestrial plants occur mainly through soil, and the uptake of organic pollutant is influenced by soil properties such as pH, temperature, water content etc. For example, in a study conducted by Carter et al. (2015), the toxicity caused by CBZ resulted in reduced photosynthesis, burnt edges in leaves, changes in plant hormones, and white spots in *Cucurbita pepo* (vegetable plants). Few studies have assessed the phytotoxic responses in aquatic plant species by analyzing toxic effect at individual level by studying their growth rate, changes in leaf area, reduction in fresh weight/plant biomass, leaf number and chlorosis. For instance, Markovic et al. (2021) reported adverse effect of drugs methotrexate, diclofenac, 17 α-ethynylestradiol and fluoxetine, on *Lemna minor* (duckweed) in a dose dependent study. They reported that effects on frond surface area and photosynthetic pigments were observed at low concentrations of drugs. Therefore, biotic and abiotic stress caused in plants may weaken the plant defense mechanism and may sometimes induce cross tolerance.

4.1 Photosynthesis and respiration

The photosynthesis in plants and cellular respiration are crucial plant metabolic processes that are essential for its survival. These metabolic processes may be affected by PPCPs depending upon the concentration and nature of pollutant (Gomes et al., 2017; Fu et al., 2019). Exposure to PPCPs may impact cellular respiration in plants by disrupting the mitochondrial function leading towards the over-production of reactive oxygen species (ROS) which negatively impacts photosystems (particularly PS-II) by interfering in efficient light energy capture and hence reduces photosynthesis (Pospíšil, 2016). Aristilde et al. (2010) reported that secondary amino group and quinolone ring of the antibiotic fluoroquinolone (ciprofloxacin) may facilitate their role in PS-II inhibition, as quinone site inhibitors, a molecule that acts as mobile electron carrier in photosynthetic electron transport chain (ETC), thereby reducing photosynthesis in spinach. Adeel et al. (2018) evaluated impact of estradiol and ethinyl estradiol (steroid estrogens, 10 mg L^{-1}), on lettuce and reported that the accumulation of estrogen was dose-dependent i.e., higher estrogen had negative impact on root growth. On the contrary at lower concentration it enhanced the photosynthetic pigments, root and shoot growth. Liu et al. (2013) also observed that higher concentration of antibiotics (>10 μg. L^{-1}) on wetland plant *Phragmites australis*, had toxic effect on leaf chlorophyll content and root activity while lower concentration (>1 μg. L^{-1}) antibiotics had hormesis effect. Similarly, Kummerová et al. (2016) conducted a study on aquatic plant *L. minor* that showed upto 50% reduction in photosynthetic pigment and decreased chlorophyll fluorescence decay values when exposed to drugs DCF and paracetamol. Gomes et al. (2017) investigated the toxic effect induced by ciprofloxacin in *L. minor* and reported drug induced toxic effect at concentration $\geq$1.05 mg. L^{-1} and hindered electron flow in respiratory ETC leading to production of hydrogen peroxide (H_2O_2) and diminished photosynthesis. Hu et al. (2021b) studied toxicity effects of ciprofloxacin in terms of chemical speciation in *Oryza sativa* (rice plant) and suggested that oxidation and cleavage of piperazine ring were the two possible transformation pathways for changes in cationic and anionic form of ciprofloxacin. They found that photosynthetic pigment contents decreased while antioxidant–enzyme activities were enhanced in pH dose-dependent manner. The adverse effect of ciprofloxacin (0.8 μg. L^{-1}–2 μg. L^{-1}) was found on maize plant causing vascular bundle shrinkage and delayed photosynthetic product transportation (Gomes et al., 2017). Likewise, several other studies reported that exposure to drugs like CBZ and antimycin A in plants (aquatic or terrestrial) may lead to the reduced chlorophyll photosynthetic activity, decreased root activity due to ROS and reduced/or discoloration of plants (Geisler et al., 2004; Ravichandran and Philip, 2022). Although, according to few studies, some antibiotics in low concentration can have hormesis effect in plant and may promote chlorophyll biosynthesis by inhibiting chlorophyllase activity that delays degradation of chlorophyll in the cell which in turn increases chlorophyll content (Liu et al., 2013). However, in majority of cases the exposure to PPCPs in general significantly reduces photosynthetic activity and cellular respiration in plants leading to the production of ROS.

4.2 Seed germination and reproduction

PPCPs may impact seed germination in several ways including delayed germination or complete inhibition of germination. In study conducted by Liu et al. (2009), seed germination in three plant species namely, rice, cucumber and *Avena sativa* (sweet oat) was inhibited in presence of sulfamethoxazole, tetracycline and chlortetracycline. Minden et al. (2017) conducted a study in which they also observed retarded seed germination in presence chlortetracycline (5 mg. L^{-1}). In addition to this, several other studies also reported negative impact of PPCPs on plant seed germination when exposed to antibiotics like less germination of *L. sativa* (lettuce) seeds was reported by Benassi-Borba et al. (2021) in presence of antibiotics. Peng et al. (2021) also observed lower seed germination rate in *Nicotiana tabacum* when exposed to triclosan ($\leq$2 mg. L^{-1}). Similarly, in a pot experiment conducted by Mukhtar et al. (2020) for 4 months, they reported adverse effect of antibiotics (10 mg. kg^{-1}) on seedling growth (root/shoot length) of rice plant. In one of the studies, the reason for stunted growth in the spinach seedlings was impaired synthesis of new chloroplast and mitochondria as estimated by Aristilde et al. (2010). The application of triclosan >30 mg. kg^{-1} was found to inhibit seed germination and plant growth (*O. sativa, T. aestivum, C. sativus* and *S. herbacea*) root elongation with EC50 values for root elongation in *O. sativa* being 57 mg. kg^{-1}. However, some studies revealed that seed germination was not influenced by PPCPs. An et al. (2009) studied biochemical responses on agricultural crop wheat when exposed to triclosan and galaxolide (concentration 50–250 mg. L^{-1}) and reported inhibited root and shoot growth of seedlings whereas less toxic effect was seen in seed germination. Li et al. (2023) studied impact of antibiotics (oxytetracycline and enrofloxacin) on *T. aestivum* and reported that seed germination was not affected by the antibiotic although reduction in root length, leaf surface area and fresh weight was observed when the experimental concentration of the pharmaceuticals were more than 10 mg. L^{-1}. Similarly, Hillis et al. (2011) found in their study that plant germination was not affected by antibiotics upto the concentration of 10 mg. L^{-1} in three plant species namely, *L. sativa*, *Daucus carota* (carrot) and *Medicago sativa* (alfalfa). PPCPs may also interfere in the plant reproduction (both asexual and sexual) cycle by inducing delayed flowering and fruit development and reducing pollen or ovule viability. For example, in a study conducted by Amy-Sagers et al. (2017) on mixotrophic plant *L. minor* significant reduction in plant asexual reproduction, root shrinkage at 323 nmol. L^{-1} and daily growth rate was observed in presence of fluoxetine. However, more research is needed in this direction.

4.3 Hormones/enzyme activity

The uptake of PPCPs in plants may disrupt signaling pathways of plant hormones by either blocking or mimicking like hormones, thereby preventing them from binding to specific receptor. In some cases, PPCPs might interfere in the production of plant hormones. For instance, ciprofloxacin interferes in the auxin production which hampers the root elongation in plants and results in suppressed seedling emergence (Gomes et al., 2020). Organic chemicals can sometimes interact with enzymes responsible for hormone synthesis and inhibit hormone production. In addition to this, imbalances in the hormones due to hindrance in their signaling pathways can lead to physiological changes like seed dormancy and stomatal closure etc. The pathways get interrupted by upregulation or downregulation of genes responsible for production of hormones like auxin, ABA, ethylene, cytokinins, gibberellins and jasmonic acid. Abscisic acid (ABA) is one of the plant hormones which is found in several plant regulation processes like photosynthesis and antioxidant defense system whereas ethylene plays crucial role in plant defense response. One of the studies in tobacco plants have suggested that imbalances in auxin occurred due to downregulation of *22D-like* and *AUX22-like* gene, hindering auxin signaling pathway when exposed to higher concentration of triclosan (Peng et al., 2021). In response to antibiotic stress, wheat plant showed difference in roots and shoot ABA content, authors explained that these changes may be associated with changes in ABA biosynthesis and can act as chemical signal in development of roots and its elongation (Li et al., 2023). Some of the studies have suggested that higher concentration of fluoroquinolone antibiotics inhibit DNA gyrases enzyme activity (chloroplast-specific) in plants like *Phragmite australis* and reported downregulation of gene encoding chlorophyllase suggesting destruction of photosynthetic system (Liu et al., 2013; Peng et al., 2021).

4.4 Growth and development

The growth and development of plants species gets adversely affected on PPCPs exposure and show developmental abnormalities in terms of stunted root/shoot growth, reduced biomass, visible signs of toxicity like yellowing of leaves, necrosis etc., as nutrients uptake and transport essential for maintaining overall health of plants gets compromised (Sunyer-Caldú et al., 2023; Singh et al., 2025). In a study conducted by Pino et al. (2016), toxicity caused by 15 nonsteroidal antiinflammatory drugs (NSAIDs) was tested on *L. sativa* seeds and reported inhibitory effect on hypocotyl and radicle elongation. They observed visible lethal effects on radicle and hypocotyl including necrotic tips, red-orange spots, underdeveloped root hairs, yellow radicles etc. Likewise, Singh et al. (2025) reported visible toxic effects such as yellowing of leaves, stunted growth in *A. calamus* during the treatment of SDS (personal care product) from wastewater in their long-term study of 144 days in VFCWs. PPCPs in plants also indirectly aid in metals and nutrients uptake essential for plant development. However, low or high uptake of these metals can damage plants and crop yield. Peng et al. (2021) observed decreased taproot length, fresh and dry root weight, lateral root numbers in tobacco plant in presence of triclosan. Any deviation from the nutritional composition of the plant may also lead to abnormal functioning and development in plants. For example, the treatment of antibiotics oxytetracycline and chlortetracycline in plants lead to decreased macronutrients (Ca, Mg, N, and K) content due to formation of macronutrients and antibiotic complexes for example, Ca-antibiotic complex (Batchelder, 1982). In addition to this, many drugs contain ZnO_2, it has been observed that Zn-induced stress leads to decreased biomass production, chlorosis and accelerates necrosis (Raina et al., 2024). Mukhtar et al. (2020) reported decreased macronutrient and micronutrient (phosphorus and iron), and protein and carbohydrate concentration in rice when treated with antibiotics (10 mg. kg^{-1}) due to increased oxidative stress leading to cell redox imbalance in the plant. Gomes et al. (2020) reported anomalies in leaf anatomy, plant biomechanical properties and decreased yield of maize kernel in presence of ciprofloxacin. The increased concentration of sulfonamide antibiotics like sulfadimethoxine (50–450 mg. L^{-1}) induced decrease in fresh mass production of *Hordeum vulgare* (barley), *Pistia stratiotes* (water lettuce), *L. minor*, and *Azolla filiculoides* (Michelini et al., 2013; Forni et al., 2006). Electrolyte leakage could result in loss of membrane functionality due to loss of K^+ ions and could be an indicator of injury caused by stress or senescence. Similarly, several other researchers have reported phytotoxic effect in plant development and physiology like deformities in plant tissues (shortened primary root and more lateral roots), redox homeostasis and inhibition of growth (Michelini et al., 2013; Christou et al., 2016).

4.5 Oxidative stress and genotoxicity

Exposure to variety of PPCPs may trigger the production of ROS and NO (reactive nitrogen species) as response in plant. ROS are produced as the toxic by-products of aerobic metabolism and serves as cell signals that triggers stress–response (plant defense mechanism) in plants (Liu et al., 2013; Christou et al., 2016). ROS are highly reactive molecules that contains oxygen and its overproduction can cause oxidative stress, as ROS damages and disrupts essential cellular processes which leads to cell death. Some of the examples of ROS are superoxide radical ($O_2^{\cdot}$) hydroxyl radical ($^{\cdot}OH$), H_2O_2 etc. The major sites for ROS production are chloroplast, mitochondria and peroxisomes while other sites include cytosol, cell wall and endoplasmic reticulum. The exposure to PPCPs in plants may lead to cellular damage like lipid peroxidation where overproduced ROS may damage lipid membranes and cause them to become leaky. For instance, Osma and Elveren (2020) found negative effect of PPCPs on *Triticum aestivum* (wheat) in terms of lipid peroxidation and electrolyte leakage, indicating toward cellular injury due to plants response to stress by producing higher H_2O_2. Some studies suggests that in addition to oxidative stress response and antioxidant enzyme activity, drugs like acetaminophen and DCF also simultaneously uncovered negative effect of drug-derived nitrosative stress in plants (Bartha et al., 2014; Christou et al., 2016). In a study conducted by Christou et al. (2016), on a forage crop alfalfa, they studied its abiotic stress response in nutrient solution spiked with different classes of pharmaceutical compounds namely, DCF, SMX, trimethoprim and 17 α-ethynylestradiol (individually and its mixture) for 50 days. They noted elevated levels of H_2O_2 and NO (indicator of nitrosative stress), in leaves and roots of the plants when exposed to cocktail of pharmaceuticals particularly, DCF and SMX exhibited increased H_2O_2 and NO levels in root. In general, plant activates it defense system against ROS by producing antioxidant enzymes like superoxide dismutase (SOD), catalase (CAT), glutathione peroxidase, ascorbate peroxidase (APX) and glutathione S-transferases (GSTs) that acts as ROS scavenger (Xu et al., 2010; Wei et al., 2023). Among these SOD scavenges $O_2^{\cdot}$, CAT and peroxidase scavenges H_2O_2 (Vardhini and Anjum, 2015; Peng et al., 2021). Liu et al. (2013) reported decreased CAT and SOD activity due to toxicity caused by the antibiotic and increase in peroxidase activity in the leaves of wetland plant *Phragmite australis* when veterinary antibiotics (sulfamethazine, ciprofloxacin HCl, oxytetracycline HCl) dose was increased. Therefore, elevated ROS levels lead to increased antioxidant enzyme activity to prevent oxidative damage and maintain cellular homeostasis. PPCPs may deplete cofactors like glutathione and NADPH that are essential for antioxidant enzymes which indirectly affects the enzyme activity and further leads to oxidative stress.

Lipid peroxidation is an important indicator of the cellular damage which is induced by malondialdehyde (MDA) production (Sun et al., 2018). The level of lipid peroxidation can be used as biomarker to assess the degree of cellular damage. The action of MDA includes cell membrane damage leading toward leaky membranes and harms the fluidity of membrane by interacting with proteins via intermolecular and intramolecular cross-linking. In a study by Karki and Philip (2024), they reported accumulation of ROS leading to higher oxidative stress in two wetland plants *Chrysopogon zizanioides* (Vetiver) and *Colocasia esculenta* (Taro) when exposed to metronidazole, naproxen and methylparaben (5–10 mg. L^{-1}) for 28 days. Further, they measured lipid peroxidases as MDA, cell membrane integrity (Evan's blue) and electrolyte leakage in plants and found that their concentration decreased as PPCPs in nutrient solution decreased, suggesting oxidative stress induced by PPCPs impaired plasma membrane integrity and elevated lipid peroxidation. Similarly, Sun et al. (2018) reported increase in lipid oxidation by 30%–92% after exposing the cucumber plant to acetaminophen (5 mg. L^{-1}), however authors observed that the damage was less to the plant, the reason might be that the tolerance toward drugs among plant may be plant specific or different GSH-acetaminophen conjugation ability. The dose-dependent exposure of estrogen (steroid) in the leaves of lettuce showed elevated amount of lipid oxidation and ROS accumulation as reported by Adeel et al. (2018). Genotoxicity in plants refers to the damage caused to the plants' genetic material (at subcellular levels mainly to plants DNA). Peng et al. (2021) in addition to seed germination rate and chlorophyll estimation, also performed ROS, MDA and transcriptome analysis, and reported the overproduction of $O_2^{\cdot}$ and H_2O_2 in tobacco cells that affected plant growth.

5. Conclusion and future prospects

Several studies have provided evidences for the indiscriminate discharge of PPCPs from WWTPs to the water sources and in soil metric which eventually gets adsorbed on the soil surface. Additionally, application of biosolids as fertilizers in the agricultural fields pave the way for PPCPs to get into the soil and ground water sources. The sorption of organic pollutants on soil occurs via interactions like hydrophobic interactions, weak Vander Waals forces, electrostatic interactions, complexation, hydrogen bonding, etc. Bioavailability, sorption capacity, and subsequent uptake of the PPCPs depends on the various physiochemical factors of organic pollutants including hydrophobicity, water solubility, half-lives, persistence, molecular size, pH, ionization properties. The plant specific factors like transpiration contributes in the translocation and bioaccumulation of these organic pollutant. The PPCPs enters in the plants through root hairs and foliar uptake mainly by passive diffusion and via symplastic movement. The translocation and accumulation of PPCPs from root–shoot that occurs via xylem, which mainly depends on transpiration and environmental factors like wind speed, relative humidity, etc. Phytotoxicity caused due to long term exposure and accumulation of PPCPs in plants results in slow growth rate, changes in leaf area, reduction in fresh weight/plant biomass, leaf number and chlorosis. Photosynthetic activity and cellular respiration in plants are significantly reduced by PPCPs leading to the production of ROS. Plant growth in terms of seed germination, decreased taproot length, fresh and dry root weight, lateral root numbers, stunted shoot growth in plants is affected negatively when exposed to most of the PPCPs. However, studies related to the phytotoxic responses on reproduction cycle of flowering plants are limited.

PPCPs hinders nutrient uptake of the plant which may lead to abnormal functioning and developmental patterns such as early signs of aging, necrosis and decreased yield of agricultural crops. Various dose-dependent studies emphasized on elevated oxidative stress in plants in terms of H_2O_2, O_2, NO content among others when exposed to PPCPs. Additionally, lipid peroxidation is an indicator of the cellular damage induced by MDA production. Although plant activate their defense mechanism in response to oxidative stress, persistent higher concentration of PPCPs leads to overproduction of ROS and supersedes the production of antioxidant enzymes such SOD, CAT, APX etc., leading to phytotoxicity and genotoxicity in plants. Future research should focus on addressing key research gaps like impact of PPCPs on reproduction of plants, more long terms studies are needed to study the extent to which these chemical compounds can cause phytotoxicity in plants. Additionally, more plant specific investigations are needed as most of the existing studies are on common plants (model plants) and does not provide the whole scenario. In addition to this researcher should focus on exploring effective and sustainable methods to mitigate phytotoxic responses.

References

Adeel, M., Yang, Y.S., Wang, Y.Y., Song, X.M., Arslan Ahmad, M., Rogers, H.J., 2018. Uptake and transformation of steroid estrogens as emerging contaminants influence plant development. Environ. Pollut. 243, 1487–1497. https://doi.org/10.1016/j.envpol.2018.09.016.

Al-Baldawi, I.A., Mohammed, A.A., Mutar, Z.H., Abdullah, S.R.S., Jasim, S.S., Almansoory, A.F., Ismail, N.'I., 2021. Application of phytotechnology in alleviating pharmaceuticals and personal care products (PPCPs) in wastewater: source, impacts, treatment, mechanisms, fate, and SWOT analysis. J. Clean. Prod. 319, 128584. https://doi.org/10.1016/j.jclepro.2021.128584.

Al-Farsi, R.S., Ahmed, M., Al-Busaidi, A., Choudri, B.S., 2017. Translocation of pharmaceuticals and personal care products (PPCPs) into plant tissues: a review. Emerging Contam. 3 (4), 132–137. https://doi.org/10.1016/j.emcon.2018.02.001.

Alhalabi, A.M., Meetani, M.A., Shabib, A., Maraqa, M.A., 2024. Sorption of pharmaceutically active compounds to soils: a review. Environ. Sci. Eur. 36 (1), 161.

Al-Khazrajy, O.S.A., Ed Bergström, A.B.A.B., 2018. Factors affecting the dissipation of pharmaceuticals in freshwater sediments. Environ. Toxicol. Chem. 37 (3), 829–838.

Amy-Sagers, C., Reinhardt, K., Larson, D.M., 2017. Ecotoxicological assessments show sucralose and fluoxetine affect the aquatic plant, Lemna minor. Aquat. Toxicol. 185, 76–85. https://doi.org/10.1016/j.aquatox.2017.01.008.

An, J., Zhou, Q., Sun, Y., Xu, Z., 2009. Ecotoxicological effects of typical personal care products on seed germination and seedling development of wheat (*Triticum aestivum* L.). Chemosphere 76 (10), 1428–1434. https://doi.org/10.1016/j.chemosphere.2009.06.004.

Aristilde, L., Melis, A., Sposito, G., 2010. Inhibition of photosynthesis by a fluoroquinolone antibiotic. Environ. Sci. Technol. 44 (4), 1444–1450. https://doi.org/10.1021/es902665n.

Azad, A., Liu, H., 2024. Pharmaceutical and personal care products in recycled water for edible crop irrigation: understanding the occurrence, crop uptake, and water quantity effects. Agric. Water Manag. 304, 109047. https://doi.org/10.1016/j.agwat.2024.109047.

Bartha, B., Huber, C., Schröder, P., 2014. Uptake and metabolism of diclofenac in Typha latifolia–how plants cope with human pharmaceutical pollution. Plant Sci. 227, 12–20.

Bartrons, M., Peñuelas, J., 2017. Pharmaceuticals and personal-care products in plants. Trends Plant Sci. 22 (3), 194–203. https://doi.org/10.1016/j.tplants.2016.12.010.

Batchelder, A.R., 1982. Chlortetracycline and oxytetracycline effects on plant growth and development in soil systems. J. Environ. Qual. 11, 675–678. https://doi.org/10.2134/jeq1982.00472425001100040023x.

Ben Mordechay, E., Tarchitzky, J., Chen, Y., Shenker, M., Chefetz, B., 2018. Composted biosolids and treated wastewater as sources of pharmaceuticals and personal care products for plant uptake: a case study with carbamazepine. Environ. Pollut. 232, 164–172. https://doi.org/10.1016/j.envpol.2017.09.029.

Benassi-Borba, L., Dal'Lin, C.M.P., Testolin, R.C., Vieira, N.M.B., Corrêa, C.V.T., Bianchi, I., Barwinski, M.J.B., Radetski, C.M., Somensi, C.A., 2021. Assessment of phytotoxicity and impact on the enzymatic activity of soil microorganisms caused by veterinary antibiotics used in Brazilian farms. J. Environ. Sci. Health B 56 (7), 675–684.

Briggs, G.G., Bromilow, R.H., Evans, A.A., 1982. Relationships between lipophilicity and root uptake and translocation of non-ionised chemicals by barley. Pestic. Sci. 13 (5), 495–504.

Carter, L.J., Williams, M., Martin, S., Kamaludeen, S.P.B., Rai, S.K., 2018. Sorption, plant uptake and metabolism of benzodiazepines. Sci. Total Environ. 628, 18–25.

Carter, L.J., Harris, E., Williams, M., Ryan, J.J., Kookana, R.S., Boxall, A.B.A., 2014. Fate and uptake of pharmaceuticals in soil–plant systems. J. Agric. Food Chem. 62 (4), 816–825. https://doi.org/10.1021/jf404282y.

Carter, L.J., Williams, M., Böttcher, C., Kookana, R.S., 2015. Uptake of pharmaceuticals influences plant development and affects nutrient and hormone homeostases. Environ. Sci. Technol. 49 (20), 12509–12518. https://doi.org/10.1021/acs.est.5b03468.

Chang, P.-H., Jiang, W.-T., Li, Z., Kuo, C.-Y., Wu, Q., Jean, J.-S., Lv, G., 2016. Interaction of ciprofloxacin and probe compounds with palygorskite PFl-1. J. Hazard Mater. 303, 55–63. https://doi.org/10.1016/j.jhazmat.2015.10.012.

Chaturvedi, V., Kumar, A., 2010. Isolation of sodium dodecyl sulfate degrading strains from a detergent polluted pond situated in Varanasi city, India/Sodyum dodesil sülfat yikan suslarin Varanasi sehri, Hindistan'da deterjan kirliligi olan göletten izolasyonu. J. Cell Mol. Biol. 8 (2), 103.

Chefetz, B., Mualem, T., Ben-Ari, J., 2008. Sorption and mobility of pharmaceutical compounds in soil irrigated with reclaimed wastewater. Chemosphere 73 (8), 1335–1343. https://doi.org/10.1016/j.chemosphere.2008.06.070.

Christou, A., Antoniou, C., Christodoulou, C., Hapeshi, E., Stavrou, I., Michael, C., Fatta-Kassinos, D., Fotopoulos, V., 2016. Stress-related phenomena and detoxification mechanisms induced by common pharmaceuticals in alfalfa (Medicago sativa L.) plants. Sci. Total Environ. 557, 652–664.

Clarke, R.M., Cummins, E., 2015. Evaluation of "classic" and emerging contaminants resulting from the application of biosolids to agricultural lands: a review. Hum. Ecol. Risk Assess. 21 (2), 492–513.

Dey, S., Bano, F., Malik, A., 2019. Pharmaceuticals and Personal Care Product (PPCP) Contamination—A Global Discharge Inventory Pharmaceuticals and Personal Care Products: Waste Management and Treatment Technology. Elsevier, pp. 1–26.

Diaz-Cruz, M.S., Lopez de Alda, M.J., Barcelo, D., 2003. Environmental behavior and analysis of veterinary and human drugs in soils, sediments and sludge. TrAC, Trends Anal. Chem. 22 (6), 340–351.

Dodgen, L.K., Ueda, A., Wu, X., Parker, D.R., Gan, J., 2015. Effect of transpiration on plant accumulation and translocation of PPCP/EDCs. Environ. Pollut. 198, 144–153.

Du, R., Duan, L., Zhang, Q., Wang, B., Huang, J., Deng, S., Yu, G., 2023. Analysis on the attenuation characteristics of PPCPs in surface water and their influencing factors based on a compilation of literature data. Water Res. 242, 120203. https://doi.org/10.1016/j.watres.2023.120203.

Fang, Y.U., Karnjanapiboonwong, A., Chase, D.A., Wang, J., Morse, A.N., Anderson, T.A., 2012. Occurrence, fate, and persistence of gemfibrozil in water and soil. Environ. Toxicol. Chem. 31 (3), 550–555.

Forni, C., Patrizi, C., Migliore, C., 2006. Soil and Water Pollution Monitoring, Protection and Remediation Floating Aquatic Macrophytes as a Decontamination Tool for Antimicrobial Drugs. Springer.

Franklin, A.M., Williams, C., Andrews, D.M., Watson, J.E., 2022. Sorption and desorption behavior of four antibiotics at concentrations simulating wastewater reuse in agricultural and forested soils. Chemosphere 289, 133038. https://doi.org/10.1016/j.chemosphere.2021.133038.

Fu, Q., Malchi, T., Carter, L.J., Li, H., Gan, J., Chefetz, B., 2019. Pharmaceutical and personal care products: from wastewater treatment into agro-food systems. Environ. Sci. Technol. 53 (24), 14083–14090. https://doi.org/10.1021/acs.est.9b06206.

Geisler, D.A., Johansson, F.I., Svensson, Å.S., Rasmusson, A.G., 2004. Antimycin A treatment decreases respiratory internal rotenone-insensitive NADH oxidation capacity in potato leaves. BMC Plant Biol. 4, 1–10.

Goldstein, M., Shenker, M., Chefetz, B., 2014. Insights into the uptake processes of wastewater-borne pharmaceuticals by vegetables. Environ. Sci. Technol. 48 (10), 5593–5600. https://doi.org/10.1021/es5008615.

Gomes, M.P., Almeida Gonçalves, C., , Júlio César Moreira de Brito, Souza, A.M., Cruz, F.V.da S., Bicalho, E.M., Cunha Figueredo, C., Garcia, Q.S., 2017. Ciprofloxacin induces oxidative stress in duckweed (Lemna minor L.): implications for energy metabolism and antibiotic-uptake ability. J. Hazard Mater. 328, 140–149. https://doi.org/10.1016/j.jhazmat.2017.01.005.

Gomes, M.P., Rocha, D.C., , Júlio César Moreira de Brito, Tavares, D.S., Marques, R.Z., Soffiatti, P., Sant'Anna-Santos, B.F., 2020. Emerging contaminants in water used for maize irrigation: economic and food safety losses associated with ciprofloxacin and glyphosate. Ecotoxicol. Environ. Saf. 196, 110549. https://doi.org/10.1016/j.ecoenv.2020.110549.

Hena, S., Gutierrez, L., Croué, J.-P., 2021. Removal of pharmaceutical and personal care products (PPCPs) from wastewater using microalgae: a review. J. Hazard Mater. 403, 124041. https://doi.org/10.1016/j.jhazmat.2020.124041.

Hillis, D.G., Fletcher, J., Solomon, K.R., Sibley, P.K., 2011. Effects of ten antibiotics on seed germination and root elongation in three plant species. Arch. Environ. Contam. Toxicol. 60, 220–232.

Hu, X., Xie, H., Zhuang, L., Zhang, J., Hu, Z., Liang, S., Feng, K., 2021a. A review on the role of plant in pharmaceuticals and personal care products (PPCPs) removal in constructed wetlands. Sci. Total Environ. 780, 146637. https://doi.org/10.1016/j.scitotenv.2021.146637.

Hu, Y., Habibul, N., Hu, Y.-Y., Meng, F.-L., Sheng, G.-P., 2021b. Chemical speciation of ciprofloxacin in aqueous solution regulates its phytotoxicity and uptake by rice (Oryza sativa L.). Sci. Total Environ. 771, 144787.

Jayampathi, T., Atugoda, T., Jayasinghe, C., 2019. Uptake and Accumulation of Pharmaceuticals and Personal Care Products in Leafy Vegetables Pharmaceuticals and Personal Care Products: Waste Management and Treatment Technology. Elsevier, pp. 87–113.

Jena, G., Dutta, K., Daverey, A., 2023. Surfactants in water and wastewater (greywater): environmental toxicity and treatment options. Chemosphere 341, 140082. https://doi.org/10.1016/j.chemosphere.2023.140082.

Karki, B.K., Philip, L., 2024. Fate of pharmaceuticals and personal care products like metronidazole, naproxen, and methylparaben and their effect on the physiological characteristics of two wetland plants. Chem. Eng. J. 483, 149180. https://doi.org/10.1016/j.cej.2024.149180.

Karnjanapiboonwong, A., Suski, J.G., Shah, A.A., Cai, Q., Morse, A.N., Anderson, T.A., 2011. Occurrence of PPCPs at a wastewater treatment plant and in soil and groundwater at a land application site. Water, Air, Soil Pollut. 216, 257–273.

Keerthanan, S., Jayasinghe, C., Biswas, J.K., Vithanage, M., 2021. Pharmaceutical and Personal Care Products (PPCPs) in the environment: plant uptake, translocation, bioaccumulation, and human health risks. Crit. Rev. Environ. Sci. Technol. 51 (12), 1221–1258.

Kinney, C.A., Heuvel, B.V., 2020. Brian Vanden Heuvel, Translocation of pharmaceuticals and personal care products after land application of biosolids. Environ. Pollut.: Biosolids 14, 23–30. https://doi.org/10.1016/j.coesh.2019.11.004.

Kočárek, M., Kodešová, R., Vondráčková, L., Golovko, O., Fér, M., Klement, A., Nikodem, A., Jakšík, O., Grabic, R., 2016. Simultaneous sorption of four ionizable pharmaceuticals in different horizons of three soil types. Environ. Pollut. 218, 563–573. https://doi.org/10.1016/j.envpol.2016.07.039.

Kumar, R., Whelan, A., Cannon, P., Sheehan, M., Reeves, L., Antunes, E., 2023. Occurrence of emerging contaminants in biosolids in northern Queensland, Australia. Environ. Pollut. 330, 121786. https://doi.org/10.1016/j.envpol.2023.121786.

Kummerová, M., Zezulka, Š., Babula, P., Tříska, J., 2016. Possible ecological risk of two pharmaceuticals diclofenac and paracetamol demonstrated on a model plant Lemna minor. J. Hazard Mater. 302, 351–361. https://doi.org/10.1016/j.jhazmat.2015.09.057.

Kurade, M.B., Ha, Y.-H., Xiong, J.-Q., Govindwar, S.P., Jang, M., Jeon, B.-H., 2021. Phytoremediation as a green biotechnology tool for emerging environmental pollution: a step forward towards sustainable rehabilitation of the environment. Chem. Eng. J. 415, 129040. https://doi.org/10.1016/j.cej.2021.129040.

Kuroda, K., Murakami, M., Oguma, K., Muramatsu, Y., Takada, H., Takizawa, S., 2012. Assessment of groundwater pollution in Tokyo using PPCPs as sewage markers. Environ. Sci. Technol. 46 (3), 1455–1464.

Li, L., Li, T., Liu, Y., Li, L., Huang, X., Xie, J., 2023. Effects of antibiotics stress on root development, seedling growth, antioxidant status and abscisic acid level in wheat (*Triticum aestivum* L.). Ecotoxicol. Environ. Saf. 252, 114621. https://doi.org/10.1016/j.ecoenv.2023.114621.

Limmer, M.A., Burken, J.G., 2014. Plant translocation of organic compounds: molecular and physicochemical predictors. Environ. Sci. Technol. Lett. 1 (2), 156–161.

Liu, F., Ying, G.-G., Tao, R., Zhao, J.-L., Yang, J.-F., Zhao, L.-F., 2009. Effects of six selected antibiotics on plant growth and soil microbial and enzymatic activities. Spec. Iss. Sect. Ozone Mediterran. Ecol. Plants People Probl. 157 (5), 1636–1642. https://doi.org/10.1016/j.envpol.2008.12.021.

Liu, L., Liu, Y.-hong, Liu, C.-xiang, Wang, Z., Dong, J., Zhu, G.-fu, Huang, X., 2013. Potential effect and accumulation of veterinary antibiotics in Phragmites australis under hydroponic conditions. Ecol. Eng. 53, 138–143. https://doi.org/10.1016/j.ecoleng.2012.12.033.

Markovic, M., Neale, P.A., Nidumolu, B., Kumar, A., 2021. Combined toxicity of therapeutic pharmaceuticals to duckweed, Lemna minor. Ecotoxicol. Environ. Saf. 208, 111428. https://doi.org/10.1016/j.ecoenv.2020.111428.

Michelini, L., La Rocca, N., Rascio, N., Ghisi, R., 2013. Structural and functional alterations induced by two sulfonamide antibiotics on barley plants. Plant Physiol. Biochem. 67, 55–62. https://doi.org/10.1016/j.plaphy.2013.02.027.

Miller, E.L., Nason, S.L., Karthikeyan, K.G., Pedersen, J.A., 2016. Root uptake of pharmaceuticals and personal care product ingredients. Environ. Sci. Technol. 50 (2), 525–541.

Minden, V., Deloy, A., Volkert, A.M., Leonhardt, S.D., Pufal, G., 2017. Antibiotics impact plant traits, even at small concentrations. AoB Plants 9, plx010. https://doi.org/10.1093/aobpla/plx010.

Mojiri, A., Zhou, J.L., Ratnaweera, H., Rezania, S., Mansoureh Nazari, V., 2022. Pharmaceuticals and personal care products in aquatic environments and their removal by algae-based systems. Chemosphere 288, 132580. https://doi.org/10.1016/j.chemosphere.2021.132580.

Moreno, L., Lanusse, C., Purslow, P.P., 2017. Chapter 23—Veterinary Drug Residues in Meat-Related Edible Tissues Woodhead Publishing Series in Food Science, Technology and Nutrition. Woodhead Publishing, pp. 581–603. https://doi.org/10.1016/B978-0-08-100593-4.00024-2.

Mukherjee, S., Das, A., Chowdhury, A., Biswas, A., Roy, S., Majumdar, S., Paul, S., 2024. Harmful effect of personal care products on ecosystem and the possible alternative approach. Biocatal. Agric. Biotechnol. 57, 103065. https://doi.org/10.1016/j.bcab.2024.103065.

Mukhtar, A., Manzoor, M., Gul, I., Zafar, R., Jamil, H.I., Khan Niazi, A., Ali, M.A., Park, T.J., Arshad, M., 2020. Phytotoxicity of different antibiotics to rice and stress alleviation upon application of organic amendments. Chemosphere 258, 127353. https://doi.org/10.1016/j.chemosphere.2020.127353.

Nishi, I., Kawakami, T., Onodera, S., 2008. Monitoring of triclosan in the surface water of the Tone Canal, Japan. Bull. Environ. Contam. Toxicol. 80, 163–166.

Okoye, C.O., Sunday Okeke, E., Okoye, K.C., Echude, D., Andong, F.A., Chukwudozie, K.I., Ukamaka Okoye, H., Damian Ezeonyejiaku, C., 2022. Occurrence and fate of pharmaceuticals, personal care products (PPCPs) and pesticides in African water systems: a need for timely intervention. Heliyon 8 (3), e09143. https://doi.org/10.1016/j.heliyon.2022.e09143.

Osma, E., Elveren, M., 2020. Exposure to pharmaceuticals affects nutrient. Environ. Eng. Manag. J. 19 (4), 667–674.

Pandey, M., Tirkey, A., Tiwari, A., Lee, S.S., Dubey, R., Kim, K.H., Pandey, S.K., 2022. The environmental significance of contaminants of concern in the soil–vegetable interface: sources, accumulation, health risks, and mitigation through biochar. Sustainability 14 (21), 14539.

Peng, D., Wang, W., Liu, A., Zhang, Y., Li, X., Wang, G., Jin, C., Guan, C., Ji, J., 2021. Comparative transcriptome combined with transgenic analysis reveal the involvement of salicylic acid pathway in the response of Nicotiana tabacum to triclosan stress. Chemosphere 270, 129456. https://doi.org/10.1016/j.chemosphere.2020.129456.

Pérez, D.J., Rodrigo Lombardero, L., Joseph Doucette, W., 2023. Influence of exposure time, physicochemical properties, and plant transpiration on the uptake dynamics and translocation of pharmaceutical and personal care products in the aquatic macrophyte Typha latifolia. Sci. Total Environ. 896, 165107. https://doi.org/10.1016/j.scitotenv.2023.165107.

Pérez, D.J., Menone, M.L., 2024. Effects and Translocation of Pharmaceuticals and Personal Care Products into Aquatic Plants. Springer, Berlin, Heidelberg, pp. 1–20. https://doi.org/10.1007/698_2024_1187.

Pérez-Lucas, G., Navarro, S., 2024. How pharmaceutical residues occur, behave, and affect the soil environment. J. Xenobiot. 14 (4), 1343–1377.

Pino, M.R., Muñiz, S., Val, J., Navarro, E., 2016. Phytotoxicity of 15 common pharmaceuticals on the germination of Lactuca sativa and photosynthesis of Chlamydomonas reinhardtii. Environ. Sci. Pollut. Control Ser. 23, 22530–22541.

Pospíšil, P., 2016. Production of reactive oxygen species by photosystem II as a response to light and temperature stress. Front. Plant Sci. 7, 1950.

Raina, P., Kumari, A., Sharma, G., Jasrotia, A., Bhardwaj, A., Raj, P., Bala, R., Kaur, R., 2024. Phytotoxicity, Cytotoxicity and Genotoxicity of Pharmaceutical Products along with Their Transport and Fate Emerging Contaminants: Sustainable Agriculture and the Environment. Elsevier, India, pp. 131–148. https://doi.org/10.1016/B978-0-443-18985-2.00005-5.

Ravichandran, M.K., Philip, L., 2022. Fate of carbamazepine and its effect on physiological characteristics of wetland plant species in the hydroponic system. Sci. Total Environ. 846, 157337.

Schaffer, M., Licha, T., 2015. A framework for assessing the retardation of organic molecules in groundwater: implications of the species distribution for the sorption-influenced transport. Sci. Total Environ. 524–525, 187–194. https://doi.org/10.1016/j.scitotenv.2015.04.006.

Scheidegger, A.M., Sparks, D.L., 1996. A critical assessment of sorption-desorption mechanisms at the soil mineral/water interface. Soil Sci. 161 (12), 813–831.

Seth, R., Singh, P., Mohan, M., Singh, R., Singh Aswal, R., 2013. Monitoring of phenolic compounds and surfactants in water of Ganga Canal, Haridwar (India). Appl. Water Sci. 3, 717–720.

Sharma, B.M., Bečanová, J., Scheringer, M., Sharma, A., Bharat, G.K., Whitehead, P.G., Klánová, J., Nizzetto, L., 2019. Health and ecological risk assessment of emerging contaminants (pharmaceuticals, personal care products, and artificial sweeteners) in surface and groundwater (drinking water) in the Ganges River Basin, India. Sci. Total Environ. 646, 1459–1467. https://doi.org/10.1016/j.scitotenv.2018.07.235.

Singh, S., Pant, A., Dutta, K., Rani, R., Vithanage, M., Daverey, A., 2024. Phytoremediation of pharmaceuticals and personal care products using the constructed wetland. Environ. Chem. Ecotoxicol. 6, 104–116. https://doi.org/10.1016/j.enceco.2024.04.001.

Singh, S., Kumar Suman, S., Dutta, K., Daverey, A., 2025. Comparison of Canna indica and Acorus calamus for surfactant removal in biochar augmented constructed wetlands. Environ. Chem. Ecotoxicol. 7, 130–140. https://doi.org/10.1016/j.enceco.2024.11.003.

Smith, D.A., Beaumont, K., Maurer, T.S., Di, L., 2017. Relevance of half-life in drug design: miniperspective. J. Med. Chem. 61 (10), 4273–4282.

Sun, C., Dudley, S., Trumble, J., Gan, J., 2018. Pharmaceutical and personal care products-induced stress symptoms and detoxification mechanisms in cucumber plants. Environ. Pollut. 234, 39–47. https://doi.org/10.1016/j.envpol.2017.11.041.

Sunyer-Caldú, A., Quintana, G., Silvia Diaz-Cruz, M., 2023. Factors driving PPCPs uptake by crops after wastewater irrigation and human health implications. Environ. Res. 237, 116923. https://doi.org/10.1016/j.envres.2023.116923.

Trapp, S., Shi, J., Zeng, L., 2023. Generic model for plant uptake of ionizable pharmaceuticals and personal care products. Environ. Toxicol. Chem. 42 (4), 793–804.

Vardhini, B.V., Anjum, N.A., 2015. Brassinosteroids make plant life easier under abiotic stresses mainly by modulating major components of antioxidant defense system. Front. Environ. Sci. 2, 67.

Veiga-Gómez, M., Nebot, C., Miranda, J.M., Vázquez, B., Verdes, S., Quintela, L., Cepeda, A., Manuel Franco, C., 2019. Consumption of pharmaceuticals by dairy cows via watering throught: uncontroled intake. Agric. Ecosyst. Environ. 280, 95–101. https://doi.org/10.1016/j.agee.2019.04.024.

Vo, H.-N.-P., Bui, X.-T., Nguyen, T.-M.-H., Koottatep, T., Bandyopadhyay, A., 2018. Insights of the removal mechanisms of pharmaceutical and personal care products in constructed wetlands. Curr. Pollut. Rep. 4, 93–103.

Walters, E., McClellan, K., Halden, R.U., 2010. Occurrence and loss over three years of 72 pharmaceuticals and personal care products from biosolids–soil mixtures in outdoor mesocosms. Water Res. 44 (20), 6011–6020. https://doi.org/10.1016/j.watres.2010.07.051.

Wang, Z., Li, X., Li, Y., Liu, H., Lin, C.S.K., Sun, J., Wang, Q., 2024. Unveiling the occurrence and ecological risks of triclosan in surface water through meta-analysis. Environ. Pollut. 361, 124901. https://doi.org/10.1016/j.envpol.2024.124901.

Wei, H., Tang, M., Xu, X., 2023. Mechanism of uptake, accumulation, transport, metabolism and phytotoxic effects of pharmaceuticals and personal care products within plants: a review. Sci. Total Environ. 892, 164413. https://doi.org/10.1016/j.scitotenv.2023.164413.

Williams, C.F., McLain, J.E.T., 2012. Soil persistence and fate of carbamazepine, lincomycin, caffeine, and ibuprofen from wastewater reuse. J. Environ. Qual. 41 (5), 1473–1480.

Xu, S.-C., Yong-ping, L.I., Jin, H.U., Guan, Y.-jing, Zheng, Y.-ye, Shui-jin, Z.H.U., 2010. Responses of antioxidant enzymes to chilling stress in tobacco seedlings. Agric. Sci. China 9 (11), 1594–1601.

Xu, Y., Yu, X., Xu, B., Peng, D., Guo, X., 2021. Sorption of pharmaceuticals and personal care products on soil and soil components: influencing factors and mechanisms. Sci. Total Environ. 753, 141891. https://doi.org/10.1016/j.scitotenv.2020.141891.

Ying, G.-G., Yu, X.-Y., Rai, S., 2007. Kookana, Biological degradation of triclocarban and triclosan in a soil under aerobic and anaerobic conditions and comparison with environmental fate modelling. Environ. Pollut. 150 (3), 300–305. https://doi.org/10.1016/j.envpol.2007.02.013.

Yu, Y., Liu, Y., Wu, L., 2013. Sorption and degradation of pharmaceuticals and personal care products (PPCPs) in soils. Environ. Sci. Pollut. Control Ser. 20, 4261–4267.

Yuan, X., Li, S., Hu, J., Yu, M., Li, Y., Wang, Z., 2019. Experiments and numerical simulation on the degradation processes of carbamazepine and triclosan in surface water: a case study for the Shahe Stream, South China. Sci. Total Environ. 655, 1125–1138. https://doi.org/10.1016/j.scitotenv.2018.11.290.

Zeng, Y., Zhang, Y., Zhang, H., Wang, J., Lian, K., Ai, L., 2022. Uptake and transport of different concentrations of PPCPs by vegetables. Int. J. Environ. Res. Publ. Health 19 (23), 15840.

Zhang, C., Feng, Y., Yuan-wang, L., Chang, H.-qing, Zhao-jun, L., Jian-ming, X., 2017. Uptake and translocation of organic pollutants in plants: a review. J. Integr. Agric. 16 (8), 1659–1668. https://doi.org/10.1016/S2095-3119(16)61590-3.

Zhang, Q.-Q., Ying, G.-G., Chen, Z.-F., Zhao, J.-L., Liu, Y.-S., 2015. Basin-scale emission and multimedia fate of triclosan in whole China. Environ. Sci. Pollut. Control Ser. 22, 10130–10143.

Zhang, Y., Geißen, S.-U., Gal, C., 2008. Carbamazepine and diclofenac: removal in wastewater treatment plants and occurrence in water bodies. Chemosphere 73 (8), 1151–1161. https://doi.org/10.1016/j.chemosphere.2008.07.086.

Zhao, M., Ma, J., Li, Mo, Zhang, Y., Jiang, B., Zhao, X., Huai, C., Shen, L., Zhang, N., He, L., 2021. Cytochrome P450 enzymes and drug metabolism in humans. Int. J. Mol. Sci. 22 (23), 12808.

Zind, H., Mondamert, L., Blancart Remaury, Q., Cleon, A., , Nathalie KARPEL VEL LEITNER, Labanowski, J., 2021. Occurrence of carbamazepine, diclofenac, and their related metabolites and transformation products in a French aquatic environment and preliminary risk assessment. Water Res. 196, 117052. https://doi.org/10.1016/j.watres.2021.117052.

CHAPTER 9

Phytotoxicity of organic nanoparticles

R. Siranjeevi[1] and R. Susmitha[2]
[1]Department of Chemistry, Saveetha School of Engineering, Saveetha Institute of Medical and Technical Sciences, Saveetha University, Chennai, Tamil Nadu, India; [2]Department of Biotechnology, Saveetha School of Engineering, Saveetha Institute of Medical and Technical Sciences, Saveetha University, Chennai, Tamil Nadu, India

1. Introduction

1.1 Definition and overview of organic nanoparticles

Organic nanoparticles consist of organic materials that exist between 1 and 100 nm in either one or multiple dimensions. The formation of nanostructures occurs from various organic substances that include lipids together with polymers, proteins, dendrimers, and small organic molecules (Prerna and Ratan, 2021; Astruc, 2020; Barhoum et al., 2022; Khan et al., 2022). The critical benefits of organic nanoparticles over inorganic particles include their better biocompatibility as well as their capacity to degrade along with their ability to adjust chemical structure which render them beneficial for biomedical implementations. Organic nanoparticles can be produced through two main methodologies known as top-down procedures along with bottom-up approaches (Khan and Hossain, 2022; Salem and Fouda, 2021). Scientists use three main top-down methods for producing nanoparticles: the three methods include nanoprecipitation and emulsification–solvent evaporation followed by milling. Self-assembly of molecular elements by noncovalent hydrogen bonding and hydrophobic and π–π stacking interactions produces structures resembling natural biological self-assembly mechanisms (Mekuye and Abera, 2023; Patil et al., 2021; Sekoai et al., 2019). Organic nanoparticles display exceptional behavioral traits due to their nanoscale size which alters manifestation of both surface-to-volume ratio and quantum effects and environmental responsiveness against bulk materials. Organic nanoparticle properties are controlled through accurate modifications made to their basic components while precise adjustments are possible in dimensions and structural shape and surface protective layers (Freitas de Freitas et al., 2018; Lismont et al., 2017). Organic nanoparticles serve various industries through drug delivery and biosensing applications alongside imaging capabilities and catalytic functions and electronic applications and environmental remediation strategies because of their controlled payload delivery mechanisms (Störmer et al., 2017; Loste et al., 2019; Bedlovičová et al., 2020; Carrasco, 2018).

Phytotoxicity of Organic Pollutants
ISBN 978-0-443-33110-7,
https://doi.org/10.1016/B978-0-443-33110-7.00015-8

1.2 Importance of studying phytotoxicity

The detrimental effects of compounds on plant biological activities during their life cycle represent research that affects agriculture and environmental protection alongside human health prospects. The development of secure agrochemicals requires complete phytotoxic effect understanding because unnecessary herbicide or fungicide or insecticide toxicity against plants creates crop damage leads to yield reduction and major economic damages (Pan and Chu, 2016; Mathur et al., 2016). The evaluation of industrial pollutants together with heavy metals and emerging contaminants on native vegetation as well as ecosystem stability and biodiversity preservation happen through the conduct of environmental phytotoxicity studies. Scientific evaluations of phytotoxicity shed light on the defensive strategies which plants employ together with their biological reactions to stress conditions (Wang et al., 2019). Plants that encounter phytotoxic substances display specific signs which include chlorosis, necrosis together with growth reductions and reproductive system changes. Various visible symptoms emerge from complex cellular and molecular changes including oxidative stress together with membrane disruption and photosynthetic inhibition and genotoxicity. The evaluation of these mechanisms enables scientists to create effective contamination biomarkers as well as develop strong plant species and protective techniques for crop defense. Research on phytotoxicity provides essential knowledge needed for phytoremediation technology implementation because plants used in this process extract and remove or detoxify contaminants from the environment (Forti et al., 2020). Scientists learn to select hyperaccumulator plant species as soil or water pollutant extractors through studying different plant responses to toxic compounds. Phytotoxicity screening operates as a global regulatory standard to evaluate new products because it ensures thorough safety assessment for market authorization which safeguards agricultural sustainability and environmental stability across generations (Werrie et al., 2020).

2. Properties and classification of organic nanoparticles

2.1 Types of organic nanoparticles (polymeric, lipid-based, carbon-based, etc.)

The extensive investigation of organic nanoparticles occurs through polymeric nanoparticles which unveil structures generated from natural and synthetic polymer methods. Natural polymer-based nanoparticles produced from chitosan and alginate and gelatin materials exhibit ideal biodegradability and biocompatibility levels which make them optimal for biomedical usages (Ren et al., 2022). The controlled physicochemical properties and degradation kinetics of surface functionality come from synthetic polymer nanoparticles based on poly(lactic-co-glycolic acid) (PLGA), polyethylene glycol (PEG), and polyethylenimine (PEI). Polymeric nanoparticles exhibit remarkable adjustable properties that support their use in controlled drug delivery systems as well as tissue engineering applications and diagnostic imaging functions. Different lipid-based nanoparticles include liposomes along with three classes of systems such as solid lipid nanoparticles (SLNs), nanostructured lipid carriers (NLCs) and lipid nanoemulsions (Bhatia, 2016). The phospholipid bilayers surrounding the aqueous core structure of liposomes enable efficient delivery of compounds, which include water-soluble and water-insoluble components. The solid lipids used to make SLNs and the combination of solid and liquid lipids in NLCs enhance stability and control release rates. Lipid nanoparticle technology has transformed nucleic acid therapy through its essential contribution to mRNA vaccination and its natural biocompatibility which makes them widely usable in pharmaceutical product development (Trivedi et al., 2022).

The carbon-based organic nanoparticles comprise carbon dots and two additional types, namely graphene quantum dots and fullerenes, which embody distinct optical, electrical, and mechanical behavior properties. The nanoparticulate dots composed of carbon known as carbon dots measure below 10 nm and exhibit superior photoluminescence properties plus low toxicity combined with exceptional chemical durability. The combination of distinctive properties makes carbon-based organic nanoparticles valuable agents in bioimaging together with sensory and catalytic and optoelectronic applications. Nanostructures made from graphene exhibit superior electronic conductivity together with mechanical strength while having extensive surface area and demonstrate practical use for electronics and energy storage as well as environmental cleanup applications because of their two-dimensional carbon atom hexagonal lattice structure (Lombardo et al., 2020). The groups of protein-based nanoparticles and dendrimers belong to significant organic nanoparticle categories which demonstrate unique characteristics. Protein nanoparticles that contain albumin or gelatin or engineered protein cages exist for exact molecular structuring while retaining natural biological recognition properties. The advantageous characteristics render these materials especially useful in pharmaceutical applications including drug delivery and vaccine development. Each dendrimer possesses precise controls over architecture, size, shape and surface functional properties in its highly branched three-dimensional structure for applications in host–guest chemistry and gene transfection and nanomedicine (Lombardo et al., 2020). Organic nanoparticles in each category demonstrate specific advantages but also come with associated limitations that researchers focus on enhancing through continuous advancements directed toward developing applications in different scientific fields.

2.2 Physicochemical properties influencing phytotoxicity

The phytotoxic behavior of organic nanoparticles depends fundamentally on their physical properties while particle dimension stands out as the essential determinant factor. Microscopic particles of 50 nm or smaller size show an improved capacity to break through plant cell barriers and vascular tissues and cellular boundaries leading to cell internalization along with possible toxic outcomes (Hechmi et al., 2020). The size of nanoparticles controls their absorption rates so smaller particles penetrate cells better than larger ones within biology. When particles decrease in size their specific surface area increases producing enhanced biological system interactions that result in amplified phytotoxic responses because of increased surface reactive availability. The relationship between nanoparticles and plant surfaces strongly depends on their surface charge measurement known as zeta potential (Young et al., 2016). Positively charged nanoparticles show enhanced adherence to negative plant cell walls because of electrostatic attraction which aids their uptake together with toxicity development. The negative electrostatic charge on some particles could create electrostatic repulsion between them and decrease the internalization process. The combination of surface properties and chemical structure determines phytotoxicity because it controls both particle stability and particle aggregation extent as well as their distinct chemical interactions with cellular compounds (Pan and Chu, 2016). The partitioning distributions of nanoparticles inside plant tissues depend on their hydrophobicity/hydrophilicity ratios because hydrophobic particles tend to concentrate in lipid membranes rather than hydrophilic particles that tend to remain in watery compartments.

Structural properties shaping size form and crystalline nature determine the harmful environmental effects of organic nanoparticles on plants. Nanorods and nanofibers because of their elongated structure generate different mechanical forces on cellular elements than spherical-shaped nanoparticles do. The release rate of toxic components from nanoparticles depends on their chemical stability together with their dissolution kinetics because unstable nanoparticles may lose their components which produce toxicity differently than the complete nanostructure (Coelho et al., 2018). Nanoparticles that heap up or congeal when exposed to environmental or biological media would transform their actual size and surface properties as well as their bioavailability which then changes their phytotoxic outcomes. The way organic nanoparticles maintain presence in environmental scenarios together with their systematic changes determines the duration of their phytotoxic properties. Biodegradable nanoparticles demonstrate harmful effects on plants in their first stages but eventually break down into nonhazardous substances that prevent long-term toxic results. Persistent nanoparticles have the tendency to build up within soil and plant tissues leading to long-lasting effects on the ecosystem. Three primary environmental factors that control nanoparticle properties are soil pH, organic matter content, ionic strength, and humidity which affect stability alongside aggregation state and surface properties (Zhao et al., 2019). The precise manner in which intrinsic physicochemical properties and surrounding environmental elements work together shapes the availability rates of organic nanoparticles in plants as well as the uptake pathways and harmful effects which plants experience. Therefore, extensive testing is essential for proper assessment of organic nanoparticle environmental consequences.

2.3 Synthesis and functionalization approaches

Scientists create organic nanoparticles through two fundamental approaches which they distinguish as top-down and bottom-up methods. Nanoscale reduction of bulk materials occurs through mechanical processes which include high-pressure homogenization together with wet milling and ultrasonication. The methods provide scalability but their size control along with uniformity remains limited. The construction of nanoparticles from individual molecular components happens through molecular self-assembly mechanisms that make up bottom-up approaches (Deshmukh and Niederberger, 2017). The methods used for bottom-up synthesis include nanoprecipitation as well as emulsification–solvent evaporation and supercritical fluid technology and template-directed synthesis. Control of particle size distribution together with morphology and internal structure is superior in bottom-up approaches but they face obstacles when producing at industrial scales. Organic nanoparticle functionalization establishes a vital process which manipulates their physical characteristics along with biological behavior and application-related functional abilities. The surface modification techniques involve covalent plus noncovalent approaches to bind desired ligands onto nanoparticles through surface reactive functional groups and electrostatic and hydrogen bond interactions and hydrophobic forces. The layer-by-layer assembly approach enables researchers to add oppositely charged polyelectrolyte layers one by one in order to develop complex nanostructures that deliver desirable surface attributes (Li et al., 2021). The functionalization approaches have multiple benefits for nanoparticle chemistry since they offer different benefits related to stability as well as control over ligand density while keeping core structures intact.

The transformation of nanoparticles gets more advanced through techniques that let them adapt their properties by detecting particular stimuli like pH changes or temperature variations and light signals or enzyme activity. The specialized nanoparticle design presents capabilities for managed discharge together with precision delivery while delivering increased selectivity to many applications. Engineers in nanoparticle science explore biomimetic functionalization while incorporating natural substances such as biomacromolecules and cell membrane fragments to improve their work. Nanoparticles receiving biomimetic functionalization achieve integration of synthetic strength from man-made nanoparticles with natural biological system abilities that enhance both immunological tolerance and compatibility between human cells. Functional and synthesized organic nanoparticles need multiple analytical tools to validate their structural composition as well as their functional capabilities (Abu-Dief and Abdel-Fatah, 2018). DLS and NTA together measure the hydrodynamic size and distribution while zeta potential evaluates surface charge of the components. The analysis of synthesized organic nanoparticles utilizes three microscopic methods which include transmission electron microscopy (TEM), scanning electron microscopy (SEM) and atomic force microscopy

(AFM). These techniques show details of structure and surface characteristics. The successful functionalization validation primarily depends on Fourier-transform infrared spectroscopy (FTIR), nuclear magnetic resonance (NMR) and X-ray photoelectron spectroscopy (XPS) because these spectroscopic methods detect distinctive chemical markers. Through characterizing methods like these researchers ensure their synthetic nanoparticles match application requirements and generate significant data for building structure–property relationships.

2.4 Applications in agriculture and industry

The deployment of organic nanoparticles brought about revolutionary changes to agricultural methods when used for advanced delivery systems of agrochemicals. Polymeric and lipid nanoparticles containing pesticides and herbicides and fungicides enable controlled extended delivery through which applications can be reduced without compromising efficacy. The targeted delivery mechanism maintains environmental safety by reducing off-target effects alongside the prevention of chemical contamination therefore it increases stability of active ingredients through protection from environmental hazards including UV radiation and microbial activity together with hydrolysis (Thakur et al., 2018). These nanoformulations help decrease conventional agrochemical application rates substantially by up to 80% for certain products thereby providing an alternative to address chemical overuse problems which still maintains their protective performance. Nanofertilizers demonstrate significant transformation through the encapsulation of essential plant nutrients within organic nanoparticles or by conjugating them to such nanoparticles. The systems possess the ability to release nutrients when particular environmental stimuli or plant biological events occur thus matching nutrient supply with crop growth needs stage-by-stage. The excellent performance of chitosan-based nanoparticles which contain nitrogen and phosphorus and potassium shows promising results for improving nutrient utilization rates while solving the problematic water body nutrient runoff that causes eutrophication. Nanoparticles serve as an effective micronutrient delivery system because they assist in solving deficiency problems in difficult-to-fertilize soil conditions where traditional methods fail due to soil fixation limitations (Lorenzo et al., 2021).

Precision farming obtained actual-time observation of environmental parameters of plant health, together with soil properties and water, using organic nanoparticles which energize high-level nanosensors. Carbon dots incorporating recognition factors attain ultrahigh sensitivity to monitor both the content of soil moisture and soil nutrients as well as pathogens within the soil system. Smart farming technologies based on data enable users to optimize resource inputs and program their farm interventions through maximized decision-making systems. When nanosensors are combined with automated fertilizer control and irrigation systems they create self-sustaining agricultural control systems that utilize real-time environmental feedback to conserve resources and yield more crops.

Organic nanoparticles are used for various applications in catalysis industry because their large surface area coupled with their tunable surface properties impart them greater benefits over conventional catalysts. Applications of catalytic-center-functionalized dendrimers and polymeric nanoparticles provide increased reaction selectivity along with recoverability potential along with enhanced reactions under milder conditions compared to traditional homogeneous or heterogeneous catalysts. The nanocatalysts function as technology enablers by reducing energy requirements while generating less waste and providing improved target selectivity in the production of chemical processes (Bahrulolum et al., 2021). The recoverability of the nanocatalysts via magnetic separation or membrane filtration offers an industrial solution to addressing the age-old problem of catalyst separation.

The packaging industry found significant advantages through adding organic nanoparticles to polymeric materials for making improved food packaging solutions which show antimicrobial and oxygen-scavenging along with biodegradable properties. Essential oils loaded within chitosan nanoparticles serve as natural food pathogen inhibitors which enhance product shelf life using no synthetic preservatives. Oxygen-scavenging nanoparticles made from cyclodextrin complexes operate inside packaging headspace to extract oxygen thus protecting food components from oxidation-caused breakdown. Research has fulfilled consumer needs for waste reduction as well as minimal synthetic additives through methods which guard safety and maintain food quality in distribution networks. Environmental remediation shows quick expansion as a domain which employs functionalized organic nanoparticles as efficient pollutants adsorbents that clean up pollutants found in water and soil systems. Polymers based on cyclodextrin show outstanding chemical binding capacities for pesticide compounds along with industrial solvents as well as pharmaceutical residue particles as organic pollutants while chitosan nanocomposites achieve heavy metal removal by chelation-based interactions (Ali et al., 2021). The nanoparticles' large surface area and plenty of functional groups allow them to remove pollutants from concentrations that standard remediation systems cannot reach. Nanoparticles with stimulus-triggered capabilities provide controlled release of trapped contaminants so adsorbents become reusable and harmful substances reach adequate concentrations for safe disposal which resolves the main issue of environmental pollution in agricultural and industrial sectors.

3. Mechanisms of interaction with plants

3.1 Uptake and translocation pathways in plants

Plant systems receive organic nanoparticles by means of multiple entry points but soil-derived root uptake functions as the main pathway. Organic nanoparticles first encounter the rhizodermis at the root surface until they find their path either through apoplastic movement (cell walls and intercellular spaces) or symplastic movement (via plasmodesmata connections between cells). A plant cell wall functions as an entry barrier because it selectively permits entry of particles that match its pore dimension scope (5–20 nm) yet this size recognition system shows species-specific variability during the plant development cycle. Plasma membranes accept nanoparticles through endocytosis-like processes which include clathrin-mediated endocytosis together with caveolae-dependent endocytosis and macropinocytosis as validated cellular uptake pathways (Schwab et al., 2016). External penetration of foliar-applied nanoparticles occurs through the plant stomata and hydathodes or open wounds in the plant epidermal layers. Plants mainly transport organic nanoparticles across their systems through vascular tissues. Transpiration pull within the xylem system transports nanoparticles upward from plant roots toward aerial tissues and phloem tissue conducts nanoparticles both upward and downward following source-to-sink directional patterns. Plant-based vascular translocation shows a direct relationship between nanoparticle transportation and the combination of particle dimensions and surface electrical charge and their water-loving and water-repellent properties. Size-exclusion manifests at three vital points in plant tissues: the Casparian strip in roots and the plasmodesmata along with sieve plates in phloem elements (ad). The surface charge plays an essential role in cellular element interaction because positively charged nanoparticles tend to stay longer within negatively charged cell walls yet neutral or negatively charged particles show higher mobility in plant tissues.

The internalized nanoparticles move through multiple cellular pathways which decide their final placement inside the cell and their resulting biological effects. Nanoparticles reside in endosomal compartments after endocytosis until these compartments either fuse with lysosomes or vacuoles at which point degradation or sequestration processes might occur. Nanoparticles that successfully escape endosomes can reach the cytosol for interactions with different organelles such as mitochondria and chloroplasts and nuclei. The accumulation patterns for organelles depend on both chemical attributes of nanoparticles and modifications made to their surfaces (Feng et al., 2021). Particular types of nanoparticles tend to accumulate selectively in target tissues or organelles which opens up opportunities for delivering agrochemicals to their sites of action and revealing unintended biological interactions that affect phytotoxicity.

3.2 Cellular and molecular interactions

When nanoparticles enter the plant system, they react with cells in ways which impact biological process at their core level. The interactions between cell membranes and nanoparticles mainly depend on membrane-contacting surface features that involve electric charge and hydrophobicity along with functional groups. The affinity of positively charged nanoparticles elevates toward negatively charged membranous bilayers because they can weaken membrane stability through various processes like membrane deformation and pore creation or lipid extraction. Membrane interactions between nanoparticles lead to the deterioration of both membrane fluidity characteristics and permeability which disrupts cellular compartmentalization and ion homeostasis (Erb and Reymond, 2019). The absorbed nanoparticles establish interactions within internal cells that particularly affect mitochondria and chloroplasts since they perform essential energy generation and photosynthetic functions). When accumulation happens in these specific organelles it disrupts electron transport chains which impairs ATP synthesis at the same time it generates reactive oxygen species (ROS) that create oxidative damage that spreads throughout the cell (Ma et al., 2016).

At the molecular level, organic nanoparticles induce intricate cascades of signaling and stress responses. Oxidative stress is a main response mechanism wherein ROS generation induced by nanoparticles overpowers cellular antioxidant defenses, leading to oxidative harm to lipids, proteins, and nucleic acids. Oxidative damage initiates stress-response pathways such as mitogen-activated protein kinase (MAPK) cascades, calcium signaling, and transcriptional reprogramming via stress-sensitive transcription factors. In parallel, nanoparticle interaction with cellular proteins can change protein conformation, inhibit enzymatic activity, and affect important cellular processes like photosynthesis, respiration, and cell division. Direct or indirect interaction with DNA could cause genotoxicity through mechanisms like strand breakage, base alteration, or adduct formation on the DNA, with potential for inducing mutations or genomic instability if repair processes are saturated (Hasanuzzaman et al., 2018).

The cellular response to nanoparticle exposure occurs through adaptive mechanisms with the intent to counteract possible injury. Upregulation of antioxidant defense mechanisms, such as increased synthesis of enzymatic antioxidants (superoxide dismutase, catalase, peroxidases) and nonenzymatic antioxidants (glutathione, ascorbate, flavonoids), is an early defense response. Activation of detoxification pathways, including glutathione S-transferase-mediated conjugation, allows for nanoparticle sequestration or modification. Autophagic mechanisms can sequester nanoparticles or damaged organelles inside autophagosomes for degradation or compartmentalization in vacuoles (Andersen et al., 2018). At the tissue level, cell wall lignification and suberization can create additional barriers to nanoparticle penetration. These multilayered defense mechanisms together determine plant resistance to nanoparticle exposure, with the balance between adaptive responses and cellular damage ultimately determining whether the outcome is successful acclimation or phytotoxicity (Kőszegi and Poór, 2016).

3.3 Role of surface chemistry and size on phytotoxicity

Surface chemistry is a critical factor in the phytotoxicity profile of organic nanoparticles and controls their interaction with biological interfaces and cellular constituents.

Surface-decorated functional groups control essential properties such as surface charge, hydrophobicity, and reactivity, thus controlling adsorption patterns to plant surfaces and cellular uptake efficiencies. Positively charged nanoparticles generally show increased interactions with negatively charged plant cell membranes and walls, potentially enhancing cellular internalization while at the same time causing more membrane disruption and cytotoxicity (Jain et al., 2017). Surface functionalization with targeted ligands, such as targeting moieties or polymer coatings like polyethylene glycol (PEG), can significantly alter these interactions—PEGylation, for example, tends to decrease phytotoxicity by lowering protein adsorption and cell interactions and increasing circulation time. In addition, the surface moieties' chemical stability also has a critical influence on nanoparticle transformation routes in plant systems, where labile surface groups can release toxic constituents that can contribute to secondary mechanisms of toxicity (Appah et al., 2020).

Particle size has far-reaching effects on phytotoxicity through a variety of mechanisms, with diminishing particle size commonly showing increased phytotoxic capability. The reciprocal relationship between particle size and specific surface area also directly relates to enhanced reactivity, as a smaller particle size presents a greater percentage of the constituent atoms on the surface, increasing interactions with biological systems. Threshold-dependent uptake efficiency is supported by the selective internalization of particles under critical thresholds—e.g., usually 50 nm for cell uptake and 5–20 nm for cell wall penetration in the majority of plant species. In addition, smaller particles are able to enter different cellular compartments, with particles less than 5 nm even being able to penetrate nuclear pores to directly engage with genetic material (Liu et al., 2019). This differential subcellular distribution is particularly important in shaping the specific cellular constituents being exposed to nanoparticle interactions and thus the resultant phytotoxicity profile.

Interrelationship between size and surface chemistry creates synergistic phenomena that cannot be anticipated by monitoring either parameter singly. For example, reducing the particle size characteristically enhances the effect of the surface charge so that small highly charged particles possess disproportionately increased cell interactions with disproportionately increased cell numbers compared to bigger counterparts with identical surface chemistry. In addition, size-dependent changes in nanoparticle curvature change the spatial organization and accessibility of surface functional groups, which may change their reactivity and biological interactions. Environmental conditions further regulate these properties, and variations in pH, ionic strength, and organic matter levels may change surface characteristics through processes such as protonation/deprotonation, corona formation, and aggregation (Pereira et al., 2018). It is critical to understand this intricate interplay between surface chemistry, size, and environmental conditions in order to create predictive models of phytotoxicity and design organic nanoparticles with reduced environmental impact or maximized agricultural performance, depending on the desired application scenario.

3.4 Bioaccumulation and biotransformation of organic nanoparticles

Organic nanoparticle bioaccumulation in plant systems occurs with a sophisticated kinetics involving a variety of factors such as exposure concentration, duration, and the physicochemical properties of the nanoparticles. When absorbed, the nanoparticles are found to concentrate in certain tissues, usually highest in root tissues with decreasing gradients to aerial structures. This kind of distribution corresponds to the barrier nature of the Casparian strip and endodermal cells limiting translocation to vascular tissue (Uddin et al., 2020). Tissue-specific deposition is also affected by nanoparticle properties, with smaller, hydrophilic particles typically having greater mobility and capacity for systemic distribution than larger, hydrophobic counterparts that tend to be localized at sites of entry. PLGA nanoparticles exhibit gradual decomposition because they are biodegradable formulations so their persistence lasts longer than unbreakable structural nanoparticles that can easily transfer along food chains through biomagnification. Plants transform organic nanoparticles through biotransforming processes that modify their physical nature and biological properties. The changes can happen through abiotic processes, such as hydrolysis, photolysis, and oxidation, and through enzyme-catalyzed processes, including the plant detoxification systems. Phase I biotransformation will typically include oxidation, reduction, or hydrolysis reactions that introduce or reveal functional groups, enhancing polarity and reactivity. The body performs these drug metabolism responses through enzyme systems including cytochrome P450 monooxygenases together with peroxidases and carboxylesterases (Xiao et al., 2020). The second step in detoxification functions through Phase II reactions that combine endogenous compounds such as glutathione and glucuronic acid and amino acids with activated intermediates to achieve more substantial water solubility and vacuole or apoplastic space compartmentalization. These biotransformation reactions proceeding sequentially can metabolize initially toxic nanoparticles into relatively innocuous metabolites, but can instead create reactive intermediates of increased toxicity than parent forms (Banu et al., 2021).

The interaction between biotransformation and bioaccumulation ultimately decides the ecological impact and environmental fate of organic nanoparticles. Different levels of susceptibility are created because plants have varying biotransformation capacities that enable select species to detoxify environmental contaminants better than others. The successful application of phytoremediation requires selection of hyperaccumulator plant species which contain powerful detoxification machinery for environmental cleanup purposes. Biotransformation affects the bioavailability of nanotechnology-associated compounds used in either controlled agrochemical delivery or the change of environmental pollutants' bioactivity (Lencioni et al., 2016). Delivery of ecological risk assessments and development of environmentally positive nanoformulations requires complex analytical methods including mass spectrometry-based metabolomics and stable isotope tracing and high-end imaging capabilities for tracking nanoparticles and their transformation products within plant tissues.

4. Methods for assessing phytotoxicity

4.1 In vitro and in vivo phytotoxicity assessment

Laboratory conditions enable testing methods to assess early the potential destructive effects of organic nanoparticles on plant systems. Laboratory methods focused on cell and tissue culture deliver distinctive advantages through enhanced reproducibility and higher throughput capability together with deeper mechanistic understanding (Mendes et al., 2021). Single cell suspensions isolated from various plant tissues permit speedy evaluation of important cell attributes such as viability along with membrane condition and metabolic state through MTT, neutral red uptake, and trypan blue exclusion experiments. Research on callus cultures offers a way to connect between cellular-level experiments and full-plant research by studying tissues with cell-to-cell relationships and structure similar to plant parts. In vitro germination of seeds and early seedling growth tests are another universally used technique wherein germination, radicle growth, and development of hypocotyl are utilized as sensitive indices of phytotoxicity (Mendes et al., 2021). These techniques complement each other by enabling quick screening of various formulations of nanoparticles within concentration ranges and exposure times with the identification of potential candidates to be further analyzed in vivo.

In vivo testing gives detailed information on phytotoxicity in more realistic exposure scenarios and during full plant growth cycles. Hydroponic growth systems offer advantages like precise control of exposure concentrations, absence of soil–nanoparticle interactions and easy root tissue access for analysis. Cultivation in soil better mimics intricate environmental conditions and includes the incorporation of factors like soil adsorption, microbial interaction and aging processes that have major impacts on nanoparticle bioavailability. Irrespective of cultivation regime, phytotoxicity testing generally involves more than one endpoint across a range of biological scales (Lima et al., 2022). This analysis tracks plant height followed by leaf area measurements and biomass evaluations in combination with evaluations of photosynthetic efficiency and transpiration rate assessments and nutrient uptake results and developmental measurements of flowering times and yield features. The microscopic level analysis involves examination of cellular structures through electron microscopy in combination with tissue histology evaluation and tests for effectiveness detection of reactive oxygen species and membrane evaluation using cytochemical methods. Scientists can determine the harmful impacts that nanoparticle exposure causes through the analysis of cellular phytotoxic mechanisms and biological processes at both molecular and biochemical levels. Antioxidant evaluation in biological systems depends on enzyme activities like superoxide dismutase and catalase and peroxidases together with nonenzyme components such as glutathione and ascorbate and stress indicators through malondialdehyde and protein carbonyls measurements. Genomic methods based on transcriptomics, proteomics, and metabolomics provide global information on changes in gene expression patterns, protein content, and metabolite levels, respectively, to allow the establishment of adverse outcome pathways connecting molecular initiating events to adverse phenotypic effects. Standardization activities are more and more concentrating on establishing consistent methodologies between laboratories, such as reference nanoparticle materials, model plant species choices, and harmonized reporting formats to improve comparability between studies (Seepe et al., 2020). This systematized method is enabling regulatory decision-making processes as well as development of predictive computational models that ultimately could minimize requirements for animal tests while maximizing phytotoxicity predictions.

4.2 Physiological and biochemical biomarkers

Physiological biomarkers offer noninvasive, early signals of nanoparticle-induced stress in plants, usually before the onset of visible morphological symptoms. The photosynthetic parameters which use chlorophyll fluorescence measurements to track photosynthetic machinery performance include maximum quantum yield (Fv/Fm) as well as effective quantum yield (ΦPSII) and nonphotochemical quenching (NPQ). These parameters render sensitive indicators for detecting any photosynthetic damage. Physiological stress causes modifications in gas exchange parameters because CO_2 assimilation rate alongside stomatal conductance and transpiration rates serve as indicators (Mostofa et al., 2017). Water relations parameters like relative water content, water potential, and hydraulic conductivity indicate effects on plant water status that often occur in advance of growth inhibition. Nutrient translocation and uptake patterns, determined by elemental analysis of various plant tissues, may reflect interference with mechanisms of nutrient homeostasis. Studies of these physiological biomarkers enable integrated functional plant status assessment and the advantage of measuring them repeatedly on single plants enables long-term tracking of stress build-up and stress recovery (Ferreira et al., 2018).

The study of biochemical biomarkers reveals changes at the molecular and subcellular level which result from nanoparticle exposure with added information about toxicity mechanisms.

Oxidative stress markers are the most common used biochemical markers, such as direct measurement of the level of reactive oxygen species (ROS) by fluorescent probes or spectrophotometric assays, and measurement of oxidative damage products like malondialdehyde (lipid peroxidation), protein carbonyls (protein oxidation), and 8–hydroxy–2′–deoxyguanosine (DNA oxidation). Antioxidant defense variables, such as enzymic activities (catalase, ascorbate peroxidase, glutathione reductase, superoxide dismutase) and nonenzymic antioxidant levels (glutathione, ascorbate, phenolic compounds), represent the protection mechanisms of the plant. Phytohormone profiling sheds light on stress signaling networks with changes in levels of abscisic acid, ethylene, salicylic acid, and jasmonic acid signaling individual stress response pathways (Nakazato et al., 2018). Physical assessments of metabolism reveal widespread changes through analyses of main metabolites (sugars, amino acids, organic acids) and additional products (phenylpropanoids, terpenoids, alkaloids). Biochemical biomarkers alongside physiological variables contribute to building adverse outcome pathways which connect initiation events to organism-level adverse outcomes while generating crucial data for risk assessment and safer-by-design strategies of organic nanoparticles (Shahid et al., 2021).

4.3 Genotoxicity and oxidative stress markers

Research evaluating genotoxicity of organic nanoparticles in plants uses various interconnected methodologies which detect multiple forms of DNA damage. The single-cell gel electrophoresis commonly known as the comet assay delivers sensitive measurements of DNA strand breaks and alkali-labile sites through analysis of tail length and tail moment and DNA percentage in tail. The micronucleus test detects chromosomal fragments resulting from clastogenic activity and absent whole chromosomes within the cell division process caused by aneugenic activity. The assessment of chromosome abnormalities identifies both structural abnormalities and breaks as well as gaps, rings and translocations after arresting cells in metaphase (Zeyad et al., 2019). TUNEL assays identify DNA fragmentation typical of programmed cell death mechanisms. Molecular analysis of DNA adducts by 32P-postlabeling or mass spectrometry identifies covalent modifications on DNA bases. These genotoxicity markers are usually followed by measurement of DNA repair mechanisms, such as measurement of repair enzyme activity and gene expression patterns of enzymes participating in various repair pathways. Through epigenetic modifications that include DNA methylation patterns with histone modifications and noncoding RNA expression scientists obtain supplemental data about inheritable changes which could endure after direct DNA damage (Tabassum et al., 2021).

The existence of oxidative stress serves as a primary pathway triggering genotoxicity alongside phytotoxicity of organic nanoparticles thus research must examine markers of oxidative harm and antioxidants thoroughly. Spatiotemporal ROS detection becomes possible through fluorescent probes which include 2′,7′-dichlorodihydrofluorescein diacetate (DCFH–DA) for hydrogen peroxide analysis and dihydroethidium (DHE) for superoxide detection. The assessment method for membrane damage utilizes malondialdehyde (MDA) by two techniques: TBARS assay and HPLC-based techniques. Protein carbonyl content together with advanced oxidation protein products reveal modifications to amino acids that lead to impaired protein functionality (Mosa et al., 2018). The assessment approach for antioxidant systems includes monitoring enzymatic elements like superoxide dismutase and catalase and ascorbate peroxidase and glutathione reductase and glutathione S–transferase along with nonenzymatic elements such as glutathione and ascorbate and tocopherols and carotenoids and phenolic compounds. Glutathione redox state (GSH/GSSG ratio) is also a sensitive biomarker for disruptions in cellular redox balance. Moreover, the analysis of oxidative stress-responsive gene expression, encompassing antioxidant enzymes and transcriptional regulators like WRKY and members of the bZIP family of proteins, indicates regulatory networks of oxidative stress response (Praveen et al., 2020). The combination between oxidative stress markers and genotoxicity tests enables researchers to understand relationships that connect ROS production to cellular injury and mutagenic effects.

4.4 Imaging and analytical techniques for NP detection in plants

Through microscopy users can observe essential details regarding organic nanoparticle spatial distribution in plant tissues and cellular structures. TEM enables nanometer-level imaging of separate nanoparticles within cellular ultrastructure while SEM and EDX surface imaging allows surface visualization with elemental analysis potential. Using CLSM researchers can explore fluorescent-tagged nanoparticle positioning throughout intact tissues in order to trace their translocation paths. The use of advanced light microscopy methods such as super-resolution microscopy surpasses conventional diffraction limits to obtain resolution measurements near 20 nm. Light sheet fluorescence microscopy offers quick observations of whole seedlings using light that harms cells to a minimum. Two-photon excitation microscopy enables researchers to study thick samples through deeper levels because of its enhanced penetration capabilities. The identification of nanoparticles relies on spectroscopic imaging through Raman microspectroscopy and Fourier-transform infrared (FTIR) imaging because they contain signature vibrational signatures that can identify nanoparticles without using fluorescent markers. Correlative light and electron microscopy (CLEM) utilizes unique strengths from different imaging technologies to provide a complete multi-scale visual analysis (Yilmaz et al., 2021).

Scientific methodologies analyze how nanoparticles penetrate cells as well as establish where they go and how their structure changes. ICP–MS serves as an element composition detector but nanoparticle separation from dissolved materials requires additional analytical techniques. Single particle ICP–MS (sp-ICP–MS) overcomes the limitation since individual nanoparticle events are detected and particle concentration, as well as size distribution information, is reported. Field-flow fractionation (FFF) with different detectors allows size-separation of nanoparticles from intricate plant matrices. Chromatographic methods such as high-performance liquid chromatography (HPLC) and size exclusion chromatography (SEC) allow separation according to chemical properties and hydrodynamic size, respectively. Mass spectrometry-based methods, especially liquid chromatography–tandem mass spectrometry (LC–MS/MS), allow identification and quantification of organic nanoparticles and their transformation products with good sensitivity and specificity (del Real et al., 2022). Hyphenated methods coupling separation techniques with multiple detection platforms (e.g., FFF–ICP–MS–UV–MALS) enable exhaustive characterization of nanoparticle properties and concentrations in matrices of biological complexity. Sample preparation is a main challenge for any analytical strategy, involving the sensitive optimization to ensure nanoparticle integrity while isolating them from plant tissues without inducing artifacts (Xie et al., 2024).

5. Factors influencing phytotoxicity

5.1 Influence of size, shape, and surface charge

Physical properties of nanoparticles—i.e., size, shape, and surface charge—have important roles to play in the behavior of nanoparticles in biological systems. Size influences cellular uptake, biodistribution, and clearance mechanisms directly; nanoparticles less than 10 nm are able to penetrate biological barriers but can be cleared rapidly by the kidneys, whereas those between 10 and 100 nm exhibit ideal circulation times and tumor accumulation through the enhanced permeability and retention effect (Quah et al., 2015). Nanosized particles with spherical shapes achieve uniform distribution patterns and their cellular uptake behaves consistently but elongated or irregularly formed particles may extend circulation duration because they resist recognition by macrophages or connect efficiently with cellular receptors through their shape proportions. The surface charge stands as the biological property which most plainly affects interactions because it manages electrostatic forces between nanoscale substances and biological structures (Liu et al., 2019). Positively charged particles enter cells at higher rates through electrostatic attraction to the cell membrane but experience more toxicity while their blood circulation is shortened. Negatively charged particles tend to have longer circulation times but can encounter diminished cellular uptake (Gikas et al., 2022). Neutral or weakly negative surface charges tend to be the best compromise between circulation time and eventual cell interaction, while zwitterionic surface modification may form a hydration layer to inhibit protein adsorption and consequent immune recognition and thus extend circulation half-life significantly (Dong et al., 2020).

5.2 Role of nanoparticle concentration and dose response

The therapeutic effectiveness together with toxicological risk and biological response of nanoparticles in biological systems comes from their concentration levels. Nanoparticles exist in two distinct behavior patterns: the minimal effects at lower concentrations stem from tissue noninteractions yet intermediate levels enable the achievement of optimal therapeutic outcomes together with reduced adverse side effects. Elevated concentrations often induce concentration-dependent toxicity mechanisms such as oxidative stress, inflammatory reactions, and membrane disruption, the exact threshold varying significantly depending on nanoparticle composition, surface properties, and the particular biological environment (Xiong et al., 2021). The dose-biological response relationship of nanoparticle concentration is not typically simple linear, but rather exhibits intricate dose-response curves that can encompass hormetic effects (useful at low doses and toxic at high doses) or threshold effects in which biological responses are minimal until a limiting concentration is achieved.

Of similar significance to concentration is delivered dose, which considers the actual amount of nanoparticles delivered to target tissue and not the dosed amount. Delivered dose factors that are considered encompass nanoparticle stability in bodily fluids, opsonization and subsequent uptake by the mononuclear phagocyte system, formation of a protein corona that changes surface characteristics, and the general microenvironmental conditions at the site of targeting.

Multiple criteria must be considered to build robust dose-response relationships because pharmacokinetics at different times in combination with biodistribution patterns affect how the starting dose concentration affects biological responses (Yan and Chen, 2019). Sophisticated analytical approaches to detect nanoparticles in tissues along with multiple concentration points and suitable time points are needed for precise characterization of analytic relationships between nanoparticle concentration and biological effects.

5.3 Impact of plant species and growth stage

Plants demonstrate extraordinary differences in nanomaterial reactions because of their distinctive fundamental variations in root architecture together with their cell wall structures and membrane transport systems as well as their defense physiological operations (Wang et al., 2022). Plants in the Cereal group react better to nanoparticles compared to the lettuce and spinach group yet their relationship with mycorrhizal fungi varies according to fungal type and nanomaterial properties (John et al., 1972). The distribution of nanomaterials within plants shows species-dependent trends because some plants retain them in their roots but others enable their movement to aboveground components and these patterns stem from natural evolutionary adaptations rather than grouping by taxonomic classification. The process of plant development affects their interaction with nanomaterials because seedlings have lower protection mechanisms and detoxification capacities during the growth stage. Young plants normally increase their nanomaterial uptake rates and reveal enhanced physiological activity after exposure whereas mature plants develop tolerance through structural defenses such as thick cuticles and strong endodermal barriers and fully functioning stress response processes (Calvelo Pereira, Monterroso and Macías, 2010). The life cycle of reproduction introduces complex interactions because nanomaterials directly contact reproductive tissues while modifying resource distribution and hormonal processes that impact flowering behavior and pollen fertility as well as seed development. Nanomaterial assessment in agroecology requires life-cycle testing because their impact varies across different developmental stages creating the potential for misleading results from testing only a single growth point (Zilkah and Gressel, 1977).

5.4 Interaction with soil and environmental factors

The sophisticated nature of soil creates fundamental changes in nanoparticle behavior which happens through aggregation dissolution and surface modification effects. Nanomaterials establish stable bindings with dirt organic substance through its multiple functional groups due to soil organic matter being an essential interaction site alongside clay minerals and metal oxides. These interactions are strongly modulated by soil pH, which impacts surface charge of both particles and soil constituents; ionic strength, which sets the thickness of electrical double layers and aggregation kinetics; and redox conditions, which can modify some nanomaterials by oxidation or reduction reactions (Jośko and Oleszczuk, 2013). Therefore, nanoparticles entering soil environments hardly preserve their initial physicochemical characteristics, rather being subject to dynamic changes that condition their mobility, bioavailability, and final ecological effect.

Environmental parameters such as temperature, humidity, and microbial activity add complexity to nanoparticle action in terrestrial systems. The equilibrium of adsorption-desorption and reaction rates change according to temperature variations while soil moisture levels determine which processes occur within solid phases versus solution suspension. The effects of microbial populations include direct internalization of nanomaterials in addition to sustaining particles through exopolymeric substances and through metabolic activities leading to the breakdown of particles. Seasonal differences amplify these forces, inducing temporal trends in nanoparticle mobility and bioavailability that need to be taken into account when assessing their environmental persistence and potential ecological effects in representative exposure conditions (Magdziak et al., 2015).

6. Case studies and experimental findings

6.1 Effects on germination, growth, and yield

Exposure to nanomaterials has a significant impact on plant growth starting from germination, where responses vary from stimulation to inhibition based on particle type, concentration, and seed properties. Metal nanoparticles tend to improve germination by enhancing water absorption and germination-related enzyme mobilization, whereas carbon nanomaterials tend to accelerate or retard emergence based on concentration and surface modification (Soltani and Soltani, 2015). Seeds with micropyles or naturally permeable seed coats tend to be more sensitive to nanomaterial action, with responses seen to include changes in germination timing, distorted radicle emergence patterns, and changes to early seedling development that can define growth trajectories for the remainder of the plant's life (Blackburn and Boutin, 2003).

Vegetative growth and total yield are compounded effects of nanomaterial influence on several systems of physiology. At intermediate concentration, some nanomaterials may increase biomass accumulation by augmented nutrient uptake, photosynthetic performance, or stress resistance, whereas at high levels they characteristically cause oxidative stress, membrane disruption, and metabolic disturbances which are reflected as decreased growth rate and lowered yield (Dhakshanamoorthy et al., 2010). During reproductive development plants demonstrate extreme sensitivity because nanomaterial exposure causes alterations to flowering onset and affects pollen fertility and both fertilization and seed developmental processes as well as seed composition and germination capacity. The complex reactions to nanomaterials demand complete phenotypic measurements combining outward plant growth features alongside fundamental biological mechanisms for comprehensive characterization of nanomaterials' influence throughout a plant's complete life period (De Souza et al., 2014).

6.2 Influence on photosynthesis and metabolic pathways

Interactions between nanomaterials and photosynthetic machinery reveal a dose-dependent dichotomy: low doses might increase light harvesting efficiency by means of improved electron transport rates and increased chlorophyll biosynthesis, while higher doses usually compromise the operation of the photosystem by means of direct physical damage to thylakoid membranes, displacement of vital metal cofactors, and production of reactive oxygen species that oxidize photosynthetic pigments (Nikiforova et al., 2004). Metal-based nanoparticles are shown to be especially challenging, having the capability to act as artificial electron donors that raise rates of photosynthesis transiently prior to finally disrupting the delicate redox equilibrium essential for sustained carbon fixation. Impact often ripples far beyond immediate primary photochemical reactions to shape downstream pathways of carbon assimilation, efficiency of photorespiration, and regulation by stomata, engendering full-blown modifications to plant energy metabolism that remain several days past single exposure (Xu and Sullivan, 2010).

Wider metabolic dysregulation results from nanomaterial interaction with essential enzymes, signaling factors, and cellular systems of compartmentalization. Carbon primary metabolism most commonly reveals altered resource distribution among glycolysis, the tricarboxylic acid cycle, and other carbohydrate use alternative pathways, whereas secondary metabolism commonly reveals augmented defense compounds like phenolics, alkaloids, and terpenoids after nanomaterial treatment. Most striking are effects on nitrogen metabolism, wherein metal-based nanomaterials have been shown to interfere with nitrogen-reducing, assimilating, and amino acid biosynthetic enzymes, causing cascading effects on protein production and overall nutritional value (Dusenge et al., 2019). The nanomaterial-induced metabolic reprogramming often mimics abiotic stress response patterns, implying the activation of shared signaling pathways and defense mechanisms that reflect evolutionary strategies against environmental stresses.

6.3 Comparative phytotoxicity of different organic nanoparticles

Carbon-based organic nanoparticles have unique phytotoxicity profiles that are highly correlated with their structural features and surface properties. Fullerenes generally display moderate toxicity due to their capacity to produce reactive oxygen species under light, whereas carbon nanotubes display more severe effects due to their high aspect ratio allowing physical penetration of cell membranes and interference with water transport systems in plant vasculature (García-Gómez et al., 2018). Carbon dots tend to have the least phytotoxicity of all carbon nanomaterials and usually show hormetic behavior with growth stimulation at low concentrations but inhibition at elevated doses. Nanoparticles derived from polymers have the widest variety of phytotoxicity profiles, including biodegradable forms such as polylactic-co-glycolic acid, which in most cases presents minimal harmful effects, whereas residual surfactant- or monomer-containing nondegradable polymers may initiate extensive developmental disorders and metabolic disfunction.

Organic nanoparticle phytotoxicity may be drastically changed by surface functionalization, which affects surface charge, hydrophobicity, and protein corona formation. Carbon nanotubes undergo carboxylation, which enhances their hydrophilicity and minimizes cellular internalization, to reduce their phytotoxicity (Kummerová et al., 2008). In contrast, amino-functionalization often increases toxicity due to more favorable electrostatic interactions with negatively charged phytocell components. Polymer nanoparticles exhibit particularly intricate structure-activity relationships, where toxicity is influenced by parameters like molecular weight, cross-linking extent, rates of degradation, and the exposed plant organs. Importantly, phytotoxicity based on growth measurements alone often underestimates effects since organic nanoparticles can cause profound biochemical as well as physiological perturbations such as changes in gene expression, enzyme inhibition, and membrane integrity disruption even when macroscopic effects on growth are minimal (Song et al., 2013).

6.4 Synergistic and antagonistic effects with other contaminants

Exposure patterns are complex when organic nanoparticles interact with other environmental contaminants; this interaction has the potential to drastically change phytotoxic reactions. Several independent pathways lead to synergistic interactions, where the cumulative impact is larger than the sum of the individual contributions. Through a mechanism called the "Trojan horse" effect, organic nanoparticles have the potential to enhance the bioavailability and cell absorption of contaminants such as heavy metals or persistent organic pollutants by acting as transporters (Cedergreen and Nazir, 2014). Because nanoparticles alter membrane permeability, they may be able to absorb more cocontaminants than would be possible with limited penetration. On top of that, nanoparticle-generated reactive oxygen species may either make prooxidant pollutants more hazardous or transform normally harmless chemicals into harmful metabolic byproducts. On the flip side, nanoparticle adsorption, which reduces pollutant bioavailability; competitive uptake processes; or the upregulation of detoxification systems by nanoparticles seeking to offer cross-protection against two or more stressors are all potential antagonistic interactions that could lead to sub additive total effects. The presence of such patterns of interaction is highly dependent on the features of nanoparticles, the qualities of contaminants, the exposure, and the species of plants. This results in a high-dimensional parameter space, which makes it difficult to do predictive modeling (Coors and De Meester, 2008).

Environmental impacts caused by these interactions include ecological context in a broader sense and extend beyond the more apparent phytotoxicity concerns. It is possible that the geographical spread of pollution may be enhanced if nanoparticle-mediated contaminant transport encouraged pollutant mobility across environmental domains. Consequently, trophic transfer and biomagnification processes within food webs may be affected by altered bioaccumulation patterns. Although complex contaminant mixtures will make remediation more difficult, engineered nanoparticles that have been specifically designed to sequester contaminants but are biocompatible with plants represent exciting opportunities to enhance phytoremediation. These interactions provide opportunities and threats from a remedial perspective. Combination toxicity concerns, not only those pertaining to individual pollutants, need to be taken into account by risk assessment systems (Khanpour-Alikelayeh and Partovinia, 2021). Additionally, these systems should include ideas such as concentration addition for similarly behaving contaminants and independent action for compounds with diverse mechanisms of action. For the purpose of systematically defining these intricate interaction patterns and developing predictive capabilities to facilitate regulatory decision-making, it will be essential to incorporate sophisticated experimental designs such as factorial and response surface methodologies, along with the statistical models that correspond to these designs.

7. Environmental implications and risks

7.1 Fate and transport in soil and water systems

Organic nanoparticles' environmental fate is dictated by complex interplay between ambient matrices and particle characteristics as they move through soil and water systems. Size, hydrophobicity, surface charge, and other physicochemical parameters are among the most important factors that influence the mobility of nanoparticles in soil systems. Soil texture, organic matter concentration, ionic strength, and pH are further factors. The electrostatic interactions between charged nanoparticles and soil colloids have a significant impact on retention. Soil components with a negative charge tend to sorb positively charged particles more effectively (Chacón et al., 2022). The incorporation of NOM into nanoparticles produces NOM coronas, which may improve retention via hydrophobic interactions or steric stabilization, leading to increased mobility. This further adds to the complexity of the matter. Factors in the environment, such as the intensity of rainfall, soil moisture dynamics, and the dynamics of freeze-thaw cycles, influence transport behaviors, which in turn release particles that were confined during high-flow episodes. Whether nanoparticles settle at application sites or go deeper into soil profiles to affect groundwater systems is determined by the combined result of these complicated processes. Surface water chemistry plays a crucial role in controlling the stability and movement of nanoparticles in aquatic systems. Aggregation behavior and sedimentation rates of nanoparticles are controlled by parameters such as dissolved organic carbon content, hardness, ionic strength, and pH. One of the most important destiny processes that determines persistence and bioavailability is heteroaggregation with naturally existing colloids. The properties and composition of biodegradable organic nanoparticles are subject to transformation processes such as hydrolysis, photolysis, and microbial degradation. These transformation processes result in the formation of transformation products that may exhibit different environmental behaviors and biological activities compared to the parent particles. The environmental hazards that are associated with these environmental fate patterns are complex. Some examples of these hazards include the potential disruption of soil microbial communities, which are essential for the cycling of nutrients; bioaccumulation in aquatic life, which has implications for trophic transfer; and disruption of ecosystem services, such as water filtration and carbon sequestration. As a result, a comprehensive environmental risk assessment requires the incorporation of exposure characterization across environmental compartments, as well as hazard evaluation that encompasses a number of ecological receptors and endpoints (Batley et al., 2013). This is complemented by analytical procedures that have been rigorously validated and are able to differentiate between nanoparticles and background material, as well as transformation products in the intricately structured environmental matrices.

7.2 Impact on soil microbial communities

Soil environments that have organic nanoparticles added to them may disrupt the delicate ecological balance that is necessary for the delivery of important ecosystem services. A large number of soil microbes are significantly affected by these nanoparticles. There are direct processes of toxicity that include membrane disruption (Simonin and Richaume, 2015). This occurs when nanoparticles contact with membranes, which causes cellular integrity to become unstable. This leads to the leaking of cytoplasmic contents, which ultimately results in the death of the cell. Nanoparticles are responsible for the generation of reactive oxygen species, which are responsible for causing damage to essential biological components such as proteins, lipids, and nucleic acids. Oxidative stress is another significant cause. In addition to their direct toxicity, nanoparticles have the potential to alter the habitat of microorganisms by altering the physicochemical qualities of soil, such as the porosity, water holding capacity, and availability of nutrients (Hull et al., 2012). These alterations have the ability to restructure the organization of the microbial community, often at the cost of bacterial groups that are tolerant of the changes while suppressing fungal groups that are sensitive to the changes. These alterations in community structure are often associated with functional deficits in significant soil processes such as the breakdown of organic matter, the cycling of nitrogen, and the sequestration of carbon. The effects on plant-nutrient-dependent microbes including mycorrhizal fungi and nitrogen-fixing bacteria remain the most important ecological outcome of nanoparticle releases into soil.

The soil microbial community displays complex and context-sensitive patterns when reacting to exposure with organic nanoparticles. Subject to concentration change effects most biological systems display the pattern known as hormetic response curves. Low exposure doses initiate hormetic stimulation of microbial activity through hormesis effects before higher doses induce inhibitory impacts. Community adjustment occurs after initial changes through genetic modifications along with biological alterations and evolutionary changes which lead communities toward populations adapted for tolerance. The described elements define the temporal dynamics of observed systems. These reactions are fundamentally shaped by the soil characteristics that include texture combined with organic matter counts and pH along with baseline microbial diversity. The level of resistance against nanoparticle disturbances increases with higher organic matter content and greater diversity. The evaluation of impacts now uses advanced molecular techniques including metagenomic sequencing as well as metatranscriptomics and functional gene arrays which superseded traditional culture-based methodologies. Used techniques enable researchers to properly analyze both structural and functional response patterns completely (Girardi et al., 2011). These sophisticated analytical methods are increasingly disclosing subtle but potentially impactful effects on microbial interactions and ecosystem processes that may occur even in the absence of acute toxicity, underlining the necessity for integral assessment frameworks that can capture both short-term impacts as well as long-term ecological ramifications of organic nanoparticles in soil systems.

7.3 Potential risks to food safety and human health

The agricultural usage of these organisms creates substantial food safety concerns because it establishes potential food chain transfer mechanisms which could result in human exposure. The nanoparticles enter crop plants that receive treatment through soil–root absorption after they are cultivated in treated soil. As a result, consumption by humans will offer a direct pathway for exposure to these compounds. Due to variations in translocation channels and tissue barriers, leafy vegetables often exhibit higher concentrations of nanoparticles than fruits or grains. The degree of accumulation varies quite a bit across different plant species and different kinds of nanoparticles. Food crop storage and processing operations may diminish the availability of organic nanoparticles in addition to modifying their biological activity levels (Nauta et al., 2008). The nanoparticles experience three different transformation paths that involve breakdown outlooks or surface alteration or integration by food matrix components. Food crops that contain organic nanoparticles will undergo chosen transformations after being incorporated. Human gastrointestinal exposure triggers multiple chemical processes between digestive fluids and dietary ingredients and intestinal bacteria to establish how much substance becomes ready for absorption. Realistic exposure assessments must account for the entire food production chain because these techniques show that initial ambient concentration assessments are insufficient.

The health-related effects of organic nanoparticles remain ill-defined but various genuine health problems may develop due to their application. Smaller devices smaller than 100 nm have increased capacity to cross and damage the intestinal epithelium thus enabling distribution throughout the body. This represents the main health concern regarding intestinal barrier translocation. Immunological syndromes such as inflammatory episodes and allergic reactions alongside immunosuppression can occur during immunological engagement after absorption. Oral particles demonstrate higher interaction capability with immune cells found inside lymphoid tissue that connects to the gut. The Trojan horse mechanism represents an advanced type of process that can augment exposure to additional agents. The use of organic nanoparticles presents an opportunity to deliver additional contaminants such as cocontaminants together with agricultural pesticides or bacterial infections. Research progress depends on using real-life oral consumption approaches in safety research frameworks and building precise gastrointestinal models and developing sophisticated analysis systems for nanomaterial detection in complex food products and human tissue samples. Standardized regulatory procedures must clearly aim at managing food chain nanomaterials for the agricultural and food production sectors to achieve full safety assessments and effective risk management (Gizaw, 2019).

7.4 Regulatory frameworks and risk assessment approaches

Regulatory environments that control organic nanoparticles are inconsistent between countries, which is a tribute to the rapid speed of technical progress in comparison to the rate at which policy is being developed. The Registration, Evaluation, Authorization, and Restriction of Chemicals (REACH) regulation is the most comprehensive framework that exists inside the European Union. Specific nanomaterials regulations exist under this legislation because authorities need specialized evaluations of both risk factors and exposure levels. The United States of America applies regulatory policies based on individual sectors (Jones, 2001). Under this policy the Environmental Protection Agency (EPA) controls nanoparticles utilizing Toxic Substances Control Act (TSCA) combined with Federal Insecticide Fungicide and Rodenticide Act (FIFRA) to handle agricultural applications. In addition, organizations such as the Organization for Economic Cooperation and Development (OECD) and the International Organization for Standardization (ISO) have developed technical standards and recommendations for the purpose of coordinating the requirements for characterization and the methods of testing. In spite of these developments, there are still significant regulatory challenges that need to be addressed. These challenges include different definitions of nanomaterials within frameworks, limited endorsement of test guideline validation of organic nanoparticles, and insufficient post-market monitoring requirements that would deliver long-term environmental impacts (Jardine et al., 2003).

Traditional chemical assessments use nano-specific evaluation methods to screen specific hazards that appear during the risk evaluation of organic nanoparticles. The scope of exposure measurement moving beyond concentration analysis has included adding factors that measure particle counts alongside surface areas along with distribution patterns. The measures provide better accuracy for bioavailability assessment and biological interaction determination because they were specifically integrated into testing methods. The current risk evaluation systems rely on testing multiple biological impacts among trophic levels but place more emphasis on toxicological assessments that unfold over time even though their ecological effects remain subtle. The combination of high-throughput screening with analytical methods coupled to computer models speeds up hazard testing operations which lowers the need to perform animal tests in chemical assessments. Future positive regulatory science changes will incorporate Modern analytical frameworks which operate using contrary outcome pathways (AOPs) to connect molecular triggers to results that affect organisms and populations and safe-by-design methods as integration standards throughout development. The advancement of technology occurs through framework methods to secure environmental security alongside human safety security (Rand and Zeeman, 1998).

8. Strategies for mitigating phytotoxicity

8.1 Surface modification and biodegradability enhancement

Surface modification stands as a primary technique which minimizes both organic nanoparticle toxicity toward plants and defends their effectiveness levels. PEG conjugation through the polyethylene glycol (PEG) process leads to the development of a hydrophilic neutral outer layer. The protective surface decreases biological membrane and protein interactions which decreases both cell uptake and subsequent toxicity effects. The underlying phytotoxic effects decrease simultaneously with the extended circulation time from this "stealth" effect. Polysaccharide coatings consisting of chitosan, alginate and dextran create surfaces which minimize harmful biological reactions (Ghorbani et al., 2024). Surface coatings provide benefits that encompass both improved stability during digestion and adjustable material breakdown patterns. Surface modification with zwitterionic charges creates powerful hydration layers because positive and negative charges stay adjacent on the surface. The multiple layers successfully block the binding of proteins and interfere with cellular union. The enhancement of biocompatibility through biomimetic approaches involves the use of natural-based coatings for cell membranes and protein coronas. Surface ligand density alongside chain length together with spatial arrangement significantly impacts the performance of these changes thus requiring precise optimization to obtain the desired combination between lower phytotoxicity and saved operational capacity for target applications.

The process of metabolizing organic nanoparticles into nonhazardous components becomes possible due to improved biodegradability while preventing environmental persistence and plant tissue accumulation of substances. Biodegradation control under physiological and environmental conditions can be achieved through proper linkage of hydrolytically active units onto the nanoparticle framework. Three chemical linkages exist for these organic nanoparticles: esters along with amides and disulfides. Addition of components that contain sensitive areas to enzymes represents one method to increase breakdown patterns with enzymes alongside introducing peptide sequences which digestive proteases recognize or glycosidic linkages which glycosidases can break. Nanoparticles break down at specified rates and degrees due to controlled modifications between unstable and stable polymers within their structures (Bartucca et al., 2022).

This allows for the optimization of the length of persistence to be tailored to the individual application. Through the incorporation of reactive groups such as acid-cleaveable acetals, reduction-sensitive disulfides, or photolecavable moieties, it is possible to use environmental cues such as changes in pH, redox, and light in order to initiate degradation. The cumulative effect of these strategies is to transform persistent nanostructures into a substance that can be digested or expelled by plants, or that may be destroyed in various ways by soil microbes. In order to guarantee that these mitigation strategies are successful and to eliminate any impacts that were not expected, it is necessary to conduct a comprehensive characterization of them.

Surface modification procedures need to prove that the coating maintains its integrity along with consistent delivery of uniformity and stability when there are modifications in relevant environmental conditions. Protection and toxic potential can increase due to unidentified interactions that stem from using unstable coating systems improperly. Each environmental condition and duration requires complete definitions regarding the biodegradation patterns. Every main degradation product and intermediate requires testing of their toxic components and their toxicity evaluations. Life cycle assessment methodology performs comprehensive studies on nanoparticle environmental effects during all their life stages. An essential condition needs that the implementation process must maintain operational capabilities for intended purposes at standard levels (Wu et al., 2024). The process of safety improvement consists of applying precise matches between safety advancements and operational performance maintenance ranges. Better efficient methods enable the creation of green nanoparticles with the help of advanced analysis tools that merge predictive modeling systems along with structure-activity relationship methods. This is accomplished via repeated cycles of design that are influenced by mechanistic understanding rather than empirical trial and error.

8.2 Controlled release and eco-friendly formulations

The modern formulation science is confronted with a dual challenge: the provision of active substances precisely where and when they are required, while simultaneously minimizing their impact on the environment. Release prediction systems found within controlled-release technologies evolve based on specific stimulus reactions from pH alterations or enzymatic activities to initiate timed carrier material deterioration. Such distribution systems boost outcome performance while minimizing active pharmaceutical ingredient requirements because they use controlled release methods. The delivery schedules consist of delayed delivery as a single event lasting up to hours or days and pulsatile release which occurs at predetermined time intervals. The environmental concerns have pushed developers of formulas toward using plant-based excipients and biodegradable carriers together with friendly manufacturing procedures (Singh et al., 2020). Chitosan, alginate, and cellulose derivatives are examples of biopolymers that have been shown to be environmentally friendly alternatives to synthetic polymers. These biopolymers biodegrade in a natural environment without causing any damage, while yet maintaining their high level of performance. By using supercritical fluid technology with solvent-free processing one can decrease environmental impacts through waste elimination and manufacturing energy conservation. The integration of controlled release and environmentally friendly method created a fundamental change in formulation research across sectors which includes agriculture, consumer goods and medicines (Zuma et al., 2023). Formulators achieve the best performance results along with minimal environmental impact when they develop biodegradable matrices which actively respond to environmental triggers to encapsulate active components. Such sustainable measures fulfill current customer requirements for green products and additionally anticipate impending government regulations for environmentally friendly formulation approaches across every sector (Kumar et al., 2022).

8.3 Use of natural antioxidants and protective agents

Natural antioxidants together with protective agents serve as the basic conceptual framework which drives modern formulation science. The wide applications benefit from this multifunctionality which extends across food management and cosmetics development as well as medication implementation. The components from plant extracts and essential oils with microorganism metabolism capabilities can deactivate free radicals while protecting the product integrity through electron donation and metal chelation alongside enzyme facilitation mechanisms. Molecular stability and matrix interaction and synergism potential among these bioactives should receive proper attention when blending them together for strategic formulation purposes. By using this process both the protection level improves and the necessary amount of dosage becomes minimized (Ragab et al., 2025).

The success of the intended application heavily relies on modern encapsulating technologies which protect sensitive antioxidants by blocking deteriorative elements while enabling controlled substance release at exact locations. Three successful methods exist for improving bioavailability and duration of action of polyphenols carotenoids and tocopherols – these are liposomal encapsulation and cyclodextrin complexation and solid lipid nanoparticles. These delivery systems offer protection to protective agents and simultaneously increase the functional properties of the agents through their controlled deliveries triggered by environmental and physiological signals (Vijayanand et al., 2025). The outlined deficiencies of natural antioxidants demand systems-based formulation techniques for their compensation. One main disadvantage of these protective agents includes the unstable potency combined with the risk of adding scent and color and the expense associated with each usage. These tactics become essential for developing laboratory discoveries into marketable products. The utilization of chemical modification approaches for stability enhancement works together with complementary ingredient systems and computational modeling allows prediction of antioxidant performance in realistic operational conditions. The combined approach enables natural protection agents to succeed as synthetic replacements in market while fulfilling consumer standards regarding product performance and taste quality and clean-label positioning in an increasingly competitive marketplace (Poornima et al., 2023).

8.4 Sustainable practices for safe nanoparticle use in agriculture

A thorough plan exists for the deployment of agricultural nanotechnology, as it provides high crop yields with minimal environmental impacts. Research on nanomaterials should follow their entire lifecycle from production to soil contact to plant uptake and decomposition or accumulation to establish proper nanoparticle procedures. Plants together with microbes and enzymes remain the top choices for sustainable green synthesis pathways. Sustainable precursor nanoparticles prepared by green synthesis methods function as well as traditional synthetics by using renewable materials and proper engineering processes that minimize environmental damage (Sharma et al., 2022).

Agricultural safety depends on delivery techniques which both reduce substance dispersion from targets and optimize beneficial effects toward living organisms and their tissues. A method of enhanced utilization efficiency at reduced necessary amounts exists through the use of polymer-coated nanoparticles which deploy nutrients or pesticides after environmental signals activate them. A test of rhizosphere pH changes or detection of plant exudates serves as environmental signals that could initiate nanoparticle release. The development of multi-functional nanoplatforms which tackle multiple agricultural issues together such as fertilizer–pesticide delivery systems or particles that both improve nutrient uptake and boost plant defense systems has led to enhanced sustainability advancements (Okeke et al., 2023). The sustainable use of nanomaterials requires well-defined stewardship frameworks as their fundamental base. The frameworks contain subrecipes for performing soil checks at regular intervals and specific requirements for nanoscale–scale interventions which specify biodegradation requirements and protect operating personnel. The agricultural industry should set up nanomaterial recycling processes with complete feedback systems to recover nanomaterials whenever possible. The construction of nanoparticles includes specific decomposition processes which produce nontoxic degradation outputs. The combined implementation of nanotechnology revolutionary powers enables the agricultural sector to utilize preventive safety measures which protect local ecosystems and human health generation after generation (Omran and Baek, 2022). The strategy requires immediate productivity sacrifices to protect environmental sustainability in the long run.

9. Future perspectives and research directions

The combination of environmentally safe nanotechnology in agriculture creates a new frontier which could reshape future innovations during the upcoming decades. The development of innovative biofabrication methods that use plant metabolites and microbial systems and enzymatic techniques decreases environment damage from nanoparticle creation while also improving biocompatibility. These techniques establish veritably closed-loop processes utilizing naturally occurring reducing agents and stabilizers found in agricultural waste products. The designed systems convert agricultural waste products to nanomaterial precursors to achieve higher agricultural productivity levels. Future research should tackle the process scale-up while maintaining control of nanoparticle shape and surface chemistry on both technical and environmental levels. These are crucial criteria that determine both the usefulness and the environmental destiny of the nanoparticles. The integration of multi-omics methods into frameworks for evaluating phytotoxicity will result in a significant transformation of our understanding of the interactions between nanoparticles and plants on both the cellular and molecular levels.

Research using transcriptomic and proteomic and metabolomic methods enables scientists to detect sub-lethal consequences that develop before they become visualized harm. Standard endpoints of toxicity assessments cannot identify small abnormalities found through these studies of biological pathways. The integration of Raman microscopy and synchrotron radiation analysis together with these technologies allows scientists to create detailed profiles of nanoparticle translocation as well as their transformation and biological effects across different parts of plant tissues. The issue that lies ahead is to standardize these methodologies over a wide range of plant species and nanomaterials in order to provide results that are comparable for the purposes of meta-analysis and regulatory processes.

The combination of machine learning techniques and quantitative structure-activity relationship (QSAR) models enables predictive modeling to produce a highly beneficial result by speeding up the development of sustainable nanomaterials. These theoretical computational models permit virtual forecasting of how nanomaterials impact the environment and human safety through their capability to align nanoparticles' physical characteristics with their biological responses. The methods rely on existing experimental information for their operation. New nanomaterials can be screened through high-throughput screening before synthesis because of advancing these models. The predictive models consider the influence of both soil composition and climatic aspects as well as microbial populations. The safety evaluation process will require fewer resources and time as well as pinpoint promising development targets through this approach. The development of agricultural nanotechnology demands parallel advancement of its associated policies and ethical framework.

The adoption of an adaptive governance model becomes necessary because it instructs how to balance innovative methods with carefully considered decisions. Worker safety together with environmental protection and food safety will exist through regulatory measures that emerge from farmer and consumer involvement with science groups and legislative bodies. This approach guarantees protective measures for technological feasibility together with social standards. The implementation of sustainable governance demands clear labeling frameworks and worldwide regulatory standardization as well as user-friendly databases for nanomaterial information and environmental reaction data to operate efficiently. Sustainable agricultural systems with better productivity practice can be created through proper implementation of fundamental factors at current technological thresholds for the industry. Satellite companies must responsibly use nanotechnology innovations for transformation while fulfilling their duty to preserve public integrity and ecological preservation.

10. Conclusion

The agricultural application of nanotechnology produces numerous environmental sustainability impacts and delivers universal food security standards. Nanomaterials developed for specific purposes demonstrate potential to improve plant nutrient uptake performance and provide tolerance against stress and reduce the need for agricultural chemicals in crop cultivation. Nanomaterials demonstrate better environmental achievements than traditional farming practices. Multiple agricultural problems find their optimal solution through accurate delivery systems connected to biodegradable nanocarriers. Different nanotechnology-based innovations create revolutionary changes in agriculture by establishing sustainable operational frameworks. Research about nanomaterial behavior dimensions in agricultural environments still remains uncertain while scientists require more data about nanomaterial effects on ecosystems and food chain accumulation. Scientific assessment of how nanomaterials change in their structure and spread in soil as well as their microorganism interaction patterns is limited by present-day experimental constraints. Fundamental knowledge gained through these processes enables the modeling of environmental paths and toxic outcomes of nanomaterials in the environment. Evaluating agricultural safety recommendations and performance measures becomes complicated due to the extensive diversity between climates and soils and farming systems. Various elements mutually influence one another since they depend on their interconnected operational characteristics. The need exists with urgency for framework-based risk assessment of nanotech in agriculture combined with standardized tests because of these outlined issues. Research must direct efforts to engineer biodegradable nanomaterials which can naturally decompose through man-made processes and establish enduring protocols for detecting nanomaterial longevity through agricultural years as well as create measurement protocols that link environmental risks to agricultural outputs. Material scientists alongside agronomists and ecologists and regulatory professionals need to collaborate for transforming laboratory findings into scalable field applications and proper regulation purposes. The focus of ordinary research funds should be manufacturing cost-efficient nanomaterials at a large scale so farmers in developing nations and smallholders can more readily utilize them. The revolution potential of agricultural nanotechnology science can be realized through combined collaboration between disciplinary experts who focus on protecting human health and environmental safety across present and future human generations.

References

Abu-Dief, A.M., Abdel-Fatah, S.M., 2018. Development and functionalization of magnetic nanoparticles as powerful and green catalysts for organic synthesis. Beni-Suef University Journal of Basic and Applied Sciences 7 (1), 55–67. https://doi.org/10.1016/j.bjbas.2017.05.008.

Ali, S.S., Al-Tohamy, R., Koutra, E., Moawad, M.S., Kornaros, M., Mustafa, A.M., Mahmoud, Y.A.G., Badr, A., Osman, M.E.H., Elsamahy, T., Jiao, H., Sun, J., 2021. Nanobiotechnological advancements in agriculture and food industry: applications, nanotoxicity, and future perspectives. Sci. Total Environ. 792. https://doi.org/10.1016/j.scitotenv.2021.148359.

Andersen, E.J., Ali, S., Byamukama, E., Yen, Y., Nepal, M.P., 2018. Disease resistance mechanisms in plants. Genes 9 (7). https://doi.org/10.3390/genes9070339.

Appah, S., Jia, W., Ou, M., Wang, P., Asante, E.A., 2020. Analysis of potential impaction and phytotoxicity of surfactant-plant surface interaction in pesticide application. Crop Prot. 127, 104961. https://doi.org/10.1016/j.cropro.2019.104961.

Astruc, D., 2020. Introduction: nanoparticles in catalysis. Chem. Rev. 120 (2), 461–463. https://doi.org/10.1021/acs.chemrev.8b00696.

Bahrulolum, H., Nooraei, S., Javanshir, N., Tarrahimofrad, H., Mirbagheri, V.S., Easton, A.J., Ahmadian, G., 2021. Green synthesis of metal nanoparticles using microorganisms and their application in the agrifood sector. J. Nanobiotechnol. 19 (1). https://doi.org/10.1186/s12951-021-00834-3.

Banu, A.N., Kudesia, N., Raut, A.M., Pakrudheen, I., Wahengbam, J., 2021. Toxicity, bioaccumulation, and transformation of silver nanoparticles in aqua biota: a review. Environ. Chem. Lett. 19 (6), 4275–4296. https://doi.org/10.1007/s10311-021-01304-w.

Barhoum, A., García-Betancourt, M.L., Jeevanandam, J., Hussien, E.A., Mekkawy, S.A., Mostafa, M., Omran, M.M., Abdalla, M.S., Bechelany, M., 2022. Review on natural, incidental, bioinspired, and engineered nanomaterials: history, definitions, classifications, synthesis, properties, market, toxicities, risks, and regulations. Nanomaterials 12 (2). https://doi.org/10.3390/nano12020177.

Bartucca, M.L., Cerri, M., Del Buono, D., Forni, C., 2022. Use of biostimulants as a new approach for the improvement of phytoremediation performance—a review. Plants 11 (15), 1946. https://doi.org/10.3390/plants11151946.

Batley, G.E., Kirby, J.K., McLaughlin, M.J., 2013. Fate and risks of nanomaterials in aquatic and terrestrial environments. Acc. Chem. Res. 46 (3), 854–862. https://doi.org/10.1021/ar2003368.

Bedlovičová, Z., Strapáč, I., Baláž, M., Salayová, A., 2020. A brief overview on antioxidant activity determination of silver nanoparticles. Molecules 25 (14), 3191. https://doi.org/10.3390/molecules25143191.

Bhatia, S., 2016. Nanoparticles Types, Classification, Characterization, Fabrication Methods and Drug Delivery Applications. Springer Nature, pp. 33–93. https://doi.org/10.1007/978-3-319-41129-3_2.

Blackburn, L.G., Boutin, C., 2003. Subtle effects of herbicide use in the context of genetically modified crops: a case study with glyphosate (Roundup®). Ecotoxicology 12 (1–4), 271–285. https://doi.org/10.1023/A:1022515129526.

Calvelo Pereira, R., Monterroso, C., Macías, F., 2010. Phytotoxicity of hexachlorocyclohexane: effect on germination and early growth of different plant species. Chemosphere 79 (3), 326–333. https://doi.org/10.1016/j.chemosphere.2010.01.035.

Carrasco, S., 2018. Metal-organic frameworks for the development of biosensors: a current overview. Biosensors 8 (4), 92. https://doi.org/10.3390/bios8040092.

Cedergreen, N., Nazir, A., 2014. Quantifying synergy: a systematic review of mixture toxicity studies within environmental toxicology. PLoS One 9 (5), e96580. https://doi.org/10.1371/journal.pone.0096580.

Chacón, L., Reyes, L., Rivera-Montero, L., Barrantes, K., 2022. Transport, Fate, and Bioavailability of Emerging Pollutants in Soil, Sediment, and Wastewater Treatment Plants: Potential Environmental Impacts. Elsevier BV, pp. 111–136. https://doi.org/10.1016/b978-0-323-85160-2.00020-2.

Coelho, J.J., Prieto, M.L., Dowling, S., Hennessy, A., Casey, I., Woodcock, T., Kennedy, N., 2018. Physical-chemical traits, phytotoxicity and pathogen detection in liquid anaerobic digestates. Waste Manag. 78, 8–15. https://doi.org/10.1016/j.wasman.2018.05.017.

Coors, A., De Meester, L., 2008. Synergistic, antagonistic and additive effects of multiple stressors: predation threat, parasitism and pesticide exposure in Daphnia magna. J. Appl. Ecol. 45 (6), 1820–1828. https://doi.org/10.1111/j.1365-2664.2008.01566.x.

De Souza, A., García, D., Sueiro, L., Gilart, F., 2014. Improvement of the seed germination, growth and yield of onion plants by extremely low frequency non-uniform magnetic fields. Sci. Hortic. 176, 63–69. https://doi.org/10.1016/j.scienta.2014.06.034.

del Real, A.E.P., Mitrano, D.M., Castillo-Michel, H., Wazne, M., Reyes-Herrera, J., Bortel, E., Hesse, B., Villanova, J., Sarret, G., 2022. Assessing implications of nanoplastics exposure to plants with advanced nanometrology techniques. J. Hazard Mater. 430. https://doi.org/10.1016/j.jhazmat.2022.128356.

Deshmukh, R., Niederberger, M., 2017. Mechanistic aspects in the formation, growth and surface functionalization of metal oxide nanoparticles in organic solvents. Chem. Eur J. 23 (36), 8542–8570. https://doi.org/10.1002/chem.201605957.

Dhakshanamoorthy, D., Selvaraj, R., Chidambaram, A., 2010. Physical and chemical mutagenesis in Jatropha curcas L. to induce variability in seed germination, growth and yield traits. Rom. J. Biol. Plant Biol. 55 (2), 113–125.

Dong, G., Nkoh, J.N., Hong, Z.n., Dong, Y., Lu, H.l., Yang, J., Pan, X.y., Xu, R.k., 2020. Phytotoxicity of Cu2+ and Cd2+ to the roots of four different wheat cultivars as related to charge properties and chemical forms of the metals on whole plant roots. Ecotoxicol. Environ. Saf. 196. https://doi.org/10.1016/j.ecoenv.2020.110545.

Dusenge, M.E., Duarte, A.G., Way, D.A., 2019. Plant carbon metabolism and climate change: elevated CO2 and temperature impacts on photosynthesis, photorespiration and respiration. New Phytol. 221 (1), 32–49. https://doi.org/10.1111/nph.15283.

Erb, M., Reymond, P., 2019. Molecular interactions between plants and insect herbivores. Annu. Rev. Plant Biol. 70 (1), 527–557. https://doi.org/10.1146/annurev-arplant-050718-095910.

Feng, R.W., Wang, L.Z., Yang, J.G., Zhao, P.P., Zhu, Y.M., Li, Y.P., Yu, Y.S., Liu, H., Rensing, C., Wu, Z.Y., Ni, R.X., Zheng, S.A., 2021. Underlying mechanisms responsible for restriction of uptake and translocation of heavy metals (metalloids) by selenium via root application in plants. J. Hazard Mater. 402, 123570. https://doi.org/10.1016/j.jhazmat.2020.123570.

Ferreira, P.A.A., Lourenzi, C.R., Tiecher, T., Tiecher, T.L., Ricachenevsky, F.K., Brunetto, G., Giachini, A.J., Soares, C.R.F.S., 2018. Physiological, Biochemical Changes, and Phytotoxicity Remediation in Agricultural Plant Species Cultivated in Soils Contaminated with Copper and Zinc Plants under Metal and Metalloid Stress: Responses, Tolerance and Remediation. Springer Singapore, Brazil, pp. 29–76. https://doi.org/10.1007/978-981-13-2242-6_2.

Forti, J.C., Loretti, G.H., Tadayozzi, Y.S., de Andrade, A.R., 2020. A phytotoxicity assessment of the efficiency 2,4-D degradation by different oxidative processes. J. Environ. Manag. 266. https://doi.org/10.1016/j.jenvman.2020.110588.

Freitas de Freitas, L., Varca, G.H.C., Gabriel Dos Santos Batista, J., Benévolo Lugão, A., 2018. An overview of the synthesis of gold nanoparticles using radiation technologies. Nanomaterials 8 (11), 939. https://doi.org/10.3390/nano8110939.

García-Gómez, C., Obrador, A., González, D., Babín, M., Fernández, M.D., 2018. Comparative study of the phytotoxicity of ZnO nanoparticles and Zn accumulation in nine crops grown in a calcareous soil and an acidic soil. Sci. Total Environ. 644, 770–780. https://doi.org/10.1016/j.scitotenv.2018.06.356.

Ghorbani, A., Emamverdian, A., Pehlivan, N., Zargar, M., Razavi, S.M., Chen, M., 2024. Nano-enabled agrochemicals: mitigating heavy metal toxicity and enhancing crop adaptability for sustainable crop production. J. Nanobiotechnol. 22 (1). https://doi.org/10.1186/s12951-024-02371-1.

Gikas, G.D., Parlakidis, P., Mavropoulos, T., Vryzas, Z., 2022. Particularities of fungicides and factors affecting their fate and removal efficacy: a review. Sustainability 14 (7), 4056. https://doi.org/10.3390/su14074056.

Girardi, C., Greve, J., Lamshöft, M., Fetzer, I., Miltner, A., Schäffer, A., Kästner, M., 2011. Biodegradation of ciprofloxacin in water and soil and its effects on the microbial communities. J. Hazard Mater. 198, 22–30. https://doi.org/10.1016/j.jhazmat.2011.10.004.

Gizaw, Z., 2019. Public health risks related to food safety issues in the food market: a systematic literature review. Environ. Health Prev. Med. 24 (1). https://doi.org/10.1186/s12199-019-0825-5.

Hasanuzzaman, M., Oku, H., Nahar, K., Bhuyan, M.H.M.B., Mahmud, J.A., Baluska, F., Fujita, M., 2018. Nitric oxide-induced salt stress tolerance in plants: ROS metabolism, signaling, and molecular interactions. Plant Biotechnology Reports 12 (2), 77–92. https://doi.org/10.1007/s11816-018-0480-0.

Hechmi, S., Hamdi, H., Mokni-Tlili, S., Ghorbel, M., Khelil, M.N., Zoghlami, I.R., Benzarti, S., Jellali, S., Hassen, A., Jedidi, N., 2020. Impact of urban sewage sludge on soil physico-chemical properties and phytotoxicity as influenced by soil texture and reuse conditions. J. Environ. Qual. 49 (4), 973–986. https://doi.org/10.1002/jeq2.20093.

Hull, R.N., Klee, U., Bryant, D., Todd, C., 2012. Soil microbial communities and ecological risk assessment: risk assessors' perspective. Hum. Ecol. Risk Assess. 5 (4), 707–714. https://doi.org/10.1080/10807039.1999.9657735.

Jain, N., Bhargava, A., Pareek, V., Sayeed Akhtar, M., Panwar, J., 2017. Does seed size and surface anatomy play role in combating phytotoxicity of nanoparticles? Ecotoxicology 26 (2), 238–249. https://doi.org/10.1007/s10646-017-1758-7.

Jardine, C.G., Hrudey, S.E., Shortreed, J.H., Craig, L., Krewski, D., Furgal, C., Mc Coll, S., 2003. Risk management frameworks for human health and environmental risks. Journal of Toxicology and

Environmental Health - Part B: Critical Reviews 6 (6), 569–718. https://doi.org/10.1080/10937400390208608.

John, M.K., Vanlaerhoven, C.J., Chuah, H.H., 1972. Factors affecting plant uptake and phytotoxicity of cadmium added to soils. Environ. Sci. Technol. 6 (12), 1005–1009. https://doi.org/10.1021/es60071a008.

Jones, R.N., 2001. An environmental risk assessment/management framework for climate change impact assessments. Nat. Hazards 23 (2–3), 197–230. https://doi.org/10.1023/A:1011148019213.

Jośko, I., Oleszczuk, P., 2013. Influence of soil type and environmental conditions on ZnO, TiO2 and Ni nanoparticles phytotoxicity. Chemosphere 92 (1), 91–99. https://doi.org/10.1016/j.chemosphere.2013.02.048.

Khan, S., Hossain, M.K., 2022. Classification and Properties of Nanoparticles Nanoparticle-Based Polymer Composites. Elsevier, United States, pp. 15–54. https://doi.org/10.1016/B978-0-12-824272-8.00009-9.

Khan, Y., Sadia, H., Ali Shah, S.Z., Khan, M.N., Shah, A.A., Ullah, N., Ullah, M.F., Bibi, H., Bafakeeh, O.T., Ben Khedher, N., Eldin, S.M., Fadhl, B.M., Khan, M.I., 2022. Classification, synthetic, and characterization approaches to nanoparticles, and their applications in various fields of nanotechnology: a review. Catalysts 12 (11). https://doi.org/10.3390/catal12111386.

Khanpour-Alikelayeh, E., Ali, P., 2021. Synergistic and Antagonistic Effects of Microbial Co-culture on Bioremediation of Polluted Environments, 26. Springer Science and Business Media LLC, pp. 229–265. https://doi.org/10.1007/978-981-15-7455-9_10.

Kőszegi, T., Poór, M., 2016. Ochratoxin a: molecular interactions, mechanisms of toxicity and prevention at the molecular level. Toxins 8 (4), 111. https://doi.org/10.3390/toxins8040111.

Kumar, R., Kumar, N., Rajput, V.D., Mandzhieva, S., Minkina, T., Saharan, B.S., Kumar, D., Sadh, P.K., Duhan, J.S., 2022. Advances in biopolymeric nanopesticides: a new eco-friendly/eco-protective perspective in precision agriculture. Nanomaterials 12 (22). https://doi.org/10.3390/nano12223964.

Kummerová, M., Váňová, L., Krulová, J., Zezulka, Š., 2008. The use of physiological characteristics for comparison of organic compounds phytotoxicity. Chemosphere 71 (11), 2050–2059. https://doi.org/10.1016/j.chemosphere.2008.01.060.

Lencioni, G., Imperiale, D., Cavirani, N., Marmiroli, N., Marmiroli, M., 2016. Environmental application and phytotoxicity of anaerobic digestate from pig farming by in vitro and in vivo trials. Int. J. Environ. Sci. Technol. 13 (11), 2549–2560. https://doi.org/10.1007/s13762-016-1088-y.

Li, H., Chen, X., Shen, D., Wu, F., Pleixats, R., Pan, J., 2021. Functionalized silica nanoparticles: classification, synthetic approaches and recent advances in adsorption applications. Nanoscale 13 (38), 15998–16016. https://doi.org/10.1039/d1nr04048k.

Lima, L.M., Pedroza, L.S., Osório, M.I.C., Souza, J.C., Nunez, C.V., 2022. Phytotoxicity of plant extracts of Vismia japurensis cultivated in vivo and in vitro. Braz. J. Biol. 82. https://doi.org/10.1590/1519-6984.235475.

Lismont, M., Dreesen, L., Wuttke, S., 2017. Metal-organic framework nanoparticles in photodynamic therapy: current status and perspectives. Adv. Funct. Mater. 27 (14). https://doi.org/10.1002/adfm.201606314.

Liu, M., Feng, S., Ma, Y., Xie, C., He, X., Ding, Y., Zhang, J., Luo, W., Zheng, L., Chen, D., Yang, F., Chai, Z., Zhao, Y., Zhang, Z., 2019. Influence of surface charge on the phytotoxicity, transformation, and translocation of CeO 2 nanoparticles in cucumber plants. ACS Appl. Mater. Interfaces 11 (18), 16905–16913. https://doi.org/10.1021/acsami.9b01627.

Lombardo, D., Calandra, P., Pasqua, L., Magazù, S., 2020. Self-assembly of organic nanomaterials and biomaterials: the bottom-up approach for functional nanostructures formation and advanced applications. Materials 13 (5), 1048. https://doi.org/10.3390/ma13051048.

Lorenzo, V.D., Schiavi, D., Petrucci, C., Balestra, G.M., 2021. Organic nanoparticles from lignocellulosic agricultural residues for plant protection applications. In: AIP Conference Proceedings, 2416. American Institute of Physics Inc, Italy. https://doi.org/10.1063/5.0069309.

Loste, J., Lopez-Cuesta, J.M., Billon, L., Garay, H., Save, M., 2019. Transparent polymer nanocomposites: an overview on their synthesis and advanced properties. Prog. Polym. Sci. 89, 133–158. https://doi.org/10.1016/j.progpolymsci.2018.10.003.

Ma, Y., Oliveira, R.S., Freitas, H., Zhang, C., 2016. Biochemical and molecular mechanisms of plant-microbe-metal interactions: Relevance for phytoremediation. Front. Plant Sci. 7. https://doi.org/10.3389/fpls.2016.00918.

Magdziak, Z., Gąsecka, M., Goliński, P., Mleczek, M., 2015. Phytoremediation and Environmental Factors Phytoremediation: Management of Environmental Contaminants, Volume 1. Springer International Publishing, Poland, pp. 45–56. https://doi.org/10.1007/978-3-319-10395-2_4.

Mathur, S., Kalaji, H.M., Jajoo, A., 2016. Investigation of deleterious effects of chromium phytotoxicity and photosynthesis in wheat plant. Photosynthetica 54 (2), 185–192. https://doi.org/10.1007/s11099-016-0198-6.

Mekuye, B., Abera, B., 2023. Nanomaterials: an overview of synthesis, classification, characterization, and applications. Nano Select 4 (8), 486–501. https://doi.org/10.1002/nano.202300038.

Mendes, P.M., Ribeiro, J.A., Martins, G.A., Lucia, T., Araujo, T.R., Fuentes-Guevara, M.D., Corrêa, L. B., Corrêa, É.K., 2021. Phytotoxicity test in check: Proposition of methodology for comparison of different method adaptations usually used worldwide. J. Environ. Manag. 291. https://doi.org/10.1016/j.jenvman.2021.112698.

Mosa, K.A., El-Naggar, M., Ramamoorthy, K., Alawadhi, H., Elnaggar, A., Wartanian, S., Ibrahim, E., Hani, H., 2018. Copper nanoparticles induced genotoxicty, oxidative stress, and changes in superoxide dismutase (SOD) gene expression in cucumber (cucumis sativus) plants. Front. Plant Sci. 9. https://doi.org/10.3389/fpls.2018.00872.

Mostofa, M.G., Hossain, M.A., Siddiqui, M.N., Fujita, M., Tran, L.S.P., 2017. Phenotypical, physiological and biochemical analyses provide insight into selenium-induced phytotoxicity in rice plants. Chemosphere 178, 212–223. https://doi.org/10.1016/j.chemosphere.2017.03.046.

Nakazato, R.K., Esposito, M.P., Cardoso-Gustavson, P., Bulbovas, P., Pedroso, A.N.V., de Assis, P.I.L.S., Domingos, M., 2018. Efficiency of biomonitoring methods applying tropical bioindicator plants for assessing the phytoxicity of the air pollutants in SE, Brazil. Environ. Sci. Pollut. Control Ser. 25 (20), 19323–19337. https://doi.org/10.1007/s11356-018-2294-6.

Nauta, M.J., Fischer, A.R.H., Van Asselt, E.D., De Jong, A.E.I., Frewer, L.J., De Jonge, R., 2008. Food safety in the domestic environment: the effect of consumer risk information on human disease risks. Risk Anal. 28 (1), 179–192. https://doi.org/10.1111/j.1539-6924.2008.01012.x.

Nikiforova, V.J., Gakière, B., Kempa, S., Adamik, M., Willmitzer, L., Hesse, H., Hoefgen, R., 2004. Towards dissecting nutrient metabolism in plants: a systems biology case study on sulphur metabolism. J. Exp. Bot. 55 (404), 1861–1870. https://doi.org/10.1093/jxb/erh177.

Okeke, E.S., Nweze, E.J., Ezike, T.C., Nwuche, C.O., Ezeorba, T.P.C., Nwankwo, C.E.I., 2023. Silicon-based nanoparticles for mitigating the effect of potentially toxic elements and plant stress in agroecosystems: a sustainable pathway towards food security. Sci. Total Environ. 898. https://doi.org/10.1016/j.scitotenv.2023.165446.

Omran, B.A., Baek, K.H., 2022. Control of phytopathogens using sustainable biogenic nanomaterials: recent perspectives, ecological safety, and challenging gaps. J. Clean. Prod. 372. https://doi.org/10.1016/j.jclepro.2022.133729.

Pan, M., Chu, L.M., 2016. Phytotoxicity of veterinary antibiotics to seed germination and root elongation of crops. Ecotoxicol. Environ. Saf. 126, 228–237. https://doi.org/10.1016/j.ecoenv.2015.12.027.

Patil, N., Bhaskar, R., Vyavhare, V., Dhadge, R., Khaire, V., Patil, Y., 2021. Overview on methods of synthesis of nanoparticles. Int. J. Curr. Pharmaceut. Res. 11–16. https://doi.org/10.22159/ijcpr.2021v13i2.41556.

Pereira, S.P.P., Jesus, F., Aguiar, S., de Oliveira, R., Fernandes, M., Ranville, J., Nogueira, A.J.A., 2018. Phytotoxicity of silver nanoparticles to Lemna minor: surface coating and exposure period-related effects. Sci. Total Environ. 618, 1389–1399. https://doi.org/10.1016/j.scitotenv.2017.09.275.

Poornima, R., Dhevagi, P., Ramya, A., Agathokleous, E., Sahasa, R.G.K., Ramakrishnan, S., 2023. Protectants to ameliorate ozone-induced damage in crops – a possible solution for sustainable agriculture. Crop Prot. 170. https://doi.org/10.1016/j.cropro.2023.106267.

Praveen, A., Pandey, C., Khan, E.nullah, Panthri, M., Gupta, M., 2020. Silicon-mediated genotoxic alterations in Brassica juncea under arsenic stress: a comparative study of biochemical and molecular markers. Pedosphere 30 (4), 517–527. https://doi.org/10.1016/s1002-0160(17)60435-1.

Prerna, D.A., Ratan, G., 2021. Nanoparticles: an overview. Drugs and cell therapies in haematology 10 (1), 1487–1497.

Quah, B., Musante, C., White, J.C., Ma, X., 2015. Phytotoxicity, uptake, and accumulation of silver with different particle sizes and chemical forms. J. Nanoparticle Res. 17 (6). https://doi.org/10.1007/s11051-015-3079-1.

Ragab, G.A., Saad-Allah, K.M., Nessem, A.A., 2025. Mitigating arsenate-induced phytotoxicity in fenugreek seedlings using garlic extract: insights into photosynthesis, arsenate uptake, antioxidative machinery and ultrastructure. J. Soil Sci. Plant Nutr. https://doi.org/10.1007/s42729-025-02386-z.

Rand, G.M., Zeeman, M.G., 1998. Ecological risk assessment approaches within the regulatory framework. Hum. Ecol. Risk Assess. 4 (4), 853–886. https://doi.org/10.1080/10807039891284857.

Ren, R., Lim, C., Li, S., Wang, Y., Song, J., Lin, T.W., Muir, B.W., Hsu, H.Y., Shen, H.H., 2022. Recent advances in the development of lipid-, metal-, carbon-, and polymer-based nanomaterials for antibacterial applications. Nanomaterials 12 (21). https://doi.org/10.3390/nano12213855.

Salem, S.S., Fouda, A., 2021. Green synthesis of metallic nanoparticles and their prospective biotechnological applications: an overview. Biol. Trace Elem. Res. 199 (1), 344–370. https://doi.org/10.1007/s12011-020-02138-3.

Schwab, F., Zhai, G., Kern, M., Turner, A., Schnoor, J.L., Wiesner, M.R., 2016. Barriers, pathways and processes for uptake, translocation and accumulation of nanomaterials in plants - critical review. Nanotoxicology 10 (3), 257–278. https://doi.org/10.3109/17435390.2015.1048326.

Seepe, H.A., Lodama, K.E., Sutherland, R., Nxumalo, W., Amoo, S.O., 2020. In vivo antifungal activity of south african medicinal plant extracts against fusarium pathogens and their phytotoxicity evaluation. Plants 9 (12), 1–21. https://doi.org/10.3390/plants9121668.

Sekoai, P.T., Ouma, C.N.M., du Preez, S.P., Modisha, P., Engelbrecht, N., Bessarabov, D.G., Ghimire, A., 2019. Application of nanoparticles in biofuels: an overview. Fuel 237, 380–397. https://doi.org/10.1016/j.fuel.2018.10.030.

Shahid, M., Khan, M.S., Ahmed, B., Syed, A., Bahkali, A.H., 2021. Physiological disruption, structural deformation and low grain yield induced by neonicotinoid insecticides in chickpea: a long term phytotoxicity investigation. Chemosphere 262. https://doi.org/10.1016/j.chemosphere.2020.128388.

Sharma, S., Shree, B., Aditika, Sharma, A., Irfan, M., Kumar, P., 2022. Nanoparticle-based toxicity in perishable vegetable crops: molecular insights, impact on human health and mitigation strategies for sustainable cultivation. Environ. Res. 212 (Pt A), 113168. https://doi.org/10.1016/j.envres.2022.113168.

Simonin, M., Richaume, A., 2015. Impact of engineered nanoparticles on the activity, abundance, and diversity of soil microbial communities: a review. Environ. Sci. Pollut. Control Ser. 22 (18), 13710–13723. https://doi.org/10.1007/s11356-015-4171-x.

Singh, A., Dhiman, N., Kar, A.K., Singh, D., Purohit, M.P., Ghosh, D., Patnaik, S., 2020. Advances in controlled release pesticide formulations: Prospects to safer integrated pest management and sustainable agriculture. J. Hazard Mater. 385. https://doi.org/10.1016/j.jhazmat.2019.121525.

Soltani, E., Soltani, A., 2015. Meta-analysis of seed priming effects on seed germination, seedling emergence and crop yield: Iranian studies. Int. J. Plant Prod. 9 (3), 413–432.

Song, U., Jun, H., Waldman, B., Roh, J., Kim, Y., Yi, J., Lee, E.J., 2013. Functional analyses of nanoparticle toxicity: a comparative study of the effects of TiO2 and Ag on tomatoes (Lycopersicon esculentum). Ecotoxicol. Environ. Saf. 93, 60–67. https://doi.org/10.1016/j.ecoenv.2013.03.033.

Störmer, A., Bott, J., Kemmer, D., Franz, R., 2017. Critical review of the migration potential of nanoparticles in food contact plastics. Trends Food Sci. Technol. 63, 39–50. https://doi.org/10.1016/j.tifs.2017.01.011.

Tabassum, Jeena, A.S., Pandey, D., 2021. Metal Induced Genotoxicity and Oxidative Stress in Plants, Assessment Methods, and Role of Various Factors in Genotoxicity Regulation Induced Genotoxicity and Oxidative Stress in Plants. Springer Nature, India, pp. 133–149. https://doi.org/10.1007/978-981-16-2074-4_5.

Thakur, S., Thakur, S., Kumar, R., 2018. Bio-nanotechnology and its role in agriculture and food industry. J. Mol. Genet. Med. 12 (324), 1747.

Trivedi, R., Arora, S., Lamptey, R.N.L., Chaulagain, B., Singh, J., Layek, B., 2022. A Summarized View of Lipid, Polyplex, Inorganic, and Carbon-Based Nanotherapeutics for Hepatocellular Carcinoma Treatment Nanotherapeutics for the Treatment of Hepatocellular Carcinoma. Bentham Science Publishers, United States, pp. 248–279. https://doi.org/10.2174/9789815039740122010008.

Uddin, M.N., Desai, F., Asmatulu, E., 2020. Engineered nanomaterials in the environment: bioaccumulation, biomagnification and biotransformation. Environ. Chem. Lett. 18 (4), 1073–1083. https://doi.org/10.1007/s10311-019-00947-0.

Vijayanand, M., Issac, P.K., Velayutham, M., Deepak, P., Thiyagarajulu, N., Alam, M.W., Guru, A., 2025. A sustainable solution: mitigating aquatic herbicide contamination through natural product interventions. Aquac. Int. 33 (2). https://doi.org/10.1007/s10499-025-01823-1.

Wang, Q., Li, C., Wang, Y., Que, X., 2019. Phytotoxicity of graphene family nanomaterials and its mechanisms: a review. Front. Chem. 7. https://doi.org/10.3389/fchem.2019.00292.

Wang, G., Yang, Y., Kong, Y., Ma, R., Yuan, J., Guoxue, L., 2022. Key factors affecting seed germination in phytotoxicity tests during sheep manure composting with carbon additives. J. Hazard Mater. 421, 126809. https://doi.org/10.1016/j.jhazmat.2021.126809.

Werrie, P.Y., Durenne, B., Delaplace, P., Fauconnier, M.L., 2020. Phytotoxicity of essential oils: opportunities and constraints for the development of biopesticides. A review. Foods 9 (9). https://doi.org/10.3390/foods9091291.

Wu, F., Wang, Z., Li, X., Pu, Q., Wu, Y., Cao, N., Wang, X., 2024. Molecular design of environment-friendly amide herbicide substitutes with high efficacy, low phytotoxicity and medication safety. J. Hazard Mater. 463, 132858. https://doi.org/10.1016/j.jhazmat.2023.132858.

Xiao, B., Wang, X., Yang, J., Wang, K., Zhang, Y., Sun, B., Zhang, T., Zhu, L., 2020. Bioaccumulation kinetics and tissue distribution of silver nanoparticles in zebrafish: the mechanisms and influence of natural organic matter. Ecotoxicol. Environ. Saf. 194, 110454. https://doi.org/10.1016/j.ecoenv.2020.110454.

Xie, H., Wei, C., Wang, W., Chen, R., Cui, L., Wang, L., Chen, D., Yu, Y.-L., Bai, L., Li, Y.-F., 2024. Screening the phytotoxicity of micro/nanoplastics through non-targeted metallomics with synchrotron radiation X-ray fluorescence and deep learning: Taking micro/nano polyethylene terephthalate as an example. J. Hazard Mater. 463, 132886. https://doi.org/10.1016/j.jhazmat.2023.132886.

Xiong, T., Zhang, S., Kang, Z., Zhang, T., Li, S., 2021. Dose-dependent physiological and transcriptomic responses of lettuce (lactuca sativa L.) to copper oxide nanoparticles—insights into the phytotoxicity mechanisms. Int. J. Mol. Sci. 22 (7), 3688. https://doi.org/10.3390/ijms22073688.

Xu, C., Sullivan, J.H., 2010. Reviewing the technical designs for experiments with ultraviolet-b radiation and impact on photosynthesis, DNA and secondary metabolism. J. Integr. Plant Biol. 52 (4), 377–387. https://doi.org/10.1111/j.1744-7909.2010.00939.x.

Yan, A., Chen, Z., 2019. Impacts of silver nanoparticles on plants: a focus on the phytotoxicity and underlying mechanism. Int. J. Mol. Sci. 20 (5), 1003. https://doi.org/10.3390/ijms20051003.

Yilmaz, M., Yilmaz, A., Karaman, A., Aysin, F., Aksakal, O., 2021. Monitoring chemically and green-synthesized silver nanoparticles in maize seedlings via surface-enhanced Raman spectroscopy (SERS) and their phytotoxicity evaluation. Talanta 225, 121952. https://doi.org/10.1016/j.talanta.2020.121952.

Young, B.J., Rizzo, P.F., Riera, N.I., Torre, V.D., López, V.A., Molina, C.D., Fernández, F.E., Crespo, D.C., Barrena, R., Komilis, D., Sánchez, A., 2016. Development of phytotoxicity indexes and their correlation with ecotoxicological, stability and physicochemical parameters during passive composting of poultry manure. Waste Manag. 54, 101–109. https://doi.org/10.1016/j.wasman.2016.05.001.

Zeyad, M.T., Kumar, M., Malik, A., 2019. Mutagenicity, genotoxicity and oxidative stress induced by pesticide industry wastewater using bacterial and plant bioassays. Biotechnol. Rep. 24, e00389. https://doi.org/10.1016/j.btre.2019.e00389.

Zhao, X., Xi, B., He, X., Li, D., Tan, W., Zhang, H., Wang, X., Yang, C., 2019. The impacts of metal ions on phytotoxicity mediate by microbial community during municipal solid waste composting. J. Environ. Manag. 242, 153–161. https://doi.org/10.1016/j.jenvman.2019.04.063.

Zilkah, S., Gressel, J., 1977. Cell cultures vs. whole plants for measuring phytotoxicity I. The establishment and growth of callus and suspension cultures; definition of factors affecting toxicity on calli. Plant Cell Physiol. 18 (3), 641–655.

Zuma, M., Arthur, G., Coopoosamy, R., Naidoo, K., 2023. Incorporating cropping systems with eco-friendly strategies and solutions to mitigate the effects of climate change on crop production. J. Agric. Food Res. 14, 100722. https://doi.org/10.1016/j.jafr.2023.100722.

CHAPTER 10

Petroleum hydrocarbons and phytotoxicity

Muhammad Imran Khan and Qammar Farooq
Institute of Soil and Environmental Sciences, University of Agriculture, Faisalabad, Pakistan

1. Introduction

Several sectors utilize petroleum hydrocarbons (PHCs) as a raw material for producing an extensive variety of goods. It is expected that PHC usage would have approached from 85 million barrels to 107 million barrels by the end of 2030 (Zoghi and Mafigholami, 2023). Globally, approximately 8.8 million metric tons (MMT) of oil are released into the aquatic environment (Dadrasnia and Agamuthu, 2013), around 90% of the oil spills in the marine environments are directly related to anthropogenic activities. The fast urbanization and mechanization in the world increased the extraction and consumption of PHCs and their derivatives in this century as tenfold as higher as previous century, and it also increased the chances of their spills in water environment causing toxic impacts on living organisms (Behera et al., 2020; Siddiqi et al., 2022). The PHCs can also enter the terrestrial environments through natural and anthropogenic activities and these may include extraction, refining, transportation, and underground leakage (Varjani, 2017; Sharma, 2022).

Severally reported PHCs incidents and their pollution around the globe make this issue of prime importance. Globally, the total volume of natural and anthropogenic oil spills is expected to be further increased in near future (Gardi et al., 2015; Adipah, 2019). In Nigeria, around 9–13 million barrels were spilled into the soil from 1958 to 2018 (Ite et al., 2013, 2018). Similarly, 60 million barrels of oil were spilled during the Gulf War, contaminating an area of around 49 km^2 (Diab, 2008). There are several spill incidents releasing substantial amounts of PHCs in the aquatic or terrestrial environment. For instance, in 2003, the Tasman Spirit tanker incident released 31,000 L (195 barrels) of crude oil in the Arabian Sea (Kahkashan et al., 2019) and in 2011 and 2017, NATO oil tanker accidents in Chaman and Ahmedpur Sharqia, Pakistan spilling 377 and 314 barrels of gasoline, respectively on roads (Manzoor, 2020). In addition to these accidental releases, oil-refining industries and automobile oil workshops are the other sources of hydrocarbon pollution, resulting in a huge amount of land contaminated with PHCs. Furthermore, in Europe alone, more than 3.0 million sites are reported to be contaminated by PHCs and related compounds (Adipah, 2019).

Most of the PHCs are persistent in nature due to their stability in various environmental conditions and are categorized as important environmental contaminants

Phytotoxicity of Organic Pollutants
ISBN 978-0-443-33110-7,
https://doi.org/10.1016/B978-0-443-33110-7.00008-0

(Song et al., 2021; Uddin et al., 2021). This ultimately decreases agricultural productivity and causes ecological imbalance (Saeki et al., 2009). The PHCs toxicity depends on its composition, and most of them cause severe short- and long-term impacts. The presence of PHCs in soil physically alters the soil structure, reduces porosity, and increases soil compaction, resulting in reduced plant growth and development, and this reduction could be further exacerbated due to inappropriate soil pH, poor nutrient assimilation and reduced microbial activities. To overcome PHCs contamination in agricultural land, a variety of remediation approaches including physical excavation followed by landfilling, incineration, soil vapor extraction, photovolataization, thermal treatments, chemical oxidation and stabilization, and biological remediation and phytoremediation, have been adopted throughout the world (Chen et al., 2020; Khan et al., 2020; Ali et al., 2023, 2024b). However, physical and chemical methods, due to their higher cost, labor requirement, by-product formation and eco-toxic nature, are not considered appropriate for the alleviation of phytotoxicity and removal of PHCs from the soil and water environments. For instance, the physical excavation is a costly, labor-intensive, and time-consuming process (Adipah, 2019). Although chemical remediation is preferred by government authorities and scientists due to the speedy recovery of soil or aquatic environments, this method also causes harmful impacts to terrestrial organisms and beneficial soil microbes by producing toxic end-products. In chemical stabilization, binding agents like lime and/or cement are added to the soil to reduce the mobility of pollutants despite their complete removal (Ortega-Calvo et al., 2013), thus, this also does not provide permanent elimination of pollutants from the site.

On the other hand, bioremediation is an ex-situ or in-situ technique in which microbes and plants are used for the breaking down of contaminants in soil or aqueous environments. In in-situ bioremediation, microbes and plants are applied on polluted sites to reduce the existing levels of PHCs contamination; however, the ex-situ bioremediation involves the excavation of the contaminated soil and their suitable treatment at a certified treatment facility. The bioremediation of hydrocarbons can be affected by several biotic, abiotic and environmental factors such as nutrients, moisture, and temperature (Sayed et al., 2021). Phytoremediation is a method of using plants to remove or break down inorganic and organic contaminants from the soil or water medium in a comparatively cost-effective and eco-friendly way. In phytoremediation, the pollutants are either degraded in the root zone or can be assimilated by plants and transported to their above-ground biomass, where they can be volatilized or decomposed (Chen et al., 2020; Khan et al., 2020). Suitable phytotoxicity alleviating and remediation strategies should be adopted for the clean-up of PHCs contaminated soil and its subsequent use for agriculture farming because soil pollution by PHCs is a global concern due to the scarcity of farmland and the need for the production of more food on existing land to overcome food crises (Ali et al., 2023). This chapter provides comprehensive information on PHCs, their composition, sources, fate, phytotoxicity, and remediation.

2. Petroleum hydrocarbons

Crude oil is a blackish flammable substance that is produced naturally below the earth's surface by the decomposition or breakdown of dead or decaying animals by microbes under high temperatures and pressure in a limited supply of oxygen (Mohanta et al., 2024). Crude oil has been classified as "light or heavy" based on its density (Tang et al., 2012). Similarly, the presence of a higher amount of sulfur in crude oil produces "sour crude oil", while with a lower sulfur content, it is recognized as "sweet crude oil". Normally, PHCs contain carbon and hydrogen as their main ingredients; however, they also contain a substantial amount of sulfur, nitrogen, and oxygen. They are the main ingredients in several well-known fuels, including motor oil, kerosene, diesel, petrol and various types of jet fuel. Moreover, PHCs also contain other organic compounds in addition to hydrocarbons, some of which are organometallic, particularly those that are complex forms of nickel and vanadium (Chen et al., 2020). The detail of PHCs composition is given below.

2.1 Physicochemical properties of PHCs

The PHCs are organic compounds made up of C and H and contain residues of nitrogen, sulfur, and some heavy metals in refined products. Saturated and aromatic fractions of PHCs constituted 80% of crude oil, while the remaining 20% comprised resins and asphaltenes (Liu et al., 2020; Kuppusamy et al., 2020). The majority of PHCs, such as gasoline and kerosene, are transparent or pale liquids that may be easily vapourized and burnt. Others, such as mineral oil and asphalt, exhibit a viscous or dark-colored liquid or semi-solid state; they do not undergo evaporation without significant heating (Selvam and Thatheyus, 2018). The VOCs (volatile organic compounds), which include BTEX (benzene, toluene, ethylbenzene, and xylene) and their alkylated derivatives, are less water-soluble and less volatile than the aliphatic fractions of PHCs (Bongiorno et al., 2022). The compounds obtained by destructive distillation of crude oil, such as PAHs, phenols, phthalates, and pyridine, are included in semi-volatile fractions of PHCs (Butkovskyi et al., 2017; McLaughlin et al., 2020). However, resins, the high molecular weight compounds containing numerous functional groups with nitrogen, sulfur, oxygen, and trace amounts of metals (Kuppusamy et al., 2020; Abouee Mehrizi et al., 2019), are generally polar; that's why they can only be dissolved in polar solutions. Similarly, like resins, asphaltenes also contain various functional groups and are colloidally dispersed in aromatics and saturated fractions of PHCs (Varjani, 2017). More information on the solubility and reactivity of different PHCs is provided below.

2.1.1 Molecular weight and solubility

The amount of a solute dissolved in a particular solvent is referred to as its solubility. According to Amodu et al. (2013), the extent of microbial uptake and hydrocarbon breakdown is influenced by the hydrocarbons' water solubility. Temperature, the chemical composition of the contaminants, and the polarity of the material all have an impact on the kinetic and equilibrium conditions, which affect how quickly a specific molecule dissolves in water from a petroleum oil-polluted environment (Varjani and Upasani, 2017). The pollutants' high-water solubility may harm the bacteria that break down PHCs. It is a fact that lower molecular weight hydrocarbons are degraded more readily in water than those with greater molecular weights (Amodu et al., 2013). For example, although the shorter-chain alkanes with high water solubility can easily be degraded by microbes, they become more dangerous for microbes that break down PHCs. Due to their intermediate water solubility, microorganisms can easily biodegrade the intermediate-length n-alkanes (C10– C25) from various petroleum fractions. However, due to their hydrophobic nature, long-chain alkanes (C25–C40) are not readily soluble in water and are hence difficult to degrade via microbial actions. Branched-chain alkanes and cycloalkanes break down more slowly than their linear alkane counterparts. Furthermore, aromatics become easier to dissolve in water as the amount of alkylation in benzenes and PAHs decreases (Varjani and Upasani, 2017). In contrast, the highly condensed aromatics, cycloparaffinic structures, and asphaltic materials are extremely resistant to microbial degradation because of their higher molecular weight and boiling temperatures (Amodu et al., 2013).

2.1.2 Bioreactivity of hydrocarbons

As PHCs are hydrophobic, they must dissolve before being up taken or broken down by microorganisms (Sajna et al., 2015). The absorption from mineral and clay sorbents observed in the polluted soils influences the solubility and subsequent mineralization of PHCs.

Surfactants are frequently utilized for enhancing the surface area and wettability of hydrophobic materials to improve the breakdown of PHCs (Saeki et al., 2009). As a result, surfactants can accelerate the microbial degradation of pollutants. The initial phase of the decomposition of aliphatic and aromatic hydrocarbons generally involves oxidation processes facilitated by cell surface-attached oxygenases (Rosenberg and Rosenberg, 1985). Microbes can interact with hydrophobic PHCs in two ways: Biosurfactant-mediated access, where cells come in connection with emulsified hydrocarbons, or interfacial access, when cells come into direct contact with the hydrocarbons (Kaczorek et al., 2011; Ron and Rosenberg, 2014). It has been observed that hydrophobic fimbriae, fibrils, outer membrane lipids, proteins, and other small cell-surface compounds, such as gramicidin S and prodigiosin, may aid in the attachment of contaminants to microbial cells. However, adhesion can be inhibited by anionic exopolysaccharides and other bacterial capsules, for instance, *Acinetobacter* sp. RAG-1 grow strongest on C16 when there are fimbriae on the cell surface. A bacterial strain may adhere to hydrophobic surfaces with the aid of these fimbriae, which the cell can subsequently absorb and metabolize (Rosenberg and Rosenberg, 1985). The surfactants could encapsulate hydrophobic hydrocarbons, enable the complex to pass through cell membranes and carry out cytoplasmic metabolism. Most bacteria that break down PHCs produce and disperse biosurfactants with a range of chemical characteristics that allow hydrophobic materials to be emulsified easily (Ron and Rosenberg, 2014).

2.2 PHCs and their chemical composition

Crude oil is transported to refineries after being extracted from the earth, where it undergoes processing to produce a wide range of products. PHCs primarily consist of carbon and hydrogen in different proportions; however, they also contain trace amounts of sulfur, oxygen, nitrogen and heavy metals (Binhazzaa, 2024). Different types and constituents of PHCs are shown in Fig. 10.1.

Most of the constituents of crude oil are saturated hydrocarbons without double bonds. They are divided into cycloalkanes and alkanes paraffin according to their chemical makeup. In aromatic hydrocarbons, one or more aromatic rings are commonly replaced with alkyl groups (Meckenstock et al., 2016). In contrast to the saturated and aromatic fractions, resin and asphaltenes are distinguished by their incorporation of polar non-hydrocarbons (Xu et al., 2024). Because resins and asphaltenes include a considerable number of sulfur, nitrogen, and oxygen atoms, their carbon structures are extremely complex and mainly unknown (Hsu and Robinson, 2024). The distinct chemical behavior of a component determines how quickly it decomposes. The interior portion

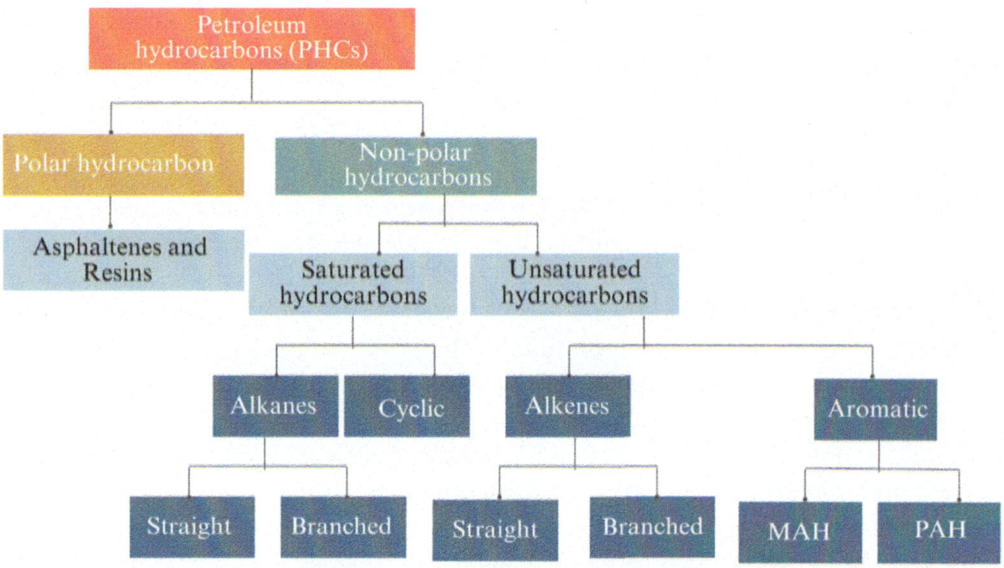

Figure 10.1 ***Types and constituents of petroleum hydrocarbons.*** Chemical composition of PHCs. MAH: Monoaromatic hydrocarbons; PAH: Polycyclic aromatic hydrocarbons.

of the oil is composed of asphaltenes, which have a greater molar mass component (Varjani and Upasani, 2017). The physicochemical properties of PHCs are given in Table 10.1.

2.2.1 Aliphatic

N-alkanes, cyclo-alkanes, naphthenes, iso-alkanes, steranes, and terpenes are typical examples of aliphatic hydrocarbons with linear or branching open-chain structures that can be both saturated or unsaturated (Rahman et al., 2003). Low molecular weight aliphatic hydrocarbons (C8–C16), medium molecular weight aliphatic hydrocarbons (C17–C28), high molecular weight aliphatic hydrocarbons (greater than C28), and gaseous alkanes are the four molecular weight groups of n-alkanes (Varjani and Upasani, 2017).

2.2.2 Aromatics

The hydrocarbons having ringed structures are commonly referred to as aromatic hydrocarbons, which include monocyclic aromatic hydrocarbons (MAHs) like BTEX (benzene, toluene, ethylbenzene, and xylenes) and polycyclic aromatic hydrocarbons (PAHs) (Farhadian et al., 2008). The phrase "low molecular weight" or "light PAHs" refers to PAHs having cyclic rings that form multiple hexagon chains with double bonds, such as phenanthrene, two-ringed naphthalene, and three-ringed anthracene. PAHs having four or more rings are considered as heavy or high molecular weight hydrocarbons. Benzo[a]pyrene, having five rings, fluoranthene, and pyrene and chrysenes with four rings are a few examples (Wilkes et al., 2016).

2.2.3 Resins

Resins comprise a range of polar functional groups, including nitrogen (N), sulfur (S), and oxygen (O), as well as trace metals, including nickel (Ni), vanadium (V), and iron (Fe). Non-crystalline substances known as resins dissolve in oil (Ugochukwu et al., 2013). The aromatic compounds that make up resins have long alkyl chains and are soluble in n-heptane and n-pentane. These substances have a chemical similarity to surface-active substances found in crude petroleum and serve as peptizers (Jada and Salou, 2002).

2.2.4 Asphaltenes

Polar functional groups are common in asphaltene-like resins. Large, complex, dark-brown molecules distribute colloidal throughout them, making them saturated and aromatic (Ugochukwu et al., 2013). Asphaltenes are viscous and have a large molecular weight. The aforementioned compounds' stability in light aromatic hydrocarbons like benzene and toluene accounts for their partial resistance to biodegradation. Polycyclic groups that are commonly replaced with alkyl groups are what define these hydrocarbons. Peptidizing agents such as resins maintain asphaltenes in suspension, increasing the shelf life of crude oil (Varjani and Upasani, 2017).

Table 10.1 Physiochemical properties of PHCs.

Compound	Gasoline (C_8H_{18})	Diesel	Light crude	Heavy crude	Intermediate fuel oil	Heavy fuel oil	References
CAS number	8006–61-9	68,476–30-2	8002–05-9	-	-	-	ATSDR (1999); Marathon Petroleum (2012)
Molecular weight (g/mol)	114.2285	150–250	300–1000	300–1000	300–1000	300–1000	ATSDR (1999); Jewell et al. (1974); Boduszynski et al. (1981)
Viscosity at 15°C (mPa)	0.5	6.5	5–50	50–50,000	1,000–15,000	10,000–50,000	Dupuis and Francisco (2015)
Density (g/cm^3)	0.7–0.8	0.84	0.78–0.88	0.88–1.00	0.94–0.99	0.96–1.04	Dupuis and Francisco (2015)
Physical state	Liquid	Liquid	Liquid	Liquid	Liquid	Liquid	ATSDR (1999)
Color	Colorless to pale yellow	Colorless to brown	Black	Brown or black	Black	Black	ATSDR (1999); U.S. EPA (2024)
Organic solvent solubility	Absolute alcohol, ether, chloroform, benzene	Dimethyl sulfoxide (DMSO), N-Methylpyrrolidone (NMP), and Dimethyl formamide (DMF)	Hexane	Aromatic, chlorinated and ketones	-	-	ATSDR (1999); Archer and Stevens (1977)
Henry's law constant at 20°C	4.8×10^{-4} m^3/mol	5.9×10^{-5}–7.4×10^{-5}	-	-	-	-	ATSDR (1999), Air force (1989)
Octanol-water partitioning coefficient (K_{ow})	2.13–4.87	3.3–7.06	-	-	-	-	ATSDR (1999)
Vapor pressure (kPa)	61.9 at 60°C 69.0 at 56°C 79.0 at 51°C 93.0 at 47°C 103 at 41°C	2.82–3.51	-	101	>0.68	-	ATSDR (1999)
Flash point (°C)	−46	58	10–30	15–30	25–30	25–35	Dupuis and Francisco (2015)

2.3 Sources of PHCs

The water and soil environments are highly polluted by the different crude oil processing, transportation and storage activities of PHCs (Song et al., 2021; Uddin et al., 2021). Throughout the world, it is common for crude oil to spill during offshore oil drilling, transportation, and transfer to onshore and cause pollution. The complicated composition of crude oil or PHCs, which includes alkanes, aromatic hydrocarbons, resins, and asphaltenes, is a fundamental issue and matter of concern for the environment. Because of their pollutant nature, chemical characteristics, and environmental mobility, they are extremely harmful to living organisms (Fatima, 2024). It is still quite difficult to comprehend their fate in the environment. To assess and estimate the pollutant's behavior and chemical characteristics as well as its short- and long-term effects on the polluted environment, PHCs must be classified (Keramea et al., 2021). A detailed description of the sources of PHCs is given below.

The now widely accepted notion that petroleum is made from the remains of living things that were integrated into the sediments during their deposition is supported by geological and chemical evidence. The discovery that a broad boiling range of hydrocarbon distillate components are optically active is one piece of evidence supporting this theory. Additionally, several hydrocarbons with similar structures or extremely large pieces of the skeletal structures of particular compounds that are extensively distributed across plants and animals have been extracted from petroleum in recent years (Mair, 1964). As crucial parts of organic molecules, hydrocarbons have received a lot of attention due to their carcinogenic, mutagenic, and poisonous characteristics. Hydrocarbons are frequently reported to be carried to the sediments in aquatic environments because of their low water solubility and hydrophobic characteristics. The relative contributions of various sources, such as biogenic, diagenetic, petrogenic, and pyrogenic, are reflected in the hydrocarbons found in sediments (Zhang et al., 2007).

There are several examples of anthropogenic origins of PHCs from wide-point sources (such as manufacturing processes and incinerators) impacted by partial combustion. Smaller point sources (such as jet aircraft exhausts, vehicle pollution, tobacco, wood-burning stove smoke, and cigar smoke) include scattered sources. The other main sources of PHCs environmental pollution are petroleum oil leaks and tarry or creosote waste products (Haider et al., 2021). The PHCs are naturally extracted from fossil fuels and are frequently used for commercial purposes (Adelaja, 2015). Environmental pollution can occur from oil spills, unintentional leaks from petrol and oil tankers, and leaks that occur during loading, transportation, storage, and distribution. Human activities are also responsible for over 90% of the 1.7–8.8 million metric tons (MMT) of PHCs released into the environment each year (Tiwari, 2024). Production, transportation, storage, refinement, and consumption are frequently the first steps in the environmental movement of these PHCs, which then spread to the air, water, soil, and entire terrestrial and aquatic ecosystems. Fig. 10.2 illustrates the main pathway through which PHCs enter and contaminate the soil.

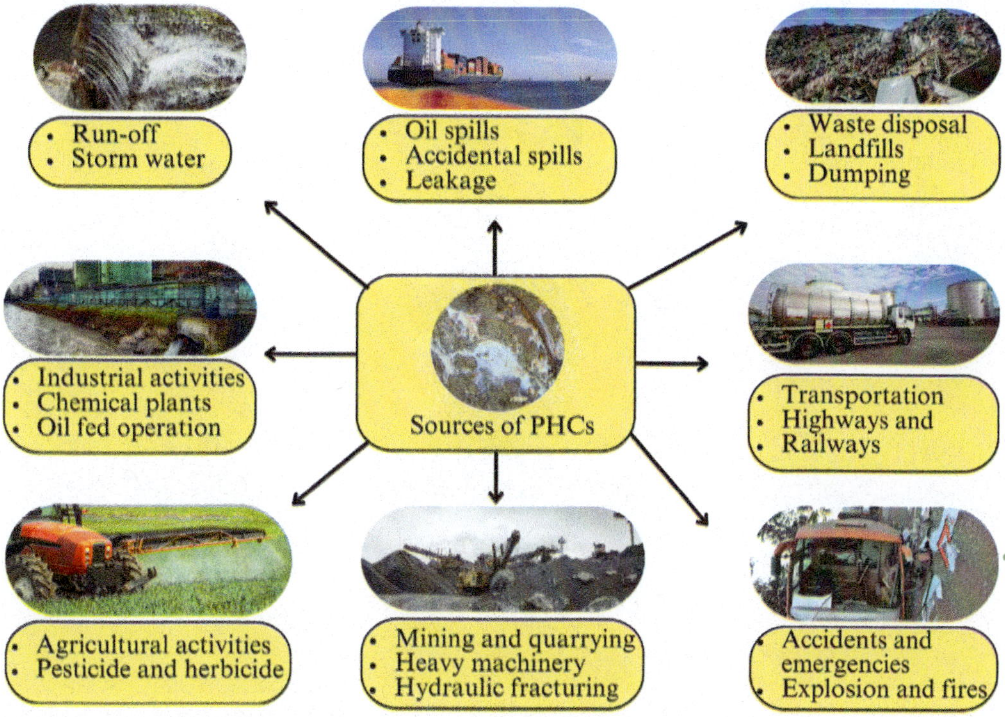

Figure 10.2 ***Sources of soil contamination by petroleum hydrocarbons.*** Different anthropogenic sources of PHCs.

2.4 Fate and dynamics of PHCs

A complex group of pollutants, PHCs are usually discharged into the surroundings as a mixture of PAHs, branched alkanes, n-alkanes, and cyclo-alkanes. With molecules ranging in size and structure from C1 (methane) to C40+, these diverse PHCs interact with the environment in different ways, frequently based on their molecular weight and characteristics (Truskewycz et al., 2019). In marine environments, distinctive hydrocarbon fractions of spilt oil are systematically subjected to evaporation, oxidation, dissolution (dissolved oxygen incorporates with oil and generates water-soluble compounds), accumulation (as persistent residues), and biological degradation by microbes. Oil floats are less dense than water when an oil spill occurs (Kuppusamy et al., 2020). The spill can also pollute the air since the most volatile hydrocarbons start to volatilize right after the oil spills (Ahmed and Fakhruddin, 2018).

After their release into the soil, the complex PHCs mixture separates into individual PHCs based on chemical properties (Fig. 10.3). The PHCs compounds having lower molecular weights, such as BTEX compounds (benzene, toluene, ethylbenzene, and the xylenes), are naturally occurring constituents of gasoline and diesel and are highly mobile in environments and more volatile and easily leach to the groundwater than

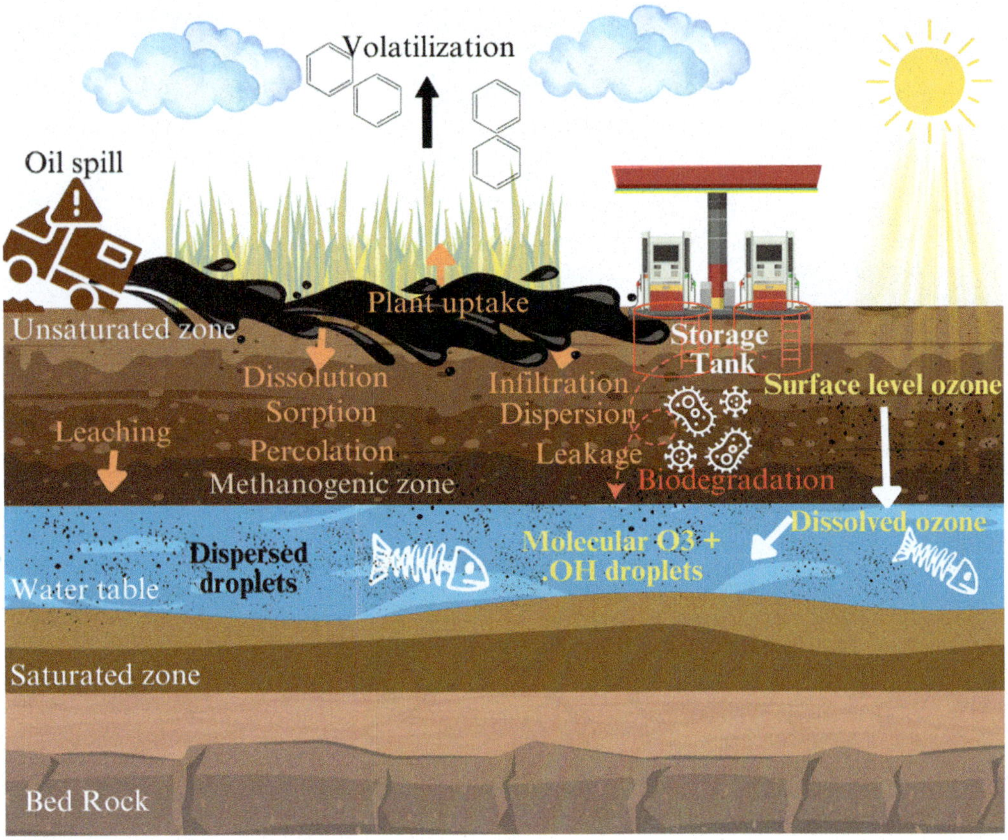

Figure 10.3 ***The fate of petroleum hydrocarbons in the environment.*** The possible fate of PHCs in the environment.

PHCs compounds of higher molecular weight (Adipah, 2019). After entering the environment, PHCs undergo several physical, chemical and biological processes (Fig. 10.3). PHCs can adsorb to organic materials and soil particles, volatilize to the atmosphere, and dissolve in water during the weathering process (Ossai et al., 2020) (Fig. 10.3). A broad variety of multiphase processes affect PHCs in soil, such as mass transfer between phases (volatilization, solution, and sorption), transport (diffusion, advection, and dispersion) between and within phases (aqueous and non-aqueous liquid, gas, and soil solids), and other natural attenuation processes like biodegradation, plant uptake, and metabolism (Balseiro-Romero et al., 2018).

3. Toxicity of PHCs

Hydrocarbon spills via natural or anthropogenic sources contaminate soil and water bodies, affecting the primary producers, such as land and aquatic plants, consumers, and decomposers. These contaminants reduce photosynthetic rates, impede growth, and eventually cause death (Haider et al., 2021). Since the primary producers are a baseline of any food chain, their loss could have an adverse negative impact on higher trophic levels and the entire food chain system. PHCs can also destroy symbiotic relationships among a variety of organisms through bioaccumulation and biomagnification. PHCs also disrupt the growth and reproduction cycles in living beings and cause changing behavior and eating patterns. These impacts lead toward environmental disturbances, declining biodiversity and populations, and impacts on human livelihoods (Sampaio et al., 2021).

3.1 Toxic effects of PHCs on soil biota

The PHCs affect the physical and chemical properties of soil, thereby affecting soil fertility and reducing plant yield. The PHCs also reduce soil porosity as they absorb organic matter from the soil and squeeze it together in the soil, resulting in the blockage of the soil pores and reduction of moisture content in the soil (Majeed et al., 2025). Water and oxygen deficiencies and a shortage of available forms of nitrogen and phosphorus are the main changes in soil properties due to contamination with PHCs. Furthermore, it has been reported that soil pollution by PHCs changes the physicochemical characteristics of topsoil and subsoil (Grifoni et al., 2020; Haider et al., 2021). PHC-contaminated soil may contaminate underground water, reducing water use and resulting in financial loss, environmental issues, and a decline in soil agriculture output. The toxicity of PHCs puts humans, animals, plants, and microorganisms in hazardous and vulnerable conditions (Mohammadi et al., 2020). The large quantities of PHCs in the soil can reduce or alter the diversity of the soil microbial environment and population (Gao et al., 2022) by preventing microbial growth, metabolism and activities beneficial to the plants. The only microorganisms that can be stimulated to grow are those that can adapt to the environment contaminated by high quantities of PHCs (Xu et al., 2018). Nevertheless, high PHCs concentrations raise the soil's carbon-to-nitrogen ratio, which enhances competition among microorganisms for nitrogen sources and promotes the quick growth of nitrogen-fixing bacteria (Ma et al., 2023), which can significantly contribute in improving soil fertility and subsequent plant productivity. Furthermore, since oil metabolites impact earthbound animals, they ought to reach the body through ingestion, inhalation, or retention (Smith et al., 2007). The primary way that most terrestrial organisms are exposed to PHCs in environments is by consuming hydrocarbons-contaminated soils (Khan et al., 2013). Therefore, terrestrial invertebrates such earthworms are generally exposed to the hydrocarbons in soil (Khan et al., 2011, 2012a, 2012b) and compared to higher-order consumers, species that ingest a lot of sediment-related invertebrates are typically and more seriously at risk of exposure to PHCs (Brooks et al., 2009).

3.2 Phytotoxicity of PHCs to plants

Although different plant species exhibit different levels of phytotoxicity of PHCs, higher concentrations of PHCs have been proven to be harmful to a wide variety of vegetation, including crop plants. Plant growth and development are significantly impacted by soil aeration and oil-clogged water infiltration, mainly reducing the root lengths, biomasses and leaf areas of many crops, including sunflower, maize, wheat, chickpea and mungbean (Steliga and Kluk, 2020; Ali et al., 2023, 2024a). Even after leaching and removing salts from the contaminated soil, Al-Joumaa (2009) found that barley seeds could not emerge after 14 days of exposure to the polluted soil. Deformations of leaf colors, H_2O_2-accumulation, oxidative stress, and even plant cell death are the effects of PHCs pollution (Al-Joumaa, 2009). There have been reports of delays in maize cell expansion, root-shoot growth inhibition, and leaf length reduction (Al-Joumaa, 2009; Steliga and Kluk, 2020). However, different plant species have varying levels of PHCs phytotoxicity. For example, certain crop species, such as wheat, rye, chickpea, and oats, can be used for phytoremediation, and they can withstand and accumulate these contaminants to a considerable degree (Al-Joumaa, 2009; Haider et al., 2021; Ali et al., 2023). More detail on the specific phytotoxic impacts of PHCs on specific attributes of the plant is provided below.

3.2.1 Phytotoxic effects of PHCs on seed germination and plant growth

The PHCs demonstrate considerable impacts on seed germination, seedling emergence and plant biomasses (Henner et al., 1999; Haider et al., 2021; Ali et al., 2024a). Some photo-induced byproducts of PHCs are more toxic than the actual compound because of their high-water solubility (Henner et al., 1999; Lin et al., 2020). Seed germination is considered a physiological index to examine the toxic impact of contaminants on plants. Although impacts on seed germination mainly depend upon the species of plant, type, concentration and solubility of hydrocarbons, the amount of organic matter and texture of soil are also very important factors that need to be considered in the degree of phytotoxicity of an organic pollutant (Czerniawska-Kusza et al., 2006; Korade and Fulekar, 2009). For example, low molecular weight hydrocarbons (LMW-HC) are found to be more toxic than heavy molecular weight hydrocarbons (HMW-HC) (Henner et al., 1999). Likewise, phytotoxicity could be higher in the early stages than in aged soil due to the loss of volatile compounds like LMW-HC (Adam and Duncan, 2002).

Recent studies reported that plant response against the stress of toxic chemicals such as PHCs and heavy metals is concentration- or dose-dependent (hormesis) (Calabrese and Blain, 2009). Hormesis is a biphasic dose, in which a low dose can stimulate while a high dose can cause inhibition (Calabrese, 2018). In many plants, hormesis can optimize and enhance many physiological processes, changes in signaling pathways, and chlorophyll content, resulting in increases in seed germination, early flowering and ultimately growth of the plants (Jalal et al., 2021). Due to the many cellular responses toward contaminants, plants build up resistance against other stresses such as plant pathogens.

In general, higher amounts of PHCs and related compounds cause detrimental impacts on plant biomass and crop yield (Kaur et al., 2017; Ali et al., 2023, 2024a). It is reported that due to the PHCs, it causes shorter root weight and length as compared to the normal soil (Maila and Cloete, 2002; Ali et al., 2024a). PHCs contaminants can disrupt cell membranes, lower the capacity of water retention, change the cell expansion processes, alter metabolism and disturb the action of the hormone (i.e., auxin) (Maila and Cloete, 2002). The presence of PHC contaminants can also cause oxidative damage and reduce the expansion of cotyledons, leading to a reduced yield of grains (Achuba, 2006; Bona et al., 2011). The effects of PHCs not only depend upon the chemical characteristics of the compound or tolerance index of the plant but also depend on soil and environmental conditions and how indigenous microbes interact with the root exudate and degrade the contaminant (Afegbua and Batty, 2018; F. Hussain et al., 2018; Martin et al., 2014; Ali et al., 2023).

3.2.2 Phytotoxic effects of PHCs on physiology

The PHCs are known to cause significant phytotoxicity to the physiological processes of plants (Haider et al., 2021; Ali et al., 2023, 2024a). For example, PHCs can induce phytotoxicity by lowering the chlorophyll content, inhibiting RuBisCO activity, and impairing photosystem II and electron transport rate due to the accumulation of reactive oxygen species (ROS). It also disrupts the TCA cycle and nitrogen metabolism by altering amino acid synthesis and enhancing nitrate reductase activity, reducing NADH synthesis and sugars to overcome the stress. Secondary metabolism is also adversely impacted by PHCs, leading to changes in polyphenol and antioxidant production. PHCs block water and nutrient uptake, reducing plant growth and biomass, and their phytotoxicity can be assessed through plant bioassays, biochemical tests, and advanced analytical techniques. The detail of the phytotoxic effects of PHCs on the physiological processes of plants is given below.

3.2.2.1 Phytotoxic effects of PHCs on plant metabolism and photosystems

The toxicity of PHCs can adversely affect plant metabolism and photosystems. Some adverse impacts and changes in plant physiological processes influenced by PHCs are shown in Fig. 10.4. PHCs cause a reduction in chlorophyll in both C3 and C4 plants, suggesting that plants are in extremely stressful conditions due to PHCs (Sivaram et al., 2020). It can inhibit carboxylation RuBisCO activity, decrease the photosynthetic rate and inhibit photosystem II activity, blocking the electron flow in photosystems II to I. This inhibition of electron flow is due to the mineralization of the D1 protein, which is due to the accumulation of ROS in hydrocarbon-exposed plants (Sivaram et al., 2018). As described earlier PHCs affect membrane permeability with the production of ROS, which causes a change in PSII (Kreslavski et al., 2017).

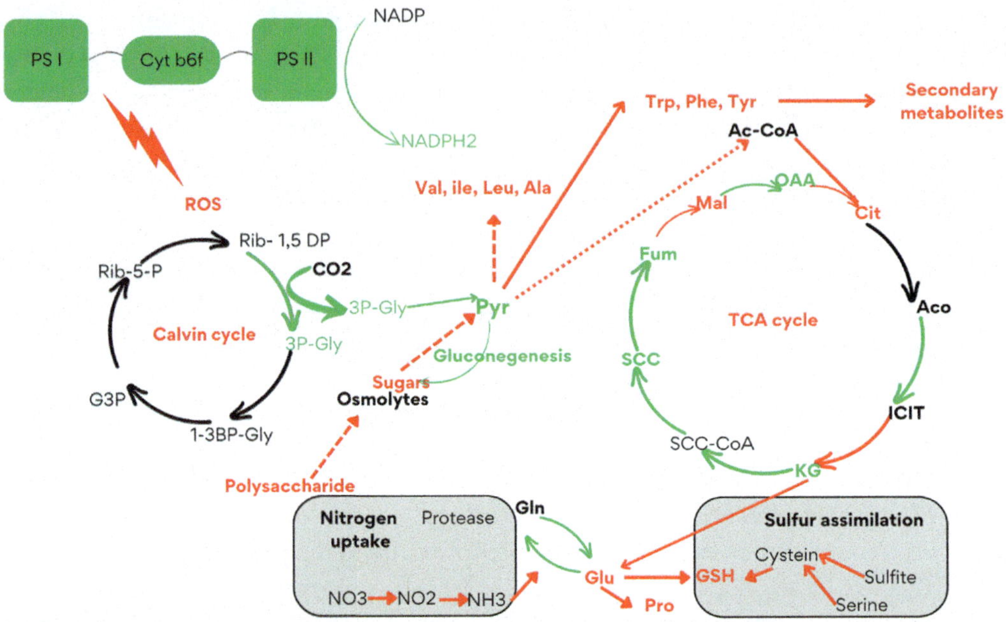

Figure 10.4 ***Different metabolic processes impacted by petroleum hydrocarbons.*** Enzymes and chemicals that are stimulated in the presence of contaminants are shown in red, whereas those that undergo inhibition are shown in green. Cit: Citrate; Aco: Aconitate; Rib-5-P: ribulose 5-phosphate; PSII: Phosphosystem II; 3P-Gly: 3-Phosphoglycerate; Pyr: Pyruvate; Mal: Malate; Ac-CoA: Acetyl-CoA; KG: α-ketoglutarate; G3P: Glyceraldehyde-3-phosphate; ICIT: Isocitrate; SCC-CoA: Succinyl-CoA; SCC: succinate; Cytb6f: Cytochrome *b*6f; OAA: Oxaloacetate; Fum: Fumarate; one to three BPGly: 1,3-Bisphosphoglycerate; Rib-1,5DP: Ribulose-1,5-bisphosphate.

3.2.2.2 Phytotoxic effects of PHCs on carbon metabolism

The PHCs can adversely impact the carbon metabolism in plants. For instance, phenanthrene (from PAH class of hydrocarbons) exposure to wheat affects the tricarboxylic acid cycle (TCA) (Shen et al., 2015). It generates the TCA intermediate concentration, enhancing malate and citrate while lowering α-ketoglutarate, oxaloacetate, pyruvate, fumarate, and succinate. The accumulation of malate and citrate is because of the induced expression of the pyruvate dehydrogenase, dihydrolipoyllysine-residue succinyltransferase, fumarate hydratase and ketoglutarate dehydrogenase and the inhibition of NADH synthesis, isocitrate dehydrogenase, GTP formation and malate dehydrogenase. Succinate reductase regulates the succinyl-CoA exposure synthase and the respiratory chain. Furthermore, it has been reported that phenanthrene can significantly cause alteration in the TCA cycle due to a decrease in pyruvate concentration (Zhan et al., 2015). Metabolic enzymes that are down-regulated in the presence of phenanthrene include NADH-forming enzyme glyceraldehyde-3-phosphate dehydrogenase and the adenosine kinase, which are involved in the formation of ATP (Shen et al., 2015). There is an inhibition of energy forming process reported i.e., NADPH and ATP in plant cells. In the presence of PHCs, inositol galactoside, sucrose, melibiose, and galactose are activated, which enhances the content of raffinose, D-mannose, galactinol, melibiose, and D-galactose in plant tissues. As mentioned earlier, the prevention of electron transport rate and photosynthetic activity also causes a decrease in water content and nutrient use efficiency in plants. Thus, the accumulation of sugar and its derivatives can demand a relatively higher amount of energy to stand stress conditions. Moreover, they can also play a role as an osmolyte protection of cellular structure to regulate osmoregulation under stress conditions caused by stress (Sivaram et al., 2019). Furthermore, it is reported that the PHCs cause a lowering of the content of hydroxypyruvate and certain amino acids, which are known to affect the gluconeogenic pathway (Sivaram et al., 2019; Xia et al., 2021).

3.2.2.3 Phytotoxic effects of PHCs on nitrogen metabolism

Plants facing PHCs stress can significantly influence nitrogen metabolism by enhancing the activity of nitrate reductase, and this enzyme is well known to play a crucial role in nitrate assimilation through the reduction of nitrate to ammonia (Zhan et al., 2015). This enhancement in NH_4 and the regulation of glutamate dehydrogenase (GDH_2) under PHCs stress could be described by the increase in the levels of glutamic acid (Xia et al., 2021) and the reduction in the levels of α-ketoglutarate (Shen et al., 2015). Proline gives a non-enzymatic response to ROS, which is derived from glutamate and accumulated under stress conditions (Sivaram et al., 2019; Xia et al., 2021). From pyruvate, branched-chain amino acids, including L-valine, L-tryptophan, L-tyrosine and D-phenylalanine, L-leucine and L-isoleucine can significantly upregulate content in plant cell tissues under PHCs stress (Sivaram et al., 2019; Xia et al., 2021) (Fig. 10.4). These amino acids are known to be precursors in important pathways for the biosynthesis of secondary metabolites. In phenanthrene stress, sulfur assimilation from sulfite, could enhance cysteine synthesis and serine transformation (Xia et al., 2021) (Fig. 10.4).

Nitrate absorption is facilitated by cysteine, which is a powerful antioxidant (Erdal and Turk, 2016). Glutathione antioxidant precursors are glutamate, glycine and cysteine (Kohlmeier, 2003). These enzymes are upregulated in the presence of PHCs and are involved in the glutathione cycle. For example, glutamate-cysteine ligase is involved in synthesis, glutathione reductase is involved in recycling, and glutathione-S-transferase is involved in the transfer of xenobiotics (Xia et al., 2021). Plant exposure to PHCs and aminoacyl-tRNA biosynthesis is involved in the biosynthesis of proteins. It could be due to the antioxidative enzyme protein stress and DNA repair enzymes (Xia et al., 2021).

3.2.2.4 Phytotoxic effects of PHCs on secondary metabolism

In recent studies, it has been reported that plant polyphenols are alleviated in response to PHCs stress (Fig. 10.4). It could be possible due to the stimulation of phenylalanine ammonia-lyase (PAL) and an increase in the concentration of precursor tryptophan, phenylalanine and tyrosine. Tyrosine is mineralized into 4-hydroxyphenylpyruvate dioxygenase. Tocopherols such as plastoquinone and vitamin E have precursor turinic acid and improve resistance against plant stress. Numerous secondary metabolites, such as auxins, alkaloids, glucosinolates and antitoxins, increase aromatic compound biosynthesis (Xia et al., 2021). In the phenylpropanoid pathway, phenylalanine is involved in the synthesis of several antioxidants and phytochemicals. It has been shown that there is a significant positive relationship between polyphenolic metabolism intensity and contaminant stress. Moreover, a significant reduction in phenolic and similar compounds such as flavanol, cinnamic alcohol, cinnamic aldehyde and cinnamic acid has been reported in plants exposed to higher concentrations of PHCs. Lipid peroxidation has also been reported under PHCs stress. Enhanced lipid contents, such as 13-hydroperoxy-9, 9-hydroxy-10E,12Z,15Z-octadecatrienoic acid, 11-octadecadienoic acid 13-HPODE, 14,15-dehydrocrepenynic acid, palmitaldehyde, 15Z-trienoic acid octadeca-11E,13E, and α-linolenic acid, have been observed in the plants exposed to the PHCs.

3.2.3 Mechanism of phytotoxicity of PHCs

In plants, PHCs toxicity causes abiotic stress, and this eventually enhances the synthesis of ROS and H_2O_2 that trigger the degradation of enzymes, cellular membranes, and nucleic acids that minimize plant growth and reduce the yield (Tariq et al., 2021). In addition, it has been reported that PHCs can block the pore spaces and obstruct the water and nutrient uptake from the soil, ultimately reducing the root, shoot length, and overall biomass of the plants (Odukoya et al., 2019; Ahmad et al., 2021). Similarly, the enzymatic activities and photosynthetic pigments were found to be reduced when plants were subjected to higher concentrations of PHCs (Song et al., 2011; Odjegba and Badejo, 2013; Al-Hawash et al., 2018). Another study revealed that PHCs toxicity could minimize chlorophyll synthesis and reduce photosynthesis in plants (Kaya et al., 2019). In addition, PHCs toxicity can also block the stomata and reduce the transpiration rate, stomatal conductance, and plant biomass (Odukoya et al., 2019) and can disturb the carbon-associated physiological mechanisms in plants (Naidoo and Naidoo, 2016). More recently, it has been found that the PHCs toxicity could reduce seedling emergence, growth, and yield attributes of mungbean plants mainly by interrupting the water osmosis, resulting in the reduced water uptake and lower moisture content in leaves (Cevher-Keskin et al., 2018; Ali et al., 2020). Phytotoxicity of PHCs can be accessed via plants, chemical and biochemical testing and advanced analysis; however, it is extremely challenging to properly characterize and identify the large variety of harmful chemicals released from PHCs. To investigate and assess harmful compounds from wastewater, plant bioassays have become widely used. Toxicants can be identified in toxicity studies by determining how sensitive plants are to various substances. Several plant bioassays can be used to screen and track environmental contaminants and assess how various chemicals or pollutants affect seed germination, physiology, growth, development and yield of plants (Table 10.2).

4. Phytotoxicity-mitigating strategies for PHCs

PHCs can significantly impact not only soil and ecosystem health but also plant growth and overall productivity by altering oxidative stress, lowering nutrient availability, and altering microbial communities (Ehis-Eriakha et al., 2024). Various strategies have been developed to mitigate the PHCs stresses on plants, including agronomic practices and biochemical and biotechnological approaches. These strategies help to enhance plant tolerance, improve soil structure, alleviate pollutant stress, and increase the degradation and removal of PHCs from polluted environments.

4.1 Role of agronomic practices in alleviating phytotoxicity of PHCs

Several agronomic practices, such as crop selection, genetic engineering, precision agriculture, irrigation and fertilizer management, are considered useful in reducing PHCs stresses on plants by mitigating overall phytotoxicity and contamination of pollutants. A brief description of these practices is given below.

4.1.1 Crop selection and genetic engineering

In recent years, several crop plants were successfully tested and reported to be efficiently growing under pollutant stress and removing hydrocarbons. These crop plants include wheat (Shahsavari et al., 2013), carpet grass (Tara et al., 2014), ryegrass (I. Hussain et al., 2018), Bermuda grass (Iqbal et al., 2019), mungbean (Ali et al., 2020) and maize (Ali et al., 2021). These plants, in association with applied microbes, were found to be successfully grown and increased the PHCs degradation in soil. Genetic engineering has the potential to modify microbes and plants to improve metabolic pathways that can break down a larger variety of pollutants (Vidali, 2001). Plants that have undergone genetic alteration may be more resistant to pollutants such as hydrocarbons and have better water absorption capacities and phytoremediation potential. More recently, it has been reported that transgenic plants could be another option that would be helpful in phytoremediation of pollutants (Ugwoha et al., 2020).

4.1.2 Irrigation and fertilizer management

PHCs serve as a carbon source for most of the soil microbes; however, polluted soils are typically deficient of important microbes and plant nutrients such as nitrogen (N), phosphate (P), potassium (K), sulfur (S) and trace elements (Ali et al., 2020). Therefore, essential nutrients can be added to soil to achieve a mass ratio of roughly 120:10:1 for C: N: P., which is also approximately the proportion of these nutrients in overall biomass. Biostimulation is the process of adding nutrients, and it can be done with

Table 10.2 Phytotoxicity assessment of PHCs.

Sr. No.	Plant species	Source of contamination	Contamination level	Parameters assessed	Key findings	References
1	Bush grass (*Calamagrostis epigejos*)	Crude oil	(0, 10 and 40 kg hm^{-2})	Enzymatic activity	Significant increase in hydrogen peroxidase, acid phosphatase, N-acetylglucosaminidase	Wei et al. (2025)
2	Rough cocklebur (*Xanthium strumarium*), Oats (*Avena sativa*), Wild carrot (*Daucus carota*), andHoney wort (*Cerinthe major*)	Diesel	(0%, 2%,5% and 5.0%)	Germination rate, root, shoot length/weight	Significant inhibition in growth (50%)	Dib and Ahmed (2024)
3	Greater club rush (*Scirpus grossus*)	Crude oil sludge BTEX	37,554 mg kg^{-1}	Stem, root weight/ length	Significant inhibition in growth (45%)	Sharuddin et al. (2024)
4	Water hyacinth (*Eichhornia crassipes*)	Diesel	2400 mg L^{-1}	Root/shoot length, fresh and dry weight, SPAD, PAR, Ft, YII, ETR, RWC	Significant inhibition in growth (62%), physiology (22%)	Ali et al. (2024b)
5	Wheat (*Triticum aestivum*)	Diesel	12,000 mg kg^{-1}	Root/shoot length, fresh and dry weight, Chlorophyll *a*,*b*,*c*, RWC,MSI, proline, antioxidant,	Significant inhibition in growth (18%–37%), physiology (14%–37%)	Ali et al. (2024a)
6	Chickpea (*Cicer arietinum*)	Diesel	12,000 mg kg^{-1}	Root/shoot length, fresh and dry weight, Chlorophyll *a*,*b*,*c*, RWC,MSI, proline, antioxidant,	Significant inhibition in growth (13%–29%), (12%–43%)	Ali et al. (2023)
7	Sunflower (Helianthus *annus*)	Poly aromatic hydrocarbons phenanthrene, anthracene, fluoranthene, and pyrene	10, 50 and 100 mg kg^{-1}	Enzymatic activity	Significant increase	Zhang et al. (2023)

8	Alfalfa (*Medicago sativa*), Tall fescue (*Festuca arundinacea*), and ryegrass (*Lolium perenne*)	Crude oil, petroleum ether	0.5 g kg^{-1}, 2.0 g kg^{-1}	Growth and physiological	Significant inhibition (32%)	Yuan et al. (2023)
9	Millet (*Sorghum bicolor* L.)	PHCs	0×10^4, 1.0×10^4, 2.5×10^4, 2.0×10^4 mg kg^{-1}	Root and shoot fresh and dry biomass	Significant inhibition	Ma et al. (2023)
10	Rice (*Oryza sativa*)	polycyclic aromatic hydrocarbons (PAH)	1 mg mL^{-1}	Chlorophyll *a*, *b*, leaf subcellular structure	Significant inhibition in physiology (32%)	Hu et al. (2023)
11	Maize (*Zea mays*)	TPH, PAH	2640 mg kg^{-1}, 6.84 mg kg^{-1}	Germination, morphological characters	Significant inhibition at higher contamination level	Wojtowicz et al. (2023)
12	Wheat (*Triticum aestivum*), ryegrass (*Lolium multiforum*), white clover (*Trifolium repens*), maize (*Zea mays*)	TPH	4%	Germination and fresh biomass	Significant reduction in germination rate alfalfa (15.1%), white clover (24.5), and biomass reduction in wheat (78%) and ryegrass (77%) white clover (62%) alfalfa (56%)	Yousaf et al. (2022)
13	Italian ryegrass (*Loklium multiflorum*)	PHCs	3.8%	Chlorophyll content, photosynthesis rate	Reduction in growth and physiological parameters	Hussain et al. (2022)
14	Barley	PHCs	27 g kg^{-1}	Biomass and physiological parameter	Decline in fresh weight of root (27%) and shoot (80%)	Kuzina et al. (2021)
15	Physic nut (*Jatropha curcas* (JC), and Khus	TPH, and TOG	22,666.67 ± 1166.67 mg kg^{-1} and 31,333.33 ± 2833.34 mg kg^{-1}	Agronomic and chemical	JC- good alternate mean	Nero (2021)

Continued

Table 10.2 Phytotoxicity assessment of PHCs.—cont'd

Sr. No.	Plant species	Source of contamination	Contamination level	Parameters assessed	Key findings	References
	grass (*Vetiveria zizanioides*)					
16	Longstamen rice (*Oryza longistaminata*), common wild sorghum (*Sorghum arundinaceum*), mexican sunflower (*Tithonia diversifolia*), and jaragua (*Hyparrhenia rufa*)	TPH	75 g kg^{-1}	Agronomic, chemical and physiological	Reduced mean percentage of TPH	Ruley et al. (2020)
17	Mung bean (*Vigna radiate*)	Diesel	5000 mg kg^{-1}	Agronomic, physiological and chemical	Decline in root length by (175%) and shoot length by (138%) of the plant. Also decrease the photosynthetic activity by (185%), stomatal conductance by (199%), transpiration rate by (205%) and chlorophyll contents by (57%)	Ali et al. (2020)
18	Maize (*Zea mays*)	PHCs, oil spills	252 mg kg^{-1} 7859 mg kg^{-1}	Agronomic and chemical	Decline in plant biomass by (41.2%)	Grifoni et al. (2020)
19	White mangrove (*Leguncularia recemosa*)	Crude oil	2.5, 5, 7,5, 10 g kg^{-1}	Agronomic, chemical, and biochemical	Decrease the plant biomass by (83.6%) and increase the oxidative stress in plant	Moradi et al. (2020)
20	Tall fescue (*F. arundinacea*)	PHCs	20 kg aliquot	Physiological, chemical, and eco toxicological tests	Significant decrease in soil toxicity	Steliga and Kluk (2020)
21	Boreau (*Hylotelephium spectabile*)	PHCs	65,986 mg kg^{-1}	PHCs determination, quantification, and sequencing	Degradation rate increased upto 37.6%–53.3%, PHC degrading bacteria rate increased due to AlkB and PAH-RHD_a genes	Cheng et al. (2019)

22	*Willow* (Salix spp.)	Waste Oil Pit	200,000 mg kg^{-1}	Contaminants quantification, biological tests, risk assessment	High survival rates, fungus showed future bioremediation ventures, biological test indicated significant decrease	Robichaud et al. (2019)
23	*Rice (Oryza sativa L.)*	Oxyflourfen	10,000 mg kg^{-1}	Chemical	Reduce the oxyflourfen by 63% and enhance the growth and biomass of soybean	Wu et al. (2019)
24	Common cordgrass (*Spartina anglica)*	PHCs	10, 30, and 50 g kg^{-1}	Agronomic, enzymatic, chemical and physiological	Significant increase	Zhen et al. (2019)
25	Italian ryegrass (*Lolium multiflorum*)	Crude oil	3.4%	Agronomic, chemical and physiological	Significant contribution in interaction with organic amendments	Hussain et al. (2018)
26	Perennial ryegrass (*Lolium perenn*)	TPH	91.73 ± 11.12 [gKg^{-1} D.M.]	Chemical and biochemical	Not significant increase	Masu et al. (2018)
27	Bread wheat *(Triticum aestivum)*	PAHs	>1000 mg kg^{-1}	Analytical and biochemical	Treatment of PAHs contents in soil by (70%)	Kong et al. (2018)
28	Desert wheatgrass (Agropyron desertorum) and bermuda grass (Cynodon spp.)	PHCs	0%, 2%, 4%, 12%	Agronomic and chemical	Cynodon spp performed better than Agropyron desertorum	Saraeian et al. (2018)
29	Ryegrass (*Lolium multiflorum Lam*)	PAHs	11.85 mg kg^{-1}	Biological, chemical and enzymatic	Lower the sorption of phenanthrene by (82.7%) and enhance the microbial biomass	Li et al. (2018)
30	Milk thistle (*Silybum marianum)*	Gas oil	0, 5, 10, 20, 40, and 80 g kg^{-1}	Growth parameters	Decline the seedling height (63%), Leaf Area (62.1%) and Plant biomass (54.1%)	Zamani et al. (2018)
31	Iridaceae species (*Iris dichotoma* Pall. and *Iris lactea* Pall.)	PHC contaminated saline-alkali soil	10,000 mg kg^{-1} 20,000 mg kg^{-1} 40,000 mg kg^{-1}	Biomass and chemical	Most effective for soils contaminated with 40,000 PHCs concentration	Cheng et al. (2017)
32	Ryegrass (*Lolium perenne)*	PHCs	500 g	Chemical and biochemical	Significant increase	Płociniczak et al. (2017)
33	Switchgrass (*Panicum virgatum*)	Motor oil and TPH	13.9 mL	Agronomic and chemical	Significant effect	McIntosh et al. (2017)

Continued

Table 10.2 Phytotoxicity assessment of PHCs.—cont'd

Sr. No.	Plant species	Source of contamination	Contamination level	Parameters assessed	Key findings	References
34	Striped Rattlepod *(Crotalaria pallida)*	Crude oil	10,000 g kg^{-1} - 100,000 g kg^{-1}	Chemical	Plant performed best at 60,000 ppm, after that the uptake decreased with increasing concentration.	Baruah et al. (2016)
35	Bermuda Grass *(Cynodon dactylon* (L.) Pers.)	PHCs	7.5%	Agronomic, chemical and physiological	Decrease crude oil up to 46.7% in low fertilizer level and 38.2% in high fertilizer level	Basumatary and Bordoloi (2016)
36	Mimosa (*Mimosa pudica.*), zinnia (*Zinnia elegans*), Gazania (*Gazania rigens*), and cypress vine (*Ipomoea quamoclit*)	Oil	2%	Agronomic, biochemical, and chemical	Significant increase in dehydrogenase activity	Ikeura et al. (2016)
37	Cock's foot (*Dactylis glomerata*)	PAHs and diesel fuel	100, 500, and 1000 mg/kg d m.	Agronomic, chemical and antioxidant	Significant increase in soil properties	Gałązka and Gałązka (2015)
38	Purple conflower ((*E. purpurea*), pulse beetle (*Callosobruchus chinensis*), fawn lily (*Erythronium oregonum*), India date (*Fire Phoenix*), and alfalfa (*Medicago sativa* L.)	PAHs	228–398 mg kg^{-1}	Chemical and agronomic	*Medicago sativa* L. and *Fire Phoenix* had significant effect on degradation	Xiao et al. (2015)
39	Lettuce (*Lactuca sativa)* and sweet potatoes (*Ipomoea batatas*)	Diesel	0–30 mL kg^{-1} 0–6 mL kg^{-1}	Agronomic, physiological and chemical	Decline in plant growth (10%), reduction in root development (64.8%)	Fatokun and Zaharare (2015)

mineral fertilizers, preferably fertilizers made of organic materials (i.e., farmyard manure, compost, activated sludge, etc.) (Minela and Zlata, 2025). Water is also one of the limiting factors for normal plant growth and development. The unavailability of an appropriate amount of water in the soil may enhance the adverse impacts of pollutants on plants and impair the normal activities of beneficial microbes. Since microorganisms require a sufficient amount of water to grow and develop, biodegradation cannot take place in soil that has less than a specific moisture content, while anaerobic processes predominate in soil that is saturated with water (Minela and Zlata, 2025).

4.1.3 Pesticide selection and application

The use of chemicals in agriculture is crucial to overcome damages caused by pests, diseases and weeds. There are many types of pesticides, e.g., organochlorine (Khan et al., 2024) and organophosphate (Farooq et al., 2023; Khan et al., 2025), which are potentially harmful to the environment. The application of such pesticides could enhance the phytotoxicity of other pollutants (e.g., hydrocarbons) through synergistic adverse impacts on plants. Biopesticides are certain kinds of pesticides that are produced from natural materials, including plants, microorganisms (bacteria, fungi, viruses, and protozoa), and animals, and are known to be effective in controlling pests in an eco-friendly way (Srijita Dutta, 2015). Biopesticides can be produced from living things, phytochemicals, microbial compounds, or metabolites (Chengala and Singh, 2017). *Pyrethrin, Datura metel, Mirabilis jalapa, Targetes minuta, Hyptis suaveolens, Lantana camara, Ryania speciose, Azadirachta indica*, and *Allium sativum* L. are examples of botanical pesticides that have been utilized effectively to control diseases and pests of common beans (*Phaseolus vulgaris* L.). These biopesticides are reported to be very effective in controlling bruchid beetle, aphids, bollworms, armyworms, cabbage loppers, potato aphids, thrips, and caterpillars (Karani et al., 2017).

4.1.4 Precision agriculture and organic farming

Precision agriculture (PA) is now considered a sustainable approach mainly due to its role in balancing environmental stewardship and agricultural productivity (Nath, 2024). The objective of sustainable development is to increase agricultural output while reducing negative environmental impacts. The utilization of cutting-edge technologies such as the internet, global positioning systems, geographic information systems, sensors, drones, and machine learning in PA has made it possible (Sun et al., 2024) to be effective in increasing crop yield and reducing environmental impacts and ecological footprints. Reduced application of fertilizers and pesticides through PA has environmental and human health advantages and thus could further improve agricultural sustainability and input efficiency (Shah and Razaq, 2022). Organic farming is an agricultural system that focuses on the use of natural and non-synthetic agricultural inputs such as green manure, farmyard manure, and compost in integration with other suitable farming techniques such as crop rotation and mixed cropping. Organic farming helps to increase soil carbon storage and decreases water pollution by pesticides by limiting nutrient leaching and soil erosion (Raimondo et al., 2021). The organic manures harboring beneficial microbes promote plant growth and mitigate environmental pollution (Khan et al., 2025). Organic manures also alleviate the phytotoxicity of pollutants by reducing their bioavailability (due to adsorption or sorption of pollutants) to plants. In short, organic farming, by minimizing synthetic inputs, could improve ecosystem health and play a significant role in sustainable agriculture (Rees, 2021).

4.2 Role of soil treatment in alleviating phytotoxicity of PHCs

The PHCs waste can be handled by numerous treatment methods, including physical, thermal, chemical, and biological approaches. Adsorption (Fu et al., 2021), membrane technology (Ahmad et al., 2021), landfill disposal (Butkovskyi et al., 2017; Mohammadi et al., 2020), thermal desorption (Al-Obaidi et al., 2017), Soil vapor extraction (Zhang and Zhu, 2024), air sparging (Koolivand et al., 2020), chemical oxidation (Checa-Fernandez et al., 2021), ultrasonic extraction, soil washing and solvent extraction (Ossai et al., 2020), and application of surfactants (Di Nino, 2017) and nanomaterials (Chen et al., 2025) are the most commonly used physicochemical methods for the soil treatment and removal of PHCs from the contaminated soils. However, these physicochemical techniques require higher cost, more labor consumption and aggravated risks to the environment due to the formation of toxic byproducts (Xu et al., 2018; Brzeszcz et al., 2020; Cui et al., 2020). From the last few years, biological options using microbes and plants, such as bioremediation and/or phytoremediation, are extensively used for the reclamation of PHCs polluted environments (Koshlaf and Ball, 2017; Kumari et al., 2018; Wu et al., 2020). Phytoremediation, using plants only or in combination with microbes, is believed to be a suitable method to clean PHCs contaminated environments due its low cost, high removal percentage, no secondary harmful by-products, and higher public acceptance, making phytoremediation one of the best remediation strategies (Suja et al., 2014; Xu et al., 2018; Almansoory et al., 2019; Cui et al., 2020). Details of these methods and strategies are given below.

4.2.1 Biological approaches for soil treatment

Bioremediation is one of the most eco-friendly, publicly acceptable and cost-effective technologies widely used for the remediation of organic pollutants (Ye et al., 2017; Koolivand et al., 2020; Ambaye et al., 2022). In bioremediation, various living organisms (particularly microbes and plants) are employed with different strategies for the removal of pollutants of concern. Combining plants with microbes further enhances the breakdown of pollutants. These plant or microbes-mediated treatment strategies not only remove the toxic chemicals from the polluted environment but also reduce the overall toxicity of the system. Potential bioremediation strategies for the removal of PHCs and related compounds are given below.

4.2.1.1 Microbial-mediated approaches for soil treatment

Microbial-mediated approaches such as Rhizodegradation or rhizoremediation is an in-situ plant-assisted biodegradation that takes place in the rhizosphere (Garrido-Sanz et al., 2019; Koolivand et al., 2020). Using metabolic or enzymatic processes, microbes transform or mineralize organic compounds during biodegradation, producing less hazardous byproducts that are then integrated into organic cycles that occur naturally (Haritash, 2020) (Table 10.3). The natural habitat around the 1–2 mm root zone containing 10–100 times more microbial population than the remaining soil is known as the rhizosphere (Naeem and Qazi, 2020). This rhizosphere area enhances the breakdown of contaminants because of enormous microbial counts (Rostami and Azhdarpoor, 2019). The plant roots secrete exudates such as amino acids, nucleotides, sugars, sterols, flavanone, and other organic acids to attract the microbes in the rhizosphere (Iqbal et al., 2019; Hassan et al., 2019). These exudates are composed of carbon and sometimes have structural similarities with organic contaminants, which is why microbes use them as a carbon source to proliferate themselves (Rohrbacher and St-Arnaud, 2016; Correa-García et al., 2018; Kumar et al., 2021). Similarly, the microbes provide nutrients by lowering the pH, water, and vitamins to the plants in addition to the breakdown of contaminants and providing resistance against pests, which becomes useful for productive plant growth (Desoky et al., 2020; Zhang et al., 2020). Biosurfactants are essential for the biodegradation of PHCs because they reduce the force that repels microorganisms from hydrocarbons and encourages interaction and mixing amongst them (Rizzo et al., 2018). Microbes participating in PHCs remediation have a high potential for producing biosurfactants, which also have a significant impact on the composting process (Varjani and Upasani, 2017). The microbial degradation of aliphatic and aromatic hydrocarbons is illustrated in Fig. 10.5. As described below, natural attenuation or microbial degradation of PHCs in the contaminated sites can be improved by adding nutrients (biostimulation), oxygen (bioventing), and engineered or specialized microbes (bioaugmentation).

LMW; Low molecular weight, HMW; High molecular weight, PAH; polycyclic aromatic hydrocarbons, TLC; Thin layer chromatogram, TPH: total petroleum hydrocarbons.

4.2.1.2 Soil treatment through biostimulation and bioaugmentation

Biostimulation, involving the addition of macronutrients such as nitrogen and phosphorous, is considered the simplest and economical bioremediation process (Soghandi and Salimi, 2024). In biostimulation, the growth of indigenous microbes is stimulated by adding limiting nutrients to severely polluted sites for rapid degradation of harmful and toxic contaminants such as PHCs. However, the concentration of added nutrients must be controlled, as it may cause a nutrient imbalance in the microflora. Bioaugmentation is a process in which exogenous or indigenous microbes (isolated single

Table 10.3 Table microbial degradation of petroleum hydrocarbons.

Sr. No.	Nature of study	Duration of experiment (days)	Source of spiking	Concentration of PHCs	PHCs degradation (%)	Bacterial sp. used in MC	Microbial identification tools	References
1.	Lab study	30	Crude Oil	-	65.5%	*Shewanella putrefaciens*	16S rRNA gene	Li et al. (2024)
2.	Pot trial, chickpea plant	90	Diesel	1.5%–3.0%	74% and 80%	BC1 (*Alcaligenes faecalis*, *Alcaligenes* sp., *Achromobacter denitrificans*, *Sphingobacterium spiritivorum*) BC2 (*Sphingobacterium spiritivorum*, *Achromobacter xylosoxidans*, *Stenotrophomonas* sp., *Alcaligenes faecalis* and *Stenotrophomonas rhizophila*)	16S rRNA gene	Ali et al. (2024c)
3.	Pot trial, wheat plant	90	Diesel	12,000 mg kg^{-1}	71% and 78%	BC1 (*Alcaligenes faecalis*, *Alcaligenes* sp., *Achromobacter denitrificans*, *Sphingobacterium spiritivorum*) BC2 (*Sphingobacterium spiritivorum*, *Achromobacter xylosoxidans*, *Stenotrophomonas* sp., *Alcaligenes faecalis* and *Stenotrophomonas rhizophila*)	16S rRNA gene	Ali et al. (2024a)
4.	Pot trial, water hyacinth	30	Diesel	2400 mg L^{-1}	72%	*Alcaligenes faecalis*, *Alcaligenes* sp. *Achromobacter denitrificans*, *Sphingobacterium spiritivorum*	16S rRNA gene	Ali et al. (2024b)
5.	Lab study	60	Benzo (a) pyrene	100 mg L^{-1}	75%	*Pseudomonas*		Goveas et al. (2023)
	Pot trial, Windmill grass, Harsh hakea	30	Diesel and GTX engine oil	1.5%	60%			Hoang et al. (2022)
6.	Pot trial, maize plant	90	Diesel	10,000 mg kg^{-1}	77%	*Bacillus* sp. MN54		Ali et al. (2021)
7.	Incubation study	28	Crude oil	10,000 $mgkg^{-1}$	78%	Four bacterial strains belonging to genera *Pseudomonas* and *Acinetobacter* (MC)	16S rRNA gene amplicon sequencing	Diallo et al. (2021)

8.	Pot trial, maize plant	90	Diesel	10,000 $mgkg^{-1}$	77%	*Bacillus* sp. MN54	Previously identified	Ali et al. (2021)
9.	Pot trial, mungbean plant	90	Diesel	5000 $mg\ kg^{-1}$	85%	*Enterobacter* sp. MN17 and biochar		Ali et al. (2020)
10.	Pot trial, mungbean plant	90	Diesel	5000 $mgkg^{-1}$	85%	*Enterobacter* sp. MN17 and biochar	Previously identified	Ali et al. (2020)
11.	Incubation study	60	Naturally contaminated soil	-	Aliphatic 86% and PAHs 70%	*Aquamicrobium, Enterococcus, Brevundimonas, and Alkaliphilus* (MC)	*Illumina MiSeq-Based Sequencing*	Brzeszcz et al. (2020)
12.	Incubation study	10	Crude oil	10,000 $mg\ kg^{-1}$	~100%	*Nocardia soli* Y48 and *Rhodococcus erythropolis* YF28-1 strains of *AcinetobacterPseudomonas*		Yang et al. (2019)
13.	Incubation study	26	Crude oil	10,000 $mg\ kg^{-1}$	~85%	*Bacillus, Pseudomonas,* Rhizobium		Xia et al. (2019)
14.	Incubation study with maize plant	60	Naturally contaminated soil	50 $mg\ kg^{-1}$ PAH	Phenanthrene 65% and pyrene 80%	*Pseudomonas*		Cavalcanti et al. (2019)
15.	Incubation study	-	Sediment sample	10,000 $mg\ kg^{-1}$	-	*Alcanivorax* and *Marinobacter* genera		Barbato et al. (2019)
16.	Incubation study	~450	Diesel, biodiesel	100,000 $mg\ kg^{-1}$	~90%	*Sphingobacteria*		Woźniak-Karczewska et al. (2019)
17.	Incubation study	16 d for LMW and HMW, PHCs	PAH polluted soil	PAH 97.63 $mg\ kg^{-1}$	LMW 52% and HMW 35% in 16 days and	*Pseudomonas, Methylobacillus, Nocardioides,* Methylophilaceae, *Achromobacter, Psedudoxanthomonas*		Lu et al. (2019)
18.	Incubation study	7	Crude oil	10,000 $mg\ kg^{-1}$	61%	*Alcanivorax*		Djahnit et al. (2019)
19.	Incubation study	30	Oily sludge	2,500–12,500 w/v	96% reduction in TPH	*Shewanallachilikensis, Bacillus firmus,* and *Halomonas hamiltonii*		Suganthi et al. (2018)

Continued

Table 10.3 Table microbial degradation of petroleum hydrocarbons.—cont'd

Sr. No.	Nature of study	Duration of experiment (days)	Source of spiking	Concentration of PHCs	PHCs degradation (%)	Bacterial sp. used in MC	Microbial identification tools	References
20.	Incubation study	8	Benzo[a]pyrene	50 mgL^{-1}	41%	*Aeromonas, acidaminiphila Ochrobactrum anthropi, salmonicid, Stenotrophomonas*, and *a salmonicida* (MC)	Identified was done by a Biolog automated system 2009	Aziz et al. (2018)
21.	Pot trial, wheat	120	Oil polluted soil	20,000 $mgkg^{-1}$	56.14%	*Pseudomonas*	-	Kumar et al. (2018)
22.	Incubation study	45	Crude oil	Naphthalene (10.0 mgL^{-1}), fluorene (1.9 mgL^{-1}), phenanthrene (3.5 mgL^{-1}) and benzo(b) fluoranthene (6.5 mgL^{-1})	Naphthalene (89.1%), Fluorene (63.8%), Phenanthrene 81% and benzo(b) fluoranthene (72.8%)	*Pseudomonas aeruginosa, Stenotrophomonas maltophilia, Ochrobactrum anthropi, Microbacterium esteraromaticum* and *Pseudomonas mendocina*	16S rRNA gene sequences were aligned and BLAST against the 16S rRNA reference database EzBioCloud	Kumari et al. (2018)
23.	Pot trial, rye grass	75	Crude oil	34,000 $mgkg^{-1}$	85% degradation with compost, biochar and microbial consortia	*Pseudomonas, Actinobacter, Stenotrophomonas*	-	Hussain et al. (2018a)
24.	Incubation study	20	Diesel	Formaldehyde (20,000 mgL^{-1})	90%	*Pseudomonas* A6, *Joostella* A8 and *Alcanivorax* A53	-	Rizzo et al. (2018)
25.	Incubation study	60	Light crude oil	TPH in Refinery reservoir soil (212 $mgkg^{-1}$) and Petroleum station soil (142 $mgkg^{-1}$)	92% of TPH	*Pseudomonas, Acinetobacter*	Illumina 16S rRNA gene sequencing	Omrani et al. (2018)

26.	Incubation study	22	PAH	Naphthalene, phenanthrene, anthracene and pyrene (250 mgL^{-1})	>30%	*Acidobacteria, Actinobacteria, Bacteroidetes*	16S rRNA gene analysis	Piubeli et al. (2018)
27.	Incubation study	35	Crude oil	20,000 $mgkg^{-1}$	84.2%	*Bacillus*	16s rRNA sequencing	Patowary et al. (2017)
28.	Field scale study, cowbeen	90	Crude oil	~12.5 mg g^{-1} soil	87.5%	Two bacterial strains, (*Bacillus* sp.) and two fungal strains, (*Geomyces* sp.)	16S rRNA gene sequence analyses	Maddela et al. (2017)
29.	Incubation study	90	Artificially TPH spiked soil	50,000 $mgkg^{-1}$	Bioaugmentation (86%), landfarming (70%) and natural attenuation (57%)	Genera *Pseudomonas, Acetobacter, Achromobacter*	Partial 16S rRNA gene sequences	Guarino et al. (2017)
30.	Incubation study	14	Crude oil	1000 mg kg-1 of a 27 mixtures of Phe, Pyr and BaP (1:1:1)	Phenanthrene (92%), Pyrene 64%, benzo(a) pyrene (65%)	*Bacillus, Pseudomonas, Enterobacter, Streptomyces*	-	Zafra et al. (2017)
31.	Incubation study	7	Crude oil	10,000 $mgkg^{-1}$	85%	*Bacillus*	16S rRNA gene through high throughput sequencing	Tao et al. (2017)
33.	Incubation study	7	Naphthalene, phenanthrene, pyrene and crude oil	Naphthalene (10 $mgkg^{-1}$), Phenanthrene (1000 $mgkg^{-1}$), pyrene (1000 $mgkg^{-1}$)and Crude oil (7000 $mgkg^{-1}$)	Naphthalene (80%), phenanthrene (30%), pyrene (56%) and crude oil (48%)	*Pseudomonas, Bacillus, Ochrobactrum* sp.	Molecular biological identification	Shen et al. (2015)

Continued

Table 10.3 Table microbial degradation of petroleum hydrocarbons.—cont'd

Sr. No.	Nature of study	Duration of experiment (days)	Source of spiking	Concentration of PHCs	PHCs degradation (%)	Bacterial sp. used in MC	Microbial identification tools	References
34.	Field study	~480	Dehydrated oil sludge	46,000 $mgkg^{-1}$	Petrofilic consortia (46%) and biosurfactants 85%	*Bacillus*, *Pseudomonas*, *Azobacter* and *Acinetobacter*	-	Helmy et al. (2015)
35.	Closed microcosm and open microcosm	90	Oily sludge	15,000 $mgkg^{-1}$ and 60,000 $mgkg^{-1}$	84% in closed and 79% in open microcosm	*Bacillus* *Cibi*, *Bacillus megaterium*, *Pseudomonas aeruginosaStenotrophomonas acidaminiphila*, and *Bacillus cereus*	Gene partial sequencing 16S rRNA	Cerqueira et al. (2014)
36.	Incubation study	30	Petrol, diesel and engine oil	Naturally oil-contaminated sites	100%	32 positive oil-degrading isolates of *Bacillus* and *Pseudomonas* sp.	-	Shankar et al. (2014)
37.	Microcosm tank experiment and Field Experiment	70	Crude oil contaminated soil, Natural condition	10,000 $mgkg^{-1}$	79% degradation in the microcosm experiment and 97% TPH degradation in field experiment	*AcinetobacterPseudomonas*	Previously identified	Suja et al. (2014)
38.	Incubation study	30	Naturally contaminated soil with TPH	63.4 $mgkg^{-1}$	94%	Genera *Actinobacteria*, *Alpha Proteobacteria* and *gamma-proteobacteria* and *Bacteroidetes* phyla	16S rRNA and Denaturing Gradient Gel Electrophoresis (DGGE) fingerprints	Gargouri et al. (2014)
39.	Incubation study	40	Oily sludge	10,000 $mgkg^{-1}$	Aliphatic fraction (90.7%) and aromatic fractions (51.8%)	*Bacillus* *Cibi*, *Bacillus megaterium*, *Pseudomonas aeruginosaStenotrophomonas acidaminiphila*, and *Bacillus cereus*	16S rRNA partial sequence	Cerqueira et al. (2011)

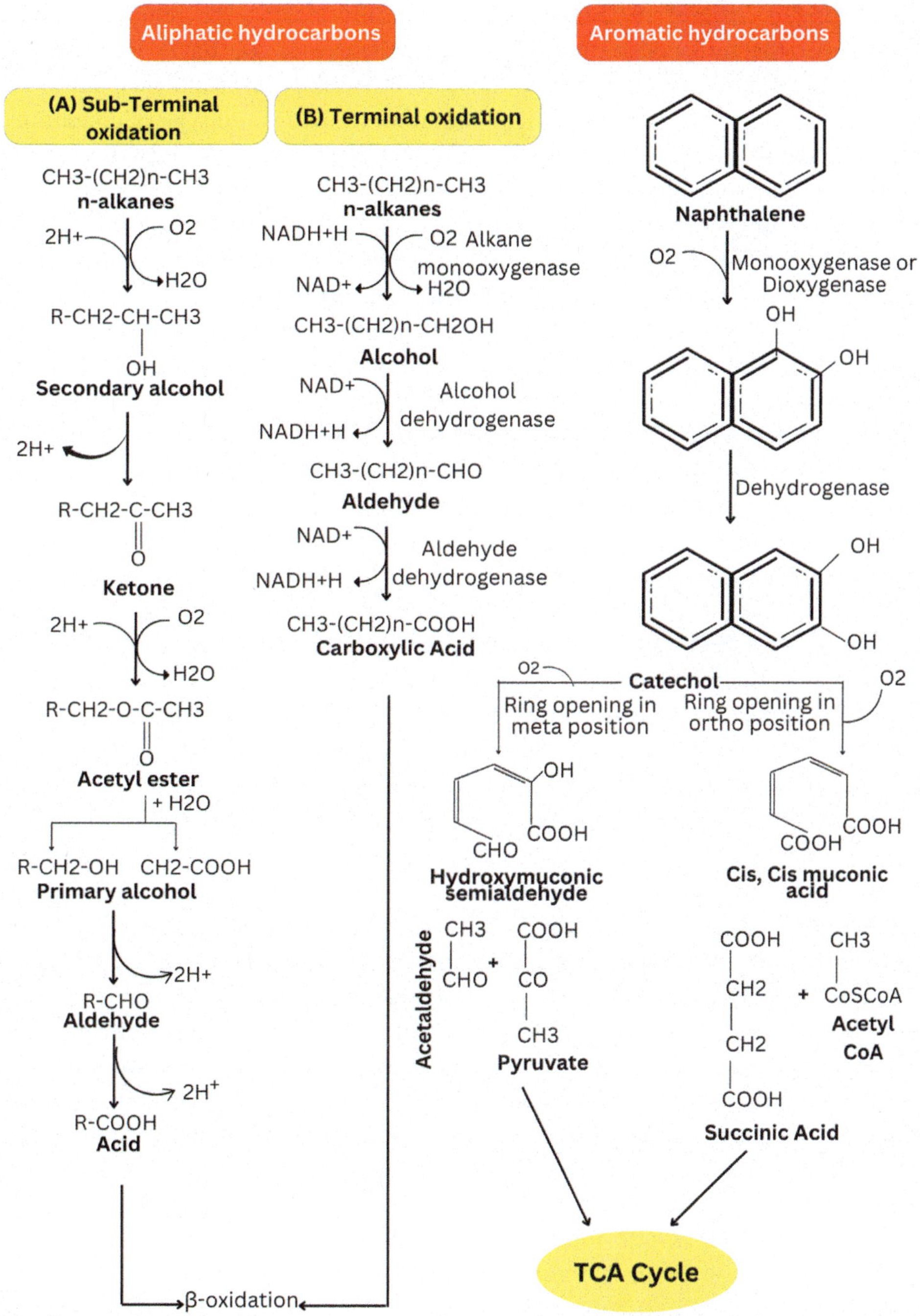

Figure 10.5 ***Microbial degradation pathways of aliphatic and aromatic hydrocarbons.*** The biodegradation pathways of PHCs.

strains or mixed culture) are added to soil (typically lacking the required pollutant-degraders) for the mineralization of PHCs (Alegbeleye et al., 2017; Vasilyeva et al., 2020). Certain factors are involved and need to be considered in the adaptation of the bioaugmentation strategy. For example, in addition to nature and type of contaminants, bioaugmentation is adopted if inadequate numbers of hydrocarbon-degrading microbes are present in the contaminated soil and the cost of bioaugmentation is lower than the alternate methods (Woźniak-Karczewska et al., 2019). Similarly, the successful implication of bioaugmentation not only requires the knowledge of contaminants type and levels but also the knowledge of suitable microbes for successful PHCs degradation (Suja et al., 2014; Varjani and Upasani, 2019). Bioaugmentation has become a major PHCs remediation strategy over the years and has been reported by several scientists around the world (Varjani, 2017; Ite and Ibok, 2019; Sayed et al., 2021; Haque et al., 2022). Bento et al. (2005) showed a four-time increase in the microbial communities, and around 75% and 72% degradation of light PHCs C12–C23 and heavy PHCs, respectively, was observed in bioaugmented soil rather than 45% PHCs degradation under natural attenuation. Gargouri et al. (2014) showed that soil contaminated with PHCs is bioaugmented with suitable microbes that cause 96% PHCs degradation after the end of the experiment. Ayotamuno et al. (2006) also reported 41%–53% in 14 d and 64%–85% in 42 d PHCs degradation with an enormous increase in colony-forming units CFU after the inoculation of microbial consortia in oil sludge contaminated soil.

The capacity of microbial inoculants to persist in PHC-polluted soil is critical for the success of the bioaugmentation technique (Varjani and Upasani, 2019). Bioaugmentation is affected by several factors including biotic such as quickly grow, easy to cultivate, internal competition for carbon and other nutrients between strains, tolerance to xenobiotics, and abiotic such as temperature, pH, humidity, organic matter, aeration, availability, concentration and type of contaminant, nutrients availability, soil physiochemical properties, soil type and amount and composition of the root exudates (Cycoń et al., 2017; Varjani and Upasani, 2019). In addition to bacteria, fungi can also be used as bioaugmenting agents. Mycoremediation is a method of bioremediation used to decontaminate the area by fungi. In mycoremediation, the mycelium secretes the two main building blocks of plant fiber, including extracellular enzymes and acids, which break down lignin and cellulose. These organic compounds are structurally similar to many organic contaminants in the long chain of carbon and hydrogen, making fungi successful pollutant degraders.

4.2.1.3 Plant-mediated soil treatment for PHCs

In plant-mediated soil treatment, also called phytoremediation, plants and their associated microbes are used to accumulate, sequester, degrade, stabilize, and detoxify toxic environmental pollutants. Phytoremediation techniques include several strategies for the removal of environmental contaminants from soil or water, including phytoextraction or phytoaccumulation, phytodegradation or phytotransformation, phytovoltalization, rhizodegradation or rhizoremediation, phytostabilization, rhizofiltration, and phytostimulation. For example, in the phytoaccumulation (also called phytoextraction) process, plants uptake and accumulate PHCs into their roots and shoots based on their composition and their availability in the root zone. Similarly, in rhizodegradation, contaminants are degraded in the rhizosphere employing microbial activity, which is enhanced by the presence of plant roots (also called phytostimulation). In rhizoremediation of PHCs, rhizosphere effect is employed to increase the abundance and activity of certain rhizospheric microbes, which in turn degrade or metabolize PHCs. In the phytostabilization process, chemical compounds produced by the plants or extensive roots immobilize contaminants by limiting their leaching from the soil. In phytotransformation (also called phytodegradation), pollutants are taken up by the plant and are degraded or transformed into a less or no toxic form in or outside the plant tissues. In the phytovolatilization process, plants take up water containing organic contaminants and transpire or release them into the air through their leaves. Thus, phytoremediation is a suitable technique for detoxification and removal of PHCs from both water and soil environments.

4.2.1.4 Microbe-assisted phytoremediation and plant growth promotion

The resistance of hydrocarbon pollutants in soil or water to microbial degradation is affected by the type, molecular weight, and quantity of benzene rings present in it (Ossai et al., 2020). In most cases, naphthalene is easily biodegradable, whereas PAHs having five or more benzene rings are degraded slowly (Aguelmous et al., 2019). Likewise, the PHCs fractions from C-10 to C-24 are easily biodegradable, while branched and long chains are difficult to degrade (Chen et al., 2020). After spillage, most of the volatile fractions evaporated quickly, and the remaining aliphatic fractions could easily reach the groundwater (Sayed et al., 2021). In general, aerobic biodegradation takes much less time than anaerobic biodegradation (Abbasian et al., 2015). Several factors can affect the bioremediation process (Fig. 10.6).

Several plant species from legumes and grasses have already been identified as suitable candidates for the successful removal of organic pollutants from soils. However, cereals or grasses are considered the most viable phytoremediation options due to their fibrous root system and short growth periods. Similarly, leguminous crops also have the potential to germinate under limited nutrient supply and can fix atmospheric nitrogen via nodules formation (Haider et al., 2021). Therefore, phytoremediation of PHCs using leguminous crops could be a more effective approach than cereals due to their ability to grow under limited nutrient supply. Various plants such as Chinese cabbage (*Brassica chinesis* L.), mustard (*Brassica parachinensis* L.), spinach (*Spinacea oleracia* L.), eggplant (*Solanum melongena* L.), broccoli (*Brassica oleracea* L.), raddish (*Raphanus sativus* L.) water spinach (*Ipomoeaaquatica* L.) chillies (*Capsicumannum* L.) amaranth (*Amaranthus tricolor*

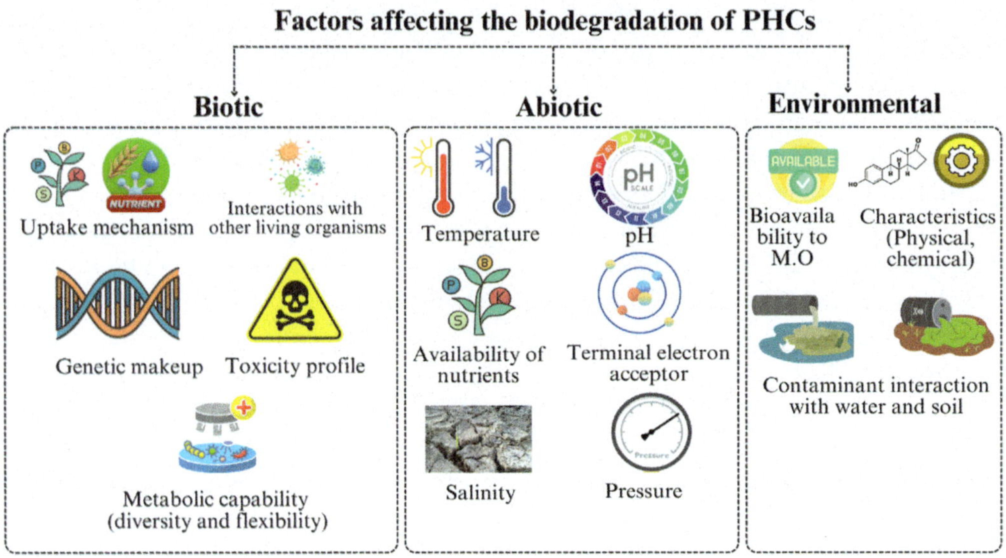

Figure 10.6 ***Factors affecting microbial degradation of PHCs.*** Biotic and abiotic factors impacting PHCs biodegradation.

L.), Italian ryegrass (*Lolium multiflorum* L.) common bean (*Phaseolus vulgaris* L.) and soybean (*Glycine* max L.) have already been successfully used for phytoremediation of PHCs.

As mentioned above, the remediation of PHCs can be improved by strengthening the rhizosphere of plants (Zhang et al., 2020; Ali et al., 2024a). More recently, it has been reported that the bacterial cultures have the ability to substantially mitigate the phytotoxicity of PHCs to cereals as well as leguminous plants and improve plant growth and productivity under PHCs contamination (Ali et al., 2024a, 2024b). Microbial activities such as the siderophore formation, N-fixation, P-solubilization, and ACC-deaminase activities further facilitate and improve the plant's health under the stress of PHCs contamination (Iqbal et al., 2019; Ali et al., 2024a). In turn, the plants promote bacterial activities in the root zone by providing aeration to aerobic microns and secreting higher amounts of organic acids, mucilage, hormones, sugars, amino acids, and vitamins (Rafique et al., 2015; Iqbal et al., 2019). The secretion of root exudates also lowers the pH of the rhizosphere and results in the enhancement of the bioavailability of nutrients as well as pollutants (Ali et al., 2024a). Due to the structural similarity of root exudates to PHCs, the microbes utilize PHCs as a carbon source, increasing microbial biomass and subsequently improving PHCs degradation (Guo et al., 2017; Ali et al., 2024b). Furthermore, microbial association with plants promote plant physiology and growth (e.g., stomatal conductance, chlorophyll pigments, evapotranspiration, root/shoot lengths and biomass of plants), eventually leading to the improvement in the rhizodegradation of PHCs (Iqbal et al., 2019; Ali et al., 2024a).

5. Conclusions

The PHCs are pollutants of grave environmental concern mainly due to their toxicity to all living organisms, including plants. The PHCs significantly impair soil quality, disturb terrestrial life, and reduce seed germination, physiology, growth and development of plants, resulting in the substantial reduction in the overall crop productivity elevating the food security crisis. Rhizoremediation or microbes-assisted phytoremediation using native plant species and indigenous microbial culture is an effective approach for the remediation of PHCs and the alleviation of their phytotoxicity to the crop plants. Furthermore, among several available options for treating PHCs-contaminated soil, rhizoremediation is technologically simpler, cost-effective, ecologically safer, socially acceptable, and aesthetically pleasant. However, there is still insufficient information on the availability of suitable PHCs-degrading microbial cultures and the selection of plants. These gaps can be filled through innovative research and collaborative efforts that can significantly advance the knowledge and practical implementation of microbe-assisted phytoremediation for PHCs-contaminated soil.

References

Abbasian, F., Lockington, R., Mallavarapu, M., Naidu, R., 2015. A comprehensive review of aliphatic hydrocarbon biodegradation by bacteria. Appl. Biochem. Biotechnol. 176, 670–699.

Abouee Mehrizi, E., Kermani, M., Farzadkia, M., Esarfili, A., Ghorbanian, M., 2019. Study of improvement of bioremediation performance for the degradation of petroleum hydrocarbons in oily sludge by a chemical pretreatment strategy. J. Mater. Cycles Waste Manag. 21, 1052–1062.

Achuba, F., 2006. The effect of sublethal concentrations of crude oil on the growth and metabolism of cowpea (*Vigna unguiculata*) seedlings. Environmentalist 26, 17–20.

Adam, G., Duncan, H., 2002. Influence of diesel fuel on seed germination. Environ. Pollut. 120, 363–370.

Adelaja, O., 2015. Bioremediation of Petroleum Hydrocarbons Using Microbial Fuel Cells. University of Westminster. A doctoral dissertation.

Adipah, S., 2019. Introduction of petroleum hydrocarbons contaminants and its human effects. Journal of Environmental Science and Public Health 3, 1–9.

Afegbua, S.L., Batty, L.C., 2018. Effect of single and mixed polycyclic aromatic hydrocarbon contamination on plant biomass yield and PAH dissipation during phytoremediation. Environ. Sci. Pollut. Control Ser. 25, 18596–18603.

Aguelmous, A., El Fels, L., Souabi, S., Zamama, M., Hafidi, M., 2019. The fate of total petroleum hydrocarbons during oily sludge composting: a critical review. Rev. Environ. Sci. Biotechnol. 18, 473–493.

Ahmad, A., Khan, W.U., Shah, A.A., Yasin, N.A., Ali, A., Rizwan, M., Ali, S., 2021. Dopamine alleviates hydrocarbon stress in *Brassica oleracea* through modulation of physio-biochemical attributes and antioxidant defense systems. Chemosphere 270, 128633.

Ahmed, F., Fakhruddin, A., 2018. A review on environmental contamination of petroleum hydrocarbons and its biodegradation. Int. J. Environ. Sci. Nat. Resour. 11, 1–7.

Air Force, 1989. Gasoline. In: The Installation Restoration Program Toxicology Guide. Volume 4. Contract No. DE-AC05840R21400. Wright-Patterson Air Force Base, OH. Document no. 65-1-65-46.

Alegbeleye, O.O., Opeolu, B.O., Jackson, V.A., 2017. Polycyclic aromatic hydrocarbons: a critical review of environmental occurrence and bioremediation. Environ. Manag. 60, 758–783.

Al-Hawash, A.B., Dragh, M.A., Li, S., Alhujaily, A., Abbood, H.A., Zhang, X., Ma, F., 2018. Principles of microbial degradation of petroleum hydrocarbons in the environment. Egypt. J. Aquat. Res. 44, 71–76.

Ali, M.H., Khan, M.I., Amjad, F., Khan, N., Seleiman, M.F., 2024c. Improved chickpea growth, physiology, nutrient assimilation and rhizoremediation of hydrocarbons by bacterial consortia. BMC Plant Biol. 24 (1), 984.

Ali, M.H., Khan, M.I., Naveed, M., Tanvir, M.A., 2024a. Microbe-assisted rhizodegradation of hydrocarbons and growth enhancement of wheat plants in hydrocarbons contaminated soil. Int. J. Environ. Sci. Technol. 21 (3), 3169–3184.

Ali, M.H., Sattar, M.T., Khan, M.I., Naveed, M., Rafique, M., Alamri, S., Siddiqui, M.H., 2020. Enhanced growth ofmungbean and remediation of petroleum hydrocarbons by *Enterobacter* sp. MN17 and biochar addition in diesel contaminated soil. Appl. Sci. 10 (23), 8548.

Ali, M.H., Muzaffar, A., Khan, M.I., Farooq, Q., Tanvir, M.A., Dawood, M., Hussain, M.I., 2024b. Microbes-assisted phytoremediation of lead and petroleum hydrocarbons contaminated water by water hyacinth. Int. J. Phytoremediation 26, 405–415.

Ali, M.H., Khan, M.I., Naveed, M., Tanvir, M.A., 2023. Microbe-assisted rhizoremediation of hydrocarbons and growth promotion of chickpea plants in petroleum hydrocarbons-contaminated soil. Sustainability 15, 6081.

Ali, M.H., Khan, M.I., Bashir, S., Azam, M., Naveed, M., Qadri, R., Bashir, S., Mehmood, F., Shoukat, M.A., Li, Y., 2021. Biochar and *Bacillus* sp. MN54 assisted phytoremediation of diesel and plant growth promotion of maize in hydrocarbons contaminated soil. Agronomy 11, 1795.

Al-Joumaa, 2009. Effect of pollution with petroleum on some soil characteristics and plant growth. Arab Univ. J. Agric. Sci. 17 (1), 267–275.

Almansoory, A.F., Hasan, H.A., Abdullah, S.R.S., Idris, M., Anuar, N., Al-Adiwish, W.M., 2019. Biosurfactant produced by the hydrocarbon-degrading bacteria: characterization, activity and applications in removing TPH from contaminated soil. Environ. Technol. Innov. 14, 100347.

Al-Obaidi, Sudad, H., Guliaeva, N.I., et al., 2017. Thermal adsorption processing of hydrocarbon residues. J. Sci. Technol. Res. 4 (1), 2349–4344.

Ambaye, T.G., Chebbi, A., Formicola, F., Prasad, S., Gomez, F.H., Franzetti, A., Vaccari, M., 2022. Remediation of soil polluted with petroleum hydrocarbons, and their reuse for agriculture: recent progress, challenges, and perspectives. Chemosphere, 133572.

Amodu, O.S., Ojumu, T.V., Ntwampe, S.K.O., 2013. Bioavailability of high molecular weight polycyclic aromatic hydrocarbons using renewable resources. Environmental Biotechnology-New Approaches and Prospective Applications 171.

Archer, W.L., Stevens, V.L., 1977. Comparison of chlorinated, aliphatic, aromatic, and oxygenated hydrocarbons as solvents. Ind. Eng. Chem. Prod. Res. Dev. 16 (4), 319–325.

ATSDR. Toxicological Profile for Total Petroleum Hydrocarbons (TPH). Atlanta (GA): Agency for Toxic Substances and Disease Registry (US). 1999. Table E-6a, Chemical Identity of Mineral Oila. Available from: https://www.ncbi.nlm.nih.gov/books/NBK595976/table/appe.tab16/.

Ayotamuno, J.M., Kogbara, R.B., Egwuenum, P.N., 2006. Comparison of corn and elephant grass in the phytoremediation of a petroleum-hydrocarbon-contaminated agricultural soil in Port Harcourt, Nigeria. J. Food Agric. Environ. 4, 218.

Aziz, A., Agamuthu, P., Alaribe, F.O., Fauziah, S.H., 2018. Biodegradation of benzo [a] pyrene by bacterial consortium isolated from mangrove sediment. Environ. Technol. 39, 527–535.

Balseiro-Romero, M., Monterroso, C., Casares, J.J., 2018. Environmental fate of petroleum hydrocarbons in soil: review of multiphase transport, mass transfer, and natural attenuation processes. Pedosphere 28, 833–847.

Barbato, M., Mapelli, F., Crotti, E., Daffonchio, D., Borin, S., 2019. Cultivable hydrocarbon degrading bacteria have low phylogenetic diversity but highly versatile functional potential. Int. Biodeterior. Biodegrad. 142, 43–51.

Baruah, P., Deka, S., Baruah, P.P., 2016. Phytoremediation of crude oil-contaminated soil employing *Crotalaria pallida* Aiton. Environ. Sci. Pollut. Control Ser. 23, 10595–10603.

Basumatary, B., Bordoloi, S., 2016. Phytoremediation of crude oil-contaminated soil using Cynodon dactylon (L.) Pers. Phytoremediation. Management of Environmental Contaminants 4, 41–51.

Behera, I.D., Basak, G., Kumar, R.R., Sen, R., Meikap, B.C., 2020. Treatment of petroleum refinery sludge by petroleum degrading bacterium Stenotrophomonas pavanii IRB19 as an efficient novel technology. J. Environ. Sci. Health Part A 56, 226–239.

Bento, F.M., Camargo, F.A., Okeke, B.C., Frankenberger, W.T., 2005. Comparative bioremediation of soils contaminated with diesel oil by natural attenuation, biostimulation and bioaugmentation. Bioresour. Technol. 96, 1049–1055.

Binhazzaa, Z. 2024. Hydrocarbons Are the Backbone of Sustainable Civilization.

Boduszynski, M.M., McKay, J.F., Latham, D.R., 1981. Composition of heavy ends of a Russian petroleum. J. Am. Chem. Soc. 26, 865–880.

Bona, C., Rezende, I.M.D., Santos, G.D.O., Souza, L.a.D., 2011. Effect of soil contaminated by diesel oil on the germination of seeds and the growth of Schinus terebinthifolius Raddi (Anacardiaceae) seedlings. Braz. Arch. Biol. Technol. 54, 1379–1387.

Bongiorno, D., Ceraulo, L., Di Stefano, V., Avellone, G., Indelicato, S., 2022. Total petroleum hydrocarbons (TPH) determination in surfactant stabilized marine water emulsions. ARKIVOC (Gainesville, FL, U. S.) 0–0.

Brooks, A.C., Gaskell, P.N., Maltby, L.L., 2009. Importance of prey and predator feeding behaviors for trophic transfer and secondary poisoning. Environ. Sci. Technol. 43, 7916–7923.

Brzeszcz, J., Kapusta, P., Steliga, T., Turkiewicz, A., 2020. Hydrocarbon removal by two differently developed microbial inoculants and comparing their actions with biostimulation treatment. Molecules 25, 661–672.

Butkovskyi, A., Bruning, H., Kools, S.A., Rijnaarts, H.H., Van Wezel, A.P., 2017. Organic pollutants in shale gas flowback and produced waters: identification, potential ecological impact, and implications for treatment strategies. Environ. Sci. Technol. 51, 4740–4754.

Calabrese, E.J., 2018. Hormesis: path and progression to significance. Int. J. Mol. Sci. 19, 2871.

Calabrese, E.J., Blain, R.B., 2009. Hormesis and plant biology. Environ. Pollut. 157, 42–48.

Cavalcanti, T.G., De Souza, A.F., Ferreira, G.F., Dias, D.S.B., Severino, L.S., Morais, J.P.S., De Sousa, K. A., Vasconcelos, U., 2019. Use of agro-industrial waste in the removal of phenanthrene and pyrene by microbial consortia in soil. Waste Biomass Valoriz. 10, 205–214.

Cerqueira, A., Souza, P., Marques, M., 2014. Effects of direct and alternating current on the treatment of oily water in an electroflocculation process. Braz. J. Chem. Eng. 31, 693–701.

Cerqueira, V.S., Hollenbach, E.B., Maboni, F., Vainstein, M.H., Camargo, F.A., Maria Do Carmo, R.P., Bento, F.M., 2011. Biodegradation potential of oily sludge by pure and mixed bacterial cultures. Bioresour. Technol. 102, 11003–11010.

Cevher-Keskin, B., Selçukcan-Erol, Ç., Yüksel, B., Ertekin, Ö., Yıldızhan, Y., Onarıcı, S., Kulen, O., Memon, A.R., 2018. Comparative transcriptome analysis of *Zea mays* in response to petroleum hydrocarbon stress. Environ. Sci. Pollut. Control Ser. 25, 32660–32674.

Checa-Fernandez, A., Santos, A., Romero, A., Dominguez, C.M., 2021. Application of chelating agents to enhance fenton process in soil remediation: A review. Catalysts 11 (6), 722.

Chen, W., Kong, Y., Li, J., Sun, Y., Min, J., Hu, X., 2020. Enhanced biodegradation of crude oil by constructed bacterial consortium comprising salt-tolerant petroleum degraders and biosurfactant producers. Int. Biodeterior. Biodegrad. 154, 105047.

Chen, Y.A., Liu, P.W.G., Whang, L.M., Wu, Y.J., Cheng, S.S., 2020. Effect of soil organic matter on petroleum hydrocarbon degradation in diesel/fuel oil-contaminated soil. J. Biosci. Bioeng. 129, 603–612.

Cheng, L., Zhou, Q., Yu, B., 2019. Responses and roles of roots, microbes, and degrading genes in rhizosphere during phytoremediation of petroleum hydrocarbons contaminated soil. Int. J. Phytoremediation 21 (12), 1161–1169.

Chen, X., Wang, S., Lai, H., Deng, L., Zhong, Q., et al., 2025. Nano-biotechnology in soil remediation: use of nanomaterials to promote plant growth and stress tolerance. Nanomaterials 15 (22), 1743.

Cheng, L., Wang, Y., Cai, Z., Liu, J., Yu, B., Zhou, Q., 2017. Phytoremediation of petroleum hydrocarbon-contaminated saline-alkali soil by wild ornamental Iridaceae species. Int. J. Phytoremediation 19 (3), 300–308.

Chengala, L., Singh, N., 2017. Botanical pesticides—a major alternative to chemical pesticides: a review. Int. J. Life Sci. 5, 722–729.

Correa-García, S., Pande, P., Séguin, A., St-Arnaud, M., Yergeau, E., 2018. Rhizoremediation of petroleum hydrocarbons: a model system for plant microbiome manipulation. Microb. Biotechnol. 11, 819–832.

Cui, J., Chen, H., Sun, M., Wen, J., 2020. Comparison of bacterial community structure and function under different petroleum hydrocarbon degradation conditions. Bioproc. Biosyst. Eng. 43, 303–313.

Cycoń, M., Mrozik, A., Piotrowska-Seget, Z., 2017. Bioaugmentation as a strategy for the remediation of pesticide-polluted soil: a review. Chemosphere 172, 52–71.

Czerniawska-Kusza, I., Ciesielczuk, T., Kusza, G., Cichoń, A., 2006. Comparison of the Phytotoxkit microbiotest and chemical variables for toxicity evaluation of sediments. Environ. Toxicol.: Int. J. 21, 367–372.

Dadrasnia, A., Agamuthu, P., 2013. Diesel fuel degradation from contaminated soil by Dracaena reflexa using organic waste supplementation. J. Jpn. Petrol. Inst. 56, 236–243.

Desoky, E.-S.M., Saad, A.M., El-Saadony, M.T., Merwad, A.-R.M., Rady, M.M., 2020. Plant growth-promoting rhizobacteria: potential improvement in antioxidant defense system and suppression of oxidative stress for alleviating salinity stress in *Triticum aestivum* (L.) plants. Biocatal. Agric. Biotechnol. 30, 101878.

Di Nino, M., 2017. Quantitative Analysis of Petroleum Hydrocarbons (PHC) in Soils and Chemically Augmented Plant Growth Promoting Rhizobacteria Enhanced Phytoremediation Systems. University of Waterloo.

Diab, E.A., 2008. Phytoremediation of oil contaminated desert soil using the rhizosphere effects. Global J. Environ. Res. 2, 66–73.

Diallo, M., Vural, C., Cay, H., Ozdemir, G., 2021. Enhanced biodegradation of crude oil in soil by a developed bacterial consortium and indigenous plant growth promoting bacteria. J. Appl. Microbiol. 130, 1192–1207.

Dib, D., Ahmed, D.S.A., 2024. Comparative study of the diesel fuel contamination effects of different types of soils on the growth and germination of four plant species. Braz. Arch. Biol. Technol. 67, e24230728.

Djahnit, N., Chernai, S., Catania, V., Hamdi, B., China, B., Cappello, S., Quatrini, P., 2019. Isolation, characterization and determination of biotechnological potential of oil-degrading bacteria from Algerian centre coast. J. Appl. Microbiol. 126, 780–795.

Dupuis, A., Francisco, U.M., 2015. A Literature Review on the Aquatic Toxicology of Petroleum Oil: An Overview of Oil Properties and Effects to Aquatic Biota. Canadian Science Advisory Secretariat, Vancouver, BC, USA.

Ehis-Eriakha, C.B., S.E. Akemu and D.O. Osofisan. 2024. Harnessing rhizospheric microbes for mitigating petroleum hydrocarbon toxicity.

Erdal, S., Turk, H., 2016. Cysteine-induced upregulation of nitrogen metabolism-related genes and enzyme activities enhance tolerance of maize seedlings to cadmium stress. Environ. Exp. Bot. 132, 92–99.

Farhadian, M., Vachelard, C., Duchez, D., Larroche, C., 2008. In situ bioremediation of monoaromatic pollutants in groundwater: a review. Bioresour. Technol. 99, 5296–5308.

Farooq, Q., Khan, M.I., Ali, M.H., Bashir, S., Azam, M., Qadri, R., Ali, B., Dawood, M., Hussain, M.I., 2023. A bacterial consortium and L-tryptophan boosted glyphosate removal, mungbean growth and physiology in contaminated soil. Arabian J. Geosci. 16, 653.

Fatima, T., 2024. Environmental chemistry and pollution control. Chemistry: The Modern Approach 187.

Fatokun, K., Zharare, G.E., 2015. Influence of diesel contamination in soil on growth and dry matter partitioning of *Lactuca sativa* and *Ipomoea batatas*. J. Environ. Biol. 36 (5), 1205.

Fu, X., Shi, K., Xue, J.L., Chen, C., Bai, Y., et al., 2021. Systematic adsorption process of petroleum hydrocarbon by immobilised petroleum-degradation bacteria system in degradation pathways. Pet. Sci. 18 (5), 1543–1550.

Gałązka, A.N.N.A., Gałązka, R., 2015. Phytoremediation of polycyclic aromatic hydrocarbons in soils artificially polluted using plant-associated-endophytic bacteria and *Dactylis glomerata* as the bioremediation plant. Pol. J. Microbiol. 64 (3), 1.

Gao, H., Wu, M., Liu, H., Xu, Y., Liu, Z., 2022. Effect of petroleum hydrocarbon pollution levels on the soil microecosystem and ecological function. Environ. Pollut. 293, 118511.

Gardi, C., Panagos, P., Van Liedekerke, M., Bosco, C., De Brogniez, D., 2015. Land take and food security: assessment of land take on the agricultural production in Europe. J. Environ. Plann. Manag. 58, 898–912.

Gargouri, B., Karray, F., Mhiri, N., Aloui, F., Sayadi, S., 2014. Bioremediation of petroleum hydrocarbons-contaminated soil by bacterial consortium isolated from an industrial wastewater treatment plant. J. Chem. Technol. Biotechnol. 89, 978–987.

Garrido-Sanz, D., Redondo-Nieto, M., Guirado, M., Pindado Jiménez, O., Millán, R., Martin, M., Rivilla, R., 2019. Metagenomic insights into the bacterial functions of a diesel-degrading consortium for the rhizoremediation of diesel-polluted soil. Genes 10, 456.

Goveas, L.C., Selvaraj, R., Vinayagam, R., Sajankila, S.P., Pugazhendhi, A., 2023. Biodegradation of benzo (a) pyrene by Pseudomonas strains, isolated from petroleum refinery effluent: degradation, inhibition kinetics and metabolic pathway. Chemosphere 321, 138066.

Grifoni, M., Rosellini, I., Angelini, P., Petruzzelli, G., Pezzarossa, B., 2020. The effect of residual hydrocarbons in soil following oil spillages on the growth of *Zea mays* plants. Environ. Pollut. 265, 114950.

Guarino, C., Spada, V., Sciarrillo, R., 2017. Assessment of three approaches of bioremediation (Natural Attenuation, Landfarming and Bioagumentation-Assistited Landfarming) for a petroleum hydrocarbons contaminated soil. Chemosphere 170, 10–16.

Guo, M., Gong, Z., Miao, R., Rookes, J., Cahill, D., Zhuang, J., 2017. Microbial mechanisms controlling the rhizosphere effect of ryegrass on degradation of polycyclic aromatic hydrocarbons in an aged-contaminated agricultural soil. Soil Biol. Biochem. 113, 130–142.

Haider, F.U., Ejaz, M., Cheema, S.A., Khan, M.I., Zhao, B., Liqun, C., Salim, M.A., Naveed, M., Khan, N., Núñez-Delgado, A., 2021. Phytotoxicity of petroleum hydrocarbons: sources, impacts and remediation strategies. Environ. Res. 197, 111031.

Haque, S., Srivastava, N., Pal, D.B., Alkhanani, M.F., Almalki, A.H., Areeshi, M.Y., Naidu, R., Gupta, V. K., 2022. Functional microbiome strategies for the bioremediation of petroleum-hydrocarbon and heavy metal contaminated soils: a review. Sci. Total Environ., 155222

Haritash, A., 2020. A comprehensive review of metabolic and genomic aspects of PAH-degradation. Arch. Microbiol. 202, 2033–2058.

Hassan, M.K., Mcinroy, J.A., Kloepper, J.W., 2019. The interactions of rhizodeposits with plant growth-promoting rhizobacteria in the rhizosphere: a review. Agriculture 9, 142.

Helmy, Q., Laksmono, R., Kardena, E., 2015. Bioremediation of aged petroleum oil contaminated soil: from laboratory scale to full scale application. Procedia Chem. 14, 326–333.

Henner, P., Schiavon, M., Druelle, V., Lichtfouse, E., 1999. Phytotoxicity of ancient gaswork soils. Effect of polycyclic aromatic hydrocarbons (PAHs) on plant germination. Org. Geochem. 30, 963–969.

Hoang, S.A., Lamb, D., Sarkar, B., Seshadri, B., Lam, S.S., Vinu, A., Bolan, N.S., 2022. Plant-derived saponin enhances biodegradation of petroleum hydrocarbons in the rhizosphere of native wild plants. Environ. Pollut. 313, 120152.

Hsu, C.S., Robinson, P.R., 2024. Classification and characterization. In: Petroleum Science and Technology: Upstream/Midstream. Springer.

Hu, J., Chen, J., Wang, W., Zhu, L., 2023. Mechanism of growth inhibition mediated by disorder of chlorophyll metabolism in rice (Oryza sativa) under the stress of three polycyclic aromatic hydrocarbons. Chemosphere 329, 138554.

Hussain, F., Khan, A.H.A., Hussain, I., Farooqi, A., Muhammad, Y.S., Iqbal, M., Arslan, M., Yousaf, S., 2022. Soil conditioners improve rhizodegradation of aged petroleum hydrocarbons and enhance the growth of *Lolium multiflorum*. Environ. Sci. Pollut. Control Ser. 1–13.

Hussain, F., Hussain, I., Khan, A.H.A., Muhammad, Y.S., Iqbal, M., Soja, G., Reichenauer, T.G., Zeshan, Yousaf, S., 2018. Combined application of biochar, compost, and bacterial consortia with Italian ryegrass enhanced phytoremediation of petroleum hydrocarbon contaminated soil. Environ. Exp. Bot. 153, 80–88.

Hussain, I., Puschenreiter, M., Gerhard, S., Schöftner, P., Yousaf, S., Wang, A., Syed, J.H., Reichenauer, T.G., 2018. Rhizoremediation of petroleum hydrocarbon-contaminated soils: improvement opportunities and field applications. Environ. Exp. Bot. 147, 202–219.

Ikeura, H., Kawasaki, Y., Kaimi, E., Nishiwaki, J., Noborio, K., Tamaki, M., 2016. Screening of plants for phytoremediation of oil-contaminated soil. Int. J. Phytoremediation 18 (5), 460–466.

Iqbal, A., Arshad, M., Karthikeyan, R., Gentry, T.J., Rashid, J., Ahmed, I., Schwab, A.P., 2019. Diesel degrading bacterial endophytes with plant growth promoting potential isolated from a petroleum storage facility. 3 Biotech 9, 35.

Ite, A.E., Ibok, U.J., 2019. Role of plants and microbes in bioremediation of petroleum hydrocarbons contaminated soils. International Journal of Environmental Bioremediation and Biodegradation 7, 1–19.

Ite, A.E., Harry, T.A., Obadimu, C.O., Asuaiko, E.R., Inim, I.J., 2018. Petroleum hydrocarbons contamination of surface water and groundwater in the Niger Delta region of Nigeria. Journal of Environment Pollution and Human Health 6, 51–61.

Ite, A.E., Ibok, U.J., Ite, M.U., Petters, S.W., 2013. Petroleum exploration and production: Past and present environmental issues in the Nigeria's Niger Delta. Am. J. Environ. Protect. 1, 78–90.

Jada, A., Salou, M., 2002. Effects of the asphaltene and resin contents of the bitumens on the water–bitumen interface properties. J. Petrol. Sci. Eng. 33, 185–193.

Jalal, A., De Oliveira Junior, J.C., Ribeiro, J.S., Fernandes, G.C., Mariano, G.G., Trindade, V.D.R., Dos Reis, A.R., 2021. Hormesis in plants: physiological and biochemical responses. Ecotoxicol. Environ. Saf. 207, 111225.

Jewell, D.M., Albaugh, E.W., Davis, B.E., Ruberto, R.G., 1974. Integration of chromatographic and spectroscopic techniques for the characterization of residual oils. Industrial and Engineering Chemistry Fundamental 13, 278–282.

Kaczorek, E., Moszyńska, S., Olszanowski, A., 2011. Modification of cell surface properties of Pseudomonas alcaligenes S22 during hydrocarbon biodegradation. Biodegradation 22, 359–366.

Kahkashan, S., Wang, X., Ya, M., Chen, J., Wu, Y., Cai, Y., Saleem, M., Inam, A., Aftab, J., 2019. Evaluation of marine sediment contamination by polycyclic aromatic hydrocarbons along the Karachi coast, Pakistan, 11 years after the Tasman Spirit oil spill. Chemosphere 233, 652–659.

Karani, A., E. Mbega and P. Ndakidemi. 2017. Botanical pesticides in management of common bean pests: importance and possibilities for adoption by small-scale farmers in Africa.

Kaur, N., Erickson, T.E., Ball, A.S., Ryan, M.H., 2017. A review of germination and early growth as a proxy for plant fitness under petrogenic contamination—knowledge gaps and recommendations. Sci. Total Environ. 603, 728–744.

Kaya, C., Okant, M., Ugurlar, F., Alyemeni, M.N., Ashraf, M., Ahmad, P., 2019. Melatonin-mediated nitric oxide improves tolerance to cadmium toxicity by reducing oxidative stress in wheat plants. Chemosphere 225, 627–638.

Keramea, P., Spanoudaki, K., Zodiatis, G., Gikas, G., Sylaios, G., 2021. Oil spill modeling: a critical review on current trends, perspectives, and challenges. J. Mar. Sci. Eng. 9, 181.

Khan, M.I., Cheema, S.A., Shen, C.F., Zhang, C.K., Tang, X.J., Zaffar, M., Chen, X.C., Chen, Y.X., 2011. Assessment of pyrene bioavailability in soil by mild hydroxypropyl-β-cyclodextrin extraction. Arch. Environ. Contam. Toxicol. 60, 107–115.

Khan, M.I., Cheema, S.A., Zhang, C.K., Shen, C.F., Tang, X.J., Chen, Y.X., 2012a. Assessment of phenanthrene bioavailability in aged and unaged soils by mild extraction. Environ. Monit. Assess. 184, 549–559.

Khan, M.I., Cheema, S.A., Tang, X.J., Shen, C.F., Park, J.H., Chen, Y.X., 2013. A battery of bioassays for the evaluation of phenanthrene biotoxicity in soil. Arch. Environ. Contam. Toxicol. 65, 47–55.

Khan, M.I., Cheema, S.A., Shen, C.F., Tang, X.J., Park, J.H., Chen, Y.X., 2012b. Biotoxicity assessment of pyrene in soil by using a battery of biological assays. Arch. Environ. Contam. Toxicol. 63, 503–512.

Khan, M.I., Cheema, S.A., Anum, S., et al., 2020. Phytoremediation of agricultural pollutants. In: Shmaefsky, B. (Ed.), Phytoremediation. Concepts and Strategies in Plant Sciences. Springer, Cham, pp. 27–81.

Khan, M.I., Yoo, K., Schwab, L., Kümmel, S., Nijenhuis, I., 2024. Characterization of anaerobic biotransformation of hexachlorocyclohexanes by novel microbial consortia enriched from channel and river sediments. J. Hazard Mater. 476, 135198.

Khan, M.I., Farooq, Q., Ali, M., Ali, M.H., Naveed, M., 2025. Integrated effects of microbial culture and nitrogen application on phytoremediation, physiology and growth of maize in glyphosate-contaminated soil. Int. J. Phytoremediation 1–13.

Kohlmeier, M., 2003. Minerals and trace elements. *Nutrient Metabolism*. Academic Press, New York.

Kong, L., Gao, Y., Zhou, Q., Zhao, X., Sun, Z., 2018. Biochar accelerates PAHs biodegradation in petroleum-polluted soil by biostimulation strategy. J. Hazard Mater. 343, 276–284.

Koolivand, A., Saeedi, R., Coulon, F., Kumar, V., Villaseñor, J., Asghari, F., Hesampoor, F., 2020. Bioremediation of petroleum hydrocarbons by vermicomposting process bioaugmentated with indigenous bacterial consortium isolated from petroleum oily sludge. Ecotoxicol. Environ. Saf. 198, 110645.

Korade, D., Fulekar, M., 2009. Effect of organic contaminants on seed germination of *Lolium multiflorum* in soil. Biol. Med. 1, 28–34.

Koshlaf, E., Ball, A.S., 2017. Soil bioremediation approaches for petroleum hydrocarbon polluted environments. AIMS Microbiology 3, 25.

Kreslavski, V., Brestic, M., Zharmukhamedov, S., Lyubimov, V.Y., Lankin, A., Jajoo, A., Allakhverdiev, S., 2017. Mechanisms of inhibitory effects of polycyclic aromatic hydrocarbons in photosynthetic primary processes in pea leaves and thylakoid preparations. Plant Biol. 19, 683–688.

Kumar, G., Bhatt, P., Lal, S., 2021. Phytoremediation: A Synergistic Interaction between Plants and Microbes for Removal of Petroleum Hydrocarbons. Soil Contamination-Threats and Sustainable Solutions, IntechOpen.

Kumar, V., Almomin, S., Al-Aqeel, H., Al-Salameen, F., Nair, S., Shajan, A., 2018. Metagenomic analysis of rhizosphere microflora of oil-contaminated soil planted with barley and alfalfa. PLoS One 13, e0202127.

Kumari, S., Regar, R.K., Manickam, N., 2018. Improved polycyclic aromatic hydrocarbon degradation in a crude oil by individual and a consortium of bacteria. Bioresour. Technol. 254, 174–179.

Kuppusamy, S., Maddela, N.R., Megharaj, M., Venkateswarlu, K., 2020. An overview of total petroleum hydrocarbons. Total Petroleum Hydrocarbons 1–27.

Kuzina, E., Rafikova, G., Vysotskaya, L., Arkhipova, T., Bakaeva, M., Chetverikova, D.Y., Kudoyarova, G., Korshunova, T., Chetverikov, S., 2021. Influence of hydrocarbon-oxidizing bacteria on the growth, biochemical characteristics, and hormonal status of barley plants and the content of petroleum hydrocarbons in the soil. Plants 10, 1745.

Li, X., Song, Y., Wang, F., Bian, Y., Jiang, X., 2018. Combined effects of maize straw biochar and oxalic acid on the dissipation of polycyclic aromatic hydrocarbons and microbial community structures in soil: a mechanistic study. J. Hazard Mater. 364, 325–331.

Li, Y., Liu, Y., Guo, D., Dong, H., 2024. Differential degradation of petroleum hydrocarbons by *Shewanella putrefaciens* under aerobic and anaerobic conditions. Front. Microbiol. 15, 1389954.

Lin, L., Jiang, W., Chen, L., Xu, P., Wang, H., 2020. Treatment of produced water with photocatalysis: recent advances, affecting factors and future research prospects. Catalysts 10 (8), 924.

Liu, H., Gao, H., Wu, M., Ma, C., Wu, J., Ye, X., 2020. Distribution characteristics of bacterial communities and hydrocarbon degradation dynamics during the remediation of petroleum-contaminated soil by enhancing moisture content. Microb. Ecol. 1–10.

Lu, C., Hong, Y., Liu, J., Gao, Y., Ma, Z., et al., 2019. A PAH-degrading bacterial community enriched with contaminated agricultural soil and its utility for microbial bioremediation. Environ. Pollut. 251, 773–782.

Ma, D., Xu, J., Zhou, J., Ren, L., Li, J., Zhang, Z., Xia, J., Xie, H., Wu, T., 2023. Using sweet sorghum varieties for the phytoremediation of petroleum-contaminated salinized soil: a preliminary study based on pot experiments. Toxics 11, 208.

Maddela, N.R., Scalvenzi, L., Venkateswarlu, K., 2017. Microbial degradation of total petroleum hydrocarbons in crude oil: a field-scale study at the low-land rainforest of Ecuador. Environ. Technol. 38, 2543–2550.

Maila, M.P., Cloete, T.E., 2002. Germination of Lepidium sativum as a method to evaluate polycyclic aromatic hydrocarbons (PAHs) removal from contaminated soil. Int. Biodeterior. Biodegrad. 50, 107–113.

Mair, B.J., 1964. Terpenoids, fatty acids and alcohols as source materials for petroleum hydrocarbons. Geochem. Cosmochim. Acta 28, 1303–1321.

Majeed, B.K., Shwan, D.M., Rashid, K.A., 2025. A review on environmental contamination of petroleum hydrocarbons; its effects and remediation approaches. Environ. Sci. Process. Impacts 27 (3), 526–548.

Manzoor, A.F., 2020. Core competencies of truck drivers responding to emergencies during transportation of hazardous materials. Journal of Health and Pollution 10 (27), 200909.

Marathon Petroleum. 2012. Crude oil - Light: Safety data sheet. Retrieved data 27-02-2025 https://www.marathonpetroleum.com/content/documents/Operations/Western_SDS/Crude_Oil-Light.

Martin, B.C., George, S.J., Price, C.A., Ryan, M.H., Tibbett, M., 2014. The role of root exuded low molecular weight organic anions in facilitating petroleum hydrocarbon degradation: current knowledge and future directions. Sci. Total Environ. 472, 642–653.

Masu, S., Cojocariu, L., Grecu, E., Morariu, F., Bordean, D.M., et al., 2018. Lolium perenne—a phytoremediation option in case of total petroleum hydrocarbons polluted soils. Rev. Chim. 69 (5), 1110–1114.

McIntosh, P., Schulthess, C.P., Kuzovkina, Y.A., Guillard, K., 2017. Bioremediation and phytoremediation of total petroleum hydrocarbons (TPH) under various conditions. Int. J. Phytoremediation 19 (8), 755–764.

Mclaughlin, M.C., Borch, T., Mcdevitt, B., Warner, N.R., Blotevogel, J., 2020. Water quality assessment downstream of oil and gas produced water discharges intended for beneficial reuse in arid regions. Sci. Total Environ. 713, 136607.

Meckenstock, R.U., Boll, M., Mouttaki, H., Koelschbach, J.S., Cunha Tarouco, P., Weyrauch, P., Dong, X., Himmelberg, A.M., 2016. Anaerobic degradation of benzene and polycyclic aromatic hydrocarbons. J. Mol. Microbiol. Biotechnol. 26, 92–118.

Minela, C., Zlata, I.S., 2025. The possibility of applying the bioremediation process on soil contaminated with petroleum hydrocarbons. International Journal of Science and Research Archive 14 (2), 170–176.

Mohammadi, L., Rahdar, A., Bazrafshan, E., Dahmardeh, H., Susan, M.a.B.H., Kyzas, G.Z., 2020. Petroleum hydrocarbon removal from wastewaters: a review. Processes 8, 447.

Mohanta, S., Pradhan, B., Behera, I.D., 2024. Impact and remediation of petroleum hydrocarbon pollutants on agricultural land: a review. Geomicrobiol. J. 41, 345–359.

Moradi, B., Kissen, R., Maivan, H.Z., Hashtroudi, M.S., Sorahinobar, M., Sparstad, T., Bones, A.M., 2020. Assessment of oxidative stress response genes in *Avicennia marina* exposed to oil contamination–Polyphenol oxidase (PPOA) as a biomarker. Biotechnol. Rep. 28, e00565.

Naeem, U., Qazi, M.A., 2020. Leading edges in bioremediation technologies for removal of petroleum hydrocarbons. Environ. Sci. Pollut. Control Ser. 27, 27370–27382.

Naidoo, G., Naidoo, K., 2016. Uptake of polycyclic aromatic hydrocarbons and their cellular effects in the mangrove Bruguiera gymnorrhiza. Mar. Pollut. Bull. 113, 193–199.

Nath, S., 2024. A vision of precision agriculture: Balance between agricultural sustainability and environmental stewardship. Agron. J. 116, 1126–1143.

Nero, B.F., 2021. Phytoremediation of petroleum hydrocarbon-contaminated soils with two plant species: *Jatropha curcas* and *Vetiveria zizanioides* at Ghana Manganese Company Ltd. Int. J. Phytoremediation 23 (2), 171–180.

Odjegba, V.J., Badejo, J.O., 2013. Crude oil induced oxidative stress in Capsicum annum L. Nat. Sci. 11, 46–50.

Odukoya, J., Lambert, R., Sakrabani, R., 2019. Understanding the impacts of crude oil and its induced abiotic stresses on agrifood production: a review. Horticulturae 5, 47.

Omrani, R., Spini, G., Puglisi, E., Saidane, D., 2018. Modulation of microbial consortia enriched from different polluted environments during petroleum biodegradation. Biodegradation 29, 187–209.

Ortega-Calvo, J., Tejeda-Agredano, M., Jimenez-Sanchez, C., Congiu, E., Sungthong, R., Niqui-Arroyo, J., Cantos, M., 2013. Is it possible to increase bioavailability but not environmental risk of PAHs in bioremediation? J. Hazard Mater. 261, 733–745.

Ossai, I.C., Ahmed, A., Hassan, A., Hamid, F.S., 2020. Remediation of soil and water contaminated with petroleum hydrocarbon: a review. Environ. Technol. Innov. 17, 100526.

Patowary, K., Patowary, R., Kalita, M.C., Deka, S., 2017. Characterization of biosurfactant produced during degradation of hydrocarbons using crude oil as sole source of carbon. Front. Microbiol. 8, 27–293.

Piubeli, F.A., Dos Santos, L.G., Fernandez, E.N., Da Silva, F.H., Durrant, L.R., Grossman, M.J., 2018. The emergence of different functionally equivalent PAH degrading microbial communities from a single soil in liquid PAH enrichment cultures and soil microcosms receiving PAHs with and without bioaugmentation. Pol. J. Microbiol. 67, 365–375.

Płociniczak, T., Fic, E., Pacwa-Płociniczak, M., Pawlik, M., Piotrowska-Seget, Z., 2017. Improvement of phytoremediation of an aged petroleum hydrocarbon-contaminated soil by *Rhodococcus erythropolis* CD 106 strain. Int. J. Phytoremediation 19 (7), 614–620.

Rafique, H., Asghar, H., Zahir, Z., Shahbaz, M., 2015. Evaluation of plant growth promoting bacteria for inducing stress tolerance in plants against petroleum hydrocarbons. Pakistan J. Agric. Sci. 52.

Rahman, K.S., Rahman, T.J., Kourkoutas, Y., Petsas, I., Marchant, R., Banat, I., 2003. Enhanced bioremediation of n-alkane in petroleum sludge using bacterial consortium amended with rhamnolipid and micronutrients. Bioresour. Technol. 90, 159–168.

Raimondo, M., Caracciolo, F., Nazzaro, C., Marotta, G., 2021. Organic farming increases the technical efficiency of olive farms in Italy. Agriculture 11, 209.

Rees, W.E., 2021. Achieving Sustainability: Reform or Transformation? the Earthscan Reader in Sustainable Cities. Routledge.

Rizzo, C., Rappazzo, A.C., Michaud, L., De Domenico, E., Rochera, C., Camacho, A., Giudice, A.L., 2018. Efficiency in hydrocarbon degradation and biosurfactant production by Joostella sp. A8 when grown in pure culture and consortia. J. Environ. Sci. 67, 115–126.

Robichaud, K., Stewart, K., Labrecque, M., Hijri, M., Cherewyk, J., Amyot, M., 2019. An ecological microsystem to treat waste oil contaminated soil: using phytoremediation assisted by fungi and local compost, on a mixed-contaminant site, in a cold climate. Sci. Total Environ. 672, 732–742.

Rohrbacher, F., St-Arnaud, M., 2016. Root exudation: the ecological driver of hydrocarbon rhizoremediation. Agronomy 6, 19.

Ron, E.Z., Rosenberg, E., 2014. Enhanced bioremediation of oil spills in the sea. Curr. Opin. Biotechnol. 27, 191–194.

Rosenberg, M., Rosenberg, E., 1985. Bacterial adherence at the hydrocarbon-water interface. Oil Petrochem. Pollut. 2, 155–162.

Rostami, S., Azhdarpoor, A., 2019. The application of plant growth regulators to improve phytoremediation of contaminated soils: a review. Chemosphere 220, 818–827.

Ruley, J.A., Amoding, A., Tumuhairwe, J.B., Basamba, T.A., Opolot, E., Oryem-Origa, H., 2020. Enhancing the phytoremediation of hydrocarbon-contaminated soils in the Sudd Wetlands, South Sudan, using organic manure. Appl. Environ. Soil Sci. 2020 (1), 4614286.

Saeki, H., Sasaki, M., Komatsu, K., Miura, A., Matsuda, H., 2009. Oil spill remediation by using the remediation agent JE1058BS that contains a biosurfactant produced by Gordonia sp. strain JE-1058. Bioresour. Technol. 100, 572–577.

Sajna, K.V., Sukumaran, R.K., Gottumukkala, L.D., Pandey, A., 2015. Crude oil biodegradation aided by biosurfactants from *Pseudozyma* sp. NII 08165 or its culture broth. Bioresour. Technol. 191, 133–139.

Sampaio, G.R., Guizellini, G.M., Da Silva, S.A., De Almeida, A.P., Pinaffi-Langley, A.C.C., Rogero, M. M., De Camargo, A.C., Torres, E.A., 2021. Polycyclic aromatic hydrocarbons in foods: biological effects, legislation, occurrence, analytical methods, and strategies to reduce their formation. Int. J. Mol. Sci. 22, 6010.

Saraeian, Z., Haghighi, M., Etemadi, N., HajAbbasi, M.A., Afyuni, M., 2018. Phytoremediation effect and growth responses *of Cynodon* spp. and *Agropyron desertorum* in a petroleum-contaminated soil. Soil Sediment Contam.: Int. J. 27 (5), 393–407.

Sayed, K., Baloo, L., Sharma, N.K., 2021. Bioremediation of total petroleum hydrocarbons (TPH) by bioaugmentation and biostimulation in water with floating oil spill containment booms as bioreactor basin. Int. J. Environ. Res. Publ. Health 18, 2226.

Selvam, A.D.G., Thatheyus, A.J., 2018. Microbial degradation of petroleum hydrocarbons: an overview. Microbial Action on Hydrocarbons 485–503.

Shah, F.M., Razaq, M., 2022. From agriculture to sustainable agriculture: Prospects for improving pest management in industrial revolution 4.0. In: Handbook of Smart Materials, Technologies, and Devices: Applications of Industry 4.0. Springer.

Shahsavari, E., Adetutu, E.M., Anderson, P.A., Ball, A.S., 2013. Tolerance of selected plant species to petrogenic hydrocarbons and effect of plant rhizosphere on the microbial removal of hydrocarbons in contaminated soil. Water Air Soil Pollut. 224, 1–14.

Shankar, S., Kansrajh, C., Dinesh, M., Satyan, R., Kiruthika, S., Tharanipriya, A., 2014. Application of indigenous microbial consortia in bioremediation of oil-contaminated soils. Int. J. Environ. Sci. Technol. 11, 367–376.

Sharma, P., 2022. Microbial communication during bioremediation of polyaromatic hydrocarbons. Syst. Microbiol. Biomanuf. 1–15.

Sharuddin, S.S.N., Abdullah, S.R.S., Hasan, H.A., Othman, A.R., 2024. Rhizobacterial-assisted phytoremediation for accelerated petroleum-hydrocarbon removal in crude-oil sludge. Sci. Total Environ. 954, 176189.

Shen, T., Pi, Y., Bao, M., Xu, N., Li, Y., Lu, J., 2015. Biodegradation of different petroleum hydrocarbons by free and immobilized microbial consortia. Environ. Sci. Process. Impacts 17, 2022–2033.

Siddiqi, H., Mishra, A., Maiti, P., Behera, I.D., Meikap, B., 2022. In-situ and ex-situ co-pyrolysis studies of waste biomass with spent motor oil: Elucidating the role of physical inhibition and mixing ratio to enhance fuel quality. Bioresour. Technol. 358, 127364.

Sivaram, A.K., Logeshwaran, P., Lockington, R., Naidu, R., et al., 2019. Low molecular weight organic acids enhance the high molecular weight polycyclic aromatic hydrocarbons degradation by bacteria. Chemosphere 222, 132–140.

Sivaram, A.K., Logeshwaran, P., Lockington, R., Naidu, R., Megharaj, M., 2020. The impact of low molecular weight organic acids from plants with C3 and C4 photosystems on the rhizoremediation of polycyclic aromatic hydrocarbons contaminated soil. Environ. Technol. Innov. 19, 100957.

Sivaram, A.K., Logeshwaran, P., Subashchandrabose, S.R., Lockington, R., Naidu, R., Megharaj, M., 2018. Comparison of plants with C3 and C4 carbon fixation pathways for remediation of polycyclic aromatic hydrocarbon contaminated soils. Sci. Rep. 8 (1), 2100.

Smith, P.N., Cobb, G.P., Godard-Codding, C., Hoff, D., Mcmurry, S.T., Rainwater, T.R., Reynolds, K. D., 2007. Contaminant exposure in terrestrial vertebrates. Environ. Pollut. 150, 41–64.

Soghandi, B., Salimi, F., 2024. Study on amendment of rapeseed meal, soybean meal, and NPK fertilizer as biostimulants in bioremediation of diesel-contaminated soil by autochthonous microorganisms. Soil Sediment. Contam. 33 (4), 430–450.

Song, H., Wang, Y.-S., Sun, C.-C., Wu, M.-L., Peng, Y.-L., Deng, C., Li, Q.P., 2011. Effects of polycyclic aromatic hydrocarbons exposure on antioxidant system activities and proline content in Kandelia candel. Oceanol. Hydrobiol. Stud. 40, 9–18.

Song, X., Wu, X., Song, X., Shi, C., Zhang, Z., 2021. Sorption and desorption of petroleum hydrocarbons on biodegradable and nondegradable microplastics. Chemosphere 273, 128553.

Srijita Dutta, S.D. 2015. Biopesticides: an ecofriendly approach for pest control.

Steliga, T., Kluk, D., 2020. Application of Festuca arundinacea in phytoremediation of soils contaminated with Pb, Ni, Cd and petroleum hydrocarbons. Ecotoxicol. Environ. Saf. 194, 110409.

Suganthi, S.H., Murshid, S., Sriram, S., Ramani, K., 2018. Enhanced biodegradation of hydrocarbons in petroleum tank bottom oil sludge and characterization of biocatalysts and biosurfactants. J. Environ. Manag. 220, 87–95.

Suja, F., Rahim, F., Taha, M.R., Hambali, N., Razali, M.R., Khalid, A., Hamzah, A., 2014. Effects of local microbial bioaugmentation and biostimulation on the bioremediation of total petroleum hydrocarbons (TPH) in crude oil contaminated soil based on laboratory and field observations. Int. Biodeterior. Biodegrad. 90, 115–122.

Sun, C., Fontanesi, G., Canberk, B., Mohajerzadeh, A., Chatzinotas, S., Grace, D., Ahmadi, H., 2024. Advancing UAV communications: a comprehensive survey of cutting-edge machine learning techniques. IEEE Open Journal of Vehicular Technology 5, 825–854.

Tang, J., Lu, X., Sun, Q., Zhu, W., 2012. Aging effect of petroleum hydrocarbons in soil under different attenuation conditions. Agric. Ecosyst. Environ. 149, 109–117.

Tao, K., Liu, X., Chen, X., Hu, X., Cao, L., Yuan, X., 2017. Biodegradation of crude oil by a defined co-culture of indigenous bacterial consortium and exogenous Bacillus subtilis. Bioresour. Technol. 224, 327–332.

Tara, N., Afzal, M., Ansari, T.M., Tahseen, R., Iqbal, S., Khan, Q.M., 2014. Combined use of alkane-degrading and plant growth-promoting bacteria enhanced phytoremediation of diesel contaminated soil. Int. J. Phytoremediation 16, 1268–1277.

Tariq, M., Shah, A.A., Yasin, N.A., Ahmad, A., Rizwan, M., 2021. Enhanced performance of bacillus megaterium OSR-3 in combination with putrescine ammeliorated hydrocarbon stress in Nicotiana tabacum. Int. J. Phytoremediation 23, 119–129.

Tiwari, G.N. Energy, Ecology and Environment. 2024. Assessed on 28-02-2024.

Truskewycz, A., Gundry, T.D., Khudur, L.S., Kolobaric, A., Taha, M., Aburto-Medina, A., Ball, A.S., Shahsavari, E., 2019. Petroleum hydrocarbon contamination in terrestrial ecosystems—fate and microbial responses. Molecules 24, 3400.

Uddin, S., Fowler, S.W., Saeed, T., Jupp, B., Faizuddin, M., 2021. Petroleum hydrocarbon pollution in sediments from the Gulf and Omani waters: status and review. Mar. Pollut. Bull. 173, 112913.

Ugochukwu, U.C., Jones, M.D., Head, I.M., Manning, D.A., Fialips, C.I., 2013. Compositional changes of crude oil SARA fractions due to biodegradation and adsorption on colloidal support such as clays using Iatroscan. Environ. Sci. Pollut. Control Ser. 20, 6445–6454.

Ugwoha, E., Amah, V.E., Oweh, G.O., 2020. Bioremediation of crude oil contaminated soil using pig droppings and bone char. Journal of Advances in Biology and Biotechnology 23, 13–24.

Varjani, S., Upasani, V.N., 2019. Influence of abiotic factors, natural attenuation, bioaugmentation and nutrient supplementation on bioremediation of petroleum crude contaminated agricultural soil. J. Environ. Manag. 245, 358–366.

Varjani, S.J., 2017. Microbial degradation of petroleum hydrocarbons. Bioresour. Technol. 223, 277–286.

Varjani, S.J., Upasani, V.N., 2017. A new look on factors affecting microbial degradation of petroleum hydrocarbon pollutants. Int. Biodeterior. Biodegrad. 120, 71–83.

Vasilyeva, G., Kondrashina, V., Strijakova, E., Ortega-Calvo, J.-J., 2020. Adsorptive bioremediation of soil highly contaminated with crude oil. Sci. Total Environ. 706, 135739.

Vidali, M., 2001. Bioremediation. an overview. Pure Appl. Chem. 73, 1163–1172.

Wei, Y., Zhu, Y., Yang, L., Chen, C., Yue, M., Mao, Z., Wang, Y., Li, Q., Li, Y., Lv, J., 2025. Effects of oil pollution on the growth and rhizosphere microbial community of Calamagrostis epigejos. Sci. Rep. 15, 1278.

Wilkes, H., Buckel, W., Golding, B.T., Rabus, R., 2016. Metabolism of hydrocarbons in n-alkane-utilizing anaerobic bacteria. J. Mol. Microbiol. Biotechnol. 26, 138–151.

Wojtowicz, K., Steliga, T., Kapusta, P., Brzeszcz, J., 2023. Oil-contaminated soil remediation with biodegradation by autochthonous microorganisms and phytoremediation by maize (*Zea mays*). Molecules 28, 6104.

Woźniak-Karczewska, M., Lisiecki, P., Białas, W., Owsianiak, M., Piotrowska-Cyplik, A., Wolko, Ł., Ławniczak, Ł., Heipieper, H.J., Gutierrez, T., Chrzanowski, Ł., 2019. Effect of bioaugmentation on long-term biodegradation of diesel/biodiesel blends in soil microcosms. Sci. Total Environ. 671, 948–958.

Wu, C., Liu, X., Wu, X., Dong, F., Xu, J., Zheng, Y., 2019. Sorption, degradation and bioavailability of oxyfluorfen in biochar-amended soils. Sci. Total Environ. 658, 87–94.

Wu, M., Ma, C., Wang, D., Liu, H., Zhu, C., Xu, H., 2020. Nutrient drip irrigation for refractory hydrocarbon removal and microbial community shift in a historically petroleum-contaminated soil. Sci. Total Environ. 713, 136331.

Xia, M., Fu, D., Chakraborty, R., Singh, R.P., Terry, N., 2019. Enhanced crude oil depletion by constructed bacterial consortium comprising bioemulsifier producer and petroleum hydrocarbon degraders. Bioresour. Technol. 282, 456–463.

Xia, C., Liu, Q., Wang, L., Tang, J., et al., 2021. Enhanced degradation of petroleum hydrocarbons in soil by FeS@ BC activated persulfate and its mechanism. Sep. Purif. Technol. 282, 120060.

Xiao, N., Liu, R., Jin, C., Dai, Y., 2015. Efficiency of five ornamental plant species in the phytoremediation of polycyclic aromatic hydrocarbon (PAH)-contaminated soil. Ecol. Eng. 75, 384–391.

Xu, X., Liu, W., Tian, S., Wang, W., Qi, Q., Jiang, P., Gao, X., Li, F., Li, H., Yu, H., 2018. Petroleum hydrocarbon-degrading bacteria for the remediation of oil pollution under aerobic conditions: a perspective analysis. Front. Microbiol. 9, 2885.

Xu, Y., Yan, G., Hou, B., Su, K., Zhang, F., Liu, B., Li, S., Zhao, S., Liu, Y., 2024. Different effects of resins and asphaltenes concentration of crude oil on sandstone wettability. Fuel 370, 131825.

Yang, R., Liu, G., Chen, T., Li, S., An, L., Zhang, G., Li, G., Chang, S., Zhang, W., Chen, X., 2019. Characterization of the genome of a *Nocardia* strain isolated from soils in the Qinghai-Tibetan Plateau that specifically degrades crude oil and of this biodegradation. Genomics 111, 356–366.

Ye, S., Zeng, G., Wu, H., Zhang, C., Dai, J., Liang, J., Yu, J., Ren, X., Yi, H., Cheng, M., 2017. Biological technologies for the remediation of co-contaminated soil. Crit. Rev. Biotechnol. 37, 1062–1076.

Yousaf, U., Khan, A.H.A., Farooqi, A., Muhammad, Y.S., Barros, R., Tamayo-Ramos, J.A., Iqbal, M., Yousaf, S., 2022. Interactive effect of biochar and compost with Poaceae and Fabaceae plants on remediation of total petroleum hydrocarbons in crude oil contaminated soil. Chemosphere 286, 131782.

Yuan, L., Wu, Y., Fan, Q., Li, P., Liang, J., Liu, Y., Ma, R., Li, R., Shi, L., 2023. Remediating petroleum hydrocarbons in highly saline–alkali soils using three native plant species. J. Environ. Manag. 339, 117928.

Zafra, G., Absalón, Á.E., Anducho-Reyes, M.Á., Fernandez, F.J., Cortés-Espinosa, D.V., 2017. Construction of PAH-degrading mixed microbial consortia by induced selection in soil. Chemosphere 172, 120–126.

Zamani, S., Ghasemnezhad, A., Ebrahimi, S., Fathi, M., 2018. Evaluation the growth potential of artichoke (*Synara scolymus* L.) and Milk thistle (*Sylibum marianum* L.) in petroleum-contaminated Soil. Journal of Chemical Health Risks 8 (1).

Zhang, L., Shi, K., Yue, L., 2007. Chemical characteristics and pollution sources of petroleum hydrocarbons and PAHs in sediments from the Beiluohe River, Northern China. Environ. Geol. 53, 307–315.

Zhang, X., Su, C., Liu, X., Liu, Z., Liang, X., Zhang, Y., Feng, Y., 2020. Effect of plant-growth-promoting rhizobacteria on phytoremediation efficiency of Scirpus triqueter in pyrene-Ni co-contaminated soils. Chemosphere 241, 125027.

Zhan, X., Yuan, J., Yue, L., Xu, G., Hu, B., Xu, R., 2015. Response of uptake and translocation of phenanthrene to nitrogen form in lettuce and wheat seedlings. Env. Sci. Pollut. Res. 22 (8), 6280–6287.

Zhang, X., Chi, G., Xia, H., Huang, X., Deng, Y., Huang, C., Zhao, H., 2023. Deciphering the Helianthus annus rhizosphere soil biodiversity under petroleum hydrocarbon compounds contamination. Process Saf. Environ. Prot. 178, 795–806.

Zhang, S., Zhu, L., 2024. Predicting removal efficiency of organic pollutants by soil vapor extraction based on an optimized machine learning method. Sci. Total Environ. 927, 172438.

Zhen, M., Chen, H., Liu, Q., Song, B., Wang, Y., Tang, J., 2019. Combination of rhamnolipid and biochar in assisting phytoremediation of petroleum hydrocarbon contaminated soil using *Spartina anglica*. J. Environ. Sci. 85, 107–118.

Zoghi, P., Mafigholami, R., 2023. Optimisation of soil washing method for removal of petroleum hydrocarbons from contaminated soil around oil storage tanks using response surface methodology. Sci. Rep. 13, 15457.

CHAPTER 11

Endocrine-disrupting chemicals and phytotoxicity: Impacts on plant growth and ecosystem health

Doris Flore Ngakeu[1], Ricardo Santos Silva[2], Thiago Abdo de Souza[2] and Lucia Helena Gomes Coelho[1]

[1]Center for Engineering, Modeling, and Applied Social Sciences, Federal University of ABC, Santo Andre, Brazil; [2]Center for Natural and Human Sciences, Federal University of ABC, Santo Andre, Brazil

1. Endocrine-disrupting chemicals (EDCs)

The World Health Organization (WHO) defines endocrine-disrupting chemicals (EDCs) as exogenous substances or mixtures that can alter one or more functions of the endocrine system, causing adverse effects in intact organisms, their offspring, or (sub)populations (Damstra et al., 2002). The United States Environmental Protection Agency (EPA) complements this definition by considering EDCs as exogenous compounds capable of interfering with the synthesis, secretion, transport, metabolism, receptor binding, or elimination of endogenous hormones, thereby impairing the endocrine and homeostatic systems (Kabir et al., 2015; Lauretta et al., 2019; Mnif et al., 2011). Each endocrine hormone has a distinct chemical composition and shape, and the hormone-binding process is referred to as the "lock-and-key" mechanism, in which each hormone (key) is specific to a receptor (lock). This mechanism is enabled by the complementary shape of receptors to their respective hormones, allowing hormone-receptor activation. Endocrine-disrupting chemicals (EDCs) are hazardous because they can mimic or block natural hormones from binding to their receptors, preventing normal hormonal signaling. As a result, receptors may be activated inappropriately (Ismanto et al., 2022). The group of molecules identified as endocrine disruptors is highly heterogeneous. It can be divided into synthetic chemicals (from anthropogenic activities, such as 17α-ethinylestradiol (EE2), pesticides like atrazine, phthalates, alkylphenol ethoxylate surfactants, nonylphenol (NP), dioxins, coplanar polychlorinated biphenyls (PCBs), paraben hydroxybenzoate derivatives, bisphenol A (BPA), polycyclic aromatic hydrocarbons (PAHs), and organotins), which are used as industrial solvents/lubricants, plasticizers, pharmaceuticals, and flame retardants, and natural substances (e.g., estrone (E1), 17β-estradiol (E2), and estriol (E3); natural androgens such as testosterone; phytosterols like (Pironti et al., 2021) β-sitosterol; and isoflavonoids such as daidzein) (Raun Andersen et al., 2002; Kojima et al., 2004; Pironti et al., 2021; Vinggaard et al., 2000; Zoeller et al., 2012). In 2013, the World Health Organization emphasized that exposure

Phytotoxicity of Organic Pollutants
ISBN 978-0-443-33110-7,
https://doi.org/10.1016/B978-0-443-33110-7.00006-7

to EDCs poses a significant concern for wildlife and human health. It urged policymakers to take action to regulate human and environmental exposure to these substances (Pironti et al., 2021). It is well established that continuous exposure to EDCs can lead to serious transgenerational health effects in humans and wildlife if proper precautions are not taken. Therefore, understanding the sources, pathways, and risks associated with exposure is crucial to preventing both short- and long-term health consequences (Tijani et al., 2016). The primary sources of EDCs, as shown in Fig. 11.1, include industrial and agricultural activities, where these compounds are released through manufacturing processes, surface runoff, and the use of pesticides and herbicides (Kumari et al., 2024). Wastewater and sewage systems also represent major sources, as EDCs are not completely removed by conventional treatment plants, resulting in the contamination of aquatic environments (*Endocrine Disrupting Chemicals in Freshwater: Monitoring and Regulating Water Quality*, 2023). Consumer products, such as plastics and personal care items, contribute to EDC pollution when improperly disposed of (Kumari et al., 2024). In addition, atmospheric deposition is another pathway of dispersion, as EDCs can be transported via air currents and deposited in regions far from their original sources, impacting even remote ecosystems (Dosis and Kamarianos, 2007).

Sources of EDC exposure are not only diverse but also vary over time and by region, depending on national regulations and bans. Historical cases of contamination involving PCBs and dioxins have demonstrated clear links between specific chemicals and endocrine dysfunction in both humans and animals. However, the most common form of exposure today is chronic and widespread, often involving complex mixtures of pollutants, particularly in industrialized areas, where EDCs can contaminate soil, water, and

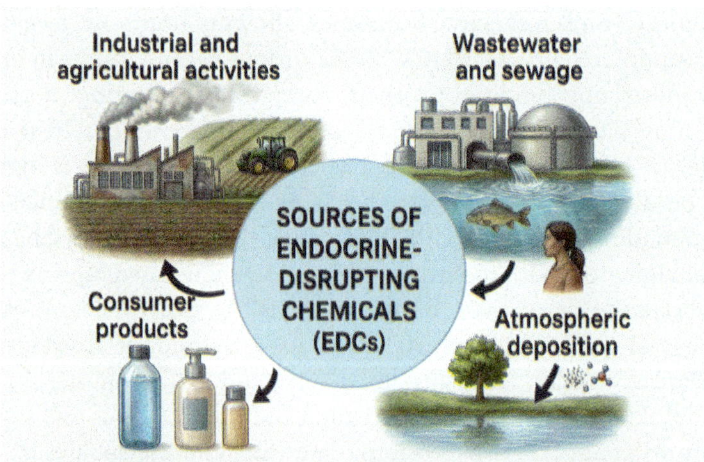

Figure 11.1 Main sources of endocrine-disrupting chemicals (EDCs) in the environment.

the food chain. Exposure may occur through ingestion, inhalation, or direct contact with contaminated environments, with workers in industries such as pesticides and chemical manufacturing being especially vulnerable to reproductive and endocrine-related health effects (Diamanti-Kandarakis et al., 2009). EDCs can enter aquatic environments through various pathways, such as the discharge of untreated or insufficiently treated wastewater and the release of pesticide residues from agricultural activities. Fish and wildlife may be directly exposed, while humans can be affected through the consumption of contaminated water and marine products (Pironti et al., 2021). Investigating the phytotoxic effects of EDCs is crucial for understanding their broader ecological and agricultural impacts.

These compounds are environmentally persistent, can bioaccumulate through the food chain, and may interfere with plant growth and development, ultimately affecting crop productivity and ecosystem health. Understanding these effects is essential for developing strategies to mitigate their impact and safeguard both natural ecosystems and agricultural systems (Mahdi et al., 2023). Ecologically, EDCs disrupt hormonal balance in wildlife, leading to reproductive and developmental abnormalities (Dosis and Kamarianos, 2007). In agricultural settings, these compounds may impair plant growth and essential physiological processes, such as photosynthesis, through root uptake or atmospheric deposition, negatively impacting both crop yield and quality (Kumari et al., 2024). Moreover, continuous human exposure to EDCs through contaminated food and water is a serious public health concern, as these substances are linked to various endocrine-related disorders (*Endocrine Disrupting Chemicals in Freshwater: Monitoring and Regulating Water Quality*, 2023). Although the harmful effects of EDCs on wildlife and human health are well documented, there is still a pressing need for more comprehensive research into their long-term ecological impacts, particularly on plants. Investigating the phytotoxic effects of EDCs is essential for a comprehensive understanding of their impact on ecosystems and agriculture, and it can inform the development of effective regulatory measures, public health policies, and phytoremediation strategies aimed at mitigating their environmental persistence and toxicity (Mahdi et al., 2023; Pridemore, 2021).

As shown in Fig. 11.2, EDCs can be transported to plants through water, soil, and air, leading to their bioaccumulation in plant tissues. Such exposure can disrupt plant growth at both macroscopic and microscopic levels, affecting critical processes such as seed germination, photosynthesis, and nutrient uptake (Mahdi et al., 2023). The disruption of phytoestrogen signaling by EDCs may interfere with essential symbiotic relationships, such as nitrogen fixation in leguminous plants, which are vital for soil fertility and plant health (Chemicals, 1993). EDCs can also alter soil microbial activity and modify the soil's physicochemical properties, triggering cascading effects on plant health and agricultural productivity (Chaturvedi et al., 2023).

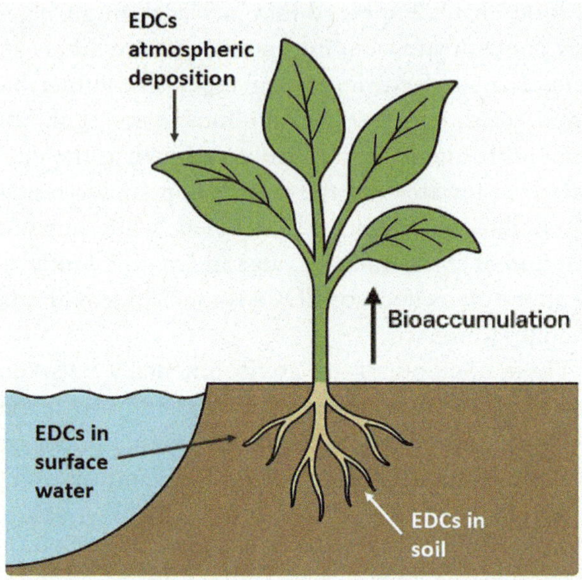

Figure 11.2 EDCs transport to plants through water, soil, and air, leading to their bioaccumulation in plant tissues.

The presence of EDCs in agricultural soils, for example, can lead to reduced crop yields and pose a threat to food safety due to the accumulation of these compounds in edible plant parts (Asenuga and Olagunju, 2023; Dosis and Kamarianos, 2007). Thus, continuous research is required to document the adverse effects of EDCs and to develop regulatory policies aimed at maintaining ecosystem health (Asenuga and Olagunju, 2023). Remediation strategies, including phytoremediation and microbial remediation, are currently being explored to mitigate the environmental impact of EDCs, emphasizing environmentally friendly and cost-effective approaches (Chaturvedi et al., 2023). Although much attention is often given to the direct effects of EDCs on human and animal health, their impact on plant systems and ecosystems is equally significant. Understanding these phytotoxic effects is vital for developing comprehensive environmental protection strategies and ensuring sustainable agricultural practices. In light of their widespread occurrence and ecological relevance, the following sections will explore the main classes of endocrine-disrupting chemicals, their phytotoxic mechanisms, and current phytoremediation strategies aimed at mitigating their environmental impact.

2. Main classes of EDCs and their interactions with plants

2.1 Pharmaceuticals and PCPs, and hormones

Pharmaceuticals, Personal Care Products (PCPs), antibiotics, and hormones (synthetic or natural) present in the environment can negatively impact plant growth and development. These compounds, often released through wastewater or contaminated irrigation water, can be absorbed by plants, leading to various adverse effects like inhibited germination, reduced biomass, and even altered plant physiology. The extent of the impact depends on the specific target compound, its concentration, and the plant species. In the next sections, we will detail the impacts of some of the main classes of these compounds.

2.1.1 Pharmaceuticals and PCPs

Pharmaceuticals and personal care products (PCPs) include analgesics and anti-inflammatories, anti-diabetics, anti-epileptics, anti-estrogenics, anti-protozoals, antiseptics, lipid regulators, diuretics, medications for treating erectile dysfunction and pulmonary arterial hypertension, psychiatric drugs and antidepressants, psycho-stimulants, veterinary and human antibiotics, β-blockers, X-ray and contrast media, cosmetics and personal care products, surfactants, and phytosanitary products. Table 11.1 presents some examples of pharmaceuticals and personal care products (PCPs) commonly detected in the influent and effluent of wastewater treatment plants.

Pharmaceuticals comprise a diverse group of biologically active substances widely used in both human and veterinary medicine. Beyond their conventional therapeutic applications, certain pharmaceutical compounds are also employed in agriculture, including their use in animal feed to promote growth and, more recently, in formulations aimed at enhancing plant development. To consolidate and evaluate current knowledge on the occurrence and dynamics of pharmaceuticals in the soil–plant system, a comprehensive literature review was conducted. These substances, along with their transformation products (metabolites), typically enter the terrestrial environment through several anthropogenic pathways (Pawłowska and Biczak, 2024). One of the most common routes involves the agricultural application of treated wastewater (effluents) and sewage sludge (biosolids), which often contain residual pharmaceuticals originating from domestic households, healthcare institutions, and industrial sources. An equally significant but sometimes overlooked pathway is the use of organic fertilizers, such as animal manure and slurry, which can be heavily contaminated with veterinary drugs administered to livestock (Petrie et al., 2015).

Once introduced into the soil environment, the fate, transport, and bioavailability of pharmaceutical compounds are influenced by multiple interacting factors. These include the physicochemical properties of the soil, such as pH, texture, organic matter content, and microbial activity, as well as the chemical nature of the pharmaceutical itself, including its solubility, persistence, and tendency to adsorb to soil particles. These variables collectively determine the degree to which pharmaceuticals remain in the soil, degrade over time, or become available for plant uptake (Pawłowska and Biczak, 2024).

Recognition that pharmaceuticals are released into, and occur in, soil environments has raised concerns over the potential for plants to take up these compounds (Wu et al.,

Table 11.1 Examples of pharmaceuticals and personal care products (PCPs) found in the influent and effluent samples from wastewater treatment plants.

Family of contaminant of emergent concern	Examples
Analgesics and anti-inflammatory	Codein, diclofenac, fenoprofen, ibuprofen, indomethacine, ketoprofen, ketorolac, paracetamol, phenylbutazone, naproxen, clofibric acid
Anti-diabetics	Metformin
Anti-epileptics	Carbamazepine, 4-aminoantipyrine, antipyrin, codein, diclofenac
Anti-estrogenics	Tamoxifen
Anti-histaminics	Dephenhyldramine
Anti-protozoals	Quinacrine dihydrochloride
Antiseptics	Triclosan, chlorophene
Lipid regulators	Acebutolol, atenolol, atorvastatin, bezafibrate, fenofibric acid, gemfibrozil
Diuretics	Furosemide, hydrochlorothiazide, amidotrizoic acid, diatrizoate, iotalamic acid
Medication used to treat erectile dysfunction and pulmonary arterial hypertension	Sildenafil
Psychiatric drugs and antidepressants	Diazepan, fluoxetin
Psycho-stimulants	Caffeine, paraxanthin
β-blockers	Celiprolol, metoprolol, propanolol, sotalol, timolol
X-ray and contrast media	Iopromide, iopamidol, iohexol, diatrizoate
Cosmetics and personal care products	Benzophenone, galaxolide, N,N-diethyltoluamide, tonalide, triclosan, triclocarban
Surfactants	Tergitol
Phytosanitary products	Clofibric acid

Modified from Petrie, B., Barden, R. & Kasprzyk-Hordern, B. (2015). Occurrence, Fate, Removal and Assessment of Emerging Contaminants in Water in the Water Cycle (from Wastewater to Drinking Water), 72, 3–27. https://doi.org/10.1016/j.watres.2014.08.053

2012; Pullagurala et al., 2018). Several investigators have demonstrated the ability of plants to take up pharmaceuticals from their surroundings (Boxall et al., 2006; Wu et al., 2012). The observed uptake of pharmaceuticals into plants has raised concerns over potential impacts on human health (Boxall et al., 2006; Duarte et al., 2022).

Studies have demonstrated that plants are capable of absorbing pharmaceutical compounds through their roots, which can then be translocated and accumulated in different tissues. The distribution of pharmaceuticals within the plant tends to follow a general pattern, with the highest concentrations typically detected in vegetative tissues. Accumulation tends to decrease in the following order: roots > leaves > stems, while the lowest levels are usually observed in reproductive organs, such as cereal grains. This distribution pattern has important implications for food safety and environmental risk assessment, especially when edible plant parts are involved(Pullagurala et al., 2018).

Non-steroidal anti-inflammatory drugs (NSAIDs) like diclofenac and ibuprofen are frequently detected in environmental matrices. Studies have shown that diclofenac can inhibit seed germination and root elongation in plants such as *Vicia faba* and *Lactuca sativa* (Kötke et al., 2024). Diclofenac's phytotoxicity is attributed to its ability to induce oxidative stress, resulting in cellular damage (Keb et al., 2022). Antidepressants like fluoxetine (Prozac) have been detected in surface waters and can be taken up by plants. Research indicates that exposure to fluoxetine can impair seed germination and reduce biomass in *Lactuca sativa* and *Allium cepa* (Bártíková et al., 2016). The mechanism involves disruption of cellular oxidative balance and interference with root cell division (Moreira et al., 2022).

To evaluate the effects of pharmaceuticals on plant health and development, researchers commonly follow standardized protocols established by the Organisation for Economic Co-operation and Development (OECD). While these guidelines provide a useful framework, typically focusing on parameters such as germination rate, biomass production, and visible phytotoxic effects, they are inherently limited in scope. This is because OECD standards emphasize acute, growth-related endpoints and often overlook subtle physiological and molecular changes that may result from long-term, low-dose exposures. Numerous studies have highlighted that chronic exposure to trace levels of pharmaceuticals in the environment can lead to significant effects that are not captured by short-term growth assessments. These effects often stem from toxicodynamic mechanisms distinct from those observed in acute toxicity studies. For instance, while many pharmaceuticals exhibit low acute toxicity in aquatic and terrestrial ecosystems, their chronic influence can manifest through specific pharmacological interactions, particularly at the cellular and molecular levels (Khetan and Collins, 2007).

To achieve a comprehensive understanding of the environmental impact of pharmaceuticals, it is essential to move beyond conventional growth metrics and adopt a more holistic, multi-level approach (Pawłowska and Biczak, 2024). Experimental evidence indicates that the influence of these compounds on plants extends beyond biomass reduction or yield inhibition. Exposure to pharmaceutical residues and other xenobiotics can trigger oxidative stress, impair photosynthetic processes, and affect key biochemical pathways (Alkimin et al., 2019). Furthermore, some xenobiotics have been shown to alter gene expression, disrupt signaling pathways, and interfere with hormone receptor activity and hormonal balance in plants (Kötke et al., 2024; Bártíková et al., 2016). These disruptions can lead to complex physiological changes, many of which remain undetectable under standard testing protocols but may have long-term implications for plant health and ecosystem function.

Similarly, PCPs, such as cosmetics, soaps, lotions, sunscreens, fragrances, and synthetic musks, are increasingly recognized as environmental contaminants due to their widespread use and persistence in wastewater streams. Once they enter the environment through treated wastewater or the application of biosolids and reclaimed water in agriculture, these compounds may reach the soil–plant system, raising concerns about their potential effects on plant health and food safety. Plants can absorb PCPs from contaminated soils or irrigation water. Compounds such as triclosan, parabens, and UV filters have been detected in root, stem, and leaf tissues (Fernandes et al., 2024). The extent of uptake depends on the chemical properties of the compound (e.g., hydrophobicity, molecular size), plant species, and soil characteristics (Wei et al., 2023).

Some PCPs can interfere with plant development. For instance, exposure to triclosan or benzophenone compounds has been associated with reduced root elongation, altered biomass distribution, and changes in germination rates. However, these effects vary by concentration and duration of exposure (Fernandes et al., 2024). Additionally, PCPs can act as xenobiotic stressors, triggering oxidative stress in plant cells. This is characterized by increased levels of reactive oxygen species (ROS), which can damage cellular components unless counteracted by antioxidant defenses. Studies have shown that compounds like triclosan and methylparaben can alter the activity of enzymes such as superoxide dismutase (SOD) and catalase (CAT), indicating stress responses (Jiang et al., 2024).

PCPs have been found to reduce chlorophyll content and inhibit photosynthetic efficiency, which can impair energy production and growth. For example, sunscreen agents like benzophenone-3 can disrupt light absorption and energy transfer in the chloroplasts, even in higher plants (Cunha Barros et al., 2023). Some PCPs molecules mimic or block plant hormones due to their structural similarity to endogenous molecules. These endocrine-disrupting properties can interfere with signal transduction pathways, affecting processes like cell elongation, flowering, and stress adaptation (Fernandes et al., 2024).

Thus, the phytotoxic effects of these pharmaceuticals and PCPs are primarily linked to oxidative stress, characterized by the overproduction of reactive oxygen species (ROS), which damages cellular components such as lipids, proteins, and DNA (Pawłowska and Biczak, 2024). Additionally, some compounds interfere with cell division and elongation, disrupting normal growth patterns. The accumulation of pharmaceuticals in soil can impair seed germination, reduce biomass, and alter plant reproductive success, ultimately affecting plant community dynamics and ecosystem health (Kötke et al., 2024). These impacts highlight the need for improved wastewater treatment and environmental monitoring.

2.1.2 Antibiotics

Although antibiotics are primarily designed to target bacterial pathogens, their presence in environmental matrices has raised concerns regarding their non-target effects on terrestrial plants. Antibiotics enter ecosystems through wastewater discharge, agricultural runoff, and biosolids application, leading to their accumulation in soil and water bodies. Given their biological activity, antibiotics can influence plant physiological processes, growth, and development, potentially disrupting plant health and soil microbial communities. In agriculture, antibiotics are commonly added to animal feed as supplements to promote the growth of food animals. However, absorption of antibiotics in the animal gut is not complete, and as a result, substantial amounts of antibiotics are excreted in urine and feces that end up in manure. Manure is used worldwide not only as a source of plant nutrients but also as a source of organic matter to improve soil quality, especially in organic and sustainable agriculture (Kumar et al., 2005).

Research has demonstrated that certain antibiotics, such as tetracyclines, sulfonamides, and chloramphenicol, can exert phytotoxic effects, including inhibited seed germination, reduced seedling growth, and morphological abnormalities. Tetracycline and its derivatives, for example, are among the most frequently detected antibiotics in the environment. Studies indicate that tetracycline can inhibit seed germination and root elongation in various plant species such as *Lactuca sativa* and *Vicia faba* (Pan and Chu, 2016; Narciso et al., 2023). The mechanisms involve interference with cellular protein synthesis and disruption of mitochondrial function, leading to energy deficits and oxidative stress (Narciso et al., 2023). Sulfamethoxazole, a common sulfonamide, has been shown to impair seed germination and seedling vigor due to its antimicrobial activity disrupting soil microbial communities that are vital for nutrient cycling, indirectly affecting plant growth (Reichel et al., 2014). Additionally, direct phytotoxic effects include membrane damage and oxidative stress in plant tissues.

Mechanisms of Antibiotic-Induced phytotoxicity can affect plants through multiple pathways:

Disruption of Mitochondrial and Chloroplast Function: Given the evolutionary similarity between bacteria and organelles like mitochondria and chloroplasts, antibiotics such as tetracyclines and chloramphenicol can inhibit organellar protein synthesis, impairing photosynthesis and respiration (Wang et al., 2006).

Induction of Oxidative Stress: Antibiotics can generate reactive oxygen species (ROS), leading to oxidative damage of lipids, proteins, and nucleic acids. Elevated ROS levels can inhibit seed germination and seedling growth (Krupka and and Michalczyk, 2022).

Alteration of Soil Microbial Communities: Antibiotics can suppress beneficial soil microbes involved in nutrient mineralization and root symbiosis, indirectly affecting plant health and nutrient acquisition (Cycoń et al., 2019).

The accumulation of antibiotics in soil can lead to reduced plant vigor, compromised reproductive capacity, and altered seed germination rates (Cycoń et al., 2019). Moreover, the persistence of antibiotics in soil may contribute to the development of antibiotic-resistant bacteria, posing additional ecological risks. The disruption of soil microbial communities can also impair soil fertility, further impacting plant growth and ecosystem stability.

2.1.3 Hormones

Steroid hormones, including natural estrogens (such as estrone [E1], 17β-estradiol [E2], and estriol [E3]) and synthetic estrogens (like 17α-ethinylestradiol [EE2]), are increasingly recognized as environmental contaminants with potential effects on plant systems. Steroidal compounds are increasingly being detected in aquatic environments worldwide, driven by their widespread use in human and veterinary medicine, heightened awareness of their ecological relevance, and advancements in analytical detection techniques (Weizel et al., 2018). Among these, certain classes of environmental steroids - particularly synthetic and natural exogenous hormones - have emerged as significant water quality contaminants due to their potent biological activity, even at extremely low concentrations (as little as 1 ng/L), especially in aquatic organisms such as fish (Gunnarsson et al., 2019). Numerous laboratory and in situ studies have confirmed that both natural and synthetic steroid analogs can disrupt the endocrine system of exposed organisms, impairing hormone regulation and leading to adverse physiological and reproductive outcomes (Ojoghoro et al., 2021). Notably, transgenerational effects have been observed, with offspring of exposed parents exhibiting compromised reproductive capacities. The endocrine-disrupting mechanisms of these compounds include inhibition of endogenous steroid biosynthesis, interference with hormone receptor synthesis, mimicking of endogenous hormone activity, and antagonism of normal hormonal signaling (Ojoghoro et al., 2021). Reflecting their high ecological hazard, a recent risk prioritization conducted by the pharmaceutical industry identified synthetic steroidal pharmaceuticals as among the most environmentally hazardous of all pharmaceutical compounds (Gunnarsson et al., 2019).

Though they are primarily studied for their effects on animal and human health, increasing attention is being given to their presence in agroecosystems, mainly due to the land application of livestock manure, sewage sludge, and treated wastewater containing these substances. Once introduced into soils, these hormone-like compounds may persist and become bioavailable to plants, with consequences that span physiological, biochemical, and molecular domains (Kumar et al., 2005). Plants can absorb estrogens through roots, especially when irrigated with contaminated water or cultivated in soils amended with manure or biosolids. Estrogens are relatively hydrophobic and tend to adsorb to soil particles, but under certain conditions (e.g., low organic matter, acidic pH), they may become mobile and bioavailable. Studies have documented their uptake and accumulation in root and shoot tissues, although generally at low concentrations (Chen et al., 2021).

Exposure to estrogens can affect plant growth and development such as: i) **altered germination rates**, either inhibition or stimulation, depending on the concentration and plant species; ii) **root and shoot elongation abnormalities**, including swelling, reduced elongation, or increased branching; and iii) **reduced biomass production**, particularly under chronic exposure scenarios (Chen et al., 2021). In some cases, low concentrations of estrogens have shown hormesis, which is an adaptive response characterized by low-dose stimulation and high-dose inhibition leading to phytotoxicity (Erofeeva, 2022).

Estrogens can cause oxidative stress in plants by enhancing the production of reactive oxygen species (ROS). This results in increased levels of malondialdehyde (MDA), a marker of lipid peroxidation, altered activities of antioxidant enzymes such as superoxide dismutase (SOD), catalase (CAT), and peroxidases (POD), and disturbance of redox balance, potentially leading to cell damage (Chen et al., 2021). Additionally, hormones can damage signal transduction pathways. Estrogens can mimic or interfere with plant endogenous hormones like auxins, cytokinins, and gibberellins, due to structural similarities. This can lead to dysregulation of cell division and elongation, changes in gene expression related to hormone biosynthesis and signaling, and potential disruption of plant reproductive processes and secondary metabolism (Jaillais and Chory, 2010). Some studies suggest that estrogens may bind to plant hormone receptors or interact with transcription factors, although the exact mechanisms remain under investigation.

The pervasive presence of natural and synthetic steroid hormones in the environment, particularly in aquatic and agricultural systems, poses a complex and significant risk to both ecosystem and plant health. Their ability to disrupt endocrine signaling at trace concentrations underscores the urgency of advancing our understanding of their environmental fate, transport, and biological effects. While the impacts on aquatic organisms are well-documented, emerging evidence suggests that plants, too, may be susceptible to subtle physiological and molecular alterations upon exposure to these contaminants. As such, further interdisciplinary research is essential to fully elucidate the mechanisms of steroidal hormone interference in non-target organisms, including plants, and to inform the development of effective mitigation strategies. This growing body of knowledge is critical not only for protecting environmental health but also for ensuring the safety and sustainability of food production systems.

2.2 Pesticides

2.2.1 Organochlorine pesticides

Organochlorine pesticides (OCPs) are synthetic chemicals that contain chlorine and were used to control pests. They are persistent organic pollutants (POPs) that are toxic and can accumulate in the environment (Jayaraj et al., 2016). Examples of OCPs chemical structures are shown in Fig. 11.3 and include DDT (dichloro-diphenyl-trichloroethane), Dieldrin, Chlordane, Lindane, Endosulfan, Methoxychlor, Hexachlorobenzene (HCB), Mirex, Toxaphene, and Heptachlor. Table 11.2 summarizes some OCPs' names, classes, and modes of action.

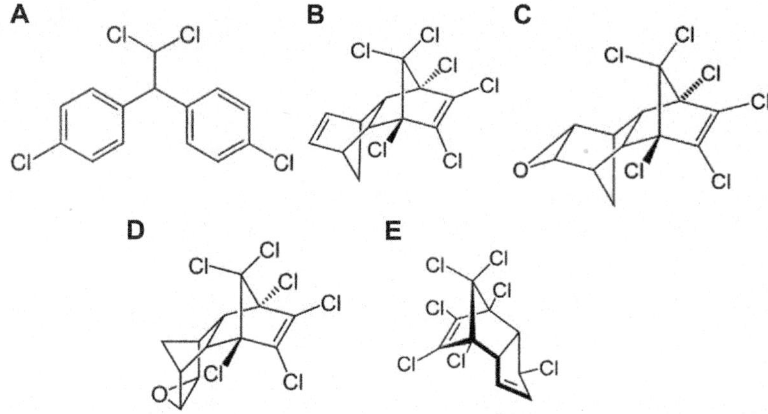

Figure 11.3 Chemical structure of some organochlorine pesticides (OCPs). *((A) From Bethesda (MD): National Library of Medicine (US), National Center for Biotechnology Information; 2004- Available from. (n.d.). NLM PubChem., (B) From Bethesda (MD): National Library of Medicine (US), National Center for Biotechnology Information; 2004- Available from. (n.d.). NLM PubChem., (C) From Bethesda (MD): National Library of Medicine (US), National Center for Biotechnology Information; 2004- Available from. (n.d.). NLM PubChem., (D) From Bethesda (MD): National Library of Medicine (US), National Center for Biotechnology Information; 2004- Available from. (n.d.). NLM PubChem., (E) From Bethesda (MD): National Library of Medicine (US), National Center for Biotechnology Information; 2004- Available from. (n.d.). NLM PubChem.)*

Due to their hydrophobic nature, OCPs tend to adsorb strongly to soil organic matter and clay minerals, reducing their mobility in the soil-water system. This limits direct plant uptake through the roots (Singh and Dileep, 2017). Despite their low bioavailability, OCPs can still induce phytotoxic effects, especially in contaminated soils with high concentrations. Generally, non-polar and undissociated OCPs pass through the cuticle, while the polar compounds enter through the aqueous route. The compounds absorbed by the plant may remain bound to the plant tissue, be stored in the vacuoles of the plant cells, form conjugates with the plant constituents, and also undergo degradation processes in the plant material (Chandra et al., 2021). The detection of OCPs in the leaf samples of paddy (*Oryza sativa*), tomato (*Lycopersicon esculentum*), tapioca (*Manihot esculenta*), and brinjal (*Solanum melongena*) could be attributed to phyllotaxy, wherein the largest portion of a leaf is exposed to the foliar pesticide spray. None of the OCP residues were found in fruit/grain/vegetable samples of paddy, tomato, tapioca, and brinjal; all the OCPs detected were found either in the stem or leaves of the four crop/vegetable plants. Although these plant parts are generally not consumed by humans, the chemical residues may enter the environment through disposal or reach the animals by direct ingestion, as the leaves are a major fodder source for cattle and other livestock (Chandra et al., 2021). Thus, while the hydrophobicity of OCPs limits their translocation in plants, they can still induce phytotoxic effects, especially in highly contaminated environments.

Table 11.2 Main OCPs' names, classes, and modes of action.

Compound	Chemical class	Mode of action
DDT (dichloro-diphenyl-trichloroethane)	Organochlorine Inseticide	Non-systemic stomach and contact action. Sodium channel modulator.
DDD (dichlorodiphenyldichloroethane)	Organochlorine Inseticide, metabolite	Non-systemic stomach and contact action.
Aldrin ((1R,4S,4aS,5S,8R,8aR)-1,2,3,4,10,10-Hexachloro-1,4,4a,5,8,8a-hexahydro-1,4:5,8-dimethanonaphthalene)	Organochlorine Inseticide	Central nervous system stimulant. GABA-gated chlorine channel antagonist. Also stomach and contact toxin.
Dieldrin ((1aR,2R,2aS,3S,6R,6aR,7S,7aS)−3,4,5,6,9,9-hexachloro-1a,2,2a,3,6,6a,7,7a-octahydro-2,7:3,6-dimethanonaphtho[2,3-b] oxirene)	Organochlorine Inseticide	Central nervous system stimulant. GABA-gated chlorine channel antagonist. Contact and stomach action.
Endrin ((1aR,2R,2aR,3R,6S,6aS,7S,7aS)-3,4,5,6,9,9-Hexachloro-1a,2,2a,3,6,6a,7,7a-octahydro-2,7:3,6-dimethanonaphtho[2,3-b] oxirene)	Organochlorine Inseticide, Avicide, Rodenticide	Broad-spectrum, with contact and stomach action. Is a chloride channel-blocking agent.
Heptachlor (1,4,5,6,7,8,8-Heptachloro-3a,4,7,7a-tetrahydro-1H-4,7-methanoindene)	Organochlorine Inseticide	Persistent, non-systemic contact and stomach poison with some fumigation action. Is a chloride channel-blocking agent.

Modified from Jayaraj et al. (2016). Review Article. Organochlorine pesticides, their toxic effects on living organisms and their fate in the environment. Interdisciplinary Toxicology, 9(3–4), 90–100. https://doi.org/10.1515/intox-2016-0012.

Understanding the interaction between OCPs, soil, and plant physiology is essential for assessing their risks and exploring phytoremediation strategies.

2.2.2 Glyphosate

Glyphosate (IUPAC name: N-(phosphonomethyl)glycine), with chemical structure shown in Fig. 11.4) is a herbicide that kills weeds and grasses by inhibiting a plant

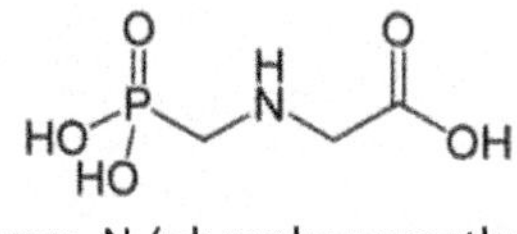

Figure 11.4 Glyphosate (IUPAC name: N-(phosphonomethyl)glycine) chemical structure. *(From Bethesda (MD): National Library of Medicine (US), National Center for Biotechnology Information; 2004-Available from. (n.d.). NLM PubChem.)*

enzyme. It's one of the most widely used herbicides worldwide (NPIC - National Pesticide Information Center, 2024). Glyphosate is applied directly to plant foliage and works as a systemic herbicide, which means it moves from treated leaves to other plant parts, including the roots. It is a non-selective compound, so it can be used to control most weeds, including grasses, sedges, and broadleaves (*NPIC - National Pesticide Information Center. Glyphosate general fact sheet.*, 2024; *NPIC - National Pesticide Information Center. Glyphosate general fact sheet.*, 2024).

The direct application of Glyphosate can be toxic to plants, damaging their growth, development, and photosynthesis. The toxicity depends on the concentration, exposure time, plant species, and environmental conditions (Singh et al., 2020). The exposure of tomato (*Lycopersicon esculentum*) plants to glyphosate led to alterations in leaf water balance regulation, increasing stomatal conductance and decreasing intrinsic water use efficiency (WUEi) at higher concentrations. It also induced slight alterations in the structural integrity of cells, mainly in chloroplasts, accompanied by a loss of cell viability. Moreover, the transcriptional and biochemical control of several photosynthetic-related parameters was reduced upon glyphosate exposure. Glyphosate impacts cellular and subcellular homeostasis (by affecting chloroplast structure, reducing photosynthetic pigments, and inhibiting photosynthetic-related genes transcription), and leaf structure, but does not reduce the carbon flow on a leaf area basis. A trade-off effect in which glyphosate-induced toxicity is compensated by a higher photosynthetic activity related to glyphosate-induced dysfunction in stomatal conductance and an increase in mesophyll thickness/density, allowing the viable leaf cells to maintain their photosynthetic capacity (Soares et al., 2020) Although plant growth and development are significantly compromised by Glyphosate exposure, the observed toxicity in leaf ultrastructure, cell viability, and water use efficiency, as well as the impairments in transcriptional and biochemical regulation of photosynthetic-related components, does not seem to substantially reduce carbon flow through photosynthesis, at least in a short-term exposure (Soares et al., 2020; Cerveira Junior et al., 2020) demonstrated that low doses of glyphosate significantly reduced plant height, stem diameter, number of leaves, leaf area, and shoot dry mass, while glyphosate field doses caused strong inhibition of plant growth in both clones. Additionally, the CO_2 assimilation rate, transpiration rate, and stomatal conductance were also reduced (Cerveira Junior et al., 2020).

2.2.3 Atrazine

Atrazine (IUPAC name 6-Chloro-N2-ethyl-N4-(propan-2-yl)-1,3,5-triazine-2,4-diamine, with chemical structure shown in Fig. 11.5), a selective, post- and pre-emergent systemic triazine herbicide, is one of the most globally used pesticides to control annual weeds and grasses in crops such as corn, cotton, sorghum, sugar cane, and pineapple, as well as in landscaping (Albuquerque et al., 2020).

Atrazine is taken up from roots and transported to the shoots, where it induces phytotoxicity effects that may reduce its further translocation to shoots. *Typha latifolia* is likely able to biodegrade atrazine via different metabolic pathways (Pérez et al., 2022). In *T. latifolia*, the herbicide can be readily absorbed by roots, and is accumulated and biodegraded mainly in leaves, demonstrating the long-lasting phytotoxic effects of atrazine, and its negative correlation with the root uptake and translocation to the shoot. The highest atrazine mass was quantified in leaves, followed by roots, without neglecting the quantities found in sprouts. Atrazine was specifically designed to inhibit the photosynthesis process by blocking the electron transport chain at the photosystem II level. As shown in Fig. 11.6, this inhibition occurs because atrazine binds to the plastoquinone QB binding site (an electron acceptor), thereby interrupting the flow between photosystem II and photosystem I (Albuquerque et al., 2020).

2.3 Industrial compounds

Industrial compounds released into the environment, such as heavy metals, persistent organic pollutants, and synthetic chemicals, can exert significant phytotoxic effects on plants. These substances may interfere with essential physiological processes, including seed germination, root elongation, and photosynthesis. Exposure to toxic industrial compounds often leads to oxidative stress, cellular damage, and disrupted hormonal signaling in plants, ultimately reducing plant growth and productivity. Sources of such contamination include industrial effluents, atmospheric deposition, and the use of contaminated wastewater or sludge in agriculture. Understanding the phytotoxicity of main industrial pollutants is essential for assessing environmental risks and developing strategies for soil and water remediation.

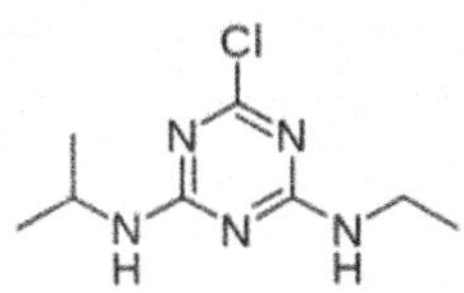

Figure 11.5 Atrazine (IUPAC name 6-chloro-N2-ethyl-N4-(propan-2-yl)-1,3,5-triazine-2,4-diamine) chemical structure. *(From Bethesda (MD): National Library of Medicine (US), National Center for Biotechnology Information; 2004- Available from. (n.d.). NLM PubChem.)*

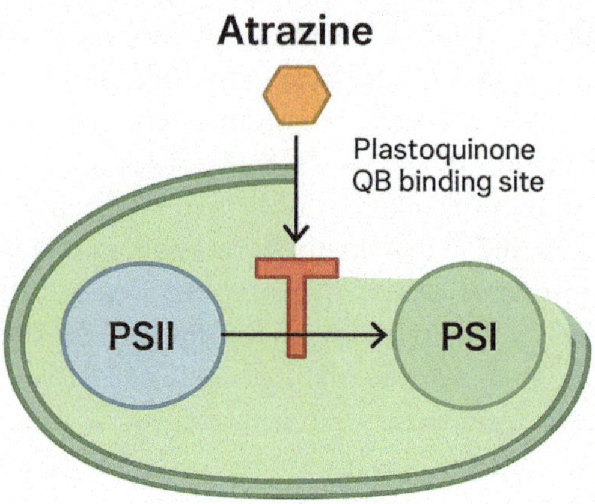

Figure 11.6 Atrazine binds to the plastoquinone QB binding site (an electron acceptor), thereby interrupting the flow between photosystemII (PSII) and photosystem I (PSI).

2.3.1 Per- and polyfluoralkyl substances (PFAS)

Per- and polyfluoroalkyl substances (PFAS) are a group of synthetic chemicals known as "forever chemicals" due to their environmental persistence. As shown in Fig. 11.7, PFAS are found in many places, including chemical production facilities, firefighting foams, food packaging, household goods, stain and water-repellent products, non-stick cookware, paints, and cosmetics (PFAS Fact Sheet).

PFAS are currently detected in numerous plant and animal species, both in terrestrial and aquatic ecosystems, ultimately affecting humans through the food chain (Teunen et al., 2021; Lal et al., 2020). The toxicity of PFAS in plants is well documented, suggesting that the onset of oxidative stress is the primary biological process induced by the presence of perfluooctanesulfonic acid (PFOS) and perfluorooctanoic acid (PFOA), whose chemical structures are shown in Fig. 11.8. This condition is characterized by an imbalance between the production and scavenging of reactive oxygen species (ROS) (Li and Li, 2021; Wang et al., 2022).

The resulting oxidative burst can alter the expression of numerous genes involved in physiological processes such as photosynthesis, hormone signaling, and energy metabolism. This can result in a reduction in growth, which is typically the most visible symptom (Zhang et al., 2022; Li and Li, 2021). Fan et al. (2020) observed a severe growth inhibition in Arabidopsis thaliana seedlings exposed to PFOA. Similarly (Li et al., 2020), demonstrated that co-exposure to PFOA and PFOS markedly affected the metabolite profiles in lettuce (*Lactuca sativa*) leaves at the tested dosages. Metabolic disorder was induced in lettuce under the co-exposure of PFOA and PFOS, which

Figure 11.7 Common sources of PFAS in the environment. *(From Mbanugo, V. S., Ojo, B. S., Lin, T. C., Huang, Y.-W., Locmelis, M. & Han, D. (2025). ACS Phys. Chem Au, 5(2), 117–133. https://doi.org/10.1021/acsphyschemau.4c00092)*

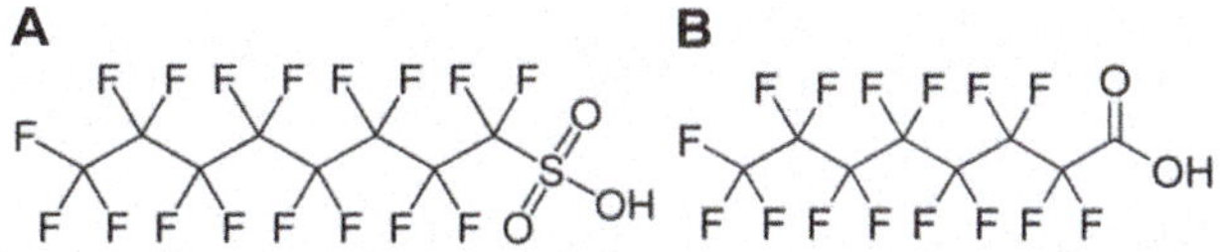

Figure 11.8 Perfluooctanesulfonic acid (PFOS) and perfluorooctanoic acid (PFOA) chemical structures. *((A) From Bethesda (MD): National Library of Medicine (US), National Center for Biotechnology Information; 2004- Available from. (n.d.). NLM PubChem. (B) From Bethesda (MD): National Library of Medicine (US), National Center for Biotechnology Information; 2004- Available from. (n.d.). NLM PubChem.)*

altered the metabolic profiles of both inorganic elements and organic compounds. Amino acids, peptides, fatty acids, lipids, and purine nucleosides were among the major reprogrammed metabolites in response to the stress induced by PFOA and PFOS co-exposure. Alterations in chlorophyll synthesis and signal transduction were observed in plants owing to the down-regulation of magnesium (Mg), iron (Fe), phytol, and abscisic acid levels. Antioxidant defense-related pathways were activated and resulted in increased levels of various phenolic antioxidants in plants (Li et al., 2020). The concerns over the phytotoxicity of PFAS have been increasing in recent years. The over-generation of ROS was reported in several plant species, such as wheat (*Triticum aestivum* L.), lettuce (*Lactuca sativa*), red chicory (*Cichorium intybus* L.), *Juncus effusus*, algae (*Scenedesmus obliquus*), and *Arabidopsis thaliana*, upon PFAS exposure (Gredelj et al., 2020; Li and Li, 2021). Excessive ROS in plants can disrupt some critical biological activities (e.g., seed germination, meristem development, photosynthesis, carbon and nitrogen metabolism, gene expression), and damage biomacromolecules (e.g., DNA, proteins, and lipids) (Choudhary et al., 2020). The accumulation of PFAS in maize plants impaired photosynthesis: the exposed plants had reduced germination and stunted growth compared to non-exposed plants (Ebinezer et al., 2022).

2.3.2 Bisphenol A (BPA)

Bisphenol A (BPA, IUPAC name 4,4'-(propane-2,2-diyl)diphenol, as chemical structure shown in Fig. 11.9) is a chemical substance that is commonly present in products that contain certain types of plastics. These products may include construction materials, electronics, food and drink packaging, biomedical equipment, and toys. In addition, BPA is also used as a base for paint and adhesive products. BPA can affect plants in a variety of ways, including inhibiting root growth and photosynthesis (Xiao et al., 2020).

The presence of BPA in water and soil may represent lethal effects on plant development in the contaminated environments. Abdelmoneim et al. (2024) demonstrated that increasing levels of BPA in the soil intensified its adverse effects on the biomass and photosynthetic pigments of tomato plants. Moreover, BPA induced osmotic stress on tomato plants by reducing soluble sugars and soluble proteins relative to the control. The soil contamination with BPA also led a decline in the levels of essential macro- and micro-elements in the tomato foliar tissues while simultaneously increasing the contents

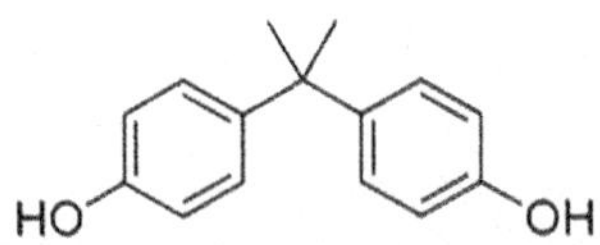

Figure 11.9 Bisphenol A (IUPAC name 4,4'-(propane-2,2-diyl)diphenol) chemical structure. *(From Bethesda (MD): National Library of Medicine (US), National Center for Biotechnology Information; 2004-Available from. (n.d.). NLM PubChem.)*

of non-essential micronutrients. Photosynthetic pigments are essential for photosynthesis and related metabolic processes in plants under environmental stress. Both chlorophyll *a* and chlorophyll *b* were adversely affected by BPA exposure, impairing the carbon assimilation capacity of the leaves. This inhibition of photosynthesis leads to a deficiency in the synthesis and accumulation of organic matter required for root development and, ultimately, the overall growth of tomato plants (Abdelmoneim et al., 2024). BPA affects plant growth and performance in a concentration-dependent manner, with low concentrations exhibiting a stimulatory effect and higher concentrations causing inhibitory effects (Xiao et al., 2020). BPA may affect the photosynthetic system, root activity, and rhizosphere microbial communities (Fu et al., 2022; Tong et al., 2021; Xiao et al., 2020). Wang et al. (2023) demonstrated that BPA's effects on soybean growth indexes occur in a concentration-dependent manner and promoted soybean growth for concentrations below 20 mg/kg (air-dry soil) and also showed that BPA had no significant impact on soybean germination rat. However, BPA treatments affected soybean growth, germination, and root nodule formation. Notably, exposure of soybean to low BPA concentrations (10 mg/kg) resulted in an increase in all growth indexes (except plant height), although these later significantly decreased under higher BPA concentrations (160 and 320 mg/kg). Moreover, soybean germination ratio and time increased with increasing BPA concentrations. On the other hand, the number and biomass of root nodules not only decreased in all treatment groups, but also had a significant negative correlation with BPA concentrations.

Germination time and nodule biomass were negatively and positively correlated with the growth indexes in soybean, respectively. Structural equation modeling results revealed that BPA reduced soybean total biomass and yield by prolonging germination time and inhibiting soybean nodule formation. Given the widespread presence of BPA in the environment, results provide invaluable insights to further understanding into the mechanisms underlying BPA's effect on crop growth (Wang et al., 2023). BPA inhibits primary root growth by inhibiting cell division and elongation through auxin accumulation in the root meristem zone (MZ), transition zone (TZ), and elongation zone (EZ) (Bahmani et al., 2020). BPA-induced root growth inhibition and auxin accumulation are mediated by ROS and ethylene (Bahmani et al., 2023).

2.3.3 Plastics by-products, microplastics, and nanoplastics

Microplastics (MPs) and nanoplastics (NPs) are by-products of plastic that can be found in the environment, food, and air. MPs are particles that are less than 5 mm in size (as shown in Fig. 11.10). They can be as small as 1 nm, which is much smaller than the width of a human hair. They can be harmful to wildlife, ecosystems, and human health. MPs and NPs negatively affected the physiological indicators of plants. The effect of MPs/NPs on plant photosynthesis was mostly negative, but had positive effects on

Figure 11.10 Macro shot of the composition of microplastics inside a laboratory Petri dish. *(From Hristov, Svetlozar. (2024).)*

the biochemical indicators of plants. The phytotoxic of MPs/NPs are related to their type, concentration and particle size (Wang et al., 2023).

Various plastics polymers commonly found in both marine and terrestrial ecosystems are polyethylene (PE), polypropylene (PP), low density polyethylene (LDPE), high density polyethylene (HDPE), polystyrene (PS), polyvinyl chloride (PVC), polyethylene terephthalate (PET), polyoxymethylene (POM), polybutylene adipate terephthalate (PBAT) and poly lactic acid (PLA). These polymers are increasingly being detected in agricultural soils and aquatic systems, with growing concern over their phytotoxic effects on crops. Polyethylene microplastics (PE-MPs) may hinder the absorption of essential mineral ions in rice plants (Tang et al., 2022). Polystyrene microplastics (PS-MPs) can accumulate on the root surface and root caps externally (Taylor et al., 2020). An increase in the water use efficiency and a general reduction in the transpiration rate are also a plant strategies and an indication of the toxicity of microplastics (MPs), as observed in maize plants (Urbina et al., 2020). Rice plants (*Oryza sativa*) exhibited dose and polymer-type-dependent responses when exposed to PS and PVC microplastics (PVC-MPs). High concentrations of both MPs (i.e., 3 mg/L) were injurious to plant growth, and PVC-MPs were more lethal than PS-MPs. The toxicity of both MPs was associated with the decline in growth, nutrient content, perturbed gas exchange attributes, and enhanced oxidative damage. Both types of MPs adversely impacted the growth and height of rice plants (Ma et al., 2022). Ma et al. (2022) also explored the potential effects of PS-MPs and PVC-MPs on the gas exchange characteristics of rice plants. Reductions in photosynthesis and stomatal conductance under PS-MPs and PVC-MPs treatments indicated the stressful situation, which may have arisen due to the MPs that the rice plants had to deal with. Although the hazards of both types of MPs were recorded in rice plants, differences in damage to attributes were evident. Microplastics also affect root architecture. Root hairs were particularly impacted, potentially limiting nutrient and water uptake capacity. Similarly, sheath formation around maize roots contributed to reduced root permeability and plant water conductivity (Ma et al., 2022; Urbina et al.,

2020). MPs decreased the uptake of phenanthrene (Phe) in soybean roots and leaves by damaging the roots. Micron-size MPs inhibited the soybean root activity and decreased the relative abundance of rhizosphere microbial community, leading to reduced uptake of Phe in soybean roots. The co-exposure of micron-sized MPs and Phe showed a higher genotoxicity compared with nano-sized MPs and Phe. MPs enhanced the toxic effects of Phe at cellular and molecular levels, and modulated the kinetics and toxicity of Phe in soybean plants. This study reveals the size effects of MPs on uptake of organic pollutants by higher plants and provides insights into their combined toxicity to plants (Xu et al., 2021). MPs could be adsorbed on the root surface of plants and reduce the water conductivity in roots, which hinders the transpiration, nutrient uptake, and growth of plants (Urbina et al., 2020). Nano-size (40 nm) and micron-size (1 μm) PS-MPs can accumulate at the root surface of wheat, especially at the root tip (Taylor et al., 2020). Crop-specific responses to exposure to NPs-PS and MPs-PS at different doses were dependent on the plant species. Italian lettuce (*Lactuca sativa*) was the most sensitive crop in terms of seed germination, with germination index reductions ranging from 18.2% to 36.0% after 7 days of exposure compared with the control system. The root growth (root dry weight, root/shoot ratio, and root length) of Italian lettuce and corn was significantly inhibited by the exposure treatment. Analyses of antioxidant enzymatic activities, lipid peroxidation, and integrated biomarker indexes confirmed that the toxic effects of nano-PS and micro-PS on crops are likely due to oxidative stress. The observed distribution of fluorescent nano-PS in the roots or germs of the tested crops suggests that nano plastics can be taken up by plants even at a very early growth stage (Gong et al., 2021).

3. Mechanisms of absorption, translocation, and metabolism of EDCs by plants

In plants, EDCs undergo metabolism through a series of biochemical transformations collectively known as xenobiotic metabolism, which include absorption, translocation, and the metabolism itself. Thus, the transformed EDCs may be sequestered in vacuoles or bound to the cell wall, limiting their mobility, preventing further transport, and reducing toxicity (Vicidomini et al., 2024).

3.1 Absorption of EDCs

Plants can absorb a variety of compounds, including water, nutrients, and organic compounds, such as EDCs. Most of the contaminants present in the plants surroundings are accumulated by roots, followed by leaves (Zhang et al., 2017), as shown in Fig. 11.11. Uptake through leaves mostly happens after intensive spraying of compounds such as agrochemicals (Tripathi et al., 2019).

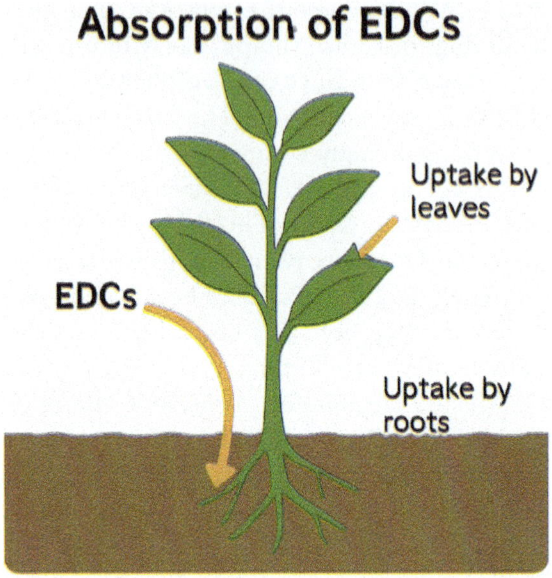

Figure 11.11 Main uptake processes of EDCs by plants.

Generally, contaminants get access to roots through passive diffusion, finally finding their way into the xylem stream (Tripathi et al., 2019). As plant systems are devoid of specifically designed transporters for the movement of xenobiotics, the plant system's transport rate is greatly influenced by their physical and chemical properties. As the process of transport is energy independent (passive) and is of a physical nature, the transport rate may be evaluated, and typical models for movement can be developed (Fujisawa et al., 2002). Passive absorption is a process that allows pollutants to enter plant cells without using energy (Zhang et al., 2017). It occurs when pollutants move from a higher concentration to a lower concentration, which is known as passive diffusion. EDCs with low molecular weight and moderate hydrophobicity can passively diffuse across root membranes (Kumar and Gupta, 2016). Some EDCs share structural similarities with plant hormones, allowing them to be accidently transported via aquaporins, nutrient transporters, or ion channels. Aquaporins (AQPs), for instance, belong to the class of major intrinsic proteins (MIPs), and have been defined as integral membrane proteins that are expressed generally in all living cells (Maurel and Chrispeels, 2001). As shown in Fig. 11.12, these proteins act as channels in cell membranes to allow water and other molecules to pass through.

The literature has not extensively addressed the potential role of AQPs as gateways facilitating the uptake of EDCs into plants. Wen et al. (2024) reported that the presence of proteins and lipids in plants influences the accumulation and distribution of PFOS and PFOA. However, the authors noted that the underlying mechanisms behind this effect remain unproven. Hacke and Laur (2016) suggest that AQP pore size determines the

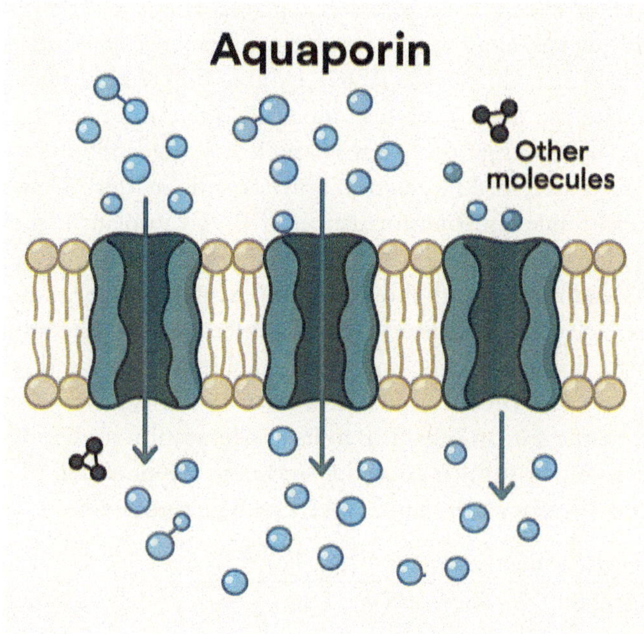

Figure 11.12 Aquaporins proteins and their water transport channels in the cell membrane.

permeability of specific molecules, regulating both their passage and transport rate through these channels. Consequently, some EDCs may be absorbed, translocated, and distributed via AQPs with pore diameters that match the compounds' length, potentially alongside other small molecules.

3.2 EDCs plant translocation and bioaccumulation

The capacity of plants to translocate contaminants from roots to shoots can be quantitatively evaluated by the translocation factor (TF), which is defined as the ratio between the contaminant concentration in aerial parts and the concentration in roots (Wang et al., 2021), according to equation:

$$TF = \frac{C_{aerial}}{C_{root}}$$

(Antoniadis et al., 2017) reported that the translocation factor (TF) reflects the plant's resistance capability, limiting the accumulation of pollutants in the roots while preventing contaminant transfer to edible aerial parts, particularly seeds. Generally, plants exhibit a TF lower than one under stress conditions. A TF greater than one indicates that the plant not only tolerates the contaminant but also utilizes the pollutant beneficially, which is a common characteristic of hyperaccumulators. Therefore, a TF greater than one is a key criterion for classifying plant species for phytoremediation (Antoniadis et al., 2017;

Chanu and Gupta, 2016). The TF is widely used to assess a plant's ability to transfer metals from roots to aerial parts. However, applying this metric to organic contaminants presents significant challenges (Zhang et al., 2017). EDCs exhibit different physicochemical behaviors in soil-plant systems. Their mobility, bioavailability, and plant uptake mechanisms vary greatly depending on factors such as hydrophobicity (Dettenmaier et al., 2009), molecular size, and interaction with organic matter (Zhang et al., 2017). Additionally, the persistence and transformation of EDCs in plant tissues further complicate the interpretation of TF values. Some compounds may undergo biotransformation or degradation, leading to metabolite formation that alters their translocation potential. Moreover, the lack of standardized analytical methods for detecting these contaminants in different plant compartments limits the reliability of TF-based assessments (Limer and Burken, 2014). The Bioaccumulation Factor (BAF) is a ratio that compares the concentration of a chemical in an organism to its concentration in the surrounding environment (typically water or soil). It is used to evaluate the potential of a substance to accumulate in biological tissues over time. BAF can be expressed by:

$$BAF = \frac{Concentration\ in\ organism\ (mg/kg)}{Concentration\ in\ environment\ (mg/L\ or\ mg/kg)}$$

A higher BAF indicates a greater tendency of the chemical to bioaccumulate, which may lead to toxic effects in organisms and biomagnification through the food chain. This can lead to direct toxic effects (e.g., cell damage, disruption of photosynthesis, endocrine interference). Certain plants with high BAFs (such as *Brassica juncea* and *Typha latifolia*) can accumulate significant amounts of contaminants in their tissues. These plants are intentionally used in phytoremediation, a green technology for cleaning up contaminated soils, water, or air. However, when used purposefully in phytoremediation, and properly managed - with controlled growth and disposal - they become powerful tools for reducing environmental contamination. Table 11.3 shows the classification of Translocation Factor and Bioaccumulation Factor (Chavan et al., 2025).

Numerous studies have provided substantial evidence to indicate that pharmaceuticals and personal care products (PPCPs) enter plants through the root hairs or stomata

Table 11.3 Classification of translocation factor (TF) and bioaccumulation factor (BAF) (Chavan et al., 2025).

	Category/Interpretation
Translocation factor <1	Low mobility
Translocation factor >1	High mobility
Bioaccumulation factor ≤ 1	Non-hyperaccumulator/non-phytoremediator
Bioaccumulation factor >1	Hyperaccumulator/Phytoremediator
Bioaccumulation factor $= 1$	Accumulator

Data from Assessing metal bioaccumulation in aquatic environments: The inverse relationship between bioaccumulation factors, trophic transfer factors and exposure concentration. (n.d.).

and cuticle on the leaf surface by passive diffusion and foliar uptake. Typically, root-to-shoot translocation of PPCPs via xylem passes through the cortex, endodermis, and Casparian strip. The uptake of PPCPs involves complex interactions among biological, chemical, and physical factors and processes (Hua Wei et al., 2023). The uptake and accumulation of PPCPs in plant tissues involve intricate biological, chemical, and physical interactions, influenced by compound properties (e.g., hydrophobicity, molecular size), soil characteristics, and plant species. Pérez et al. (2022) indicated that atrazine is taken up by plant roots, translocated into the shoot, and mainly accumulated in leaves and sprouts, independently of the plant growth habits. The accumulation of 17β-estradiol (17β-E2) in wheat roots mainly resulted from adsorption and active uptake that involved aquaporins and anion channel transport. The chlorophyll and protein contents of plants were positively correlated with the uptake of 17β-E2, whereas competitive inhibition occurred when 17β-E2 and 17α-ethynylestradiol (EE2) coexisted in the same solution (Chen et al., 2021). These findings underscore the potential risk of PPCPs bioaccumulating in edible plant parts, posing threats to ecological integrity and human health via the food chain. Moreover, understanding plant uptake dynamics is essential for developing phytoremediation strategies and for predicting the behavior of emerging pollutants in agroecosystems.

3.3 Metabolism

Once inside plant cells, EDCs undergo metabolic transformations to reduce their toxicity and facilitate detoxification and storage. A series of biochemical transformations, known as xenobiotic metabolism, occurs in three distinct phases, as shown in Fig. 11.13.

3.3.1 Phase I: xenobiotic oxidation, reduction, or hydrolysis reactions

This step is primarily catalyzed by cytochrome P450 monooxygenases (P450) to enhance the compound's reactivity and solubility (Fan and Cong, 2024). Enzymes of the P450 superfamily are responsible for the Phase I metabolism of numerous herbicides representing several classes of organic compounds, as described by Zimmerlin and Durst (1992) for wheat (*Triticum aestivum*). BPA, for example, can be hydroxylated to form more water-soluble derivatives, which are then processed in Phase II metabolism (Tian et al., 2014).

3.3.2 Phase II: Anabolic conjugation process

Phase II involves conjugation of Phase I metabolites with endogenous molecules such as glutathione, sugars, or amino acids, facilitated by transferases like glutathione S-transferases (GSTs). These modified compounds are often less biologically active and can no longer interfere with plant hormone pathways. As a result, this process enhances solubility, facilitating detoxification and subsequent excretion (Stanley, 2017). Electrophilic xenobiotics tend to spontaneously react with nucleophilic sites in

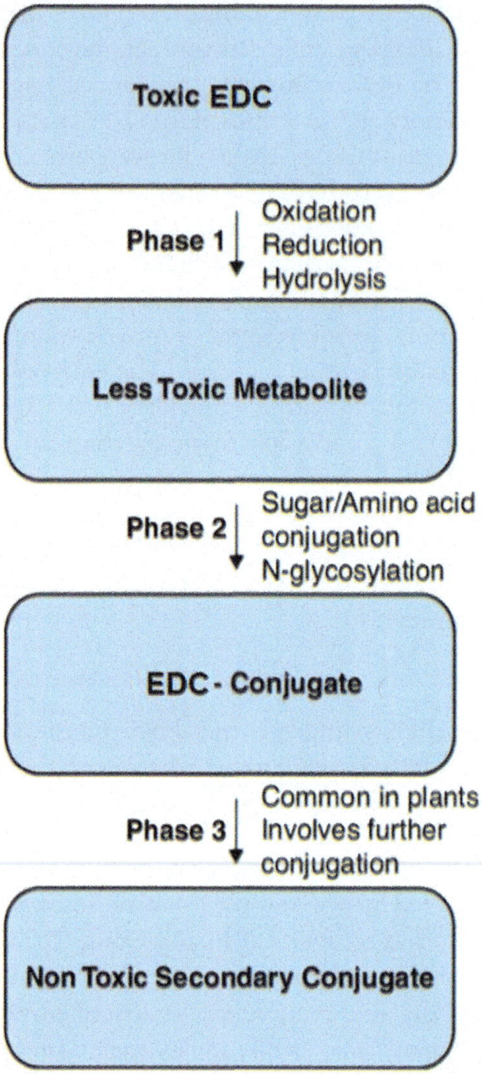

Figure 11.13 Biochemical transformation of EDCs along plants' metabolism. *(Modified from Shweta, N., Samatha, S. & Sahu, K. (2021). (pp. 73–99). https://doi.org/10.1016/C2019-0-04695-X-1)*

biomolecules, such as proteins and nucleic acids. The affinity of these reactions can be explained by the concept of hard and soft nucleophiles/electrophiles, where the sulfur in glutathione acts as a soft nucleophile, interacting with a wide range of xenobiotics (Stanley, 2017). Glutathione conjugation transforms hydrophobic compounds into more hydrophilic products, reducing their mobility and preventing uncontrolled diffusion across biomembranes (Vazquéz-Meza et al., 2023). This process decreases the biological half-life of xenobiotics, minimizes cellular exposure to toxins, and prevents their accumulation in critical tissues. However, amphiphilic metabolites derived from conjugation can inhibit cytoplasmic enzymes, justifying the strict regulation of glutathione conjugate concentrations in the cytosol (van Bladeren, 2000). In addition to glutathione conjugation, sugar transfer also plays a fundamental role in xenobiotic metabolism in plants. Reactions catalyzed by glycosyltransferases result in the formation of glycosides, which exhibit higher hydrophobicity and lower reactivity, facilitating the compartmentalization and excretion of compounds (Andreau et al., 2023). Glucose conjugation can occur directly on the xenobiotic molecule or after modifications catalyzed by P-450 monooxygenases, serving as an essential process for detoxification and storage of secondary metabolites in plants (Guengerich, 2026).

3.3.3 Phase III: transport of conjugated xenobiotics into vacuoles or extracellular compartments

Phase III involves the transport of conjugated xenobiotics from the cytosol to vacuoles or extracellular compartments. This process is primarily mediated by ATP-binding cassette (ABC) transporters, which actively sequester the conjugated molecules away from essential cellular components. By compartmentalizing these modified xenobiotics, plants effectively reduce their toxicity and prevent potential interference with cellular functions. This phase plays a critical role in ensuring the safe storage or excretion of xenobiotics, contributing to the overall detoxification process (Wu et al., 2019). The specific conjugated xenobiotics transported by most ABC transporters remain uncertain or controversial. This ambiguity arises partly because certain ABC transporters can transport a wide range of structurally diverse substrates. Therefore, identifying a single substrate does not provide a complete understanding of their function. Additionally, few studies have employed direct methods to determine the substrates of plant ABC transporters. Addressing this gap requires transport assays using isolated cells, membrane vesicles, or reconstituted liposomes, which are essential but technically demanding. Over the past decades, significant progress in these assays has been achieved through the efficient expression of proteins in heterologous systems, including plant cells and transgenic plants (Lefèvre and Boutry, 2018).

3.3.3.1 Plant secondary metabolism

Plant secondary metabolism refers to compounds that are not essential for basic metabolic processes (such as growth and reproduction) but play critical roles in defense against pathogens, herbivory, and environmental stresses. Examples of secondary metabolites include alkaloids, flavonoids, terpenoids, phenolics, and glycosides. These compounds can act as antioxidants, repellents, growth inhibitors of pathogens, or even signaling agents, which are essential for xenobiotic detoxification (Chakraborty et al., 2023). These compounds contribute to detoxification by modulating enzyme activity, strengthening antioxidant defenses, and directly interacting with xenobiotics to mitigate their toxic effects (Vicidomini et al., 2024). For example, flavonoids inhibit the production of reactive oxygen species (ROS) and enhance the activity of glutathione S-transferase (GST) (Agati et al., 2012), while terpenoids, recognized for their antioxidant properties, shield plant cells from oxidative damage caused by xenobiotics (Vicidomini et al., 2024).

4. Mechanisms of phytotoxicity of EDCs

As previously mentioned, antibiotics, fertilizers, pesticides, microplastics, perfluoroalkyl compounds, hormones, and others are among the major classes of substances currently impacting the environment (Abdulrazaq et al., 2020; Baralla et al., 2021; Steeves et al., 2024; Martín et al., 2022). Although these substances possess distinct chemical and physicochemical properties, studying their interrelations in the context of ecological contamination caused by anthropogenic activities is essential. In this context, the main point is that, due to the characteristics attributed to these classes of compounds as emerging environmental pollutants, it is inferred that they, in some way, exert negative effects on the environment in which they are deposited. Therefore, it is crucial to understand not only their origin and chemical composition but also their mechanisms of interaction with different environmental compartments such as soil, water, and air (Bayabil et al., 2022).

4.1 Changes in plant growth and development: impacts on germination, root growth and leaf development

EDCs pollutants can cause ecological imbalances, affect the health of living organisms, including humans, which is the main reason for initiating studies on these substances, and compromise the quality of natural resources. Moreover, many of these compounds exhibit environmental persistence, bioaccumulation in food chains, and the potential for long-distance transport (Bayabil et al., 2022). As a result, phytotoxicity tests are conducted using plant species to identify the effects of physiological changes in plants caused by oxidative stress induced by these substances. One of the simplest and fastest parameters to assess the phytotoxicity of a given chemical substance or class of compounds is the germination rate and root elongation (Wang et al., 2001). The use of pre-germinated seeds, as recommended by OECD (2006), allows for a more accurate analysis of the inhibitory or stimulating effects caused by contaminants present in the samples. Thus, measuring the length of the radicles becomes a sensitive and reliable parameter for detecting changes in plant development, reflecting the potential toxicity of the evaluated material. This approach is particularly useful in environmental and quality control studies, contributing to the identification of ecological risks associated with waste, effluents, and chemical compounds (OECD, 2006). According to Zucconni et al. (1985), the phytotoxicity of substances or organic waste can be assessed through seed bioassays, using analytical indices such as the Germination Index (GI). This index combines the germination rate with the root length of seedlings exposed to the test substance in comparison to a control, allowing for a more integrated analysis of the impact on plant development. The GI is calculated using the following equation:

$$GI\ (\%) = \frac{G_C \times L_C}{G_S \times L_S}$$

Where: **Gs**: number of seeds germinated in the sample; **Ls**: average root length in the sample (mm or cm); **Gc**: number of seeds germinated in the control; and **Lc**: average root length in the control (mm or cm).

In other words, the values obtained from the GI analysis allow for the determination of the degree of toxicity, ranging from low toxicity, which affects both root growth and the number of germinated seeds, to high toxicity. Thus, GI values below 80% generally indicate low phytotoxicity, while values below 50% point to high toxicity. Table 11.4 shows the relationship between the GI and the degree of phytotoxicity assessed in germination assays (Coelho et al., 2020).

Therefore, due to being a simple, sensitive, and effective method (Wang et al., 2001), this methodology has been widely used in environmental and agricultural studies to assess the effects of various classes of organic pollutant compounds on the seeds of cultivated plants, to understand the impacts that the presence of emerging pollutants may cause.

Seed germination and seedling establishment are fundamental stages in the plant life cycle and directly impact crop productivity. Traits like seedling vigor, although highly desirable in crops such as *Brassica oleracea*, rice, and oilseed rape (*Brassica napus*), are complex and influenced by polygenic, hormonal, genetic, and environmental factors. Among plant hormones, abscisic acid (ABA) and gibberellins (GA) play central antagonistic roles: ABA promotes dormancy, while GA promotes germination. Their balance, regulated through hormone synthesis, signaling, and crosstalk with other hormones like auxin, cytokinins, ethylene, and brassinosteroids, determines germination success (Nguyen et al., 2016). Recent studies highlight the modulatory role of auxin in

Table 11.4 Phytotoxicity classification based on the Germination Index (GI) (Coelho et al., 2020).

Germination index (IG)	Phytotoxicity classification
<30	Highly phytotoxic
30–60	Phytotoxic
60–80	Moderately phytotoxic
80–100	Non-phytotoxic
>100	Promotes germination

ABA-driven dormancy, where high auxin levels enhance ABA signaling via transcription factors like ABI3, reinforcing dormancy. Conversely, disruption of auxin signaling can relieve dormancy. Auxin-responsive genes such as *ARF10*, *GH3.5*, and *GH3.6* are implicated in post-germination repression of growth, especially in species like *Arabidopsis thaliana* (Nguyen et al., 2016). Recent studies have shown that even at low concentrations, EDCs can negatively affect seedling growth and development, highlighting the importance of applying seed bioassays as a complementary tool to traditional chemical analysis (Wang et al., 2001; Pan and Chu, 2016). Thus, the integration of toxicity assays with environmental data contributes to a better assessment of the ecological risks associated with contamination by emerging pollutants. Table 11.5 presents data from several studies that aimed to demonstrate the phytotoxic effects of certain classes of pollutants through seed germination bioassays.

Pan and Chu (2016), for example, conducted phytotoxicological analyses aiming to establish the relationship between the effects caused by veterinary antibiotics and the physiological and morphological responses presented by plant organisms. These analyses sought to understand how the physicochemical properties of these substances—such as solubility, lipophilicity, persistence, and soil adsorption capacity—can influence, directly or indirectly, fundamental processes of plant metabolism, such as seed germination, root growth, and nutrient absorption. In the study, it was observed that exposure to different concentrations of antibiotics resulted in significant inhibition of germination and root elongation in common crops.

Phytotoxic effects vary according to plant species and compound concentration, as phytotoxicity depends not only on the substance itself but also on the specific susceptibility of each plant. Therefore, studies must be conducted to obtain results that can

Table 11.5 Studies on phytotoxicity assays involving classes of compounds characterized as EDCs.

Reference	Pollutant class	Substance	Plant species tested
Pan and Chu (2016)	Antibiotics	Tetracycline, sulfamethazine, norfloxacin, erythromycin, chloramphenicol	*Lactuca sativa*, *Daucus carota*, *Cucumis sativus*, *Lycopersicon esculentum*
Kebrom et al. (2019)	Chemical fertilizers	Milwaukee Organic nitrogen	*Brassica oleracea* L.
Miranda et al. (2022)	Microplastics	–	*Lepidium sativum* and *Sinapis alba*
Zhou et al. (2016)	PFAS	Perfluorooctanoic acid (PFOA)	*Triticum aestivum* L.
Souza et al. (2025)	Pesticides and microplastics	Atrazine and polyethylene	*Lactuca sativa*

model the impacts caused by the introduction of these chemical compounds into the environment (Blok et al., 2019). These findings reinforce the need to consider the ecological impacts of these contaminants, particularly given their increasing deposition in the environment across various environmental matrices.

A common outcome of EDS exposure is the inhibition of seed germination, as observed in *Lactuca sativa* (lettuce) and *Triticum aestivum* (wheat) exposed to estrogens and triclosan, likely due to oxidative stress and disruption of metabolic processes essential for embryo activation (Bhardwaj, 2016). Root development is often more sensitive than shoot growth, with several studies reporting reduced root elongation, thickened root tips, and altered branching patterns in *Zea mays* (maize), *Pisum sativum* (pea), and *Cucumis sativus* (cucumber) exposed to bisphenol A and antibiotics (Bahmani et al., 2020). Such alterations impair the plant's ability to acquire water and nutrients, reducing overall fitness. Leaf development is also negatively affected, with symptoms such as decreased leaf area, chlorosis, necrosis, and altered stomatal density reported in *Arabidopsis thaliana* and *Solanum lycopersicum* (tomato) following exposure to EDS like nonylphenol and triclosan. These effects often occur at environmentally relevant concentrations and can lead to long-term impacts on photosynthesis, transpiration, and biomass accumulation (Kim et al., 2019). Therefore, the presence of EDS in agroecosystems poses a potential threat to plant health, crop productivity, and ecological balance, highlighting the need for further research and regulation.

Among the studies analyzed, there was variation in the methodologies. However, the assays similarly involved the application of increasing concentrations of the analyzed substance to observe the Germination Index (GI) for each test, aiming to identify some level of phytotoxicity according to the studied class of pollutants. In the studies by (Miranda et al., 2022), for example, the tests involving the absorption of microplastics by seeds during their early development indicated that there was little to no phytotoxic effect under the conditions studied. However, the study highlights the need for additional tests to better understand the behavior of these contaminants within plant–soil systems and their environmental dynamics. Conversely, when considering the interaction of microplastics with another class of emerging pollutants, such as pesticides, as described by (Souza et al., 2025), the presence of Atrazine (a commonly used pesticide in Brazil) adsorbed onto microplastics enhanced the phytotoxic effects, significantly inhibiting seed germination.

In the study by (Kebrom et al., 2019), collard greens (*Brassica oleracea* L.) were used as the test plant to investigate the toxicity of three commercial organic amendments: poultry manure, Milorganite (a municipal biosolid), and composted cattle manure. Seeds were exposed to aqueous extracts of the amendments, and parameters such as the Germination Index (GI) were measured. The results showed that poultry manure and Milorganite exhibited significant phytotoxic effects, with GI values near zero in concentrated extracts, suggesting the presence of salts and heavy metals such as nickel (Ni) at toxic levels. On the other hand, cattle manure did not exhibit phytotoxicity and even showed a stimulatory effect on root growth at certain dilutions.

Zhou et al. (2016) investigated the phytotoxic effects of perfluorooctanoic acid (PFOA) on wheat (*Triticum aestivum* L.) through germination and pot cultivation assays, simulating different levels of soil contamination. Seeds were exposed to increasing concentrations of PFOA to assess morphological parameters such as germination rate, root and shoot length, as well as physiological indicators such as proline content and antioxidant enzyme activity. The results showed that high concentrations ($\geq$800 mg/kg) significantly inhibited germination and seedling growth, with more severe effects observed in roots. In contrast, very low concentrations ($<$0.2 mg/kg) showed an initial stimulatory effect, suggesting a possible hormetic response. However, even at intermediate levels (200 mg/kg), PFOA caused growth reductions and physiological alterations associated with oxidative stress. The study demonstrated that PFOA phytotoxicity is related to both dose and the plant's ability to activate defense mechanisms, highlighting the importance of physiological analysis to identify sublethal impacts caused by emerging contaminants in agricultural systems.

In conclusion, although it is not possible to establish a direct correlation between the studies mentioned—given the use of different classes of compounds and various plant organisms—this work aims to demonstrate that seed germination bioassays are sensitive and effective tools for detecting phytotoxic effects caused by potentially hazardous organic compounds present in organic amendments or waste. These studies reinforce the importance of considering not only the chemical composition of materials applied to soil but also their biological impacts, especially during the early stages of plant development, which are particularly vulnerable to the presence of toxic substances. Thus, even with different methodologies, the results converge in highlighting the need for careful evaluation of the inputs used in sustainable agricultural systems.

4.2 Physiological and biochemical damage: oxidative stress, hormonal imbalance, and impact on plant defense mechanisms

The use of pesticides, antibiotics, and other emerging contaminants induces oxidative stress in plants, which is considered a critical indicator of phytotoxicity. The characterization of both oxidative stress status and the plant antioxidant defense response is considered a key aspect for evaluating the impact of substances such as PFOA on plants (Li et al., 2022). Among antibiotics, the chlorophyll *a* content in leaves decreased with increasing Oxytetracycline (OTC) level, even though the contaminant was not detected in shoots, suggesting a systemic stress response (Bao et al., 2022). Bao et al. (2022) demonstrated that, regardless of the presence of Polyethylene (PE) microplastics, OTC inhibited plant growth and triggered the stress response of wheat during seed germination and early seedling cultivation. During seed germination, OTC induced phytotoxicity according to sensitive indices (root elongation and sprout length, fresh weight, and vigor index). In addition, under combined exposure, the effect of PE depended on the OTC level. During seedling cultivation, the metabolic response was more sensitive than other physiological indices. Khan et al. (2021) found that OTC stress decreases the photosynthetic ability of Brassica rapa ssp. chinensis. Quinolone antibiotics at environmentally relevant concentrations disrupted the conformation of the photosystem II center D2 protein (psbD protein) in photosystem II (PSII) to interfere with the light reaction phase of photosynthesis, which in turn hampered the photon electron transfer and decreased the carbon sequestration capacity of rice. Additionally, they confirmed, through comparative analysis, that the photosynthetic damage by quinolone antibiotics on rice intensified with each generation as the side chain substituents of the antibiotic molecules increase in size and complexity (Li et al., 2024). Interestingly, strategies to mitigate pathogen-induced stress in plants have been explored. The results of the greenhouse experiment showed that pectin-functionalized Fe- MOF NPs (Fe-MOF-PT NPs) improved rice resistance to *R. solani* by inhibiting mycelial invasion, enhancing antioxidant enzyme activities, activating the jasmonic acid signaling pathway, and enhancing photosynthesis (Yang et al., 2024).

4.3 Dysfunctions in secondary metabolism and photosynthesis

Although all pesticides are designed to eliminate or prevent specific plant and animal species, there is growing concern and interest regarding their broader biological and physiological impacts, not only on the target organisms but also on non-target species and ecosystems. Pesticides, particularly agrochemicals such as herbicides, insecticides, and fungicides, can cause a range of unintended effects that extend beyond their initial application purpose. While these chemicals are crucial for increasing agricultural productivity and controlling pests and diseases, they also pose significant environmental and ecological risks.

One critical aspect often overlooked is the disproportionate exposure of seeds and seedlings to chemical treatments. Chemically treated seeds are frequently exposed to substantially higher concentrations of pesticides than mature plants encounter during their life cycle. This early exposure can lead to detrimental effects on seed viability, germination rate, and initial plant development. As a result, the potential benefits of pesticide application are counterbalanced by the heightened risk of phytotoxicity—damage to plant tissues caused by toxic substances.

Herbicides, in particular, function through diverse mechanisms to suppress or eliminate unwanted plant species (weeds). These mechanisms include the inhibition of essential biological processes such as photosynthesis, mitosis, cell division, enzymatic activity, root and shoot development, as well as the synthesis of DNA and proteins. Some herbicides disrupt cell membranes or induce hormonal imbalances that result in uncontrolled growth, leading to plant death. The specific mode of action varies by chemical class but often involves complex biochemical pathways.

Moreover, the use of pesticides triggers both enzymatic and non-enzymatic responses in plants. These include alterations in antioxidant defense systems, such as changes in the activity of superoxide dismutase (SOD), catalase (CAT), and peroxidases, as well as modifications in non-enzymatic antioxidants like glutathione and ascorbate. These changes may lead to oxidative stress and cellular damage, even in non-target plants. Such stress can hinder physiological processes from the very early stages of germination and vegetative growth to the reproductive phase, ultimately reducing the overall yield of agricultural products such as vegetables, fruits, and seeds. Therefore, understanding the phytotoxic potential of pesticides is essential for developing safer agricultural practices and minimizing negative impacts on crop productivity and environmental health (Yengkokpam and Mazumder, 2020).

5. Phytotoxicity assay to assess plant species for phytoremediation

Anthropogenic activities have led to widespread contamination of the environment, including thousands of sites that are severely polluted. The concentrations of the contaminants can vary from highly toxic concentrations to barely detectable concentrations that, after long-term exposure, can be detrimental to ecosystemic health (Wang et al., 2023). The growing concern about endocrine-disrupting chemicals (EDCs) in the environment has driven the search for effective strategies for their mitigation and removal (Wei et al., 2023). The cost of cleaning up contaminated sites is extremely high. In the USA for example, about $six to eight billion is spent annually in remediation efforts, with global costs in the range of $25–50 billion (Sainger et al., 2011). Engineering methods for the remediation of contaminated sites include excavation, transport, soil washing, extraction, pumping and treating of contaminated water, addition of reactants such as hydrogen peroxide or potassium permanganate, and incineration (Doty, 2008). A serious consequence of the high cost of remediation technologies is that polluted commercial properties are often abandoned rather than cleaned up (Sainger et al., 2011). Among the most promising approaches are biotechnological strategies, phytoremediation, and bioengineering to minimize the impacts of these contaminants on plant physiology (Kristanti et al., 2023).

Phytoremediation mechanisms include:

Phytoextraction: Plants absorb pollutants from the soil, water, or sediment through their roots. The pollutants are then transferred to the aboveground biomass, such as the shoots or leaves.

Phytovolatilization: Plants absorb pollutants (especially heavy metals) and transform them into less hazardous chemical species through transpiration.

Phytostabilization: Plants limit the movement of contaminants by accumulating or immobilizing them in the rhizosphere.

Rhizofiltration: Plant roots absorb and sequester metal pollutants from water.

Degradation: Rhizo-degradation and phytodegradation are mechanisms that break down pollutants

Hydraulic control: A mechanism that immobilizes pollutants through transpiration and root development.

Given the complexity of environmental contamination by EDCs, selecting an effective phytoremediation strategy requires consideration of multiple factors, such as soil characteristics, contaminant properties, plant species, and local climate conditions, as shown in Fig. 11.14. The following section discusses current phytoremediation and biotechnological approaches aimed at mitigating EDCs' pollution and enhancing ecosystem resilience.

5.1 Phytoremediation and biotechnological strategies for EDCs mitigation

Phytoremediation has emerged as a sustainable approach for the removal of EDCs from the environment. This technique uses plants to absorb, transform, or stabilize contaminants in soil and water (Kafle et al., 2022). Phytotoxicity assessment is essential to determine the feasibility of using plants in the phytoremediation of EDCs. Phytotoxicity

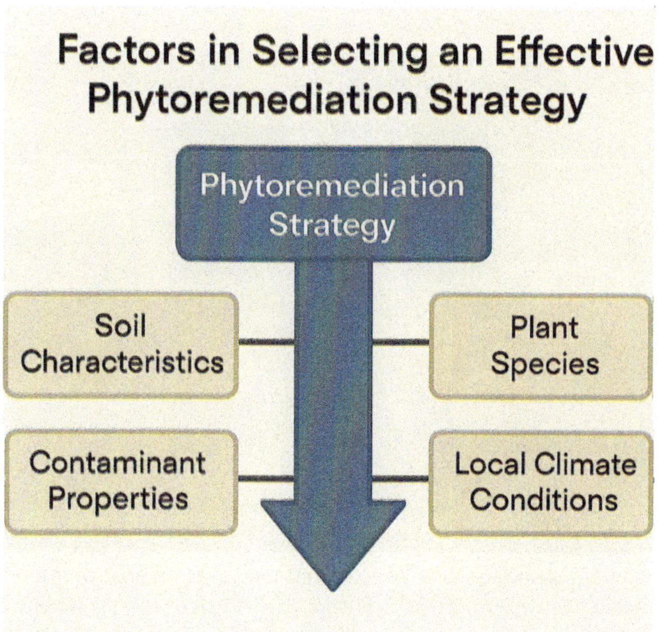

Figure 11.14 Factors affecting phytoremediation technlogies.

assays help identify the impacts of contaminants on plant physiology and evaluate the ability of plant species to survive and develop in contaminated environments (Oubohssaine and Dahmani, 2024). The main parameters analyzed in phytotoxicity assays include: Germination rate: assesses whether EDCs affect seedling emergence and early growth. - Growth index: evaluates changes in root and shoot length as indicators of contaminant toxicity. - Dry and fresh biomass: measures organic matter accumulation and the plant's ability to withstand EDCs exposure. - Biochemical analyses: quantification of ROS, photosynthetic pigments, and enzymes related to oxidative stress. - Gene expression and enzymatic activity: determine physiological and biochemical responses of plants to EDCs contamination.

The combination of these assays enables the selection of plant species that are more efficient for phytoremediation, as well as the optimization of environmental conditions to maximize EDC removal (Oubohssaine and Dahmani, 2024). The constructed wetlands (CW), for example, were a peculiar strategy, which was utilized to clean up the organic and inorganic water contaminants. The CW strategy has been noted to help eliminate a variety of EDCs with cost effectiveness, ease of operation, increased removal rate, and improved water recycling (Horner et al., 2016). Figure 11.15 shows a horizontal subsurface flow constructed wetlands as post-treatment of aerated pond effluent.

The rhizomes and biofilms present near the root in the floating mat enhance the adsorption and degradation of the organic compounds (Mimmo et al., 2014). A study reported that cucumber plants uptake the anticonvulsant drug carbamazepine found to be present

Figure 11.15 Horizontal subsurface flow constructed wetlands as post-treatment of aerated pond effluent. *(From Stopa, J. M., Benassi, R. F., F. Bueno, R. & Coelho, L. H. G. (2022). Horizontal subsurface flow constructed wetlands as post-treatment of aerated pond effluent. International Journal of Environmental Science and Technology, 19(5), 4447–4458. https://doi.org/10.1007/s13762-021-03418-8)*

in the effluent in a hydroponic culture, which was used as a water source (Shenker et al., 2010). Another study which has been taken place in Vietnam on the herbaceous plant narrow-leaf cattail (*Typha angustifolia*) reported that it has the potential to remove the pharmaceutical residues released into the aquatic environment. The study shows that the macrophyte, along with its various rhizospheric bacteria, removes the residues of the drug ibuprofen and its various analogs via phytotransformation, rhizodegradation, and phytoextraction mechanisms (Li et al., 2016). Gorito et al. (2018) designed a CW with *Phragmites australis* and treated the aquaculture effluents. They removed 36 organic micropollutants with a more than 87% efficiency rate. The combined effect of the plant (phytodegradation) and beneficial microbes (microbial degradation) resulted in the high removal rate of the contaminants in the CW system (Gorito et al., 2018). To further enhance the efficiency of phytoremediation strategies such as CW, recent research has focused on the synergistic role of plant-associated microorganisms. The following section delves into the contributions of endophytic and rhizosphere bacteria in supporting phytoremediation processes and improving EDCs degradation in contaminated environments.

5.2 Endophytic bacteria, rizhosphere bacteria, and phytoremediation

Endophytic bacteria reside within plant tissues, whereas rhizospheric bacteria inhabit the root surface and surrounding soil (Doty, 2008). Both types of microbes can enhance plant tolerance to abiotic stressors, including salinity, drought, nutrient deficiency, temperature extremes, and contamination (Ameen et al., 2024). While rhizospheric bacteria are well recognized for their role in phytoremediation, endophytes offer advantages such as reduced microbial competition and better control over colonization, as their populations are selected and maintained by the host plant (Doty, 2008). Studies have demonstrated that plants can recruit pollutant-degrading endophytes from the environment. For instance, petroleum-contaminated soils influenced the enrichment of specific endophytes capable of degrading hydrocarbons within plant roots, with this recruitment being species-dependent (Li et al., 2022). Poplar (*Populus* spp.) and willow (*Salix* spp.), commonly used in phytoremediation, harbor diverse endophytes - including Rhizobium tropici, Burkholderia, Pseudomonas, and Herbaspirillum - that contribute to nutrient acquisition and contaminant degradation (Doty, 2008). Several endophytes have demonstrated the ability to degrade pollutants such as 2,4-D, toluene, naphthalene, TNT, RDX, and HMX (Burragoni and Jeon, 2021; Germaine et al., 2006). These capabilities, combined with tolerance to contaminants and promotion of plant growth, make endophytes powerful allies in phytoremediation. Notably, engineered or naturally selected endophytic strains have been successfully inoculated into host plants to enhance their remediation efficiency. For example, inoculating Pseudomonas putida into peas enabled effective 2,4-D herbicide removal without phytotoxicity, highlighting the potential of plant–microbe partnerships to improve the phytoremediation of herbicide-contaminated soils (Germaine et al., 2006).

Moreover, rhizospheric microorganisms, including bacteria and fungi, can synergistically work with plants to mineralize or transform EDCs into less toxic compounds, boosting both plant resilience and contaminant degradation. This integrated microbial network enhances enzymatic activity and optimizes the phytoremediation process, reinforcing the importance of microbial consortia in sustainable environmental remediation strategies. Species from genera such as Pseudomonas, Bacillus, Trichoderma, and Aspergillus have been reported to mineralize or co-metabolize various EDCs, including bisphenol A, phthalates, and nonylphenol. The synergistic interaction between plants and rhizospheric microbes enhances the overall efficiency of EDC removal, making the rhizosphere a dynamic and essential zone for sustainable remediation efforts (Ameen et al., 2024).

5.3 Role of bioengineering in reducing the impact of EDCs on plant physiology

Bioengineering has been a valuable tool in mitigating the phytotoxic effects of EDCs. Advanced genetic modification and biotechnology techniques allow for the development of plants more resistant to environmental contamination, reducing oxidative stress and cellular damage caused by these compounds. Another promising biotechnological strategy involves the genetic engineering of plants to enhance their ability to absorb and metabolize emerging contaminants. The insertion of genes encoding specific enzymes, such as peroxidases and cytochromes P450, can accelerate the degradation of EDCs and minimize their adverse effects on the environment. Among the approaches employed, the following stand out:

Genetic modification to increase the expression of antioxidant proteins reduces the impact of EDCs on plant physiology. Genetic modifications can enhance the production of antioxidant enzymes such as superoxide dismutase (SOD), catalase (CAT), and glutathione peroxidase (GPX), which play critical roles in detoxifying reactive oxygen species (ROS) generated by EDC exposure. By boosting the antioxidant defense system, genetically modified plants can better withstand oxidative stress, maintain cellular integrity, and sustain normal metabolic functions even in contaminated environments. Additionally, the overexpression of genes encoding heat shock proteins (HSPs) and other stress-related proteins can further improve plant resilience by enhancing their ability to repair cellular damage and modulate stress signaling pathways. To improve the breakdown and absorption of organic pollutants by inserting genes encoding pollutant-degrading enzymes (Vijayanand et al., 2023). However, ethical considerations arise as a result of the potential transmission of altered genes to wild types, which could have unforeseen implications in natural ecosystems (Bohua et al., 2023).

Application of growth regulators to minimize the adverse effects of contaminants on plant development and reproduction. Growth regulators, such as auxins, gibberellins, cytokinins, and abscisic acid, play a crucial role in modulating plant responses to environmental stress. These compounds can enhance root development, improve nutrient uptake, and regulate stress-responsive pathways to mitigate the toxic effects of EDC exposure. By optimizing hormonal balance, growth regulators can help maintain normal growth and reproductive processes, even in contaminated environments. Studies have shown that exogenous application of plant growth regulators can increase tolerance to heavy metals and organic pollutants, making them a valuable tool in phytoremediation efforts.

Use of beneficial microorganisms that act in protection against chemical stresses, promoting the enzymatic degradation of EDCs and improving plant adaptation to contaminated environments. The genetic modification of root-associated bacteria aims at elevating their activity in the rhizosphere, which is an important component of nutrient cycling and pollution breakdown (Li et al., 2025). This alteration, however, may have an impact on soil bacteria and have a ripple effect on ecosystem functioning. Concerns regarding unexpected environmental consequences, genetic contamination, long-term repercussions on the environment and human health, as well as control over genetically modified organisms, are all part of the ethical considerations surrounding genetic changes (Louwaars and Jochemsen, 2021). To strike a balance between potential advantages and accompanying hazards, strict regulation and continual scientific evaluation are required. The integration of these biotechnological and bioengineering strategies represents a sustainable and innovative alternative for mitigating the presence of EDCs in the environment, ensuring the preservation of ecosystems and the safety of agricultural production. Continuous research in this field is essential for developing more effective and economically viable solutions for controlling contamination by endocrine-disrupting compounds. However, while improving a plant's stress tolerance by genetic alterations broadens the spectrum of appropriate species for phytoremediation, concerns regarding unintended ecological consequences and competitive advantages in natural environments remain.

6. Summary

Endocrine-Disrupting Chemicals (EDCs), including pharmaceuticals, hormones, and personal care products, significantly interfere with plant growth and development, even at low environmental concentrations. These substances disrupt hormonal signaling pathways, resulting in inhibited seed germination, altered root and shoot elongation, impaired leaf development, and reduced biomass accumulation. Species-specific responses have been observed in plants such as *Lactuca sativa*, *Zea mays*, and *Brassica napus*, with vegetative tissues often accumulating higher concentrations of contaminants than reproductive organs. In addition to these direct phytotoxic effects, EDCs can induce oxidative stress, alter gene expression, and interfere with hormone receptor activity and signal transduction pathways.

At the ecosystem level, the continuous introduction of EDCs via wastewater irrigation, biosolids, and manure disrupts soil microbial communities, nutrient cycling, and plant-soil interactions. These alterations may reduce biodiversity, compromise soil fertility, and affect food web dynamics, ultimately threatening agroecosystem sustainability and key ecosystem services. Moreover, EDCs may impair critical symbiotic relationships, such as those between plants and nitrogen-fixing bacteria or mycorrhizal fungi, further diminishing plant health and productivity.

In light of these challenges, there is an urgent need for integrated risk assessment frameworks that extend beyond traditional growth-based endpoints to include physiological, biochemical, and molecular markers. Despite growing awareness of the ecological risks posed by EDCs, current regulatory frameworks and monitoring protocols remain limited and inconsistent, particularly in low- and middle-income countries where resources for environmental surveillance are scarce. The lack of standardized analytical methods, toxicological reference values, and long-term exposure data further hinders effective policy development and environmental protection.

Given the persistence, bioaccumulation potential, and sublethal effects of many EDCs, future research should prioritize interdisciplinary approaches—combining environmental chemistry, ecotoxicology, molecular biology, and microbiome science—to better understand their fate and ecological impact. These insights are crucial for refining risk models, designing remediation strategies, and developing evidence-based regulations. Promoting the use of phytoremediation-capable plant species and reducing dependence on synthetic agrochemicals by adopting sustainable alternatives such as biopesticides are essential strategies to mitigate pollution, protect environmental health, and safeguard human well-being through a more resilient food system.

AI Disclosure

During the preparation of this work the author(s) used Chat GPT in order to To enhance the quality and efficiency of academic writing and research by improving language accuracy. After using this tool/service, the author(s) reviewed and edited the content as needed and take(s) full responsibility for the content of the publication.

References

Abdelmoneim, M., Hafez, E., Dawood, M., Hammad, S., Ghazy, M., et al., 2024. Toxicity of bisphenol A and *p*-nitrophenol on tomato plants: morpho-physiological, ionomic profile, and antioxidants/defense-related gene expression studies. Biomol. Concepts. https://doi.org/10.1515/bmc-2022-0049.

Abdulrazaq, Y., Abdulsalam, A., Rotimi, A.L., Abdulbasit, A.A., Clifford, O., Abdulsalam, O.A., Racheal, O.N., Joy, A.A., Victor, F.O., Johannes, Z.M., Bilal, M., 2020. Classification, potential routes and risk of emerging pollutants/contaminant. In *Emerging contaminants*. IntechOpen. https://doi.org/10.5772/intechopen.94447.

Agati, G., Azzarello, E., Pollastri, S., Tattini, M., 2012. Flavonoids as antioxidants in plants: location and functional significance. Plant Sci. https://doi.org/10.1016/j.plantsci.2012.07.014.

Albuquerque, F.P., de Oliveira, J.L., Moschini-Carlos, V., Fraceto, L.F., 2020. An overview of the potential impacts of atrazine in aquatic environments: Perspectives for tailored solutions based on nanotechnology. Sci. Total Environ. 700. https://doi.org/10.1016/j.scitotenv.2019.134868.

Alkimin, G.D., Daniel, D., Frankenbach, S., Serôdio, J., Soares, A.M.V.M., Barata, C., Nunes, B., 2019. Evaluation of pharmaceutical toxic effects of non-standard endpoints on the macrophyte species Lemna minor and Lemna gibba. Sci. Total Environ. 657, 926–937. https://doi.org/10.1016/j.scitotenv.2018.12.002.

Ameen, M., Mahmood, A., Sahkoor, A., et al., 2024. The role of endophytes to combat abiotic stress in plants. Plant Stress. https://doi.org/10.1016/j.stress.2024.100435.

Andreau, A., Korovic, M., Garcia-Sanz, C., et al., 2023. Enzymatic glycosylation strategies in the production of bioactive compounds. Catalyst. https://doi.org/10.3390/catal13101359.

Antoniadis, V., Levizou, E., Shaheen, S.M., Sik Ok, Y., Sebastian, A., Baum, C., Prasad, M.N.V., Wenzel, W.W., Rinklebe, J., 2017. Trace elements in the soil-plant interface: Phytoavailability, translocation, and phytoremediation–A review. Earth Sci. Rev. 171, 621–645. https://doi.org/10.1016/j.earscirev.2017.06.005.

Asenuga, E.R., Olagunju, A.S., 2023. Exposure and associated health risk of endocrine disruption compounds. African Journal of Health, Safety and Environment 4 (1), 57–66. https://doi.org/10.52417/ajhse.v4i1.332.

Bahmani, R., Kim, D., Modareszadeh, M., Hwang, S., 2023. Ethylene and ROS mediate root growth inhibition induced by the endocrine disruptor bisphenol A (BPA). Plant Physiol. Biochem. https://doi.org/10.1016/j.plaphy.2023.108212.

Bahmani, R., Kim, D.G., Modareszadeh, M., Thompson, A.J., Park, J.H., Yoo, H.H., Hwang, S., 2020. The mechanism of root growth inhibition by the endocrine disruptor bisphenol A (BPA). Environ. Pollut. 257, 113516. https://doi.org/10.1016/j.envpol.2019.113516.

Bao, Y., Pan, C., Li, D., et al., 2022. Stress response to oxytetracycline and microplastic-polyethylene in wheat (*Triticum aestivum* L.) during seed germination and seedling growth stages. Sci. Total Environ. https://doi.org/10.1016/j.scitotenv.2021.150553.

Baralla, E., Demontis, M.P., Dessì, F., Varoni, M.V., 2021. An overview of antibiotics as emerging contaminants: occurrence in bivalves as biomonitoring organisms. Animals 11 (11). https://doi.org/10.3390/ani11113239.

Barnabas Ebinezer, L., Battisti, I., Sharma, N., Ravazzolo, L., Ravi, L., Trentin, A.R., Barion, G., Panozzo, A., Dall'Acqua, S., Vamerali, T., Quaggiotti, S., Arrigoni, G., Masi, A., 2022. Perfluorinated alkyl substances affect the growth, physiology and root proteome of hydroponically grown maize plants. J. Hazard Mater. 438, 129512. https://doi.org/10.1016/j.jhazmat.2022.129512.

Bártíková, H., Podlipná, R., Skálová, L., 2016. Veterinary drugs in the environment and their toxicity to plants. Chemosphere 144, 2290–2301. https://doi.org/10.1016/j.chemosphere.2015.10.137.

Bayabil, H.K., Teshome, F.T., Li, Y.C., 2022. Emerging contaminants in soil and water. Front. Environ. Sci. 10. https://doi.org/10.3389/fenvs.2022.873499.

Bhardwaj, D., 2016. Effect of triclosan on germination and seedling growth of mustard seeds. Current Opinion in Agriculture Curr. Opin. Agric. 5 (1), 5–8. ISSN: 2322-4193.

Blok, C., Baumgarten, A., Baas, R., Wever, G., Lohr, D., Raviv, M., Lieth, J.H., Bar-Tal, A., 2019. Chapter 11 - Analytical Methods Used with Soilless Substrates. Elsevier, Boston, pp. 509–564. https://doi.org/10.1016/B978-0-444-63696-6.00011-6.

Bohua, L., Yuexin, W., Yakun, O., et al., 2023. Ethical framework on risk governance of synthetic biology. J. Biosaf. Biosecur. https://doi.org/10.1016/j.jobb.2023.03.002.

Boxall, A.B.A., Johnson, P., Smith, E.J., Sinclair, C.J., Stutt, E., Levy, L.S., 2006. Uptake of veterinary medicines from soils into plants. J. Agric. Food Chem. 54 (6), 2288–2297. https://doi.org/10.1021/jf053041t.

Burragoni, S., Jeon, J., 2021. Applications of endophytic microbes in agriculture, biotechnology, medicine, and beyond. Microbiol. Res. https://doi.org/10.1016/j.micres.2020.126691.

Chakraborty, P., Biswas, A., Dey, D., et al., 2023. Cytochrome P450 gene families: role in plant secondary metabolites production and plant defense. J. Xenobiotics. https://doi.org/10.3390/jox13030026.

Chandra, R., Sharpanabharathi, N., Anjan Kumar Prusty, B., Azeez, P.A., Kurakalva, R.M., 2021. Organochlorine pesticide residues in plants and their possible ecotoxicological and agri food impacts. Sci. Rep. 11 (1). https://doi.org/10.1038/s41598-021-97286-4.

Chanu, L., Gupta, A., 2016. Phytoremediation of lead using *Ipomoea aquatica* Forsk in hydroponic solution. Chemosphere. https://doi.org/10.1016/j.chemosphere.2016.05.001.

Chaturvedi, M., Joy, S., Gupta, R.D., Pandey, S., Sharma, S., 2023. Endocrine disrupting chemicals (EDCs): chemical fate, distribution, analytical methods and promising remediation strategies–a critical review. Environmental Technology Reviews 12 (1), 286–315. https://doi.org/10.1080/21622515.2023.2205026.

Chavan, A., Khan, A., Ansari, A., Kakde, U., 2025. Phytoremediation of mercury using cymbopogon citratus: growth performance & bioaccumulation study. J. Technol. 13, 735–746.

Chemicals, E.-D., 1993. Phytoestrogen signaling and symbiotic gene activation are disrupted by endocrine-disrupting chemicals. Environ. Health Perspect. Suppl. 112, 672–677.

Chen, X., Li, Y., Jiang, L., Hu, B., Wang, L., An, S., Zhang, X., 2021. Uptake, accumulation, and translocation mechanisms of steroid estrogens in plants. Sci. Total Environ. 753, 141979. https://doi.org/10.1016/j.scitotenv.2020.141979.

Choudhary, A., Kumar, A., Kaur, N., 2020. ROS and oxidative burst: roots in plant development. Plant Divers. 42 (1), 33–43. https://doi.org/10.1016/j.pld.2019.10.002.

Coelho, L.H.G., de Jesus, T.A., Kohatsu, M.Y., Poccia, G.T., Chicarolli, V., Helwig, K., Hunter, C., Roberts, J., Teedon, P., Pahl, O., 2020. Estrogenic hormones in são paulo waters (Brazil) and their relationship with environmental variables and sinapis alba phytotoxicity. Water, Air, Soil Pollut. 231 (4), 150. https://doi.org/10.1007/s11270-020-04477-2.

Cunha Barros, D.G., dos Santos Gonçalves do Nascimento, G.C., Okon, C., Rocha, M.B., Santo, D.E., de Lima Feitoza, L., Junior, O.V., da Silva Gonzalez, R., de Souza, D.C., Peron, A.P., 2023. Benzophenone-3 sunscreen causes phytotoxicity and cytogenotoxicity in higher plants. Environ. Sci. Pollut. Control Ser. 30 (52), 112788–112798. https://doi.org/10.1007/s11356-023-30365-3.

Cycoń, M., Mrozik, A., Piotrowska-Seget, Z., 2019. Antibiotics in the soil environment—degradation and their impact on microbial activity and diversity. Front. Microbiol. 10 (2019). https://doi.org/10.3389/fmicb.2019.00338.

Dettenmaier, M., Doucette, W., Bugbee, B., 2009. Chemical hydrophobicity and uptake by plant roots. Environ. Sci. Technol. https://doi.org/10.1021/es801751x.

Diamanti-Kandarakis, E., Bourguignon, J.P., Giudice, L.C., Hauser, R., Prins, G.S., Soto, A.M., Zoeller, R.T., Gore, A.C., 2009. Endocrine-disrupting chemicals: an endocrine society scientific statement. Endocr. Rev. 30 (4), 293–342. https://doi.org/10.1210/er.2009-0002Greece.

Dosis, I., Kamarianos, A., 2007. Environmental toxic endocrine disrupting compounds (EDCs): effects on environment, animal production and human. Journal of the Hellenic Veterinary Medical Society 58, 321–334.

Doty, S., 2008. Enhancing phytoremediation through the use of transgenics and endophytes. New Phytol. https://doi.org/10.1111/j.1469-8137.2008.02446.x.

Duarte, D.J., Oldenkamp, R., Ragas, A.M.J., 2022. Human health risk assessment of pharmaceuticals in the European Vecht River. Integrated Environ. Assess. Manag. 18 (6), 1639–1654. https://doi.org/10.1002/ieam.4588.

Zimmerlin, A., Durst, F., 1992. Aryl hydroxylation of the herbicide diclofop by a wheat cytochrome p-450 monooxygenase : substrate specificity and physiological activity. Plant Phisiol. https://doi.org/10.1104/pp.100.2.874.

Zucconi, F., Monaco, A., Forte, M., 1985. Phytotoxins during the stabilization of organic matte. Compost. Agric. Other Wastes.

Endocrine Disrupting Chemicals in Freshwater: Monitoring and Regulating Water Quality OECD Studies on Water, 2023. OECD. https://doi.org/10.1787/5696d960-en.

Erofeeva, E.A., 2022. Hormesis in plants: its common occurrence across stresses. Current Opinion in Toxicology 30, 100333. https://doi.org/10.1016/j.cotox.2022.02.006.

Fan, S., Cong, Z., 2024. Emerging strategies for modifying cytochrome P450 monooxygenases into peroxizymes. Acc. Chem. Res. https://doi.org/10.1021/acs.accounts.3c00746.

Fan, L., Tang, J., Zhang, D., Ma, M., Wang, Y., Han, Y., 2020. Investigations on the phytotoxicity of perfluorooctanoic acid in Arabidopsis thaliana. Environ. Sci. Pollut. Control Ser. 27 (1), 1131–1143. https://doi.org/10.1007/s11356-019-07018-5.

Fernandes, A.S., Bragança, I., Homem, V., 2024. Personal care products in soil-plant and hydroponic systems: uptake, translocation, and accumulation. Sci. Total Environ. 912, 168894. https://doi.org/10.1016/j.scitotenv.2023.168894.

Fu, W., Chen, X., Zheng, X., Liu, A., Wang, W., Ji, J., Wang, G., Guan, C., et al., 2022. Phytoremediation potential, antioxidant response, photosynthetic behavior and rhizosphere bacterial community adaptation of tobacco (*Nicotiana tabacum* L.) in a bisphenol A-contaminated soil. Environ. Sci. Pollut. Res. Int. https://doi.org/10.1007/s11356-022-21765-y.

Fujisawa, T., Ichise, K., Fukushima, M., Katagi, T., Takimoto, Y., 2002. Improved uptake models of nonionized pesticides to foliage and seed of crops. J. Agric. Food Chem. https://doi.org/10.1021/jf010985j.

Germaine, K., Liu, X., Cabellos, G., et al., 2006. Bacterial endophyte-enhanced phytoremediation of the organochlorine herbicide 2,4-dichlorophenoxyacetic acid. FEMS Microbiol. Ecol. https://doi.org/10.1111/j.1574-6941.2006.00121.x.

Gomes Moreira, D., Aires, A., Pereira, M.de L., Oliveira, M., 2022. Levels and effects of antidepressant drugs to aquatic organisms. Comp. Biochem. Physiol. C Toxicol. Pharmacol. 256, 109322. https://doi.org/10.1016/j.cbpc.2022.109322.

Gong, W., Zhang, W., Jiang, M., Li, S., et al., 2021. Species-dependent response of food crops to polystyrene nanoplastics and microplastics. Sci. Total Environ. https://doi.org/10.1016/j.scitotenv.2021.148750.

Gorito, A., Ribeiro, A., Gomes, C., et al., 2018. Constructed wetland microcosms for the removal of organic micropollutants from freshwater aquaculture effluents. Sci. Total Environ. https://doi.org/10.1016/j.scitotenv.2018.06.371.

Gredelj, A., Nicoletto, C., Polesello, S., Ferrario, C., Valsecchi, S., Lava, R., Barausse, A., Zanon, F., Palmeri, L., Guidolin, L., Bonato, M., 2020. Uptake and translocation of perfluoroalkyl acids (PFAAs) in hydroponically grown red chicory (Cichorium intybus L.): growth and developmental toxicity, comparison with growth in soil and bioavailability implications. Sci. Total Environ. 720, 137333. https://doi.org/10.1016/j.scitotenv.2020.137333.

Guengerich, F.P., 2026. Chapter 4.02 - The importance of biotransformation. In: Charlene, A.M. (Ed.), Comprehensive Toxicology, 4th. Elsevier. https://doi.org/10.1016/B978-0-323-95488-4.00006-1.

Gunnarsson, L., Snape, J.R., Verbruggen, B., Owen, S.F., Kristiansson, E., Margiotta-Casaluci, L., Österlund, T., Hutchinson, K., Leverett, D., Marks, B., Tyler, C.R., 2019. Pharmacology beyond the patient – the environmental risks of human drugs. Environ. Int. 129, 320–332. https://doi.org/10.1016/j.envint.2019.04.075.

Hacke, U., Laur, J., 2016. Aquaporins: channels for the molecule of life. In: Encyclopedia of Life Sciences. https://doi.org/10.1002/9780470015902.a0001289.pub2.

Horner, M., Beever, A., Johson, D., et al., 2016. Predictors of current and longer-term patterns of abundance of *American pikas* (*Ochotona princeps*) across a leading-edge protected area. PLoS One. https://doi.org/10.1371/journal.pone.0167051.

Ismanto, A., Hadibarata, T., Ayu Kristanti, R., Maslukah, L., Safinatunnajah, N., Kusumastuti, W., 2022. Endocrine disrupting chemicals (EDCs) in environmental matrices: occurrence, fate, health impact, physio-chemical and bioremediation technology. Environ. Pollut. 302, 119061. https://doi.org/10.1016/j.envpol.2022.119061.

Jaillais, Y., Chory, J., 2010. Unraveling the paradoxes of plant hormone signaling integration. Nat. Struct. Mol. Biol. 17, 642–645. https://doi.org/10.1038/nsmb0610-642.

Jayaraj, R., Megha, P., Sreedev, P., Article, R., 2016. Organochlorine pesticides, their toxic effects on living organisms and their fate in the environment. Interdiscip. Toxicol. 9 (3–4), 90–100. https://doi.org/10.1515/intox-2016-0012.

Jiang, Y., Liu, L., Jin, B., Liu, Y., Liang, X., 2024. Critical review on the environmental behaviors and toxicity of triclosan and its removal technologies. Sci. Total Environ. 932, 173013. https://doi.org/10.1016/j.scitotenv.2024.173013.

Kabir, E.R., Rahman, M.S., Rahman, I., 2015. A review on endocrine disruptors and their possible impacts on human health. Environ. Toxicol. Pharmacol. 40 (1), 241–258. https://doi.org/10.1016/j.etap.2015.06.009.

Kafle, A., Timilsina, A., Gautam, A., et al., 2022. Phytoremediation: mechanisms, plant selection and enhancement by natural and synthetic agents. Environ. Adv. https://doi.org/10.1016/j.envadv.2022.100203.

Keb, C., Agraz Hernández, C., Gutiérrez-Alcántara, E.J., Ake-Canché, B., Tirado-Torres, D., Ruiz-Hernández, J., Sarabia-Alcocer, B., Gómez-Solano, M.I., Moo-Chim, D.J., Lopez-Gutierrez, T.J., Perez-Balan, R., 2022. Acute phytotoxicity of four common pharmaceuticals on the germination and growth of Lactuca sativa L. Appl. Ecol. Environ. Res. 20, 3737–3746. https://doi.org/10.15666/aeer/2005_37373746.

Kebrom, T.H., Woldesenbet, S., Bayabil, H.K., Garcia, M., Gao, M., Ampim, P., Awal, R., Ali, F., 2019. Evaluation of phytotoxicity of three organic amendments to collard greens using the seed germination bioassay. Environ. Sci. Pollut. Control Ser. 26 (6), 5454–5462. https://doi.org/10.1007/s11356-018-3928-4.

Khan, Y., Ali, B., Zhang, S., et al., 2021. Effects of antibiotics stress on growth variables, ultrastructure, and metabolite pattern of *Brassica rapa* ssp. Chinensis. Sci. Total Environ. https://doi.org/10.1016/j.scitotenv.2021.146333.

Khetan, S.K., Collins, T.J., 2007. Human pharmaceuticals in the aquatic environment: a challenge to green chemistry. Chem. Rev. 107 (6), 2319–2364. https://doi.org/10.1021/cr020441w.

Kim, D., Il Kwak, J., An, Y.-J., 2019. Physiological response of crop plants to the endocrine-disrupting chemical nonylphenol in the soil environment. Environ. Pollut. 251, 573–580. https://doi.org/10.1016/j.envpol.2019.04.101.

Kojima, H., Katsura, E., Takeuchi, S., Niiyama, K., Kobayashi, K., 2004. Screening for estrogen and androgen receptor activities in 200 pesticides by in vitro reporter gene assays using Chinese hamster ovary cells. Environ. Health Perspect. 112 (5), 524–531. https://doi.org/10.1289/ehp.6649.

Kötke, D., Gandrass, J., Bento, C.P.M., Ferreira, C.S.S., Ferreira, A.J.D., 2024. Occurrence and environmental risk assessment of pharmaceuticals in the Mondego river (Portugal). Heliyon 10 (15), e34825. https://doi.org/10.1016/j.heliyon.2024.e34825.

Kristanti, A., Tirtalistyani, R., Tang, Y., et al., 2023. Phytoremediation mechanism for emerging pollutants: a Rreview. Trop. Aqua.Soil Pollut. https://doi.org/10.53623/tasp.v3i1.222.

Krupka, M., Piotrowicz-Cieślak, Dariusz, J., Michalczyk, 2022. Effects of antibiotics on the photosynthetic apparatus of plants. J. Plant Interact. 17 (1), 96–104. https://doi.org/10.1080/17429145.2021.2014579.

Kumar, K., Gupta, S., 2016. A framework to predict uptake of trace organic compounds by plants. J. Environ. Qual. https://doi.org/10.2134/jeq2015.06.0261.

Kumar, K., Gupta, S.C., Baidoo, S.K., Chander, Y., Rosen, C.J., 2005. Antibiotic uptake by plants from soil fertilized with animal manure. J. Environ. Qual. 34 (6), 2082–2085. https://doi.org/10.2134/jeq2005.0026.

Kumari, A., Rajput, V.D., Mandzhieva, S.S., Minkina, T., Hullebusch, E.v., 2024. Emerging Contaminants: Sustainable Agriculture and the Environment Emerging Contaminants: Sustainable Agriculture and the Environment. Elsevier, Japan, pp. 1–418. https://doi.org/10.1016/C2022-0-00910-7.

Lal, M., Mallavarapu, M., Naidu, R., Bahar, Md, 2020. Uptake of perfluorooctane sulfonate (PFOS) by common home-grown vegetable plants and potential risks to human health. Environ. Technol. Innovat. 19, 100863. https://doi.org/10.1016/j.eti.2020.100863.

Lauretta, R., Sansone, A., Sansone, M., Romanelli, F., Appetecchia, M., 2019. Endocrine disrupting chemicals: effects on endocrine glands. Front. Endocrinol. 10. https://doi.org/10.3389/fendo.2019.00178.

Lefèvre, F., Boutry, M., 2018. Towards identification of the substrates of ATP-binding cassette transporters. Plant Physiol. https://doi.org/10.1104/pp.18.00325.

Li, C., Chen, J., Xu, L., et al., 2024. Quinolone antibiotics inhibit the rice photosynthesis by targeting photosystem II center protein: generational differences and mechanistic insights. Environ. Sci. Technol. https://doi.org/10.1021/acs.est.4c01866.

Li, P., Li, J., 2021. Perfluorooctanoic acid (PFOA) caused oxidative stress and metabolic disorders in lettuce (Lactuca sativa) root. Sci. Total Environ. 770, 144726. https://doi.org/10.1016/j.scitotenv.2020.144726.

Li, Y., Li, C., Xin, Y., et al., 2022. Petroleum pollution affects soil chemistry and reshapes the diversity and networks of microbial communities. Ecotoxicol. Environ. Safety. https://doi.org/10.1016/j.ecoenv.2022.114129.

Li, P., Oyang, X., Xie, X., Guo, Y., Li, Z., Xi, J., Zhu, D., Ma, X., Liu, B., Li, J., Xiao, Z., 2020. Perfluorooctanoic acid and perfluorooctane sulfonate co-exposure induced changes of metabolites and defense pathways in lettuce leaves. Environ. Pollut. 256, 113512. https://doi.org/10.1016/j.envpol.2019.113512.

Li, J., Sun, J., Li, P., 2022. Exposure routes, bioaccumulation and toxic effects of per- and polyfluoroalkyl substances (PFASs) on plants: a critical review. Environ. Int. https://doi.org/10.1016/j.envint.2021.106891.

Li, Y., Sun, M., Raaijmakers, J., et al., 2025. Predicting rhizosphere-competence-related catabolic gene clusters in plant-associated bacteria with rhizoSMASH. Nat. Commun. https://doi.org/10.1038/s41467-025-63526-8.

Li, Y., Zhang, J., Zhu, G., et al., 2016. Phytoextraction, phytotransformation and rhizodegradation of ibuprofen associated with *Typha angustifolia* in a horizontal subsurface flow constructed wetland. Water Res. https://doi.org/10.1016/j.watres.2016.06.049.

Limer, M., Burken, J., 2014. Plant translocation of organic compounds: molecular and physicochemical predictors. Environ. Sci. Technol. Lett. https://doi.org/10.1021/ez400214q.

Louwaars, N., Jochemsen, H., 2021. An ethical and societal analysis for biotechnological methods in plant breeding. Agronomy. https://doi.org/10.3390/agronomy11061183.

Ma, J., Aqeel, M., Khalid, N., Nazir, A., et al., 2022. Effects of microplastics on growth and metabolism of rice (*Oryza sativa* L.). Chemosphere. https://doi.org/10.1016/j.chemosphere.2022.135749.

Mahdi, M.M., Islam, M.D., Hossain, M.A., 2023. Toxic Effects of the Endocrine Disrupter on Plants Current Developments in Biotechnology and Bioengineering: Bioremediation of Endocrine Disrupting Pollutants in Industrial Wastewater. Elsevier, Iceland, pp. 139–152. https://doi.org/10.1016/B978-0-323-91902-9.00003-1.

Martín, J., Luis Santos, J., Aparicio, I., Alonso, E., 2022. Microplastics and associated emerging contaminants in the environment: analysis, sorption mechanisms and effects of co-exposure. Trends in Environmental Analytical Chemistry 35, e00170. https://doi.org/10.1016/j.teac.2022.e00170.

Maurel, C., Chrispeels, A., 2001. A molecular entry into plant water relations. Plant Physiol . https://doi.org/10.1104/pp.125.1.135.

Mimmo, T., Del Buono, D., Terzano, R., et al., 2014. Rhizospheric organic compounds in the soil–microorganism–plant system: their role in iron availability. Eur. J. Soil Sci. https://doi.org/10.1111/ejss.12158.

Miranda, M.N., Lado Ribeiro, A.R., Silva, A.M.T., Pereira, M.F.R., 2022. Can aged microplastics be transport vectors for organic micropollutants? – Sorption and phytotoxicity tests. Sci. Total Environ. 850, 158073. https://doi.org/10.1016/j.scitotenv.2022.158073.

Mnif, W., Hassine, A.I.H., Bouaziz, A., Bartegi, A., Thomas, O., Roig, B., 2011. Effect of endocrine disruptor pesticides: a review. Int. J. Environ. Res. Publ. Health 8 (6), 2265–2303. https://doi.org/10.3390/ijerph8062265.

Narciso, A., Barra Caracciolo, A., De Carolis, C., 2023. Antibiotics. Overview of direct and indirect effects of antibiotics on terrestrial organisms 12 (9). https://doi.org/10.3390/antibiotics12091471.

Nguyen, T., Obermeier, C., Friedt, W., Abrams, S., Snowdon, R., 2016. Disruption of germination and seedling development in brassica napus by mutations causing severe seed hormonal imbalance. Front. Plant Sci. 7. https://doi.org/10.3389/fpls.2016.00322.

Damstra, T., Barlow, S., Bergman, A., Kavlock, R., 2002. Global Assessment of the State-of-Science of Endocrine. Disruptors.

NPIC - National Pesticide Information Center. Glyphosate general fact sheet 2024. https://npic.orst.edu/factsheets/glyphogen.html.

OECD, 2006. Test No. 208: Terrestrial Plant Test: Seedling Emergence and Seedling Growth Test. OECD Publishing, Paris. https://doi.org/10.1787/9789264070066-en.

Ojoghoro, J.O., Scrimshaw, M.D., Sumpter, J.P., 2021. Steroid hormones in the aquatic environment. Sci. Total Environ. 792, 148306. https://doi.org/10.1016/j.scitotenv.2021.148306.

Oubohssaine, M., Dahmani, I., 2024. Phytoremediation: Harnessing plant power and innovative technologies for effective soil remediation. Plant Stress. https://doi.org/10.1016/j.stress.2024.100578.

Pan, M., Chu, L.M., 2016. Phytotoxicity of veterinary antibiotics to seed germination and root elongation of crops. Ecotoxicol. Environ. Saf. 126, 228–237. https://doi.org/10.1016/j.ecoenv.2015.12.027.

Pawłowska, B., Biczak, R., 2024. Drugs in the environment - impact on plants: a review. Environ. Toxicol. Pharmacol. 111, 104557. https://doi.org/10.1016/j.etap.2024.104557.

Pérez, D.J., Doucette, W.J., Moore, M.T., 2022. Atrazine uptake, translocation, bioaccumulation and biodegradation in cattail (Typha latifolia) as a function of exposure time. Chemosphere 287. https://doi.org/10.1016/j.chemosphere.2021.132104.

Petrie, B., Barden, R., Kasprzyk-Hordern, B., 2015. A review on emerging contaminants in wastewaters and the environment: current knowledge, understudied areas and recommendations for future monitoring. Occurrence, fate, removal and assessment of emerging contaminants in water in the water cycle (from wastewater to drinking water) 72, 3–27. https://doi.org/10.1016/j.watres.2014.08.053 (PFAS Fact Sheets).

PFAS Fact Sheet. PFAS Fact Sheet. https://pfas-1.itrcweb.org/fact-sheets/.

Pironti, C., Ricciardi, M., Proto, A., Massimiliano Bianco, P., Montano, L., Motta, O., 2021. Endocrine-disrupting compounds: an overview on their occurrence in the aquatic environment and human exposure. Water 13 (10), 1347. https://doi.org/10.3390/w13101347.

Pridemore, C., 2021. The Effects of Endocrine Disrupting Chemicals on Plants.

Raun Andersen, H., Vinggaard, A.M., Høj Rasmussen, T., Gjermandsen, I.M., Cecilie Bonefeld-Jørgensen, E., 2002. Effects of currently used pesticides in assays for estrogenicity, androgenicity, and aromatase activity in vitro. Toxicol. Appl. Pharmacol. 179 (1), 1–12. https://doi.org/10.1006/taap.2001.9347.

Reddy Pullagurala, V.L., Rawat, S., Adisa, I.O., Hernandez-Viezcas, J.A., Peralta-Videa, J.R., Gardea-Torresdey, J.L., 2018. Plant uptake and translocation of contaminants of emerging concern in soil. Sci. Total Environ. 636, 1585–1596. https://doi.org/10.1016/j.scitotenv.2018.04.375.

Reichel, R., Radl, V., Rosendahl, I., Albert, A., Amelung, W., Schloter, M., Sören, T.-B., 2014. Soil microbial community responses to antibiotic-contaminated manure under different soil moisture regimes. Appl. Microbiol. Biotechnol. 98. https://doi.org/10.1007/s00253-014-5717-4.

Roberto Cerveira Junior, W., da Costa, Y.K.S., Antonio Carbonari, C., Oscar Duke, S., da Costa Aguiar Alves, P.L., Bianco de Carvalho, L., 2020. Growth, morphological, metabolic and photosynthetic responses of clones of eucalyptus to glyphosate. For. Ecol. Manag. 470–471, 118218. https://doi.org/10.1016/j.foreco.2020.118218.

Sainger, P., Dhankhar, R., Sainger, M., et al., 2011. Assessment of heavy metal tolerance in native plant species from soils contaminated with electroplating effluent. Ecotoxicol. Environ. Safety. https://doi.org/10.1016/j.ecoenv.2011.07.028.

Shenker, M., Harush, D., Ben-Ari, J., et al., 2010. Uptake of carbamazepine by cucumber plants – A case study related to irrigation with reclaimed wastewater. Chemosphere. https://doi.org/10.1016/j.chemosphere.2010.10.052.

Singh, T., Dileep, K., et al., 2017. Phytoremediation of organochlorine pesticides: concept, method, and recent developments. Int. J. Phytoremediation 19 (9), 834–843. https://doi.org/10.1080/15226514.2017.1290579.

Singh, S., Kumar, V., Gill, J.P.K., Datta, S., Singh, S., Dhaka, V., Kapoor, D., Wani, A.B., Dhanjal, D.S., Kumar, M., Harikumar, S.L., Singh, J., 2020. Herbicide glyphosate: toxicity and microbial degradation. Int. J. Environ. Res. Publ. Health 17 (20), 1–18. https://doi.org/10.3390/ijerph17207519.

Soares, C., Pereira, R., Martins, M., Tamagnini, P., Serôdio, J., Moutinho-Pereira, J., Cunha, A., Fidalgo, F., 2020. Glyphosate-dependent effects on photosynthesis of Solanum lycopersicum L.—an ecophysiological, ultrastructural and molecular approach. J. Hazard Mater. 398, 122871. https://doi.org/10.1016/j.jhazmat.2020.122871.

Souza, L., Teran, F., Cuba, R., Chaves, A., Silva, K., 2025. Interaction of microplastics with emerging organic pollutants: a study on atrazine adsorption and phytotoxicity. Toxics 13, 257. https://doi.org/10.3390/toxics13040257.

Stanley, L.A., 2017. Chapter 27 - Drug metabolism. In: Simone, B., Rupika, D. (Eds.), Pharmacognosy. https://doi.org/10.1016/B978-0-12-802104-0.00027-5.

Steeves, K., Cahill, L.S., Jobst, K.J., 2024. Emerging perfluoroalkyl substances in environmental waters revealed by non-targeted screening. Current Opinion in Environmental Science & Health 37, 100531. https://doi.org/10.1016/j.coesh.2024.100531.

Tang, M., Huang, Y., Zhang, W., Fu, T., Zeng, T., Huang, Y., Yang, X., 2022. Effects of microplastics on the mineral elements absorption and accumulation in hydroponic rice seedlings (*Oryza sativa* L.). Bull. Environ. Contam. Toxicol. https://doi.org/10.1007/s00128-021-03453-8.

Taylor, S., Pearce, C., Saguinet, K., Hu, D., Chrisler, W., Kim, W., Wang, Z., Flury, M., 2020. Polystyrene nano-A nd microplastic accumulation at *Arabidopsis* and wheat root cap cells, but no evidence for uptake into roots. Environ. Sci. Nano. https://doi.org/10.1039/d0en00309c.

Teunen, L., Bervoets, L., Belpaire, C., De Jonge, M., Groffen, T., 2021. PFAS accumulation in indigenous and translocated aquatic organisms from Belgium, with translation to human and ecological health risk. Environ. Sci. Eur. 33 (1), 39. https://doi.org/10.1186/s12302-021-00477-z.

Thomas Zoeller, R., Brown, T.R., Doan, L.L., Gore, A.C., Skakkebaek, N.E., Soto, A.M., Woodruff, T. J., Vom Saal, F.S., 2012. Endocrine-disrupting chemicals and public health protection: a statement of principles from the endocrine society. Endocrinology 153 (9), 4097–4110. https://doi.org/10.1210/en.2012-1422.

Tian, Y.-S., Jin, X.-F., Fu, X.-Y., et al., 2014. Microarray analysis of differentially expressed gene responses to Bisphenol A in *Arabidopsis*. J. Toxicol. Sci. https://doi.org/10.2131/jts.39.671.

Tijani, J.O., Fatoba, O.O., Babajide, O.O., Petrik, L.F., 2016. Pharmaceuticals, endocrine disruptors, personal care products, nanomaterials and perfluorinated pollutants: a review. Environ. Chem. Lett. 14 (1), 27–49. https://doi.org/10.1007/s10311-015-0537-z.

Tong, T., Li, R., Chai, M., Wang, Q., Yang, Y., Xie, S., et al., 2021. Metagenomic analysis of microbial communities continuously exposed to Bisphenol A in mangrove rhizosphere and non-rhizosphere soils. Sci. Total Environ. https://doi.org/10.1016/j.scitotenv.2021.148486.

Tripathi, S., Singh, V., Srivastava, P., et al., 2019. Phytoremediation of organic pollutants: Current status and future directions, In: Abatement of Environmental Pollutants: Trends and Strategies. Elsevier. https://doi.org/10.1016/B978-0-12-818095-2.00004-7.

Urbina, M., Correa, F., Aburto, F., Ferrio, J., 2020. Adsorption of polyethylene microbeads and physiological effects on hydroponic maize. Sci. Total Environ. https://doi.org/10.1016/j.scitotenv.2020.140216.

van Bladeren, 2000. Glutathione conjugation as a bioactivation reaction. Chem. Biol. Interact. https://doi.org/10.1016/s0009-2797(00)00214-3.

Vazquéz-Meza, H., Vilchis-Landeros, M., Vazquéz-Carrada, M., et al., 2023. Cellular compartmentalization, glutathione transport and its relevance in some pathologies. Antioxidants (Basel). https://doi.org/10.3390/antiox12040834.

Vicidomini, C., Palumbo, R., Moccia, M., Roviello, G., 2024. Oxidative processes and xenobiotic metabolism in plants: mechanisms of defense and potential therapeutic implications. J. Xenobiot. https://doi.org/10.3390/jox14040084.

Vijayanand, M., Ramakrishnan, A., Subramanian, R., et al., 2023. Polyaromatic hydrocarbons (PAHs) in the water environment: a review on toxicity, microbial biodegradation, systematic biological advancements, and environmental fate. Environ. Res. https://doi.org/10.1016/j.envres.2023.115716.

Vinggaard, A., Hnida, C., Breinholt, V., Larsen, J.C., 2000. Screening of selected pesticides for inhibition of CYP19 aromatase activity in vitro. Toxicol. Vitro 14 (3), 227–234. https://doi.org/10.1016/s0887-2333(00)00018-7.

Wang, K., Huang, C., Tang, J., Liu, S., Miao, Y., Liu, Y., Wu, C., et al., 2023. Bisphenol A affects soybean growth by inhibiting root nodules and germination. Water Air Soil Pollut. https://doi.org/10.1007/s11270-023-06097-y.

Wang, F., Li, X., Yu, S., et al., 2021. Chemical factors affecting uptake and translocation of six pesticides in soil by maize (*Zea mays* L.). J. Hazard Mater. https://doi.org/10.1016/j.jhazmat.2020.124269.

Wang, X., Sun, C., Gao, S., Wang, L., Han, S., 2001. Validation of germination rate and root elongation as indicator to assess phytotoxicity with cucumis sativus. Chemosphere 44 (8), 1711–1721. https://doi.org/10.1016/S0045-6535(00)00520-8.

Wang, R., Z Liu, T.-, Wang, T., 2006. The fate of antibiotics in environment and its ecotoxicology: a review. Acta Ecol. Sin. 26, 265–270.

Wang, T., Wang, S., Shao, S., et al., 2023. Toxic effects of microplastics and nanoplastics on plants: a global meta-analysis. Environ. Pollut. https://doi.org/10.1016/j.envpol.2023.122593.

Wang, T.-T., Wang, S., Shao, S., Wang, X.-D., Wang, D.-Y., Liu, Y.-S., Ge, C.-J., Ying, G.-G., Chen, Z.-B., 2022. Perfluorooctanoic acid (PFOA)-induced alterations of biomolecules in the wetland plant Alisma orientale. Sci. Total Environ. 820, 153302. https://doi.org/10.1016/j.scitotenv.2022.153302.

Wei, H., Tang, M., Xu, X., 2023. Mechanism of uptake, accumulation, transport, metabolism and phytotoxic effects of pharmaceuticals and personal care products within plants: a review. Sci. Total Environ. 892, 164413. https://doi.org/10.1016/j.scitotenv.2023.164413.

Weizel, A., Schlüsener, M.P., Dierkes, G., Ternes, T.A., 2018. Occurrence of glucocorticoids, mineralocorticoids, and progestogens in various treated wastewater, rivers, and streams. Environ. Sci. Technol. 52 (9), 5296–5307. https://doi.org/10.1021/acs.est.7b06147.

Wen, B., Wu, Y., Zhang, Z., et al., 2024. Public perception on human exposure risk: A case study on endocrine disrupting compounds in the environment. Ecotoxicol. Environ. Saf. https://doi.org/10.1016/j.ecoenv.2023.115830.

Wu, C., Chakrabarty, S., Jin, M., et al., 2019. Insect ATP-Binding Cassette (ABC) transporters: roles in xenobiotic detoxification and Bt insecticidal activity. Int. J. Mol. Sci. https://doi.org/10.3390/ijms20112829.

Wu, C., Spongberg, A.L., Witter, J.D., Maruthi Sridhar, B.B., 2012. Transfer of wastewater associated pharmaceuticals and personal care products to crop plants from biosolids treated soil. Ecotoxicol. Environ. Saf. 85, 104–109. https://doi.org/10.1016/j.ecoenv.2012.08.007.

Xiao, C., Wang, L., Zhou, Q., Huang, X., et al., 2020. Hazards of bisphenol A (BPA) exposure: a systematic review of plant toxicology studies. J. Hazard. Mater. https://doi.org/10.1016/j.jhazmat.2019.121488.

Xu, G., Liu, Y., Yu, Y., 2021. Effects of polystyrene microplastics on uptake and toxicity of phenanthrene in soybean. Sci. Total Environ. https://doi.org/10.1016/j.scitotenv.2021.147016.

Yang, L., Chen, H., Du, P., et al., 2024. Inhibition mechanism of *Rhizoctonia solani* by pectin-coated iron metal-organic framework nanoparticles and evidence of an induced defense response in rice. J. Hazard. Mater. https://doi.org/10.1016/j.jhazmat.2024.134807.

Yengkokpam, P., Mazumder, P., 2020. Phytotoxicity of malathion (PM) and tatafen (PTF) towards *Solanum melongena* L. *cv. Longai*: a case study. Plant Phys. Rep. https://doi.org/10.1007/s40502-019-00498-0.

Zhang, C., Feng, Y., Liu, Y., et al., 2017. Uptake and translocation of organic pollutants in plants: a review. J. Integr. Agric. https://doi.org/10.1016/S2095-3119(16)61590-3.

Zhang, P., Sun, L., Liu, F., Gao, Q., Wang, R., Ju, Q., Xu, J., 2022. Perfluorooctanoic acid and perfluorooctane sulfonic acid inhibit plant growth through the modulation of phytohormone signalling pathways: evidence from molecular and genetic analysis in Arabidopsis. Sci. Total Environ. 851, 158287. https://doi.org/10.1016/j.scitotenv.2022.158287.

Zhou, L., Xia, M., Wang, L., Mao, H., 2016. Toxic effect of perfluorooctanoic acid (PFOA) on germination and seedling growth of wheat (Triticum aestivum L.). Chemosphere 159, 420–425. https://doi.org/10.1016/j.chemosphere.2016.06.045.

CHAPTER 12

Impacts of microplastics on plants: Mechanisms of toxicity and detection methods

Alice Pita-Barbosa[1], Oussama Bouaicha[2], Júlia Melo Henrique[3], Andreia Neves Fernandes[3], Crislaine Bertoldi[4], Walter Ruggeri Waldman[5], Raphael Tiziani[2, 6], Felipe Klein Ricachenevsky[7] and Tanja Mimmo[2, 6]

[1]Centre for Coastal, Limnological and Marine Studies, Litoral Norte Campus, Federal University of Rio Grande do Sul, Imbé, Brazil; [2]Faculty of Agricultural, Environmental and Food Sciences, Free University of Bozen-Bolzano, Bolzano, Italy; [3]Institute of Chemistry, Federal University of Rio Grande do Sul, Porto Alegre, Brazil; [4]Department of Earth and Environmental Sciences, Faculty of Science, Lund, Sweden; [5]Centre for Science and Technology for Sustainability, Sorocaba Campus, Federal University of São Carlos, São Carlos, Brazil; [6]Competence Centre of Plant Health, Free University of Bozen-Bolzano, Bolzano, Italy; [7]Institute of Biosciences, Federal University of Rio Grande do Sul, Porto Alegre, Brazil

1. Introduction

Plastic pollution has become one of the most pervasive and long-lasting environmental challenges because of human interference on our planet (Barnes et al., 2009). Among the most concerning forms of this contaminant are microplastics (MPs), a term first proposed 20 years ago by Thompson et al. (2004) to refer to solid plastic particles smaller than 5 mm in size, which are composed of polymers, functional additives, and other intentionally and unintentionally added chemicals (Hartmann et al., 2022). The continuous fragmentation of MPs by various physical, chemical, and biological processes can further break down MPs into even smaller particles with colloidal properties and sizes between 1 and 1000 nm, named nanoplastics (NPs) (Gigault et al., 2018). As a matter of standardization, this chapter will consider the expression micro- and nanoplastics (MNP) to refer to the plastic contaminants smaller than 5 mm, and the word nanoplastic (NP) when this category was used as a description by the chosen literature.

Plastics are complex materials. All MNP are made of: 1) the polymer as a solid matrix, 2) the additives, chemicals dispersed intentionally inside the matrix to tune the plastic properties (Hahladakis et al., 2018), 3) the degradation products, produced by the several processes during and after the useful lifetime of a piece of plastic (Waldman and Rillig, 2020), 4) contaminants on MP surface, either pollutants and other substances coming from the environment (Endo et al., 2013), or the biofilms grown on its surface (Luo et al., 2022). Each of these four categories can have a multitude of chemicals, increasing the complexity of the MPs found in the environment and interacting with the biota components.

Phytotoxicity of Organic Pollutants
ISBN 978-0-443-33110-7,
https://doi.org/10.1016/B978-0-443-33110-7.00005-5

Early studies focused on MNP pollution in marine and other aquatic environments, and only in 2012, the question "Why have MPs not been studied in soils and terrestrial systems?" was first raised (Rillig, 2012). Seven years later, the first hypothesis for MNP effects on plants was proposed and tested by the same author and his research group (De Souza Machado et al., 2018a,b; Rillig et al., 2019). The potential effects of different MP types were tested (including beads, fibers, and fragments) on a terrestrial plant-soil model, and changes were observed in plant biomass, tissue elemental composition, root traits, and soil microbial activities. They also showed that plant responses were dependent on plastic type and proposed that, although direct effects must occur, changes in soil structure and water dynamics may impact plant traits and function. During the same period, Qi et al. (2018) exposed wheat (*Triticum aestivum* L.) to contaminated soil with different plastic mulch films (biodegradable or polyethylene) and revealed that both have negative effects on vegetative and reproductive growth. From 2018 on, the number of publications about the phytotoxicity of plastic pollution in plants (especially crop species) rapidly increased, reaching about 50 publications in 2021 (Hartmann et al., 2022).

Although being an uncharted issue, during the last decades MNP have been accumulating in terrestrial ecosystems (Büks and Kaupenjohann, 2020; Rillig, 2012), especially in agricultural fields (Zang et al., 2020), which are particularly vulnerable to plastic contamination due to degradation of agricultural plastics such as mulch films, irrigation systems, greenhouse coverings, as well as fertilization with sewage sludge. The forecast models predict a potential doubling of MP concentrations in the environment by 2040 due to rising plastic production and the degradation of existing plastic pollution (Lau et al., 2020). According to Hartmann et al. (2022), the main studied plastics, concerning phytotoxicity, are (in decrescent order): polystyrene (PS), polyethylene (PE), low-density polyethylene (LDPE), high-density polyethylene (HDPE), polypropylene (PP), polyvinyl chloride (PVC), polyester (PES), polyethylene terephthalate (PET), starch-based biodegradable polymers (Bio), polyamide (PA), polylactic acid (biodegradable polymer; PLA), polyurethane (PU), polymethylmethacrylate (PMMA), and polycarbonate (PC).

We highlight that most studies on MNP accumulation and toxicity in vegetative and reproductive organs were mostly performed under strictly controlled conditions using standard plastics and non-realistic plastic concentrations–several times higher than the actual concentrations found in the environment (Büks and Kaupenjohann, 2020). More experiments under environmentally realistic conditions are required to better understand this process under natural conditions. Besides, a unique polymer type was often used as a contaminant (mainly polystyrene), generally with a size higher than 100 nm (Hartmann et al., 2022). However, investigations on MNP phytotoxicity are recent—-starting in 2016 (Qi et al., 2018)—and the outputs indicate a wide and paradoxical range of results, indicating negative and positive effects. This variation is likely coming from the wide range of experimental conditions, varying from plant species to its growth stage or physiological maturity to the MNP preparation, using different degradation levels, considering differently the presence or the type of additives, with no recommended standards for the size of the particulate, or its concentration—often unrealistic ones. To be able to properly address the issue in a limited space, this chapter will delve mainly into the representative portion of the literature describing the negative results.

2. Uptake, translocation, and accumulation in plant tissues

In roots, plastic particles seem to accumulate mainly in the root cap, in the subapical meristem, and in the differentiation zone (Li et al., 2020a). The presence of NPs and small MPs has been widely recorded inside the root: both in the cortex and inside the stele (Liu et al., 2022). Li et al. (2020a) proposed two apoplastic pathways to explain how MNP enter roots: by crack-entry mode, where the emergence of lateral roots in the main root creates cracked areas, and through regions with high cell division rates, such as the root cap and the apical meristem. The uptake of MNP is increased under higher transpiration rates since the movement close to the rhizosphere occurs through mass flow (Li et al., 2020a). The movement of MNP toward the stele occurs via apoplast. Most of the particles accumulate in the cortex, around the endodermis (Li et al., 2020a), due to the presence of the Casparian strip, which acts as a selective filter in the roots (Geldner, 2013). However, to access the stele in mature root regions, the Casparian strip barrier is transposed through the passage cells, which lack the Casparian strip (Li et al., 2020a).

Once inside the stele, plastic particles can be translocated to shoot through the xylem stream, reaching the leaves and reproductive organs (Dong et al., 2022; Li et al., 2020a). Oliveri Conti et al. (2020) quantified MNP in several fruits and vegetables to better access possible human dietary intakes. The most contaminated fruit was apple (*Malus domestica* (Suckow) Borkh.), while carrot (*Daucus carota* L.) was the most contaminated vegetable. Besides the accumulation in fruits, NP can also be parentally transferred, as shown by Kim et al. (2024). The authors exposed pea plants (*Pisum sativum* L., synonym of *Lathyrus oleraceus* Lam.) to fluorescent NP beads (200 nm) in soil and the harvested fruits presented NP in the tissues. A subsequent generation of plants was replanted in uncontaminated soil. This generation presented fluorescence in the cell wall of the vascular bundles, indicating parental transferring.

Small plastic particles are also widespread in the air and, because of that, Brahney et al. (2020) proposed the term 'plastic rain'. The authors evaluated the rates of MP deposition in several remote U.S. conservation areas and found deposition rates of 132 plastics per square meter per day, which accounts for more than 1000 metric tons of plastic deposition to U.S. protected lands annually. The ubiquitous presence of MPs in the air plus its easy transportation has widespread implications for vegetation, since leaves are an important sink of airborne MPs (Jiao et al., 2024), intercepting the particles and acting as a natural barrier (Huang et al., 2022). The stomata are the main micromorphological structure driving MPs interception and there is a positive correlation between the stomata pore area and MPs abundance (Jiao et al., 2024), but other structures such as trichomes and the cuticle also trape MPs on the leaf surface (Lian et al., 2021). The electrostatic interactions also affect this process, since positively charged MPs tend to associate more with negatively charged leaf surfaces, while negatively charged MPs are often repealed (Sun et al., 2020). MPs intercepted by leaves can access internal tissues through the stomata. The particles can then access the phloem and reach other sink tissues of the plant, including reproductive organs such as fruits and seeds (Sun et al., 2021).

3. Mechanisms of phytotoxicity and plant responses to microplastic pollution

The direct effects of MNP on plants result from their interaction with plant surfaces and internal structures (Bouaicha et al., 2025). Larger particles primarily damage root surfaces by abrasion and adhesion to the epidermis, accumulating in the cortex and obstructing water and nutrient uptake (Bosker et al., 2019; Lian et al., 2020; Wu et al., 2020). In contrast, smaller particles, such as NP, can be absorbed by plant cells, potentially altering molecular and metabolic processes (De Souza Machado et al., 2018a). Indirect effects involve alterations in soil physicochemical properties and the rhizosphere, thereby impacting plant growth and development. In this topic, we explore some of the direct and indirect effects on plants (Fig. 12.1).

3.1 Toxicity of microplastic pollution to plants

Seed germination is a crucial stage in plant development and crop yield. Large plastic particles can delay this process by accumulating in the seed coat, blocking pores, and reducing water uptake, as observed in garden cress (*Lepidium sativum* L.) exposed to 4.8 μm polystyrene MPs (Bosker et al., 2019). It was suggested that the water deficit induced by MPs hinders starch hydrolysis and increases reactive oxygen species production, inhibiting germination and reducing seedling vigor. Germination inhibition by MPs has been reported in several species, such as ryegrass (*Lolium perenne* L.) exposed to polyamide (PA) fibers (>2 mm) (Boots et al., 2019), lettuce treated with MNP (Gong et al., 2021) and even non-crop native species (Lozano et al., 2025). Additionally, plastic debris can cause seedling malformations and reduced root growth, likely due to cytotoxic effects and impaired energy production (Bosker et al., 2019; Li et al., 2021b). Conversely, exposure to NP improved wheat germination, root elongation, and seed vigor. This positive effect was attributed to metabolic changes in energy production pathways, including the tricarboxylic acid cycle and starch metabolism (Bouaicha

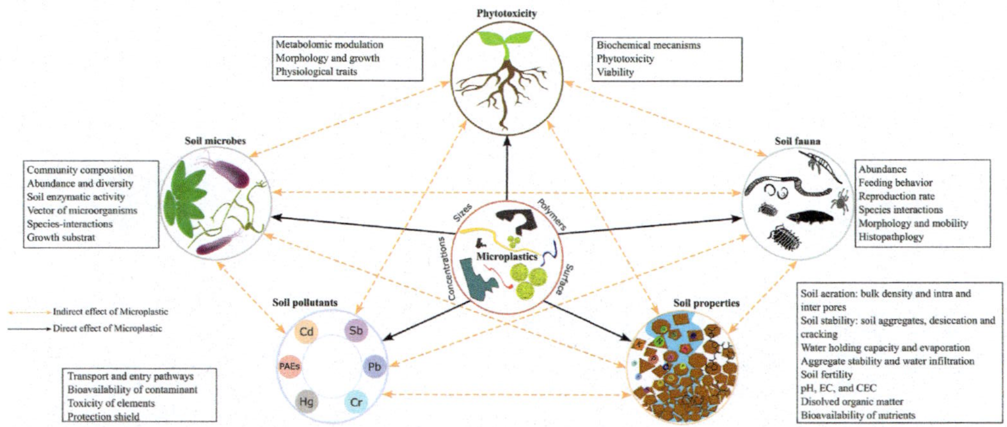

Figure 12.1 ***Schematic representation of the potential alterations promoted by micro- and nanoplastics on plants, soil physical and chemical properties, and soil fauna***. The solid and dashed lines represent the direct and indirect effects, respectively.

et al., 2022b; Lian et al., 2020). Overall, most germination studies indicate that increased plastic concentration inhibits seed germination, posing long-term risks to agricultural systems. As a result, delayed germination and seedling malformations can compromise plant growth and crop yields, especially in annual species (Rice and Dyer, 2001).

MNP have significant impacts on crop growth and yield, particularly affecting edible plant parts under higher concentrations, representing risks to food safety (Kim et al., 2024; Oliveri Conti et al., 2020). Exposure to MNP has been linked to reduced root biomass in carrot and maize (Dong et al., 2021, 2022; Urbina et al., 2020; Fu et al., 2022), decreased bulb development in spring onion (De Souza Machado et al., 2019), and lower dry weight, biomass, and height in lettuce (Gao et al., 2021; Lian et al., 2021). Similarly, Chinese cabbage exhibits reduced fresh weight (Yang et al., 2021), while rice and wheat show a decline in grain number and biomass (Qi et al., 2018; Wu et al., 2020). Under high MNP concentrations, Arabidopsis, maize, onion, and rice exhibit significant biomass reduction and decreased root and shoot length (Dong et al., 2020; Giorgetti et al., 2020; Sun et al., 2020; Urbina et al., 2020). However, some plants display contrasting responses. For instance, rice seedlings exposed to NP show shorter main roots but increased lateral root density, likely due to altered root elongation and nutrient uptake (Zhou et al., 2021a). Conversely, MNP exposure in lettuce and wheat promotes root growth, potentially as an adaptive response to stress and limited nutrient availability (Li et al., 2020a; Lian et al., 2020). These findings raise concerns about the long-term implications of plastic contamination in agricultural soils, where plastic accumulation is expected to increase in the coming decades (Büks and Kaupenjohann, 2020). For example, the extent of MNP effects varies depending on the experimental setup, particularly between soil-based and hydroponic systems. In soil, root responses to MPs exposure often resemble those induced by water deficit, including increased elongation, reduced diameter, larger surface area, and a higher root-to-shoot biomass ratio (Khalil et al., 2020). Additionally, MPs contamination negatively affects shoot development, leading to reductions in leaf length, fruit biomass, and tiller production across multiple species (Boots et al., 2019; De Souza Machado et al., 2019; Dong et al., 2021; Li et al., 2020a; Lian et al., 2021; Lozano and Rillig, 2020; Pignattelli et al., 2020; Qi et al., 2018; Sun et al., 2020). In hydroponic conditions, MNP interactions with plant tissues tend to be more direct, as these particles remain in constant contact with roots rather than interacting with soil components.

Reduced photosynthetic rates were observed in several species in response to MNP exposure, including maize, lettuce, and rice (Dong et al., 2020; Gao et al., 2019; Urbina et al., 2020; Fu et al., 2022). This occurs by stomatal and non-stomatal limitations. The first is a consequence of reduced tissue hydration due to the blockage of the cell wall and aquaporin channels by the NMP particles, as well as by changes in soil characteristics, negatively affecting water absorption (Boots et al., 2019; De Souza Machado et al., 2019; Dong et al., 2020; Lian et al., 2020; Urbina et al., 2020). Conversely, NP (approx.

100 nm) enhanced gas exchange in wheat at low concentrations, but reduced it at higher levels (Lian et al., 2020). Beyond stomatal effects, MNP also impairs photosynthesis non-stomatically (Li et al., 2020b; Pignattelli et al., 2020; Wang et al., 2022; Wu et al., 2020) by decreasing RuBisCO synthesis as a consequence of oxidative stress (Gao et al., 2019). The chlorophyll content is also affected by NMP, with reductions and increases, depending on species and conditions (Dong et al., 2020; Gao et al., 2019; Lian et al., 2021; Liu and Shen, 2024; Yang et al., 2021). Additionally, a high concentration of plastic particles can disrupt photosystem II functioning, reducing electron transport rates in thylakoids and maximum quantum efficiency (Dong et al., 2020; Gao et al., 2019; Li et al., 2020b; Pehlivan and Gedik, 2021).

The effects of MNP on growth and photosynthesis are related to oxidative stress, which is evidenced by elevated levels of malondialdehyde (Gong et al., 2021), a lipid peroxidation byproduct, along with reactive oxygen species (ROS) such as superoxide anion, hydroxyl radical, and hydrogen peroxide. Additionally, antioxidant enzymes such as catalase, peroxidase, and superoxide dismutase show increased activity in plants exposed to MNP (Bouaicha et al., 2026; Gao et al., 2021; Gong et al., 2021; Jiang et al., 2019; Li et al., 2020b; Maity et al., 2020; Wu et al., 2020; Zhou et al., 2021a). The rise in ROS and enzymes suggests excessive ROS production and insufficient scavenging system, which can lead to oxidative damage in plants. The exact pathways of MNP-induced oxidative stress remain unclear, but it is suggested that the mechanical damage to root surfaces caused by MP abrasion, leaching of absorbed chemical compounds from MNP (Lozano et al., 2025), water deficit due to soil structure alterations (De Souza Machado et al., 2019; Wan et al., 2019), and photosynthesis disruption (Dong et al., 2020) are some possible ways of action.

MNP concentration is of paramount importance when we evaluate plastic phytotoxicity. Most of the plant responses we describe in this chapter are related to exposure to high concentrations of particles, generally higher than found in nature. Beyond particle size and concentration, the shape of plastic particles also affects the extent of damage they cause. Sharper particles, such as fragments, can inflict more harm on root and cell surfaces compared to fibers or spheres, for example (Rozman et al., 2021). Additionally, the chemical composition and the electrical change of polymers also influence phytotoxicity. Lettuce seedlings exposed to 200 nm polystyrene coated with different charge groups (-OH, –COOH, $–SO_3H$, and-NH_2) showed different degrees of damage, with PS–NH_2 promoting the worst changes. This was attributed to its positive electrical charge, which easily interacts with negatively charged cell walls, accumulating in root surface and binding sites of the plasma membrane, hindering the absorption of cationic elements (Xu et al., 2022). The effect of the electrical charge was also shown in arabidopsis exposed to PS-NH_2 (positive) and PS-COOH (negative). The positively charged PS accumulated at relatively low levels in the root tips, promoting a higher accumulation of reactive oxygen species and growth inhibition than the negatively charged ones, which in turn were more internalized–often detected in the apoplast and xylem–but less harmful (Sun et al., 2020).

3.2 Toxicity of microplastic pollution in soil and rhizosphere and the consequences to plants

MPs can also affect plants by acting on the different rhizosphere and soil components rather than directly contacting the plant. The rhizosphere is one of the key interfaces between plants and their environment, with intensive root-induced physical, chemical, and biological processes (Zhang et al., 2022). It is the narrow soil region directly surrounded and influenced by root activity and associated soil microorganisms (Philippot et al., 2013). The rhizosphere is one of the most complex ecosystems on earth, being inhabited by various organisms including nematodes, fungi, bacteria, and arthropod herbivores (Venturi and Keel, 2016). The study of plastic phytotoxicity in the rhizosphere has gained increasing attention due to the widespread environmental distribution of MNP (Chen et al., 2022). These contaminants in the rhizosphere can disrupt soil structure and water retention, affect microbial diversity and abundance, and absorb or release pollutants that impact nutrient bioavailability and uptake (Bouaicha et al., 2022a). Understanding these complex interactions is essential to assess the long-term impacts of plastic contamination on terrestrial ecosystems and crop productivity. Once MPs enter the soil system, they interact in complex ways that create challenges in understanding their effects, due to the variety of soil types, the MPs' properties, and the different study conditions (Bouaicha et al., 2022a). Predicting how soil properties change in the presence of MPs is therefore difficult, but it has been demonstrated that MPs pollution significantly alters several soil characteristics, including texture, aggregate size, pore structure, moisture retention capability, water stability, and bulk density (De Souza Machado et al., 2018b; Fang et al., 2025; Hanif et al., 2024). These changes can negatively impact root dynamics and growth, leading to potential phytotoxicity.

Bulk density and water dynamics, including moisture retention and evaporation, play crucial roles in soil aeration, penetration, and root growth (Yu et al., 2024). MPs (including PS fibers and PP) lower bulk density and enhance soil water retention capacity, by creating water flow channels and cracks, facilitating evaporation and infiltration (Wan et al., 2019; Yu et al., 2023). On the contrary, Zhang et al. (2019) found that the addition of plastics to soil did not significantly affect its saturated hydraulic conductivity or bulk density. These findings are closely related to the structure and stability of soil pores, which are profoundly influenced by microplastics (Sepehrnia et al., 2024; Xing et al., 2024). Introducing MPs substantially alters the interconnected pore network's structure, morphology, composition, and connectivity. Due to MPs' physicochemical characteristics (e.g., size, charge, and hydrophobicity) and soil pores' shape, MPs can act as blockers of crucial interconnection. In the more severe scenario, pore sizes can diminish and even disappear entirely (Li et al., 2024a), potentially affecting the water balance in plants.

Even though the overall structure of soil pores differs between textures, MPs still affect pore morphology and number independently of soil texture, but to a different extent. For instance, Dike and Apte (2024) found that sandy soil is most affected by MP contamination, followed by clay soil and silt soil regarding water dynamic and dry density. Similarly, MPs in sandy soils rapidly reduced or eliminated connected pores, while loamy soils could maintain a certain degree of pore connectivity, probably due to soil particle sizes and surface area (Wang et al., 2024). Indeed, sand has the biggest particle size with a lower water-holding capacity (Bao et al., 2024), rendering it less exposed to pores blocking. Thus, in sandy soils MPs move across soil profiles more easily, increasing their bioavailability to plant roots and microorganisms and increasing the probability of groundwater contamination. Additionally, aggregates as pivotal indicators for assessing soil quality are affected by MPs (Fang et al., 2025).

Soil aggregates serve as a sink for MPs within the soil ecosystem, with a higher abundance in macroaggregates (>5 mm fractions) than in microaggregates (Zhang et al., 2022). Significantly, HDPE and PLA MPs decreased the stability of soil aggregates, by altering the proportions of aggregate fractions (Han et al., 2024; Liu et al., 2022). Due to MP characteristics, such as shape, flexibility, and crimpness, a higher abundance of aggregates is expected to be observed in the presence of MPs (Lehmann et al., 2022), highlighting the participation of MPs in the formation of soil aggregates. Nevertheless, the neutral effects of MPs on soil aggregation were also reported (Liang et al., 2021; Yu et al., 2023). Despite this, the phytotoxicity of MPs will depend on the specific needs and preferences of the crops. Well-aerated soil with moderate bulk density and water-holding capacity, characterized by an optimal number, size, and structure of pores, would be suitable for crops such as tomatoes (Tracy et al., 2013). In contrast, crops such as rice would prefer denser soil with higher water-holding capacity (El Baroudy et al., 2014). Additionally, tuber crops would thrive better in lighter soil with low density compared to dense clay soil (Chow et al., 2007).

MPs potentially alter the chemical properties of soil and the rhizosphere, reducing soil fertility (Hasan and Tarannum, 2025). Soil chemical factors such as pH, electrical conductivity, organic matter, cation exchange capacity, redox potential, and nutrient and toxic content are important factors for vegetation growth and development, where any malfunctioning and alteration could lead to phytotoxicity (George et al., 2012). Depending on the soil and MP polymer, the presence of MPs can increase or decrease soil pH (Boots et al., 2019; Joos and De Tender, 2022). For example, PE has been shown to lower pH in acidic soils and increase it in alkaline soils (Dissanayake et al., 2022). On the contrary, Yuan et al. (2023) found that MPs did not affect the soil pH in saline soil, even when they decreased the electrical conductivity and increased the soil organic matter. In addition (Ma et al., 2023), found that PE MPs reduced the contents of dissolved organic matter and increased the cation exchange capacity, suggesting that MPs interfere with the decomposition of organic matter and disrupt nutrient cycling in the soil (Goyal et al., 2024). The previously reported chemical properties highly affect nutrient profiling and heavy metals (Alkaç and Günes, 2024; Liu et al., 2025). As mentioned, MP polymers have different chemical compositions and charges, which may lead to varying effects on soil components. MPs can release metals such as Cu, Zn, and Mn—commonly used as plastic additives into the soil (Taghizadeh Rahmat Abadi et al., 2024). At the same time, MPs have the potential to adsorb and mobilize elements such as B, Cl, Cu, Fe, Mn, Mo, P, and Zn, influencing their bioavailability and stoichiometry in the soil (Pita-Barbosa et al., 2025). This process enhances their transfer across environmental systems and within soil profiles (Fang et al., 2022; Wang et al., 2024).

Beyond these effects on elemental content and mobility, MPs also influence nitrogen (N), carbon (C), and phosphorus (P) dynamics in the soil (Hou et al., 2023; Yin et al., 2023)). The impact on the N cycle has been linked to MPs's effects on processes like mineralization, nitrification, and denitrification (Hanif et al., 2024). For instance, low density polyethylene (LDPE, 2%–7%) alters functional groups involved in the N cycle (Rong et al., 2021). Regarding C, MPs primarily composed of C are considered a major external source of C in soil (Rillig, 2018), influencing the C cycle and sequestration (Shen et al., 2023) and thus affecting the C/N ratio (Moharana et al., 2024). Changes in soil nutrient content may impact their availability and accessibility, especially given their interconnected synergy and antagonistic relationships (Rietra et al., 2017; Xie et al., 2021). Such disruptions can trigger plant responses by altering root morphology and potentially causing phytotoxicity. Similarly, MPs can absorb and interact with heavy metals such as Cd, Zn, Ni, Pb, Ag, Hg, Fe, Sb, and Mn (Wang et al., 2022). While sorption may initially decrease bioavailability to plants, prolonged low desorption could lead to toxic accumulation in the rhizosphere (Shirkhorshidi et al., 2023). Besides, MPs exhibit a high surface area-to-volume ratio and hydrophobic properties, enabling them to sorb various chemical substances, including pesticides and antibiotics, depending on the type of MPs, chemical composition, and environmental conditions (Kinigopoulou et al., 2022). This sorption capacity can significantly impact agrochemical efficacy, as MPs may bind pesticide (e.g., herbicide) molecules, making them unavailable to target plants. As a result, it reduces pesticide effectiveness, leading to increased application rates and heightened soil and ecosystem contamination risks (Brochado et al., 2024; Lajmanovich et al., 2022). Such disruptions can trigger plant responses by altering root morphology and potentially causing phytotoxicity.

MPs have been shown to interfere with the beneficial relationships between plants and soil microorganisms, such as mycorrhizal fungi and nitrogen-fixing bacteria (Ansari et al., 2025). This interference can occur directly through the decomposition of MPs in the soil and the release of additives (Pedram Jarf et al., 2024). Numerous studies have reported that MPs alter microbial community composition, diversity, and activity in the soil and the rhizosphere (Ng et al., 2021). MPs have been found to decrease the richness and diversity of bacterial communities while enriching specific bacterial taxa (Wang et al., 2022). For instance, PLA and PE MPs reduce bacterial diversity in rice and pakchoi (*Brassica rapa* L. subsp. *chinensis*) rhizosphere, altering the dominant bacterial species' community structure (Lai et al., 2025; Liu et al., 2025). However, in *Pennisetum alopecuroides* (L.) Spreng (a synonym of *Cenchrus alopercuroides* (L.). Thunb.) soil, MPs significantly increased the alpha diversity of eukaryotes while decreasing in bacterial communities (Li and Xiao, 2023). This suggests that the impact of MPs on soil microbial communities depends on several factors, including crop species. The type and concentration of MPs are also critical factors influencing their effects on rhizosphere microorganisms. For instance, PP MPs increased crop mycorrhizal infection rates, while PET MPs decreased (De Souza Machado et al., 2019). Additionally, both polyvinyl chloride (PVC) and PE MPs at concentrations greater than 7% affect the alpha diversity of fungal communities. However, for bacterial communities, only PVC MPs decreased it (Fan et al., 2022). In contrast, PVC MPs at 0.1% and 1% concentrations did not significantly impact bacterial communities' overall diversity and composition (Yan et al., 2021).

MPs play a unique role as novel niches for soil microbes, fostering the formation of distinct microbial communities depending on factors such as soil type, pollution level, and the presence of additives (Zhou et al., 2021b). Some studies indicated that MPs can promote the growth of arbuscular mycorrhizal fungi (AMF) (De Souza Machado et al., 2019; Jiang et al., 2024). Similarly, PE MPs have been found to promote the growth of specific bacterial taxa associated with stress responses. Additionally, MPs have been shown to increase the abundance of Proteobacteria, Actinobacteria, and Ascomycota while decreasing the abundance of Acidobacteria, Basidiomycota, and Chytridiomycota (Fan et al., 2022). Furthermore, PE and PVC NPs can shift microbial communities from Gram-positive to Gram-negative bacteria (Zang et al., 2020) suggesting that MPs affect the soil microbes differently, as backed by Bouaicha et al. (2024), who found that PE MP pollution differentially influences the growth and sporulation of *Trichoderma viride* Pers. and *Fusarium solani* (Mart.) Sacc., two Ascomycota species commonly found in the rhizosphere. Moreover, MPs can act as carriers for microbial colonies, potentially introducing new species. For example (Luo et al., 2023), demonstrated that microplastics transferred into the soil were accompanied by the transfer of green, fluorescent *Escherichia coli* E (GFP-E. coli). This finding raises concerns about the risks of spreading multi-resistant species, including plant soil-borne pathogens (Trojan et al., 2024). Such effects could disrupt complex ecosystems such as soil and rhizosphere, ultimately hindering plant growth and several mechanisms by which MPs can disrupt bacterial and fungal communities, including promoting, reducing, suppressing, or introducing new species into the soil. As a result, MPs potentially alter networking adaptations, inducing the formation of specialized microbial networks adapted to metabolize MPs depending on MPs type, size, and concentration (Ren et al., 2020). For example, soil bio-MPs have been shown to create more complex microbial networks than those with PE MPs (Cheng et al., 2024). Similarly, PS MPs at 5% concentration (<149 μm) display a more complex ecological network, while PS at 10% concentration (<149 μm) reduces network complexity with more negative interactions (Liu et al., 2025). PLA MPs have been found to create a more complex bacterial network than PE MPs and control soils, though in fungal communities, MPs slightly reduce network complexity (Qiu et al., 2024). Interestingly, higher temperature and MP diversity in soil amplify microbial networks, resulting in higher network connectivity and species interactions than individual MP types and control conditions (Wu et al., 2024).

In contrast to the extensive research on soil microbes, the impact of MPs on soil fauna remains less documented. However, studies have shown that MPs negatively affect the composition and abundance of soil fauna. For example, Lin et al. (2020) found that LDPE MPs at concentrations of 5–15 g per m^2 significantly decreased the abundance of microarthropods (oribatid mites, Dipteran larvae, Lepidopteran larvae, and Hymenoptera) and nematodes (omnivores, predators, and plant feeders) compared to the control. Soil fauna is crucial in maintaining soil health and biodiversity (Zhang et al., 2022). Therefore, the observed effects on their composition and abundance could hinder plant growth and increase the likelihood of phytotoxicity. Rather than that, MP ingestion by soil fauna may impair several functions, including inducing false satiety, weak nutrition supply, oxidative stress, and intestinal damage (Zhang et al., 2022). MPs also hinder the activity, survival, growth, and reproductive rates of soil fauna, posing potential risks to plant health (Chang et al., 2024; Wang et al., 2024).

Soil enzymes are critical indicators of soil health, often responding to environmental stressors such as pollution earlier than other soil properties. They play a vital role in organic matter decomposition and nutrient cycling (e.g., amylase, β-glucosidase, cellulase, chitinase, dehydrogenase, phosphatase, protease, and urease) (Li et al., 2025). Some studies suggest that MPs can enhance enzymatic activity. For instance, meta-analyses have shown that MPs can boost the activity of soil urease and leucine aminopeptidase (Feng et al., 2022; Liu et al., 2023; Xiang et al., 2025). Similarly, PA, PE, and PES MPs increased the activity of fluorescein diacetate hydrolase (De Souza Machado et al., 2019). Additionally, urease and acid phosphatase activities increased dramatically in the presence of PVC, LDPE, PA, PS, and PBS MPs (Fei et al., 2020; Guo et al., 2024; Huang et al., 2019). In contrast, other studies have reported inhibitory effects of MPs on soil enzymatic activities. For example, enzymes such as β-D-glucosidase, cellobiosidase, N-acetyl-β-glucosaminidase, urease, dehydrogenase, and phosphatase were negatively affected by MPs (Guo et al., 2024; Khan and Sikder, 2024; Liang et al., 2021; Lozano et al., 2021; Liang et al., 2021, 2021; Lozano et al., 2021, 2021). Interestingly, some studies have found no significant effect of MPs on certain soil enzymatic activities (De Souza Machado et al., 2019; Fei et al., 2020; Liang et al., 2021; Xiang et al., 2025). This further underscores the complexity of MP interactions with soil enzymes, suggesting that the effects are highly context-dependent, as well as on polymer type and enzyme category (Zhao et al., 2021). Likewise, MPs' size also plays a crucial role. For example, medium-sized PVC MPs (1500 μm) significantly enhanced soil invertase activity, while larger particles (4500 μm) had no effect (Gao et al., 2024). Similarly, smaller PE MPs (300 μm) reduced the activity of β-1,4-N-acetylglucosaminidase, whereas larger particles (600 μm) did not (Ma et al., 2023). These findings indicate that smaller MP particles tend to have a more pronounced effect on soil enzyme activities, likely due to their greater surface area and reactivity. From another perspective, the duration of MP exposure also influences their impact on soil enzymes. Short-term exposure often results in more pronounced effects, which may diminish over extended periods (Zhao et al., 2021). This could be due to microbial adaptation to the new environmental conditions or the inactivation or degradation of MPs and their additives over time, leading to a reactivation of enzyme activities (Kim et al., 2020; Yi et al., 2021). In summary, the impact of MPs (MPs) on soil enzyme activities is highly variable and influenced by several factors, including the type of polymer, the size of the MPs, and the duration of exposure. Further research would enhance our understanding of MP pollution's implications for soil enzyme activities, health, and overall phytotoxicity.

4. Ecological impacts of microplastic pollution

Several studies have demonstrated that MPs affect plant performance; however, most of this knowledge has been derived from experiments using crop species. Little is known about the ecological effects of MPs, particularly their impact on plant community productivity and structure. In this topic, we bring some examples of MNP pollution as a potential risk to native species and natural ecosystems.

MNPs influence plant species dominance (Lozano and Rillig, 2020), with invasive alien species generally outperforming native ones (Li et al., 2024b; Lozano et al., 2025; Lozano and Rillig, 2020; Zhang et al., 2024), raising concerns for biodiversity conservation. Native plant species, often adapted to specific ecological niches, tend to be more vulnerable to environmental changes. In contrast, invasive species, characterized by high adaptability, resilience, high root-to-shoot ratio, and rapid growth rates (Grotkopp and Rejmánek, 2007), may have a competitive advantage in MP-polluted soils, leading to suppression of native plant growth (Zhang et al., 2024). For instance, an experiment with seven species cultivated in MP-contaminated soil showed that invasive and allelopathic species became dominant, while species that could potentially facilitate the establishment of others experienced reduced biomass (Lozano and Rillig, 2020). Additionally, another study demonstrated that invasive alien species exhibited enhanced competitiveness over native species under pronounced dry-wet cycle conditions (Zhang et al., 2024), suggesting that alien species would take advantage of the actual climate change and MP pollution scenario.

Mangrove ecosystems have been recognized as important sinks for MP pollution, with mangrove plants playing a crucial role in the interception and accumulation of these pollutants (Deng et al., 2021). The accumulation and distribution of MPs in mangrove sediments are strongly influenced by plant species and, more specifically, root morphology. The unique root structures of mangroves, called pneumatophores, promote the trapping of MPs from terrestrial and marine sources, as demonstrated by Duan et al. (2021). The authors compared the concentration of MPs in three monospecific mangrove stands dominated by *Sonneratia caseolaris*(L.) Engl., *Kandelia obovata*Sheue, H.Y.Liu & J.W.H.Yong, and *S. apetala*Buch. -Ham. They found that MPs abundance positively relates to pneumatophores density, resulting in a lower concentration of MPs in the forest interior, when this density is high. This is the first report of how plant species affect the spatial distribution of MPs in mangrove sediments, providing useful information for environmental processes of MPs in coastal wetlands (Duan et al., 2021).

Another highly plastic-polluted habitat is the coastal dunes. In a recent study, Bitton et al. (2025) evaluated the effects of MNP exposure on seed germination of *Cutandia maritima* L. (Barbey), a native dune species, in a greenhouse experiment. No effect was found on plants submitted to chronic exposure to pristine and weathered MPs. However, weathered HDPE MPs significantly reduced seed germination after exposure of mother plants, suggesting that epigenetic modifications might have affected the embryo. These findings highlight the vulnerability of coastal vegetation to microplastic pollution and the potential disruption of dune ecosystem regeneration.

MPs also accumulate and affect marine grasses. Seagrass beds, which serve as nearshore marine ecosystems, are also important targets for plastic contamination (Li et al., 2023). In *Cymodocea nodosa*(Ucria) Asch., MPs negatively affected growth by increasing leaf loss rate and oxidative stress, and by decreasing photosynthesis, which makes this emergent pollutant a novel threat to seagrass ecosystems (Menicagli et al., 2022). On the other hand, the seagrasses also act as ecosystem service providers, helping to remove MPs from the ocean by transporting them away from the meadows to the coastline as the so-called aegagropilae or "neptune balls", made of aggregates of vegetal fibers intertwined by seawater motion. This process was nicely described for the species *Posidonia oceanica*(L.) Delile by Sanchez-Vidal et al. (2021).

5. Plastic detection in plants and the environment

Plastic pollution significantly impacts ecosystems across the globe, encompassing air, land, and marine environments. Similar to climate change and persistent organic pollutants, plastics exemplify the extensive and pervasive human influence on the environment (Dong et al., 2022). To assess and understand this impact, studying the detection and characterization of plastics on meso-, micro-, and nanoscale levels is crucial. Common analytical techniques include Fourier-transform infrared spectroscopy (FTIR) and Raman spectroscopy. However, emerging methods such as optical photothermal infrared spectroscopy (O-PTIR), and chromatographic techniques have shown considerable potential for identifying and characterizing MNP (Xie et al., 2024). Herein we will highlight a few techniques to enable MNP detection in soil-plant systems.

5.1 Challenges for the detection and quantification of micro- and nanoplastics in the environment and in plant tissues

Identifying MNP in environmental samples is challenging due to their heterogeneity in morphology and chemical nature (Blevins et al., 2021; Kang et al., 2020; Maes et al., 2017; Tympa et al., 2021). Additionally, size plays a role in the MNP pervasiveness, being the smaller particles (1–1000 nm) more pervasive within terrestrial ecosystems, including incorporation into plant and animal tissues. Nanoparticles, in particular, can penetrate intracellular environments, where their impacts on the metabolism of living organisms are more pronounced and observable, being transported from soil to plants through root uptake, eventually reaching stems, leaves, and even fruits (Azeem et al., 2023).

Azeem et al. (2021) reviewed the analytical challenges of determining MNP in soil and plant matrices. The authors identified the need for a universal, cost-effective, and efficient analytical method as a key obstacle in this field. The study also emphasized the critical importance of sample treatment prior to any physical or chemical characterization, precisely because of the size of the plastic particles involved in the matrix. In soil-plant systems, it is already expected that NPs will be found, mainly due to the transportation of these particles from the roots to the leaves of the plant. For that, sample preparation must be tailored to the specific matrix. Moreover, matrices with high organic content generally require density separation, filtration, and digestion to remove organic matter aggregated with MPs.

The primary challenge in these scenarios lies in identifying particles smaller than 1 μm, as submicron-sized plastics can potentially be internalized within plant tissues. Plants exhibit high selectivity for nanoparticle size, further complicating the detection of these internalized plastics. In addition, NPs tend to form homo- or heteroaggregates or undergo physicochemical changes (such as oxidation or adsorption of biomolecules) within plant tissues, which can alter their properties and make it difficult to distinguish between plastic particles and other nanoparticles present in the plant (Hua et al., 2023). Exposure to diverse physical, chemical, and biological stressors, complicates their subsequent detection and characterization (Tirkey and Upadhyay, 2021).

5.2 Techniques for the detection and determination of micro- and nanoplastics

FTIR and Raman spectroscopies identify polymer types by detecting their unique vibrational spectra. These techniques are non-destructive and enable inferring degradation levels (Mogha and Shin, 2023; Gomes et al., 2024). FTIR is particularly effective for identifying particles larger than 20 μm (Chen et al., 2020), whereas Raman can identify particles of around 10 μm. Contamination can interfere with the results for both techniques, so sample preparation is key for good outputs (Bialkowski et al., 2019) . Despite the greater range for smaller particles, Raman can be strongly influenced by fluorescence, so color and some additives can prevent the usage of this technique.

O-PTIR is also a spectroscopic technique in the infrared region used to investigate the thermal and optical properties of materials, especially on a microscopic scale. It is based on the absorption of infrared radiation by the sample, which generates local heating due to the conversion of light energy into heat (Katzenmeyer et al., 2013). This temperature increase alters the optical properties of the material or its surroundings, such as the refractive index, and these changes are monitored using laser beams. The main advantage of O-PTIR is its high resolution, which can reach up to 500 nm without light scattering effects and fluorescence interference. These features make the analysis easier and more practical than other spectroscopic techniques, such as FTIR and Raman. In addition, it is considered a non-destructive technique, depending on the energy level used, and allows both solid and liquid samples to be studied (Adhikari et al., 2020; Martín-Gómez et al., 2024).

Although physical methods such as microscopy and spectroscopy are widely used to identify and characterize MNP, chromatography offers specific advantages for more detailed analysis, especially concerning the chemical composition of these materials, such as additives. Plastics often contain chemical compounds like plasticizers, stabilizers, and pigments that can leach into the environment. Techniques such as gas chromatography (GC) and high-performance liquid chromatography (HPLC) can detect and quantify the leached compounds, providing a comprehensive understanding of the environmental impact of MPs (Jiménez-Skrzypek et al., 2021). As aforementioned, MNPs must be prepared to reduce the contamination, usually by digestion processes (Biver et al., 2018; Picó and Barceló, 2020).

In several analyses, pyrolysis combined with gas chromatography and mass spectrometry (Py-GC-MS) are used (Jin et al., 2022). The plastic particles are heated to high temperatures, breaking down into unique smaller compounds that the chromatography technique can identify. This technique enables the precise identification of polymer types and allows for mass quantification, although the latter is highly dependent on the detection limit of the equipment used. Hybrid methods combining chromatography with spectroscopic techniques, such as FTIR or Raman, enhance the detection and differentiation of plastic particles within complex environmental matrices (Kabir et al., 2023). However, using this technique can be time-consuming, expensive, and require a well-trained operator.

Advanced microscopy techniques, such as scanning electron microscopy (SEM) and transmission electron microscopy (TEM) provide detailed images of the shape, roughness, and morphological features of MNP (Azeem et al., 2023). The SEM is more suitable for surface studies and 3D analysis on a larger scale, while TEM is used for investigations on an atomic and subnanometric level. Although they are simple analyses, the sample preparation process can be time-consuming (Claessens et al., 2013).

5.3 Use of fluorescent dyes in investigations

Environmental samples have plenty of particulates made of several different substances. MPs share similar size, shape, and color with many of them. Textile fibers can be confounded with vegetal and animal fibers (Suaria et al., 2020), fragments can be confounded with pieces of insect and plant debris, and plastic films can be confused with vegetable skins, epidermal cuticles, and detached biofilm layers. Due to this similarity among the different MPs and several other particulates, visual inspection, although straightforward and convenient, has an error rate of up to 70% when applied to micro- and nanometric particles, raising concerns about its reliability in quantification. To overcome this issue, fluorescent dyes have emerged as a promising and innovative approach for detecting MPs in environmental samples, particularly within terrestrial ecosystems (Sturm et al., 2021). Fluorescent dyes function by selectively binding to the surfaces of plastic materials. When exposed to light at specific wavelengths, these dyes emit fluorescence, thereby facilitating the visualization of MNP under fluorescence microscopy or other light-based detection systems (Nam et al., 2024). Since different polymers exhibit distinct surface properties stemming from their unique chemical structures and manufacturing processes, the selection of an appropriate dye is critical and depends on the type of polymer present in the sample matrix (Pencik et al., 2023). Nile Red, a widely used fluorescent dye, is particularly effective due to its affinity for hydrophobic surfaces, characteristic of most polymers (Sancataldo et al., 2020; Shruti et al., 2022).

In environmental samples, fluorescent dyes can detect MNP at low concentrations, ranging from 0.02 μg/mL to 10 μg/mL (Lozano et al., 2021). They are effective even for particles smaller than the detection limits of traditional methods, rendering such particles visible under ultraviolet light or fluorescence microscopes and enabling their detection in complex matrices such as sediments, biosolids, and soils (Blevins et al., 2021). Compared to conventional methods like SEM/TEM, FTIR, Raman spectroscopy, and chromatography, fluorescent dyes offer the advantage of relatively rapid analysis. While sample preparation is still required, visual detection becomes almost instantaneous following dye application (Ho and Masura, 2024). This approach, although, has notable limitations like some dyes binding to other hydrophobic, non-plastic materials, such as organic or lipid particles, potentially leading to false positives (Maes et al., 2017).

6. Future perspectives

For less than 10 years, the phytotoxicity of MNP have been studied and demonstrated that these particles can interfere with key physiological and metabolic processes in plants. However, the extent of this interference remains insufficiently understood, particularly in terms of how MNP may impact plant reproduction and overall productivity considering the chronic exposure. As such, future studies must refine our understanding of these mechanisms, exploring the intricate ways in which plastics affect plants at the cellular and molecular levels.

The discrepancy between laboratory conditions and real-world scenarios also raises concerns about the relevance and applicability of experimental findings to natural environments. The concentrations of MNP observed in agricultural and horticultural fields tend to be relatively low—up to 13,000 units per kilogram of dry soil or 4.5 mg per kilogram of dry soil (Büks and Kaupenjohann, 2020)—compared to the experimental studies, which often work with much higher concentrations to assess phytotoxicity (Hartmann et al., 2022). Given that, long-term experiments using environmentally relevant concentrations, that incorporate a variety of soil types, plant species, and plastic particle characteristics are essential to truly evaluate the long-term phytotoxic effects of MPs and MNPs. Additionally, the absence of effective methods for detecting and tracking MPs in natural environments also impedes the practical application of laboratory findings, limiting field interventions.

In addition to experimental efforts, it is critical to focus on how plastic inputs into agricultural fields can be reduced. Sustainable agricultural practices, such as the use of biodegradable materials and alternative plastic-free technologies, offer potential solutions to mitigate the growing risks associated with MNP pollution. Future research must explore the potential for integrating such practices into agricultural systems, ensuring that strategies for minimizing plastic input are both practical and effective. This focus on prevention will complement experimental studies, helping to create a more sustainable framework for addressing the environmental challenges posed by plastic pollution.

References

Adhikari, S., Spaeth, P., Kar, A., Baaske, M.D., Khatua, S., Orrit, M., 2020. Photothermal microscopy: imaging the optical absorption of single nanoparticles and single molecules. ACS Nano 14 (12), 16414–16445. https://doi.org/10.1021/acsnano.0c07638.

Alkaç, O.S., Günes, M., 2024. The effects of compost and fertilization applications on plant growth, development and nutrient content in leaves in cut tulip cultivation. J. Agric. Fac. Gaziosmanpasa Univ. https://doi.org/10.55507/gopzfd.1576758.

Ansari, I., Arora, C., Verma, A., Mahmoud, A.E.D., El-Kady, M.M., Rajarathinam, R., Verma, D.K., Mahish, P.K., 2025. A critical review on biological impacts, ecotoxicity, and health risks associated with microplastics. Water, Air, Soil Pollut. 236 (2), 88. https://doi.org/10.1007/s11270-024-07731-z.

Azeem, I., Adeel, M., Ahmad, M.A., Shakoor, N., Jiangcuo, G.D., Azeem, K., Ishfaq, M., Shakoor, A., Ayaz, M., Xu, M., Rui, Y., 2021. Uptake and accumulation of nano/microplastics in plants: a critical review. Nanomaterials 11 (11), 2935. https://doi.org/10.3390/nano11112935.

Azeem, I., Shakoor, N., Chaudhary, S., Adeel, M., Zain, M., Ahmad, M.A., Li, Y., Zhu, G., Shah, S.A.A., Khan, K., Khan, A.A., Xu, M., Rui, Y., 2023. Analytical challenges in detecting microplastics and nanoplastics in soil-plant systems. Plant Physiol. Biochem. 204, 108132. https://doi.org/10.1016/j.plaphy.2023.108132.

Bao, X., Bao, Z., Shen, J., Chen, X., Wu, X., Cui, H., 2024. Dynamic properties of sand with different contents of clay. Powder Technol. 431, 119070. https://doi.org/10.1016/j.powtec.2023.119070.

Barnes, D.K.A., Galgani, F., Thompson, R.C., Barlaz, M., 2009. Accumulation and fragmentation of plastic debris in global environments. Phil. Trans. Biol. Sci. 364 (1526), 1985–1998. https://doi.org/10.1098/rstb.2008.0205.

Bialkowski, S.E., Astrath, G.C., Proskurnin, M.A., 2019. Photothermal spectroscopy methods. John Wiley & Sons.

Bitton, N.K., Zucker, I., Gruntman, M., 2025. Microplastic exposure reduces seed germination in a coastal plant. Sci. Total Environ. 972, 179098. https://doi.org/10.1016/j.scitotenv.2025.179098.

Biver, T., Bianchi, S., Carosi, M.R., Ceccarini, A., Corti, A., Manco, E., Castelvetro, V., 2018. Selective determination of poly(styrene) and polyolefin microplastics in sandy beach sediments by gel permeation chromatography coupled with fluorescence detection. Mar. Pollut. Bull. 136, 269–275. https://doi.org/10.1016/j.marpolbul.2018.09.024.

Blevins, M., Allen, H., Colson, B., Cook, A.-M., Greenbaum, A., Hemami, S., Hollmann, J., Kim, E., LaRocca, A., Markoski, K., Miraglia, P., Mott, V., Robberson, W., Santos, J., Sprachman, M., Swierk, P., Tate, S., Witinski, M., Kratchman, L., Michel, A., 2021. Field-portable microplastic sensing in aqueous environments: a perspective on emerging techniques. Sensors 21 (10), 3532. https://doi.org/10.3390/s21103532.

Boots, B., Russell, C.W., Green, D.S., 2019. Effects of microplastics in soil ecosystems: above and below ground. Environ. Sci. Technol. 53 (19), 11496–11506. https://doi.org/10.1021/acs.est.9b03304.

Bosker, T., Bouwman, L.J., Brun, N.R., Behrens, P., Vijver, M.G., 2019. Microplastics accumulate on pores in seed capsule and delay germination and root growth of the terrestrial vascular plant Lepidium sativum. Chemosphere 226, 774–781. https://doi.org/10.1016/j.chemosphere.2019.03.163.

Bouaicha, O., Mimmo, T., Tiziani, R., Praeg, N., Polidori, C., Lucini, L., Vigani, G., Terzano, R., Sanchez-Hernandez, J.C., Illmer, P., Cesco, S., Borruso, L., 2022a. Microplastics make their way into the soil and rhizosphere: a review of the ecological consequences. Rhizosphere 22, 100542. https://doi.org/10.1016/j.rhisph.2022.100542.

Bouaicha, O., Tiziani, R., Maver, M., Lucini, L., Miras-Moreno, B., Zhang, L., Trevisan, M., Cesco, S., Borruso, L., & Mimmo, T. (2022b). Plant species-specific impact of polyethylene microspheres on seedling growth and the metabolome.

Bouaicha, O., Maver, M., Mimmo, T., Cesco, S., Borruso, L., 2024. Microplastic influences the ménage à trois among the plant, a fungal pathogen, and a plant growth-promoting fungal species. Ecotoxicol. Environ. Saf. 279, 116518. https://doi.org/10.1016/j.ecoenv.2024.116518.

Bouaicha, O., Tiziani, R., Schmid, A.E., Praeg, N., Illmer, P., Del Buono, D., Puglia, D., Fracasso, I., Borruso, L., Mimmo, T., 2026. Aerial polyethylene deposition alters strawberry physiology and rhizosphere functionality. Rhizosphere 38, 101313. https://doi.org/10.1016/j.rhisph.2026.101313.

Bouaicha, O., Trevisan, F., Tiziani, R., Brenner, M., Weckwerth, W., Onelli, E., Moscatelli, A., Mimmo, T., Borruso, L., 2025. From the shoot to the rhizosphere: The short-term cascade impact of aerial microplastic. Environ. Exp. Bot. https://doi.org/10.1016/j.envexpbot.2025.106222.

Brahney, J., Hallerud, M., Heim, E., Hahnenberger, M., Sukumaran, S., 2020. Plastic rain in protected areas of the United States. Science 368 (6496), 1257–1260. https://doi.org/10.1126/science.aaz5819.

Brochado, M.G.D.S., Magalhães, I.B., Soares, J.M., Lima, A.D.C., Bordignon, L., Guedes, A.G., Castillo, M.R., Mendes, K.F., 2024. Does microplastic contamination in agricultural soils decrease the efficiency of herbicides for weed control? Microplastics 3 (4), 771–788. https://doi.org/10.3390/microplastics3040048.

Büks, F., Kaupenjohann, M., 2020. Global concentrations of microplastics in soils – a review. SOIL 6 (2), 649–662. https://doi.org/10.5194/soil-6-649-2020.

Chang, N., Chen, L., Wang, N., Cui, Q., Qiu, T., Zhao, S., He, H., Zeng, Y., Dai, W., Duan, C., Fang, L., 2024. Unveiling the impacts of microplastic pollution on soil health: a comprehensive review. Sci. Total Environ. 951, 175643. https://doi.org/10.1016/j.scitotenv.2024.175643.

Chen, G., Fu, Z., Yang, H., Wang, J., 2020. An overview of analytical methods for detecting microplastics in the atmosphere. TrAC, Trends Anal. Chem. 130, 115981. https://doi.org/10.1016/j.trac.2020.115981.

Chen, G., Li, Y., Liu, S., Junaid, M., Wang, J., 2022. Effects of micro(nano)plastics on higher plants and the rhizosphere environment. Sci. Total Environ. 807, 150841. https://doi.org/10.1016/j.scitotenv.2021.150841.

Cheng, X., Li, X., Cai, Z., Wang, Z., Zhou, J., 2024. The structural and functional responses of rhizosphere bacteria to biodegradable microplastics in the presence of biofertilizers. Plants 13 (18), 2627. https://doi.org/10.3390/plants13182627.

Chow, T.L., Rees, H.W., Monteith, J.O., Toner, P., Lavoie, J., 2007. Effects of coarse fragment content on soil physical properties, soil erosion and potato production. Can. J. Soil Sci. 87 (5), 565–577. https://doi.org/10.4141/CJSS07006.

Claessens, M., Van Cauwenberghe, L., Vandegehuchte, M.B., Janssen, C.R., 2013. New techniques for the detection of microplastics in sediments and field collected organisms. Mar. Pollut. Bull. 70 (1–2), 227–233. https://doi.org/10.1016/j.marpolbul.2013.03.009.

De Souza Machado, A.A., Kloas, W., Zarfl, C., Hempel, S., Rillig, M.C., 2018a. Microplastics as an emerging threat to terrestrial ecosystems. Glob. Change Biol. 24 (4), 1405–1416. https://doi.org/10.1111/gcb.14020.

De Souza Machado, A.A., Lau, C.W., Till, J., Kloas, W., Lehmann, A., Becker, R., Rillig, M.C., 2018b. Impacts of microplastics on the soil biophysical environment. Environ. Sci. Technol. 52 (17), 9656–9665. https://doi.org/10.1021/acs.est.8b02212.

De Souza Machado, A.A., Lau, C.W., Kloas, W., Bergmann, J., Bachelier, J.B., Faltin, E., Becker, R., Görlich, A.S., Rillig, M.C., 2019. Microplastics can change soil properties and affect plant performance. Environ. Sci. Technol. 53 (10), 6044–6052. https://doi.org/10.1021/acs.est.9b01339.

Deng, H., He, J., Feng, D., Zhao, Y., Sun, W., Yu, H., Ge, C., 2021. Microplastics pollution in mangrove ecosystems: a critical review of current knowledge and future directions. Sci. Total Environ. 753, 142041. https://doi.org/10.1016/j.scitotenv.2020.142041.

Dike, S., Apte, S.D., 2024. Impact of microplastics on strength parameters of clayey, Sandy, silty soil: a comparative assessment. Sci. Total Environ. 946, 174161. https://doi.org/10.1016/j.scitotenv.2024.174161.

Dissanayake, P.D., Kim, S., Sarkar, B., Oleszczuk, P., Sang, M.K., Haque, M.N., Ahn, J.H., Bank, M.S., Ok, Y.S., 2022. Effects of microplastics on the terrestrial environment: a critical review. Environ. Res. 209, 112734. https://doi.org/10.1016/j.envres.2022.112734.

Dong, Y., Gao, M., Song, Z., Qiu, W., 2020. Microplastic particles increase arsenic toxicity to rice seedlings. Environ. Pollut. 259, 113892. https://doi.org/10.1016/j.envpol.2019.113892.

Dong, Y., Gao, M., Qiu, W., Song, Z., 2021. Uptake of microplastics by carrots in presence of as (III). combined toxic effects. J. Hazard Mater. 411, 125055. https://doi.org/10.1016/j.jhazmat.2021.125055.

Dong, Y., Bao, Q., Gao, M., Qiu, W., Song, Z., 2022. A novel mechanism study of microplastic and as co-contamination on indica rice (Oryza sativa L.). J. Hazard Mater. 421, 126694. https://doi.org/10.1016/j.jhazmat.2021.126694.

Duan, J., Han, J., Cheung, S.G., Chong, R.K.Y., Lo, C.-M., Lee, F.W.-F., Xu, S.J.-L., Yang, Y., Tam, N. F., Zhou, H.-C., 2021. How mangrove plants affect microplastic distribution in sediments of coastal wetlands: Case study in Shenzhen Bay, South China. Sci. Total Environ. 767, 144695. https://doi.org/10.1016/j.scitotenv.2020.144695.

El Baroudy, A.A., Ibrahim, M.M., Mahmoud, M.A., 2014. Effects of deficit irrigation and transplanting methods of irrigated rice on soil physical properties and rice yield. Soil Use Manag. 30 (1), 88–98. https://doi.org/10.1111/sum.12084.

Endo, S., Yuyama, M., Takada, H., 2013. Desorption kinetics of hydrophobic organic contaminants from marine plastic pellets. Mar. Pollut. Bull. 74 (1), 125–131. https://doi.org/10.1016/j.marpolbul.2013.07.018.

Fan, P., Tan, W., Yu, H., 2022. Effects of different concentrations and types of microplastics on bacteria and fungi in alkaline soil. Ecotoxicol. Environ. Saf. 229, 113045. https://doi.org/10.1016/j.ecoenv.2021.113045.

Fang, Y., Xu, W., Zhang, W., Guang, C., Mu, W., 2022. Microbial elimination of pyrethroids: specific strains and involved enzymes. Appl. Microbiol. Biotechnol. 106 (21), 6915–6932. https://doi.org/10.1007/s00253-022-12200-w.

Fang, S., Hua, C., Yang, J., Liu, F., Wang, L., Wu, D., Ren, L., 2025. Combined pollution of soil by heavy metals, microplastics, and pesticides: mechanisms and anthropogenic drivers. J. Hazard Mater. 485, 136812. https://doi.org/10.1016/j.jhazmat.2024.136812.

Fei, Y., Huang, S., Zhang, H., Tong, Y., Wen, D., Xia, X., Wang, H., Luo, Y., Barceló, D., 2020. Response of soil enzyme activities and bacterial communities to the accumulation of microplastics in an acid cropped soil. Sci. Total Environ. 707, 135634. https://doi.org/10.1016/j.scitotenv.2019.135634.

Feng, X., Wang, Q., Sun, Y., Zhang, S., Wang, F., 2022. Microplastics change soil properties, heavy metal availability and bacterial community in a Pb-Zn-contaminated soil. J. Hazard Mater. 424, 127364. https://doi.org/10.1016/j.jhazmat.2021.127364.

Fu, Q., Lai, J., Ji, X., Luo, Z., Wu, G., Luo, X., 2022. Alterations of the rhizosphere soil microbial community composition and metabolite profiles of Zea mays by polyethylene-particles of different molecular weights. J. Hazard Mater. 423, 127062. https://doi.org/10.1016/j.jhazmat.2021.127062.

Gao, M., Liu, Y., Song, Z., 2019. Effects of polyethylene microplastic on the phytotoxicity of di-n-butyl phthalate in lettuce (Lactuca sativa L. var. Ramosa Hort). Chemosphere 237, 124482. https://doi.org/10.1016/j.chemosphere.2019.124482.

Gao, M., Xu, Y., Liu, Y., Wang, S., Wang, C., Dong, Y., Song, Z., 2021. Effect of polystyrene on di-butyl phthalate (DBP) bioavailability and DBP-induced phytotoxicity in lettuce. Environ. Pollut. 268, 115870. https://doi.org/10.1016/j.envpol.2020.115870.

Gao, B., Gao, F., Zhang, X., Li, Y., Yao, H., 2024. Effects of different sizes of microplastic particles on soil respiration, enzyme activities, microbial communities, and seed germination. Sci. Total Environ. 933, 173100. https://doi.org/10.1016/j.scitotenv.2024.173100.

Geldner, N., 2013. The endodermis. Annu. Rev. Plant Biol. 64 (1), 531–558. https://doi.org/10.1146/annurev-arplant-050312-120050.

George, E., Horst, W., Neumann, E., 2012. Chapter 17 - Adaptation of plants to adverse chemical soil conditions. In: Matschner, P. (Ed.), Marschner's Mineral Nutrition of Higher Plants, 3rd. Academic Press, San Diego, pp. 409–472.

Gigault, J., Halle, A.T., Baudrimont, M., Pascal, P.-Y., Gauffre, F., Phi, T.-L., El Hadri, H., Grassl, B., Reynaud, S., 2018. Current opinion: what is a nanoplastic? Environ. Pollut. 235, 1030–1034. https://doi.org/10.1016/j.envpol.2018.01.024.

Giorgetti, L., Spanò, C., Muccifora, S., Bottega, S., Barbieri, F., Bellani, L., Ruffini Castiglione, M., 2020. Exploring the interaction between polystyrene nanoplastics and Allium cepa during germination: Internalization in root cells, induction of toxicity and oxidative stress. Plant Physiol. Biochem. 149, 170–177. https://doi.org/10.1016/j.plaphy.2020.02.014.

Gomes, S., Fernandes, A.N., Waldman, W.R., 2024. How to measure polymer degradation? an analysis of authors' choices when calculating the carbonyl index. Environ. Sci. Technol. 58 (17), 7609–7616. https://doi.org/10.1021/acs.est.3c10855.

Gong, W., Zhang, W., Jiang, M., Li, S., Liang, G., Bu, Q., Xu, L., Zhu, H., Lu, A., 2021. Species-dependent response of food crops to polystyrene nanoplastics and microplastics. Sci. Total Environ. 796, 148750. https://doi.org/10.1016/j.scitotenv.2021.148750.

Goyal, S., Bagri, S., Bishnoi, A., Jhajhriya, S., Kumar, S., Paul, D., & Dutta, S. (2024). A Comprehensive Review on Microplastic Impacts on Plant and Terrestrial Food Chain. vol. 4.

Grotkopp, E., Rejmánek, M., 2007. High seedling relative growth rate and specific leaf area are traits of invasive species: Phylogenetically independent contrasts of woody angiosperms. Am. J. Bot. 94 (4), 526–532. https://doi.org/10.3732/ajb.94.4.526.

Guo, W., Ye, Z., Zhao, Y., Lu, Q., Shen, B., Zhang, X., Zhang, W., Chen, S.-C., Li, Y., 2024. Effects of different microplastic types on soil physicochemical properties, enzyme activities, and bacterial communities. Ecotoxicol. Environ. Saf. 286, 117219. https://doi.org/10.1016/j.ecoenv.2024.117219.

Hahladakis, J.N., Velis, C.A., Weber, R., Iacovidou, E., Purnell, P., 2018. An overview of chemical additives present in plastics: migration, release, fate and environmental impact during their use, disposal and recycling. J. Hazard Mater. 344, 179–199. https://doi.org/10.1016/j.jhazmat.2017.10.014.

Han, L., Chen, L., Feng, Y., Kuzyakov, Y., Chen, Q., Zhang, S., Chao, L., Cai, Y., Ma, C., Sun, K., Rillig, M.C., 2024. Microplastics alter soil structure and microbial community composition. Environ. Int. 185, 108508. https://doi.org/10.1016/j.envint.2024.108508.

Hanif, M.N., Aijaz, N., Azam, K., Akhtar, M., Laftah, W.A., Babur, M., Abbood, N.K., Benitez, I.B., 2024. Impact of microplastics on soil (physical and chemical) properties, soil biological properties/ soil biota, and response of plants to it: a review. Int. J. Environ. Sci. Technol. 21 (16), 10277–10318. https://doi.org/10.1007/s13762-024-05656-y.

Hartmann, G.F., Ricachenevsky, F.K., Silveira, N.M., Pita-Barbosa, A., 2022. Phytotoxic effects of plastic pollution in crops: what is the size of the problem? Environ. Pollut. 292, 118420. https://doi.org/10.1016/j.envpol.2021.118420.

Hasan, M.M., Tarannum, M.N., 2025. Adverse impacts of microplastics on soil physicochemical properties and crop health in agricultural systems. J. Hazard. Mater. Adv. 17, 100528. https://doi.org/10.1016/j.hazadv.2024.100528.

Ho, D., Masura, J., 2024. Dyeing to Know: Optimizing Solvents for nile red fluorescence in microplastics analysis. Chemistry. https://doi.org/10.26434/chemrxiv-2024-wbf7n-v2.

Hou, X., Mu, L., Hu, X., Guo, S., 2023. Warming and microplastic pollution shape the carbon and nitrogen cycles of algae. J. Hazard Mater. 447, 130775. https://doi.org/10.1016/j.jhazmat.2023.130775.

Hua, Z., Ma, S., Ouyang, Z., Liu, P., Qiang, H., Guo, X., 2023. The review of nanoplastics in plants: detection, analysis, uptake, migration and risk. TrAC, Trends Anal. Chem. 158, 116889. https://doi.org/10.1016/j.trac.2022.116889.

Huang, Y., Zhao, Y., Wang, J., Zhang, M., Jia, W., Qin, X., 2019. LDPE microplastic films alter microbial community composition and enzymatic activities in soil. Environ. Pollut. 254, 112983. https://doi.org/10.1016/j.envpol.2019.112983.

Huang, X., Chen, Y., Meng, Y., Liu, G., Yang, M., 2022. Are we ignoring the role of urban forests in intercepting atmospheric microplastics? J. Hazard Mater. 436, 129096. https://doi.org/10.1016/j.jhazmat.2022.129096.

Jiang, X., Chen, H., Liao, Y., Ye, Z., Li, M., Klobučar, G., 2019. Ecotoxicity and genotoxicity of polystyrene microplastics on higher plant Vicia faba. Environ. Pollut. 250, 831–838. https://doi.org/10.1016/j.envpol.2019.04.055.

Jiang, M., Zhao, W., Liang, Q., Cai, M., Fan, X., Jiang, Y., Li, T., Wang, Y., Peng, C., Liu, J., 2024. Advances in physiological and ecological effects of microplastic on crop. J. Soil Sci. Plant Nutr. 24 (2), 1741–1760. https://doi.org/10.1007/s42729-024-01752-7.

Jiao, M., Wang, Y., Yang, F., Zhao, Z., Wei, Y., Li, R., Wang, Y., 2024. Dynamic fluctuations in plant leaf interception of airborne microplastics. Sci. Total Environ. 906, 167877. https://doi.org/10.1016/j.scitotenv.2023.167877.

Jiménez-Skrzypek, G., Ortega-Zamora, C., González-Sálamo, J., Hernández-Sánchez, C., Hernández-Borges, J., 2021. The current role of chromatography in microplastic research: plastics chemical characterization and sorption of contaminants. Journal of Chromatography Open 1, 100001. https://doi.org/10.1016/j.jcoa.2021.100001.

Jin, N., Song, Y., Ma, R., Li, J., Li, G., Zhang, D., 2022. Characterization and identification of microplastics using Raman spectroscopy coupled with multivariate analysis. Anal. Chim. Acta 1197, 339519. https://doi.org/10.1016/j.aca.2022.339519.

Joos, L., De Tender, C., 2022. Soil under stress: the importance of soil life and how it is influenced by (micro)plastic pollution. Comput. Struct. Biotechnol. J. 20, 1554–1566. https://doi.org/10.1016/j.csbj.2022.03.041.

Kabir, M.S., Wang, H., Luster-Teasley, S., Zhang, L., Zhao, R., 2023. Microplastics in landfill leachate: sources, detection, occurrence, and removal. Environ. Sci. Ecotechnol. 16, 100256. https://doi.org/10.1016/j.ese.2023.100256.

Kang, H., Park, S., Lee, B., Ahn, J., Kim, S., 2020. Modification of a nile red staining method for microplastics analysis: a nile red plate method. Water 12 (11), 3251. https://doi.org/10.3390/w12113251.

Katzenmeyer, A.M., Aksyuk, V., Centrone, A., 2013. Nanoscale infrared spectroscopy: improving the spectral range of the photothermal induced resonance technique. Anal. Chem. 85 (4), 1972–1979. https://doi.org/10.1021/ac303620y.

Khalil, A.M., Murchie, E.H., Mooney, S.J., 2020. Quantifying the influence of water deficit on root and shoot growth in wheat using X-ray Computed Tomography. AoB PLANTS 12 (5), plaa036. https://doi.org/10.1093/aobpla/plaa036.

Khan, T.F., Sikder, A.H.F., 2024. Microplastic can decrease enzyme activities and microbes in soil. Open J. Soil Sci. 14 (01), 1–12. https://doi.org/10.4236/ojss.2024.141001.

Kim, S.W., Waldman, W.R., Kim, T.-Y., Rillig, M.C., 2020. Effects of different microplastics on nematodes in the soil environment: tracking the extractable additives using an ecotoxicological approach. Environ. Sci. Technol. 54 (21), 13868–13878. https://doi.org/10.1021/acs.est.0c04641.

Kim, D., Kim, H., Lee, J., Choi, M.-J., Kweon, H.-S., An, Y.-J., 2024. Evidence of parental transfer of nanoplastics in pea (Pisum sativum) plants. J. Hazard Mater. 465, 133516. https://doi.org/10.1016/j.jhazmat.2024.133516.

Kinigopoulou, V., Pashalidis, I., Kalderis, D., Anastopoulos, I., 2022. Microplastics as carriers of inorganic and organic contaminants in the environment: a review of recent progress. J. Mol. Liq. 350, 118580. https://doi.org/10.1016/j.molliq.2022.118580.

Lai, S., Fan, C., Yang, P., Fang, Y., Zhang, L., Jian, M., Dai, G., Liu, J., Yang, H., Shen, L., 2025. Effects of different microplastics on the physicochemical properties and microbial diversity of rice rhizosphere soil. Front. Microbiol. 15, 1513890. https://doi.org/10.3389/fmicb.2024.1513890.

Lajmanovich, R.C., Attademo, A.M., Lener, G., Cuzziol Boccioni, A.P., Peltzer, P.M., Martinuzzi, C.S., Demonte, L.D., Repetti, M.R., 2022. Glyphosate and glufosinate ammonium, herbicides commonly used on genetically modified crops, and their interaction with microplastics: ecotoxicity in anuran tadpoles. Sci. Total Environ. 804, 150177. https://doi.org/10.1016/j.scitotenv.2021.150177.

Lau, W.W.Y., Shiran, Y., Bailey, R.M., Cook, E., Stuchtey, M.R., Koskella, J., Velis, C.A., Godfrey, L., Boucher, J., Murphy, M.B., Thompson, R.C., Jankowska, E., Castillo Castillo, A., Pilditch, T.D., Dixon, B., Koerselman, L., Kosior, E., Favoino, E., Gutberlet, J., , … Palardy, J.E., 2020. Evaluating scenarios toward zero plastic pollution. Science 369 (6510), 1455–1461. https://doi.org/10.1126/science.aba9475.

Lehmann, A., Leifheit, E.F., Feng, L., Bergmann, J., Wulf, A., Rillig, M.C., 2022. Microplastic fiber and drought effects on plants and soil are only slightly modified by arbuscular mycorrhizal fungi. Soil Ecol. Lett. 4 (1), 32–44. https://doi.org/10.1007/s42832-020-0060-4.

Li, W., Xiao, Y., 2023. Microplastics increase soil microbial network complexity and trigger diversity-driven community assembly. Environ. Pollut. 333, 122095. https://doi.org/10.1016/j.envpol.2023.122095.

Li, L., Luo, Y., Li, R., Zhou, Q., Peijnenburg, W.J.G.M., Yin, N., Yang, J., Tu, C., Zhang, Y., 2020a. Effective uptake of submicrometre plastics by crop plants via a crack-entry mode. Nat. Sustain. 3 (11), 929–937. https://doi.org/10.1038/s41893-020-0567-9.

Li, Z., Li, Q., Li, R., Zhao, Y., Geng, J., Wang, G., 2020b. Physiological responses of lettuce (Lactuca sativa L.) to microplastic pollution. Environ. Sci. Pollut. Control Ser. 27 (24), 30306–30314. https://doi.org/10.1007/s11356-020-09349-0.

Li, C., Gao, Y., He, S., Chi, H.-Y., Li, Z.-C., Zhou, X.-X., Yan, B., 2021b. Quantification of nanoplastic uptake in cucumber plants by pyrolysis gas chromatography/mass spectrometry. Environ. Sci. Technol. Lett. 8 (8), 633–638. https://doi.org/10.1021/acs.estlett.1c00369.

Li, C., Zhu, L., Li, W.-T., Li, D., 2023. Microplastics in the seagrass ecosystems: a critical review. Sci. Total Environ. 902, 166152. https://doi.org/10.1016/j.scitotenv.2023.166152.

Li, G., Tang, Y., Lou, J., Wang, Y., Yin, S., Li, L., Iqbal, B., Lozano, Y.M., Zhao, T., Du, D., 2024a. The promoting effects of soil microplastics on alien plant invasion depend on microplastic shape and concentration. Sci. Total Environ. 926, 172089. https://doi.org/10.1016/j.scitotenv.2024.172089.

Li, W., Zhao, J., Zhang, Z., Ren, Z., Li, X., Zhang, R., Ma, X., 2024b. Uptake and effect of carboxyl-modified polystyrene microplastics on cotton plants. J. Hazard Mater. 466, 133581. https://doi.org/10.1016/j.jhazmat.2024.133581.

Li, K., He, Q., Zhu, J., Wang, J., Sun, C., Tan, A., Zhao, X., Peng, Y., Huang, C., Cai, J., Wang, P., Liu, Z., 2025. Responses of microbial communities to the addition of different types of microplastics in agricultural soils. Environ. Pollut. 364, 125220. https://doi.org/10.1016/j.envpol.2024.125220.

Lian, J., Wu, J., Xiong, H., Zeb, A., Yang, T., Su, X., Su, L., Liu, W., 2020. Impact of polystyrene nanoplastics (PSNPs) on seed germination and seedling growth of wheat (Triticum aestivum L.). J. Hazard Mater. 385. https://doi.org/10.1016/j.jhazmat.2019.121620.

Lian, J., Liu, W., Meng, L., Wu, J., Chao, L., Zeb, A., Sun, Y., 2021. Foliar-applied polystyrene nanoplastics (PSNPs) reduce the growth and nutritional quality of lettuce (Lactuca sativa L.). Environ. Pollut. 280, 116978. https://doi.org/10.1016/j.envpol.2021.116978.

Liang, Y., Lehmann, A., Yang, G., Leifheit, E.F., Rillig, M.C., 2021. Effects of microplastic fibers on soil aggregation and enzyme activities are organic matter dependent. Front. Environ. Sci. 9, 650155. https://doi.org/10.3389/fenvs.2021.650155.

Lin, D., Yang, G., Dou, P., Qian, S., Zhao, L., Yang, Y., Fanin, N., 2020. Microplastics negatively affect soil fauna but stimulate microbial activity: Insights from a field-based microplastic addition experiment. Proc. Biol. Sci. 287 (1934), 20201268. https://doi.org/10.1098/rspb.2020.1268.

Liu, R., Shen, J., 2024. Effect of chronic and humped input of polystyrene microplastics on rice growth. Sci. Total Environ. 906, 167642. https://doi.org/10.1016/j.scitotenv.2023.167642.

Liu, Y., Guo, R., Zhang, S., Sun, Y., Wang, F., 2022. Uptake and translocation of nano/microplastics by rice seedlings: evidence from a hydroponic experiment. J. Hazard Mater. 421, 126700. https://doi.org/10.1016/j.jhazmat.2021.126700.

Liu, X., Li, Y., Yu, Y., Yao, H., 2023. Effect of nonbiodegradable microplastics on soil respiration and enzyme activity: a meta-analysis. Appl. Soil Ecol. 184, 104770. https://doi.org/10.1016/j.apsoil.2022.104770.

Liu, Y., Wen, Y., Cai, H., Song, X., Wang, X., Zhang, Z., 2025. Stress of polyethylene and polylactic acid microplastics on pakchoi(Brassica rapa subsp. chinensis) and soil bacteria: Biochar mitigation. J. Hazard Mater. 487, 137301. https://doi.org/10.1016/j.jhazmat.2025.137301.

Lozano, Y.M., Rillig, M.C., 2020. Effects of microplastic fibers and drought on plant communities. Environ. Sci. Technol. 54 (10), 6166–6173. https://doi.org/10.1021/acs.est.0c01051.

Lozano, Y.M., Lehnert, T., Linck, L.T., Lehmann, A., Rillig, M.C., 2021. Microplastic shape, polymer type, and concentration affect soil properties and plant biomass. Front. Plant Sci. 12, 616645. https://doi.org/10.3389/fpls.2021.616645.

Lozano, Y.M., Landt, L., Rillig, M.C., 2025. Plastic particles and their additives promote plant invasion through physicochemical mechanisms on seed germination. J. Ecol. 113 (2), 275–288. https://doi.org/10.1111/1365-2745.14476.

Luo, H., Liu, C., He, D., Xu, J., Sun, J., Li, J., Pan, X., 2022. Environmental behaviors of microplastics in aquatic systems: a systematic review on degradation, adsorption, toxicity and biofilm under aging conditions. J. Hazard Mater. 423, 126915. https://doi.org/10.1016/j.jhazmat.2021.126915.

Luo, Y., Wang, L., Cao, T., Chen, J., Lv, M., Wei, S., Lu, S., Tian, X., 2023. Microplastics are transferred by soil fauna and regulate soil function as material carriers. Sci. Total Environ. 857, 159690. https://doi.org/10.1016/j.scitotenv.2022.159690.

Ma, J., Xu, M., Wu, J., Yang, G., Zhang, X., Song, C., Long, L., Chen, C., Xu, C., Wang, Y., 2023. Effects of variable-sized polyethylene microplastics on soil chemical properties and functions and microbial communities in purple soil. Sci. Total Environ. 868, 161642. https://doi.org/10.1016/j.scitotenv.2023.161642.

Maes, T., Jessop, R., Wellner, N., Haupt, K., Mayes, A.G., 2017. A rapid-screening approach to detect and quantify microplastics based on fluorescent tagging with Nile Red. Sci. Rep. 7 (1), 44501. https://doi.org/10.1038/srep44501.

Maity, S., Chatterjee, A., Guchhait, R., De, S., Pramanick, K., 2020. Cytogenotoxic potential of a hazardous material, polystyrene microparticles on Allium cepa L. J. Hazard Mater. 385, 121560. https://doi.org/10.1016/j.jhazmat.2019.121560.

Martín-Gómez, B., Stephen Elmore, J., Valverde, S., Ares, A.M., Bernal, J., 2024. Recent applications of chromatography for determining microplastics and related compounds (bisphenols and phthalate esters) in food. Microchem. J. 197, 109903. https://doi.org/10.1016/j.microc.2024.109903.

Menicagli, V., Castiglione, M.R., Balestri, E., Giorgetti, L., Bottega, S., Sorce, C., Spanò, C., Lardicci, C., 2022. Early evidence of the impacts of microplastic and nanoplastic pollution on the growth and physiology of the seagrass Cymodocea nodosa. Sci. Total Environ. 838, 156514. https://doi.org/10.1016/j.scitotenv.2022.156514.

Mogha, N.K., Shin, D., 2023. Nanoplastic detection with surface enhanced Raman spectroscopy: present and future. TrAC, Trends Anal. Chem. 158, 116885. https://doi.org/10.1016/j.trac.2022.116885.

Moharana, T., Patnaik, A., Mishra, C.S.K., Behera, B.P., Samal, R.R., 2024. High-density polyethylene microplastics in agricultural soil: impact on microbes, enzymes, and carbon-nitrogen ratio. J. Environ. Qual. 53 (5), 711–726. https://doi.org/10.1002/jeq2.20610.

Nam, H., Gil, D., Lee, J.J., Kim, C., 2024. Dual-channel fluorescence dye: fluorescent color-dependent visual detection of microplastics and selective polyurethane. Sci. Total Environ. 912, 169219. https://doi.org/10.1016/j.scitotenv.2023.169219.

Ng, E.L., Lin, S.Y., Dungan, A.M., Colwell, J.M., Ede, S., Huerta Lwanga, E., Meng, K., Geissen, V., Blackall, L.L., Chen, D., 2021. Microplastic pollution alters forest soil microbiome. J. Hazard Mater. 409, 124606. https://doi.org/10.1016/j.jhazmat.2020.124606.

Oliveri Conti, G., Ferrante, M., Banni, M., Favara, C., Nicolosi, I., Cristaldi, A., Fiore, M., Zuccarello, P., 2020. Micro- and nano-plastics in edible fruit and vegetables. The first diet risks assessment for the general population. Environ. Res. 187, 109677. https://doi.org/10.1016/j.envres.2020.109677.

Pedram Jarf, M., Kamali, A., Khara, H., Pourang, N., Shekarabi, S.P.H., 2024. Microplastic pollution and heavy metal risk assessment in Perca fluviatilis from Anzali wetland: implications for environmental health and human consumption. Sci. Total Environ. 907, 167978. https://doi.org/10.1016/j.scitotenv.2023.167978.

Pehlivan, N., Gedik, K., 2021. Particle size-dependent biomolecular footprints of interactive microplastics in maize. Environ. Pollut. 277, 116772. https://doi.org/10.1016/j.envpol.2021.116772.

Pencik, O., Molnarova, K., Durdakova, M., Kolackova, M., Klofac, D., Kucsera, A., Capal, P., Svec, P., Bytesnikova, Z., Richtera, L., Brtnický, M., Adam, V., Huska, D., 2023. Not so dangerous? PET microplastics toxicity on freshwater microalgae and cyanobacteria. Environ. Pollut. 329, 121628. https://doi.org/10.1016/j.envpol.2023.121628.

Philippot, L., Raaijmakers, J.M., Lemanceau, P., Van Der Putten, W.H., 2013. Going back to the roots: the microbial ecology of the rhizosphere. Nat. Rev. Microbiol. 11 (11), 789–799. https://doi.org/10.1038/nrmicro3109.

Picó, Y., Barceló, D., 2020. Pyrolysis gas chromatography-mass spectrometry in environmental analysis: focus on organic matter and microplastics. TrAC, Trends Anal. Chem. 130, 115964. https://doi.org/10.1016/j.trac.2020.115964.

Pignattelli, S., Broccoli, A., Renzi, M., 2020. Physiological responses of garden cress (L. sativum) to different types of microplastics. Sci. Total Environ. 727, 138609. https://doi.org/10.1016/j.scitotenv.2020.138609.

Pita-Barbosa, A., Bouaicha, O., Tiziani, R., Chibesa, M.C., Borruso, L., Cesco, S., Mimmo, T., 2025. Combined impact of short-term phosphorus deficiency and microplastic contamination on tomato mineral elements, chlorophyll fluorescence and root exudates. Plant Physiol. Biochem. 110496. https://doi.org/10.1016/j.plaphy.2025.110496.

Qi, Y., Yang, X., Pelaez, A.M., Huerta Lwanga, E., Beriot, N., Gertsen, H., Garbeva, P., Geissen, V., 2018. Macro- and micro- plastics in soil-plant system: effects of plastic mulch film residues on wheat (Triticum aestivum) growth. Sci. Total Environ. 645, 1048–1056. https://doi.org/10.1016/j.scitotenv.2018.07.229.

Qiu, G., Wang, Q., Wang, T., Zhang, S., Song, N., Yang, X., Zeng, Y., Sun, Z., Wu, G., Yu, H., 2024. Microplastic risk assessment and toxicity in plants: a review. Environ. Chem. Lett. 22 (1), 209–226. https://doi.org/10.1007/s10311-023-01665-4.

Ren, X., Tang, J., Liu, X., Liu, Q., 2020. Effects of microplastics on greenhouse gas emissions and the microbial community in fertilized soil. Environ. Pollut. 256, 113347. https://doi.org/10.1016/j.envpol.2019.113347.

Rice, K.J., Dyer, A.R., 2001. Seed aging, delayed germination and reduced competitive ability in Bromus tectorum. Plant Ecol. 155 (2), 237–243. https://doi.org/10.1023/A:1013257407909.

Rietra, R.P.J.J., Heinen, M., Dimkpa, C.O., Bindraban, P.S., 2017. Effects of nutrient antagonism and synergism on yield and fertilizer use efficiency. Commun. Soil Sci. Plant Anal. 48 (16), 1895–1920. https://doi.org/10.1080/00103624.2017.1407429.

Rillig, M.C., 2012. Microplastic in terrestrial ecosystems and the soil? Environ. Sci. Technol. 46 (12), 6453–6454. https://doi.org/10.1021/es302011r.

Rillig, M.C., 2018. Microplastic disguising as soil carbon storage. Environ. Sci. Technol. 52 (11), 6079–6080. https://doi.org/10.1021/acs.est.8b02338.

Rillig, M.C., Lehmann, A., De Souza Machado, A.A., Yang, G., 2019. Microplastic effects on plants. New Phytol. 223 (3), 1066–1070. https://doi.org/10.1111/nph.15794.

Rong, L., Zhao, L., Zhao, L., Cheng, Z., Yao, Y., Yuan, C., Wang, L., Sun, H., 2021. LDPE microplastics affect soil microbial communities and nitrogen cycling. Sci. Total Environ. 773, 145640. https://doi.org/10.1016/j.scitotenv.2021.145640.

Rozman, U., Turk, T., Skalar, T., Zupančič, M., Korošin, N.Č., Marinšek, M., Olivero-Verbel, J., Kalčíková, G., 2021. An extensive characterization of various environmentally relevant microplastics – Material properties, leaching and ecotoxicity testing. Sci. Total Environ. 145576. https://doi.org/10.1016/j.scitotenv.2021.145576.

Sancataldo, G., Avellone, G., Vetri, V., 2020. Nile Red lifetime reveals microplastic identity. Environ. Sci. Process. Impacts 22 (11), 2266–2275. https://doi.org/10.1039/D0EM00348D.

Sanchez-Vidal, A., Canals, M., De Haan, W.P., Romero, J., Veny, M., 2021. Seagrasses provide a novel ecosystem service by trapping marine plastics. Sci. Rep. 11 (1), 254. https://doi.org/10.1038/s41598-020-79370-3.

Sepehrnia, N., Gubry-Rangin, C., Tanino, Y., Hallett, P.D., 2024. Microplastics alter soil structural stability as quantified by high-energy moisture characteristics. J. Hazard Mater. 476, 134940. https://doi.org/10.1016/j.jhazmat.2024.134940.

Shen, M., Liu, S., Hu, T., Zheng, K., Wang, Y., Long, H., 2023. Recent advances in the research on effects of micro/nanoplastics on carbon conversion and carbon cycle: a review. J. Environ. Manag. 334, 117529. https://doi.org/10.1016/j.jenvman.2023.117529.

Shirkhorshidi, B., Ghanatghestani, M.D., Moeinpour, F., Parvaresh, H., 2023. The impacts of microplastics on sorption and desorption Specifications of Iron in soil. Water, Air, Soil Pollut. 234 (6), 379. https://doi.org/10.1007/s11270-023-06395-5.

Shruti, V.C., Pérez-Guevara, F., Roy, P.D., Kutralam-Muniasamy, G., 2022. Analyzing microplastics with Nile Red: emerging trends, challenges, and prospects. J. Hazard Mater. 423, 127171. https://doi.org/10.1016/j.jhazmat.2021.127171.

Sturm, M.T., Horn, H., Schuhen, K., 2021. The potential of fluorescent dyes—comparative study of Nile red and three derivatives for the detection of microplastics. Anal. Bioanal. Chem. 413 (4), 1059–1071. https://doi.org/10.1007/s00216-020-03066-w.

Sun, X.-D., Yuan, X.-Z., Jia, Y., Feng, L.-J., Zhu, F.-P., Dong, S.-S., Liu, J., Kong, X., Tian, H., Duan, J.-L., Ding, Z., Wang, S.-G., Xing, B., 2020. Differentially charged nanoplastics demonstrate distinct accumulation in Arabidopsis thaliana. Nat. Nanotechnol. 15 (9), 755–760. https://doi.org/10.1038/s41565-020-0707-4.

Suaria, G., Aikaterini, A., Perold, V., Lee, J.R., Pierucci, A., Bornman, T.G., Aliano, S., Ryan, P.G., 2020. Microfibers in oceanic surface waters: A global characterization. Science 6, eaay8493. https://doi.org/10.1126/sciadv.aay8493.

Sun, H., Lei, C., Xu, J., Li, R., 2021. Foliar uptake and leaf-to-root translocation of nanoplastics with different coating charge in maize plants. J. Hazard Mater. 416, 125854. https://doi.org/10.1016/j.jhazmat.2021.125854.

Taghizadeh Rahmat Abadi, Z., Abtahi, B., Fathi, M.B., Mashhadi, N., Grossart, H.-P., 2024. Size, shape, and elemental composition as predictors of microplastic surface erosion. J. Hazard Mater. 476, 134961. https://doi.org/10.1016/j.jhazmat.2024.134961.

Thompson, R.C., Olsen, Y., Mitchell, R.P., Davis, A., Rowland, S.J., John, A.W.G., McGonigle, D., Russell, A.E., 2004. Lost at Sea: where is All the plastic? Science 304 (5672), 838. https://doi.org/10.1126/science.1094559.

Tirkey, A., Upadhyay, L.S.B., 2021. Microplastics: an overview on separation, identification and characterization of microplastics. Mar. Pollut. Bull. 170, 112604. https://doi.org/10.1016/j.marpolbul.2021.112604.

Tracy, S.R., Black, C.R., Roberts, J.A., Mooney, S.J., 2013. Exploring the interacting effect of soil texture and bulk density on root system development in tomato (Solanum lycopersicum L.). Environ. Exp. Bot. 91, 38–47. https://doi.org/10.1016/j.envexpbot.2013.03.003.

Trojan, M., Koutný, M., Brtnický, M., Holátko, J., Zlámalová Gargošová, H., Fojt, J., Procházková, P., Kalčíková, G., Kučerík, J., 2024. The interaction of microplastics and microbioplastics with soil and a comparison of their potential to spread pathogens. Appl. Sci. 14 (11), 4643. https://doi.org/10.3390/app14114643.

Tympa, L.-E., Katsara, K., Moschou, P.N., Kenanakis, G., Papadakis, V.M., 2021. Do microplastics enter our food chain via root vegetables? a Raman based spectroscopic study on raphanus sativus. Materials 14 (9), 2329. https://doi.org/10.3390/ma14092329.

Urbina, M.A., Correa, F., Aburto, F., Ferrio, J.P., 2020. Adsorption of polyethylene microbeads and physiological effects on hydroponic maize. Sci. Total Environ. 741, 140216. https://doi.org/10.1016/j.scitotenv.2020.140216.

Venturi, V., Keel, C., 2016. Signaling in the rhizosphere. Trends Plant Sci. 21 (3), 187–198. https://doi.org/10.1016/j.tplants.2016.01.005.

Waldman, W.R., Rillig, M.C., 2020. Microplastic research should embrace the complexity of secondary particles. Environ. Sci. Technol. 54 (13), 7751–7753. https://doi.org/10.1021/acs.est.0c02194.

Wan, Y., Wu, C., Xue, Q., Hui, X., 2019. Effects of plastic contamination on water evaporation and desiccation cracking in soil. Sci. Total Environ. 654, 576–582. https://doi.org/10.1016/j.scitotenv.2018.11.123.

Wang, X., Xing, Y., Lv, M., Zhang, T., Ya, H., Jiang, B., 2022. Recent advances on the effects of microplastics on elements cycling in the environment. Sci. Total Environ. 849, 157884. https://doi.org/10.1016/j.scitotenv.2022.157884.

Wang, Z., Li, J., Qu, Z., Ayurzana, B., Zhao, G., Li, W., 2024. Effects of microplastics on the pore structure and connectivity with different soil textures: based on CT scanning. Environ. Technol. Innovat. 36, 103791. https://doi.org/10.1016/j.eti.2024.103791.

Wu, X., Liu, Y., Yin, S., Xiao, K., Xiong, Q., Bian, S., Liang, S., Hou, H., Hu, J., Yang, J., 2020. Metabolomics revealing the response of rice (Oryza sativa L.) exposed to polystyrene microplastics. Environ. Pollut. 266, 115159. https://doi.org/10.1016/j.envpol.2020.115159.

Wu, H., Gao, T., Hu, A., Wang, J., 2024. Network complexity and stability of microbes enhanced by microplastic diversity. Environ. Sci. Technol. https://doi.org/10.1021/acs.est.3c08704 acs.est.3c08704.

Xiang, Y., Yao, B., Peñuelas, J., Sardans, J., Nizzetto, L., Li, R., Liu, Y., Luo, Y., Räty, M., Long, J., Li, Y., 2025. Microplastic effects on soil nitrogen cycling enzymes: a global meta-analysis of environmental and edaphic factors. J. Hazard Mater. 484, 136677. https://doi.org/10.1016/j.jhazmat.2024.136677.

Xie, K., Cakmak, I., Wang, S., Zhang, F., Guo, S., 2021. Synergistic and antagonistic interactions between potassium and magnesium in higher plants. The Crop Journal 9 (2), 249–256. https://doi.org/10.1016/j.cj.2020.10.005.

Xie, J., Gowen, A., Xu, W., Xu, J., 2024. Analysing micro- and nanoplastics with cutting-edge infrared spectroscopy techniques: a critical review. Anal. Methods 16 (15), 2177–2197. https://doi.org/10.1039/D3AY01808C.

Xing, X., Jing, X., Zhao, F., Jiao, S., Su, L., Yu, M., Zhao, L., 2024. Dry-wet alternation and microplastics particle size effects on and contributions to soil water and soil pore properties. Eur. J. Soil Sci. 75 (3), e13522. https://doi.org/10.1111/ejss.13522.

Xu, Z., Zhang, Y., Lin, L., Wang, L., Sun, W., Liu, C., Yu, G., Yu, J., Lv, Y., Chen, J., Chen, X., Fu, L., Wang, Y., 2022. Toxic effects of microplastics in plants depend more by their surface functional groups than just accumulation contents. Sci. Total Environ. 833, 155097. https://doi.org/10.1016/j.scitotenv.2022.155097.

Yan, Y., Chen, Z., Zhu, F., Zhu, C., Wang, C., Gu, C., 2021. Effect of polyvinyl chloride microplastics on bacterial community and nutrient status in two agricultural soils. Bull. Environ. Contam. Toxicol. 107 (4), 602–609. https://doi.org/10.1007/s00128-020-02900-2.

Yang, M., Huang, D.-Y., Tian, Y.-B., Zhu, Q.-H., Zhang, Q., Zhu, H.-H., Xu, C., 2021. Influences of different source microplastics with different particle sizes and application rates on soil properties and growth of Chinese cabbage (Brassica chinensis L.). Ecotoxicol. Environ. Saf. 222, 112480. https://doi.org/10.1016/j.ecoenv.2021.112480.

Yi, M., Zhou, S., Zhang, L., Ding, S., 2021. The effects of three different microplastics on enzyme activities and microbial communities in soil. Water Environ. Res. 93 (1), 24–32. https://doi.org/10.1002/wer.1327.

Yin, M., Yan, B., Wang, H., Wu, Y., Wang, X., Wang, J., Zhu, Z., Yan, X., Liu, Y., Liu, M., Fu, C., 2023. Effects of microplastics on nitrogen and phosphorus cycles and microbial communities in sediments. Environ. Pollut. 318, 120852. https://doi.org/10.1016/j.envpol.2022.120852.

Yu, Y., Battu, A.K., Varga, T., Denny, A.C., Zahid, T.M., Chowdhury, I., Flury, M., 2023. Minimal impacts of microplastics on soil physical properties under environmentally relevant concentrations. Environ. Sci. Technol. 57 (13), 5296–5304. https://doi.org/10.1021/acs.est.2c09822.

Yu, C., Mawodza, T., Atkinson, B.S., Atkinson, J.A., Sturrock, C.J., Whalley, R., Hawkesford, M.J., Cooper, H., Zhang, X., Zhou, H., Mooney, S.J., 2024. The effects of soil compaction on wheat seedling root growth are specific to soil texture and soil moisture status. Rhizosphere 29, 100838. https://doi.org/10.1016/j.rhisph.2023.100838.

Yuan, Y., Zu, M., Li, R., Zuo, J., Tao, J., 2023. Soil properties, microbial diversity, and changes in the functionality of saline-alkali soil are driven by microplastics. J. Hazard Mater. 446, 130712. https://doi.org/10.1016/j.jhazmat.2022.130712.

Zang, H., Zhou, J., Marshall, M.R., Chadwick, D.R., Wen, Y., Jones, D.L., 2020. Microplastics in the agroecosystem: are they an emerging threat to the plant-soil system? Soil Biol. Biochem. 148, 107926. https://doi.org/10.1016/j.soilbio.2020.107926.

Zhang, G.S., Zhang, F.X., Li, X.T., 2019. Effects of polyester microfibers on soil physical properties: Perception from a field and a pot experiment. Sci. Total Environ. 670, 1–7. https://doi.org/10.1016/j.scitotenv.2019.03.149.

Zhang, Y., Zhang, X., Li, X., He, D., 2022. Interaction of microplastics and soil animals in agricultural ecosystems. Current Opinion in Environmental Science & Health 26, 100327. https://doi.org/10.1016/j.coesh.2022.100327.

Zhang, R., Guo, Y., Lai, Y., Zhao, T., Li, C., Yan, Z., Wang, Y., Rillig, M.C., 2024. Microplastics promote the invasiveness of invasive alien species under fluctuating water regime. J. Appl. Ecol. 61 (9), 2281–2293. https://doi.org/10.1111/1365-2664.14726.

Zhao, T., Lozano, Y.M., Rillig, M.C., 2021. Microplastics increase soil ph and decrease microbial activities as a function of microplastic shape, polymer type, and exposure time. Front. Environ. Sci. 9, 675803. https://doi.org/10.3389/fenvs.2021.675803.

Zhou, C.-Q., Lu, C.-H., Mai, L., Bao, L.-J., Liu, L.-Y., Zeng, E.Y., 2021a. Response of rice (Oryza sativa L.) roots to nanoplastic treatment at seedling stage. J. Hazard Mater. 401, 123412. https://doi.org/10.1016/j.jhazmat.2020.123412.

Zhou, J., Gui, H., Banfield, C.C., Wen, Y., Zang, H., Dippold, M.A., Charlton, A., Jones, D.L., 2021b. The microplastisphere: biodegradable microplastics addition alters soil microbial community structure and function. Soil Biol. Biochem. 156, 108211. https://doi.org/10.1016/j.soilbio.2021.108211.

CHAPTER 13

Phytotoxicity of organic amendments and wastes

Minh-Ky Nguyen
Faculty of Environment and Natural Resources, Nong Lam University, Ho Chi Minh City, Viet Nam

1. Introduction

Organic amendments such as manure (particularly green manure), straws, sludge/biogas, and other organic solid waste are commonly suggested in agricultural practices to enhance soil nutrient levels and boost crop yields (Luan et al., 2019; Zhong et al., 2010). Additionally, compost used in agriculture is increasingly acknowledged as a sustainable soil amendment. These organic sources supply crops with essential nutrients and promote soil health by altering its physical, biological, and chemical profiles. For example, the application of composted crop residues and municipal solid waste has been demonstrated to enhance structural stability and improve soil fertility in agricultural soils. Nevertheless, organic soil amendments can introduce unwanted hazards, including toxic heavy metal(loid)s, salts, or organic contaminants, which may slow plant growth and lessen yields (Kebrom et al., 2019). Sewage biosolids and animal manure, which are among the most common organic amendments, involve important components that can be hazardous to human health as well as to other animals and plants (Goss et al., 2013). The growing amount of waste presents a significant challenge for managing the health and environment (Zhou et al., 2023). Concerning toxic elements such as metal(loid)s, their accumulation in soils is frequently cited as a significant potential risk, especially with the use of biosolid waste compost. Fertilizer amendments can affect metal mobility; it is essential to consider their transfer to crops and the associated toxicity (Battaglia-Brunet et al., 2021). Therefore, the phytotoxicity of soil amendments and commercial organic fertilizers and wastes should be assessed prior to their application to agricultural soils in order to mitigate risks to crops. Determining metal bioavailability for plants is crucial, especially when these plants are cultivated for consumption. These contaminants typically have prolonged half-lives, are highly mobile within plant-soil systems, could be accumulated in the food chain, and display moderate levels to high toxicity (O'Connor et al., 2022).

The application of compost in agriculture is anticipated to increase due to the growing emphasis on sustainable practices to enhance soil health and promote carbon sequestration. Yet, they can give potential risks. As an example, compost can contain toxic pollutants, including heavy metal(loid)s, pesticides, pathogens, pharmaceuticals,

Phytotoxicity of Organic Pollutants
ISBN 978-0-443-33110-7,
https://doi.org/10.1016/B978-0-443-33110-7.00004-3

and microplastics, which may adversely affect plant health and soil quality (Ahsan et al., 2023; Hrčka et al., 2024; Le et al., 2023; Okori et al., 2024). Phytotoxicity from compost refers to the negative effects on plant growth resulting from high salt concentrations, nutrient imbalances, pathogens, contaminants, or incomplete decomposition. Also, both fresh and aged municipal solid waste has been found to contain several organic elements, including decane, hexanal, two-ethyl, 2,2-dimethyl, 3,6-dimethyl, dodecane, undecane, hexane, oxalic acid, 3,4-bis(1,1-dimethyl ethyl)-2,2,5,5-tetramethyl, 2,7,10-trimethyl, and butyl neopentyl ester, etc (Sharma and Kumar, 2021). The findings regarding organic pollutants indicated that certain carcinogenic and endocrine-disrupting chemicals (EDCs) present in the aged municipal solid waste samples indirectly and/or directly impacted living forms.

Although organic amendments may be mature and stable, they can still include phytotoxic chemicals from the original materials that negatively impact crop/plant growth and yield losses. The occurrence of various complex compounds in the digestate, including micro- and macro-nutrients, metal(loid) ions, and organic micropollutants, resulted in ecotoxicological effects/interactions ranging from "synergism to antagonism", complicating the identification of the root cause of toxicity. It is necessary to understand plant responses to phytotoxicity resulting from organic amendments and to develop effective management strategies. This chapter aimed to assess the phytotoxicity of organic amendments and waste materials. This highlights the immediate necessity for suitable, environmentally sustainable, and long-term treatment solutions to reduce phytotoxicity and mitigate potential human and environmental health risks.

2. Characteristics and impacts of organic amendments and wastes

2.1 Characteristics

Globally, the amount of organic waste produced is significant. Recent statistics indicate that approximately 1000 million tons of agricultural waste are generated worldwide annually, making it one of the critical sources of organic waste (Karić et al., 2022; Peng et al., 2023). This encompasses a variety of materials, including crop residues (e.g., husks, straws, and stalks), animal manure, food processing by-products, and other organic by-products generated from agricultural activities. Agricultural waste can generate several toxic contaminants, i.e., pesticides, pathogens, heavy metal(loid)s, and antibiotics, which pose risks to both environmental and human beings (Nguyen et al., 2023; Shu et al., 2025). It is estimated that around 2000 million tons of municipal (solid) waste are formed globally, with approximately 33% of this waste not being collected by municipal authorities (Hoornweg and Bhada-Tata, 2012). It is reported that 46% of municipal solid waste consists of organic waste, which consists of food scraps, green/yard waste (e.g., grass, leaves, and brush), and wood, etc. (Seadi et al., 2013) In developing countries, biodegradable materials, mainly composed of organic material, represent 40–70 wt% of urban solid waste (Wei et al., 2017). The 1300 million tons of food waste generated yearly is equivalent to about 1400 million hectares of usable agricultural area (Paritosh et al., 2017; Sahoo et al., 2024; Sharma et al., 2021).

Municipal, domestic, and agricultural waste can include metals, plastics, and pharmaceuticals, along with microbes and both organic and inorganic chemicals from sources such as solid amendments, fertilizers, compost, sludge, wastewater, and others (Hoang et al., 2024; Nguyen et al., 2022, 2024; Goss et al., 2013; Urra et al., 2019). (Table 13.1) reports the pollutant levels that can be found in organic amendments and wastes. For instance, potentially phytotoxic chemicals, including zinc, nickel, copper, and salts, were found to be at higher levels in milorganite and chicken manure than in dairy manure (Kebrom et al., 2019). The elevated levels of these elements may contribute to the high phytotoxicity observed in solid waste compost, as similar findings have been reported in compost made from swine manure and sewage sludge (Miaomiao et al., 2009). Compost produced from urban solid waste and plant residues exhibited phytotoxicity, and its heavy metal(loid) content exceeded regulatory limits (Siles-Castellano et al., 2020).

Overall, organic amendments, composts, and commercial organic fertilizers are prepared for immediate application and are suitable for direct use in agricultural soils that do not need any additional processing (Kebrom et al., 2019; Araji et al., 2001). However, it is crucial to investigate their potential phytotoxic effects in their application for the cultivation of crops and plants. These highlight the necessity for further consideration of the phytotoxicity of organic soil amendments to make certain their safe application in fruit/vegetable crop cultivation systems.

2.2 Potential impacts

As mentioned, in spite of the many advantages, the agricultural use of composted wastes and raw materials can also lead to negative ecological impacts (Diacono and Montemurro, 2011). Organic fertilizers, composts, and organic soil amendments may pose risks to the environment and humans (Kebrom et al., 2019). The organic amendment-induced phytotoxicity may arise from its impact on seed germination and/or root growth. Agricultural waste often displays high phytotoxicity associated with the presence of substances such as organic acids, phenols, ammonia/nitrogen, and heavy metal(loid)s (Cui et al., 2024). Also, the content of heavy metal(loid)s in compost is typically more than the average level found in soil, which raises the potential of metal accumulation with compost applications repeatedly (Zhang et al., 2006). It is primarily linked to the dangerous agents present in these organic matters, which can enter the food web as plants absorb them from the soil.

Challenges associated with the application of organic amendments stem from the occurrence of heavy metal(loid)s, and hazardous organic compounds (such as pesticides), and could pose a risk to human health (Fig. 13.1) (Alengebawy et al., 2021; Goss et al.,

Table 13.1 Pollutant levels from organic amendments and wastes globally.

Location	Wastes/sources	Contaminants	Levels	Remarks	References
Spain	Organic waste composts	Cd, Cu, Zn, cr, Ni, Pb, As	0.06–0.61 to 30.84–140.48 mg kg^{-1}	Vegetable residues and urban solid waste-based compost were phytotoxic	Siles-Castellano et al. (2020)
Lithuania	Food waste composts	Cd, Cu, Zn, cr, Ni, Pb, Hg	0.00–0.19 to 85–108 mg kg^{-1}	Heavy metals in compost	Višniauskė et al. (2018)
UK	Domestic food waste	Cd, As, Cu, Zn, cr, Ni, Pb	<0.01–33 mg kg^{-1}	Potentially toxic elements	Zhang et al. (2012)
Europe	Composts	• Hexachlorocyclohexane • Pentachlorobenzen	• 0.55–0.78 μg kg^{-1} • 0.039–0.33 μg kg^{-1}	Cause negative impacts on the soil quality	Beníšek et al. (2015)
Global	Composts	Microplastics	Thousands of items kg^{-1}	A pathway for microplastics to enter the environment	Le et al. (2023)
Japan	Bio-composts	PAHs	27–210 μg kg^{-1}	Polluted to a low level	Takigami et al. (2010)
Switzerland	Compost + digestate	• PFAS • PCDD/Fs • PCBs	• 6.3 μg kg^{-1} • 3.2 ng-TEQ kg^{-1} • 0.4–6.8 ng–TEQ kg^{-1}	• Organic pollutants (> levels of background soils) • The introduction of organic contaminants into the soil	Brändli et al. (2007b), Brändli et al. (2007)
Sweden	Household waste	• Chlorpyrifos • Endosulfan-α • Pentachloroaniline	• 3–3.4 μg kg−1 • 3–5.8 μg kg−1 • 0.7–1.2 μg kg−1	Were found in the biodegradable fraction	Hellström et al. (2011)

Italy	Municipal solid waste	PAHs	• 1120–1670 μg kg^{-1}	Persistent organic pollutants	Beggio et al. (2019)
US	Organic waste composts	• PFBA • PFBS • PFOA	• 0.15–12.04 μg kg−1 • 0–7.63 μg kg−1 • 0.04–10.31 μg kg−1	The distribution and challenges related to PFAS	Choi et al. (2019)
Norway	Organic wastes/ feedstocks	PAHs, PCBs, and PCDD/Fs	PAHs (0.38±0.01 to 5.05 ± 0.09 mg kg^{-1}), PCBs (7.6 ± 0.6 to 20.7 ± 0.6 μg kg^{-1}), and PCDD/Fs (1.2 ± 0.1 to 8.3 ± 0.2 ng TEQ kg^{-1})	• Contaminated organic waste • Has the potential to be toxic	Sørmo et al. (2024)
Scotland	Sludge, compost, and manure-amended soils	PAHs	• 331 μg kg−1 • 378 μg kg−1 • 223 μg kg−1	• Increase in soil PAHs (in soils amended with organic wastes) • Minor health risks associated with PAHs	Ma et al. (2024)

per- and polyfluoroalkyl substances (PFAS), polycyclic aromatic hydrocarbons (PAHs); polychlorinated biphenyls (PCBs), polychlorinated dibenzo-p-dioxins and dibenzofurans (*PCDD/Fs*); *perfluorobutanoic acid* (*PFBA*), perfluorobutane sulfonate (PFBS); *perfluorooctanoic acid* (*PFOA*), toxicity equivalent (*TEQ*).

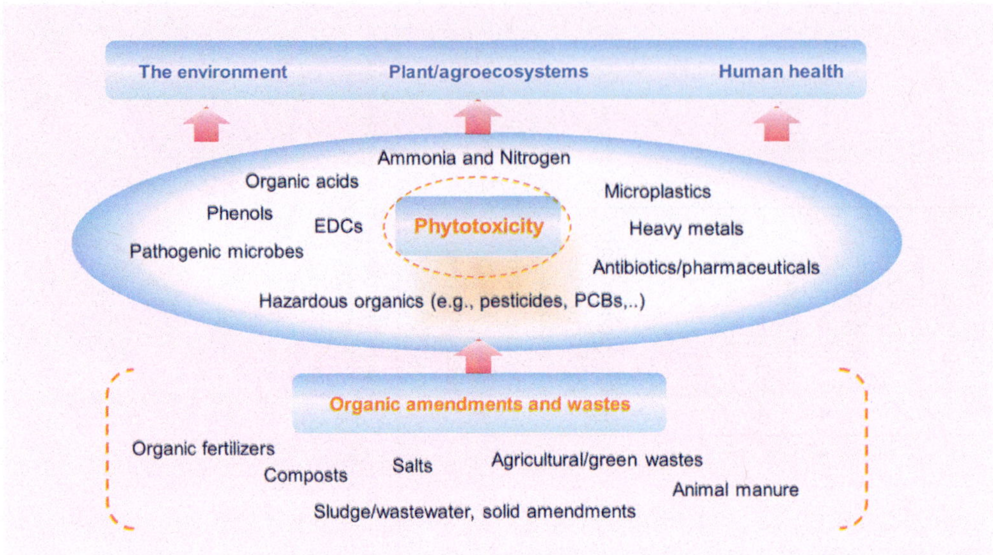

Figure 13.1 Challenges and potential impacts originated from organic amendments and wastes.

2013). Heavy metal(loid)s and pesticides rank among the most critical environmental toxicants that pose a threat significantly (Alengebawy et al., 2021). Organic amendments serve as natural carbon sources for microbes, including those that may be pathogenic to humans. Amendments derived from feces may also contain metals, antibiotics used for disease treatment or prophylaxis, as well as sex and growth hormones that persist in the environment (Goss et al., 2013). Thus, risks to ecosystems and human health from antibiotics, pathogens, and hormonal substances that exist in organic amendments—transferred via the food chain/web and/or sources supply for drinking water—must be mitigated prior to land application (Smith, 2009; Urra et al., 2019).

Nowadays, with increasing application rates, heavy metal(loid)s, hazardous agents, or salts in composts and organic soil amendments may accumulate to toxic concentrations, inhibiting crop/plant development and reducing yield (Fig. 13.2). (Chotsaeng et al., 2018) studied the various aldehydes-caused phytotoxicity, e.g., furans, and found that furans substantially restricted the Chinese amaranth's germination. The presence of unstable organic matter is likely to lead to phytotoxic effects in the final products (Cesaro et al., 2019). A poorly executed composting process can lead to immature compost or insufficiently stabilized organic matter, which may adversely influence the soil environment and plant development, be a disease agent, and lead to crop damage due to phytotoxicity (Cui et al., 2017). The phytotoxicity of organic soil amendments may also arise from synergistic effects, even if the level of each potentially harmful substance is below the threshold that could typically trigger toxic effects (Barral and Paradelo, 2011).

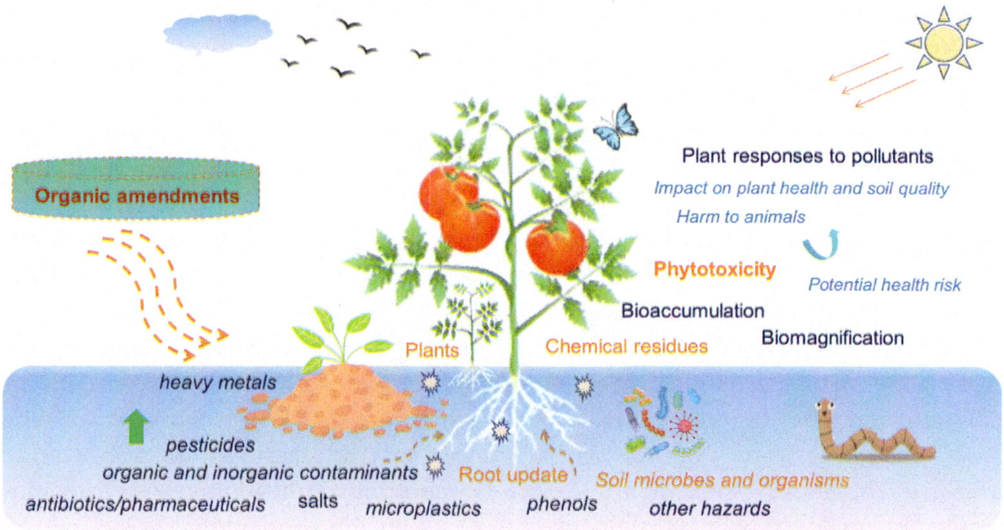

Figure 13.2 Phytotoxicity and potential impacts on soil-plant ecosystems.

The persistence of recent pollutants poses a significant ecological concern, as they are capable of contaminating cropland, surrounding water sources, and the food network. Heavy metal(loid)s are recognized as toxic to living organisms at high levels and are regarded as emerging pollutants in aquatic and terrestrial ecosystems (Tran et al., 2024). They can enter food waste sources via polluted farmland or agricultural fields and persist throughout the food supply chain. Moreover, microplastics are frequently found in various food wastes and compost (Le et al., 2023). It can be accumulated to thousands of items per kilogram in compost, presenting an emerging threat to agroecosystems. A wide variety of unknown and known dangerous contaminants, such as agrochemicals like pesticides, are frequently present in food wastes related to persistent residues in soil environments and foods (Nguyen et al., 2020). OCPs, known as "persistent organochlorine pesticides", strongly adsorb to soils, can be found in high concentrations in agricultural soils, and could be taken up by various plants. With regards to commercial composts and organic waste for a nutrient-boosting soil amendment, resulting in the loads and leachability of perfluoroalkyl acids (PFAAs) were analyzed, ranging from 28.7 to 75.9 μg kg^{-1}(Choi et al., 2019). Additionally, digestates and waste composts can involve elevated concentrations of polychlorinated biphenyls (PCBs) (Brändli et al., 2007). For example, it was found that sewage sludge-originated composts were highly contaminated with PCBs, with levels totaling 37.8 μg kg^{-1}(Višniauskė et al., 2018). The harmful influences, prolonged persistence, and rising concentrations in the ecosystems pose a severe risk to food safety and security.

3. Solutions for limiting phytotoxicity of organic amendments and wastes

Composting food waste and agricultural wastes using suitable solutions can provide sources of organic fertilizers as soil amendments. Also, green waste holds significant potential for use as a growing medium; however, it often exhibits strong phytotoxicity (Yang et al., 2022). Mitigating or eliminating phytotoxicity is essential for effectively using different organic wastes in agricultural sectors. The methods investigated include commonly used techniques—(1) "additive-based composting", (2) "heat treatment", (3) "washing treatment", (4) "aging"—as well as a recently introduced approach, and (5) "ammonium incubation" (Fig. 13.3) (Noor et al., 2024; Barthod et al., 2018; Cui et al., 2024). As an illustration, additive-mediated composting involves using specific additives to enhance the composting process, which can break down organic material transformed into stable, humus-like substances via the action of microbes under managed conditions (Tran et al., 2023). Composting with additives—is the most frequently employed approach for reducing or eliminating agricultural waste-induced phytotoxicity (Tran et al., 2022). Diverse additives can improve compost humification and effectively mitigate phytotoxicity (Nguyen et al., 2023). Biochar has become increasingly popular as a recent compost additive. Adding biochar to compost can effectively decrease phytotoxic effects by enhancing the compost mixture's physical and chemical properties. Nguyen et al. (2022) mentioned that biochar contributed to better

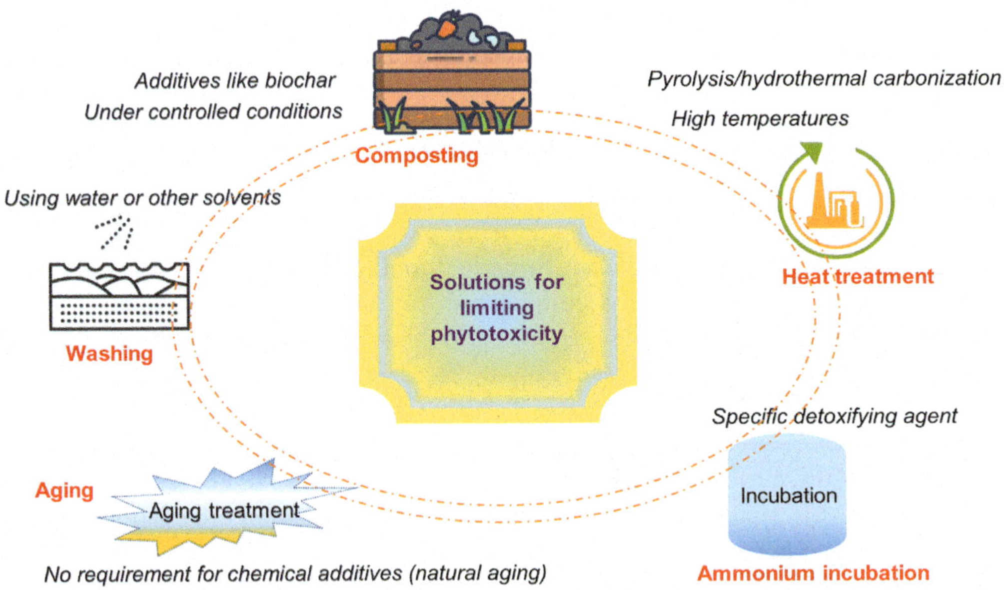

Figure 13.3 Solutions for limiting phytotoxicity induced by organic amendments.

pore structure in compost materials, enhancing oxygen accessibility rates, eliminating toxic compounds, and improving overall compost quality. Most importantly, in the scope of a circular economy, these approaches could serve as an ideal alternative to agrochemicals, promoting more environmentally friendly and sustainable agricultural practices (Pecorini et al., 2020).

As the heat treatment process occurs, phytotoxic components in the material are volatilized, condensed, and thermally degraded, leading to a substantial reduction in phytotoxicity (Cui et al., 2024). Subsequent to the hydrothermal carbonization and/or pyrolysis of municipal sewage sludge, the bioavailability of heavy metal(loid)s was diminished, resulting in decreased phytotoxicity (Alipour et al., 2021). Similarly, washing is a technique that uses water and/or other liquid solvents to reduce the contaminant level from materials, thereby improving quality and their applicability (Ahsan et al., 2023; Feng et al., 2020). A washing approach has also been employed to remediate organic waste from agriculture in order to neutralize its phytotoxic effects (Intani et al., 2019). Despite the fact that chemical washing is efficient and allows for the selection of reagents suitable to specific pollutants, it can be expensive.

As one of the other practical approaches(Chemetova et al., 2019), revealed that fresh Acacia melanoxylon bark exhibited a strong phytotoxicity. Interestingly, post-aging treatment, the phenolic substances were completely removed, eliminating their inhibition of celery germination. As a matter of fact, the advantages of natural aging are evident: it is cost-effective, easy to implement, and does not require chemical additives (Cui et al., 2024). During natural aging, microbial activity and environmental exposure help degrade these phytotoxic compounds, making the material safer for plants and reducing the risk of plant growth. Nonetheless, natural aging requires much time, can produce inconsistent outcomes, and is restricted by the supply of feedstocks or raw materials. These factors pose challenges for industrial production towards high quality (Qi et al., 2018). Further, drawing on these findings recently(Zhou et al., 2023), proposed a novel method for diminishing the phytotoxic effects of green waste known as the ammonium incubation process. The core concept of this method involves a targeted chemical detoxifying agent reacting with the phytotoxic substances in the material, thereby reducing their level or activity and ultimately eliminating or mitigating phytotoxicity. Throughout the incubation process of green/organic wastes with ammonium, complex non-biological and biological reactions involving ammonium salts, acidic organic substances (organic acids), and other toxic substances can occur, leading to a reduction in their phytotoxicity. This solution is characterized by energy-saving, minimal pollution, and promising prospects for industrial application. Given these findings, upcoming strategies should enhance phytotoxicity removal approaches to reduce material loss and processing expenses while avoiding secondary contamination.

4. Conclusions and outlooks

Organic amendments like compost and manure are commonly used to improve soil fertility; however, if not adequately processed, they can release phytotoxic compounds that hinder seed germination and root development and detrimental effects that can have on plant growth. Organic amendments and wastes can exhibit different levels of phytotoxicity depending on their composition and stage of decomposition. Wastes from agricultural or industrial processes may contain hazardous substances that can negatively impact neighboring plants, highlighting the importance of carefully managing organic waste applications in agricultural systems. Thus, while advantageous in many ways, certain organic amendments can result in the accumulation of toxic elements or pathogens in the soil, emphasizing the need for effective management and monitoring of these risks.

Additional research is necessary to assess the phytotoxicity of organic amendments in soils to make certain their safe application in fruit and vegetable crop cultivation systems. Enhance toxicology studies concerning soil, plant, and animal health. Assessing the phytotoxicity of organic wastes is essential for sustainable agricultural practices, as incorporating non-composted or inadequately treated wastes can result in decreased crop yields and deterioration of soil health. Additionally, it is crucial to investigate emerging microcontaminants in compost derived from food waste and digestate to confirm food security and mitigate future risks to humans. The quantity of pathogens released into the ecosystems is crucial for conducting a risk assessment regarding the use of organic amendments. Further, expanded research is required to evaluate the impacts of plastics in waste, as well as the extent of pesticides, PAHs, and PFAS contamination in organic wastes and their byproducts. Effective management of these wastes and their products is essential before their application in the environment.

Furthermore, circular sustainability represents a promising approach that emphasizes an economic and environmental strategy focused on the efficient use of resources and waste reduction through a closed-loop system. In contrast to local and traditional agriculture, which relied on circular economy frameworks, modern agriculture can now reintroduce these wastes into the production chain for new purposes (Jimenez-Lopez et al., 2020). Proper composting processes can transform organic waste into stable, nutrient-rich amendments that minimize phytotoxicity. Additionally, implementing closed-loop systems in agriculture can improve nutrient cycling and further reduce phytotoxicity. Ensuring the thorough decomposition of organic materials helps eliminate harmful pathogens and phytotoxic compounds, resulting in a safer product for soil application. Using organic waste along with biotechnologies offers an eco-friendly solution for remediating toxic-contaminated soils, such as PAHs (Omoni et al., 2024). The enhancement of soils contaminated by emerging micropollutants through bioaugmentation not only aids in remediation but also enhances the soil's nutrients. Combined organic amendments and biotechnologies prospects as efficient and environmentally friendly solutions for treating contaminated sites. Thus, these critical solutions based on circular sustainability are expected to significantly reduce phytotoxicity from organic amendments by promoting practices that minimize harmful effects on plant growth while enhancing soil health and resource efficiency.

References

Ahsan, W.A., Hussain, A., Lin, C., Nguyen, M.K., 2023. Biodegradation of different types of bioplastics through composting—a recent trend in green recycling. Catalysts 13 (2). https://doi.org/10.3390/catal13020294.

Alengebawy, A., Abdelkhalek, S.T., Qureshi, S.R., Wang, M.Q., 2021. Heavy metals and pesticides toxicity in agricultural soil and plants: ecological risks and human health implications. Toxics 9 (3), 1–34. https://doi.org/10.3390/toxics9030042.

Alipour, M., Asadi, H., Chen, C., Rezaei Rashti, M., 2021. Bioavailability and eco-toxicity of heavy metals in chars produced from municipal sewage sludge decreased during pyrolysis and hydrothermal carbonization. Ecol. Eng. 162, 106173. https://doi.org/10.1016/j.ecoleng.2021.106173.

Araji, A.A., Abdo, Z.O., Joyce, P., 2001. Efficient use of animal manure on cropland – economic analysis. Bioresour. Technol. 79 (2), 179–191. https://doi.org/10.1016/s0960-8524(01)00042-6.

Barral, M.T., Paradelo, R., 2011. A review on the use of phytotoxicity as a compost quality indicator. Dyn. Soil Dyn. Plant 5 (2), 36–44.

Barthod, J., Rumpel, C., Dignac, M.F., 2018. Composting with additives to improve organic amendments. A review. Agron. Sustain. Dev. 38 (2). https://doi.org/10.1007/s13593-018-0491-9.

Battaglia-Brunet, F., Le Guédard, M., Faure, O., Charron, M., Hube, D., Devau, N., Joulian, C., Thouin, H., Hellal, J., 2021. Influence of agricultural amendments on arsenic biogeochemistry and phytotoxicity in a soil polluted by the destruction of arsenic-containing shells. J. Hazard Mater. 409, 124580. https://doi.org/10.1016/j.jhazmat.2020.124580.

Beggio, G., Schievano, A., Bonato, T., Hennebert, P., Pivato, A., 2019. Statistical analysis for the quality assessment of digestates from separately collected organic fraction of municipal solid waste (OFMSW) and agro-industrial feedstock. Should input feedstock to anaerobic digestion determine the legal status of digestate? Waste Management 87, 546–558. https://doi.org/10.1016/j.wasman.2019.02.040.

Beníšek, M., Kukučka, P., Mariani, G., Suurkuusk, G., Gawlik, B.M., Locoro, G., Giesy, J.P., Bláha, L., 2015. Dioxins and dioxin-like compounds in composts and digestates from European countries as determined by the in vitro bioassay and chemical analysis. Republic Chemosphere 122, 168–175. https://doi.org/10.1016/j.chemosphere.2014.11.039.

Brändli, R.C., Bucheli, T.D., Kupper, T., Furrer, R., Stahel, W.A., Stadelmann, F.X., Tarradellas, J., 2007a. Organic pollutants in compost and digestate.: Part 1. Polychlorinated biphenyls, polycyclic aromatic hydrocarbons and molecular markers. J. Environ. Monit. 9 (5), 456–464. https://doi.org/10.1039/b617101j.

Brändli, R.C., Kupper, T., Bucheli, T.D., Zennegg, M., Huber, S., Ortelli, D., Müller, J., Schaffner, C., Iozza, S., Schmid, P., Berger, U., Edder, P., Oehme, M., Stadelmann, F.X., Tarradellas, J., 2007b. Organic pollutants in compost and digestate.: Part 2. Polychlorinated dibenzo-p-dioxins, and -furans, dioxin-like polychlorinated biphenyls, brominated flame retardants, perfluorinated alkyl substances, pesticides, and other compounds. J. Environ. Monit. 9 (5), 465–472. https://doi.org/10.1039/b617103f.

Cesaro, A., Conte, A., Belgiorno, V., Siciliano, A., Guida, M., 2019. The evolution of compost stability and maturity during the full-scale treatment of the organic fraction of municipal solid waste. J. Environ. Manag. 232, 264–270. https://doi.org/10.1016/j.jenvman.2018.10.121.

Chemetova, C., Quilhó, T., Braga, S., Fabião, A., Gominho, J., Ribeiro, H., 2019. Aged acacia melanoxylon bark as an organic peat replacement in container media. J. Clean. Prod. 232, 1103–1111. https://doi.org/10.1016/j.jclepro.2019.06.064.

Choi, Y.J., Kim Lazcano, R., Yousefi, P., Trim, H., Lee, L.S., 2019. Perfluoroalkyl acid characterization in U.S. municipal organic solid waste composts. Environ. Sci. Technol. Lett. 6 (6), 372–377. https://doi.org/10.1021/acs.estlett.9b00280.

Chotsaeng, N., Laosinwattana, C., Charoenying, P., 2018. Inhibitory effects of a variety of aldehydes on amaranthus tricolor L. and echinochloa crus-galli (L.) Beauv. Molecules 23 (2), 471. https://doi.org/10.3390/molecules23020471.

Cui, H.Y., Zhao, Y., Chen, Y.N., Zhang, X., Wang, X.Q., Lu, Q., Jia, L.M., Wei, Z.M., 2017. Assessment of phytotoxicity grade during composting based on EEM/PARAFAC combined with projection pursuit regression. J. Hazard Mater. 326, 10–17. https://doi.org/10.1016/j.jhazmat.2016.09.059.

Cui, W., Qi, B., Liu, J., Chen, J., Qi, Z., Zhou, W., 2024. Phytotoxicity removal technologies for agricultural waste as a growing media component: a review. Agronomy 14 (1), 40. https://doi.org/10.3390/agronomy14010040.

Diacono, M., Montemurro, F., 2011. Long-term effects of organic amendments on soil fertility. Sustainable Agriculture 2, 761–786. https://doi.org/10.1007/978-94-007-0394-0_34.

Feng, W., Zhang, S., Zhong, Q., Wang, G., Pan, X., Xu, X., Zhou, W., Li, T., Luo, L., Zhang, Y., 2020. Soil washing remediation of heavy metal from contaminated soil with EDTMP and PAA: properties, optimization, and risk assessment. J. Hazard Mater. 381, 120997. https://doi.org/10.1016/j.jhazmat.2019.120997.

Goss, M.J., Tubeileh, A., Goorahoo, D., 2013. A review of the use of organic amendments and the risk to human health. Adv. Agron. 120, 275–379. https://doi.org/10.1016/B978-0-12-407686-0.00005-1.

Hellström, A., Nilsson, M.L., Kylin, H., 2011. Current-use and organochlorine pesticides and polychlorinated biphenyls in the biodegradable fraction of source separated household waste, compost, and anaerobic digestate. Bull. Environ. Contam. Toxicol. 86 (1), 60–64. https://doi.org/10.1007/s00128-010-0147-1.

Hoang, V.H., Nguyen, M.K., Hoang, T.D., Ha, M.C., Huyen, N.T.T., Bui, V.K.H., Pham, M.T., Nguyen, C.M., Chang, S.W., Nguyen, D.D., 2024. Sources, environmental fate, and impacts of microplastic contamination in agricultural soils: a comprehensive review. Sci. Total Environ. 950. https://doi.org/10.1016/j.scitotenv.2024.175276.

D. Hoornweg, P. Bhada-Tata, What a waste: a global review of solid waste management. (2012).

Hrčka, M., Hřebečková, T., Hanč, A., Grasserová, A., Cajthaml, T., 2024. Changes in the content of emerging pollutants and potentially hazardous substances during vermi/composting of a mixture of sewage sludge and moulded pulp. Environ. Pollut. 348, 123736. https://doi.org/10.1016/j.envpol.2024.123736.

Intani, K., Latif, S., Islam, MdS., Müller, J., 2019. Phytotoxicity of corncob biochar before and after heat treatment and washing. Sustainability 11 (1), 30. https://doi.org/10.3390/su11010030.

Jimenez-Lopez, C., Fraga Corral, M., Carpena, M., García-Oliveira, P., Echave, J., Pereira, A.G., Lourenço-Lopes, C., Prieto, M.A., Simal-Gandara, J., 2020. Agriculture waste valorisation as a source of antioxidant phenolic compounds within a circular and sustainable bioeconomy. Food Funct. 11 (6), 4853–4877. https://doi.org/10.1039/d0fo00937g.

Karić, N., Maia, A.S., Teodorović, A., Atanasova, N., Langergraber, G., Crini, G., Ribeiro, A.R.L., Đolić, M., 2022. Bio-waste valorisation: agricultural wastes as biosorbents for removal of (in)organic pollutants in wastewater treatment. Chem. Eng. J. Adv. 9. https://doi.org/10.1016/j.ceja.2021.100239.

Kebrom, T.H., Woldesenbet, S., Bayabil, H.K., Garcia, M., Gao, M., Ampim, P., Awal, R., Fares, A., 2019. Evaluation of phytotoxicity of three organic amendments to collard greens using the seed germination bioassay. Environ. Sci. Pollut. Control Ser. 26 (6), 5454–5462. https://doi.org/10.1007/s11356-018-3928-4.

Le, V.R., Nguyen, M.K., Nguyen, H.L., Lin, C., Rakib, M.R.J., Thai, V.A., Le, V.G., Malafaia, G., Idris, A.M., 2023. Organic composts as a vehicle for the entry of microplastics into the environment: a comprehensive review. Sci. Total Environ. 892. https://doi.org/10.1016/j.scitotenv.2023.164758.

Luan, H., Gao, W., Huang, S., Tang, J., Li, M., Zhang, H., Chen, X., 2019. Partial substitution of chemical fertilizer with organic amendments affects soil organic carbon composition and stability in a greenhouse vegetable production system. Soil Tillage Res. 191, 185–196. https://doi.org/10.1016/j.still.2019.04.009.

Ma, X., Cooper, P., Shi, H., Osprey, M., Riach, D., Paton, G.I., Xiong, Q., Zhou, X., Zhang, Z., 2024. Temporal trends of polycyclic aromatic hydrocarbons in soils amended with sludge, compost, and manure in a Scotland pasture: an 8-year field experiment. Environ. Pollut. 360, 124614. https://doi.org/10.1016/j.envpol.2024.124614.

Miaomiao, H., Wenhong, L., Xinqiang, L., Wu, D., Guangming, T., 2009. Effect of composting process on phytotoxicity and speciation of copper, zinc and lead in sewage sludge and swine manure. Waste Management 29 (2), 590–597. https://doi.org/10.1016/j.wasman.2008.07.005.

Nguyen, M.K., Lin, C., Hoang, H.G., Bui, X.T., Ngo, H.H., Le, V.G., Tran, H.T., 2023. Investigation of biochar amendments on odor reduction and their characteristics during food waste co-composting. Sci. Total Environ. 865. https://doi.org/10.1016/j.scitotenv.2022.161128.

Nguyen, M.K., Hadi, M., Lin, C., Nguyen, H.L., Thai, V.B., Hoang, H.G., Vo, D.V.N., Tran, H.T., 2022. Microplastics in sewage sludge: Distribution, toxicity, identification methods, and engineered technologies. Chemosphere 308. https://doi.org/10.1016/j.chemosphere.2022.136455.

Nguyen, M.K., Lin, C., Bui, X.T., Rakib, M.R.J., Nguyen, H.L., Truong, Q.M., Hoang, H.G., Tran, H. T., Malafaia, G., Idris, A.M., 2024. Occurrence and fate of pharmaceutical pollutants in wastewater: Insights on ecotoxicity, health risk, and state–of–the-art removal. Chemosphere 354. https://doi.org/10.1016/j.chemosphere.2024.141678.

Nguyen, T.T., Rosello, C., Bélanger, R., Ratti, C., 2020. Fate of residual pesticides in fruit and vegetable waste (FVW) Processing. Foods 9 (10), 1468. https://doi.org/10.3390/foods9101468.

Noor, R.S., Shah, A.N., Tahir, M.B., Umair, M., Nawaz, M., Ali, A., Ercisli, S., Abdelsalam, N.R., Ali, H. M., Yang, S.H., Ullah, S., Assiri, M.A., 2024. Recent trends and advances in additive-mediated composting technology for agricultural waste resources: a comprehensive review. ACS Omega 9 (8), 8632–8653. https://doi.org/10.1021/acsomega.3c06516.

O'Connor, J., Mickan, B.S., Siddique, K.H.M., Rinklebe, J., Kirkham, M.B., Bolan, N.S., 2022. Physical, chemical, and microbial contaminants in food waste management for soil application: a review. Environ. Pollut. 300. https://doi.org/10.1016/j.envpol.2022.118860.

Okori, F., Lederer, J., Komakech, A.J., Schwarzböck, T., Fellner, J., 2024. Plastics and other extraneous matter in municipal solid waste compost: a systematic review of sources, occurrence, implications, and fate in amended soils. Environ. Adv. 15, 100494. https://doi.org/10.1016/j.envadv.2024.100494.

Omoni, V.T., Bankole, P.O., Semple, K.T., Ojo, A.S., Ibeto, C., Okekporo, S.E., Harrison, I.A., 2024. Enhanced remediation of polycyclic aromatic hydrocarbons in soil through fungal delignification strategy and organic waste amendment: a review. Indian J. Microbiol. https://doi.org/10.1007/s12088-024-01383-7.

Paritosh, K., Kushwaha, S.K., Yadav, M., Pareek, N., Chawade, A., Vivekanand, V., 2017. Food waste to energy: an overview of sustainable approaches for food waste management and nutrient recycling. BioMed Res. Int. 2017. https://doi.org/10.1155/2017/2370927.

Pecorini, I., Peruzzi, E., Albini, E., Doni, S., Macci, C., Masciandaro, G., Iannelli, R., 2020. Evaluation of MSW compost and digestate mixtures for a circular economy application. Sustainability 12 (7), 3042. https://doi.org/10.3390/su12073042.

Peng, X., Jiang, Y., Chen, Z., Osman, A.I., Farghali, M., Rooney, D.W., Yap, P.S., 2023. Recycling municipal, agricultural and industrial waste into energy, fertilizers, food and construction materials, and economic feasibility: a review. Environ. Chem. Lett. 21 (2), 765–801. https://doi.org/10.1007/s10311-022-01551-5.

Qi, D., Miao, A., Cao, J., Wang, W., Chen, W., Pang, S., He, X., Ma, C., 2018. Study on the effects of rapid aging technology on the aroma quality of white tea using GC–MS combined with chemometrics: in comparison with natural aged and fresh white tea. Food Chem. 265, 189–199. https://doi.org/10.1016/j.foodchem.2018.05.080.

Sahoo, A., Dwivedi, A., Madheshiya, P., Kumar, U., Sharma, R.K., Tiwari, S., 2024. Insights into the management of food waste in developing countries: with special reference to India. Environ. Sci. Pollut. Control Ser. 31 (12), 17887–17913. https://doi.org/10.1007/s11356-023-27901-6.

Seadi, Al, Owen, T., Hellström, N.E., Kang, H., 2013. Source Separation of MSW. IEA Bioenergy.

Sharma, P., Gaur, V.K., Sirohi, R., Varjani, S., Hyoun Kim, S., Wong, J.W.C., 2021. Sustainable processing of food waste for production of bio-based products for circular bioeconomy. Bioresour. Technol. 325. https://doi.org/10.1016/j.biortech.2021.124684.

Sharma, P., Kumar, S., 2021. Characterization and phytotoxicity assessment of organic pollutants in old and fresh municipal solid wastes at open dump site: a case study. Environ. Technol. Innov. 24, 101938. https://doi.org/10.1016/j.eti.2021.101938.

Shu, Y., Li, D., Tong, X., Zhao, K., Zhou, L., Li, F., 2025. Antibiotics-heavy metals combined pollution in agricultural soils: sources, fate, risks, and countermeasures. Green Energy Environ. 10 (5), 869–897. https://doi.org/10.1016/j.gee.2024.07.007.

Siles-Castellano, A.B., López, M.J., López-González, J.A., Suárez-Estrella, F., Jurado, M.M., Estrella-González, M.J., Moreno, J., 2020. Comparative analysis of phytotoxicity and compost quality in industrial composting facilities processing different organic wastes. J. Clean. Prod. 252. https://doi.org/10.1016/j.jclepro.2019.119820.

Smith, S.R., 2009. Organic contaminants in sewage sludge (biosolids) and their significance for agricultural recycling. Philos. Trans. R. Soc. A Math. Phys. Eng. Sci. 367 (1904), 4005–4041. https://doi.org/10.1098/rsta.2009.0154.

Sørmo, E., Krahn, K.M., Flatabø, G.Ø., Hartnik, T., Arp, H.P.H., Cornelissen, G., 2024. Distribution of PAHs, PCBs, and PCDD/Fs in products from full-scale relevant pyrolysis of diverse contaminated organic waste. J. Hazard Mater. 461. https://doi.org/10.1016/j.jhazmat.2023.132546.

Takigami, H., Suzuki, G., Sakai, S.I., 2010. Screening of dioxin-like compounds in bio-composts and their materials: chemical analysis and fractionation-directed evaluation of AhR ligand activities using an in vitro bioassay. J. Environ. Monit. 12 (11), 2080–2087. https://doi.org/10.1039/c0em00200c.

Tran, H.T., Nguyen, M.K., Hoang, H.G., Hutchison, J.M., Vu, C.T., 2022. Composting and green technologies for remediation of phthalate (PAE)-contaminated soil: Current status and future perspectives. Chemosphere 307. https://doi.org/10.1016/j.chemosphere.2022.135989.

Tran, H.T., Bolan, N.S., Lin, C., Binh, Q.A., Nguyen, M.K., Luu, T.A., Le, V.G., Pham, C.Q., Hoang, H.G., Vo, D.V.N., 2023. Succession of biochar addition for soil amendment and contaminants remediation during co-composting: a state of art review. J. Environ. Manag. 342. https://doi.org/10.1016/j.jenvman.2023.118191.

Tran, T.K., Huynh, L., Nguyen, H.L., Nguyen, M.K., Lin, C., Hoang, T.D., Hung, N.T.Q., Nguyen, X. H., Chang, S.W., Nguyen, D.D., 2024. Applications of engineered biochar in remediation of heavy metal(loid)s pollution from wastewater: Current perspectives toward sustainable development goals. Sci. Total Environ. 926. https://doi.org/10.1016/j.scitotenv.2024.171859.

Urra, J., Alkorta, I., Garbisu, C., 2019. Potential benefits and risks for soil health derived from the use of organic amendments in agriculture. Agronomy 9 (9), 542. https://doi.org/10.3390/agronomy9090542.

Višniauskė, I., Barčauskaitė, K., Bakšienė, E., Mažeika, R., 2018. Evaluation of contamination levels of different types of composts and their suitability for usage in agriculture. Zemdirbyste-Agriculture 105 (3), 211–220. https://doi.org/10.13080/z a.2018.105.027.

Wei, Y., Li, J., Shi, D., Liu, G., Zhao, Y., Shimaoka, T., 2017. Environmental challenges impeding the composting of biodegradable municipal solid waste: a critical review. Resour. Conserv. Recycl. 122, 51–65. https://doi.org/10.1016/j.resconrec.2017.01.024.

Yang, R., Chen, X., Zhang, D., Wang, H., Zhou, W., Lin, W., Qi, Z., 2022. Steam-exploded pruning waste as peat substitute: physiochemical properties, phytotoxicity and their implications for plant cultivation. Int. J. Environ. Res. Publ. Health 19 (9), 5328. https://doi.org/10.3390/ijerph19095328.

Zhang, M., Heaney, D., Henriquez, B., Solberg, E., Bittner, E., 2006. A four-year study on influence of biosolids/msw cocompost application in less productive soils in alberta: nutrient dynamics. Compost Sci. Util. 14 (1), 68–80. https://doi.org/10.1080/1065657x.2006.10702265.

Zhang, Y., Banks, C.J., Heaven, S., 2012. Anaerobic digestion of two biodegradable municipal waste streams. J. Environ. Manag. 104, 166–174. https://doi.org/10.1016/j.jenvman.2012.03.043.

Zhong, W., Gu, T., Wang, W., Zhang, B., Lin, X., Huang, Q., Shen, W., 2010. The effects of mineral fertilizer and organic manure on soil microbial community and diversity. Plant Soil 326 (1–2), 511–522. https://doi.org/10.1007/s11104-009-9988-y.

Zhou, W., Liao, J., Zhou, B., Yang, R., Lin, W., Zhang, D., Wang, H., Qi, Z., 2023. Rapidly reducing phytotoxicity of green waste for growing media by incubation with ammonium. Environ. Technol. Innov. 31, 103136. https://doi.org/10.1016/j.eti.2023.103136.

CHAPTER 14

Biodegradation and natural attenuation of organic pollutants in environments

Modupe Stella Ayilara[1], Ajakaye Abisola Eunice[2], Bartholomew Saanu Adeleke[1,3], Saheed Adekunle Akinola[1,4], Akinlolu Olalekan Akanmu[1] and Olubukola Oluranti Babalola[1]

[1]Food Security and Safety Focus Area, Faculty of Natural and Agricultural Sciences, North-West University, Mmabatho, South Africa; [2]Department of Public Health, University of Suffolk, Microbiology Programme, Department of Biological Sciences, School of Science, Ipswich, United Kingdom; [3]Agagu University of Science and Technology, Okitipupa, Ondo, Nigeria; [4]Department of Microbiology and Parasitology, School of Medicine and Pharmacy, College of Medicine and Health Sciences, University of Rwanda, Butare, Rwanda

1. Introduction

Environmental pollutants derived from human-made and natural sources make up a significant organic pollutant category (Chaitanya et al., 2024). The diverse contaminants, which include petroleum hydrocarbons along with PAHs and pesticides in addition to pharmaceuticals and industrial solvents and dyes, present serious dangers to ecosystems and public health because they exhibit characteristics of toxicity, remain persistent in the environment as well as accumulate within biological systems (Adeleke et al., 2025). Organic pollutants are widespread in terrestrial and aquatic environments due to the ever-growing industrial and agricultural activities, indiscriminate waste disposal practices, and urban runoff. Organic pollutants are persistent in the environment and trigger damaging effects on ecosystems. They result in contamination of groundwater and surface water and have negative impacts on biodiversity and human health.

Organic pollutants are conventionally managed through incineration methods, physical removal techniques and chemical oxidation procedures, which prove to be expensive, need substantial energy input and produce additional pollutant compounds (Kumari and Kumar, 2023). The biological degradation techniques along with natural attenuation, represent effective methods for managing organic contaminants since they provide eco-friendly and economical solutions. The microbial process of biodegradation transforms organic materials into non-toxic, simpler products, which lead to the complete breakdown of carbon dioxide combined with water and microbial biomass (Bharti et al., 2020). The breakdown occurs through microorganisms utilizing organic pollutants as their carbon source under specific environmental conditions.

Natural attenuation includes multiple passive processes that occur in situ without human handling to reduce contaminant concentration, toxic potential, and movement in

Phytotoxicity of Organic Pollutants
ISBN 978-0-443-33110-7,
https://doi.org/10.1016/B978-0-443-33110-7.00016-X

the environment (Ayilara et al., 2023). Natural attenuation involves the presence of microorganisms and abiotic methods, which include chemical transformation, dilution, sorption, and volatilization (Bharti et al., 2020). Natural attenuation provides an effective remediation solution when correctly controlled and understood, as it works best for complex or vast contaminated sites where conventional remediation alternatives are not feasible (Nie et al., 2024). A reliable outcome of natural attenuation requires a full comprehension of the polluted sites' geochemical, biological and hydrological conditions. In recent years, due to the attention that has been on the use of microbes and sustainable approaches to control organic pollutants, scientists have improved their capability to predict, measure and detect degradation processes in different ecosystems using molecular biology, microbial ecology and environmental monitoring. Therefore, this article presents comprehensive information on the applications, mechanisms, and factors affecting the application of biodegradation and natural attenuation for organic pollution control. This article also proposes how to maximally utilize biodegradation and natural attenuation procedures for organic pollution management.

2. Mechanisms of biodegradation

The process of biodegradation occurs through an active relationship between microbial enzymatic functions, environmental conditions, and contaminant chemical characteristics (Fig. 14.1). Research on these pathways enables scientists to improve predictive

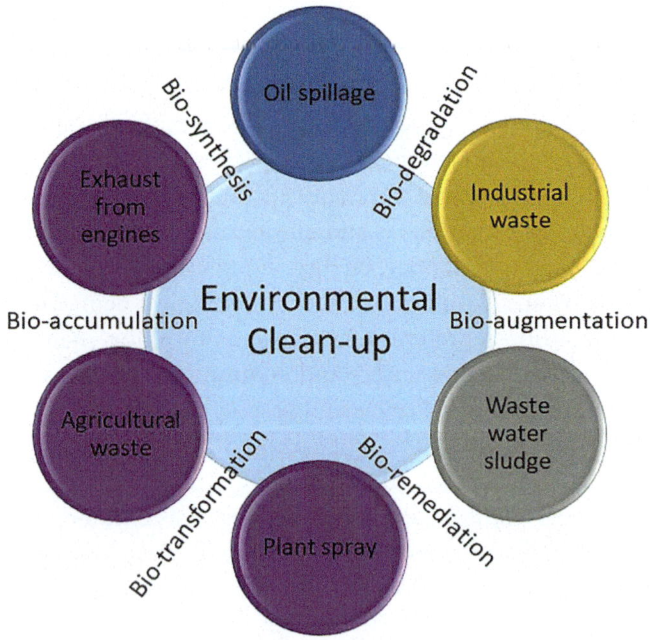

Figure 14.1 Mechanism of remediation by enhanced natural attenuation.

capabilities and the scalability of biological degradation methods, which have become vital for environmentally sustainable management. Organic pollutants undergo microbial breakdown as a complex process which creates products that become nontoxic or less toxic compounds in the environment through a process called biodegradation (Singh et al., 2021). The process relies mainly on microorganisms' metabolic functions, where the organisms use pollution materials as energy and carbon sources. The nature of microbial consortia, the type of pollutant and the environmental condition determine both the biodegradation path and efficiency as this process can occurs either in aerobic and anaerobic conditions (Cao et al., 2022). The knowledge of these mechanisms enables more beneficial application of biodegradation-based remediation methods and supports predictions for organic contaminants remediation.

Oxygen functions as the terminal electron acceptor in aerobic conditions to permit oxygen-dependent enzymes such as monooxygenases and dioxygenases increase their activity (Mrnjavac et al., 2024). These enzymes facilitate the breakdown of complex organic compounds, which include aromatic and aliphatic hydrocarbon compounds, while inserting oxygen atoms throughout the molecule. During PAH degradation ring-hydroxylating dioxygenases create hydroxyl group additions that enhance both solubility and reactivity features in the substance. The enzymatic breakdown of intermediates continues until all molecules mineralize into water, carbon dioxide, and other biomass products. The biodegradation process through aerobic methods happens quickly with a total breakdown yet its effectiveness on the level of oxygen levels across the area where the degradation is taking place (Bher et al., 2022).

On the other hand, the anaerobic biodegradation process is carried out in the absence of oxygen, where microbes use other electron acceptors, including pollutants, nitrates and carbon dioxide. This process often occurs in environments that are void of oxygen, such as in the groundwater and underground soil. This process of degradation is always slower than the aerobic process, and the degradation is often incomplete, although they are advantageous in anoxic zones where the aerobic process is ineffective (Huang et al., 2022). There are different types of anaerobic degradation, namely methanogenesis, reductive dehalogenation and fermentation (Wang et al., 2021a; Yuan et al., 2021); De (Marines et al., 2023).

An important factor in biodegradation is the enzymatic specificity and adaptive capabilities of the microbes involved. Microbes obtain catabolic genes through horizontal gene transfer mechanisms and develop the ability to break down synthetic chemicals, including both xenobiotics and new hazard materials (Brito, 2021). Plasmid genes generally contain degradative enzyme code that helps microbial populations quickly acquire new metabolic abilities through the spread of these genes (Mohapatra and Phale, 2021). The degradation process often demands synergetic efforts from several microbial species, which perform their functions in succession in a process termed syntrophic degradation.

Environmental factors are crucial in determining the rate and pathways through which biodegradation takes place. The survival and activities of microbes in the environment are affected by environmental conditions, including nutrients, temperature, and moisture (Kebede et al., 2021). The accessibility of pollutants to microorganisms serves as a significant factor in biodegradation. Soil particles and non-aqueous phases strongly adsorb hydrophobic organic compounds, reducing their susceptibility to biodegradation. However, to beat this challenge, biosurfactant production or physical agitation serves as an alternative strategy to increase bioavailability when these limitations exist.

Microbial degradation mechanisms have benefited significantly from applying modern molecular and omics-based scientific methods during the past few years. Three molecular techniques, metagenomics, proteomics and transcriptomics, have been used to reveal the gene expression, microbial community structure and functional abilities of the microbes that take part in degradation processes. These recent discoveries furnish crucial data about metabolic resistance, fundamental genes, and regulatory sequences, thus allowing for a better bioremediation method design.

3. Natural attenuation processes

Natural attenuation is the naturally occurring physical, biological and environmental processes which decrease the contaminant's volume, concentration, toxicity and mobility. The ecosystems possess the inherent ability to naturally self-purify, and this is usually used to manage distinct environments such as groundwater and soil pollution, most especially because of the high cost of biodegradation and when the biodegradation through this means is infeasible. It should, however, be noted that adopting natural attenuation for building sustainable passive remediation solutions relies on strong site monitoring alongside risk assessment. Natural attenuation can be characterized into biodegradation, sorption, dilution, volatilization and chemical transformation.

Biodegradation, which is the metabolic process through which indigenous microorganisms degrade pollutants, remains the main natural and most efficient attenuation method. Microbes with aerobic and anaerobic capabilities use pollutants either to generate energy or to co-metabolize pollutants while processing alternative substrates. The process enables complex contaminants to disintegrate into multiple safer end products, including biomass, carbon dioxide and water. Biodegradation is a method that is used to bioremediate pollutants arising from the pharmaceutical sectors as well as those from pesticides, organic and inorganic solvents and hydrocarbons (Ayilara and Babalola, 2023).

Sorption refers to the process by which contaminants bind to the soil materials, organic substances, sediments, and thereby decreasing their circulation potential and susceptibility to biological activity (Zamulina et al., 2022). Sorption can either be in the form of absorption, where the pollutants bind to the surface of the remediation materials or adsorption, where pollutants are absorbed into the matrics of the remediation materials (Hu et al., 2023). Through the sorption remediation process, hydrophobic organic compounds including heavy metals and polycyclic aromatic hydrocarbons (PAHs) become immobilized in polluted environments, which blocks their movement into groundwater and deep into the soil (Duffy et al., 2025).

Dilution is a physical remediation process that deals with the mixing of contaminants with clean surface water and groundwater to reduce the levels of contaminants in the affected environment (Douglas and Pandey, 2025). In this type of remediation, the mass of contaminants remains unchanged through dilution but the process lowers their quantities at particular sites which reduces the danger of contact with pollutants. The mechanism plays a vital role in water systems such as aquifers, which exhibit rapid flow velocities alongside hydraulic connection pathways.

Volatilization refers to the process where pollutants are moved from water or solid contents into atmospheric gases and vapor forms (Ke et al., 2021). Majorly, this method is usually used for volatile organic compounds such as chlorinated solvents, toluene and benzene. Volatilization remediation method is affected by the properties of the soil, temperature and vapor pressure of the environment. It should however be noted that although this remediation method removes pollution from the atmosphere, it could lead to secondary pollution in the atmosphere and could also transfer the pollutants elsewhere.

Chemical transformation involves different abiotic processes that transform pollutants both chemically and toxicologically (Ahmad et al., 2025). Examples of this remediation method include photolysis, oxidation-reduction reactions as well as hydrolysis (Cui et al., 2025; Banayan Esfahani et al., 2022). A notable instance of chemical transformation is the situation where trichlorethylene transforms through abiotic reductive dechlorination through iron-bearing minerals to produce benign end products. The process of chemical transformation occurs naturally or with environmental factor-induced catalysis that includes pH levels, reactive compounds and redox potentials. Examples of studies showing different natural attenuation process is shown in Table 14.1 below.

4. Monitoring natural attenuation in situ

To achieve successful natural attenuation remediation, one must establish comprehensive onsite testing procedures that show how inherent system processes reduce environmental threats. A well-designed monitoring system remains essential to gain regulatory approval and public acceptance for natural attenuation techniques because the passive process extends across long periods. The monitoring process for in situ evaluation includes regular systematic data collection that shows natural attenuation results alongside potential indications for corrective measures.

Table 14.1 Studies showing different natural attenuation processes.

S/N	Location	Type of waste/ Contaminant	Natural attenuation mechanism	Year	Reference
1	Beijing, China	Industrial dye	Microbial degradation	2010	Anastasi et al. (2010)
2	Rhine Valley, Germany	Mono- and poly-aromatic hydrocarbons	Microbial degradation	2007	Morasch et al. (2007)
3	Queensland, Australia	Coal tar (PAHs)	Sorption, microbial degradation	2009	Megharaj et al. (2011)
4	Mumbai, India	Textile dyes	Phytoremediation, sorption	2016	Aravantinou and Manariotis (2016)
5	Paris, France	Polluted water	Biodegradation using ozonization	2018	Guo et al. (2018)
6	Lab based research	Arsenic (As)	Sorption (composts/ biochars from municipal solid waste)		Lima et al. (2022)
7	Cold marine environments	Petroleum oil	Biodegradation (cold-adapted microbes)	2017	Brakstad et al. (2017)
8	Labrador Sea, Canada	Diesel and crude oil	Biodegradation (cold-adapted microbial communities)	2021	Murphy et al. (2021)
9	Not specified (laboratory study)	Medium-chain alkanes (C8–C16)	Biodegradation (alkB gene-mediated)	2023	Fenibo et al. (2023)
10	Niger Delta, Nigeria	Crude oil (spill scenario)	Microbial community bioremediation potential	2023	Muhammad (2023)
11	Marine sediments (Lebanon)	Petroleum hydrocarbons	Aerobic biodegradation + biostimulation	2020	Hamdan and Salam (2020)
12	Oslo, Norway	PCBs (industrial waste)	Anaerobic dechlorination	2008	Isa et al. (2008)
13	Netherlands	Strontium-90 (^{90}Sr)	Sorption/desorption kinetics	2020	Zang et al. (2020)

The success of natural attenuation in situ depends fundamentally on the sampling of groundwater wells that are constructed at different spots of the contaminant zones. The installed wells serve as observation points to track pollution levels with time to create plume boundary maps and observe rebound or migration events (Hellal et al., 2021;

White et al., 2021). Continuous records collected from sampled wells help experts detect patterns of pollutant concentration reduction and plume stabilization as well as pollutant natural depletion as a result of degradation and dilution.

Monitoring the success of natural attenuation activity also includes bioindicator organism identification and evaluation of the microbes that are involved in the biodegradation process (Ezeah et al., 2023). Different technologies, including quantitative polymerase chain reaction (qPCR) and metagenomic sequencing, have been used in recent years to identify key functional gene expressions which codes for enzymes such as oxygenases or dehalogenases and document how microbial communities change over a period of time (Mohan et al., 2021; Thippeswamy et al., 2021). The molecular-level tools determine the effectiveness and role of microbial populations in natural attenuation.

The evaluation of redox parameters plays a key role in natural attenuation since they illustrate the geochemical environment that facilitates different degradation pathways. The analysis of water parameters consisting of methane, dissolved oxygen, iron (II), nitrate and sulfate helps identify current redox settings and active microbial respiration types that occur during the natural attenuation process. For example, the presence of methane and ferrous iron indicates anaerobic degradation processes, but aerobic degradation occurs while on the other hand, nitrate or oxygen appears in the system (Tang et al., 2021). The knowledge of these factors allows for confirmation regarding the sustained favorable conditions for biodegradation progress.

Stable isotope analysis is an advanced method that identifies in situ biodegradation (Nelson et al., 2022). The changes in some element's isotopic ratio can be used to infer the type and extent of degradation that happens in an environment. The types of elements that are used in this context include chlorine, carbon and hydrogen. The fractionation patterns of stable isotopes function as distinct identifying marks for microbial processes, which provide additional proof of the degradation of pollutants in an environment. The long-term activity of contaminants in an environment can also be done using predictive models like geochemical modeling (Matys Grygar et al., 2023). This modeling system unites microbial, chemical and hydrological data to perform simulations about contaminant movement and breakdown. These models enable scientists to predict different situations and determine the cleanup duration, which helps decision-makers assess potential.

5. Environmental factors affecting biodegradation and attenuation

Biodegradation processes and natural attenuation mechanisms of organic pollutants in the environment experience tremendous changes based on multiple environmental conditions. The breakdown, availability and mobility of contaminants depend heavily on environmental factors, which also ultimately modify microbial community structure and function. Organic pollutant remediation through natural attenuation requires a thorough understanding of such environmental variables. The main environmental factors that affect this process include soil characteristics, redox reactions, water table and hydrology, temperature and seasonal variations and microbial community structure. Soil characteristics, including the texture of the soil and the organic matter component, are fundamental in determining the biodegradation of contaminants in the soil (Wang et al., 2021b). The fraction of different components of the soil (the sandy, clayey and silty content of the soil), determines the water retention ability and permeability in the soil (Wang et al., no date), which consequently affect the oxygen diffusion and the mobility of microbes in the soil. For example, soil with a sandy texture promotes aeration and fast water infiltration, enabling aerobic degradation; conversely, clayey soils restrict oxygen penetration, resulting in anaerobic environments.

Heterotrophic microorganisms obtain their energy from soil organic matter while the matter also creates surfaces that can adsorb hydrophobic organic pollutants (Li and Wei, 2022). The pollutant sorption process has a dual effect on degradation; on the one hand, it can enhance it, while on the other hand, it can restrict it; that is, it can either make the pollutant unavailable or prevent the microorganisms from gaining access to the toxic pollutants. Similarly, the electron transfer function of humic substances in soil supports necessary redox reactions that aid biodegradation processes. The water table depth and the hydrology determine how biodegradation progresses in an environment (Wang et al., 2023). Soils found in vadose regions that have access to atmospheric oxygen enable aerobic biodegrading activities ((Sookhak Lari et al., 2024)), while saturated zones primarily encourage anaerobic processes because oxygen diffusion is restricted. Changes in the water table levels introduce oxygen into normally oxygen-depleted areas, where aerobic microbial processes begin to take place, resulting in a sequential breakdown of complex compounds.

Factors such as groundwater flow speed, hydraulic relationships and recharge levels determine how contaminants, nutrients and microorganisms move within the environment (Jin et al., 2022; Salowsky et al., 2021). High flow velocities might disperse pollutants from the microbial environment and reduce toxic substance concentrations. In contrast, when groundwater remains stagnant and flows slowly, it enables degradation products to build up to form stable redox zones that support extended natural attenuation processes.

Another environmental factor that determines which microbial processes can occur at contaminated sites is redox potential (Brumovský et al., 2021). Different microorganisms use different electron acceptors to degrade pollutants, but these acceptors exist based on the current redox environment. The terminal electron acceptor in high redox conditions is oxygen, which enables aerobic degradation and offers faster and more complete breakdown of pollutants like petroleum hydrocarbons. When redox potential values drop, microbes usually shift to alternative electron acceptors, including iron, sulfate and nitrate, before finally using carbon dioxide for methanogenesis (Jakus et al., 2021; Pang and Wang, 2021; Zheng et al., 2025). The anaerobic degradation routes represent essential processes through which microorganisms break down the recalcitrant environmental pollutants, including chlorinated solvents. Proper redox gradients enable microbial functions, which are required for the continuous operation of natural attenuation processes.

Temperature greatly influences both microbial metabolism and enzymatic activity, which drive the degradation mechanisms (Qu et al., 2023). The activity of microorganisms and degradation speeds increase with temperature until they reach an optimal range that typically spans from 20°C to 35°C for most mesophilic organisms (Rahman et al., 2023). Unfortunately, enzyme denaturation along with microbial growth inhibition occurs at extremely high temperatures, but low temperatures result in biochemical reactions and diffusion reduction.

Dry and wet seasons have effects on the patterns of biodegradation process. Biodegradation becomes progressively slow or stops completely at various times in temperate zones since the microbial activity reduces while water mobility decreases and soil moisture levels shift. Freeze-thaw cycles cause two damaging effects: they destroy microbial habitats and they damage cell membranes through lysis. However, season warming as well as increased rainfall trigger microbial growth and enhance nutrients availability, which promote pollutant diffusion, consequently boosting the rate of pollutant degradation (Ayilara et al., 2020).

6. Microbial community structure during degradation

The types, nature and the ability of microbes to survive in an environment is crucial in determining the potential of microbial biodegradation in a polluted site. Complex pollutant breakdown depends on which degraders or consortia inhabit the contaminated site since microorganisms display different abilities to metabolize pollution. A notable instance of this is the ability of *Pseudomonas* sp., *Dehalococcoidessp.* and *Sphingomonas* sp. to break down hydrocarbons and also eliminate chlorinated compounds (Yu et al., 2021; Ruikar and Pawar, 2022).

The type of relationshsips that exist among and between microbes also determines the potential of microbial biodegradation., examples of such relationship is syntrophy, where many organisms come together to degrade a particular pollutant which none of them can degrade alone (Harindintwali et al., 2023). Another example of such relationship is competition, where multiple organisms require the same pollutant as the source of nutriemt, which eventually reduces the availability of the pollutant in the environment. Also, microbial communities gain adaptive abilities to degrade contamination through horizontal gene transfer and mobile genetic elements, such as plasmids that encode catabolic genes.

In all, environmental stressors like pollutants that are toxic to microorganisms, lack of nutrients for microbial growth and survival, and fluctuating redox conditions modify the composition of microbial communities, resulting in inefficient breakdown of pollutants. Hence, the knowledge of microbial populations found in contaminated environments is very important for the prediction of natural attenuation success and to determine and guide on the type of bioaugmentation or biostimulation for specific sites or pollution remediation.

7. Challenges and limitations

The implementation of biodegradation and natural attenuation techniques for polluted environment cleanup encounters multiple obstacles that inhibit their wide-scale usage. Examples of these include the characteristics of specific pollutants, rules established by governmental agencies, social beliefs of the populace and unpredictable outcomes from scientific simulation and monitoring, and these are discussed in details below. A proper understanding of these factors would go a long way in enhancing the potential of these approaches as sustainable strategies for pollution remediation. A major issue associated with the application of natural attenuation is the fact that some specific organic pollutants, which cannot be biologically broken down through degradation persistently remain in the environment (Akhtar et al., 2023). For example, polycyclic aromatic hydrocarbons, polyfluoroalkyl substances, polychlorinated biphenyls, per-polyfluoroalkyl substances and some specific pesticides have complex structures and hydrophobic characteristics and they are also sometimes resistant to degrading enzymes (Chen et al., 2023; Renella et al., 2025). The chemicals tend to remain trapped in sediment layers and organic material compounds, which reduces accessibility to microorganisms and delays their breakdown processes. These chemicals maintain their presence in the environment because they survive for extended durations, thus creating enduring ecological and health concerns. In cases where these compounds manage to undergo partial degradation, the process often results in toxic intermediate breakdown products whose risks surpass those of initial pollutant chemicals.

The process of natural attenuation typically requires a long time to significantly reduce the concentration of a pollutant, especially when it involves recalcitrant pollutants, or if the environment is cold or lacks nutrition and oxygen (Liu et al., 2022). Although specific hydrocarbon and organic pollutants require weeks or a few months to degrade, others take a very long time to degrade to a level of safe concentrations. Many times this timelines does not usually align with the timeline designed by regulatory bodies or expectation of the populace. Also, degradation processes often have varying durations across different seasons and environmental conditions which causes the remediation timeline to lengthen and makes it unpredictable when cleanup goals will be accomplished.

The results from monitored natural attenuation heavily depend on obtaining precise data, accurate interpretation and reliable simulation models (Alozie et al., 2022). Unfortunately, the continuous uncertainties that persist during the monitoring process and the simulation modeling pose a challenge to the adoption of the technology. The natural heterogeneity and dynamism of environmental systems create barriers toward obtaining representative samples when tracking the subtle changes during natural attenuation.

The flow patterns of groundwater become intricate and change with time, which eventually alters the way contaminants are transported. Analytical methods cannot detect low levels of transformation products and contaminants bound to other molecules or compounds that might not be captured by modeling. Moreover, modeling processes require multiple assumptions alongside specific site-based parameters, which are always very challenging to quantify. Model-to-field observation mismatches can produce incorrect assessments of attenuation effectiveness which results either in too high or too low predictions of remediation progress. The unpredictability results in delayed crucial decisions, while it might force decision-makers to use excessive or inefficient remediation techniques.

Additionally, regulatory acceptance and public perception remain a huge challenge to adopting natural attenuation as regulatory bodies might resist approving natural attenuation techniques for remediation due to its extended durations, passive nature and the movement of contaminant to groundwater within an aquifer (Azuazu et al., 2023). Nevertheless, regulators require extra measures or backup remediation plans when site-specific proof about degradation and risk reduction cannot be demonstrated sufficiently.

The general public forms its opinions based on what it can see and observe as a quick pollution remediation technique (Zhang et al., 2022). Affected communities usually find it unacceptable to leave pollution unattended while relying on natural processes, especially when sensitive receptors like drinking water and residential areas are concerned. The public trust must be built while scientists explain natural attenuation methods through monitoring and the enforcement of safety guidelines to achieve stakeholder support.

8. Advances and future perspectives

Modern scientific innovation now gives researchers a better perception of microbial communities involved in remediation, improved knowledge about pollutant behaviors and new combined remediation solutions. Advancements in environmental biotechnology need to focus on interdisciplinary innovation combined with enhanced monitoring and policy coordination to make the most of natural attenuation processes as environmental challenges increase in complexity and emerging pollutants rise.

Relevant scientific progress in recent times includes using genomic and metagenomic technologies to understand microbial communities involved in pollutant breakdown in terms of diversity, population fluctuations and their functional attributes (Mohan et al., 2021; Thippeswamy et al., 2021). Current traditional culture-based techniques struggle to capture environmental microbes because they only detect limited fractions of them. The use of next-generation sequencing together with metagenomics enables researchers to acquire complete microbial DNA profiles directly from environmental samples, thus allowing them to detect functional genes responsible for specific degradation routes from all microorganisms including those that cannot be cultured.

These technology tools help scientists locate functional biomarkers such as reductive dehalogenases and monooxygenases, which reveal microbial biodegradation. Equally, technologies such as metatranscriptomics and advanced bioinformatics are being used in recent years to monitor gene expression patterns, which result from environmental changes to give researchers live data about microbial activities (Sharma et al., 2022). New sequencing technologies that improve speed and reduce costs open opportunities for using these methods in regular site monitoring practices to study how microbes function in contaminated environments. Bioremediation technologies face new obstacles from the worldwide rise of emerging pollutants, which include pharmaceuticals, endocrine-disrupting chemicals and personal care products. Despite the fact that the quantity of these chemicals is minute in the environment, they cause significant environmental and health-related problems. The structure of these new pollutants is not known by the typical bacterial metabolic pathways, affecting their natural detoxification by microbes. Synthetic biology, metabolic engineering, and adaptive evolution provide scientists with the tools needed to create or identify microbes that effectively degrade resistant compounds. Scientists are using modern identification and engineering techniques to create bacterial and fungal strains that effectively break down persistent pollutants like per- and polyfluoroalkyl substances, bisphenol A.

Further studies should focus on combining biodegradation systems with phytoremediation technologies and chemical oxidation techniques to enhance environmental remediation outcomes. Phytoremediation incorporates plants to degrade or remove contaminants, and when used together with microbes in the rhizosphere zone (plant root system), can leverage the plant-microbe interaction to improve pollutant remediation. Plants release organic substances to activate microbial populations, which in succession minimize toxicity toward plants and boost their overall growth and development.

Chemical oxidation processes involving Fenton's reaction alongside ozonation and persulfate treatments reduce pollutant concentrations while lowering their toxicity, enabling bioavailability and microbial degradation of intermediates (Han et al., 2023). This remediation approach relies on abiotic and biotic factors to create a stronger and more efficient cleanup solution. Further studies should investigate the combinations of these methods through experimental tests alongside predictive modeling systems. Good policy establishment and proper regulatory platforms are part of the main elements determining how natural attenuation gets implemented in real-life situations. Although efficiency has been established, natural attenuation is generally considered an alternative after other steps fail. Therefore, the implementation of natural attenuation requires improved regulatory standards, which should include standardized evaluation methods to make it an acceptable first-remediation option. Environmental regulation that applies adaptive management principles, which depend on ongoing monitoring to inform decision-making, can enhance the effective and sustainable use of attenuation strategies. Building capacity with stakeholders, educating them, and fostering public involvement are essential to increasing acceptance of these strategies for implementation. Countries working to meet international sustainability targets create an emerging chance to establish natural attenuation as the fundamental principle of nature-based remediation methods.

9. Conclusion

Biodegradation and natural attenuation are sustainable methods to address extensive organic pollutant contamination affecting ground and water bodies. The intrinsic abilities of microorganisms, together with environmental conditions, allow the transformation of harmful organic compounds into end-products that pose less of a hazard to the environment while enabling ecological recovery. Understanding the potential of biodegradation from bacteria-based metabolic pathways through enzymes like oxygenases and dehydrogenases enables scientists to understand ecosystem resistance while developing sustainable removal methods.

The effectiveness of these natural processes, however, is governed by a range of environmental factors, including pH, redox potential, temperature, soil composition, and the structure of microbial communities. Aerobic and anaerobic degradation routes each offer distinct advantages depending on the pollutant type and environmental conditions, and numerous examples demonstrate successful biodegradation of hydrocarbons, chlorinated solvents, pesticides, and other organic contaminants. Monitoring these processes in situ through advanced techniques such as stable isotope analysis, redox profiling, and metagenomics ensures that natural attenuation strategies remain safe, effective, and scientifically robust.

Despite their promise, natural attenuation and biodegradation face several limitations. Recalcitrant pollutants, extended remediation timelines, uncertainties in predictive modeling, and public skepticism continue to challenge their broader adoption. Nevertheless, ongoing advances in microbial genomics, synthetic biology, and integrated remediation technologies are opening new pathways for expanding the utility of these natural processes. Combining biodegradation with phytoremediation, chemical oxidation, or engineered bioreactors offers innovative solutions for complex and mixed-pollutant scenarios. Looking forward, there is a critical need to embed natural attenuation more formally within regulatory frameworks, guided by sound science, reliable monitoring, and transparent risk communication. As environmental policies evolve to support sustainable and cost-effective remediation, biodegradation and natural attenuation are poised to play increasingly central roles in the global response to organic pollution—bridging the gap between scientific discovery, environmental stewardship, and societal well-being.

References

Ahmad, Z., Hassan Abbasi, M., Naz, N., Sheikh, N., Khawar, M.B., 2025. Migration and Chemical Transformation of Pollutants in the Marine Ecosystem. Springer Science and Business Media LLC, pp. 161–200. https://doi.org/10.1007/978-3-031-75713-6_7.

Akhtar, S., Mohsin, A., Riaz, A., Mohsin, F., 2023. Worldwide efficiency of bioremediation techniques for organic pollutants in soil: a brief review. Geosfera Indonesia 8 (1), 102. https://doi.org/10.19184/geosi.v8i1.30875.

Alozie, E., Abdulkarim, A., Abdullahi, I., Usman, A.D., Faruk, N., Olayinka, I.F.Y., Adewole, K.S., Oloyede, A.A., Chiroma, H., Sowande, O.A., Olawoyin, L.A., Garba, S., Imoize, A.L., Musa, A., Adediran, Y.A., Taura, L.S., 2022. A review on rain signal attenuation modeling, analysis and validation techniques: advances, challenges and future direction. Sustainability 14 (18). https://doi.org/10.3390/su141811744.

Anastasi, A., Prigione, V., Varese, G.C., 2010. Industrial dye degradation and detoxification by basidiomycetes belonging to different eco-physiological groups. J. Hazard Mater. 177 (1–3), 260–267. https://doi.org/10.1016/j.jhazmat.2009.12.027.

Aravantinou, A.F., Manariotis, I.D., 2016. Effect of operating conditions on chlorococcum sp. growth and lipid production. J. Environ. Chem. Eng. 4 (1), 1217–1223. https://doi.org/10.1016/j.jece.2016.01.028.

Ayilara, M.S., Babalola, O.O., 2023. Bioremediation of environmental wastes: the role of microorganisms. Frontiers in Agronomy 5. https://doi.org/10.3389/fagro.2023.1183691.

Ayilara, M., Olanrewaju, O., Babalola, O., Odeyemi, O., 2020. Waste management through composting: challenges and potentials. Sustainability 12 (11), 4456. https://doi.org/10.3390/su12114456.

Ayilara, M.S., Adeleke, B.S., Adebajo, M.T., Akinola, S.A., Fayose, C.A., Adeyemi, U.T., Gbadegesin, L. A., Omole, R.K., Johnson, R.M., Edhemuino, M., Ogundolie, F.A., Babalola, O.O., 2023. Remediation by enhanced natural attenuation; an environment-friendly remediation approach. Front. Environ. Sci. 11. https://doi.org/10.3389/fenvs.2023.1182586.

Banayan Esfahani, E., Asadi Zeidabadi, F., Zhang, S., Mohseni, M., 2022. Photo-chemical/catalytic oxidative/reductive decomposition of per- and poly-fluoroalkyl substances (PFAS), decomposition mechanisms and effects of key factors: a review. Environmental Science: Water Research and Technology 8 (4), 698–728. https://doi.org/10.1039/d1ew00774b.

Bharti, R.K., Srivastava, S., Thakur, I.S., 2020. Sequestration of Carbon Dioxide by Microorganism and Production of Value Added Product Environmental Microbiology and Biotechnology: Volume 2: Bioenergy and Environmental Health. Springer Singapore, India, pp. 235–249. https://doi.org/10.1007/978-981-15-7493-1_11.

Bher, A., Mayekar, P.C., Auras, R.A., Schvezov, C.E., 2022. Biodegradation of biodegradable polymers in mesophilic aerobic environments. Int. J. Mol. Sci. 23 (20). https://doi.org/10.3390/ijms232012165.

Brakstad, O.G., Lofthus, S., Ribicic, D., Netzer, R., 2017. Biodegradation of Petroleum Oil in Cold Marine Environments Psychrophiles: From Biodiversity to Biotechnology, second ed. Springer International Publishing, Norway, pp. 613–644. https://doi.org/10.1007/978-3-319-57057-0_27.

Brito, I.L., 2021. Examining horizontal gene transfer in microbial communities. Nat. Rev. Microbiol. 19 (7), 442–453. https://doi.org/10.1038/s41579-021-00534-7.

Brumovský, M., Oborná, J., Lacina, P., Hegedüs, M., Sracek, O., Kolařík, J., Petr, M., Kašlík, J., Hofmann, T., Filip, J., 2021. Sulfidated nano-scale zerovalent iron is able to effectively reduce in situ hexavalent chromium in a contaminated aquifer. J. Hazard Mater. 405, 124665. https://doi.org/10.1016/j.jhazmat.2020.124665.

Cao, Z., Yan, W., Ding, M., Yuan, Y., 2022. Construction of microbial consortia for microbial degradation of complex compounds. Front. Bioeng. Biotechnol. 10. https://doi.org/10.3389/fbioe.2022.1051233.

Chaitanya, M.V.N.L., Arora, S., Pal, R.S., Ali, H.S., Haj, B.M.E., Logesh, R., 2024. Assessment of Environmental Pollutants for Their Toxicological Effects of Human and Animal Health Organic Micro Pollutants in Aquatic and Terrestrial Environments. Springer Nature, India, pp. 67–85. https://doi.org/10.1007/978-3-031-48977-8_4.

Chen, J.C., Baumert, B.O., Li, Y., Li, Y., Pan, S., Robinson, S., Rubbo, B., Costello, E., He, J., Hampson, H., 2023. Associations of per-and polyfluoroalkyl substances, polychlorinated biphenyls, organochlorine pesticides, and polybrominated diphenyl ethers with oxidative stress markers: a systematic review and meta-analysis. Environ. Res. 239.

Cui, N., Ge, L., Halsall, C., Niu, J., Zheng, J., Zhang, P., 2025. Dissociation-dependent kinetics and distinct pathways for direct photolysis and •OH/SO4•− radical dominated photodegradation of ionizable antiviral drugs in aquatic systems. Water Res. 279, 123452. https://doi.org/10.1016/j.watres.2025.123452.

Douglas, C., Pandey, P., 2025. Nitrate and bacterial loads in dairy cattle drinking water and potential treatment options for pollutants—a review. Appl. Sci. 15 (6), 3017. https://doi.org/10.3390/app15063017.

Duffy, A.M., Lockyear, K.T., Hughes, M.T., Garimella, S., 2025. Investigation of a novel combined ab- and ad- sorption based thermal energy storage and upgrade system. Appl. Therm. Eng. 258. https://doi.org/10.1016/j.applthermaleng.2024.124520.

Ezeah, C., Esonye, C., Azuama, O.C., 2023. Detoxification of Sewage Sludge by Natural Attenuation Sustainable Treatment and Management of Sewage Sludge. CRC Press, Nigeria, pp. 277–302. https://doi.org/10.1201/9781003354765-20.

Fenibo, E.O., Selvarajan, R., Abia, A.L.K., Matambo, T., 2023. Medium-chain alkane biodegradation and its link to some unifying attributes of alkB genes diversity. Sci. Total Environ. 877. https://doi.org/10.1016/j.scitotenv.2023.162951.

Guo, Y., Wang, H., Wang, B., Deng, S., Huang, J., Yu, G., Wang, Y., 2018. Prediction of micropollutant abatement during homogeneous catalytic ozonation by a chemical kinetic model. Water Res. 142, 383–395. https://doi.org/10.1016/j.watres.2018.06.019.

Hamdan, H.Z., Salam, D.A., 2020. Microbial community evolution during the aerobic biodegradation of petroleum hydrocarbons in marine sediment microcosms: effect of biostimulation and seasonal variations. Environ. Pollut. 265. https://doi.org/10.1016/j.envpol.2020.114858.

Han, M., Wang, H., Jin, W., Chu, W., Xu, Z., 2023. The performance and mechanism of iron-mediated chemical oxidation: advances in hydrogen peroxide, persulfate and percarbonate oxidation. J. Environ. Sci. 128, 181–202. https://doi.org/10.1016/j.jes.2022.07.037.

Harindintwali, J.D., Xiang, L., Wang, F., Chang, S.X., Zhao, Z., Mei, Z., Jia, Z., Jiang, X., Zhu, Y.g., Tiedje, J.M., 2023. Syntrophy of bacteria and archaea in the anaerobic catabolism of hydrocarbon contaminants. Crit. Rev. Environ. Sci. Technol. 53 (13), 1331–1357. https://doi.org/10.1080/10643389.2022.2134702.

Hellal, J., Joulian, C., Urien, C., Ferreira, S., Denonfoux, J., Hermon, L., Vuilleumier, S., Imfeld, G., 2021. Chlorinated ethene biodegradation and associated bacterial taxa in multi-polluted groundwater: Insights from biomolecular markers and stable isotope analysis. Sci. Total Environ. 763, 142950. https://doi.org/10.1016/j.scitotenv.2020.142950.

Hu, Y., Liu, L., Xu, K., Song, Y., Jing, J., Zhang, H., Feng, J., 2023. Efficiency analysis of sorption-enhanced method in steam methane reforming process. Carbon Resour. Convers. 6 (2), 132–141. https://doi.org/10.1016/j.crcon.2022.12.004.

Huang, L., Lu, Z., Xie, T., Wang, L., Mo, C., 2022. Nitrogen and phosphorus removal by coupling Anaerobic ammonia oxidation reaction with algal-bacterial symbiotic system. J. Environ. Chem. Eng. 10 (6), 108905. https://doi.org/10.1016/j.jece.2022.108905.

Isa, M.H., Ibrahim, N., Aziz, H.A., Adlan, M.N., Sabiani, N.H.M., Zinatizadeh, A.A.L., Kutty, S.R.M., 2008. Removal of chromium (VI) from aqueous solution using treated oil palm fibre. J. Hazard Mater. 152 (2), 662–668. https://doi.org/10.1016/j.jhazmat.2007.07.033.

Jakus, N., Blackwell, N., Osenbrück, K., Straub, D., Byrne, J.M., Wang, Z., Glöckler, D., Elsner, M., Lueders, T., Grathwohl, P., Kleindienst, S., Kappler, A., 2021. Nitrate removal by a novel Lithoautotrophic nitrate-reducing, iron(II)-Oxidizing culture enriched from a Pyrite-Rich Limestone aquifer. Appl. Environ. Microbiol. 87 (16), 1–15. https://doi.org/10.1128/AEM.00460-21.

Jin, H., Huang, Y., Bense, V.F., Ma, Q., Marchenko, S.S., Shepelev, V.V., Hu, Y., Liang, S., Spektor, V. V., Jin, X., Li, X., Li, X., 2022. Permafrost degradation and its hydrogeological impacts. Water (Switzerland) 14 (3). https://doi.org/10.3390/w14030372.

Ke, C., Ma, X., Zheng, W., Chen, L., Tang, Y., 2021. The volatilization of heavy metals during combustion of polyvinyl chloride after hydrothermal carbonization. J. Clean. Prod. 285, 124825. https://doi.org/10.1016/j.jclepro.2020.124825.

Kebede, G., Tafese, T., Abda, E.M., Kamaraj, M., Assefa, F., 2021. Factors influencing the bacterial bioremediation of hydrocarbon contaminants in the soil: mechanisms and impacts. J. Chem. 2021. https://doi.org/10.1155/2021/9823362.

Kumari, P., Kumar, A., 2023. Advanced oxidation process: a remediation technique for organic and non-biodegradable pollutant. Results Surf. Interfaces 11, 100122. https://doi.org/10.1016/j.rsurfi.2023.100122.

Li, Y., Wei, M., 2022. Evaluation on adsorption capacity of low organic matter soil for hydrophobic organic pollutant. J. Environ. Chem. Eng. 10 (3), 107561. https://doi.org/10.1016/j.jece.2022.107561.

Lima, J.Z., Ferreira da Silva, E., Patinha, C., Durães, N., Vieira, E.M., Rodrigues, V.G.S., 2022. Sorption of arsenic by composts and biochars derived from the organic fraction of municipal solid wastes: kinetic, isotherm and oral bioaccessibility study. Environ. Res. 204. https://doi.org/10.1016/j.envres.2021.111988.

Liu, N., Wei, S., Yang, Y., Li, S., Sun, F., Gao, J., 2022. Seismic attenuation estimation using an enhanced log spectral ratio method. IEEE Geoscience and Remote Sensing Letters 19. https://doi.org/10.1109/LGRS.2022.3140343.

Marines, I.C., Bella, Trapani, Giustra, M.G., Calabrisotto, L.S., Lucchina, P.G., Quatrini, Viviani, G., 2023. Degradation of 1, 2-dichloroethane in real polluted groundwater by using enriched bacterial consortia in aerobic and anaerobic laboratory-scale conditions. Int. Biodeterior. Biodegrad. 183.

Matys Grygar, T., Elznicová, J., Tůmová, Š., Kylich, T., Skála, J., Hron, K., Álvarez-Vázquez, M.Á., 2023. Moving from geochemical to contamination maps using incomplete chemical information from long-term high-density monitoring of Czech agricultural soils. Environ. Earth Sci. 82 (1). https://doi.org/10.1007/s12665-022-10692-3.

Megharaj, M., Ramakrishnan, B., Venkateswarlu, K., Sethunathan, N., Naidu, R., 2011. Bioremediation approaches for organic pollutants: a critical perspective. Environ. Int. 37 (8), 1362–1375. https://doi.org/10.1016/j.envint.2011.06.003.

Mohan, S., Chigadannavar, P.S., Quadri, Z., Dey, P., Gola, D., Chauhan, N., Bharti, R.K., 2021. Microbial Community Diversity in a Wastewater Treatment Plant Wastewater Treatment Reactors: Microbial Community Structure. Elsevier, India, pp. 373–385. https://doi.org/10.1016/B978-0-12-823991-9.00013-7.

Mohapatra, B., Phale, P.S., 2021. Microbial degradation of naphthalene and substituted naphthalenes: metabolic diversity and genomic insight for bioremediation. Front. Bioeng. Biotechnol. 9. https://doi.org/10.3389/fbioe.2021.602445.

Morasch, B., Höhener, P., Hunkeler, D., 2007. Evidence for in situ degradation of mono-and polyaromatic hydrocarbons in alluvial sediments based on microcosm experiments with 13C-labeled contaminants. Environ. Pollut. 148 (3), 739–748. https://doi.org/10.1016/j.envpol.2007.01.029.

Mrnjavac, N., Nagies, F.S.P., Wimmer, J.L.E., Kapust, N., Knopp, M.R., Trost, K., Modjewski, L., Bremer, N., Mentel, M., Esposti, M.D., Mizrahi, I., Allen, J.F., Martin, W.F., 2024. The radical impact of oxygen on prokaryotic evolution—enzyme inhibition first, uninhibited essential biosyntheses second, aerobic respiration third. FEBS (Fed. Eur. Biochem. Soc.) Lett. 598 (14), 1692–1714. https://doi.org/10.1002/1873-3468.14906.

R.A. Muhammad, The impact of an oil spill on microbial communities (and their bioremediation potential); a Niger Delta case study. (2023),

Murphy, S.M.C., Bautista, M.A., Cramm, M.A., Hubert, C.R.J., 2021. Diesel and crude oil biodegradation by cold-adapted microbial communities in the labrador sea. Appl. Environ. Microbiol. 87 (20), 1–21. https://doi.org/10.1128/AEM.00800-21.

Ndubuisi Azuazu, I., Sam, K., Campo, P., Coulon, F., 2023. Challenges and opportunities for low-carbon remediation in the Niger Delta: towards sustainable environmental management. Sci. Total Environ. 900, 165739. https://doi.org/10.1016/j.scitotenv.2023.165739.

Nelson, T.F., Baumgartner, R., Jaggi, M., Bernasconi, S.M., Battagliarin, G., Sinkel, C., Künkel, A., E Kohler, H.-P., McNeill, K., Sander, M., 2022. Biodegradation of poly (butylene succinate) in soil laboratory incubations assessed by stable carbon isotope labelling. Nat. Commun. 13 (1).

Nie, J., Wang, Q.m., Han, L.j., Li, J.s., 2024. Synergistic remediation strategies for soil contaminated with compound heavy metals and organic pollutants. J. Environ. Chem. Eng. 12 (4). https://doi.org/10.1016/j.jece.2024.113145.

Pang, Y., Wang, J., 2021. Various electron donors for biological nitrate removal: a review. Sci. Total Environ. 794, 148699. https://doi.org/10.1016/j.scitotenv.2021.148699.

Qu, R., Liu, G., Yue, M., Wang, G., Peng, C., Wang, K., Gao, X., 2023. Soil temperature, microbial biomass and enzyme activity are the critical factors affecting soil respiration in different soil layers in Ziwuling Mountains, China. Front. Microbiol. 14. https://doi.org/10.3389/fmicb.2023.1105723.

Rahman, M.F., Manogaran, M., Muhammad Yakasai, H., Abubakar, A., 2023. Activation energy, temperature coefficient and Q10 value estimations of the growth of an SDS-degrading bacterium. Journal of Environmental Microbiology and Toxicology 11 (2), 57–63. https://doi.org/10.54987/jemat.v11i2.892.

Renella, G., Carletti, A., Masi, 2025. Sustainable management of poly-and perfluoroalkyl substances (PFASs)-contaminated areas: tackling a wicked environmental problem. Sustainability 17 (2).

Ruikar, A., Pawar, H.S., 2022. Diversity and Interaction of Microbes in Biodegradation. Informa UK Limited, pp. 185–213. https://doi.org/10.1201/9781003354147-12.

Saanu Adeleke, B., Amuzu, P., Tosin Alawiye, T., Akinola, S.A., Ayilara, M.S., Fasusi, O.A., Erinoso, S. M., Fathi Abd_Allah, E., 2025. Roles of Nanobioinoculants in Remediation of Heavy Metals from Soil and Water System. Elsevier BV, pp. 221–251. https://doi.org/10.1016/b978-0-443-22285-6.00009-4.

Salowsky, H., Schäfer, W., Schneider, A.L., Müller, A., Dreher, C., Tiehm, A., 2021. Beneficial effects of dynamic groundwater flow and redox conditions on natural Attenuation of mono-, poly-, and NSO-heterocyclic hydrocarbons. J. Contam. Hydrol. 243. https://doi.org/10.1016/j.jconhyd.2021.103883.

Sharma, P., Singh, S.P., Iqbal, H.M.N., Tong, Y.W., 2022. Omics approaches in bioremediation of environmental contaminants: an integrated approach for environmental safety and sustainability. Environ. Res. 211. https://doi.org/10.1016/j.envres.2022.113102.

Singh, G., Singh, A., Singh, P., Shukla, R., Tripathi, S., Mishra, V.K., 2021. The Fate of Organic Pollutants and Their Microbial Degradation in Water Bodies Pollutants and Water Management: Resources, Strategies and Scarcity. wiley, India, pp. 210–240. https://doi.org/10.1002/9781119693635.ch9.

Sookhak Lari, K., Davis, G.B., Rayner, J.L., Bastow, T.P., 2024. Advective and diffusive gas phase transport in vadose zones: Importance for defining vapour risks and natural source zone depletion of petroleum hydrocarbons. Water Res. 255. https://doi.org/10.1016/j.watres.2024.121455.

Tang, Y., Yang, L., Zhang, M., Xiong, P., Liu, L., Bao, Y., Zhao, Z., 2021. Link between characteristics of Fe(III) oxides and critical role in enhancing anaerobic methanogenic degradation of complex organic compounds. Environ. Res. 194, 110498. https://doi.org/10.1016/j.envres.2020.110498.

Thippeswamy, M., Rajasrerlatha, V., Shubha, D., Niveditha, B., 2021. Metagenomics and future perspectives in discovering pollutant degrading enzymes from soil microbial communities. Recent Developments in Applied Microbiology and Biochemistry: Elsevier 257–267.

C. Wang, S.Y Li, X.J He, Q. Chen, H. Zhang, X.Y Liu, Improved prediction of water retention characteristic based on soil gradation and clay fraction. Geoderma. 404 .

Wang, D., Yi, N., Wang, Y., Yang, J., Fu, Q., Liu, X., Qi, Y., Cai, Z., Ye, J., Liu, Y., Wang, Q., Ni, B.-J., 2021a. Triclosan degradation in sludge anaerobic fermentation and its impact on hydrogen production. Chem. Eng. J. 421, 129948. https://doi.org/10.1016/j.cej.2021.129948.

Wang, L., Cheng, Y., Naidu, R., Bowman, M., 2021b. The key factors for the fate and transport of petroleum hydrocarbons in soil with related in/ex situ measurement methods: an overview. Front. Environ. Sci. 9. https://doi.org/10.3389/fenvs.2021.756404.

Wang, J., Zhen, J., Hu, W., Chen, S., Lizaga, I., Zeraatpisheh, M., Yang, X., 2023. Remote sensing of soil degradation: progress and perspective. Int. Soil Water Conserv. Res. 11 (3), 429–454. https://doi.org/10.1016/j.iswcr.2023.03.002.

White, S.A., Conrad, S.R., Woodrow, R.L., Tucker, J.P., Wong, W.W., Cook, P.M., Sanders, C.J., Wadnerkar, P.D., Davis, K.L., Holloway, C.J., Santos, I.R., 2021. Natural attenuation of large anthropogenic nitrate loads in a subtropical stream revealed by δ15N and δ18O. J. Hydrol. 598. https://doi.org/10.1016/j.jhydrol.2021.126077.

Yu, F., Li, Y., Wang, H., Peng, T., Wu, Y.R., Hu, Z., 2021. Microbial debromination of hexabromocyclododecanes. Appl. Microbiol. Biotechnol. 105 (11), 4535–4550. https://doi.org/10.1007/s00253-021-11095-3.

Yuan, J., Li, S., Cheng, J., Guo, C., Shen, C., He, J., Yang, Y., Hu, P., Xu, J., He, Y., 2021. Potential role of methanogens in microbial reductive dechlorination of organic chlorinated pollutants in situ. Environ. Sci. Technol. 55 (9), 5917–5928. https://doi.org/10.1021/acs.est.0c08631.

Zamulina, I.V., Gorovtsov, A.V., Minkina, T.M., Mandzhieva, S.S., Burachevskaya, M.V., Bauer, T.V., 2022. Soil organic matter and biological activity under long-term contamination with copper. Environ. Geochem. Health 44 (2), 387–398. https://doi.org/10.1007/s10653-021-01044-4.

Zang, J., Wang, J., Han, X., Yao, H., Fu, B., Zhao, J., Li, B., Chen, J., 2020. Sorption and desorption of Sr onto a rough single fractured granite. J. Contam. Hydrol. 228, 103558. https://doi.org/10.1016/j.jconhyd.2019.103558.

Zhang, X., Luo, M., Deng, S., Long, T., Sun, L., Yu, R., 2022. Field study of microbial community structure and dechlorination activity in a multi-solvents co-contaminated site undergoing natural attenuation. J. Hazard Mater. 423 (Pt A), 127010. https://doi.org/10.1016/j.jhazmat.2021.127010.

Zheng, L., Yue, H., Wang, X., Guan, X., Shen, Y., 2025. Nano-zerovalent iron (nZVI) shifts microbial nitrate respiration from denitrification to dissimilatory nitrate reduction to ammonium (DNRA) for the treatment of low-C/N wastewater. ACS ES&T Water 5 (4), 1683–1693. https://doi.org/10.1021/acsestwater.4c01065.

CHAPTER 15

Phytoremediation of organic pollutants: Mechanisms, applications, and challenges

Nadeem Iqbal[1, 2], Midhat Rahim[1], Maria Khan[1], Laiba Arif[1], Hadia Riaz[1], Malik Tahir Hayat[1], Asma Imtiaz[1], Qaisar Mahmood[1, 3], Nazneen Bangash[4], Shamim Akhtar[5] and Zhangtao Li[6]

[1]Department of Environmental Sciences, COMSATS University Islamabad, Abbottabad, Pakistan; [2]State Key Laboratory of Geohazard Prevention and Geoenvironment Protection, College of Ecology and Environment, Chengdu University of Technology, Chengdu, Sichuan, P.R. China; [3]Department of Biology, College of Science, University of Bahrain, Sakhir, Bahrain; [4]Department of Biosciences, COMSATS University Islamabad, Islamabad, Pakistan; [5]Department of Botany, Rawalpindi Women University, Rawalpindi, Pakistan; [6]Key Laboratory of Recycling and Eco-Treatment of Waste Biomass of Zhejiang Province, School of Environment and Natural Resources, Zhejiang University of Science and Technology, Hangzhou, China

1. Introduction

Phytoremediation is a known sustainable technology that uses plants for the removal, degradation, and immobilization of the pollutants that are present in the environment (Bakhshayeshan-Agdam et al., 2024). This process deals majorly with both organic and inorganic pollutants; the former contains pesticides, chlorinated solvents, polychlorinated biphenyls (PCBs), and polycyclic aromatic hydrocarbons (PAHs). These toxic chemicals are released into the environment through different agricultural, urban, and industrial activities, posing adverse effects on living organisms. Organic pollutants have considerable human and environmental concerns due to their toxicity, bioaccumulation, and persistence in the environment (Khan et al., 2022). Compared to conventional methods, such as chemical treatments, excavation, and thermal desorption, phytoremediation is more cost-effective and environmentally friendly. Phytoremediation is considered a green technology as it reduces operational costs and lessens disturbance to target sites than energy-extensive traditional processes (Xu et al., 2017). Phytoremediation also has the benefit of carbon sequestration, soil stabilization, and habitat restoration. This process underlines the economic and ecological benefits by improving biodiversity and recovering soil health and the ecosystem (Wang et al., 2020). Therefore, phytoremediation emerges as a promising technology to treat contaminated sites to enhance ecosystem functioning. Phytoremediation comprises different mechanisms, such as phytovolatilization, phytotransfromation, phytostabilization, phytodegradation, and phytoextraction, which have their unique characteristics, applications, and specificity for a diverse range of pollutants.

Phytotoxicity of Organic Pollutants
ISBN 978-0-443-33110-7,
https://doi.org/10.1016/B978-0-443-33110-7.00009-2

The efficiency of phytoremediation for pollutant removal mainly depends on several factors, including environmental factors, soil properties, pollutant concentrations, bioavailability of pollutants, and plant species (Yan et al., 2020). Plants absorb hydrophobic organic pollutants passively, while plants use active transport for the absorption of hydrophilic organic pollutants. Thereafter, pollutants are stored, degraded through plant metabolism, and translocated to different plant tissues (Pandey et al., 2016). Transgenic plants have shown huge potential for the remediation of organic pollutants by enhancing their removal capabilities and adaptability to surrounding polluted environments. Advancements in this area can provide plants with enhanced degradation efficiencies of the contaminants through optimization of the process and selection of appropriate plant species with distinct biochemical processes for remediation purposes (Farooq et al., 2022). However, the effectiveness of phytoremediation can be achieved through deploying innovative technologies such as transgenic plants and genetically modified microbes to enhance the success rate of this technique. Further, the use of microbial stimulation, bioaugmentation, genetic engineering, and chelating agents can significantly enhance the overall efficiency of phytoremediation (Yin et al., 2024). Hence, phytoremediation can be used as an effective and efficient technology to treat contaminated sites with organic pollutants to improve ecosystem functioning.

This chapter highlights the role of phytoremediation and its various techniques for the removal of organic pollutants from contaminated sites. Further, it explores the key mechanisms along with key factors responsible for the removal efficiency of phytoremediation. Furthermore, the practical application and advantages of phytoremediation have been discussed in detail, followed by the limitations or key challenges of this process. Overall, this chapter provides current insights into the understanding of phytoremediation and its potential application for environmental cleanup and sustainable ecosystem management to mitigate the toxic effects posed by organic pollutants.

2. Types of organic pollutants remediated by phytoremediation

2.1 Persistent organic pollutants (POPs)

POPs are harmful chemicals that negatively impact the environment on a global scale. They are commonly found in air and dust, entering the human body through inhalation or ingestion and causing toxic effects (Groffen et al., 2021). POPs resist breakdown through physical, chemical, or biological processes. Due to their stability and low reactivity, they are widely used in various applications, including flame retardants, fire suppressants, heat transfer agents, surfactants, cosmetics, and pesticides (Megson et al., 2016). POPs are highly resistant to photolytic, chemical, and biological breakdowns, making their elimination from the environment difficult. Processing further increases their stability and semi-volatility, enabling them to persist for long periods and travel great distances before settling in the soil. Their buildup in food chains and environmental circulation contributes to rising air pollution (Adebusuyi et al., 2022). POPs, particularly pesticides, are immunotoxins to humans and interfere with thyroid hormone production by reducing biosynthesis. Many pesticide chemicals belong to the organochlorine group, which is known for its harmful effects on human health and long-term environmental persistence. These chemicals include aldrin, chlordane, chlordecone, dicofol, dieldrin, DDT (1,1′-(2,2,2-Trichloroethane-1,1-diyl)bis(4-chlorobenzene)), endosulfan and its isomers, endrin, heptachlor, hexachlorobenzene (HCB), α-hexachlorocyclohexane, β-hexachlorocyclohexane, lindane (γ-hexachlorocyclohexane), mirex, pentachlorobenzene, pentachlorophenol (including its salts and esters), perfluoro octane sulfonyl fluoride, perfluoro octane sulfonic acid (along with its salts), and toxaphene (Hernández et al., 2020). Thus, eliminating these toxic pollutants is crucial. Phytoremediation, along with microbial interactions, is an efficient and sustainable approach for removing POPs from soil, air, and water in a cost-effective manner (Anerao et al., 2022). POPs in industrial chemicals include decabromodiphenyl ether, hexabromobiphenyl, hexabromocyclododecane, hexabromobiphenyl ether, heptabromodiphenyl ether, hexachlorobenzene, hexachlorobutadiene, pentachlorobenzene, PCBs, polychlorinated naphthalene, perfluorooctanoic acid (along with its salts and related compounds), short-chain chlorinated paraffins, tetrabromobisphenol ether, heptabromodiphenyl ether, perfluoro octane sulfonic acid (including its salts), and perfluoro octane sulfonyl fluoride. These chemicals are extensively utilized in industrial manufacturing processes. For example, PCBs were widely applied in industrial lubricants and coolants for electrical equipment such as transformers and capacitors (Iqbal et al., 2022). Additionally, perfluorooctanoic acid has been used in producing consumer goods with resistance to heat, grease, oil, stains, and water (Al-Mulali et al., 2015; Bakirtas and Akpolat, 2018). Phytoremediation utilizes plants to restore polluted soil and groundwater, offering a cost-effective solution for removing harmful contaminants, including toxic PCBs (Aken et al., 2010).

Polychlorinated dibenzo-p-dioxins (PCDDs) and polychlorinated dibenzofurans (PCDFs), commonly known as dioxins, are toxic and persistent organic pollutants. They are byproducts of industrial activities, combustion, and certain natural processes. Dioxins can be released into the atmosphere from both human activities and natural events such as volcanic eruptions, forest fires, and spontaneous combustion (Salamanca et al., 2016). Natural sources release only a small fraction of dioxins compared to human activities. Key industrial contributors include the manufacturing of chlorine-based chemicals, paper pulp production, and the creation of herbicides and pesticides (Mudhoo et al., 2013). The burning of organic materials, including municipal waste, wood, and fossil fuels, can release dioxins into the environment through smoke, emissions, and waste incineration. Various bioremediation methods have been developed to address dioxin contamination, such as biosurfactants, composting, and phytoremediation. Aerobic composting has shown effectiveness in reducing dioxin levels, with removal rates ranging from

65% to 85%. Phytoremediation, particularly using crops like zucchini, cucumber, and wheat, plays a crucial role in dioxin removal through root uptake and transpiration. Moreover, biosurfactants like rhamnolipids and sophorolipids have been successfully utilized to treat dioxin-contaminated soil by improving dioxin bioavailability and enhancing microbial activity (Tran et al., 2024). Herbicides play a vital role in weed control and are the most widely used pesticides worldwide. However, their residues in water, soil, air, and non-target organisms present major challenges to both agriculture and environmental sustainability. Phytoremediation is an eco-friendly solution aimed at reducing the adverse effects of herbicide residues. Remediating plants are classified into three main categories: herbaceous species, trees, and aquatic macrophytes. This method has been shown to eliminate at least 50% of herbicide residues from the environment. The Fabaceae family is among the most commonly used herbaceous and tree species for herbicide remediation. Triazines are the most frequently identified group of herbicides across different plant types. The primary phytoremediation mechanisms for herbicide removal involve extraction and accumulation (Barroso et al., 2023).

2.2 Petroleum hydrocarbons

Benzene, toluene, ethylbenzene, and xylene (BTEX) are volatile organic compounds commonly associated with petroleum products. Their environmental impact raises concern due to their toxicity and ability to contaminate soil and groundwater. Implementing efficient remediation methods is essential to reduce their harmful effects. Phytoremediation offers an eco-friendly solution by utilizing plants to absorb, transfer, stabilize, or degrade contaminants present in soil and water (Ojimelukwe et al., 2021).

BTEX are persistent organic pollutants that are highly toxic and carcinogenic, posing serious risks to both the environment and human health. While BTEX contamination is often studied with surface water and above-ground pollution, its impact on deep soil and groundwater is frequently neglected (Li et al., 2024). BTEX are prominent organic contaminants in soil and groundwater, commonly originating from industrial processes and daily activities. Among them, monocyclic aromatic hydrocarbons such as benzene, toluene, and ethylbenzene are highly prevalent, making up 46%–91% of the total BTEX concentration in the environment. Their presence significantly affects soil and groundwater quality (Xu et al., 2024). The adsorption, transport, transformation, and toxicity of BTEX influence the functioning of plants, animals, and microbial communities in soil and groundwater, leading to decreased biodiversity and disrupted ecosystem processes. As a result, the potential health hazards posed by BTEX should not be overlooked. These compounds are volatile, carcinogenic, and poorly soluble in water, making them particularly concerning from a toxicological standpoint. They can irritate the skin and mucous membranes and can enter the central nervous system, causing damage to various bodily systems. Prolonged exposure to BTEX has been associated with serious health conditions, including bone marrow disorders, lymphomas, and leukemia (Caniani et al., 2021). BTEX compounds can be transported, transformed, bound, or broken down within plant tissues, with phytoremediation proving to be an effective method for their removal. After the phytoremediation process, organic acids were identified as the primary byproducts of benzene and toluene degradation. These findings highlight phytoremediation as a sustainable and environmentally friendly approach for eliminating BTEX contaminants.

2.3 Volatile organic compounds (VOCs)

Various volatile organic compounds (VOCs), including hydrogen cyanide, acetaldehyde, aromatic amines, ethylene oxide, formaldehyde, vinyl chloride, 1,3-butadiene (a hazardous gas), polycyclic aromatic hydrocarbons (PAHs), and BTEX (benzene, toluene, ethylbenzene, and xylene), are known to be harmful to human health, with many classified as toxic and carcinogenic (Knoell et al., 2020). VOCs are regularly released through various human activities, including combustion, painting, spray drying, and furniture production. Given that individuals spend nearly 80% of their time indoors, exposure to these pollutants can have a significant impact on health. Indoor pollutant concentrations are generally much higher than those in outdoor environments (Saramak, 2019). VOCs pose a serious concern and require effective indoor remediation methods. Several technologies have been introduced for their removal, such as fabrication filters, electrostatic filtration, and gravity deposition. However, the high maintenance costs of these systems make their widespread use in homes and offices difficult (Xiong et al., 2020). Air phytoremediation utilizes plants to remove air pollutants, making it a suitable option for indoor and semi-indoor environments. This approach effectively eliminates VOCs while being cost-effective, environmentally sustainable, and safe for human health. It is a widely recognized method due to its low maintenance requirements and eco-friendly nature (Chaudhary and Rathore, 2019). Plants have natural structures like stomata, cuticles, trichomes, and aqueous pores on their leaves and stems that facilitate the absorption of VOCs from the air. They also contain enzymes that neutralize harmful compounds, helping them cope with pollution stress. Various plant species have shown significant effectiveness in removing benzene from the air. This makes air phytoremediation a promising approach for improving air quality and reducing pollution (Setsungnern et al., 2017). Air phytoremediation is a time-intensive process that requires a significant number of plants for effectiveness. To improve its efficiency, systems like botanical biofilters and living walls have been designed to enhance air circulation and pollutant removal (Permana et al., 2022). Integrating air phytoremediation with optimized airflow can be an effective approach to addressing VOC contamination, reducing both the duration and the number of plants required for passive air purification.

2.4 Pharmaceutical and personal care products (PPCPs)

PPCPs refer to a broad category of substances used for healthcare, cosmetic purposes, and livestock treatment. The pharmaceutical segment includes antibiotics, pain relievers, diuretics, non-steroidal anti-inflammatory drugs (NSAIDs), antiseptics, lipid-lowering agents, and β-blockers. On the other hand, personal care products (PCPs) consist of items like cosmetics, sunscreens, and other skincare formulations (Shahid et al., 2021). The annual pharmaceutical consumption is estimated to surpass several hundred kilotons (Ogunwole et al., 2021). The significant demand for PPCPs inevitably leads to considerable environmental pollution. This raises concerns, as these substances can enter the environment through various pathways, such as excretion, bathing, household waste disposal, and livestock activities. These pollutants often flow into water bodies and ultimately reach sewage treatment plants (Saxena et al., 2021). Even at minimal concentrations, these pollutants present serious ecological and health risks, such as infertility and reproductive disorders in aquatic organisms, along with the development of antibiotic-resistant genes in bacteria. In aquatic ecosystems, PPCPs can interfere with the endocrine system of higher organisms like fish, contribute to microbial resistance, and accumulate in plants and animals. Due to their widespread occurrence in the environment, the removal of these contaminants is crucial to prevent potential harm (García et al., 2020; Hossein et al., 2023).

Pharmaceuticals and personal care products (PPCPs) are widely distributed contaminants that persist in the environment and can cause ecological harm even at extremely low concentrations. Their tendency to accumulate and remain in ecosystems poses risks to human health and aquatic life over time. Several treatment methods, such as sedimentation, membrane filtration, advanced oxidation, activated sludge, and membrane bioreactors, can remove 50%–95% of these pollutants. However, these techniques come with challenges, including high costs, chemical usage, and membrane fouling. Phytoremediation, which relies on plant absorption, movement, and breakdown of contaminants, offers a more sustainable approach. Certain plants, such as *Canna indica* and *Phragmites australis*, have been found to effectively remove 70%–90% of antibiotics from the environment (Singh et al., 2024a).

2.5 Other emerging pollutants

Nanoparticles (NPs) are materials measuring between 1 and 100 nm in size and can be composed of metals, metal oxides, or carbon-based substances (Sengul and Asmatulu, 2020). NPs exhibit unique properties that make them beneficial for a wide range of applications, including their use in consumer products like sunscreens and electronics (Khan et al., 2019). However, the release of NPs into the environment, including air, water, and soil, can pose ecological risks. Their toxicity is influenced by factors such as size, shape, chemical composition, and surface charge (Rajput et al., 2018). Several remediation techniques, both physical and chemical, have been implemented to reduce contamination (Song et al., 2017). However, these methods are often labor-intensive, time-consuming, and costly, making them less effective for long-term environmental pollution control and remediation. Phytoremediation offers an alternative approach that utilizes plants for pollutant removal. As integral parts of ecosystems, plants play a key role in the uptake, accumulation, and transport of elements, including metal-based nanoparticles (Kumari et al., 2018).

Harmful chemicals in water adversely impact the aquatic environment by blocking light penetration, which disrupts the photosynthesis process in aquatic plants (Abbas et al., 2016). The *Salvinia molesta* plant demonstrates a high efficiency in removing toxic dyes and is increasingly being recognized as a valuable solution in various industries (Al-Baldawi et al., 2020). In recent years, advanced technologies such as ultrasonic treatment, electrochemical methods, photochemical processes, and membrane separation, whether used individually or in combination, have gained significant attention for their effectiveness in removing toxic dyes from wastewater (Arslan et al., 2016). A range of chemical, physical, and biological methods are employed to remove color dyes. Chemical oxidation techniques include H_2O_2, cavitation, ozone, and the Fenton process, while physical and biological approaches involve adsorption, filtration, and the utilization of plants, fungi, bacteria, and algae (Holkar et al., 2016). Green technology using plants and adsorption methods are widely employed for dye removal, with notable advancements in this field in recent years. Phytoremediation, a plant-based approach, is an emerging area of study that focuses on using plants to eliminate dyes from textile sources (Bharathiraja et al., 2018). Experiments have demonstrated that biomass adsorbents possess diverse adsorption capacities, operational conditions, and application methods, making them highly effective and reusable. Plant-derived materials, such as leaves, peels, and other components, have shown significant efficiency in dye removal (Kadhom et al., 2020).

Microplastics have spread throughout the environment, making their removal essential. Various strategies have been suggested to address microplastic pollution, including physically extracting mismanaged plastic waste and using microorganisms and biocatalysts for biodegradation. Phytoremediation is another potential method for plastic removal, though it has received little attention. Plants can act as natural interceptors and temporary reservoirs for microplastics by facilitating their deposition, adsorbing and trapping them in the root zone, promoting clustering on roots, and enabling their uptake, translocation, and accumulation in different plant parts (Tang, 2023).

3. Mechanisms of phytoremediation of organic pollutants

Phytoremediation incorporates numerous discrete mechanisms, each suitable for diverse types of pollutants and species of plants. The nature of organic contaminants (e.g., their

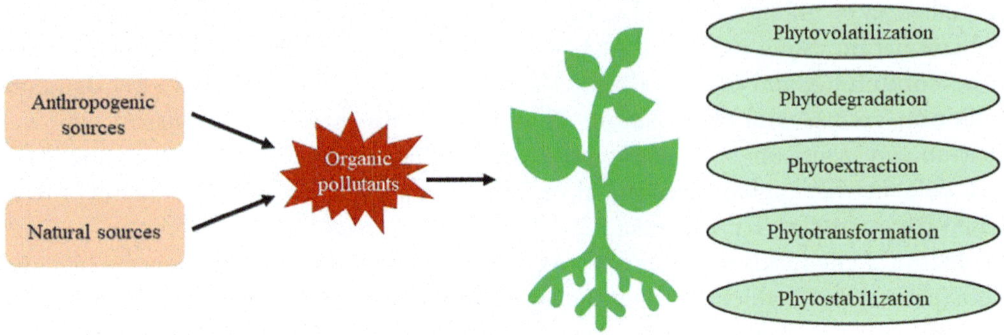

Figure 15.1 Various mechanisms of phytoremediation of organic pollutants.

hydrophobicity, solubility, and molecular structure), the environmental conditions (e.g., soil type, pH, temperature), and the plant species selected help to define the success of phytoremediation. Different plants that are hyper-accumulators, grasses, aquatic plants, and trees have shown the ability to remove the toxins from the contaminated sites. Modern progressions in molecular biology have enhanced phytoremediation efficiency. Genetic engineering has helped to develop transgenic plants that have the ability to metabolize, absorb, and cleanse organic pollutants. For instance, genetically modified poplar trees (*Populus* spp.) have shown improved removal of trichlorethylene (TCE), which is a known organic pollutant usually found underground (Doty et al., 2007). This section offers a complete summary of the mechanisms involved in phytoremediation of pollutants, including phytodegradation, phytostabilization, phytoextraction, phytovolatilization, and phytotransformation (Fig. 15.1). Each mechanism has a separate role in removing organic contamination, and understanding these mechanisms is important for the selection of suitable plant species and optimization of strategies involved in the remediation.

3.1 Phytodegradation

Phytodegradation, also called phytotransformation, is a vital bioremediation mechanism where plants use enzymes to degrade the organic pollutants; they convert the pollutants into less toxic, simpler, and more manageable forms. Through the action of specific enzymes that carry out hydrolysis, oxidation, reduction, and conjugation reactions, this process mainly occurs in the tissues of plants. Phytodegradation is known as a feasible and sustainable technology for eliminating environmental contamination, mostly in soil and groundwater polluted by chemicals that are organic in nature (Li et al., 2021). The enzymes present in the plants play a vital role in the removal and transformation of numerous organic pollutants. The enzymes involved in phytodegradation are mainly peroxidases, cytochrome P450 monooxygenases, and dehydrogenases (Malik et al., 2022). These enzymes facilitate a series of biochemical reactions that further the breakdown of persistent organic pollutants (POPs) like PAHs, herbicides, and pesticides. The oxidative degradation of hydrophobic and recalcitrant organic compounds occurs mainly due to the presence of Cytochrome P450 Monooxygenases enzymes. These enzymes make the non-polar pollutants more hydrophilic by inducing the oxygen atom into them, thus making it easier for the plants or microorganisms to process. Research has proven that cytochrome P450 enzymes can oxidize hydrocarbons, reducing their concentration in contaminated soils by up to 70% over 12 weeks. These hydrocarbons include benzo[a]pyrene, a known carcinogenic PAH (Bukowska et al., 2022). Studies on plant species like *Arabidopsis thaliana* and *Populus tremula* have shown a major increase in the regulation of cytochrome P450 enzymes when subjected to xenobiotic exposure, improving the phytodegradation of complex pollutants.

Peroxidases and dehydrogenases also play a critical role in speeding up the breakdown of aromatic hydrocarbons and phenolic compounds. Peroxidases have been reported to rupture the aromatic rings of pollutants, which leads to detoxification. In a study by Aguilar-Zarate et al. (2018), *Zea mays* (corn) plants showed an increase in peroxidase activity when they interacted with atrazine, which is a known herbicide; peroxidase converted the atrazine into a less toxic form. Moreover, dehydrogenases are known to reduce the pollutants, for example, nitroaromatic compounds, converting them into less toxic derivatives and facilitating further degradation (Wiatrowska et al., 2021). Phytodegradation is extensively used in the remediation of soil and water bodies polluted with organic pollutants. Agricultural fields exposed to pesticides and herbicides or industrial sites with historical contamination by hydrocarbons and solvents are more relevant areas for the application of phytodegradation with diffuse pollution. For instance, *Populus deltoides*, a fast-growing tree species, has been widely used in phytoremediation to degrade chlorinated solvents in groundwater, reaching the degradation rate of trichlorethylene (TCE) as high as 97% in field trials. Additionally, phytodegradation is gaining attention in urban and industrial settings, where contamination by organic pollutants such as benzene, toluene, and xylene (BTX) is predominant (Sriprapat et al., 2014). Studies show that plant-based systems can knowingly decrease BTX levels in soil and groundwater, with reductions as low as 60% within 6 months of planting phytoremediation species like willows and poplars.

3.2 Phytoextraction

Phytoextraction is a phytoremediation approach in which plants absorb organic pollutants from the soil through their root systems and then translocate them to the aerial parts of the plant, like the stems and leaves. This process efficiently removes pollutants from contaminated soil or water, mainly hydrophilic and moderately hydrophobic organic compounds (Cassina et al., 2011). The effectiveness of phytoextraction depends on the plant species, root architecture, pollutant bioavailability, and the physicochemical properties of the contaminant. The first step in phytoextraction includes pollutant absorption by plant roots. This process depends on a number of factors, including plant species, soil conditions, root structure, and pollutant characteristics. Pollutants must either be dissolved in the soil solution or adhere to the root surfaces for successful uptake. Phytoextraction depends on pollutant bioavailability, which defines the success of the process. The bioavailability of these pollutants is improved by the presence of chelating agents or soil amendments that improve their solubility (de Souza et al., 2023). Plant species with widespread and deep root systems are suitable for phytoextraction as they can target pollutants in both the upper and lower soil layers. Plants like poplar (*Populus* spp.) and willow (*Salix* spp.) are deep-rooted and are often used in phytoextraction projects. As pollutants get absorbed by the roots, they are then transported to the aerial parts of the plants, such as stems, leaves, and shoots, through the xylem.

The plant vascular system controls the translocation of pollutants. The efficiency of this movement is because of the chemical properties of the pollutant, such as molecular size, polarity, and hydrophobicity (Kristanti et al., 2021). Pollutants with more polarity and lower molecular weights are easily transported through the xylem of the plant. Due to their solubility in water and compatibility with xylem transport, herbicides such as glyphosate and 2,4-D, which are moderately hydrophilic, are readily translocated within plants. On the contrary, hydrophobic compounds like PAHs and PCBs accumulate in root tissues because of their low solubility and slow translocation rates. Once the pollutants reach the aerial parts, they may either get sequestered in the plant tissues or undergo metabolic transformation. Sequestration takes place when pollutants are stored in plant vacuoles or bound to cellular structures, hence reducing their toxicity (Ramborger et al., 2021). Phytoextraction is chiefly suitable for the remediation of soils and water that are contaminated with pollutants that are hydrophilic organic pollutants, comprising of industrial solvents, herbicides, and aromatic hydrocarbons. The process was previously applied in both agricultural and industrial settings, where soil is repeatedly exposed to chemical pollutants.

3.3 Phytovolatilization

Phytovolatilization is a process in which plants absorb organic pollutants from the soil or water through their root systems, change these pollutants into volatile compounds, and release them into the atmosphere through the stomata. This method is more suitable for the remediation of organic compounds that are volatile and semi-volatile (VOCs and SVOCs) in nature, such as hydrocarbons, halogenated solvents, and some methylated metals (Cheng et al., 2024). Although this process has the potential to remove soil and groundwater contamination, its consequences on air quality and environmental safety are to be considered. Initially, in the phytovolatilization process, the uptake of organic pollutants by plant roots from the soil or water takes place. In effective phytovolatilization, the pollutants first get changed into volatile forms that can evaporate into the atmosphere. This transformation usually takes place because of the plant's enzymatic and metabolic pathways. The absorbed pollutants are metabolized in the plant's root or shoot tissues. Major pollutants exposed to phytovolatilization include chlorinated solvents, like perchlorethylene (PCE), trichlorethylene (TCE), and heavy metals that are volatile, like mercury (Hg) and selenium (Se) (de Souza et al., 2023). Metabolic transformations in plants occur through enzymatic activity, frequently including oxidation and reduction reactions. For example, TCE is oxidized into compounds that are less toxic, such as trichloroethanol and trichloroacetic acid, which are then easily volatilized.

Certain plant enzymes have a crucial role in converting non-volatile pollutants into volatile forms. Cytochrome P450 monooxygenases, glutathione S-transferases (GSTs), and peroxidases are major enzymes that are involved in the metabolism of pollutants. Cytochrome P450 is responsible for the catalyzing oxidative reactions that can change TCE into volatile compounds. Likewise, GSTs help to conjugate pollutants to glutathione, enabling further metabolism (Akram et al., 2021). The species of plants affect their ability to volatilize the pollutants, where hyper-accumulators or transgenic plants show a better performance during this process. After the metabolization of organic pollutants, they are changed into volatile compounds, they are then transported from the roots to the aerial parts of the plant, chiefly through the xylem. The volatile compounds are then released into the atmosphere through the stomata in the leaves. Poplar trees (*Populus* spp.) are frequently used in phytoremediation due to their ability to uptake and volatilize organic pollutants like TCE (Rabideau et al., 2022). Field trials have shown that poplar trees planted over TCE-contaminated groundwater sites can remove significant amounts of contaminant through phytovolatilization. In one study, it was estimated that approximately 90% of the absorbed TCE was volatilized through the leaves, with the remainder being transformed into less toxic byproducts (Newman et al., 1999).

The physicochemical properties, like vapor pressure, molecular weight, and polarity of the plants, affect the movement of volatile compounds through the vascular system. Pollutants having higher vapor pressure, like benzene, toluene, and xylene (BTX compounds), are moved easily to the leaves, where they are volatilized (Cai et al., 2021). Compounds having lower vapor pressure might face slower transport or limitation in volatilization rates. Still, in many cases, plant metabolism may transform these compounds into more volatile products, facilitating their movement through the plant. Stomata have small pores that are microscopic on the leaf's surface. They act as the main channels through which the release of volatile compounds into the atmosphere occurs. The release of pollutants is due to a passive process, including the vapor pressure gradient between the internal plant tissues and the outer environment. In the case of TCE, studies revealed that volatilization rates rise expressively because of the temperature and stomatal conductance (Cheng et al., 2024). However, different species have varied efficiency of stomatal release, which depends on the density and behavior of the stomata and other specific features of the involved volatile compounds. Phytovolatilization is a strategy that shows effectiveness in remediating soil and groundwater contamination with volatile organic compounds, but has effects on air quality and can leave possible ecological footprints. The volatilized pollutants are changed into less toxic forms, but still, when released into the atmosphere, they can cause potential risk to the atmosphere.

3.4 Phytostabilization

Phytostabilization is a key phytoremediation method that uses plants to immobilize the organic pollutants in soil, stopping their migration to the groundwater or the adjacent ecosystems (Malik et al., 2022). Unlike the other phytoremediation methods that are used to decrease or degrade contaminants, phytostabilization concentrates on the containment, lowering the mobility and the bioavailability of the pollutants. This technique is remarkably beneficial in stabilizing the contaminated soils in the areas where degradation or removal procedures are either very expensive or impractical, as in the arid and semi-arid areas where water is limited. At the core of phytostabilization is the capability of the plants to act as a natural barrier, reducing the movement of pollutants in the soil. Plants with long and fibrous root systems, including certain grasses and perennial species, physically trap the contaminants inside the root zone, blocking their movement into deeper layers of soil or other water sources (Begum et al., 2021). This containment is also supported by the combination of physical, chemical, and biological interactions linking the plant roots, particles of the soil, and the organic pollutants. Plants that have properly developed and extensive root systems make a physical network system that binds the soil particles, lowering the permeability of the soil and reducing erosion. This network system limits the lateral and vertical movement of the contaminants while preventing the pollutants from migrating into the surrounding areas or being carried away by the surface runoff of water. For instance, in the case of contaminated soils that contain persistent organic pollutants (POPs), including the polycyclic aromatic hydrocarbons (PAHs) and the polychlorinated biphenyls (PCBs), grasses having deep roots can efficiently trap the compounds in the root zone, lowering their potential for leaching (Gogoi and Sarma, 2023).

Along with the physical containment, the adsorption of the pollutants onto soil particles has a key role in immobilization. Plants excrete root exudates, such as organic acids, amino acids, and enzymes, which interact with soil minerals along with organic matter. These exudates increase the adsorption of the pollutants onto the soil colloids, such as clays and organic matter, so they can further stabilize them (Yaqoob et al., 2024). Organic pollutants, including the PAHs and PCBs, which are hydrophobic, bind readily to the soil organic matter because of their lipophilic nature. This binding decreases their bioavailability, which makes it harder for the pollutants to travel through the soil or be taken up by the other organisms. Phytostabilization not only has a role in immobilizing the pollutants, but it also increases the overall health of the soil, making an environment with a lower risk of pollutant migration. By increasing the soil structure, promoting useful microbial activity, and lowering erosion, phytostabilization upholds a healthier and resilient soil ecosystem that is not likely to be prone to contaminant leaching and groundwater contamination (Nie et al., 2024). Phytostabilization is specifically used to manage organic pollutants in the soil where traditional remediation techniques, including excavation or chemical treatments, are not viable. It is broadly applied in areas contaminated with persistent organic pollutants (POPs), such as PAHs, PCBs, and specified pesticides, which are involved in degrading or eradicating from the environment. These contaminants, because of their hydrophobic nature, bind strongly to the soil particles and are very prone to accumulating in the environment, presenting long-term risks to the ecosystem and human health.

3.5 Phytotransformation

Phytotransformation, also known as phytometabolism, is an important phytoremediation procedure whereby the plants absorb the organic pollutants and transform them into less toxic and more manageable forms through a major series of enzymatic metabolic reactions (Begum et al., 2021). Different from phytoextraction or phytostabilization, which either eliminate or immobilize pollutants, phytotransformation directly functions within the metabolic systems of the plant to change the chemical structure of the contaminants. This transformation decreases their toxicity or enables their further degradation by the plants and microorganisms. Phytotransformation is specifically effective for the remediation of organic pollutants, including pesticides, industrial solvents, and herbicides, providing a biologically sustainable solution for polluted environments (Park and Oh, 2023). When the organic pollutants are absorbed by the plant roots, they are usually translocated to various parts of the plant, where they usually undergo a metabolic transformation. This includes a variety of biochemical procedures, including oxidation, reduction, and conjugation, which are catalyzed by the key enzymes inside plant cells.

One primary mechanism of phytotransformation includes oxidation-reduction (redox) reactions, which are carried by enzymes, including the cytochrome P450 monooxygenases. These enzymes add oxygen to the organic pollutants, often resulting in the formation of the hydroxyl groups that improve the polarity and the solubility of the contaminants (Pathak et al., 2024). For instance, pesticides, including atrazine, can go through oxidation in plant tissues, resulting in the formation of more soluble metabolites, which are easier to degrade further or excrete. Likewise, reduction reactions might remove or decrease the specified functional groups in the pollutants, making them less toxic and more reactive toward further degradation. Another important metabolic procedure in phytotransformation is conjugation, in which the enzymes, including the glutathione S-transferases (GSTs), support the attachment of the polar groups to pollutants (Chunduri et al., 2024). GSTs catalyze the pollutants' conjugation with glutathione, a tripeptide that has a key role in detoxification. Through conjugation, pollutants are converted into more water-soluble molecules, making it easier to sequester in vacuoles or to metabolize further. For instance, the herbicide 2,4-dichlorophenoxyacetic acid (2,4-D) could be conjugated with sugars or amino acids in plant tissues, forming compounds that are relatively less harmful and much easier to transport or store (Pinto et al., 2023). Once the pollutants are transformed well into polar or less harmful metabolites, plants sequester these compounds in vacuoles, which are specialized organelles that store waste products or transport them in the rhizosphere. In a few cases, these transformed metabolites are stored in aerial parts, including the leaves, stems, and roots, where they are grouped into a non-toxic form. For instance, the conjugated metabolites of the pesticides can be stored inside the vacuoles of plant cells, where these metabolites are rendered inert and isolated from other cellular functions. This sequestration procedure decreases the potential toxicity to the plant itself and stops metabolites from interfering with plant growth or development.

The final goal of phytotransformation is to convert toxic organic pollutants into less harmful or entirely benign compounds. This transformation might result in various effects depending on the nature of the pollutant and the specific metabolic paths involved. In several cases, phytotransformation results in the detoxification of toxic pollutants. For instance, organic solvents, including trichlorethylene (TCE), are metabolized by the plants into less toxic compounds, such as trichloroethanol and trichloroacetic acid (AL Mamun et al., 2024). In some cases, plants can fully mineralize the organic pollutants, breaking them into the basic components, including carbon dioxide, water, and other harmless byproducts. This full degradation happens when the pollutants go through complete oxidation inside the metabolic system of the plant. For instance, plants have been demonstrated to mineralize simple organic pollutants, including phenol or benzene, into carbon dioxide through metabolic oxidation, efficiently removing the contaminant from the environment (Treesubsuntorn and Thiravetyan, 2012). In some cases where complete degradation is not usually possible, phytotransformation usually results in the formation of intermediate products that are either less toxic than the original pollutant or more prone to further degradation. Such intermediate products could be stored in the plant tissues or excreted in the rhizosphere, where the soil microbes can continue the degradation process. For instance, pollutants, including the polycyclic aromatic hydrocarbons (PAHs), might be partially metabolized into simpler aromatic compounds, which are easier for rhizosphere microorganisms to degrade. By producing these intermediate metabolites, the phytotransformation environment supports more thorough remediation.

4. Key plant species used in phytoremediation

Aquatic, terrestrial, and decorative plants are all utilized in phytoremediation. Phytoremediation is a sustainable method of cleaning up contaminated regions since these plants can extract harmful compounds from soil and water. Soil pollution by heavy metals has recently been recognized as one of the major environmental problems worldwide (Dong et al., 2020). The development of heavy metal buildup in agricultural soils may seriously endanger the health of microorganisms, animals, and plants (Ullah et al., 2023). Long-term wastewater irrigation exposure increased the buildup of heavy metals in the soil and greatly increased their transfer to the food chain (Nawaz et al., 2021). Hyperaccumulator plants' beneficial characteristics, such as their large biomass allocation, deep-rooted root system, and ability to extract water-soluble heavy metals, work together to enable them to access and accumulate higher concentrations of heavy metals (Patil et al., 2023). Usually, plants manage pollutants without harming topsoil, preserving its fertility and usefulness. By adding organic matter, they might increase soil fertility.

4.1 Hyperaccumulators and tolerant species

In phytoremediation, hyper-accumulators and tolerant plant species are essential, especially when it comes to the removal of heavy metals and organic contaminants. Plant species that can absorb and store abnormally high quantities of pollutants—often in their shoots—without becoming hazardous are known as hyperaccumulators.A variety of metals can be extracted and concentrated in the upper sections of hyperaccumulator plants. These species can grow and live in soils with abnormally high concentrations of heavy metals, and they can accumulate these metals up to 1%–2% of their total dry weight (Nedjimi, 2021; Shrivastava et al., 2019). These plants have special physiological and biochemical processes that enable them to withstand, absorb, or break down contaminants (Sarma, 2011). Well-known hyperaccumulators that can accumulate significant amounts of heavy metals include *Phyllanthus balgooyi, Alyssum cassium, Pycnandra acuminata, and Agrostis capillaris.* Through the xylem, hyper-accumulators effectively transport, disperse, and sequester HM to shoots without posing any phytotoxic risks to themselves (van Der Ent et al., 2018). From annual herbs to perennial shrubs and trees, the 721 species of hyperaccumulators are found in 52 families and 130 genera, including the prominent *families* Brassicaceae, Fabaceae, Euphorbiaceae, Asteraceae, Phyllanthaceae, Lamiaceae, and Scrophulariaceae (Reeves et al., 2018).Edible crops should not be used for phytoremediation, however, because heavy metals can accumulate in plant components that humans or animals consume, potentially entering the food chain and posing a risk to human health. Therefore, choosing non-edible hyperaccumulators is essential for effective and secure heavy metal phytoremediation (Skuza et al., 2022).

4.2 *Populus* spp.

Numerous studies have also been conducted on the beneficial effects of poplar trees on soil and underwater environments. Their naturally well-designed root structure, which absorbs a lot of water, is their secret. Because of their rapid growth, broad root systems, and capacity to absorb and break down organic contaminants like hydrocarbons, poplar trees are frequently utilized in phytoremediation. Populus species have been used to clean up contaminated groundwater and soils, especially those containing polycyclic aromatic hydrocarbons (PAHs) and chlorinated solvents (Huslina et al., 2024). According to a study from the National Institute of Environmental Health Sciences, hybrid poplars are better able to withstand organic contaminants like chlorinated solvents, like trichlorethylene or the well-known carcinogenic carbon tetrachloride (95% of material eliminated). Additionally, according to a study from the National Institute of Environmental Health Sciences, hybrid poplars are better able to withstand organic contaminants like chlorinated solvents, like trichlorethylene or the well-known carcinogenic carbon tetrachloride (95% of material eliminated).

4.3 *Brassica* spp.

The capacity of mustard plants to collect heavy metals, such as lead (Pb) and cadmium (Cd), is well-documented (Kaur et al., 2017). Many plant species belonging to the genera *Brassica*, *Rhamphospermum*, and *Sinapis* within the Brassicaceae family, also known as the mustard family, are considered potential candidates to play a role in the remediation of organic pollutants. A lot of veggies are grown from mustard plants, which may have been domesticated as early as 6000 years ago. Because of their high biomass production and efficient pollutant uptake, these plants are commonly employed in soils contaminated by organic pollutants. One significant method that has emerged is phytoremediation using the botanical genus *Brassica*. The phytoremediation potential of mustard cultivars is strongly influenced by their genetic composition (Singh et al., 2024a, 2024b). A number of plant species of the *Brassica* genus are suitable for use in phytoremediation because they can withstand the harmful effects of organic pollutants. *Brassica* species can perform HMs phytoremediation through physiological mechanisms, such as phytovolatilization (Kumari et al., 2020), phytostabilization (Yadav et al., 2018), and phytoextraction (Sarwar et al., 2017).

4.4 *Salix* spp. (Willows)

Willow species have demonstrated a great deal of promise in soil stabilization and organic pollution remediation, especially in hydrocarbons and phenolic chemicals. Because of their vast root systems, willows can draw contaminants out of the groundwater and soil (Bardos et al., 2011). Although water-loving plants enhance landscapes, their value extends beyond their esthetic appeal. Additionally, they have a more intriguing application in phytoremediation: their roots have shown greater vitality than Brassicaceae and function even in environments contaminated by diesel fuel and other mixed heavy metals.

4.5 *Phragmites australis* (common reed)

In artificial wetlands, the common reed is a widespread plant that is utilized for the phytoremediation of organic pollutants and wastewater. Through interactions with the rhizosphere, it is very effective at breaking down nutrients, hydrocarbons, and other organic pollutants (Jeelani et al., 2020). New wetland species like *Phragmites australis*, a genuinely global grass that thrives in wetlands, produce a lot of biomass and offer vital ecosystem services (Eller et al., 2017). *P. australis* has been extensively utilized for the phytoremediation of areas that are co-contaminated with organic and metal contaminants, according to compelling evidence.

5. Role of genetic engineering in enhancing phytoremediation

Through the creation of plants with improved capacities for pollution breakdown and uptake, genetic engineering has paved the way for new developments in phytoremediation. Scientists have been able to increase plants' capacity to withstand and digest pollutants by adding foreign genes. Adding bacterial genes that encode enzymes that can break down contaminants is a common genetic alteration technique (Koźmińska et al., 2018). For example, it has been demonstrated that plants modified with bacterial genes expressing dehalogenases may break down halogenated organic chemicals, including trichlorethylene (TCE) (Meagher, 2000). Likewise, the accumulation of heavy metals in plants has improved due to the overexpression of metal transporter genes (Kurade et al., 2021). By introducing the genes that increase the metabolic pathways included in a pollutant transformation, scientists are trying to create plants that can volatilize a wide range of contaminants. For example, plants that express bacterial genes for the degradation of chlorinated hydrocarbons are now being grown to increase the volatilization of pollutants. These innovations are enabling more efficient and better phytovolatilization applications.

6. Applications of phytoremediation

Phytoremediation is a sustainable environmental management strategy that utilizes plants to treat heavy metal, organic, and radioactive contamination, among other forms of pollution. It improves soil health and restores fertility by efficiently breaking down contaminants, including pesticides, heavy metals, and petroleum hydrocarbons. This approach works particularly well in areas close to oil extraction sites and agricultural settings affected by excessive chemical use (Yang et al., 2022). Aquatic ecosystems can be restored and water quality improved by using phytoremediation to remove organic contaminants. Wetlands that have been planted with plant species, for instance, can efficiently filter contaminants out of contaminated water.Phytoremediation is also very beneficial in urban and industrial settings, where it can clean up brownfield areas and abandoned sites that have been contaminated with organic pollutants and industrial waste, leading to redevelopment and improving the quality of life in urban areas. Additionally, phytoremediation can improve air quality by lowering the amount of airborne volatile organic compounds (VOCs) through absorption by plants and degradation (Diarra et al., 2022).Moreover, phytoremediation has several drawbacks. The process can take an extended period, and high pollution concentrations can harm plants. Selecting the right plant species and doing a thorough site assessment are essential for successful implementation. Despite these drawbacks, phytoremediation is an economical and sustainable method of tackling a range of environmental pollution issues.

6.1 Phytoremediation of organic pollutant-contaminated soils

On a global scale, soil is acknowledged as the third essential element of the environment, following air and water. Because 95% of the food that humans eat comes from the soil, maintaining fertile soil is vital to life. It is essential to plan for the preservation of healthy soil to guarantee human survival. When pollutants are introduced into the soil, the quality of the soil changes, and environmental problems arise (Chen et al., 2009). Numerous environmental problems have arisen because of pollution caused by human activity and the discharge of contaminants into the environment. The pollution of soil with pesticide-laden soil and petroleum compounds in oil-rich areas and near oil refineries is one prominent issue among them (Zhu et al., 2020). Although crude oil is one of the most important energy sources in the world, the complex procedures involved in its production, transportation, consumption, and disposal greatly contribute to widespread environmental degradation. The infiltration, seepage, and leakage of petroleum pollutants and derivatives into the soil are some of the most significant soil contaminants in nations with large oil resources, facilities, refineries, and continuous exploration efforts (Jaskulak et al., 2020).

Hydrocarbons from petroleum are another contaminant that enters the land through crude oil. Globally, petroleum hydrocarbons constitute the main cause of environmental contamination (Jung et al., 2017). Petroleum hydrocarbon-induced soil contamination is a serious issue in regions where oil is extracted. The environmental effects of hydrocarbon pollution have received a lot of attention in recent decades (Perera, 2018). When petroleum hydrocarbons are present in soil, they can negatively impact the biological activities of soil microorganisms, which can result in toxicity. Soil quality may suffer because of this problem. Additionally, these substances may have detrimental effects on the soil's chemical composition (Roy et al., 2013). The blockage of pore spaces between soil particles caused by petroleum hydrocarbons raises soil temperature and reduces the number of helpful microorganisms. Furthermore, these contaminants hinder plants' ability to absorb nutrients, which eventually leads to plant death (Igoni, 2018). Recent years have seen the development and application of novel techniques for the bioremediation of soil contaminants, with promising outcomes (Guarino et al., 2017; Wang et al., 2019). There are several physical and chemical techniques, including solvent-based extraction and burning, to deal with HMs contamination in the soil. However, because of their high prices and insufficient pollution clearance, these methods are less effective. Nowadays, biological techniques are frequently used to remove pollutants from soil (García-Sánchez et al., 2018).

Agriculture, environmental health, and human well-being are all impacted by the management and disposal of outdated pesticides as well as the cleanup of pesticide-contaminated soil. Due to an inadequate understanding of appropriate planning, storage, and use, pesticide consumption and management have a history of issues. Due to the numerous advantages of using pesticides, their manufacture and application have a long history (Tarla et al., 2020). Phytoremediation processes involve the transpiration stream driving the bioavailable pesticides to be absorbed by plant roots together with water and then transported to shoots via the xylem (Fan et al., 2020, 2021). Pesticides travel through cell walls and membranes before arriving at different plant tissues or volatilizing outside of the plants. Plant detoxification and the biotransformation of xenobiotic compounds are generally said to involve three stages: hydroxylation by cytochrome P450 monooxygenases, conjugation by glutathione-S-transferase, and sequestration as bound residues in plant tissues (Del Buono et al., 2020). Through mechanisms like phytostimulation and phytodegradation, plants indirectly improve their surroundings by releasing exudates that allow the bacteria they are connected with to break down pesticides (Singh and Singh, 2017).

6.2 Phytoremediation of water contaminated with organic pollutants

Water is a valuable natural resource that is essential to human civilization and the survival of agriculture. Over 70% of the earth's crust is made up of water, and most of that water has been heavily exploited and highly contaminated or damaged because of human activity (Obinna and Ebere, 2019). Providing clean water to the entire world is one of the current issues brought on by the growing global population (Ashraf et al., 2021; Chen et al., 2018). There are several possible sources of contaminated water, such as pathogens, inorganic substances, organic material, and macroscopic pollutants, which are the four main groups into which they can be divided. Even in treated water, organic and inorganic pollutants are the most prevalent (Karthigadevi et al., 2021; Pi et al., 2018). Certain contaminants are treated using physical, chemical, and electrochemical methods, but they might not be reduced. Some methods for removing metals from the soil include adsorption, leaching, and the use of nano-sorbents. Therefore, remediation systems can be improved by emerging technologies such as bioremediation, which uses plants to remove some of the most common toxins in water (Hua et al., 2012). When a contaminating substance is discharged into water bodies and reaches a critical limit of accumulation in the system, the aquatic environment becomes contaminated. In general, there are two types of sources of water pollution: point sources and non-point sources (Li and Wu, 2019). There are many different types of contaminants, including both organic and inorganic ones. Hexachlorobenzene (HCB), polybrominated diphenyl ethers (PBDEs), polychlorinated biphenyls (PCBs), polycyclic aromatic hydrocarbons (PAHs), and pesticides such as lindane and dichloro diphenyl trichloroethane (DDT) are examples of organic pollutants (Weber et al., 2011). Plants and the microbial communities that live in the rhizosphere can have a variety of effects on organic contaminants, such as immobilizing, storing, transforming, degrading, or mineralizing the pollutants into less harmful forms. However, the exact compounds and the environmental conditions in which the plant is growing will affect these effects (Kushwaha et al., 2015).

Rhizodegradation, phytodegradation, transformation, mineralization, and phytostabilization are the phyto-technologies that are in charge of eliminating organic pollutants (Agarwal et al., 2020). However, in the case of inorganic pollutants, radionuclides, metalloids, and heavy metals can accumulate and be removed in the plant's roots and shoots. The main phyto-technologies used to reduce inorganic pollutants are enzymatic modification, phytoextraction, rhizofiltration, phytoaccumulation, and phytovolatilization (Agarwal et al., 2020; Kushwaha et al., 2015). Plants can store contaminants within their vacuoles or sequester certain cations and anions in their cell walls (Rascio and Navari-Izzo, 2011). The low energy requirements, floating mat, natural bio-filtration, and rich wetland plant life of Floating Treatment Wetlands (FTWs) have made them a unique treatment solution for petroleum hydrocarbon-contaminated water (Colares et al., 2020). To eliminate contaminants, the roots of plants that dangle into the body of water serve as a biological filter. Additionally, these plant roots offer a lot of surface area for bacteria to grow and for biofilms to form (Afzal et al., 2019). On the other hand, in FTWs, organic pollutants found in petroleum-contaminated water are broken down by bacteria that break down hydrocarbons and develop either inside or outside the roots. Additionally, the vegetated plant's aerial parts provide the biota in the FTWs system with a pleasing appearance (Rehman et al., 2019).

The wetlands' ability to purify water is now acknowledged as a desirable alternative for wastewater treatment because of its low cost, ease of use, and capacity to handle several pollutants. Constructed wetlands (CWs) are designed to take advantage of many of the same processes that occur in natural wetlands but do so within a more controlled environment (Vymazal 2010). Phytoremediation of constructed wetlands has been used to improve the quality of contaminated waters by acting as a sink for various contaminants discharged from sewage, industrial and agricultural wastewater, landfill leachate, and storm water runoff (Jomjun et al., 2010). Constructed wetlands are demarcated and constructed systems that imitate the natural processes of wetland vegetation, soils, and the microbes that live in symbiotic interaction with them in order to remediate wastewater in a more controlled environment. Many of the same mechanisms seen in natural wetlands can be exploited in constructed wetlands (CWs), but in a more regulated setting (Vymazal 2010). Using artificial wetlands as a sink for different pollutants released from sewage, industrial and agricultural wastewater, landfill leachate, and storm water runoff, phytoremediation has been employed to improve the quality of contaminated waters (Jomjun et al., 2010). The purpose of artificial wetlands is to remediate wastewater in a more controlled setting by creating systems that mimic the natural processes of wetland vegetation, soils, and the bacteria that coexist with them (Vymazal 2007). They are intended to enhance and maximize the natural processes that take place in wetlands (Vymazal 2005). The components of created wetlands include water, a filtering medium, microorganisms, and vegetation. Wastewater is permitted to flow through filter media, where physical and biochemical processes facilitate infiltration and the removal of pollutants and toxins with the help of microbes and vegetation. Constructed wetlands can remove pollutants and need little energy input as well as low operating and maintenance expenses (Kadlec et al., 2000).

6.3 Phytoremediation of air

One economical and ecologically favorable method of assisting in the reduction of air pollution is phytoremediation. Based on the metabolic processes of plants, the mechanism of phytoremediation in the air involves the absorption and destruction of airborne contaminants (Kim et al., 2020). The surface of plants above ground, primarily the stomata of leaves, is where pollutants are primarily absorbed (Kulshreshtha et al., 2009). When foreign pollutants are broken down by plant metabolites such as enzymes (catalase, dehalogenase, etc.), the degradation of pollutants continues (Rachmadiarti et al., 2019). Therefore, throughout their growth, plants can help purify the environment by removing harmful substances like metals and organic compounds as well as greenhouse gases like carbon dioxide (CO_2) and methane (CH_4) (Wei and Wang, 2020). Phytoremediation plants can be categorized according to how well they work indoors or outside. Small shrubs and herbs that grow well indoors are typically considered relevant indoor plants. Experiments in sealed, fixed-volume containers can be used to evaluate these plants' ability to adsorb contaminants. Conversely, most relevant outdoor plants are tall trees or big shrubs, and their ability to adsorb contaminants of concern through their leaves or roots is often how purification effects are assessed.

Through direct and indirect impacts, volatile organic compounds (VOCs) can seriously endanger human health, the environment, climate change, and ambient air quality. Industrial production activities have been identified since 2000 as the primary anthropogenic source of volatile organic compounds (VOCs) emissions due to the growth of urban industrialization; they account for more than half of all VOC emissions from all anthropogenic sources (Li et al., 2022). Because of the activities of building occupants, emissions from building materials, and volatile chemical products (VCPs), such as consumer goods like paints, dry cleaners, household solvents, personal care products, fragrances, and cleaning products, as well as other commercial and industrial goods like oil and gas solvents, coatings, printing ink, pesticides, adhesives, and sealants, VOCs can be easily produced inside buildings (Gkatzelis et al., 2020). Numerous detrimental health impacts, such as skin and eye irritation and trouble breathing, are linked to high levels of indoor volatile organic compounds. The risk of lung cancer, asthma, sick building syndrome, and other building-related illnesses rises with prolonged exposure to volatile organic compounds (VOCs) (Tran et al., 2020).

Depending on the kind of contaminant, different plants have varied ways of eliminating it from the air, particularly volatile organic compounds (VOCs). The contaminants are usually taken up by the plant's stems or leaves, processed by plant cells through an enzymatic pathway, and then broken down by microorganisms in the rhizosphere (Pettit et al., 2019). Additionally, the phyllosphere and rhizosphere of plants include microbial communities that interact with one another to remove air contaminants. It has been reported that endophytes, which are microorganisms that live on the surface of leaves and within their bodies, can detoxify a portion of airborne contaminants that have been adsorbed or absorbed through transformation, degradation, or sequestration. In order to further detoxify the plant, the residual contaminants are then carried to the soil and come into touch with the microorganisms that reside in the roots and rhizosphere (Lee et al., 2020).

6.4 Phytoremediation in urban and industrial areas

Social development has, unfortunately, been accompanied by irreversible natural resource destruction. Significant volumes of trash are produced because of the quest for a higher level of life, which harms the soil ecosystem. The situation is made worse by the growth of urban areas, which encroach on formerly agricultural and forested areas (Qin et al., 2019; Shabbir et al., 2020). A promising method for dealing with the various and complicated contaminants found in urban soil is phytoremediation. Organic pollutants, such as pesticides, which are mostly sourced from agricultural land, are the main cause of soil pollution in urban areas (Antoniadis et al., 2021; Kumar et al., 2022; Qin et al., 2021). It is essential to comprehend the origins of pollutants in urban soil to address soil pollution and guarantee a healthy environment for residents. Both man-made and natural disaster-induced pollutants are possible. Nonetheless, the "three wastes" produced by manufacturers, direct sewage discharge, pesticide use, high metal ore smelting, burning fossil fuels, and leaching from municipal waste landfills have combined to cause most of the anthropogenic urban soil contamination (Sharma, 2021).

Organic or inorganic contaminants, or sometimes both, are the cause of brownfield contamination, which is typically found in both the soil and the groundwater (Thornton et al., 2008). Brownfields are considered to have a high potential for reclamation after in-situ green technologies are implemented because of their unique physical characteristics and nature, which include low concentrations and a heterogeneous distribution of contaminants restricted to shallow soil layers. In situations where pollution levels are minimal, natural vegetation may eliminate or stabilize contaminants without the need for human assistance. Plant metabolism, natural soil microbial activity, light exposure, wind, and humidity are some of the naturally occurring processes that contribute to soil reclamation in this process, which is called natural attenuation. While certain brownfield land sites have been successfully stabilized by the promotion of natural attenuation, the technique is particular to contaminants and is best suited for petroleum hydrocarbons (Hartley et al., 2012; Megharaj and Naidu, 2017). One environmentally friendly biological method for treating industrial effluents is phytoremediation. Pollutants negatively impact plant growth and development, crops are not used because of the potential for pollutants to enter the food chain, plants typically accumulate a single pollutant, whereas industrial effluents contain multiple pollutants, and the phytoremediation process alone is slower. In this regard, microorganisms that can mobilize, solubilize, sequester, volatilize, accumulate, and detoxify contaminants include bacteria, which are prevalent in soil and water. However, because there is food and shelter available, there are a lot more bacteria in the rhizosphere. More effective than phytoremediation, this bacterial-plant combination detoxifies and accumulates industrial toxins (Danyal et al., 2023).

7. Factors affecting phytoremediation efficiency

A variety of plant-based remediation techniques, including phytoextraction, phytostabilization, phytoimmobilization, rhizofiltration, and phytovolatilization, are included in phytoremediation (Alkorta et al., 2004; Vithanage et al., 2012). Phytoextraction is the most important of these methods since it works based on metal translocation to shoots and bioconcentration in the plant species above-ground portions (Peuke and Rennenberg 2005; Vithanage et al., 2012). However, the effectiveness of phytoextraction (and other phytoremediation techniques) is influenced by several interconnected environmental factors (Fig. 15.2). In general, they can be defined as elements that affect organisms' lives and are necessary for them to function properly in both land and aquatic environments. The most common classification of environmental factors separates

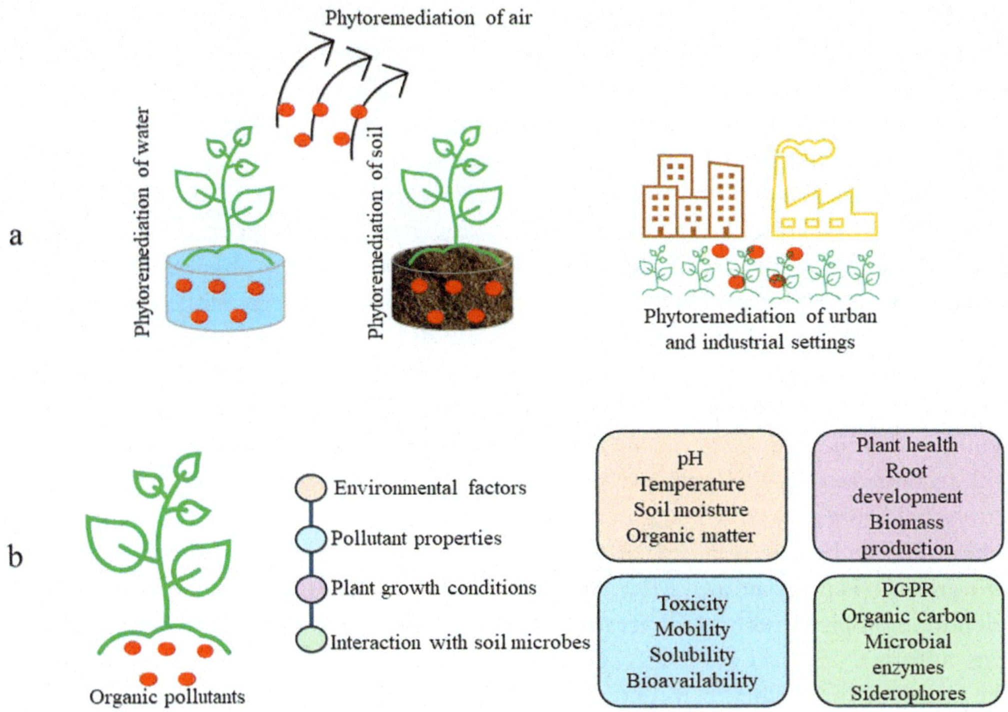

Figure 15.2 Overview of plant-based remediation and factors affecting organic pollutant degradation. (a) Phytoremediation of different environmental compartments, including water, soil, air, and urban/industrial settings (b) Factors affecting phytoremediation efficiency.

them into two primary categories: biotic (living) and abiotic (nonliving) factors (Schulze et al., 2005). While the abiotic group includes weather (rain, isolation, temperature, cloud, snow), rocks, oceans, rivers, and anthropogenic factors (all the things that are products of human activity, such as organic and/or inorganic pollutants), the biotic group includes plants, animals, protists, fungi, and bacteria. Since the factors' characteristic traits are interdependent, changing one of them can change the system (Beard et al., 2005). However, both internal plant factors and external environmental factors affect the phytoextraction potential (Seth et al., 2012).

7.1 Environmental factors

The most crucial environmental component for plant growth and development is soil. It must be noted, nevertheless, that soil is, in fact, diverse and lacks a regular, predictable structure; as a result, one area of soil may differ greatly from another nearby in terms of its physical, chemical, and biological characteristics. The amount of organic and mineral matter, soil water, aeration, climate, the presence of fungi and bacteria in the soil, and animals that impede the ground are just a few of the numerous variables that affect it. In addition to having a significant impact on plant life, each of these factors also affects how well the phytoremediation process works. In addition to determining whether plant growth is stimulated or inhibited, dynamic changes in the soil environment also significantly affect the remediation process (Gonzalez et al., 2013). The type of pollution, however, has the biggest influence on the many factors listed above that affect the phytoremediation process in soil. because metals, pesticides, herbicides, fungicides, and antibiotics are examples of man-made chemicals that can contaminate soil, as well as natural aromatic and hydrocarbon compounds.

Air temperature has a major impact on plant growth, bringing up the hotly debated subject of global warming in the modern era. Global warming is associated with higher atmospheric concentrations of carbon dioxide (CO_2) and other greenhouse gases because of both natural and human-induced activity. Despite a decrease in CO_2 emissions, industrialization significantly alters the biological, chemical, and physical characteristics of the environment by raising the concentration of CO_2 in the atmosphere. It has been shown that higher atmospheric CO_2 concentrations have a beneficial impact on heavy metal phytoextraction in contaminated soil, such as in *Pteridium revolutum* and *Pinus densiflora*, by boosting biomass production or soil uptake of heavy metals (Jia et al., 2010; Kim et al., 2011; Zheng et al., 2008).

7.2 Pollutant properties

The physicochemical characteristics of the pollutants, including their toxicity, bioavailability, and solubility (Fig. 15.2), have a major impact on how effective this process is. The mobility and availability of organic pollutants for plant uptake are influenced by their solubility in water. Strong binding to soil particles by hydrophobic organic compounds (HOCs) lowers their concentration in soil water and, as a result, their bioavailability to plants. Because it functions as a sorbent, soil organic matter affects how HOCs are distributed between the soil and water phases (Aryal, 2024). Plant health can be negatively impacted by organic pollutants' toxicity, which can lower growth and metabolic activity. In order to reduce toxicity, phytoremediation uses plant processes like contaminant uptake, transformation, and sequestration. Determining the potential and constraints of phytoremediation requires an understanding of these processes (Reichenauer and Germida, 2008).The ability of organic pollutants to be absorbed and degraded by plants is determined by their bioavailability. Certain factors, such as the chemical makeup of the pollutant, the properties of the soil, and the existence of soil amendments or bacteria that promote plant growth, affect the bioavailability of pollutants. The efficiency of phytoremediation can be increased by increasing bioavailability through such interventions (Aryal, 2024).

7.3 Plant growth conditions

Numerous elements pertaining to plant growth conditions, root development, plant health, and biomass production have a substantial impact on this process's efficiency. Improving the efficiency of phytoremediation requires optimizing soil nutrients. Inorganic fertilizers that provide potassium (K), phosphorus (P), and nitrogen (N) are essential. Nitrogen (N): Encourages the production of chlorophyll and leaf growth. The development of roots and flowers depends on phosphorus (P). Potassium (K): Promotes root and stem development (Grzegórska et al., 2020). The ratio of these nutrients should be carefully controlled, though, because improper balances can have a detrimental effect on plant health and nutrient absorption. According to studies, fertilizers containing nitrogen and phosphorus greatly boost plant biomass and contaminant uptake. For example, studies on Brassica juncea showed that the highest copper uptake was achieved with nitrogen and phosphorus additions at 100 and 200 mg/kg, respectively (Grzegórska et al., 2020).

The ability of the plant to absorb and access pollutants is enhanced by a robust root system. Mycorrhizal fungi form symbiotic relationships with plant roots to expand their reach and improve the uptake of nutrients and contaminants. This association may lead to increased phytoremediation efficiency by enhancing root growth and function. Healthy plants are better at absorbing and tolerating contaminants by generating phytohormones (Zhu et al., 2023) that promote plant growth, plant growth-promoting rhizobacteria (PGPR) improve plant health, fixing nitrogen so that plants can access it more easily, making vital nutrients soluble, and generating siderophores that increase the availability of iron by sequestering it (Sahoo et al., 2024). Increased contaminant uptake is correlated with higher biomass production. More biomass from roots and shoots results from improved plant growth, which increases the number of sites available for absorbing contaminants. To increase biomass without creating nutrient imbalances that might impede phytoremediation, nutrient addition must be balanced (Zhou et al., 2021).

7.4 Interaction with soil microorganisms

Siderophores are small-molecule chelating compounds produced by bacteria, fungi, and plants, and they have a significant impact on the soil solution (Kidd et al., 2009; Rabęda et al., 2011). Phytosiderophores are the pollutant-mobilizing substances that plants' roots secrete into the rhizosphere (Lone et al., 2008), which solubilize pollutants in soil through specific mechanisms. Furthermore, plants that grow in environments with high levels of trace element contamination have low iron content in their tissue. Because they can mineralize organic matter and increase the bioavailability and efficiency of nutrient uptake, bacteria are a crucial component of soil (Domínguez-Crespo et al., 2012). Phytoremediation is significantly impacted by soil bacteria, particularly those that are members of the plant growth-promoting rhizobacteria/bacteria (PGPR/PGPB) group. The rhizosphere's bacteria contribute to the buildup of trace elements (Jing et al., 2007). Additionally, the reduction of phytotoxicity in contaminated soil is impacted by the presence of bacteria. This is because the plant and bacteria work together to create a unique arrangement whereby the plant gives the bacteria organic carbon (for example, through molecules or compounds released by the root cells), which causes the bacteria to lessen the soil's phytotoxicity. Furthermore, microbial activity leads to the degradation of contaminants because of the symbiotic relationship between bacteria and plants. Additionally, the substances that the plant roots secrete into the root zone raise metal bioavailability, which in turn raises the likelihood that bacteria will be able to perform additional phytoremediation. This process enhances plants' capacity to adjust to adverse circumstances and boosts the potential for phytoremediation through the presence of plant–bacterium symbiosis (Jing et al., 2007). Due to their close involvement in the natural carbon cycle, fungi play a significant role in the availability of trace elements in the soil environment, second only to bacteria. They contribute significantly to the distribution of energy produced by the breakdown of organic matter. They also convert proteins from plant residues into nitrogen dissolved in the acidic form, or NH^{4+} ions, which are forms that plants can access more readily. Furthermore, mushrooms are a part of the inorganic component circuit (Domínguez-Crespo et al., 2012). Additionally, the presence of plant-fungus symbiosis increases the plant's resistance to environmental stress (Khan, 2005; Gamalero et al., 2009; Rabęda et al., 2011). On the one hand, mycorrhizal fungi can greatly expand the absorbent surface area of the root system, improving plant nutrition; on the other hand, they cause a substantial increase in the buildup of metals within the cells (Liang et al., 2009; Rabęda et al., 2011).

7.5 Root systems and rhizospheric microbes

The rhizosphere and root systems play a crucial role in the effectiveness of phytoremediation. Microbial activity and the breakdown of pollutants are accelerated in the rhizosphere, a small area of soil that is impacted by root exudates and related microorganisms (Chen et al., 2019). A variety of organic substances, such as sugars, amino acids, and organic acids, are released by roots and promote the development and activity of rhizosphere microbes. These microbes are essential for converting organic contaminants into less harmful or non-toxic forms. For instance, petroleum-based pollutants can be broken down by hydrocarbon-degrading bacteria in the rhizosphere, demonstrating how phytoremediation is a cooperative plant-microbe process (Singh et al., 2024b). Pollutant uptake is greatly influenced by root structure. Pollutants in deeper soil layers and groundwater can be reached by plants with deep and wide root systems, such as *Populus* and *Salix* species. Fine, fibrous roots promote deeper interactions with the rhizosphere bacteria and provide a greater surface area for pollution absorption. Both phytostabilization and rhizodegradation result in the breakdown of pollutants, aided by root exudates and microbial enzymes are improved by these structural changes (Jeelani et al., 2020).

8. Advantages of phytoremediation

Phytoremediation offers the most suitable, eco-friendly, and cost-effective solution to treat organic contaminants (Fig. 15.3). This approach has numerous advantages compared to other conventional treatment methods and provides a better alternative for the remediation of polluted sites.

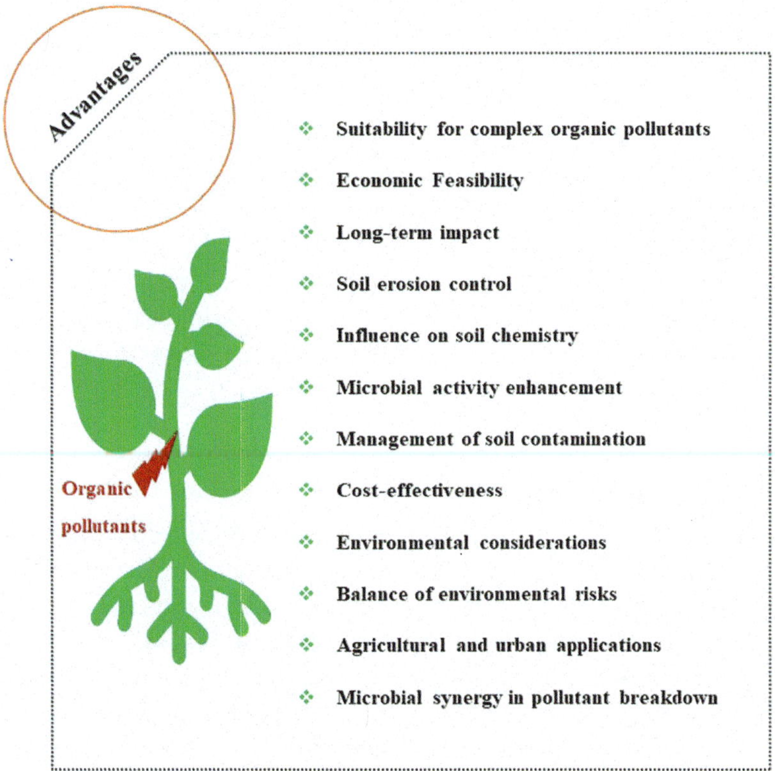

Figure 15.3 Advantages of phytoremediation.

8.1 Suitability for complex organic pollutants

Phytotransformation especially works for organic pollutants that are resistant to natural degradation, such as pesticides, herbicides, industrial solvents, and halogenated hydrocarbons. These pollutants create important environmental challenges due to their toxicity. However, the enzymatic processes inside plants can change these compounds into less toxic or more biodegradable forms, reducing their impact on ecosystems (Malik et al., 2022). Plants show that they transform chlorinated solvents into less harmful byproducts, which are more easily poorly managed by microorganisms in the surrounding soil.

8.2 Economic feasibility

As a low-cost remediation technique, phytotransformation requires fewer requirements than traditional methods. When plants are transferred to the contaminated area, they can continue to absorb and transform pollutants with minimal maintenance. This makes phytotransformation suitable for large-scale or remote sites where the cost of remediation could be very high (Riaz et al., 2022). Additionally, using native plant species adapted to local conditions can further reduce maintenance costs while ensuring the resilience and success of the remediation effort.

8.3 Long-term impact

One of the key advantages of phytotransformation is its ability for long-term site stabilization and ecological restoration. They continuously transform the pollutants into less toxic forms, plants can contribute to the natural attenuation of contaminants over time, making the slow recovery of the ecosystem (Enochs et al., 2023). This specifically has more value in areas where immediate remediation is not possible or where slow and good improvements are needed, such as in agricultural fields or industrial sites.

8.4 Soil erosion control

A primary mechanism by which phytostabilization controls the pollutant spread is by the decrease of soil erosion. Erosion is a major driver of contaminant dispersal, and it allows the pollutants to be carried away by the wind or the water, escalating the risk of contamination in the nearby ecosystems (Lee et al., 2023). Plants with deep root systems alleviate the soil, anchor it in place, and prevent it from both wind and water erosion. This is specifically important in arid and semi-arid areas, where the absence of vegetation cover can result in extreme soil erosion and intensified pollutant mobility.

8.5 Influence on soil chemistry

Plants that are involved in phytostabilization can also change the chemical properties of the soil, making conditions that additionally control the mobility of the pollutants. For instance, the organic acids released by the roots can lower the soil pH, affecting the solubility and the stability of specified organic pollutants (Li et al., 2023). Likewise, changes in the redox conditions in the rhizosphere can change the chemical speciation of the pollutants, which makes them less mobile or less bioavailable. In the contaminated soils where the redox-sensitive pollutants, including the PCBs or PAHs, are present, the capability of the plants to change the rhizosphere environment can substantially increase the efficiency of phytostabilization.

8.6 Microbial activity enhancement

The existence of plant roots also increases microbial action in the rhizosphere of the soil, making a microenvironment that plays a role in the degradation or the stabilization of the organic pollutants. While phytostabilization aims to lower the containment instead of degradation, the presence of beneficial microorganisms can further lower the pollutant bioavailability by enabling their binding to the soil particles or by altering them into less mobile forms (Manzoor et al., 2021). Rhizosphere microbes, including bacteria and fungi, interact with the root exudates and the organic matter, contributing to the stabilization of the contaminants by changing them into more stable and less dangerous byproducts.

8.7 Management of soil contamination

In areas with limited water availability, phytostabilization is usually the most suitable option to manage the contaminated soil. phytostabilization depends on the natural ability of the plants to stabilize the soils without the requirement for the additional resources (Lee et al., 2023). It is advantageous to manage large-scale contaminated sites in arid and semi-arid environments, where soil erosion and contaminated leaching are specifically prevalent.

8.8 Cost-effectiveness

Phytostabilization provides a cost-effective solution to more resource-intensive remediation approaches. The method is usually low maintenance, depending majorly on the natural growth of the vegetation to stabilize the contaminated soils. Once they're established, the plants that are used in phytostabilization need minimal input, including water or nutrients, which makes it a very sustainable long-term solution (Yadav et al., 2022). Furthermore, the use of native plant species adapted to the local soil and the climate conditions can additionally lower the maintenance costs while supporting biodiversity and restoration of the ecosystem.

8.9 Environmental considerations

When the pollutants are changed into volatile forms through this process and released into the atmosphere, this can eliminate the contamination associated with the soil and water but can pose a risk to the quality of the air (Phang et al., 2024). For example, though the volatilized forms of TCE or PCE are less toxic than their original forms, when released into the atmosphere, they can still need monitoring to confirm that their concentrations do not surpass regulatory thresholds.

8.10 Balance of environmental risks

Regardless of the potential effect on air, phytovolatilization can help to achieve a more sustainable alternative to traditional remediation methods, which are not environmentally friendly. These processes include excavation or incineration, which can also lead to atmospheric pollution. In addition, phytovolatilization volatilizes lesser quantities of pollutants when compared to the amount present in the contaminated soil or groundwater, which can result in a reduction of risk associated with the environment (Aransiola and Maddela, 2024). To mitigate the release of harmful compounds into the atmosphere, phytovolatilization can be used in combination with other phytoremediation methods, such as phytodegradation or phytoextraction.

8.11 Agricultural and urban applications

In agricultural areas, the soils contaminated by pesticides and herbicides have been remediated by phytoextraction. For instance, field trials using *Zea mays* (corn) and *Helianthus annuus* (sunflower) have demonstrated the potential for phytoextraction of atrazine, resulting in reduced soil concentrations by up to 65% over a growing season (Burken and Schnoor, 1998). In urban areas, phytoextraction has been used to remove volatile organic compounds (VOCs), such as benzene, toluene, and xylene, from contaminated soils and groundwater near industrial sites.

8.12 Microbial synergy in pollutant breakdown

The otherwise recalcitrant organic pollutants break down due to the interaction between plant roots and rhizosphere microorganisms, creating a synergistic relationship that accelerates the process. It is proven by the studies that the pollutants, such as PAHs, degrade increasingly by up to 45% due to the presence of rhizosphere microorganisms. This improvement in degradation is because the plants offer the needed nutrients for microbial proliferation, and the microbes have specific catabolic pathways for the breakdown of pollutants, which plants alone cannot degrade effectively (Wang et al., 2022). For example, the bacterium *Pseudomonas putida*, commonly found in the rhizosphere of poplar trees, can break down toluene and other volatile organic compounds (VOCs), expressively decreasing pollutant levels.

9. Challenges in phytoremediation

Photoremediation offers effective and efficient remediation of organic pollutants, especially in cases where other physical and chemical remediation methods are not practical. It's an environmentally friendly process that controls natural plant processes to reduce the pollution toxicity without the need for any chemical treatment or extensive soil excavation (Sharma et al., 2023). However, there are still some challenges that need to be dealt with to improve the efficiency of this process and its various mechanisms.

9.1 Incomplete degradation

Incomplete mineralization. Even though complete mineralization is not common for more recalcitrant pollutants, the potential for the plants to fully degrade into simpler compounds suggests a very effective remediation approach. The formation of intermediate products, which in some cases could be more toxic than their parent molecules. These harmful intermediate or end products pose a serious threat to the ecosystem and disturb the soil ecosystem, resulting in soil deterioration. The effectiveness of this process depends on several factors, including specific plant species, the nature of the pollutant, and environmental conditions (Lee et al., 2025). In a few cases, phytotransformation can lead to the formation of intermediate products that are less harmful than the original pollutant, but they still present some environmental risk. While these products can be more readily degraded by microbes, there is usually the possibility that they can gather in the soil or plant tissues, resulting in secondary contamination. Hence, it is important to monitor the area over time to ensure that phytotransformation is efficiently decreasing the overall pollutant load without generating any new risks.

9.2 Species-specific limitations

The capacity for phytotransformation varies widely among plant species. Not all plants can metabolize the same pollutants, and selecting the most suitable species is important for the success of the remediation effort (Yadav et al., 2023). For example, some plants are better at metabolizing pesticides, while others may be good at specializing in transforming industrial solvents. Selecting the right species based on the nature of the pollutant and local environmental conditions is essential for optimizing the phytotransformation process.

9.3 Environmental constraints

Phytotransformation is affected by environmental factors, including temperature, soil moisture, pH, and nutrient availability. In areas with severe climatic conditions or poor soil quality, the metabolic activity of the plants may be decreased, reducing the efficiency of the transformation procedure (Sharma et al., 2023). Moreover, pollutants with complex chemical structures or strong resistance to the metabolic breakdown may not be completely transformed by the plant enzymes, decreasing the overall effectiveness of the phytoremediation efforts.

9.4 Long-term containment

One key limitation of phytostabilization is that it does not decrease or decay the contaminants, but it immobilizes them inside the soil. This means that these pollutants persist in the soil environment, and their long-term stability should be monitored to make sure that they do not become remobilized because of environmental changes, including shifts in the soil pH, redox conditions, or erosion (Padhye et al., 2023). Regular monitoring of the soil and the groundwater is crucial to ensure that these contaminants stay immobilized and do not present a risk to human health or the ecosystem.

9.5 Potential risks of contaminated biomass

While phytostabilization aims to prevent the movement of contaminants, there is still a risk that pollutants might accumulate in the biomass of plants. In particular cases, pollutants can be taken up by the roots of plants and sequestered in the tissues of plants, which presents a threat if the biomass is consumed by the herbivores or if these plants are harvested and disposed of improperly (Khan et al., 2023). Sustainable management of plant biomass is fundamental to avoiding the spread of contaminants through the food chain or the environment.

9.6 Plant selection

The success of phytostabilization is greatly dependent on the selection of the appropriate species of plant. All plants are not capable of efficiently stabilizing organic pollutants, and the species choice must consider the factors, including the root structure, growth rate, and the tolerance to contaminants (Kafle et al., 2022). Native species are preferred because of their adaptability to the local conditions, but in specific cases, non-native or transgenic plants may suggest superior performance in stabilizing the specific pollutants. Research for the use of genetically modified plants for improved phytostabilization is ongoing, with promising results in the improvement of the immobilization of the challenging contaminants.

9.7 Pollutant bioavailability

Because of the low solubility in water and tendency to adsorb onto soil organic matter, it can be challenging to extract organic pollutants, such as polychlorinated biphenyls (PCBs), dioxins, and certain herbicides (Lee et al., 2025).

Therefore, it is obvious from several reported literature that phytoremediation has certain limitations for its successful applications, which require more scientific research to enhance its efficiency and removal capability of potential plant species to eliminate the toxic effects of organic pollutants and restore environmental health.

10. Conclusions and future perspectives

Contamination of organic pollutants has severely affected all environmental types such as soil, water, air, and residential areas and resultantly poses risks to all living organisms, including humans. The contamination of food and feed with organic contaminants has resulted in serious diseases in humans and animals, respectively. Organic pollutants are affecting soil properties and disturbing nutrient cycling and microbial activities. In this current scenario of the problem, phytoremediation offers the most suitable and economic solution to reduce the harmful effects of organic pollutants on the environment by providing different mechanisms to remove, degrade, transform, stabilize, and volatilize toxic organic compounds into less or non-toxic compounds. However, the phytoremediation of such noxious pollutants by plants is affected by specific optimal conditions that favor this process and enhance the removal efficiency of pollutants.

Moreover, the efficiency of phytoremediation of organic pollutants can be significantly increased using genetic engineering approaches to produce highly resistant plants and microbes capable of degrading these toxic pollutants. New bacterial species can be identified, and the degradation potential of already existing microbial strains can be enhanced via gene editing to promote pronounced degradation of organic pollutants during plant-microbe interactions. Further, the resistance in plants can be increased against organic pollutants, and the uptake of organic pollutants can be enhanced using specific plant species known for more biomass production and bioaccumulation.

References

Abbas, A., Al-Amer, A.M., Loui, T., Al-Marri, M.J., Nasser, M.S., Khraisheh, M., Atieh, M.A., 2016. Heavy metal removal from aqueous solution by advanced carbon nanotubes: critical review of adsorption applications. Separ. Purif. Technol. 157, 141–161.

Adebusuyi, A.T., Sojinu, S.O., Aleshinloye, A.O., 2022. The prevalence of persistent organic pollutants (POPs) in West Africa–a review. Environ. Chall. 7, 100486.

Afzal, M., Rehman, K., Shabir, G., Tahseen, R., Ijaz, A., Brix, H., 2019. Large-scale remediation of oil-contaminated water using floating treatment wetlands. npj Clean Water 2 (1), 3.

Agarwal, P., Giri, B.S., Rani, R., 2020. Unravelling the role of rhizospheric plant-microbe synergy in phytoremediation: a genomic perspective. Curr. Genom. 21 (5), 334–342.

Aguilar-Zarate, P., Wong-Paz, J.E., Buenrostro-Figueroa, J.J., Ascacio, J.A., Contreras-Esquivel, J.C., Aguilar, C.N., 2018. Ellagitannins: bioavailability, purification and biotechnological degradation. Mini Rev. Med. Chem. 18 (15), 1244–1252.

Aken, B.V., Correa, P.A., Schnoor, J.L., 2010. Phytoremediation of polychlorinated biphenyls: new trends and promises. Environ. Sci. Technol. 44 (8), 2767–2776.

Akram, M.S., Rashid, N., Basheer, S., 2021. Physiological and molecular basis of plants tolerance to linear halogenated hydrocarbons. In: Handbook of Bioremediation. Academic Press, pp. 591–602.

Al Mamun, A., Rahman, M.M., Huq, M.A., Rahman, M.M., Rana, M.R., Rahman, S.T., Khatun, M.L., Alam, M.K., 2024. Phytoremediation: a transgenic perspective in omics era. Transgenic Res. 33 (4), 175–194.

Al-Baldawi, I.A., Abdullah, S.R.S., Almansoory, A.F., Ismail, N.I., Hasan, H.A., Anuar, N., 2020. Role of *Salvinia molesta* in biodecolorization of methyl orange dye from water. Sci. Rep. 10 (1), 13980.

Alkorta, I., Hernández-Allica, J., Garbisu, C., 2004. Plants against the global epidemic of arsenic poisoning. Environ. Int. 30 (7), 949–951.

Al-Mulali, U., Ozturk, I., Lean, H.H., 2015. The influence of economic growth, urbanization, trade openness, financial development, and renewable energy on pollution in Europe. Nat. Hazards 79, 621–644.

Anerao, P., Kaware, R., kumar Khedikar, A., Kumar, M., Singh, L., 2022. Phytoremediation of persistent organic pollutants: Concept challenges and perspectives. Phytoremediation technology for the removal of heavy metals and other contaminants from soil and water 375–404.

Antoniadis, V., Shaheen, S.M., Stärk, H.J., Wennrich, R., Levizou, E., Merbach, I., Rinklebe, J., 2021. Phytoremediation potential of twelve wild plant species for toxic elements in a contaminated soil. Environ. Int. 146, 106233.

Aransiola, S. A., & Maddela, N. R. (Eds.). (2024). Phytoremediation in Food Safety: Risks and Prospects.

Arslan, S., Eyvaz, M., Gürbulak, E., Yüksel, E., 2016. A review of state-of-the-art technologies in dye-containing wastewater treatment–the textile industry case. Textile wastewater treatment 1–29.

Aryal, M., 2024. Phytoremediation strategies for mitigating environmental toxicants. Heliyon 10 (19).

Ashraf, N., Glaeser, E., Holland, A., Steinberg, B.M., 2021. Water, health and wealth: the impact of piped water outages on disease prevalence and financial transactions in Zambia. Economica 88 (351), 755–781.

Bakhshayeshan-Agdam, H., Houshani, M., Salehi-Lisar, S.Y., 2024. Arbuscular mycorrhizal fungi in plant tolerance to organic pollutants and associated food safety. In: Ahammed, G.J., Hajiboland, R. (Eds.), Arbuscular Mycorrhizal Fungi and Higher Plants. Springer, Singapore.

Bakirtas, T., Akpolat, A.G., 2018. The relationship between energy consumption, urbanization, and economic growth in new emerging-market countries. Energy 147, 110–121.

Bardos, R.P., Bone, B., Andersson-Sköld, Y., Suer, P., Track, T., Wagelmans, M., 2011. Crop-based systems for sustainable risk-based land management for economically marginal damaged land. Remed. J. 21 (4), 11–33.

Barroso, G.M., Dos Santos, E.A., Pires, F.R., Galon, L., Cabral, C.M., Dos Santos, J.B., 2023. Phytoremediation: a green and low-cost technology to remediate herbicides in the environment. Chemosphere 334, 138943.

Beard, K.H., Vogt, K.A., Vogt, D.J., Scatena, F.N., Covich, A.P., Sigurdardottir, R., Siccama, T.G., Crowl, T.A., 2005. Structural and functional responses of a subtropical forest to 10 years of hurricanes and droughts. Ecol. Monogr. 75 (3), 345–361.

Begum, M., Sarmah, B., Kandali, G.G., Kalita, S., Ojha, I., Bhagawati, R., Talukdar, L., 2021. Persistant organic pollutants in soil and its phytoremediation. Biodegradation technology of organic and inorganic pollutants 61–78.

Bharathiraja, B., Jayamuthunagai, J., Praveenkumar, R., Iyyappan, J., 2018. Phytoremediation techniques for the removal of dye in wastewater. Bioremediation: Applications for Environmental Protection and Management 243–252.

Bukowska, B., Mokra, K., Michałowicz, J., 2022. Benzo [a] pyrene—environmental occurrence, human exposure, and mechanisms of toxicity. Int. J. Mol. Sci. 23 (11), 6348.

Burken, J.G., Schnoor, J.L., 1998. Predictive relationships for uptake of organic contaminants by hybrid poplar trees. Environ. Sci. Technol. 32 (21), 3379–3385.

Cai, R., Xue, J., Tan, L., Zhao, Z., Zhang, Z., 2021. Study on physicochemical properties of high-density sludge sediment in copper mine after phytoremediation. J. Soil Sci. Plant Nutr. 21 (4), 2884–2892.

Caniani, D., Caivano, M., Mazzone, G., Masi, S., Mancini, I.M., 2021. Effect of site-specific conditions and operating parameters on the removal efficiency of petroleum-originating pollutants by using ozonation. Sci. Total Environ. 800, 149393.

Cassina, L., Tassi, E., Morelli, E., Giorgetti, L., Remorini, D., Chaney, R.L., Barbafi eri, M., 2011. Exogenous cytokinin treatments of an Ni hyper-accumulator, *Alyssum murale*, grown in a serpentine soil: implications for phytoextraction. Int. J. Phytoremediation 13, 90–101.

Chaudhary, I.J., Rathore, D., 2019. Dust pollution: its removal and effect on foliage physiology of urban trees. Sustain. Cities Soc. 51, 101696.

Chen, L., Zhou, S., Shi, Y., Wang, C., Li, B., Li, Y., Wu, S., 2018. Heavy metals in food crops, soil, and water in the Lihe River Watershed of the Taihu Region and their potential health risks when ingested. Science of the total environment 615, 141–149.

Chen, T., Liu, X., Li, X., Zhao, K., Zhang, J., Xu, J., Dahlgren, R.A., 2009. Heavy metal sources identification and sampling uncertainty analysis in a field-scale vegetable soil of Hangzhou, China. Environ. Pollut. 157 (3), 1003–1010.

Chen, Y., Palta, J.A., Wu, P., Siddique, K.H.M., 2019. Crop root systems and rhizosphere interactions. Plant Soil 439 (1–2), 1–5.

Cheng, X., Jiang, L., Liu, W., Song, X., Kumpiene, J., Luo, C., 2024. Phytoremediation of trichloroethylene in the soil/groundwater environment: progress, problems, and potential. Sci. Total Environ., 176566

Chunduri, V., Kapoor, P., Kumari, A., Kumar, A., Sharma, S., Sharma, N., Kaur, S., Garg, M., 2024. Biotechnological advancements in phytoremediation. Aquatic Contamination: Tolerance and Bioremediation 165–188.

Colares, G.S., Dell'Osbel, N., Wiesel, P.G., Oliveira, G.A., Lemos, P.H.Z., da Silva, F.P., Lutterbeck, C. A., Kist, L.T., Machado, Ê.L., 2020. Floating treatment wetlands: a review and bibliometric analysis. Sci. Total Environ. 714, 136776.

Danyal, Y., Mahmood, K., Ullah, S., Rahim, A., Raheem, G., Khan, A.H., Ullah, A., 2023. Phytoremediation of industrial effluents assisted by plant growth promoting bacteria. Environ. Sci. Pollut. Control Ser. 30 (3), 5296–5311.

de Souza, D.M., da Silva, J.D.L., Ludwig, L.D.C., Petersen, B.C., Brehm, F.A., Modolo, R.C.E., De Marchi, T.C., Figueiredo, R., Moraes, C.A., 2023. Study of the phytoremediation potential of native plant species identified in an area contaminated by volatile organic compounds: a systematic review. Int. J. Phytoremediation 25 (11), 1524–1541.

Del Buono, D., Terzano, R., Panfili, I., Bartucca, M.L., 2020. Phytoremediation and detoxification of xenobiotics in plants: herbicide-safeners as a tool to improve plant efficiency in the remediation of polluted environments. A mini review. Int. J. Phytoremediation 22 (8), 789–803.

Diarra, I., Kotra, K.K., Prasad, S., 2022. Application of phytoremediation for heavy metal contaminated sites in the South Pacific: strategies, current challenges and future prospects. Appl. Spectrosc. Rev. 57 (6), 490–512.

Domínguez-Crespo, M.A., Sánchez-Hernández, Z.E., Torres-Huerta, A.M., Negrete-Rodríguez, M.X., Conde-Barajas, E., Flores-Vela, A., 2012. Effect of the heavy metals Cu, Ni, Cd and Zn on the growth and reproduction of epigeic earthworms (*E. fetida*) during the vermistabilization of municipal sewage sludge. Water Air Soil Pollut. 223, 915–931.

Dong, W., Zhang, Y., Quan, X., 2020. Health risk assessment of heavy metals and pesticides: a case study in the main drinking water source in Dalian, China. Chemosphere 242, 125269.

Doty, S.L., James, C.A., Moore, A.L., Vajzovic, A., Singleton, G.L., Ma, C., Khan, Z., Xin, G., Kang, J. W., Park, J.Y., Meilan, R., 2007. Enhanced phytoremediation of volatile environmental pollutants with transgenic trees. Proc. Natl. Acad. Sci. 104 (43), 16816–16821.

Eller, F., Skálová, H., Caplan, J.S., Bhattarai, G.P., Burger, M.K., Cronin, J.T., Guo, W.-Y., Guo, X., Hazelton, E.L.G., Kettenring, K.M., Lambertini, C., McCormick, M.K., Meyerson, L.A., Mozdzer, T.J., Pyšek, P., Sorrell, B.K., Whigham, D.F., Brix, H., 2017. Cosmopolitan species as models for ecophysiological responses to global change: the common reed phragmites australis. Front. Plant Sci. 8, 1833.

Enochs, B., Meindl, G., Shidemantle, G., Wuerthner, V., Akerele, D., Bartholomew, A., Bulgrien, B., Davis, A., Hoyt, K., Kung, L., Molina, M., 2023. Short and long-term phytoremediation capacity of aquatic plants in Cu-polluted environments. Heliyon 9 (1).

Fan, T., Chen, X., Xu, Z., Liu, L., Shen, D., Dong, S., Zhang, Q., 2020. Uptake and translocation of triflumezopyrim in rice plants. J. Agric. Food Chem. 68 (27), 7086–7092.

Fan, T., Chen, X., Zhao, M., Wang, J., Meng, Z., Dong, S., Miao, X., Wu, Q., 2021. Uptake, translocation and subcellular distribution of chlorantraniliprole and tetrachlorantraniliprole in maize. Sci. Total Environ. 800, 149429.

Farooq, M.A., Ali, S., Yasmeen, T., 2022. Heavy metal accumulation and tolerance potential of Indian mustard (Brassica juncea L.) grown in contaminated soil. Ecotoxicol. Environ. Saf. 229 (2022). Article 113080.

Gamalero, E., Lingua, G., Berta, G., Glick, B.R., 2009. Beneficial role of plant growth promoting bacteria and arbuscular mycorrhizal fungi on plant responses to heavy metal stress. Can. J. Microbiol. 55 (5), 501–514.

García, J., García-Galán, M.J., Day, J.W., Boopathy, R., White, J.R., Wallace, S., Hunter, R.G., 2020. A review of emerging organic contaminants (EOCs), antibiotic resistant bacteria (ARB), and antibiotic resistance genes (ARGs) in the environment: increasing removal with wetlands and reducing environmental impacts. Bioresour. Technol. 307, 123228.

García-Sánchez, M., Košnář, Z., Mercl, F., Aranda, E., Tlustoš, P., 2018. A comparative study to evaluate natural attenuation, mycoaugmentation, phytoremediation, and microbial-assisted phytoremediation strategies for the bioremediation of an aged PAH-polluted soil. Ecotoxicol. Environ. Saf. 147, 165–174.

Gkatzelis, G.I., Coggon, M.M., McDonald, B.C., Peischl, J., Aikin, K.C., Gilman, J.B., Trainer, M., Warneke, C., 2020. Identifying volatile chemical product tracer compounds in US cities. Environ. Sci. Technol. 55 (1), 188–199.

Gogoi, B., Sarma, H., 2023. Rhizoremediation of persistent organic pollutants (POPs) from the soil. In: Land Remediation and Management: Bioengineering Strategies. Springer Nature Singapore, Singapore, pp. 49–78.

Gonzalez, V., Diez-Ortiz, M., Simón, M., van Gestel, C.A.M., 2013. Assessing the impact of organic and inorganic amendments on the toxicity and bioavailability of a metal contaminated soil to the earthworm Eisenia andrei. Environ. Sci. Pollut. Res. 20 (11), 8162–8171.

Groffen, T., Rijnders, J., van Doorn, L., Jorissen, C., De Borger, S.M., Oude Luttikhuis, D., de Deyn, L., Covaci, A., Bervoets, L., 2021. Preliminary study on the distribution of metals and persistent organic pollutants (POPs), including perfluoroalkylated acids (PFAS), in the aquatic environment near Morogoro, Tanzania, and the potential health risks for humans. Environ. Res. 192, 110299.

Grzegórska, A., Rybarczyk, P., Rogala, A., Zabrocki, D., 2020. Phytoremediation—from environment cleaning to energy generation—current status and future perspectives. Energies 13 (11), 2905.

Guarino, C., Spada, V., Sciarrillo, R., 2017. Assessment of three approaches of bioremediation (Natural Attenuation, Landfarming and Bioaugmentation Assistive Landfarming) for a petroleum hydrocarbons contaminated soil. Chemosphere 170, 10–16.

Hartley, W., Dickinson, N.M., Riby, P., Shutes, B., 2012. Sustainable ecological restoration of brownfield sites through engineering or managed natural attenuation? A case study from Northwest England. Ecol. Eng. 40, 70–79.

Hernández, A.F., Bennekou, S.H., Hart, A., Mohimont, L., Wolterink, G., 2020. Mechanisms underlying disruptive effects of pesticides on the thyroid function. Current Opinion in Toxicology 19, 34–41.

Holkar, C.R., Jadhav, A.J., Pinjari, D.V., Mahamuni, N.M., Pandit, A.B., 2016. A critical review on textile wastewater treatments: possible approaches. J. Environ. Manag. 182, 351–366.

Hossein, M., Asha, R., Bakari, R., Islam, N.F., Jiang, G., Sarma, H., 2023. Exploring eco-friendly approaches for mitigating pharmaceutical and personal care products in aquatic ecosystems: a sustainability assessment. Chemosphere 316, 137715.

Hua, M., Zhang, S., Pan, B., Zhang, W., Lv, L., Zhang, Q., 2012. Heavy metal removal from water/wastewater by nanosized metal oxides: a review. J. Hazard Mater. 211, 317–331.

Huslina, F., Khudur, L.S., Shah, K., Surapaneni, A., Netherway, P., Ball, A.S., 2024. Mine site restoration: the phytoremediation of arsenic-contaminated soils. Environments 11 (5), 99.

Igoni, A.H., 2018. Understanding the Niger Delta Environment for Sustainable Agricultural Mechanization. Annual Lecture/Awards Ceremony of the Nigerian Society of Engineers, Port Harcourt Branch, Rivers State, Nigeria.

Iqbal, N., Nazir, N., Numan, M., Hayat, M.T., Mahmood, Q., Zeb, B.S., Ma, B., Abbas, Z., 2022. Effects of polychlorinated biphenyls on plant growth. In: Sustainable Plant Nutrition under Contaminated Environments. Springer International Publishing, Cham, pp. 187–208.

Jaskulak, M., Grobelak, A., Vandenbulcke, F., 2020. Modelling assisted phytoremediation of soils contaminated with heavy metals–main opportunities, limitations, decision making and future prospects. Chemosphere 249, 126196.

Jeelani, N., Yang, W., Zhu, H.-L., An, S., 2020. Phytoremediation for co-contaminated soils of cadmium and pyrene using Phragmites australis (common reed). Int. J. Phytoremediation 22 (13), 1385–1395.

Jia, L., Wang, W., Li, Y., Yang, L., 2010. Heavy metals in soil and crops of an intensively farmed area: a case study in Yucheng City, Shandong Province, China. Int. J. Environ. Res. Publ. Health 7 (2), 395–412.

Jing, Y.D., He, Z.L., Yang, X.E., 2007. Role of soil rhizobacteria in phytoremediation of heavy metal contaminated soils. J. Zhejiang Univ. - Sci. B 8 (3), 192–207.

Jomjun, N., Siripen, T., Maliwan, S., Jintapat, N., Prasak, T., Somporn, C., Petch, P., 2010. Phytoremediation of arsenic in submerged soil by wetland plants. Int. J. Phytoremediation 13 (1), 35–46.

Jung, D., Kim, J.A., Park, M.S., Yim, U.H., Choi, K., 2017. Human health and ecological assessment programs for Hebei Spirit oil spill accident of 2007: status, lessons, and future challenges. Chemosphere 173, 180–189.

Kadhom, M., Albayati, N., Alalwan, H., Al-Furaiji, M., 2020. Removal of dyes by agricultural waste. Sustain. Chem. Pharm. 16, 100259.

Kadlec, R., Knight, R., Vymazal, J., Brix, H., Cooper, P., Haberl, R., 2000. Constructed Wetlands for Pollution Control: Processes, Performance, Design and Operation. IWA publishing.

Kafle, A., Timilsina, A., Gautam, A., Adhikari, K., Bhattarai, A., Aryal, N., 2022. Phytoremediation: mechanisms, plant selection and enhancement by natural and synthetic agents. Environ. Adv. 8, 100203.

Karthigadevi, G., Manikandan, S., Karmegam, N., Subbaiya, R., Chozhavendhan, S., Ravindran, B., Awasthi, M.K., 2021. Chemico-nanotreatment methods for the removal of persistent organic pollutants and xenobiotics in water–A review. Bioresour. Technol. 324, 124678.

Kaur, R., Yadav, P., Thukral, A.K., Walia, A., Bhardwaj, R., 2017. Co-application of 6-ketone type brassinosteroid and metal chelator alleviates cadmium toxicity in B. juncea L. Environ. Sci. Pollut. Control Ser. 24, 685–700.

Khan, A.G., 2005. Role of soil microbes in the rhizospheres of plants growing on trace metal contaminated soils in phytoremediation. J. Trace Elem. Med. Biol. 18 (4), 355–364.

Khan, A.H.A., Kiyani, A., Santiago-Herrera, M., Ibánez, J., Yousaf, S., Iqbal, M., Martel-Martín, S., Barros, R., 2023. Sustainability of phytoremediation: post-harvest stratagems and economic opportunities for the produced metals contaminated biomass. J. Environ. Manag. 326, 116700.

Khan, A.U., Khan, A.N., Waris, A., Ilyas, M., Zamel, D., 2022. Phytoremediation of pollutants from wastewater: a concise review. Open Life Sci. 17 (1), 488–496.

Khan, I., Saeed, K., Khan, I., 2019. Nanoparticles: properties, applications and toxicities. Arab. J. Chem. 12 (7), 908–931.

Kidd, P., Barceló, J., Bernal, M.P., Navari-Izzo, F., Poschenrieder, C., Shilev, S., Clemente, R., Monterroso, C., 2009. Trace element behaviour at the root–soil interface: implications in phytoremediation. Environ. Exp. Bot. 67 (1), 243–259.

Kim, S.H., Lee, I.S., Kang, H.J., 2011. Effects of *Pinus densiflora* on soil chemical and microbial properties in Pb-contaminated forest soil. Journal of Ecology and Environment 34 (3), 315–322.

Kim, S., Lee, C.G., Kim, Y.T., Kim, K.H., Lee, J., 2020. Effect of Pt catalyst on the condensable hydrocarbon content generated via food waste pyrolysis. Chemosphere 248, 126043.

Knoell, D.L., Smith, D., Bao, S., Sapkota, M., Wyatt, T.A., Zweier, J.L., Flury, J., Borchers, M.T., Knutson, M., 2020. Imbalance in zinc homeostasis enhances lung Tissue Loss following cigarette smoke exposure. J. Trace Elem. Med. Biol. 60, 126483.

Koźmińska, A., Wiszniewska, A., Hanus-Fajerska, E., Muszyńska, E., 2018. Recent strategies of increasing metal tolerance and phytoremediation potential using genetic transformation of plants. Plant Biotechnology Reports 12, 1–14.

Kristanti, R.A., Ngu, W.J., Yuniarto, A., Hadibarata, T., 2021. Rhizofiltration for removal of inorganic and organic pollutants in groundwater: a review. Biointerface Res. Appl. Chem. 11, 12326–12347.

Kulshreshtha, K., Rai, A., Mohanty, C.S., Roy, R.K., Sharma, S.C., 2009. Particulate pollution mitigates the ability of some plant species. Int. J. Environ. Res. 3 (1), 137–142.

Kumar, M., Bolan, N., Jasemizad, T., Padhye, L.P., Sridharan, S., Singh, L., Rinklebe, J., 2022. Mobilization of contaminants: potential for soil remediation and unintended consequences. Sci. Total Environ. 839, 156373.

Kumari, P., Rastogi, A., Shukla, A., Srivastava, S., Yadav, S., 2018. Prospects of genetic engineering utilizing potential genes for regulating arsenic accumulation in plants. Chemosphere 211, 397–406.

Kumari, S., Jamwal, R., Mishra, N., Singh, D.K., 2020. Recent developments in environmental mercury bioremediation and its toxicity: a review. Environ. Nanotechnol. Monit. Manag. 13, 100283.

Kurade, M.B., Ha, Y.H., Xiong, J.Q., Govindwar, S.P., Jang, M., Jeon, B.H., 2021. Phytoremediation as a green biotechnology tool for emerging environmental pollution: a step forward towards sustainable rehabilitation of the environment. Chem. Eng. J. 415, 129040.

Kushwaha, A., Rani, R., Kumar, S., Gautam, A., 2015. Heavy metal detoxification and tolerance mechanisms in plants: implications for phytoremediation. Environ. Rev. 24 (1), 39–51.

Lee, B.X.Y., Hadibarata, T., Yuniarto, A., 2020. Phytoremediation mechanisms in air pollution control: a review. Water, Air, Soil Pollut. 231 (8), 437.

Lee, S.H., Park, H., Kim, J.G., 2023. Current status of and challenges for phytoremediation as a sustainable environmental management plan for abandoned mine areas in Korea. Sustainability 15 (3), 2761.

Lee, Y.Y., Cho, K.S., Yun, J., 2025. Phytoremediaton strategies for Co-contaminated soils: Overcoming challenges, enhancing efficiency, and exploring future advancements and innovations. Processes 13 (1), 132.

Li, H., Gu, X., Song, J., Hui, K., Chen, G., Tan, W., Wang, H., Jiang, Y., Yuan, Y., 2024. Effects of soil-groundwater environmental factors on BTEX transport and transformation: a review. J. Environ. Chem. Eng., 113697

Li, J., Deng, S., Li, G., Lu, Z., Song, H., Gao, J., Xu, K., 2022. VOCs characteristics and their ozone and SOA formation potentials in autumn and winter at Weinan, China. Environ. Res. 203, 111821.

Li, M., He, W., Han, Z., Zhou, M., Chen, X., Li, Y., 2021. Mechanism analysis of the phytotoxicity and phytodegradation of PCBs based on the 2D-QASR model and sensitivity analysis method. J. Environ. Chem. Eng. 9 (5), 106241.

Li, N., Jiang, L., Li, X., Su, Y., 2023. Enhancing phytoremediation of arsenic-contaminated soil by agronomic practices (drip irrigation and intercropping): influence of soil organic matter. Sci. Total Environ. 891, 164463.

Li, P., Wu, J., 2019. Drinking water quality and public health. Expo. Health 11 (2), 73–79.

Liang, H.M., Lin, T.H., Chiou, J.M., Yeh, K.C., 2009. Model evaluation of the phytoextraction potential of heavy metal hyperaccumulators and non-hyperaccumulators. Environ. Pollut. 157 (6), 1945–1952.

Lone, M.I., He, Z.L., Stoffella, P.J., Yang, X.E., 2008. Phytoremediation of heavy metal polluted soils and water: progresses and perspectives. J. Zhejiang Univ. - Sci. B 9 (3), 210–220.

Malik, B., Pirzadah, T.B., Hakeem, K.R., 2022. Phytoremediation of persistent organic pollutants (POPs). In: Phytoremediation. Academic Press, pp. 415–436.

Manzoor, M., Gul, I., Manzoor, A., Kallerhoff, J., Arshad, M., 2021. Optimization of integrated phytoremediation system (IPS) for enhanced lead removal and restoration of soil microbial activities. Chemosphere 277, 130243.

Meagher, R.B., 2000. Phytoremediation of toxic elemental and organic pollutants. Curr. Opin. Plant Biol. 3 (2), 153–162.

Megharaj, M., Naidu, R., 2017. Soil and brownfield bioremediation. Microb. Biotechnol. 10 (5), 1244–1249.

Megson, D., Reiner, E.J., Jobst, K.J., Dorman, F.L., Robson, M., Focant, J.F., 2016. A review of the determination of persistent organic pollutants for environmental forensics investigations. Anal. Chim. Acta 941, 10–25.

Mudhoo, A., Thayalan, G., Muthoora, N.J., Muthoora, M.N., Oozeer, B.Z., Rago, Y.P., Ramphul, M.P., Valaydon, A.K., Kumar, S., 2013. Dioxins and furans: sources, impacts and remediation. Pollutant diseases, remediation and recycling 479–541.

Nawaz, H., Anwar-ul-Haq, M., Akhtar, J., Arfan, M., 2021. Cadmium, chromium, nickel and nitrate accumulation in wheat (Triticum aestivum L.) using wastewater irrigation and health risks assessment. Ecotoxicol. Environ. Saf. 208, 111685.

Nedjimi, B., 2021. Phytoremediation: a sustainable environmental technology for heavy metals decontamination. SN Appl. Sci. 3 (3), 286.

Newman, L.A., Wang, X., Muiznieks, I.A., Ekuan, G., Ruszaj, M., Cortellucci, R., Domroes, D., Karscig, G., Newman, T., Crampton, R.S., Hashmonay, R.A., 1999. Remediation of trichloroethylene in an artificial aquifer with trees: a controlled field study. Environ. Sci. Technol. 33 (13), 2257–2265.

Nie, J., Wang, Q.M., Han, L.J., Li, J.S., 2024. Synergistic remediation strategies for soil contaminated with compound heavy metals and organic pollutants. J. Environ. Chem. Eng., 113145

Obinna, I.B., Ebere, E.C., 2019. Phytoremediation of polluted waterbodies with aquatic plants: recent progress on heavy metal and organic pollutants. Anal Methods Environ Chem J 2, 66–104.

Ogunwole, G.A., Saliu, J.K., Osuala, F.I., Odunjo, F.O., 2021. Chronic levels of ibuprofen induce haematoxic and histopathology damage in the gills, liver, and kidney of the African sharptooth catfish (*Clarias gariepinus*). Environ. Sci. Pollut. Control Ser. 28, 25603–25613.

Ojimelukwe, A.E., Nafagha-Lawal, M.O., Lelei, K.E., Uche, A.O., Kika, E.P., Igbiri, S., Babatunde, B.B., Sikoki, F.D., 2021. Petroleum hydrocarbon pollution in the Niger Delta: human health risk assessment of BTEX in biota. Toxicology and Environmental Health Sciences 13, 65–72.

Padhye, L.P., Srivastava, P., Jasemizad, T., Bolan, S., Hou, D., Shaheen, S.M., Rinklebe, J., O'Connor, D., Lamb, D., Wang, H., Siddique, K.H., 2023. Contaminant containment for sustainable remediation of persistent contaminants in soil and groundwater. J. Hazard Mater. 455, 131575.

Pandey, S.K., Bhattacharya, T., Chakraborty, S., 2016. Metal phytoremediation potential of naturally growing plants on fly ash dumpsite of Patratu thermal power station, Jharkhand, India. Int. J. Phytoremediation 18 (1), 87–93.

Park, J.K., Oh, K., 2023. Advancements in phytoremediation research for soil and water resources: Harnessing plant power for environmental cleanup. Sustainability 15 (18), 13901.

Pathak, A., Gupta, M.K., Rabani, M.S., Tripathi, S., Pandey, S., Gupta, C., Shrivastav, M., 2024. Enzymatic approach for phytoremediation. Aquatic Contamination: Tolerance and Bioremediation 123–130.

Patil, P.M., Matkar, A.R., Patil, V.B., Gurav, R., Dhanavade, M.J., 2023. Removal of heavy metals from industrial wastewater using bioremediation approach. Modern Approaches in Waste Bioremediation: Environ. Microbiol. 377–407.

Perera, F., 2018. Pollution from fossil-fuel combustion is the leading environmental threat to global pediatric health and equity: Solutions exist. Int. J. Environ. Res. Publ. Health 15 (1), 16.

Permana, B.H., Thiravetyan, P., Treesubsuntorn, C., 2022. Effect of airflow pattern and distance on removal of particulate matters and volatile organic compounds from cigarette smoke using Sansevieria trifasciata botanical biofilter. Chemosphere 295, 133919.

Pettit, T., Irga, P.J., Torpy, F.R., 2019. The in-situ pilot-scale phytoremediation of airborne VOCs and particulate matter with an active green wall. Air Qual. Atmos. Health 12, 33–44.

Peuke, A.D., Rennenberg, H., 2005. Phytoremediation: molecular biology, requirements for application, environmental protection, public attention and feasibility. EMBO Rep. 6 (6), 497–501.

Phang, L.Y., Mingyuan, L., Mohammadi, M., Tee, C.S., Yuswan, M.H., Cheng, W.H., Lai, K.S., 2024. Phytoremediation as a viable ecological and socioeconomic management strategy. Environ. Sci. Pollut. Control Ser. 31 (38), 50126–50141.

Pi, Y., Li, X., Xia, Q., Wu, J., Li, Y., Xiao, J., Li, Z., 2018. Adsorptive and photocatalytic removal of Persistent Organic Pollutants (POPs) in water by metal-organic frameworks (MOFs). Chem. Eng. J. 337, 351–371.

Pinto, A., Azenha, M., Fidalgo, F., Teixeira, J., 2023. 2, 4-dichlorophenoxyacetic acid detoxification occurs primarily in tomato leaves by the glutathione S-transferase phi members 4 and 5. Sci. Hortic. 321, 112214.

Qin, G., Niu, Z., Yu, J., Li, Z., Ma, J., Xiang, P., 2021. Soil heavy metal pollution and food safety in China: effects, sources and removing technology. Chemosphere 267, 129205.

Qin, G., Wu, J., Zheng, X., Zhou, R., Wei, Z., 2019. Phosphorus forms and associated properties along an urban–rural gradient in Southern China. Water 11 (12), 2504.

Rabęda, I., Woźny, A., Krzesłowska, M., 2011. Bacteria and mycorrhizal fungi increase plant efficiency in phytoremediation of trace metals. Kosmos 60 (3–4), 423–433.

Rabideau, A.J., Swensen, K., Chen, Z., Santiago-Collazo, F.L., Bronner, C., Lowry, C., Gnat, R., Lockwood, T., 2022. Hydrologic evaluation of a poplar phytoextraction system. Int. J. Phytoremediation 24 (2), 145–155.

Rachmadiarti, F., Purnomo, T., Azizah, D.N., Fascavitri, A., 2019. Syzigium oleina and Wedelia trilobata for phytoremediation of lead pollution in the atmosphere. Nat. Environ. Pollut. Technol. 18 (1).

Rajput, V.D., Minkina, T., Sushkova, S., Tsitsuashvili, V., Mandzhieva, S., Gorovtsov, A., Nevidomskyaya, D., Gromakova, N., 2018. Effect of nanoparticles on crops and soil microbial communities. J. Soils Sediments 18, 2179–2187.

Ramborger, B.P., Paz, M.E.G., Kieling, K.M.C., Carriço, M.R.S., de Paula Gollino, G., Costa, M.T., Ribeiro, V.B., Folmer, V., Denardin, E.L.G., de Jesus Soares, J., Roehrs, R., 2021. Toxicological parameters of aqueous residue after using *Plectranthus neochilus* for 2, 4-D phytoremediation. Chemosphere 270, 128638.

Rascio, N., Navari-Izzo, F., 2011. Heavy metal hyperaccumulating plants: how and why do they do it? And what makes them so interesting? Plant Sci. 180 (2), 169–181.

Reeves, R.D., Baker, A.J., Jaffré, T., Erskine, P.D., Echevarria, G., van Der Ent, A., 2018. A global database for plants that hyperaccumulate metal and metalloid trace elements. New Phytol. 218 (2), 407–411.

Rehman, K., Imran, A., Amin, I., Afzal, M., 2019. Enhancement of oil field-produced wastewater remediation by bacterially augmented floating treatment wetlands. Chemosphere 217, 576–583.

Reichenauer, T.G., Germida, J.J., 2008. Phytoremediation of organic contaminants in soil and groundwater. ChemSusChem: Chemistry & Sustainability Energy & Materials 1 (8-9), 708–717.

Riaz, U., Athar, T., Mustafa, U., Iqbal, R., 2022. Economic feasibility of phytoremediation. In: Phytoremediation. Academic Press, pp. 481–502.

Roy, A.S., Yenn, R., Singh, A.K., Boruah, H.P.D., Saikia, N., Deka, M., 2013. Bioremediation of crude oil contaminated tea plantation soil using two *Pseudomonas aeruginosa* strains AS 03 and NA 108. Afr. J. Biotechnol. 12 (19).

Sahoo, R., Sow, S., Ranjan, S., Dharminder, Kumar, R., Roy, D.K., Kumar, S., Kumar, A., Srivastava, R. K., Prasad, R., Padhan, S.R., 2024. Unveiling the potential of plant growth promoting rhizobacteria (PGPR) in phytoremediation of heavy metal. Discov. Appl. Sci. 6 (6), 324.

Salamanca, M., Chandía, C., Hernández, A., 2016. Impact of forest fires on the concentrations of polychlorinated dibenzo-p-dioxin and dibenzofurans in coastal waters of central Chile. Science of the total Environment 573, 1397–1405.

Saramak, A., 2019. Comparative analysis of indoor and outdoor concentration of PM10 particulate matter on example of Cracow City Center. Int. J. Environ. Sci. Technol. 16 (11), 6609–6616.

Sarma, H., 2011. Metal hyperaccumulation in plants: a review focusing on phytoremediation technology. J. Environ. Sci. Technol. 4 (2), 118–138.

Sarwar, N., Imran, M., Shaheen, M.R., Ishaque, W., Kamran, M.A., Matloob, A., Rehim, A., Hussain, S., 2017. Phytoremediation strategies for soils contaminated with heavy metals: modifications and future perspectives. Chemosphere 171, 710–721.

Saxena, P., Hiwrale, I., Das, S., Shukla, V., Tyagi, L., Pal, S., Dafale, N., Dhodapkar, R., 2021. Profiling of emerging contaminants and antibiotic resistance in sewage treatment plants: an Indian perspective. J. Hazard Mater. 408, 124877.

Schulze, E.D., Beck, E., Müller-Hohenstein, K., 2005. Plant Ecology. Springer Science & Business Media.

Sengul, A.B., Asmatulu, E., 2020. Toxicity of metal and metal oxide nanoparticles: a review. Environ. Chem. Lett. 18 (5), 1659–1683.

Seth, C.S., Remans, T., Keunen, E., Jozefczak, M., Gielen, H., Opdenakker, K., Weyens, N., Vangronsveld, J., Cuypers, A., 2012. Phytoextraction of toxic metals: a central role for glutathione. Plant Cell Environ. 35 (2), 334–346.

Setsungnern, A., Treesubsuntorn, C., Thiravetyan, P., 2017. The influence of different light quality and benzene on gene expression and benzene degradation of *Chlorophytum comosum*. Plant Physiol. Biochem. 120, 95–102.

Shabbir, Z., Sardar, A., Shabbir, A., Abbas, G., Shamshad, S., Khalid, S., Shahid, M., 2020. Copper uptake, essentiality, toxicity, detoxification and risk assessment in soil-plant environment. Chemosphere 259, 127436.

Shahid, M.K., Kashif, A., Fuwad, A., Choi, Y., 2021. Current advances in treatment technologies for removal of emerging contaminants from water–A critical review. Coord. Chem. Rev. 442, 213993.

Sharma, P., 2021. Efficiency of bacteria and bacterial assisted phytoremediation of heavy metals: an update. Bioresour. Technol. 328, 124835.

Sharma, P., Rai, S., Gautam, K., Sharma, S., 2023. Phytoremediation strategies of plants: challenges and opportunities. Plants and their interaction to environmental pollution 211–229.

Shrivastava, M., Khandelwal, A., Srivastava, S., 2019. Heavy metal hyperaccumulator plants: the resource to understand the extreme adaptations of plants towards heavy metals. In: Plant-metal interactions. Springer International Publishing, Cham, pp. 79–97.

Singh, P.K., Yadav, J.S., Kumar, I., Kumar, U., Sharma, R.K., 2024b. Screening of mustard cultivars for phytoremediation of heavy metals contamination in wastewater irrigated soil systems. Environ. Monit. Assess. 196 (3), 321.

Singh, S., Pant, A., Dutta, K., Rani, R., Vithanage, M., Daverey, A., 2024a. Phytoremediation of pharmaceuticals and personal care products using the constructed wetland. Environ. Chem. Ecotoxicol. 6, 104.

Singh, T., Singh, D.K., 2017. Phytoremediation of organochlorine pesticides: Concept, method, and recent developments. Int. J. Phytoremediation 19 (9), 834–843.

Skuza, L., Szućko-Kociuba, I., Filip, E., Bożek, I., 2022. Natural molecular mechanisms of plant hyperaccumulation and hypertolerance towards heavy metals. Int. J. Mol. Sci. 23 (16), 9335.

Song, B., Zeng, G., Gong, J., Liang, J., Xu, P., Liu, Z., Zhang, Y., Zhang, C., Cheng, M., Liu, Y., Ye, S., 2017. Evaluation methods for assessing effectiveness of in situ remediation of soil and sediment contaminated with organic pollutants and heavy metals. Environ. Int. 105, 43–55.

Sriprapat, W., Boraphech, P., Thiravetyan, P., 2014. Factors affecting xylene-contaminated air removal by the ornamental plant *Zamioculcas zamiifolia*. Environ. Sci. Pollut. Control Ser. 21, 2603–2610.

Tang, K.H.D., 2023. Phytoremediation of microplastics: a perspective on its practicality. Industrial and Domestic Waste Management 3 (2), 90–102.

Tarla, D.N., Erickson, L.E., Hettiarachchi, G.M., Amadi, S.I., Galkaduwa, M., Davis, L.C., Pidlisnyuk, V., 2020. Phytoremediation and bioremediation of pesticide-contaminated soil. Applied Sciences 10 (4), 1217.

Thornton, I., Farago, M.E., Thums, C.R., Parrish, R.R., McGill, R.A., Breward, N., Watt, J., 2008. Urban geochemistry: research strategies to assist risk assessment and remediation of brownfield sites in urban areas. Environ. Geochem. Health 30, 565–576.

Tran, H.T., Hoang, H.G., Chacha, W.E., Mukherjee, S., Duong, T.V.H., Nguyen, N.S.H., Nguyen, K. N., Naidu, R., 2024. A review of advanced bioremediation technologies for dioxin-contaminated soil treatment: current and Future outlook. Chemosphere, 143400.

Tran, V.V., Park, D., Lee, Y.C., 2020. Indoor air pollution related human diseases, and recent trends in the control and improvement of indoor air quality. Int. J. Environ. Res. Publ. Health 17 (8), 2927.

Treesubsuntorn, C., Thiravetyan, P., 2012. Removal of benzene from indoor air by Dracaena sanderiana: effect of wax and stomata. Atmos. Environ. 57, 317–321.

Ullah, S., Naeem, A., Calkaite, I., Hosney, A., Depar, N., Barcauskaite, K., 2023. Zinc (Zn) mitigates copper (cu) toxicity and retrieves yield and quality of lettuce irrigated with cu and Zn-contaminated simulated wastewater. Environ. Sci. Pollut. Control Ser. 30 (19), 54800–54812.

van Der Ent, A., Mak, R., De Jonge, M.D., Harris, H.H., 2018. Simultaneous hyperaccumulation of nickel and cobalt in the tree Glochidion cf. sericeum (Phyllanthaceae): elemental distribution and chemical speciation. Sci. Rep. 8 (1), 9683.

Vithanage, M., Dabrowska, B.B., Mukherjee, A.B., Sandhi, A., Bhattacharya, P., 2012. Arsenic uptake by plants and possible phytoremediation applications: a brief overview. Environ. Chem. Lett. 10, 217–224.

Vymazal, J., 2005. Horizontal sub-surface flow and hybrid constructed wetlands systems for wastewater treatment. Ecol. Eng. 25 (5), 478–490.

Vymazal, J., 2007. Removal of nutrients in various types of constructed wetlands. Science of the total environment 380 (1–3), 48–65.

Vymazal, J., 2010. Constructed wetlands for wastewater treatment. Water 2 (3), 530–549.

Wang, A., Fu, W., Feng, Y., Liu, Z., Song, D., 2022. Synergetic effects of microbial-phytoremediation reshape microbial communities and improve degradation of petroleum contaminants. J. Hazard Mater. 429, 128396.

Wang, C., Gu, L., Ge, S., Liu, X., Zhang, X., Chen, X., 2019. Remediation potential of immobilized bacterial consortium with biochar as carrier in pyrene-Cr (VI) co-contaminated soil. Environ. Technol. 40 (18), 2345–2353.

Wang, J., Chen, X., Yan, W., Ning, C., Gsell, T., 2020. Both artificial root exudates and natural *Koelreuteria paniculata* exudates modify bacterial community structure and enhance phenanthrene biodegradation in contaminated soils. Chemosphere 263 (2020), 128041.

Weber, R., Watson, A., Forter, M., Oliaei, F., 2011. Persistent organic pollutants and landfills-a review of past experiences and future challenges. Waste Manag. Res. 29 (1), 107–121.

Wei, C., Wang, M., 2020. Spatial distribution of greenhouse gases (CO 2 and CH 4) on expressways in the megacity Shanghai, China. Environ. Sci. Pollut. Control Ser. 27, 31143–31152.

Wiatrowska, K., Komisarek, J., Olejnik, J., 2021. Variations in organic carbon content and dehydrogenases activity in post-agriculture forest soils: a case study in south-western Pomerania. Forests 12 (4), 459.

Xiong, Z., Gao, Y., Li, X., Ding, L., Luo, G., Yu, D., Yao, H., 2020. A novel CO2-water leaching method for AAEM removal from coal: Suppression of PM formation and release during Zhundong coal combustion. Fuel 271, 117689.

Xu, J., Zuo, R., Wu, G., Liu, J., Liu, J., Huang, C., Wang, Z., 2024. Global distribution, drivers, and potential hazards of microplastics in groundwater: a review. Sci. Total Environ., 176194

Xu, N., Wei, Z., Yang, X., Shen, Q., 2017. Biodegradation of polycyclic aromatic hydrocarbons by duckweed-associated bacteria. Chemosphere 168 (2017), 345–353.

Yadav, K.K., Gupta, N., Kumar, A., Reece, L.M., Singh, N., Rezania, S., Khan, S.A., 2018. Mechanistic understanding and holistic approach of phytoremediation: a review on application and future prospects. Ecol. Eng. 120, 274–298.

Yadav, R., Singh, G., Santal, A.R., Singh, N.P., 2023. Omics approaches in effective selection and generation of potential plants for phytoremediation of heavy metal from contaminated resources. J. Environ. Manag. 336, 117730.

Yadav, R., Singh, S., Kumar, A., Singh, A.N., 2022. Phytoremediation: a wonderful cost-effective tool. In: Cost Effective Technologies for Solid Waste and Wastewater Treatment. Elsevier, pp. 179–208.

Yan, A., Wang, Y., Tan, S.N., Mohd Yusof, M.L., Ghosh, S., Chen, Z., 2020. Phytoremediation: a promising approach for revegetation of heavy metal-polluted land. Front. Plant Sci. 11 (2020). Article 513099.

Yang, Z., Yang, F., Liu, J.L., Wu, H.T., Yang, H., Shi, Y., Liu, J., Zhang, Y.F., Luo, Y.R., Chen, K.M., 2022. Heavy metal transporters: functional mechanisms, regulation, and application in phytoremediation. Sci. Total Environ. 809, 151099.

Yaqoob, S., Zahoor, I., Kazerooni, E.A., 2024. Phytostabilization of contaminated soils. Int. J. Chem. Biochem. Sci.(IJCBS) 25, 136–153.

Yin, F., Li, J., Wang, Y., Yang, Z., 2024. Biodegradable chelating agents for enhancing phytoremediation: mechanisms, market feasibility, and future studies. Ecotoxicol. Environ. Saf. 272 (2024). Article 116113.

Zheng, J., Wang, H., Li, Z., Tang, S., Chen, Z., 2008. Using elevated carbon dioxide to enhance copper accumulation in Pteridium revolutum, a copper-tolerant plant, under experimental conditions. Int. J. Phytoremediation 10 (2), 161–172.

Zhou, K.N., Zhang, Y.Y., Wu, J.S., Dou, C.Y., Ye, Z.H., Ye, Z.Q., Fu, W.J., 2021. Integrated Fertilization Regimes Boost Heavy Metals Accumulation and Biomass of Sedum Alfredii Hance.

Zhu, Y., Wang, Y., He, X., Li, B., Du, S., October 2023. Plant growth-promoting rhizobacteria: a good companion for heavy metal phytoremediation. Chemosphere 338, 139475.

Zhu, A., Liu, J., Qiao, S., Zhang, H., 2020. Distribution and assessment of heavy metals in surface sediments from the Bohai Sea of China. Mar. Pollut. Bull. 153, 110901.

CHAPTER 16

Biochar and phytotoxicity

Md Shafiqul Islam[1], Md Arifur Rahaman[2], Zulqarnain Haider Khan[3] and Belayneh Azene[1]

[1]Yunnan Key Laboratory of Forest Ecosystem Stability and Global Change, Xishuangbanna Tropical Botanical Garden, Chinese Academy of Sciences, Mengla, China; [2]International Maize and Wheat Improvement Center (CIMMYT), Dhaka, Bangladesh; [3]State Key Laboratory for Ecological Security of Regions and Cities, Ningbo Observation and Research Station, Institute of Urban Environment, Chinese Academy of Sciences, Xiamen, China

1. Introduction

Biochar is a carbon-rich material that results from pyrolysis under oxygen-limited conditions of biomass, and is heralded as an agricultural and environmental resource due to its unique physicochemical properties (Bolan et al., 2022; Gunarathnea et al., 2022). A highly porous structure positively impacts soil quality by increasing its cation exchange capacity, water retention, and microbial environment (Ding et al., 2016; Huang et al., 2023) and is, thus, much more suited to restoring degraded soils. Adsorption by way of their large surface area and functional groups, biochar retains and slowly releases these nutrients into the soil for plant utilization (Cai et al., 2022; Gwenzi et al., 2015). However, depending on types of feedstock, production downtimes, and soil properties, their performance varies greatly. Yet, a little phytotoxic effect could occur with initial applications either via nitrogen immobilization associated with high C:N ratios (Gwenzi et al., 2015) or leaching of inhibitory compounds (Prendergast-Miller et al., 2014); hence a coordinated manner of application is necessary. Simultaneously, Biochar may influence the soil microbiology by stimulating populations of beneficial microbes, which could be involved in nutrient cycling (Srivastava et al., 2023; Yang et al., 2023). However, the microbial interactions will depend largely on the properties of biochar and the nutrients it provides (Ahmad et al., 2020). Considering such intricate biophysical and biochemical dynamics, there is a continuous requirement for refining biochar formulations and techniques to apply them such that they will benefit agricultural practices whilst minimizing the possible adverse effects.

Phytotoxicity described as undesired effects on plant growth and development, including inhibited germination, stunted growth, physiological changes, forms one main concern while handling biochar (Corrêa et al., 2020; Gale et al., 2016; Intani et al., 2019). Four main factors determining the phytotoxicity of biochar are the source of feedstock, pyrolysis variables, pH properties, and chemical composition. Feedstock choice basically sets the chemical profile for biochar, which has been found that the biochar from animal sources (say poultry litter) offers lesser phytotoxic effects when compared to biochar from plant sources owing to the entirely different distribution of

Phytotoxicity of Organic Pollutants
ISBN 978-0-443-33110-7
https://doi.org/10.1016/B978-0-443-33110-7.00014-6

nutrients and contaminants (Carril et al., 2023; Rombolà et al., 2015). Pyrolysis temperature is an important parameter affecting stability with higher temperatures, particularly above 500°C, producing biochars with lower amounts of volatile organic compounds (VOCs) and therefore lesser opportunity of these being phytotoxic compared with those produced at lower temperatures (Arán et al., 2023; Gulyás et al., 2022). Particularly through salt solubilization and ionic imbalance, the inherent alkalinity of most biochars could develop phytotoxic effects if extremes of pH are encountered (Rogovska et al., 2012; Uchimiya and Bannon, 2013). Fresh biochars may have residual organic contaminants such as polycyclic aromatic hydrocarbons (PAHs) and VOCs, which leach out and inhibit root growth and germination (Qian et al., 2016; Thomas et al., 2024). Multi-tiered interactions warrant a thorough check of biochar properties pre-application to weigh the agricultural benefits with phytotoxic potentials.

A rather intriguing duality characterizes the relationship between biochar and phytotoxicity. In some instances, the very material that improves soil fertility inhibits plant growth. This paradoxical relationship stems from the variable chemical compositions of biochar, with fresh biochars including volatile organic compounds and soluble salts causing species-specific phytotoxicity effects, mainly by disturbing plant cell functions under high pH or electric conductivity (Gale et al., 2016; Walter and Rao, 2015). This is a behavior that depends largely on the feedstock origin-probably because of their specific nutrient profiles and contaminant loads, biochars from poultry manure show less phytotoxicity in contrast to woody biomass variants (Burachevskaya et al., 2021; Park et al., 2011). Therefore, pyrolysis temperature must be taken into account as well, and it holds that the higher the temperature at which biochar is produced (>500°C), the more stable the biochar would be, thus having a lower content of phytotoxic compounds as opposed to the biochars produced at low temperatures (Dróżdż et al., 2023; Gale et al., 2016). Application rates follow a non-monotonic response curve, where moderate doses improve soil quality while excessive applications may trigger salt or metal toxicity (Jaiswal et al., 2015). The paradox is that biochar can also help minimize certain phytotoxic risks due to heavy metal immobilization in contaminated soils (Li et al., 2023b; Phillips et al., 2016), emphasizing the demanding requirements to characterize biochar properties aptly and to develop application strategies accurate to derive maximum agricultural use of biochar while minimizing the adverse effects on plant systems.

This chapter treats the double-edged nature of biochar, the positive soil amendment on the one hand and a potential phytotoxin on the other-with an intention of presenting a fair perspective of its agricultural applications. Starting with how biochar is produced and its basic properties, the discussion shifts to its ability to stimulate soil fertility, water retention, and microbial flask while implicating concerns about phytotoxicity with plant growth. Next, phytotoxicity is analyzed, listing key factors that influence the matter: the nature of feedstocks, pyrolysis conditions, and chemical characteristics that can either promote or inhibit plant growth. Then follows the exploration of the complex interplay between biochar and plants, scrutinizing application methods, soil interactions, and species-specific responses that mediate phytotoxic effects. Through the synthesis of up-to-date research, the aim of this work is to provide practical guidelines for optimizing biochar usage in agriculture so that practitioners can derive maximum phytological benefits while minimizing risks to crop health and productivity through decision-making with regard to the choice of material, processing methods, or the application techniques used in the field.

2. Source, preparation and physicochemical properties of biochar

2.1 Raw materials of biochar

Biochar, a carbon-rich material produced through biomass pyrolysis, derives its properties from feedstock selection and production conditions, with raw materials determining critical characteristics like nutrient content, pH, porosity, and stability that influence agricultural applications (Prakongkep et al., 2015; Venkatesh et al., 2022). Agricultural wastes such as rice straw, corn stalks, and sugarcane leaves are widely used due to their abundance, yielding biochar with high porosity and water retention capacity that improves soil structure (Sangsuwan et al., 2023), while high-lignin feedstocks enhance biochar's recalcitrance for long-term carbon sequestration (Arias et al., 2023). Forestry residues like wood chips and bark produce biochar with high carbon stability but potentially elevated ash content (Kalus et al., 2019), with pyrolysis temperature critically affecting properties - lower temperatures (300–500°C) preserve functional groups for nutrient exchange while higher temperatures (>600°C) increase surface area and reduce volatiles (Abdelaal et al., 2021; Gunes et al., 2015). Animal manure (e.g., poultry/cattle) provides nutrient-rich biochar containing nitrogen and phosphorus (Chen and Wang, 2023) but requires careful processing to address potential salt and contaminant issues (Chen and Wang, 2023). The resulting physicochemical properties determine biochar's soil amendment efficacy, with agricultural waste-derived biochars typically exhibiting alkaline pH (8.7-11.0) for acidic soil remediation (Prakongkep et al., 2015), porous structures that improve aeration and water retention, and functional groups (e.g., carboxyl, hydroxyl) that enhance nutrient adsorption (Kalus et al., 2019; Yang et al., 2022).

2.2 Preparation methods of biochar

Biochar is produced through distinct thermochemical processes including pyrolysis, gasification, and hydrothermal carbonization (HTC), each imparting unique properties to the final product. Pyrolysis, the most common method conducted at 300–700°C under oxygen-limited conditions, includes slow pyrolysis which yields more stable biochar (~35% yield) with enhanced carbon retention and cation exchange capacity (Brown et al., 2011; Meyer et al., 2011), and fast pyrolysis that primarily produces bio-oil with lower biochar yields (10%–20%) containing higher volatile content (Gao et al., 2014; Ghorbani et al., 2022), where higher temperatures generally increase pH, surface area and porosity while reducing hydrophilicity (Li et al., 2023a). Gasification, operating at 700–1000°C, generates syngas with biochar as a byproduct that has lower yields but remains valuable for carbon sequestration and soil applications (Khan et al., 2025; Yin et al., 2019), with adjustable nutrient content and stability through process modifications (Yin et al., 2019). HTC processes wet biomass at 180–250°C under pressure without requiring pre-drying (Enaime et al., 2020; Li et al., 2025), producing biochar with higher oxygen content but lower carbon stability compared to pyrolysis-derived biochar (Nagaraju et al., 2023).

2.3 Physical and chemical properties of biochar

The physical and chemical properties of biochar, including pH, pore structure, specific surface area (SSA), and functional groups, fundamentally determine its effectiveness as a soil amendment and carbon sequestration agent. Biochar typically exhibits an alkaline pH (6.0–11.0), influenced by feedstock and pyrolysis conditions, with wood-derived biochars often having higher pH values that benefit acidic soil remediation (Kloss et al., 2012; Phillips et al., 2016; Thomas et al., 2024). Its hierarchical pore structure, comprising micropores (<2 nm) and macropores, enhances water retention and nutrient adsorption, particularly at higher pyrolysis temperatures (>500°C) that promote porosity (Brewer et al., 2017; Uchimiya et al., 2011). SSA, critical for pollutant and nutrient adsorption, increases with elevated pyrolysis temperatures (650–850°C) due to micropore development (Li et al., 2014; Zhang et al., 2015). Surface functional groups (e.g., carboxyl, hydroxyl, carbonyl) govern biochar's reactivity, with feedstock and post-treatment modulating their abundance to improve nutrient exchange and metal adsorption (Kharel et al., 2019; Sahin et al., 2017; Wang et al., 2019).

2.4 Key factors affecting biochar properties

The properties of biochar are primarily determined by pyrolysis temperature, reaction time, and feedstock type, which collectively influence its effectiveness as a soil amendment. Higher pyrolysis temperatures (>500°C) increase carbon content and stability while enhancing surface area and microporosity (Liu et al., 2023; Xing et al., 2018), though they reduce functional groups like hydroxyl and carboxyl that aid nutrient retention (Liu et al., 2020; Pantoja et al., 2024). Reaction time affects carbonization, with longer durations improving stability but potentially causing energy inefficiency if excessive (Ferreira et al., 2019; Widowati et al., 2024). Feedstock composition plays a critical role, as lignin-rich materials (e.g., hardwoods) yield more stable biochars with higher carbon content (Arias et al., 2023; Liu et al., 2017b, 2019), while ash-rich feedstocks (e.g., manure) may increase soil salinity but provide more nutrients (Pantoja et al., 2024; Wang et al., 2019). The chemical composition of raw materials further influences biochar's adsorption characteristics, with organic waste-derived biochars exhibiting different adsorption behavior than wood-based ones (López et al., 2022; Zhou et al., 2024).

3. Effects of biochar on soil environment

3.1 Improve soil physicochemical properties

Biochar application significantly enhances soil physicochemical properties by improving pH, organic matter content, and nutrient availability. Its alkaline nature helps neutralize acidic soils, with studies demonstrating pH increases that improve nutrient availability (Laird et al., 2010; Liu et al., 2022). As a stable carbon source, biochar boosts soil organic matter, improving soil structure, water retention, and microbial activity (Hossain et al., 2020; Singh et al., 2022). The material's high surface area and cation exchange capacity enhance nutrient retention, particularly for nitrogen, phosphorus, and potassium (Banik et al., 2021; Liu et al., 2019), with corn straw-derived biochar shown to increase nutrient availability and crop yields (Tian et al., 2018). When combined with organic amendments like manure, biochar further enhances nutrient release and soil fertility (Lestiyani et al., 2023).

3.2 Adsorption and fixation of heavy metals and organic pollutants

Biochar demonstrates remarkable potential for remediating contaminated soils through its ability to adsorb and immobilize both heavy metals and organic pollutants, owing to its unique physicochemical properties including high surface area, porous structure, and diverse functional groups. For heavy metals like As, Pb, Cd, and Cu, biochar facilitates adsorption through multiple mechanisms including surface complexation, ion exchange, precipitation, and redox reactions (Islam et al., 2024; Wang et al., 2025), with modified biochars such as nitrogen-doped variants showing enhanced metal adsorption capacity due to amino groups that form stable coordination complexes (Zhang et al., 2025). Similarly, biochar effectively sorbs organic pollutants like pesticides and pharmaceuticals through hydrophobic interactions, pore filling, and π-π electron donor-acceptor mechanisms, with adsorption efficiency varying based on feedstock type and pyrolysis conditions (Shen et al., 2015; Wang et al., 2016), while chemical modifications can further optimize its pollutant removal capacity (Islam et al., 2021c; Li et al., 2020). Beyond direct adsorption, biochar contributes to long-term contaminant stabilization by forming immobile metal-biochar complexes (Islam et al., 2021a; Soria et al., 2020) and improving soil conditions like pH and cation exchange capacity that reduce metal solubility (Islam et al., 2021b; Shen et al., 2017), exemplified by its success in decreasing bioavailability of toxic elements such as cadmium and arsenic in polluted soils (Wang et al., 2020; Zhou et al., 2024). These multifunctional remediation capabilities position biochar as a sustainable solution for soil pollution mitigation.

3.3 Soil microbial community structure and activity

Biochar application significantly influences soil microbial community structure and activity through multiple mechanisms, primarily mediated by its porous structure, chemical properties, and interactions with soil components (Islam et al., 2024). The material enhances microbial diversity and abundance by providing protective microhabitats and serving as a carbon source, with studies demonstrating increased species-level diversity and metabolic potential (Islam et al., 2023; Kolton et al., 2017), while selectively enriching functional microbial groups like cellulolytic bacteria that facilitate organic matter decomposition (Halmi and Simarani, 2024). Biochar's impact on nutrient cycling manifests through stimulation of enzymatic activities critical for nutrient mobilization, including increased dehydrogenase activity that indicates enhanced microbial respiration and metabolic processes (Mierzwa-Hersztek et al., 2020; Nash et al., 2021; Tan et al., 2015), while also promoting microbial physiological processes that improve nitrogen and phosphorus availability (Mao et al., 2022). Its pollutant adsorption capacity protects microorganisms by immobilizing heavy metals and organic contaminants like PAHs, reducing their bioavailability and creating safer microhabitats within its porous structure (Han et al., 2016; Wang et al., 2024a,b). However, these effects show context-dependent variability based on biochar production parameters (e.g., pyrolysis temperature affecting organic matter bioavailability) and soil conditions, with some studies reporting reduced microbial activity in already fertile soils or with high-temperature biochars (Elzobair et al., 2016; Hardy et al., 2019; Luo et al., 2018; Mao et al., 2022), emphasizing the need for tailored biochar selection based on specific soil-microbial system requirements.

3.4 Soil moisture retention capacity

Biochar application significantly enhances soil moisture retention capacity through multiple mechanisms, particularly in coarse-textured or degraded soils, as demonstrated by numerous studies (Deka et al., 2019; Huang et al., 2021; Thao et al., 2024). The material's porous structure and high surface area create additional water adsorption sites, substantially improving water holding capacity (Shomana et al., 2020; Trifunovic et al., 2018), with feedstock selection and pyrolysis conditions playing crucial roles in determining retention efficiency - for instance, sunflower husk-derived biochars show superior water retention properties compared to other feedstocks (Licht et al., 2017; Shomana et al., 2020). Biochar modifies soil pore architecture by creating a balanced distribution of macro- and micropores (Trifunovic et al., 2018; Ulyett et al., 2014) while simultaneously enhancing soil aggregation and structural stability (Lebrun et al., 2024), collectively contributing to improved moisture retention. However, the effectiveness varies considerably depending on application rates, soil types, and biochar characteristics (Edeh and Mašek, 2022; Liu et al., 2017b; Sherman et al., 2018), with potential negative effects observed at high application rates or with hydrophobic biochars that may temporarily reduce water infiltration (Basso et al., 2013). Additionally, biochar's indirect benefits on soil pH, nutrient availability, and microbial activity (Laird et al., 2010; Martinsen et al., 2014) further support moisture retention through improved plant growth and soil structure development, highlighting the complex, context-dependent nature of biochar's impact on soil water dynamics.

4. Biochar and plant toxicity

4.1 Positive and negative effects on plant growth

Biochar application provides multiple benefits for plant growth through its positive influence on soil properties and nutrient dynamics. By enhancing the cation exchange capacity (CEC) of soils, biochar improves the retention and availability of essential nutrients such as nitrogen (N), phosphorus (P), and potassium (K), leading to increased crop yields (Jaiswal et al., 2015; Rajkovich et al., 2012). Additionally, biochar promotes soil aggregation and porosity, significantly improving water retention and root development, particularly in arid regions (Ahmad et al., 2020; Hye et al., 2025). The material also stimulates beneficial microbial communities, including arbuscular mycorrhizal fungi, which enhance nutrient uptake and overall plant performance (Prendergast-Miller et al., 2014). Furthermore, biochar's alkaline nature helps neutralize acidic soils, creating a more favorable environment for plant growth (Huang and Gu, 2019).

Despite its advantages, biochar can exhibit detrimental effects under certain conditions. New or improperly produced biochar may release phytotoxic compounds, such as short-chain carboxylic acids and phenolics, which can inhibit seed germination and early plant growth (Intani et al., 2019). Excessive application rates can disrupt nutrient availability, leading to imbalances that suppress plant development (Jaiswal et al., 2015; Zou et al., 2021). Plant responses to biochar are highly variable, with some species thriving while others experience stunted growth due to differences in biochar properties and nutrient bioavailability (Jabborova et al., 2021; Zheng et al., 2023). Additionally, biochars derived from contaminated feedstocks may introduce heavy metals or organic pollutants, posing risks to plant and soil health (Huang and Gu, 2019; Schulz et al., 2013). Thus, careful selection and application of biochar are crucial to mitigating these potential drawbacks.

4.2 Plant toxicity

4.2.1 Toxic substances in biochar

Biochar application may introduce toxic substances such as polycyclic aromatic hydrocarbons (PAHs) and heavy metals that can adversely affect plant health, particularly when produced at lower pyrolysis temperatures where PAH formation is more prevalent (Buss and Mašek, 2014; Strawn et al., 2015). These hydrophobic compounds can adsorb onto biochar surfaces, and their leachates may inhibit seed germination and root development in sensitive plant species (Buss and Mašek, 2014), while heavy metals like lead (Pb), cadmium (Cd), and arsenic (As) present in feedstock can become bioavailable depending on soil pH and biochar characteristics, potentially increasing metal uptake by plants (Butnan et al., 2016; Namgay et al., 2010). Although biochar can reduce the mobility of certain contaminants, it may inadvertently enhance the solubility of others, complicating remediation efforts (Puga et al., 2015), with high-VOC biochar leachates showing acute toxicity to plants during initial application phases (Buss and Mašek, 2014). To mitigate these risks, strategies such as chemical modification of biochar to enhance contaminant adsorption (Aziz et al., 2021; Duwiejuah et al., 2020) or co-application with organic amendments like compost (Jiang et al., 2022; Yaseen et al., 2021) have been proposed to reduce toxicity while preserving soil health benefits.

4.2.2 Soil nutrient imbalance

Biochar application can inadvertently cause soil nutrient imbalances and plant toxicity through multiple mechanisms, including excessive nutrient accumulation and introduction of harmful substances. While biochar enhances nutrient retention, over-application may lead to toxic levels of certain nutrients like potassium (K), disrupting plant growth (Biederman and Harpole, 2013; Kuttippurath et al., 2023), and its interaction with fertilizers can exacerbate nutrient leaching or induce toxicities of micronutrients like copper (Cu) and zinc (Zn) (Tsai and Chang, 2020). Furthermore, biochar produced at low pyrolysis temperatures may contain polycyclic aromatic hydrocarbons (PAHs) that inhibit seed germination and plant development (Liao et al., 2014), while feedstocks contaminated with heavy metals can increase metal bioavailability in soils, leading to plant uptake and physiological stress (Aziz et al., 2021; Liu et al., 2023). Additionally, biochar-induced reactive oxygen species (ROS) may cause oxidative damage to plants (Aziz et al., 2021), necessitating careful quality control through contaminant screening (Erdal et al., 2018), optimized application rates to prevent nutrient overload (Glaser and Lehr, 2019), and strategic combination with organic amendments to balance nutrient dynamics (Gaskin et al., 2010; Kuttippurath et al., 2023). These measures are critical for harnessing biochar's benefits while minimizing its potential adverse effects on soil-plant systems.

4.2.3 Soil salinity accumulation due to excessive application rate

Excessive biochar application can induce soil salinity accumulation through multiple pathways, primarily through the leaching of soluble cations (K^+, Ca^{2+}, Mg^{2+}) that elevate soil electrical conductivity and create osmotic stress, restricting plant water uptake (Lin et al., 2015; Sun et al., 2016; Zhang et al., 2017b). Sodium-enriched biochars or those facilitating Na^+ absorption can exacerbate salinity issues, particularly harming salt-sensitive crops (Elshaikh et al., 2018; Hazman et al., 2022; She et al., 2018), while simultaneously disrupting nutrient ratios by disproportionately increasing potassium and magnesium concentrations at the expense of nitrogen and phosphorus availability (Major et al., 2010; van Zwieten et al., 2010). These salinity-induced changes impair root development, nutrient assimilation, and microbial activity critical for nutrient cycling (Wang et al., 2022), necessitating careful management through controlled application rates (5% by weight or less) to prevent salt accumulation (Anwari et al., 2019; Wang et al., 2022; Wu et al., 2024), strategic co-application with organic amendments like compost to buffer salinity effects (Gulyás et al., 2022), and site-specific adaptations for saline-prone soils through tailored biochar selection and dosage (Elshaikh et al., 2018; Wang et al., 2024a,b).

4.3 Differences in response of different plants to biochar toxicity

Plant species exhibit markedly different responses to biochar application, with outcomes ranging from significant growth enhancement to phytotoxicity, depending on biochar characteristics and plant-specific tolerances. Annual crops generally show greater positive responses than perennials, benefiting from biochar-induced improvements in soil structure and nutrient availability due to their rapid growth rates and resource utilization efficiency (Biederman and Harpole, 2013), with soybean (Glycine max) demonstrating particular growth improvements when amended with moderately pyrolyzed organic waste-derived biochar (Adekanmbi et al., 2022). Conversely, sensitive species like wheat and radish may experience growth inhibition from certain biochars, particularly those produced from food waste or applied at high rates, due to induced salinity or nutrient imbalances (Rajkovich et al., 2012), while rice plants show unique benefits like reduced nematode infections that other crops may not experience (Huang et al., 2015). These differential responses are mediated by biochar properties - manure-derived biochars typically enhance growth through higher nutrient content, whereas high-temperature biochars may offer limited benefits due to reduced nutrient availability (Xu et al., 2023), with potential phytotoxicity arising from residual PAHs or heavy metals that affect species differently based on tolerance thresholds (Honvault et al., 2022). Soil interactions further modulate these effects, as biochar-induced changes in salinity, pH, and microbial communities create variable growing conditions; salt-tolerant species like Salicornia herbacea thrive under conditions that would stress less tolerant plants (Shin et al., 2024), while shifts in nutrient cycling mediated by biochar-altered soil microbiomes differentially impact species based on their specific nutritional requirements (Honvault et al., 2022; Qian et al., 2016). This complex interplay necessitates species-specific biochar application strategies to optimize benefits while minimizing adverse effects. Plant-specific responses to biochar application: growth enhancement versus phytotoxicity risks are provided in Table 1.

4.4 Potential risks of long-term application of biochar to plant growth

The long-term application of biochar presents several risks to plant growth, primarily through nutrient imbalances and accumulation of toxic substances. Excessive biochar use can disrupt soil nutrient dynamics, with studies showing reduced total nitrogen (N) and sulfur (S) availability in calcareous soils (Lentz and Ippolito, 2012), while the liming effect of high-carbon biochars may decrease phosphorus (P), copper (Cu), and iron (Fe) solubility, creating micronutrient deficiencies (Paymaneh et al., 2018). These nutrient imbalances are compounded by biochar's potential to introduce heavy metals (Cd, Pb, Cu) from contaminated feedstocks, which may accumulate in soils and become bioavailable over time (Ai et al., 2023; Knoblauch et al., 2021; Mu et al., 2024). Additionally, biochars produced at low pyrolysis temperatures can retain polycyclic aromatic hydrocarbons (PAHs) and other organic contaminants that impair germination and plant development (Erdal et al., 2018; Uzoma et al., 2011). Soil salinity may also increase with repeated biochar applications, particularly in sandy soils, causing osmotic stress that reduces water uptake and induces nutrient toxicity symptoms (Kuttippurath et al., 2023; Yan et al., 2019). Furthermore, prolonged biochar use can fundamentally alter soil chemistry, potentially disrupting nutrient cycling efficiency and long-term fertility (Novak et al., 2018), necessitating careful monitoring of application rates and soil conditions to mitigate these cumulative effects on plant health.

5. Key factors influencing biochar phytotoxicity

5.1 Preparation conditions

The phytotoxic potential of biochar is fundamentally determined by its preparation conditions, with pyrolysis temperature being a critical factor influencing its chemical composition and plant toxicity. Higher pyrolysis temperatures (typically above 500°C) generally reduce volatile organic compounds (VOCs) and phytotoxic residues, while lower temperatures (around 300°C) tend to retain higher levels of potentially harmful compounds that can negatively affect plant growth (Arán et al., 2023; Buss et al., 2015; Corrêa et al., 2020). Post-production treatments such as leaching or washing can mitigate phytotoxicity by removing soluble salts and organic contaminants, though excessive washing may also deplete beneficial nutrients, requiring careful optimization (Intani et al., 2019; Thomas et al., 2024; Visioli et al., 2016). The choice of feedstock equally impacts biochar's properties, with nutrient-rich materials like poultry litter typically exhibiting lower phytotoxicity and even promoting plant growth, whereas agricultural residues such as wheat straw may contain phytotoxic compounds like phenols or aromatic hydrocarbons that adversely affect plants (Corrêa et al., 2020; Jain et al., 2022; Rombolà et al., 2015). Furthermore, processing methods like co-pyrolysis with plastics or thermal treatments can modify biochar's chemical profile, potentially reducing

Table 1 Plant-specific responses to biochar application: growth enhancement vs. phytotoxicity risks.

Plant type	Positive responses to biochar	Potential phytotoxic effects	Key influencing factors	Reference
Legumes (e.g., soybean, peas)	Improved nodulation, N-fixation, and P availability	Reduced growth if biochar contains high Cu/Zn or induces N immobilization	Feedstock (avoid heavy metals), pyrolysis temperature (>500 °C preferred)	(Bolan et al., 2023; Egamberdieva et al., 2022)
Cereals (e.g., wheat, maize)	Enhanced water retention, K/Mg uptake, and yield	Germination inhibition from VOC-rich biochars (low-temperature)	Application rate (<5% w/w), aging/pre-washing of biochar	(Farhangi-Abriz et al., 2021; Han et al., 2023)
Leafy vegetables (e.g., lettuce, spinach)	Reduced heavy metal uptake in contaminated soils	Sensitivity to high pH/salinity (e.g., manure-derived biochars)	Co-application with compost, soil pH monitoring	(Qin et al., 2022)
Root crops (e.g., carrot, radish)	Improved soil porosity and root development	Stunted growth from PAHs or nutrient imbalances (high K:Na ratio)	Biochar particle size, avoidance of low-temperature pyrolysis	(Agbede, 2021; Nyambo et al., 2023)
Salt-tolerant species (e.g., barley, quinoa)	Resilience to biochar-induced salinity	Rare; may benefit from biochar's moisture retention	High-temperature biochars (>600 °C) for stability	(Abbas et al., 2022)

heavy metal toxicity through increased silicon availability or introducing mobile organic compounds that differentially affect plant species based on their sensitivity (Buss et al., 2015; Buss and Mašek, 2014; Li et al., 2023b). These factors collectively underscore the importance of tailored production approaches to minimize phytotoxic risks while preserving biochar's agronomic benefits. Key biochar properties and their effects on soil-plant systems based on pyrolysis temperature are provided in Table 2.

5.2 Application rate and application method

The application rate and method of biochar significantly influence its effects on soil-plant systems, with improper usage potentially causing phytotoxicity while optimized application can enhance soil quality. Excessive biochar application (>10% by weight) may induce nutrient imbalances, particularly elevating potassium (K) levels to concentrations that create osmotic stress and hinder water uptake (Duwiejuah et al., 2020), while also increasing soil salinity and pH to levels that reduce micronutrient availability (e.g., iron and manganese) and inhibit germination in sensitive species (Duwiejuah et al., 2020; Gale et al., 2016). Conversely, moderate application rates (5%–10% by weight) typically improve soil structure, moisture retention, and nutrient cycling without adverse effects (Biederman and Harpole, 2013; Jaiswal et al., 2017). Application methods further modulate these outcomes: basal application (pre-plant incorporation) enables uniform biochar distribution that enhances root-zone nutrient retention and microbial activity (Bird et al., 2012), though excessive rates in this method risk elevating salinity (Biederman and Harpole, 2013). Mixed application with organic amendments like compost can buffer potential toxicity by improving nutrient synchrony and microbial degradation of contaminants (Rogovska et al., 2012), but requires careful balancing of ratios to prevent nutrient oversaturation and salt leaching (Biederman and Harpole, 2013). These findings underscore the need for precision in both application rate and method to harness biochar's benefits while avoiding phytotoxic consequences.

5.3 Soil type and environmental factors

The phytotoxic potential of biochar is strongly influenced by both its production parameters and the soil environment where it is applied. The feedstock type and pyrolysis conditions fundamentally determine biochar's chemical composition, with volatile organic compounds (VOCs) from certain raw materials and low-temperature pyrolysis (<500°C) being particularly associated with phytotoxic effects (Corrêa et al., 2020; Gale et al., 2016). Optimal pyrolysis temperatures around 600°C tend to reduce phytotoxic compounds while enhancing beneficial properties (Crespo-Barreiro et al., 2023). However, these intrinsic properties interact dynamically with soil characteristics - soil pH modifies biochar's surface charge and cation exchange capacity, altering nutrient and contaminant bioavailability (Crane-Droesch et al., 2013), while soil moisture

Table 2 Key biochar properties and their effects on soil-plant systems based on pyrolysis temperature.

Feedstock type	Temperature range	Changes observed	Functional groups/Elemental composition	Reference
Corn stover; switchgrass; wood	175–950 °C	pH, CEC increase with temperature; nitrogen content decreases	H/C ratio decreases; carbon content increases with temperature	(Li et al., 2022)
Vetiver roots; wheat straw; sweet potato vine; apricot leaves	300–750 °C	pH increases with temperature; CEC not mentioned	More basic groups at high temperature; loss of acidic groups	(Sun et al., 2017)
Cottonseed hulls	200–800 °C	No mention found	Nitrogen:carbon ratio peaks at 300–400°C; varies by feedstock	(Uchimiya et al., 2011)
Pine saw dust; rice husk; food waste; poultry litter; paper sludge	350–650 °C	pH increases; CEC decreases with temperature	Aromaticity increases, polarity decreases with temperature	(Pariyar et al., 2020)
Wheat straw; poplar wood; spruce wood	400–525 °C	CEC decreases with temperature	H/C ratio: 0.46–0.40 at 525°C; more aromatic carbon	(Kloss et al., 2012)
Pine wood chips; wheat straw; wheat straw pellets	350–650 °C	pH, CEC increase with temperature (peak CEC at 450–550°C)	Stable carbon increases, labile carbon decreases with temperature	(Crombie et al., 2015)
No mention found	200–850 °C	pH, CEC increase with temperature	Aliphatic to aromatic carbon shift with temperature (NMR)	(Zhang et al., 2017a)
Vine shoots; corn stover	350–500 °C	CEC: 20.1–75.8 cmol/kg	No mention found	(Balmuk et al., 2023)

increases the mobility of phytotoxic compounds (Dhareesank et al., 2006). Post-production treatments like leaching can mitigate phytotoxicity by removing mobile organic compounds and short-chain carboxylic acids (Thomas et al., 2024), highlighting how both production choices and environmental conditions must be considered to optimize biochar's agricultural value while minimizing risks to plant health.

This complex interplay between biochar properties and soil factors creates species-specific responses that necessitate tailored application strategies. The same biochar that stimulates growth in one crop may inhibit another depending on their sensitivity to residual VOCs or changes in soil chemistry (Gale et al., 2016), while soil organic matter content and texture further modulate these interactions by affecting compound retention and microbial activity. These multidimensional relationships underscore why a comprehensive evaluation of both biochar characteristics and the target soil environment is essential for effective, non-phytotoxic implementation in agricultural systems.

5.4 Plant species and physiological characteristics

The phytotoxic effects of biochar exhibit significant variation across plant species, primarily governed by three key factors: biochar's chemical composition, plant physiological characteristics, and soil interactions. Biochars derived from contaminated feedstocks may contain phytotoxic heavy metals (Cr, Pb, As) that particularly affect sensitive species like lettuce and tomato (Kochanek et al., 2016), while lignin-derived phenolic compounds and variable nutrient content (N, P, K) further contribute to species-specific responses (Liu et al., 2017a; Waqas et al., 2018). Certain crops like maize demonstrate improved growth through enhanced nutrient availability and soil microbiology (Cong et al., 2023), while others experience altered hormonal regulation, particularly in auxin-mediated root development (Huang et al., 2021) or compromised stress resilience due to modified gene expression (Viger et al., 2015). The soil environment mediates these interactions through pH-dependent nutrient availability (Cong et al., 2023) and electrical conductivity effects, with some biochars improving water retention and salinity tolerance (Abbas et al., 2024; Akhtar et al., 2015), while microbial community shifts induced by biochar can either benefit plant nutrition or disrupt established symbiotic relationships (Jin et al., 2022). These multidimensional interactions underscore the need for species-specific evaluation of biochar effects, particularly regarding heavy metal content, nutrient composition, and soil modification potential, to optimize agricultural applications while minimizing phytotoxic risks.

6. Mechanism of biochar to alleviate plant toxicity

6.1 Mitigation mechanism for heavy metal toxicity

Biochar effectively reduces the bioavailability and toxicity of heavy metals in soils. Its porous structure and large specific surface area facilitate the immobilization of cationic heavy metals, such as cadmium (Cd) and lead (Pb), through mechanisms including adsorption, complexation, and precipitation (Guo et al., 2020; Park et al., 2011; Wang et al., 2020). Additionally, biochar's electrostatic attraction and ion exchange capabilities enhance its efficiency in capturing heavy metal ions, thereby reducing their mobility and potential uptake by plants (Guo et al., 2020; Islam et al., 2024). Studies further indicate that biochar application not only stabilizes toxic metals but also alters their chemical forms, contributing to reduced toxicity (Qiu et al., 2022; Wang et al., 2020). For example, arsenic (As) and Cd can be effectively immobilized by iron (Fe)- and manganese (Mn)-modified biochar, which binds these contaminants and prevents their uptake by plant roots (Islam et al., 2024; Wang et al., 2020).

6.2 Degradation and fixation of biochar on the toxicity of organic pollutants

Biochar plays a critical role in the degradation and immobilization of organic pollutants in soil. Its abundant functional groups enhance the adsorption of organic contaminants, reducing their bioavailability and toxicity to plants (Bahia, 2022; Cui et al., 2019). Studies demonstrate that biochar amendments significantly decrease the leaching potential and phytotoxicity of pollutants such as 2,4,6-trichlorophenol and sulfamethoxazole by enhancing their sorption and limiting their mobility in soil (Cui et al., 2021; Trinh et al., 2017). Additionally, biochar can facilitate oxidative reactions, particularly in the presence of manganese oxides, promoting both the adsorption and chemical degradation of organic pollutants (Islam et al., 2023; Tian et al., 2019).

6.3 Mitigation of plant toxicity by improving soil microbial communities

A key mechanism by which biochar mitigates plant toxicity is through its positive effects on soil microbial communities. Studies show that biochar application enhances microbial biomass and diversity, which play vital roles in organic contaminant degradation and heavy metal immobilization (Islam et al., 2024; Khan et al., 2020). This stimulation of microbial activity improves nutrient cycling and promotes the secretion of biosurfactants, aiding in the detoxification of harmful compounds and enhancing overall soil health and plant resilience (Guo et al., 2020; Islam et al., 2024). Additionally, biochar serves as a habitat for beneficial microorganisms, fostering the development of mycorrhizal networks that facilitate nutrient uptake while restricting plant absorption of toxic heavy metals (Islam et al., 2023, 2024; Khan et al., 2020).

6.4 Enhancement of soil nutrient release to mitigate toxic effects

Biochar not only reduces toxicity through contaminant immobilization but also enhances soil fertility by improving nutrient availability. When amended to soil, biochar acts as a slow-release fertilizer, gradually supplying essential nutrients such as phosphorus, thereby improving long-term soil fertility (Fang et al., 2014; Graber et al., 2010). This controlled nutrient release helps alleviate toxicity caused by nutrient imbalances, which are often aggravated by heavy metal contamination (Fang et al., 2014). The enhanced nutrient availability supports plant growth and stress tolerance in polluted soils, enabling plants to better cope with toxic conditions (Graber et al., 2010). Furthermore, biochar amendments can increase soil pH and cation exchange capacity (CEC), which promotes better nutrient uptake by plants while simultaneously reducing heavy metal toxicity (Boukar et al., 2023; Ippolito et al., 2012; Islam et al., 2024). These combined effects make biochar a valuable tool for restoring productivity to degraded soils.

7. Application of biochar in agriculture and environmental remediation

7.1 The role of biochar in sustainable agriculture

Biochar has demonstrated significant potential for improving soil fertility and crop productivity through enhanced nutrient retention and stimulation of microbial activity. Studies show that biochar application increases soil cation exchange capacity (CEC), thereby improving plant nutrient availability (Shiyal et al., 2022). Specifically, it can alleviate phosphorus (P) and potassium (K) limitations in soils, leading to increased pasture yields without negative impacts on soil respiration or nitrous oxide (N_2O) emissions (van Zwieten et al., 2019). The combination of biochar with mineral fertilizers creates synergistic effects that enhance soil microbiological activity and nutrient cycling, further boosting crop production (Islam et al., 2023; Mustafa et al., 2022).

Meta-analyses reveal that biochar's effects on crop yields vary depending on regional soil conditions, with heterogeneous benefits observed across different agricultural systems (Crane-Droesch et al., 2013). Beyond productivity gains, biochar enhances soil resilience to climate variability, supporting more sustainable agricultural practices (Koide et al., 2015). Additionally, biochar incorporation contributes to long-term soil carbon sequestration, offering a valuable strategy for climate change mitigation while improving soil quality (Lin et al., 2015; Liu et al., 2017a; Venkatesh et al., 2022).

7.2 Application of biochar in remediation of contaminated soil

Biochar demonstrates significant potential for soil remediation beyond agricultural applications, particularly in soils contaminated with heavy metals and organic pollutants. Research shows that biochar amendment effectively reduces heavy metal bioavailability in contaminated soils, thereby protecting crop health and ensuring food safety (Islam et al., 2023, 2024; Murtaza et al., 2021). This effect is attributed to biochar's porous structure and high adsorption capacity, which immobilize contaminants and promote soil rehabilitation (Islam et al., 2023, 2024; Murtaza et al., 2021).

For organic pollutants, biochar serves as an effective remediation agent by facilitating the degradation of harmful compounds (Shiyal et al., 2022). Its exceptional adsorption capacity for diverse organic contaminants makes it particularly valuable for environmental cleanup and waste management applications (Murtaza et al., 2021). These dual remediation capabilities position biochar as a versatile solution that not only improves soil quality but also reduces environmental risks associated with soil contamination.

7.3 Application of biochar in horticulture and urban greening

Biochar has emerged as a valuable organic amendment for horticulture and urban greening, demonstrating multiple benefits for soil quality and plant performance. Research indicates that biochar application significantly improves water retention and nutrient availability in urban soils, particularly in areas characterized by impervious surfaces and degraded soil conditions (Thao et al., 2024). These moisture retention properties prove especially valuable in arid environments where water scarcity limits plant growth (Nurida et al., 2023).

Beyond physical soil improvements, biochar enhances urban green space establishment by fostering root development and supporting beneficial soil microbial communities (Yadav and Ramakrishna, 2023). When incorporated into urban gardening practices, biochar contributes to broader sustainability objectives, including air quality improvement and biodiversity enhancement (Tisserant and Cherubini, 2019). These combined effects make biochar an effective tool for creating resilient urban green infrastructure.

7.4 The role of biochar in climate change mitigation

Biochar plays a pivotal role in climate change mitigation by enhancing carbon sequestration and reducing greenhouse gas emissions. When applied to agricultural soils, biochar creates a stable, long-term carbon sink by converting waste biomass into recalcitrant carbon that can persist for centuries (Crespo-Barreiro et al., 2023; Woolf et al., 2010). This carbon stabilization process effectively removes atmospheric CO_2 while simultaneously improving soil organic matter content. Beyond carbon storage, biochar significantly reduces emissions of potent greenhouse gases, particularly methane and N_2O, through improved soil aeration, optimized nitrogen cycling, and physical gas adsorption (Shrestha et al., 2023). These combined effects make biochar a valuable negative emissions technology that aligns with global climate policy objectives.

The climate benefits of biochar extend to agricultural productivity and soil health enhancement. By improving soil structure and nutrient availability, biochar application enables more efficient ecosystem functioning that further offsets agricultural emissions (Pressler et al., 2019). This dual capacity for emission reduction and yield improvement positions biochar as a key strategy for sustainable intensification of food production systems, particularly in the context of climate change adaptation (Nurida et al., 2023). The multifaceted advantages of biochar, from carbon sequestration to soil fertility enhancement, demonstrate its significant potential in global efforts to achieve net-negative emissions while maintaining agricultural productivity.

8. Future research directions and challenges

8.1 Optimization and standardization of biochar preparation process

The widespread adoption of biochar as a soil amendment requires consistent and efficient production methods. Current research emphasizes the need to standardize pyrolysis processes, as key variables including feedstock composition, temperature profiles, and reactor configurations significantly influence biochar's physical and chemical properties (Biederman and Harpole, 2013; Liu et al., 2016). Studies demonstrate that these production parameters directly affect critical biochar characteristics such as porosity, surface area, and pH, which ultimately determine its effectiveness in different agricultural applications (Tammeorg et al., 2017). Developing standardized production protocols would ensure reliable biochar quality while maintaining flexibility for adaptation to diverse soil types and farming systems (Liu et al., 2016; Tisserant and Cherubini, 2019).

Emerging technologies offer promising solutions for optimizing biochar production. Recent advances in machine learning applications enable precise tuning of pyrolysis conditions to produce biochars with specific properties tailored to particular soil remediation or crop production needs (He et al., 2024). This technological integration could revolutionize biochar manufacturing by simultaneously improving production efficiency while customizing products for targeted agricultural outcomes. Such innovations in production optimization will be crucial for scaling up biochar applications while maintaining quality standards across different geographical regions and farming contexts.

8.2 Effects of long-term biochar application on soil-plant-microbial systems

Biochar application initiates complex, evolving interactions within soil ecosystems that extend far beyond initial soil quality improvements. Research demonstrates that while biochar consistently enhances nutrient retention, water-holding capacity, and microbial activity, the magnitude and duration of these effects show significant temporal variation (Edeh et al., 2020; Palansooriya et al., 2019). This variability underscores the critical need for longitudinal studies that track not just physicochemical soil parameters, but also the adaptive responses of microbial communities and plant systems over multi-year timescales (Maslouski et al., 2025; Palansooriya et al., 2019).

A comprehensive understanding of these long-term interactions is particularly crucial for evaluating biochar's dual role in climate-smart agriculture. The evolving relationships between biochar, soil biota, and plants directly influence both carbon sequestration potential and agricultural productivity (Rittl et al., 2021; Singh et al., 2022). Such insights are becoming increasingly valuable for developing sustainable farming systems capable of withstanding climate change pressures while maintaining soil health and crop yields over decadal timescales.

8.3 Effects of biochar application under different soil and climatic conditions

The effectiveness of biochar as a soil amendment varies substantially depending on specific environmental and agronomic conditions. Studies demonstrate that biochar application can dramatically improve water retention in sandy soils, providing critical drought resilience for crops in arid regions (Vitková et al., 2021; Yu et al., 2021). However, these benefits become less evident in clay-rich or naturally fertile soils where native properties already support adequate water and nutrient retention (Blanco-Canqui, 2021). This variability highlights the need for targeted research across different agricultural systems, particularly in climate-vulnerable areas where biochar's potential to enhance soil resilience remains underexplored (Hafez et al., 2021; Michael, 2020).

Developing region-specific biochar application protocols requires synthesizing knowledge from both global research and localized field trials. Such an integrated approach would enable the creation of tailored guidelines that optimize agricultural productivity while ensuring environmental sustainability (Gurwick et al., 2013). By accounting for the complex interplay between soil characteristics, climate conditions, and farming practices, these guidelines could help maximize biochar's benefits across diverse agricultural landscapes while minimizing potential drawbacks associated with inappropriate application.

8.4 Economic and environmental feasibility analysis to promote the industrial application of biochar

The industrial-scale adoption of biochar requires a thorough evaluation of both economic viability and environmental sustainability. Existing studies reveal that despite its demonstrated agronomic benefits, widespread implementation faces barriers from substantial production and logistics costs (Jeffery et al., 2015; Spokas et al., 2012). Research efforts should focus on reconciling these short-term economic challenges with biochar's long-term environmental benefits, including greenhouse gas mitigation and improved soil quality (Fidel et al., 2019; He et al., 2017).

Developing robust economic frameworks that recognize biochar's carbon sequestration potential through carbon credit systems could significantly enhance its commercial attractiveness (Tisserant and Cherubini, 2019). A comprehensive valuation model integrating both ecological benefits and financial returns will be essential for convincing agricultural stakeholders and policymakers of biochar's dual value proposition. Such models must account for regional variations in production costs, application benefits, and potential policy incentives to accurately represent biochar's full socioeconomic potential across different implementation contexts.

9. Conclusions

One great tool with multifarious applications in agriculture and environmental sustainability is biochar, but the practicality largely depends on its phytotoxic and inhibiting properties. When used as a soil amendment, biochar can improve soil physicochemical properties, enhance nutrient retention, and foster beneficial microbial communities that support plant growth and resistance. In cases of soil contamination, biochar can immobilize heavy metals and hence serve as a medium where organic pollutants can be hydrolyzed and degraded. Nonetheless, the phytotoxic aspects of biochar underscore the need for its proper use, which could cause nutrient imbalances, salinity buildup, and even the release of toxic compounds alongside issues of inherent and extrinsic biochar toxicities.

Selection of feedstock, treatment via pyrolysis, rates of biochar application, and the reaction of soil with plants are all crucial in considering biochar efficacy and safety. High pyrolysis temperatures, defined as above 500°C, as well as post-production procedures such as leaching, are required to reduce phytotoxic compounds; moderate rates of application, in the range of 5%–10% wt., in addition to co-application with organic amendments, are conversely recommended to minimize negative effects. The variability in plant response underscores a need for customized approaches to yield optimal benefits.

Future research should prioritize the standardization of biochar production, long-term studies on soil-plant-microbial interactions, and context-specific evaluations under diverse climatic and soil conditions. Additionally, economic and environmental feasibility analyses will be critical to scaling up biochar applications industrially. By addressing these challenges, biochar can be strategically integrated into sustainable agriculture and environmental remediation, balancing its advantages with minimized risks to plant health.

References

Abbas, H.M.M., Rais, U., Sultan, H., Tahir, A., Bahadur, S., Shah, A., Iqbal, A., Li, Y., Khan, M.N., Nie, L., 2024. Residual effect of microbial-Inoculated biochar with nitrogen on rice growth and salinity reduction in paddy soil. Plants 13. https://doi.org/10.3390/plants13192804.

Abdelaal, A., Pradhan, S., AlNouss, A., Tong, Y., Al-Ansari, T., McKay, G., Mackey, H.R., 2021. The impact of pyrolysis conditions on orange peel biochar physicochemical properties for sandy soil. Waste Manag. Res. 39, 995–1004. https://doi.org/10.1177/0734242X20978456.

Adekanmbi, A.A., Oghenewiro, F., Fagbenro, J.A., Bala, A., Osunde, O.A., 2022. Effect of four biochar types and inorganic phosphate fertilizer on growth and nodulation of soybean (Glycine Max (L.) Moench). Plant Physiol. Soil Chem. 2, 24–28. https://doi.org/10.26480/ppsc.01.2022.24.28.

Agbede, T.M., 2021. Effect of tillage, biochar, poultry manure and NPK 15-15-15 fertilizer, and their mixture on soil properties, growth and carrot (Daucus carota L.) yield under tropical conditions. Heliyon 7 (6), e07391. https://doi.org/10.1016/J.HELIYON.2021.E07391.

Ahmad, M., Wang, X., Hilger, T.H., Luqman, M., Nazli, F., Hussain, A., Zahir, Z.A., Latif, M., Saeed, Q., Malik, H.A., Mustafa, A., 2020. Evaluating biochar-microbe synergies for improved growth, yield of maize, and post-harvest soil characteristics in a semi-arid climate. Agronomy 10. https://doi.org/10.3390/AGRONOMY10071055.

Ai, Y., Wang, Y., Song, L., Hong, W., Zhang, Z., Li, X., Zhou, S., Zhou, J., 2023. Effects of biochar on the physiology and heavy metal enrichment of Vetiveria zizanioides in contaminated soil in mining areas. J. Hazard. Mater. 448, 130965. https://doi.org/10.1016/j.jhazmat.2023.130965.

Akhtar, S.S., Andersen, M.N., Liu, F., 2015. Biochar mitigates salinity stress in potato. J. Agron. Crop Sci. 201, 368–378. https://doi.org/10.1111/jac.12132.

Anwari, G., Mandozai, A., Feng, J., 2019. Effects of biochar amendment on soil problems and improving rice production under salinity conditions. Adv. J. Grad. Res. 7, 45–63. https://doi.org/10.21467/ajgr.7.1.45-63.

Arán, D.S., Deza, M., Monferrán, M.V., Pignata, M.L., Harguinteguy, C.A., 2023. Use of local waste for biochar production: influence of feedstock and pyrolysis temperature on chromium removal from aqueous solutions. Integr. Environ. Assess. Manag. 19, 717–725. https://doi.org/10.1002/ieam.4643.

Arias, C.M., Da Silva, L.F.S., Soares, M.R., Forti, V.A., 2023. A bibliometric analysis on the agricultural use of biochar in Brazil from 2003 to 2021: research status and promising raw materials. Renew. Agric. Food Syst. https://doi.org/10.1017/S1742170522000412.

Aziz, H., Murtaza, G., Saleem, M.H., Ali, S., Rizwan, M., Riaz, U., Niaz, A., Abualreesh, M.H., Alatawi, A., 2021. Alleviation of chlorpyrifos toxicity in maize (Zea mays L.) by reducing its uptake and oxidative stress in response to soil-applied compost and biochar amendments. Plants 10. https://doi.org/10.3390/plants10102170.

Bahia, W., 2022. Adsorption-desorption behavior and pesticide bioavailability of biochar in soil. Sci. Insights 41, 725–731. https://doi.org/10.15354/si.22.re093.

Balmuk, G., Videgain, M., Manyà, J.J., Duman, G., Yanik, J., 2023. Effects of pyrolysis temperature and pressure on agronomic properties of biochar. J. Anal. Appl. Pyrolysis 169, 105858. https://doi.org/10.1016/j.jaap.2023.105858.

Banik, C., Koziel, J.A., Bonds, D., Singh, A.K., Licht, M.A., 2021. Comparing biochar-swine manure mixture to conventional manure impact on soil nutrient availability and plant uptake-a greenhouse study. Land 10. https://doi.org/10.3390/land10040372.

Basso, A.S., Miguez, F.E., Laird, D.A., Horton, R., Westgate, M., 2013. Assessing potential of biochar for increasing water-holding capacity of sandy soils. GCB Bioenergy 5, 132–143. https://doi.org/10.1111/gcbb.12026.

Biederman, L.A., Harpole, W.S., 2013. Biochar and its effects on plant productivity and nutrient cycling: a meta-analysis. GCB Bioenergy 5, 202–214. https://doi.org/10.1111/gcbb.12037.

Bird, M.I., Wurster, C.M., de Paula Silva, P.H., Paul, N.A., de Nys, R., 2012. Algal biochar: effects and applications. GCB Bioenergy 4, 61–69. https://doi.org/10.1111/j.1757-1707.2011.01109.x.

Blanco-Canqui, H., 2021. Does biochar improve all soil ecosystem services? GCB Bioenergy 13, 291–304. https://doi.org/10.1111/gcbb.12783.

Bolan, N., Hoang, S.A., Beiyuan, J., Gupta, S., Hou, D., Karakoti, A., Joseph, S., Jung, S., Kim, K.H., Kirkham, M.B., Kua, H.W., Kumar, M., Kwon, E.E., Ok, Y.S., Perera, V., Rinklebe, J., Shaheen, S. M., Sarkar, B., Sarmah, A.K., Singh, B.P., Singh, G., Tsang, D.C.W., Vikrant, K., Vithanage, M., Vinu, A., Wang, H., Wijesekara, H., Yan, Y., Younis, S.A., Van Zwieten, L., 2022. Multifunctional applications of biochar beyond carbon storage. Int. Mater. Rev. https://doi.org/10.1080/09506608.2021.1922047.

Bolan, S., Hou, D., Wang, L., Hale, L., Egamberdieva, D., Tammeorg, P., Li, R., Wang, B., Xu, J., Wang, T., Sun, H., Padhye, L.P., Wang, H., Siddique, K.H.M., Rinklebe, J., Kirkham, M.B., Bolan, N., 2023. The potential of biochar as a microbial carrier for agricultural and environmental applications. Sci. Total Environ. 886, 163968. https://doi.org/10.1016/j.scitotenv.2023.163968.

Boukar, H., Davy, F.K., Emmanuel, S., Martin, N., 2023. Contribution of enriched biochar to the reduction of the hydrosoluble fraction of some heavy metals in an urban soil in ngaoundere, Cameroon. Int. J. Plant Soil Sci. 35, 485–498. https://doi.org/10.9734/ijpss/2023/v35i173236.

Brewer, C.E., Hall, E.T., Schmidt-Rohr, K., Laird, D.A., Brown, R.C., Zygourakis, K., 2017. Temperature and reaction atmosphere effects on the properties of corn stover biochar. Environ. Prog. Sustain. Energy 36, 696–707. https://doi.org/10.1002/ep.12503.

Brown, T.R., Wright, M.M., Brown, R.C., 2011. Estimating profitability of two biochar production scenarios: slow pyrolysis vs fast pyrolysis. Biofuels, Bioprod. Biorefining 5, 54–68. https://doi.org/10.1002/bbb.254.

Burachevskaya, M., Mandzhieva, S., Bauer, T., Minkina, T., Rajput, V., Chaplygin, V., Fedorenko, A., Chernikova, N., Zamulina, I., Kolesnikov, S., Sushkova, S., Perelomov, L., 2021. The effect of granular activated carbon and biochar on the availability of cu and zn to hordeum sativum distichum in contaminated soil. Plants 10. https://doi.org/10.3390/plants10050841.

Buss, W., Mašek, O., 2014. Mobile organic compounds in biochar - a potential source ofcontamination - phytotoxic effects on cress seed (Lepidiumsativum) germination. J. Environ. Manage. 137, 111–119. https://doi.org/10.1016/j.jenvman.2014.01.045.

Buss, W., Mašek, O., Graham, M., Wüst, D., 2015. Inherent organic compounds in biochar-Their content, composition and potential toxic effects. J. Environ. Manage. 156, 150–157. https://doi.org/10.1016/j.jenvman.2015.03.035.

Butnan, S., Deenik, J.L., Toomsan, B., Antal, M.J., Vityakon, P., 2016. Biochar properties influencing greenhouse gas emissions in tropical soils differing in texture and mineralogy. J. Environ. Qual. 45, 1509–1519. https://doi.org/10.2134/jeq2015.10.0532.

Cai, T., Wang, Z., Guo, C., Huang, H., Chai, H., Zhang, C., 2022. Effects of biochar and manure Co-application on aggregate stability and pore size distribution of vertisols. Int. J. Environ. Res. Public Health 19. https://doi.org/10.3390/ijerph191811335.

Carril, P., Ghorbani, M., Loppi, S., Celletti, S., 2023. Effect of biochar type, concentration and washing conditions on the germination parameters of three model crops. Plants 12. https://doi.org/10.3390/plants12122235.

Chen, J., Wang, J., 2023. Synthesis of modified walnut shell biochar and its performance of cadmium adsorption. J. Geosci. Environ. Prot. 11, 317–332. https://doi.org/10.4236/gep.2023.119020.

Cong, M., Hu, Y., Sun, X., Yan, H., Yu, G., Tang, G., Chen, S., Xu, W., Jia, H., 2023. Long-term effects of biochar application on the growth and physiological characteristics of maize. Front. Plant Sci. 14, 1–15. https://doi.org/10.3389/fpls.2023.1172425.

Corrêa, L.F., Bertol, A., Ushiwata, S.Y., Reis, R. de G.E., Marimon-Junior, B.H., Vilar, C.C., 2020. Evaluation of the biochar quality using the germination index/Avaliação da qualidade de biochar por meio de índice de germinação. Brazilian J. Anim. Environ. Res. 3, 2710–2718. https://doi.org/10.34188/bjaerv3n3-178.

Crane-Droesch, A., Abiven, S., Jeffery, S., Torn, M.S., 2013. Heterogeneous global crop yield response to biochar: a meta-regression analysis. Environ. Res. Lett. 8. https://doi.org/10.1088/1748-9326/8/4/044049.

Crespo-Barreiro, A., Gómez, N., González-Arias, J., Ortiz-Liébana, N., González-Andrés, F., Cara-Jiménez, J., 2023. Scaling-up of the production of biochar from olive tree pruning for agricultural use: evaluation of biochar characteristics and phytotoxicity. Agric. For. 13. https://doi.org/10.3390/agriculture13051064.

Cui, L., Yin, C., Chen, T., Quan, G., Ippolito, J.A., Liu, B., Yan, J., Hussain, Q., 2019. Biochar immobilizes and degrades 2,4,6-trichlorophenol in soils. Environ. Toxicol. Chem. 38, 1364–1371. https://doi.org/10.1002/etc.4401.

Crombie, K., Masek, O., Cross, A., Sohi, S., 2015. Biochar – synergies and trade-offs between soilenhancing properties and C sequestration potential. GCB Bioenergy 7, 1161–1175. https://doi.org/10.1111/gcbb.12213.

Cui, L., Fan, Q., Sun, J., Quan, G., Yan, J., Hina, K., Wang, H., Zhang, Z., Hussain, Q., 2021. Changes in surface characteristics and adsorption properties of 2,4,6-trichlorophenol following Fenton-like aging of biochar. Sci. Rep. 11, 1–11. https://doi.org/10.1038/s41598-021-82129-z.

Deka, K., Hazarika, P.P., Medhi, B.K., Borah, R., Kandali, G.G., Gogoi, B., 2019. Characterization of biochars and their application rates on soil moisture retention in light textured soil of Assam. Int. Res. J. Pure Appl. Chem. 1–9. https://doi.org/10.9734/irjpac/2019/v19i230106.

Dhareesank, A., Kobayashi, K., Usui, K., 2006. Residual phytotoxic activity of pethoxamid in soil and its concentration in soil water under different soil moisture conditions. Weed Biol. Manag. 6, 50–54. https://doi.org/10.1111/j.1445-6664.2006.00195.x.

Ding, Y., Liu, Y., Liu, S., Li, Z., Tan, X., Huang, X., Zeng, G., Zhou, L., Zheng, B., 2016. Biochar to improve soil fertility. A review. Agron. Sustain. Dev. https://doi.org/10.1007/s13593-016-0372-z.

Dróżdż, D., Malińska, K., Wystalska, K., Meers, E., Robles-Aguilar, A., 2023. The influence of poultry manure-derived biochar and compost on soil properties and plant biomass growth. Materials 16. https://doi.org/10.3390/ma16186314.

Duwiejuah, A.B., Abubakari, A.H., Quainoo, A.K., Amadu, Y., 2020. Review of biochar properties and remediation of metal pollution of water and soil. J. Heal. Pollut. 10. https://doi.org/10.5696/2156-9614-10.27.200902.

Edeh, I.G., Mašek, O., 2022. The role of biochar particle size and hydrophobicity in improving soil hydraulic properties. Eur. J. Soil Sci. 73, 1–14. https://doi.org/10.1111/ejss.13138.

Edeh, I.G., Mašek, O., Buss, W., 2020. A meta-analysis on biochar's effects on soil water properties – New insights and future research challenges. Sci. Total Environ. 714. https://doi.org/10.1016/j.scitotenv.2020.136857.

Egamberdieva, D., Ma, H., Reckling, M., Omari, R.A., Wirth, S., Bellingrath-Kimura, S.D., 2022. Interactive effects of biochar, nitrogen, and phosphorous on the symbiotic performance, growth, and nutrient uptake of soybean (Glycine max l.). Agronomy 12 (27), 1–14. https://doi.org/10.3390/agronomy12010027.

Elshaikh, N.A., Zhipeng, L., Dongli, S., Timm, L.C., 2018. Increasing the okra salt threshold value with biochar amendments. J. Plant Interact. 13, 51–63. https://doi.org/10.1080/17429145.2017.1418914.

Elzobair, K.A., Stromberger, M.E., Ippolito, J.A., Lentz, R.D., 2016. Contrasting effects of biochar versus manure on soil microbial communities and enzyme activities in an Aridisol. Chemosphere 142, 145–152. https://doi.org/10.1016/j.chemosphere.2015.06.044.

Enaime, G., Baçaoui, A., Yaacoubi, A., Lübken, M., 2020. Biochar for wastewater treatment-conversion technologies and applications. Appl. Sci. https://doi.org/10.3390/app10103492.

Erdal, I., Memici, M., Dogan, A., Yaylaci, C., Ekinci, K., 2018. Effects of tomato harvest residue derived biochars obtained from different pyrolysis temperature and duration on plant growth and nutrient concentrations of corn. Eng. Rural Dev. 17, 547–553. https://doi.org/10.22616/ERDev2018.17.N137.

Fang, C., Zhang, T., Li, P., Jiang, R.F., Wang, Y.C., 2014. Application of magnesium modified corn biochar for phosphorus removal and recovery from swine wastewater. Int. J. Environ. Res. Public Health 11, 9217–9237. https://doi.org/10.3390/ijerph110909217.

Farhangi-Abriz, S., Torabian, S., Qin, R., Noulas, C., Lu, Y., Gao, S., 2021. Biochar effects on yield of cereal and legume crops using meta-analysis. Sci. Total Environ. 775, 145869. https://doi.org/10.1016/j.scitotenv.2021.145869.

Ferreira, S.D., Junges, J., Bassanesi, G.R., Lazzarotto, I.P., Osório, E., Godinho, M., 2019. Investigation of the structure of the biochar obtained by slow pyrolysis of elephant grass during its steam gasification. Chem. Eng. Technol. 42, 2546–2555. https://doi.org/10.1002/ceat.201800680.

Fidel, R.B., Laird, D.A., Parkin, T.B., 2019. Effect of biochar on soil greenhouse gas emissions at the laboratory and field scales. Soil Syst. 3, 1–18. https://doi.org/10.3390/soilsystems3010008.

Gale, N.V., Sackett, T.E., Thomas, S.C., 2016. Thermal treatment and leaching of biochar alleviates plant growth inhibition from mobile organic compounds. PeerJ. https://doi.org/10.7717/PEERJ.2385.

Gao, X., Yani, S., Wu, H., 2014. Pyrolysis of spent biomass from mallee leaf steam distillation: biochar properties and recycling of inherent inorganic nutrients. In: Energy and Fuels. American Chemical Society, pp. 4642–4649. https://doi.org/10.1021/ef501114v.

Gaskin, J.W., Speir, R.A., Harris, K., Das, K.C., Lee, R.D., Morris, L.A., Fisher, D.S., 2010. Effect of peanut hull and pine chip biochar on soil nutrients, corn nutrient status, and yield. Agron. J. 102, 623–633. https://doi.org/10.2134/agronj2009.0083.

Ghorbani, M., Amirahmadi, E., Neugschwandtner, R.W., Konvalina, P., Kopecký, M., Moudrý, J., Perná, K., Murindangabo, Y.T., 2022. The impact of pyrolysis temperature on biochar properties and its effects on soil hydrological properties. Sustain. Times 14. https://doi.org/10.3390/su142214722.

Glaser, B., Lehr, V.I., 2019. Biochar effects on phosphorus availability in agricultural soils: a meta-analysis. Sci. Rep. 9, 1–9. https://doi.org/10.1038/s41598-019-45693-z.

Graber, E.R., Harel, Y.M., Kolton, M., Cytryn, E., Silber, A., David, D.R., Tsechansky, L., Borenshtein, M., Elad, Y., 2010. Biochar impact on development and productivity of pepper and tomato grown in fertigated soilless media. Plant Soil 337, 481–496. https://doi.org/10.1007/s11104-010-0544-6.

Gulyás, M., Someus, E., Klátyik, S., Fuchs, M., Varga, Z.I., Dér, S., Fekete, G., Czinkota, I., Székács, A., Gyuricza, C., Aleksza, L., 2022. Effects of combined application of solid pyrolysis products and digestate on selected soil properties of arenosol and plant growth and composition in laboratory experiments. Agronomy 12. https://doi.org/10.3390/agronomy12061440.

Gunarathnea, V., Bandara, T., Ramanayaka, S., Vithanage, M., 2022. Benefits of biochar addition in a sustainable agriculture practice: soil nutrients dynamics, enzyme activities and plant growth. Current Scientia 25 (01), 9–25. https://doi.org/10.31357/vjs.v25i01.5921.

Gunes, A., Inal, A., Sahin, O., Taskin, M.B., Atakol, O., Yilmaz, N., 2015. Variations in mineral element concentrations of poultry manure biochar obtained at different pyrolysis temperatures, and their effects on crop growth and mineral nutrition. Soil Use Manag. 31, 429–437. https://doi.org/10.1111/sum.12205.

Guo, M., Song, W., Tian, J., 2020. Biochar-facilitated soil remediation: mechanisms and efficacy variations. Front. Environ. Sci. 8. https://doi.org/10.3389/fenvs.2020.521512.

Gurwick, N.P., Moore, L.A., Kelly, C., Elias, P., 2013. A systematic review of biochar research, with a focus on its stability in situ and its promise as a climate mitigation strategy. PLoS One 8. https://doi.org/10.1371/journal.pone.0075932.

Gwenzi, W., Chaukura, N., Mukome, F.N.D., Machado, S., Nyamasoka, B., 2015. Biochar production and applications in sub-saharan africa: opportunities, constraints, risks and uncertainties. J. Environ. Manage. https://doi.org/10.1016/j.jenvman.2014.11.027.

Hafez, E.M., Gowayed, S.M., Nehela, Y., Sakran, R.M., Rady, A.M.S., Awadalla, A., Omara, A.E.D., Alowaiesh, B.F., 2021. Incorporated biochar-based soil amendment and exogenous glycine betaine foliar application ameliorate rice (Oryza sativa l.) tolerance and resilience to osmotic stress. Plants 10. https://doi.org/10.3390/plants10091930.

Halmi, M.F.A., Simarani, K., 2024. Response of soil microbial glycoside hydrolase family 6 cellulolytic population to lignocellulosic biochar reveals biochar stability toward microbial degradation. J. Environ. Qual. 53, 546–551. https://doi.org/10.1002/jeq2.20588.

Han, S., Li, H., Rengel, Z., Du, Z., Hu, N., Wang, Y., Zhang, A., 2023. Biochar application promotes crops yield through regulating root development and the community structure of root endophytic fungi in wheat-maize rotation. Soil Tillage Res. 234, 105827. https://doi.org/10.1016/j.still.2023.105827.

Han, T., Zhao, Z., Bartlam, M., Wang, Y., 2016. Combination of biochar amendment and phytoremediation for hydrocarbon removal in petroleum-contaminated soil. Environ. Sci. Pollut. Res. 23, 21219–21228. https://doi.org/10.1007/s11356-016-7236-6.

Hardy, B., Sleutel, S., Dufey, J.E., Cornelis, J.T., 2019. The long-term effect of biochar on soil microbial abundance, activity and community structure is overwritten by land management. Front. Environ. Sci. 7, 1–14. https://doi.org/10.3389/fenvs.2019.00110.

Hazman, M.Y., El-Sayed, M.E.A., Kabil, F.F., Helmy, N.A., Almas, L., McFarland, M., El Din, A.S., Burian, S., 2022. Effect of biochar application to fertile soil on tomato crop production under saline irrigation regime. Agronomy 12, 1–17. https://doi.org/10.3390/agronomy12071596.

He, Y., Zhou, X., Jiang, L., Li, M., Du, Z., Zhou, G., Shao, J., Wang, X., Xu, Z., Hosseini Bai, S., Wallace, H., Xu, C., 2017. Effects of biochar application on soil greenhouse gas fluxes: a meta-analysis. GCB Bioenergy 9, 743–755. https://doi.org/10.1111/gcbb.12376.

He, D., Dong, Z., Zhu, B., 2024. An optimal global biochar application strategy based on matching biochar and soil properties to reduce global cropland greenhouse gas emissions: findings from a global meta - analysis and density functional theory calculation. Biochar. https://doi.org/10.1007/s42773-024-00383-6.

Honvault, N., Nobile, C., Faucon, M.P., Firmin, S., Houben, D., 2022. Direct and indirect interactions between biochar properties, plant belowground traits, and plant performance. GCB Bioenergy 14, 1254–1265. https://doi.org/10.1111/gcbb.12993.

Hossain, M.Z., Bahar, M.M., Sarkar, B., Donne, S.W., Ok, Y.S., Palansooriya, K.N., Kirkham, M.B., Chowdhury, S., Bolan, N., 2020. Biochar and its importance on nutrient dynamics in soil and plant. Biochar. Springer Singapore. https://doi.org/10.1007/s42773-020-00065-z.

Huang, L., Gu, M., 2019. Effects of biochar on container substrate properties and growth of plants—a review. Horticulturae 5, 1–26. https://doi.org/10.3390/horticulturae5010014.

Huang, W. kun, Ji, H. li, Gheysen, G., Debode, J., Kyndt, T., 2015. Biochar-amended potting medium reduces the susceptibility of rice to root-knot nematode infections. BMC Plant Biol. 15, 1–15. https://doi.org/10.1186/s12870-015-0654-7.

Huang, H., Reddy, N.G., Huang, X., Chen, P., Wang, P., Zhang, Y., Huang, Y., Lin, P., Garg, A., 2021. Effects of pyrolysis temperature, feedstock type and compaction on water retention of biochar amended soil. Sci. Rep. 11, 1–19. https://doi.org/10.1038/s41598-021-86701-5.

Huang, K., Zhang, J., Tang, G., Bao, D., Wang, T., Kong, D., 2023. Impacts and mechanisms of biochar on soil microorganisms. Plant. Soil Environ. https://doi.org/10.17221/348/2022-PSE.

Hye, F.B., Saha, N., Shamim, M., Dutta, O., 2025. Assessing suitable tree species for biochar production and their performance in enriching soil quality and plant growth of Oryza sativa L . And Acacia auriculiformis A. Cunn . 11, 32–52.

Intani, K., Latif, S., Islam, M.S., Müller, J., 2019. Phytotoxicity of corncob biochar before and after heat treatment and washing. Sustain. Times 11. https://doi.org/10.3390/su11010030.

Ippolito, J.A., Laird, D.A., Busscher, W.J., 2012. Environmental benefits of biochar. J. Environ. Qual. 41, 967–972. https://doi.org/10.2134/jeq2012.0151.

Islam, M.S., Chen, Y., Weng, L., Ma, J., Khan, Z.H., Liao, Z., Magid, A.S.I.A., Li, Y., 2021a. Watering techniques and zero-valent iron biochar pH effects on as and Cd concentrations in rice rhizosphere soils, tissues and yield. J. Environ. Sci. (China) 100. https://doi.org/10.1016/j.jes.2020.07.002.

Islam, M.S., Magid, A.S.I.A., Chen, Y., Weng, L., Arafat, M.Y., Khan, Z.H., Ma, J., Li, Y., 2021b. Arsenic and cadmium load in rice tissues cultivated in calcium enriched biochar amended paddy soil. Chemosphere 283, 131102.

Islam, M.S., Magid, A.S.I.A., Chen, Y., Weng, L., Ma, J., Arafat, M.Y., Khan, Z.H., Li, Y., 2021c. Effect of calcium and iron-enriched biochar on arsenic and cadmium accumulation from soil to rice paddy tissues. Sci. Total Environ. 785. https://doi.org/10.1016/j.scitotenv.2021.147163.

Islam, M.S., Zhu, J., Xiao, L., Khan, Z.H., Saqib, H.S.A., Gao, M., Song, Z., 2023. Enhancing rice quality and productivity: multifunctional biochar for arsenic, cadmium, and bacterial control in paddy soil. Chemosphere 342, 140157.

Islam, M.S., Deng, H., Dong, Y., Zhu, J., Gao, M., Song, Z., 2024. Improving arsenic and cadmium contaminated paddy soil health and rice quality with plant-animal-based modified biochar: a mechanistic study. J. Clean. Prod. 448, 141659.

Jabborova, D., Annapurna, K., Paul, S., Kumar, S., Saad, H.A., Desouky, S., Ibrahim, M.F.M., Elkelish, A., 2021. Beneficial features of biochar and arbuscular mycorrhiza for improving spinach plant growth, root morphological traits, physiological properties, and soil enzymatic activities. J. Fungi 7. https://doi.org/10.3390/jof7070571.

Jain, S., Yadav, V., Mishra, D., Kumar, B., Deshmukh, Y., Nigam, N., Khare, P., 2022. Utilizing three different biochars for attenuation of toxic acidic mine spoils reflected by lixiviate quality vis-a-vis phyto-toxicity on ocimum sanctum and cassia angustifolia. Glob. J. Agric. Innov. Res. Dev. 8, 49–65. https://doi.org/10.15377/2409-9813.2021.08.4.

Jaiswal, A.K., Frenkel, O., Elad, Y., Lew, B., Graber, E.R., 2015. Non-monotonic influence of biochar dose on bean seedling growth and susceptibility to Rhizoctonia solani: the "Shifted Rmax-Effect." Plant Soil 395, 125–140. https://doi.org/10.1007/s11104-014-2331-2.

Jaiswal, A.K., Elad, Y., Paudel, I., Graber, E.R., Cytryn, E., Frenkel, O., 2017. Linking the belowground microbial composition, diversity and activity to soilborne disease suppression and growth promotion of tomato amended with biochar. Sci. Rep. 7, 1–17. https://doi.org/10.1038/srep44382.

Jeffery, S., Bezemer, T.M., Cornelissen, G., Kuyper, T.W., Lehmann, J., Mommer, L., Sohi, S.P., van de Voorde, T.F.J., Wardle, D.A., van Groenigen, J.W., 2015. The way forward in biochar research: Targeting trade-offs between the potential wins. GCB Bioenergy 7, 1–13. https://doi.org/10.1111/gcbb.12132.

Jiang, S., Liu, Y., Shu, Y., 2022. Biochar and exogenous calcium assisted alleviation of Pb phytotoxicity in water spinach (Ipomoea aquatica Forsk) cultivated in Pb-spiked soil. Environ. Geochem. Health 44, 207–219. https://doi.org/10.1007/s10653-021-00977-0.

Jin, X., Bai, Y., Khashi u Rahman, M., Kang, X., Pan, K., Wu, F., Pommier, T., Zhou, X., Wei, Z., 2022. Biochar stimulates tomato roots to recruit a bacterial assemblage contributing to disease resistance against Fusarium wilt. iMeta 1. https://doi.org/10.1002/imt2.37.

Kalus, K., Koziel, J.A., Opaliński, S., 2019. A review of biochar properties and their utilization in crop agriculture and livestock production. Appl. Sci. https://doi.org/10.3390/app9173494.

Khan, S., Shah, Z., Mian, I.A., Dawar, K., Tariq, M., Khan, B., Mussarat, M., Amin, H., Ismail, M., Ali, S., Shah, T., Alamri, S., Siddiqui, M.H., Adnan, M., Romman, M., Fahad, S., Noman, A., Kamal, A., 2020. Soil fertility, n2 fixation and yield of chickpea as influenced by long-term biochar application under mung–chickpea cropping system. Sustain. Times 12, 1–14. https://doi.org/10.3390/su12219008.

Khan, W.A., Arain, M.B., Balal, S., Niaz, A., Mollahosseini, A., Soylak, M., 2025. Eco-friendly and green biochar sorbent–extraction techniques for pharmaceuticals, environmental, and food analysis—a review. J. Sep. Sci. https://doi.org/10.1002/jssc.70074.

Kharel, G., Sacko, O., Feng, X., Morris, J.R., Phillips, C.L., Trippe, K., Kumar, S., Lee, J.W., 2019. Biochar surface oxygenation by ozonization for super high cation exchange capacity. ACS Sustain. Chem. Eng. 7, 16410–16418. https://doi.org/10.1021/acssuschemeng.9b03536.

Kloss, S., Zehetner, F., Dellantonio, A., Hamid, R., Ottner, F., Liedtke, V., Schwanninger, M., Gerzabek, M.H., Soja, G., 2012. Characterization of slow pyrolysis biochars: effects of feedstocks and pyrolysis temperature on biochar properties. J. Environ. Qual. 41, 990–1000. https://doi.org/10.2134/jeq2011.0070.

Knoblauch, C., Priyadarshani, S.H.R., Haefele, S.M., Schröder, N., Pfeiffer, E.M., 2021. Impact of biochar on nutrient supply, crop yield and microbial respiration on sandy soils of northern Germany. Eur. J. Soil Sci. 72, 1885–1901. https://doi.org/10.1111/ejss.13088.

Kochanek, J., Long, R.L., Lisle, A.T., Flematti, G.R., 2016. Karrikins identified in biochars indicate post-fire chemical cues can influence community diversity and plant development. PLoS One 11, 1–19. https://doi.org/10.1371/journal.pone.0161234.

Koide, R.T., Nguyen, B.T., Skinner, R.H., Dell, C.J., Peoples, M.S., Adler, P.R., Drohan, P.J., 2015. Biochar amendment of soil improves resilience to climate change. GCB Bioenergy 7, 1084–1091. https://doi.org/10.1111/gcbb.12191.

Kolton, M., Graber, E.R., Tsehansky, L., Elad, Y., Cytryn, E., 2017. Biochar-stimulated plant performance is strongly linked to microbial diversity and metabolic potential in the rhizosphere. New Phytol. 213, 1393–1404. https://doi.org/10.1111/nph.14253.

Kuttippurath, J., Abbhishek, K., Chander, G., Dixit, S., Singh, A., Das, D., Dey, S., 2023. Biochar-based nutrient management as a futuristic scalable strategy for C-sequestration in semiarid tropics. Agron. J. 115, 2311–2324. https://doi.org/10.1002/agj2.21424.

Laird, D.A., Fleming, P., Davis, D.D., Horton, R., Wang, B., Karlen, D.L., 2010. Impact of biochar amendments on the quality of a typical Midwestern agricultural soil. Geoderma 158, 443–449. https://doi.org/10.1016/j.geoderma.2010.05.013.

Lebrun, M., Aguinaga, M., Zahid, Z., Šimek, P., Ouředníček, P., Klápště, P., Száková, J., Bímová, K.B., Jačka, L., Beesley, L., Punčochář, J., Trakal, L., 2024. Manure-biochar blends effectively reduce nutrient leaching and increase water retention in a sandy, agricultural soil: Insights from a field experiment. Soil Use Manag. 40, 1–27. https://doi.org/10.1111/sum.13135.

Lentz, R.D., Ippolito, J.A., 2012. Biochar and manure affect calcareous soil and corn silage nutrient concentrations and uptake. J. Environ. Qual. 41, 1033–1043. https://doi.org/10.2134/jeq2011.0126.

Lestiyani, A., Purnomo, E., Widyaningsih, S., Fadillah, K., 2023. Investigating the effects of bamboo biochar on fertile soil for improving the production of brassica oleraceae. Agric. For. 35, 169–180. https://doi.org/10.24246/agric.2023.v35.i2.p169-180.

Li, Y., Shao, J., Wang, X., Deng, Y., Yang, H., Chen, H., 2014. Characterization of modified biochars derived from bamboo pyrolysis and their utilization for target component (furfural) adsorption. Energy Fuels 28, 5119–5127. https://doi.org/10.1021/ef500725c.

Li, Y., Gao, L., Wang, Y., Wang, Y., Wan, S., 2020. Enhanced removal of heavy metals from water by hydrous ferric oxide-modified biochar. ACS Omega 5, 28702–28711. https://doi.org/10.1021/acsomega.0c03893.

Li, L., Long, A., Fossum, B., Kaiser, M., 2023a. Effects of pyrolysis temperature and feedstock type on biochar characteristics pertinent to soil carbon and soil health: a meta-analysis. Soil Use Manag. 39, 43–52. https://doi.org/10.1111/sum.12848.

Li, S., Xie, Y., Jiang, S., Yang, M., Lei, H., Cui, W., Wang, F., 2023b. Biochar decreases Cr toxicity and accumulation in sunflower grown in Cr(VI)-polluted soil. Toxics 11. https://doi.org/10.3390/toxics11090787.

Li, L., Wang, J., Sun, W., Peng, X., Qi, X., 2025. Production of algae-derived biochar and its application in pollutants adsorption—a mini review. Separations. https://doi.org/10.3390/separations12020032.

Liao, S., Pan, B., Li, H., Zhang, D., Xing, B., 2014. Detecting free radicals in biochars and determining their ability to inhibit the germination and growth of corn, wheat and rice seedlings. Environ. Sci. Technol. 48, 8581–8587. https://doi.org/10.1021/es404250a.

Licht, J., Smith, N., Mitchell, P., Shields, F., 2017. Impact of lignocellulosic and hemicellulosic biochar on soil moisture in low clay soils. J. Plant Nutr. Soil Sci. 180, 576–584. https://doi.org/10.1002/jpln.201700221.

Lin, X.W., Xie, Z.B., Zheng, J.Y., Liu, Q., Bei, Q.C., Zhu, J.G., 2015. Effects of biochar application on greenhouse gas emissions, carbon sequestration and crop growth in coastal saline soil. Eur. J. Soil Sci. 66, 329–338. https://doi.org/10.1111/ejss.12225.

Liu, S., Zhang, Y., Zong, Y., Hu, Z., Wu, S., Zhou, J., Jin, Y., Zou, J., 2016. Response of soil carbon dioxide fluxes, soil organic carbon and microbial biomass carbon to biochar amendment: a meta-analysis. GCB Bioenergy 8, 392–406. https://doi.org/10.1111/gcbb.12265.

Liu, C., Liu, F., Ravnskov, S., Rubæk, G.H., Sun, Z., Andersen, M.N., 2017a. Impact of wood biochar and its interactions with mycorrhizal fungi, phosphorus fertilization and irrigation strategies on potato growth. J. Agron. Crop Sci. 203, 131–145. https://doi.org/10.1111/jac.12185.

Liu, Z., Dugan, B., Masiello, C.A., Gonnermann, H.M., 2017b. Biochar particle size, shape, and porosity act together to influence soil water properties. PLoS One 12, 1–19. https://doi.org/10.1371/journal.pone.0179079.

Liu, L., Tan, Z., Gong, H., Huang, Q., 2019. Migration and transformation mechanisms of nutrient elements (N, P, K) within biochar in straw-biochar-soil-plant systems: a review. ACS Sustain. Chem. Eng. 7, 22–32. https://doi.org/10.1021/acssuschemeng.8b04253.

Liu, C., Wang, W., Wu, R., Liu, Y., Lin, X., Kan, H., Zheng, Y., 2020. Preparation of acid- and alkali-modified biochar for removal of methylene blue pigment. ACS Omega 5, 30906–30922. https://doi.org/10.1021/acsomega.0c03688.

Liu, M., Linna, C., Ma, S., Ma, Q., Song, W., Shen, M., Song, L., Cui, K., Zhou, Y., Wang, L., 2022. Biochar combined with organic and inorganic fertilizers promoted the rapeseed nutrient uptake and improved the purple soil quality. Front. Nutr. 9. https://doi.org/10.3389/fnut.2022.997151.

Liu, Y., Yao, L., Hu, B., Li, T., Tian, H., 2023. Adsorption behavior and residue degradation of triazine herbicides in soil amended with rice straw biochar. Agric. For. 13, 1–12. https://doi.org/10.3390/agriculture13071282.

López, J.E., Arroyave, C., Aristizábal, A., Almeida, B., Builes, S., Chavez, E., 2022. Reducing cadmium bioaccumulation in Theobroma cacao using biochar: basis for scaling-up to field. Heliyon 8. https://doi.org/10.1016/j.heliyon.2022.e09790.

Luo, Y., Dungait, J.A.J., Zhao, X., Brookes, P.C., Durenkamp, M., Li, G., Lin, Q., 2018. Pyrolysis temperature during biochar production alters its subsequent utilization by microorganisms in an acid arable soil. L. Degrad. Dev. 29, 2183–2188. https://doi.org/10.1002/ldr.2846.

Major, J., Rondon, M., Molina, D., Riha, S.J., Lehmann, J., 2010. Maize yield and nutrition during 4 years after biochar application to a Colombian savanna oxisol. Plant Soil 333, 117–128. https://doi.org/10.1007/s11104-010-0327-0.

Mao, Q., He, B., Ma, M., Wang, L., Koike, T., Agathokleous, E., 2022. Effects of biochar on karst lime soil nutrients, soil microbial communities and physiology of Sichuan pepper plants. Ann. Appl. Biol. 181, 357–366. https://doi.org/10.1111/aab.12787.

Martinsen, V., Mulder, J., Shitumbanuma, V., Sparrevik, M., Børresen, T., Cornelissen, G., 2014. Farmer-led maize biochar trials: effect on crop yield and soil nutrients under conservation farming. J. Plant Nutr. Soil Sci. 177, 681–695. https://doi.org/10.1002/jpln.201300590.

Maslouski, M., Eschenbach, A., Beer, C., Thomsen, S., Porada, P., 2025. Soil and vegetation responses to biochar application in terms of its feedback on carbon sequestration under different environmental conditions—LiDELS model overview. Environ. Res. Lett. 20. https://doi.org/10.1088/1748-9326/adbfa6.

Meyer, S., Glaser, B., Quicker, P., 2011. Technical, economical, and climate-related aspects of biochar production technologies: a literature review. Environ. Sci. Technol. https://doi.org/10.1021/es201792c.

Michael, P.S., 2020. Cogon grass biochar amendment and Panicum coloratum planting improve selected properties of sandy soil under humid lowland tropical climatic conditions. Biochar 2, 489–502. https://doi.org/10.1007/s42773-020-00057-z.

Mierzwa-Hersztek, M., Wolny-Koładka, K., Gondek, K., Gałązka, A., Gawryjołek, K., 2020. Effect of Coapplication of biochar and nutrients on Microbiocenotic composition, dehydrogenase activity index and chemical properties of sandy soil. Waste Biomass Valoriz. 11, 3911–3923. https://doi.org/10.1007/s12649-019-00757-z.

Mu, M., Nie, W., Zhao, S., Jiang, J., Wang, Z., Wu, Y., Zheng, L., 2024. Effects of biochar on the growth, physiology and heavy metal uptake of Trifolium pratense L. In: Pb-contaminated Soil, pp. 1–25.

Murtaza, G., Ditta, A., Ullah, N., Usman, M., Ahmed, Z., 2021. Biochar for the management of nutrient impoverished and metal contaminated soils: preparation, applications, and prospects. J. Soil Sci. Plant Nutr. 21, 2191–2213. https://doi.org/10.1007/s42729-021-00514-z.

Mustafa, A., Brtnicky, M., Hammerschmiedt, T., Kucerik, J., Kintl, A., Chorazy, T., Naveed, M., Skarpa, P., Baltazar, T., Malicek, O., Holatko, J., 2022. Food and agricultural wastes-derived biochars in combination with mineral fertilizer as sustainable soil amendments to enhance soil microbiological activity, nutrient cycling and crop production. Front. Plant Sci. 13, 1–11. https://doi.org/10.3389/fpls.2022.1028101.

Nagaraju, K., Prasad, T.N.V.K.V., Naidu, M.V.S., Chari, M.S., Ramu, Y.R., Murthy, B.R., 2023. Exploring the benefits of rice husk waste: synthesis and characterization of biochar and nanobiochar for agricultural and environmental sustainability. Int. J. Environ. Clim. Chang. 13, 715–725. https://doi.org/10.9734/ijecc/2023/v13i92292.

Namgay, T., Singh, B., Singh, B.P., 2010. Influence of biochar application to soil on the availability of As, Cd, Cu, Pb, and Zn to maize (Zea mays L.). Aust. J. Soil Res. 48, 638–647. https://doi.org/10.1071/SR10049.

Nash, J.A., Miesel, J.R., Bonito, G.M., Sakalidis, M.L., Ren, H., Warnock, D., Tiemann, L.K., 2021. Biochar restructures plant–soil–microbe relationships in a woody cropping system. Soil Sci. Soc. Am. J. 85, 2019–2039. https://doi.org/10.1002/saj2.20334.

Novak, J.M., Johnson, M.G., Spokas, K.A., 2018. Concentration and release of phosphorus and potassium from lignocellulosic- and manure-based biochars for fertilizer reuse. Front. Sustain. Food Syst. 2, 1–9. https://doi.org/10.3389/fsufs.2018.00054.

Nurida, N.L., Salma, S., Jubaedah, 2023. Corn cob biochar as agricultural waste management for increasing soil resilient and crop yield. E3S Web Conf. 467, 1–5. https://doi.org/10.1051/e3sconf/202346701003.

Nyambo, P., Motsi, H., Chiduza, C., Malobane, M.E., 2023. Biochar ageing improves soil properties, growth and yield of red radish (*Raphanus sativus*) in a Haplic Cambisol. PLoS ONE 18, 1–14. https://doi.org/10.1371/journal.pone.0288709.

Palansooriya, K.N., Wong, J.T.F., Hashimoto, Y., Huang, L., Rinklebe, J., Chang, S.X., Bolan, N., Wang, H., Ok, Y.S., 2019. Response of microbial communities to biochar-amended soils: a critical review. Biochar 1, 3–22. https://doi.org/10.1007/s42773-019-00009-2.

Pantoja, F., Beszédes, S., Gyulavári, T., Illés, E., Kozma, G., László, Z., 2024. Impact of pyrolysis temperature on the physical and chemical properties of non-modified biochar produced from banana leaves: a case study on ammonium ion adsorption. Materials 17. https://doi.org/10.3390/ma17133180.

Pariyar, P., Kumari, K., Jain, M.K., Jadhao, P.S., 2020. Evaluation of change in biochar properties derived from different feedstock and pyrolysis temperature for environmental and agricultural application. Sci. Total Environ. 713, 136433. https://doi.org/10.1016/j.scitotenv.2019.136433.

Park, J.H., Choppala, G.K., Bolan, N.S., Chung, J.W., Chuasavathi, T., 2011. Biochar reduces the bioavailability and phytotoxicity of heavy metals. Plant Soil 348, 439–451. https://doi.org/10.1007/s11104-011-0948-y.

Paymaneh, Z., Gryndler, M., Konvalinková, T., Benada, O., Borovička, J., Bukovská, P., Püschel, D., Řezáčová, V., Sarcheshmehpour, M., Jansa, J., 2018. Soil matrix determines the outcome of interaction between mycorrhizal symbiosis and biochar for Andropogon gerardii growth and nutrition. Front. Microbiol. 9, 1–16. https://doi.org/10.3389/fmicb.2018.02862.

Phillips, C.L., Trippe, K.M., Whittaker, G., Griffith, S.M., Johnson, M.G., Banowetz, G.M., 2016. Gasified grass and wood biochars facilitate plant establishment in acid mine soils. J. Environ. Qual. 45, 1013–1020. https://doi.org/10.2134/jeq2015.09.0470.

Prakongkep, N., Gilkes, R.J., Wiriyakitnateekul, W., 2015. Forms and solubility of plant nutrient elements in tropical plant waste biochars. J. Plant Nutr. Soil Sci. 178, 732–740. https://doi.org/10.1002/jpln.201500001.

Prendergast-Miller, M.T., Duvall, M., Sohi, S.P., 2014. Biochar-root interactions are mediated by biochar nutrient content and impacts on soil nutrient availability. Eur. J. Soil Sci. 65, 173–185. https://doi.org/10.1111/ejss.12079.

Pressler, Y., Hunter-Laszlo, M., Bucko, S., Covitt, B.A., Urban, S., Benton, C., Bartholomew, M., Morrison, A.J., Foster, E.J., Parker, S.D., Cotrufo, M.F., Moore, J.C., 2019. Teaching authentic soil & plant science in middle school classrooms with a biochar case study. Am. Biol. Teach. 81, 256–268. https://doi.org/10.1525/abt.2019.81.4.256.

Puga, A.P., Abreu, C.A., Melo, L.C.A., Beesley, L., 2015. Biochar application to a contaminated soil reduces the availability and plant uptake of zinc, lead and cadmium. J. Environ. Manage. 159, 86–93. https://doi.org/10.1016/j.jenvman.2015.05.036.

Qian, L., Chen, B., Chen, M., 2016. Novel alleviation mechanisms of aluminum phytotoxicity via released biosilicon from rice straw-derived biochars. Sci. Rep. 6, 1–11. https://doi.org/10.1038/srep29346.

Qin, J., Wang, X., Ying, J., Lin, C., 2022. Biochar is not durable for remediation of heavy metal-contaminated soils affected by acid-mine drainage. Toxics 10 (462), 1–15. https://doi.org/10.3390/toxics10080462.

Qiu, M., Liu, L., Ling, Q., Cai, Y., Yu, S., Wang, S., Fu, D., Hu, B., Wang, X., 2022. Biochar for the removal of contaminants from soil and water: a review. Biochar 4, 1–25. https://doi.org/10.1007/s42773-022-00146-1.

Rajkovich, S., Enders, A., Hanley, K., Hyland, C., Zimmerman, A.R., Lehmann, J., 2012. Corn growth and nitrogen nutrition after additions of biochars with varying properties to a temperate soil. Biol. Fertil. Soils 48, 271–284. https://doi.org/10.1007/s00374-011-0624-7.

Rittl, T.F., Oliveira, D.M.S., Canisares, L.P., Sagrilo, E., Butterbach-Bahl, K., Dannenmann, M., Cerri, C.E.P., 2021. High application rates of biochar to mitigate N2O emissions from a N-fertilized tropical soil under warming conditions. Front. Environ. Sci. 8, 1–5. https://doi.org/10.3389/fenvs.2020.611873.

Rogovska, N., Laird, D., Cruse, R.M., Trabue, S., Heaton, E., 2012. Germination tests for assessing biochar quality. J. Environ. Qual. 41, 1014–1022. https://doi.org/10.2134/jeq2011.0103.

Rombolà, A.G., Marisi, G., Torri, C., Fabbri, D., Buscaroli, A., Ghidotti, M., Hornung, A., 2015. Relationships between chemical characteristics and phytotoxicity of biochar from poultry litter pyrolysis. J. Agric. Food Chem. 63, 6660–6667. https://doi.org/10.1021/acs.jafc.5b01540.

Sahin, O., Taskin, M.B., Kaya, E.C., Atakol, O., Emir, E., Inal, A., Gunes, A., 2017. Effect of acid modification of biochar on nutrient availability and maize growth in a calcareous soil. Soil Use Manag. 33, 447–456. https://doi.org/10.1111/sum.12360.

Sangsuwan, P., Detraksa, J., Srimawong, P., 2023. Proximate analysis of the growth of organic grey oyster mushrooms on biochar from agricultural waste. Pakistan J. Phytopathol. 35, 83–91. https://doi.org/10.33866/phytopathol.035.01.0852.

Schulz, H., Dunst, G., Glaser, B., 2013. Positive effects of composted biochar on plant growth and soil fertility. Agron. Sustain. Dev. 33, 817–827. https://doi.org/10.1007/s13593-013-0150-0.

She, D., Sun, X., Gamareldawla, A.H.D., Nazar, E.A., Hu, W., Edith, K., Yu, S., 2018. Benefits of soil biochar amendments to tomato growth under saline water irrigation. Sci. Rep. 8, 1–10. https://doi.org/10.1038/s41598-018-33040-7.

Shen, Z., Jin, F., Wang, F., McMillan, O., Al-Tabbaa, A., 2015. Sorption of lead by Salisbury biochar produced from British broadleaf hardwood. Bioresour. Technol. 193, 553–556. https://doi.org/10.1016/j.biortech.2015.06.111.

Shen, Z., Zhang, Y., Jin, F., McMillan, O., Al-Tabbaa, A., 2017. Qualitative and quantitative characterisation of adsorption mechanisms of lead on four biochars. Sci. Total Environ. 609, 1401–1410. https://doi.org/10.1016/j.scitotenv.2017.08.008.

Sherman, L.A., Page-Dumroese, D.S., Coleman, M.D., 2018. Idaho forest growth response to post-thinning energy biomass removal and complementary soil amendments. GCB Bioenergy 10, 246–261. https://doi.org/10.1111/gcbb.12486.

Shin, H., Chun, D., Cho, I.R., Hanif, M.A., Kang, S.S., Kwac, L.K., Kim, H.G., Kim, Y.S., 2024. Systematic characterization of cow manure biochar and its effect on salicornia herbacea L. growth. Sustain. Times 16. https://doi.org/10.3390/su16083396.

Shiyal, V., Patel, V.M., Patel, H.K., Rathwa, M., Patel, P., 2022. Biochar: an emerging soil amendment for sustaining soil health and black gold for indian agriculture. J. Exp. Agric. Int 6–12. https://doi.org/10.9734/jeai/2022/v44i122072.

Shomana, T., Botha, D., Agachi, P.S., 2020. The water retention properties of biochar derived from broiler poultry litter as applied to the Botswana soil. DRC Sustain. Futur. J. Environ. Agric. Energy 1, 66–72. https://doi.org/10.37281/drcsf/1.1.9.

Shrestha, R.K., Jacinthe, P.A., Lal, R., Lorenz, K., Singh, M.P., Demyan, S.M., Ren, W., Lindsey, L.E., 2023. Biochar as a negative emission technology: a synthesis of field research on greenhouse gas emissions. J. Environ. Qual. 52, 769–798. https://doi.org/10.1002/jeq2.20475.

Singh, H., Northup, B.K., Rice, C.W., Prasad, P.V.V., 2022. Biochar applications influence soil physical and chemical properties, microbial diversity, and crop productivity: a meta-analysis. Biochar 4, 1–17. https://doi.org/10.1007/s42773-022-00138-1.

Soria, R.I., Rolfe, S.A., Betancourth, M.P., Thornton, S.F., 2020. The relationship between properties of plant-based biochars and sorption of Cd(II), Pb(II) and Zn(II) in soil model systems. Heliyon 6. https://doi.org/10.1016/j.heliyon.2020.e05388.

Spokas, K.A., Cantrell, K.B., Novak, J.M., Archer, D.W., Ippolito, J.A., Collins, H.P., Boateng, A.A., Lima, I.M., Lamb, M.C., McAloon, A.J., Lentz, R.D., Nichols, K.A., 2012. Biochar: a synthesis of its agronomic impact beyond carbon sequestration. J. Environ. Qual. 41, 973–989. https://doi.org/10.2134/jeq2011.0069.

Srivastava, P., Gadi, Y., Supriya, B., Lytand, W., Singh, B.V., Katiyar, D., 2023. Biochar's influence on soil microorganisms: understanding the impacts and mechanisms. Int. J. Plant Soil Sci. 35, 455–464. https://doi.org/10.9734/ijpss/2023/v35i183310.

Strawn, D.G., Rigby, A.C., Baker, L.L., Coleman, M.D., Koch, I., 2015. Biochar soil amendment effects on arsenic availability to mountain brome (bromus marginatus). J. Environ. Qual. 44, 1315–1320. https://doi.org/10.2134/jeq2014.11.0477.

Sun, H., He, F., Pan, Y., Zhang, Z., 2017. Effects of pyrolysis temperature and residence time on physicochemicalproperties of different biochar types. Acta Agric. Scand. B Soil Plant Sci. 67 (1), 12–22. https://doi.org/10.1080/09064710.2016.1214745.

Sun, J., He, F., Shao, H., Zhang, Z., Xu, G., 2016. Effects of biochar application on Suaeda salsa growth and saline soil properties. Environ. Earth Sci. 75. https://doi.org/10.1007/s12665-016-5440-9.

Tammeorg, P., Bastos, A.C., Jeffery, S., Rees, F., Kern, J., Graber, E.R., Ventura, M., Kibblewhite, M., Amaro, A., Budai, A., Cordovil, C.M.d.S., Domene, X., Gardi, C., Gascó, G., Horák, J., Kammann, C., Kondrlova, E., Laird, D., Loureiro, S., Martins, M.A.S., Panzacchi, P., Prasad, M., Prodana, M., Puga, A.P., Ruysschaert, G., Sas-Paszt, L., Silva, F.C., Teixeira, W.G., Tonon, G., Delle Vedove, G., Zavalloni, C., Glaser, B., Verheijen, F.G.A., 2017. Biochars in soils: towards the required level of scientific understanding. J. Environ. Eng. Landsc. Manag. 25, 192–207. https://doi.org/10.3846/16486897.2016.1239582.

Tan, X., Liu, Y., Gu, Y., Zeng, G., Hu, X., Wang, X., Hu, X., Guo, Y., Zeng, X., Sun, Z., 2015. Biochar amendment to lead-contaminated soil: effects on fluorescein diacetate hydrolytic activity and phytotoxicity to rice. Environ. Toxicol. Chem. 34, 1962–1968. https://doi.org/10.1002/etc.3023.

Thao, T., Gonzales, M., Ryals, R., Dahlquist-Willard, R., Diaz, G.C., Ghezzehei, T.A., 2024. Biochar impacts on soil moisture retention and respiration in a coarse-textured soil under dry conditions. Soil Sci. Soc. Am. J. 1919–1931. https://doi.org/10.1002/saj2.20746.

Thomas, S.C., Ruan, R., Gale, N.V., Gezahegn, S., 2024. Phytotoxicity and hormesis in common mobile organic compounds in leachates of wood-derived biochars. Biochar 6. https://doi.org/10.1007/s42773-024-00339-w.

Tian, X., Li, C., Zhang, M., Wan, Y., Xie, Z., Chen, B., Li, W., 2018. Biochar derived from corn straw affected availability and distribution of soil nutrients and cotton yield. PLoS One 13, 1–19. https://doi.org/10.1371/journal.pone.0189924.

Tian, S.Q., Wang, L., Liu, Y.L., Yang, T., Huang, Z.S., Wang, X.S., He, H.Y., Jiang, J., Ma, J., 2019. Enhanced permanganate oxidation of sulfamethoxazole and removal of dissolved organics with biochar: formation of highly oxidative manganese intermediate species and in situ activation of biochar. Environ. Sci. Technol. 53, 5282–5291. https://doi.org/10.1021/acs.est.9b00180.

Tisserant, A., Cherubini, F., 2019. Potentials, limitations, co-benefits, and trade-offs of biochar applications to soils for climate change mitigation. Land 8. https://doi.org/10.3390/LAND8120179.

Trifunovic, B., Gonzales, H.B., Ravi, S., Sharratt, B.S., Mohanty, S.K., 2018. Dynamic effects of biochar concentration and particle size on hydraulic properties of sand. L. Degrad. Dev. 29, 884–893. https://doi.org/10.1002/ldr.2906.

Trinh, B.S., Werner, D., Reid, B.J., 2017. Application of a full-scale wood gasification biochar as a soil improver to reduce organic pollutant leaching risks. J. Chem. Technol. Biotechnol. 92, 1928–1937. https://doi.org/10.1002/jctb.5219.

Tsai, C.C., Chang, Y.F., 2020. Effects of biochar to excessive compost-fertilized soils on the nutrient status. Agronomy 10. https://doi.org/10.3390/agronomy10050683.

Uchimiya, M., Bannon, D.I., 2013. Solubility of lead and copper in biochar-amended small arms range soils: influence of soil organic carbon and pH. J. Agric. Food Chem. 61, 7679–7688. https://doi.org/10.1021/jf401481x.

Uchimiya, M., Wartelle, L.H., Klasson, K.T., Fortier, C.A., Lima, I.M., 2011. Influence of pyrolysis temperature on biochar property and function as a heavy metal sorbent in soil. J. Agric. Food Chem. 59, 2501–2510. https://doi.org/10.1021/jf104206c.

Ulyett, J., Sakrabani, R., Kibblewhite, M., Hann, M., 2014. Impact of biochar addition on water retention, nitrification and carbon dioxide evolution from two sandy loam soils. Eur. J. Soil Sci. 65, 96–104. https://doi.org/10.1111/ejss.12081.

Uzoma, K.C., Inoue, M., Andry, H., Fujimaki, H., Zahoor, A., Nishihara, E., 2011. Effect of cow manure biochar on maize productivity under sandy soil condition. Soil Use Manag. 27, 205–212. https://doi.org/10.1111/j.1475-2743.2011.00340.x.

van Zwieten, L., Kimber, S., Morris, S., Chan, K.Y., Downie, A., Rust, J., Joseph, S., Cowie, A., 2010. Effects of biochar from slow pyrolysis of papermill waste on agronomic performance and soil fertility. Plant Soil 327, 235–246. https://doi.org/10.1007/s11104-009-0050-x.

van Zwieten, L., Kimber, S., Morris, S., Macdonald, L.M., Rust, J., Petty, S., Joseph, S., Rose, T., 2019. Biochar improves diary pasture yields by alleviating P and K constraints with no influence on soil respiration or N2O emissions. Biochar 1, 115–126. https://doi.org/10.1007/s42773-019-00005-6.

Venkatesh, G., Gopinath, K.A., Reddy, K.S., Reddy, B.S., Prabhakar, M., Srinivasarao, C., Kumari, V.V., Singh, V.K., 2022. Characterization of biochar derived from crop residues for soil amendment, carbon sequestration and energy use. Sustain. Times 14. https://doi.org/10.3390/su14042295.

Viger, M., Hancock, R.D., Miglietta, F., Taylor, G., 2015. More plant growth but less plant defence? First global gene expression data for plants grown in soil amended with biochar. GCB Bioenergy 7, 658–672. https://doi.org/10.1111/gcbb.12182.

Visioli, G., Conti, F.D., Menta, C., Bandiera, M., Malcevschi, A., Jones, D.L., Vamerali, T., 2016. Assessing biochar ecotoxicology for soil amendment by root phytotoxicity bioassays. Environ. Monit. Assess. 188, 1–11. https://doi.org/10.1007/s10661-016-5173-y.

Vitková, J., Šurda, P., Rončák, P., Botková, N., Zvala, A., 2021. Statistical analysis of soil water content differences after biochar application and its repeated application during 2020 growing season. Acta Hydrol. Slovaca 22, 320–325. https://doi.org/10.31577/ahs-2021-0022.02.0036.

Walter, R., Rao, B.K.R., 2015. Biochars influence sweet-potato yield and nutrient uptake in tropical Papua New Guinea. J. Plant Nutr. Soil Sci. 178, 393–400. https://doi.org/10.1002/jpln.201400405.

Wang, Z., Han, L., Sun, K., Jin, J., Ro, K.S., Libra, J.A., Liu, X., Xing, B., 2016. Sorption of four hydrophobic organic contaminants by biochars derived from maize straw, wood dust and swine manure at different pyrolytic temperatures. Chemosphere 144, 285–291. https://doi.org/10.1016/j.chemosphere.2015.08.042.

Wang, D., Li, B., Yang, H., Zhao, C., Yao, D., Chen, H., 2019. Influence of biochar on the steam reforming of biomass volatiles: effects of activation temperature and atmosphere. Energy Fuels 33, 2328–2334. https://doi.org/10.1021/acs.energyfuels.8b04412.

Wang, Y.M., Wang, S.W., Wang, C.Q., Zhang, Z.Y., Zhang, J.Q., Meng, M., Li, M., Uchimiya, M., Yuan, X.Y., 2020. Simultaneous immobilization of soil Cd(II) and As(V) by fe-modified biochar. Int. J. Environ. Res. Public Health 17, 1–12. https://doi.org/10.3390/ijerph17030827.

Wang, Z., Wang, H., Zhao, C., Yang, K., Li, Z., Yin, K., 2022. Effects of biochar on the microenvironment of saline-sodic soil and maize growth. Agronomy 12. https://doi.org/10.3390/agronomy12112859.

Wang, J., Riaz, M., Babar, S., El-Desouki, Z, Li, Y., Wang, X., Xia, X., Jiang, C., 2024a. Multiple effects of modified biochar in saline-alkali soil : reducing salinity , providing nutrition , and promoting root colonization by beneficial bacteria 1–25.

Wang, Y., Li, A., Zou, B., Qian, Y., Li, X., Sun, Z., 2024b. The combination of buchloe dactyloides engelm and biochar promotes the remediation of soil contaminated with polycyclic aromatic hydrocarbons. Microorganisms 12. https://doi.org/10.3390/microorganisms12050968.

Wang, T., Luo, X., Chen, K., Zheng, H., Song, Z., Zhao, L., Gong, J., Li, F., Liu, R., 2025. Effects of Mg-palygorskite modified biochar on the growth of sedum alfredii hance in heavy metal contaminated soil. Plants 14, 1–15. https://doi.org/10.3390/plants14050790.

Waqas, M., Shahzad, R., Hamayun, M., Asaf, S., Khan, A.L., Kang, S.M., Yun, S., Kim, K.M., Lee, I.J., 2018. Biochar amendment changes jasmonic acid levels in two rice varieties and alters their resistance to herbivory. PLoS One 13, 1–18. https://doi.org/10.1371/journal.pone.0191296.

Widowati, F.,W., Wakhid, U., Kolambani, F., 2024. Residue of biochar-organic fertilizer after one year of use on corn (Zea mays L.) plants in Alfisol. Agromix 15, 67–74. https://doi.org/10.35891/agx.v15i1.3930.

Woolf, D., Amonette, J.E., Street-Perrott, F.A., Lehmann, J., Joseph, S., 2010. Sustainable biochar to mitigate global climate change. Nat. Commun. 1, 1–9. https://doi.org/10.1038/ncomms1053.

Wu, Q., Ning, K., Liu, B., Zheng, X., Li, C., Li, X., Zhou, X., Li, J., Li, J., Zhang, C., Dong, Z., 2024. Co-application of biochars and piriformospora indica improved the quality of coastal saline soil and promoted the growth of forage. Front. Plant Sci. 15, 1–11. https://doi.org/10.3389/fpls.2024.1434097.

Xing, X., Fan, F., Jiang, W., 2018. Characteristics of biochar pellets from corn straw under different pyrolysis temperatures. R. Soc. Open Sci. 5. https://doi.org/10.1098/rsos.172346.

Xu, P., Gao, Y., Cui, Z., Wu, B., Yan, B., Wang, Y., Zaitongguli, K., Wen, M., Wang, H., Jing, N., Wang, Y., Chao, C., Xue, W., 2023. Research progress on effects of biochar on soil environment and crop nutrient absorption and utilization. Sustain. Times 15. https://doi.org/10.3390/su15064861.

Yadav, R., Ramakrishna, W., 2023. Biochar as an environment-friendly alternative for multiple applications. Sustain. Times 15. https://doi.org/10.3390/su151813421.

Yan, S., Niu, Z., Zhang, A., Yan, H., Zhang, H., He, K., Xiao, X., Wang, N., Guan, C., Liu, G., 2019. Biochar application on paddy and purple soils in southern China: soil carbon and biotic activity. R. Soc. Open Sci. 6. https://doi.org/10.1098/rsos.181499.

Yang, W., Lei, G., Quan, S., Zhang, L., Wang, B., Hu, H., Li, L., Ma, H., Yin, C., Feng, F., Jing, Y., 2022. The removal of Cr(VI) from aqueous solutions with corn stalk biochar. Int. J. Environ. Res. Public Health 19. https://doi.org/10.3390/ijerph192114188.

Yang, X., You, M., Liu, S., Sarkar, B., Liu, Z., Yan, X., 2023. Microbial responses towards biochar application in potentially toxic element (PTE) contaminated soil: a critical review on effects and potential mechanisms. Biochar. https://doi.org/10.1007/s42773-023-00255-5.

Yaseen, S., Amjad, S.F., Mansoora, N., Kausar, S., Shahid, H., Alamri, S.A.M., Alrumman, S.A., Eid, E.M., Ansari, M.J., Danish, S., Datta, R., 2021. Supplemental effects of biochar and foliar application of ascorbic acid on physio-biochemical attributes of barley (Hordeum vulgare l.) under cadmium-contaminated soil. Sustain. Times 13. https://doi.org/10.3390/su13169128.

Yin, W., Zhang, W., Zhao, C., Xu, J., 2019. Evaluation of removal efficiency of Ni(II) and 2,4-DCP using in situ nitrogen-doped biochar modified with aquatic animal waste. ACS Omega 4, 19366–19374. https://doi.org/10.1021/acsomega.9b02769.

Yu, P., Li, T., Fu, Q., Liu, D., Hou, R., Zhao, H., 2021. Effect of biochar on soil and water loss on sloping farmland in the black soil region of northeast China during the spring thawing period. Sustain. Times 13, 1–16. https://doi.org/10.3390/su13031460.

Zhang, J., Lu, F., Zhang, H., Shao, L., Chen, D., He, P., 2015. Multiscale visualization of the structural and characteristic changes of sewage sludge biochar oriented towards potential agronomic and environmental implication. Sci. Rep. 5, 1–8. https://doi.org/10.1038/srep09406.

Zhang, H., Chen, C., Gray, E.M., Boyd, S.E., 2017a. Effect of feedstock and pyrolysis temperature on properties of biochar governing end use efficacy. Biomass Bioenergy 105, 136–146. https://doi.org/10.1016/j.biombioe.2017.06.024.

Zhang, H., Voroney, R.P., Price, G.W., 2017b. Effects of temperature and activation on biochar chemical properties and their impact on ammonium, nitrate, and phosphate sorption. J. Environ. Qual. 46, 889–896. https://doi.org/10.2134/jeq2017.02.0043.

Zhang, Y., Xia, Z., Nafsun, A.I., Feng, W., 2025. Preparation of nitrogen-doped biochar and its adsorption performance for Cr6+ and Pb2+ in aqueous systems. Toxics 13. https://doi.org/10.3390/toxics13050402.

Zheng, W.L., Wang, Y.F., Mo, J., Zeng, P., Chen, J., Sun, C., 2023. Effects of biochar application and nutrient fluctuation on the growth, and cadmium and nutrient uptake of Trifolium repens with different planting densities in Cd-contaminated soils. Front. Plant Sci. 14, 1–10. https://doi.org/10.3389/fpls.2023.1269082.

Zhou, L., Chen, L., Zhang, Y., Zhang, Y., Li, Z., Yang, K., Chen, L., 2024. The adsorption characteristics of phosphorus-modified corn stover biochar on lead and cadmium. Agric. For. 14. https://doi.org/10.3390/agriculture14071118.

Zou, Z., Fan, L., Li, X., Dong, C., Zhang, L., Zhang, L., Fu, J., Han, W., Yan, P., 2021. Response of plant root growth to biochar amendment: a meta-analysis. Agronomy 11, 1–14. https://doi.org/10.3390/agronomy11122442.

Index

'*Note*: Page numbers followed by "f" indicate figures and "t" indicate tables.'

D

E

F

G

N

O

T

Printed and bound by CPI Group (UK) Ltd, Croydon, CR0 4YY
16/07/2026
02169300-0001